U.S. Customary Units and Their SI Equivalents

Quantity	U.S. Customary Units	SI Equivalent
Acceleration	ft/s^2	$0.3048 \ m/s^2$
	$in./s^2$	$0.0254 \ m/s^2$
Area	ft^2	$0.0929 \ m^2$
	in^2	$645.2 \ mm^2$
Energy	$ft \cdot lb$	$1.356 \ J$
Force	kip	$4.448 \ kN$
	lb	$4.448 \ N$
	oz	$0.2780 \ N$
Impulse	$lb \cdot s$	$4.448 \ N \cdot s$
Length	ft	$0.3048 \ m$
	in.	$25.40 \ mm$
	mi	$1.609 \ km$
Mass	oz mass	$28.35 \ g$
	lb mass	$0.4536 \ kg$
	slug	$14.59 \ kg$
	ton	$907.2 \ kg$
Moment of a force	$lb \cdot ft$	$1.356 \ N \cdot m$
	$lb \cdot in.$	$0.1130 \ N \cdot m$
Moment of inertia		
Of an area	in^4	$0.4162 \times 10^6 \ mm^4$
Of a mass	$lb \cdot ft \cdot s^2$	$1.356 \ kg \cdot m^2$
Power	$ft \cdot lb/s$	$1.356 \ W$
	hp	$745.7 \ W$
Pressure or stress	lb/ft^2	$47.88 \ Pa$
	lb/in^2 (psi)	$6.895 \ kPa$
Velocity	ft/s	$0.3048 \ m/s$
	$in./s$	$0.0254 \ m/s$
	mi/h (mph)	$0.4470 \ m/s$
	mi/h (mph)	$1.609 \ km/h$
Volume, solids	ft^3	$0.02832 \ m^3$
	in^3	$16.39 \ cm^3$
Liquids	gal	$3.785 \ L$
	qt	$0.9464 \ L$
Work	$ft \cdot lb$	$1.356 \ J$

MECHANICS OF MATERIALS

Third Edition

MECHANICS OF MATERIALS

FERDINAND P. BEER
Lehigh University

E. RUSSELL JOHNSTON, JR.
University of Connecticut

JOHN T. DEWOLF
University of Connecticut

Boston   Burr Ridge, IL   Dubuque, IA   Madison, WI   New York   San Francisco   St. Louis
Bangkok   Bogotá   Caracas   Kuala Lumpur   Lisbon   London   Madrid   Mexico City
Milan   Montreal   New Delhi   Santiago   Seoul   Singapore   Sydney   Taipei   Toronto

McGraw-Hill Higher Education 🐝

A Division of *The McGraw-Hill Companies*

The photograph on the cover shows a steel wide-flange column being tested in the five-million-pound universal testing machine at Lehigh University, Bethlehem, Pennsylvania. (Courtesy of Fritz Engineering Laboratory.)

MECHANICS OF MATERIALS

Published by McGraw-Hill, an imprint of The McGraw-Hill Companies, Inc. 1221 Avenue of the Americas, New York, NY, 10020. Copyright © 2001, 1992, 1981, by The McGraw-Hill Companies, Inc. All rights reserved. No part of this publication may be reproduced or distributed in any form or by any means, or stored in a database or retrieval system, without the prior written consent of The McGraw-Hill Companies, Inc., including, but not limited to, in any network or other electronic storage or transmission, or broadcast for distance learning.

Some ancillaries, including electronic and print components, may not be available to customers outside the United States.

This book is printed on acid-free paper.

International	3 4 5 6 7 8 9 0 VNH/VNH 0 9 8 7 6 5 4 3 2
Domestic	3 4 5 6 7 8 9 0 VNH/VNH 0 9 8 7 6 5 4 3 2

ISBN 0-07-365935-5

Publisher: *Elizabeth A. Jones*
Senior sponsoring editor: *Jonathan Plant*
Associate editor: *Debra Matteson*
Marketing manager: *Ann Caven*
Project manager: *Kimberly D. Hooker*
Lead production supervisor: *Heather Burbridge*
Director, design **BR**: *Keith McPherson*
Manager, publication services: *Ira Roberts*
Photo researcher: *Sabina Dowell*
Senior supplement coordinator: *Carol Loreth*
Senior producer/developer, media technology: *Phillip Meek*
Cover design: *Michael Warrell*
Cover photo: *Courtesy of Fritz Engineering Laboratory*
Interior design: *Maureen McCutcheon/***TECHBOOKS**
Compositor: **TECHBOOKS**/Production Manager: *E. Gail Downey*
Typeface: *10/12 New Caledonia*
Printer: *Von Hoffmann Press, Inc.*

Library of Congress Cataloging-in-Publication Data

Beer, Ferdinand Pierre, 1915–
  Mechanics of materials / Ferdinand P. Beer, E. Russell Johnston, Jr., John T. DeWolf.—
3rd ed.
      p.   cm.
    ISBN 0-07-365935-5
    1. Strength of materials.   I. Johnston, E. Russell (Elwood Russell), 1925–   II. DeWolf,
John T., 1943–   III. Title.
TA405 .B39 2002
620.1′12—dc21
                                                                                    2001030518

INTERNATIONAL EDITION ISBN 0-07-112167-6
Copyright © 2002 Exclusive rights by The McGraw-Hill Companies, Inc., for manufacture and export. This book cannot be re-exported from the country to which it is sold by McGraw-Hill. The International Edition is not available in North America.

www.mhhe.com

About the Authors

Ferd Beer and Russ Johnston are often asked how they happened to write their books together, with one of them at Lehigh and the other at the University of Connecticut.

The answer to this question is simple. Russ Johnston's first teaching appointment was in the Department of Civil Engineering and Mechanics at Lehigh University. There he met Ferd Beer, who had joined that department two years earlier and was in charge of the courses in mechanics. Born in France and educated in France and Switzerland (he holds an M.S. degree from the Sorbonne and an Sc.D. degree in the field of theoretical mechanics from the University of Geneva), Ferd had come to the United States after serving in the French army during the early part of World War II and had taught for four years at Williams College in the Williams-MIT joint arts and engineering program. Born in Philadelphia, Russ had obtained a B.S. degree in civil engineering from the University of Delaware and an Sc.D. degree in the field of structural engineering from MIT.

Ferd was delighted to discover that the young man who had been hired chiefly to teach graduate structural engineering courses was not only willing but eager to help him reorganize the mechanics courses. Both believed that these courses should be taught from a few basic principles and that the various concepts involved would be best understood and remembered by the students if they were presented to them in a graphic way. Together they wrote lecture notes in statics and dynamics, to which they later added problems they felt would appeal to future engineers, and soon they produced the manuscript of the first edition of *Mechanics for Engineers*. The second edition of *Mechanics for Engineers* and the first edition of *Vector Mechanics for Engineers* found Russ Johnston at Worcester Polytechnic Institute and the next editions at the University of Connecticut. In the meantime, both Ferd and Russ had assumed administrative responsibilities in their departments, and both were involved in research, consulting, and supervising graduate students—Ferd in the area of stochastic processes and random vibrations, and Russ in the area of elastic stability and structural analysis and design. However, their interest in improving the teaching of the basic mechanics courses had not subsided, and they both taught sections of these courses as they kept revising their texts and began writing together the manuscript of the first edition of *Mechanics of Materials*.

Their collaboration has spanned the years of the revolution in computing. The first editions of *Mechanics for Engineers* and of *Vector Mechanics for Engineers* included notes on the proper use of the slide rule. With the advent of the pocket calculator, these notes were replaced by notes on the use of calculators. Now, problems requiring the use of a computer are included in each chapter of their texts.

Ferd and Russ's contributions to engineering education have earned them a number of honors and awards. They were presented with the Western Electric Fund Award for excellence in the instruction of engineering students by their respective regional sections of the American Society for Engineering Education, and they both received the Distinguished Educator Award from the Mechanics Division of the same society. In 1991 Russ received the Outstanding Civil Engineer Award from the Connecticut Section of the American Society of Civil Engineers, and in 1995 Ferd was awarded an honorary Doctor of Engineering degree by Lehigh University.

John T. DeWolf, Professor of Civil Engineering at the University of Connecticut, joined the Beer and Johnston team as a collaborator to the second edition of *Mechanics of Materials* and, starting with this third edition, is now a co-author of that text. John holds a B.S. degree in civil engineering from the University of Hawaii and M.E. and Ph.D. degrees in structural engineering from Cornell University. His research interests are in the area of elastic stability, bridge monitoring, and structural analysis and design. He is a member of the Connecticut Board of Examiners for Professional Engineers.

Contents

5
ANALYSIS AND DESIGN OF BEAMS FOR BENDING
308

6
SHEARING STRESSES IN BEAMS AND THIN-WALLED MEMBERS
372

7
TRANSFORMATIONS OF STRESS AND STRAIN
423

8
PRINCIPAL STRESSES UNDER A GIVEN LOADING
496

9
DEFLECTION OF BEAMS
530

APPENDICES
735

PREFACE

The main objective of a basic mechanics course should be to develop in the engineering student the ability to analyze a given problem in a simple and logical manner and to apply to its solution a few fundamental and well-understood principles. This text is designed for the first course in mechanics of materials—or strength of materials—offered to engineering students in the sophomore or junior year. The authors hope that it will help instructors achieve this goal in that particular course in the same way that their other texts may have helped them in statics and dynamics.

In this text the study of the mechanics of materials is based on the understanding of a few basic concepts and on the use of simplified models. This approach makes it possible to develop all the necessary formulas in a rational and logical manner, and to clearly indicate the conditions under which they can be safely applied to the analysis and design of actual engineering structures and machine components.

Free-body diagrams are used extensively throughout the text to determine external or internal forces. The use of "picture equations" will also help the students understand the superposition of loadings and the resulting stresses and deformations.

It is expected that students using this text will have completed a course in statics. However, Chap. 1 is designed to provide them with an opportunity to review the concepts learned in that course, while shear and bending-moment diagrams are covered in detail in Secs. 5.2 and 5.3. The properties of moments and centroids of areas are described in Appendix A; this material can be used to reinforce the discussion of the determination of normal and shearing stresses in beams (Chaps. 4, 5, and 6).

The first four chapters of the text are devoted to the analysis of the stresses and of the corresponding deformations in various structural members, considering successively axial loading, torsion, and pure bending. Each analysis is based on a few basic concepts, namely, the conditions of equilibrium of the forces exerted on the member, the relations existing between stress and strain in the material, and the conditions imposed by the supports and loading of the member. The study of each type of loading is complemented by a large number of examples, sample problems, and problems to be assigned, all designed to strengthen the students' understanding of the subject.

The concept of stress at a point is introduced in Chap. 1, where it is shown that an axial load can produce shearing stresses as well as normal stresses, depending upon the section considered. The fact that stresses depend upon the orientation of the surface on which they are computed is emphasized again in Chaps. 3 and 4 in the cases of torsion and pure bending. However, the discussion of computational techniques—such as Mohr's circle—used for the transformation of stress at a point is delayed until Chap. 7, after students have had the opportunity to solve problems involving a combination of the basic loadings and have discovered for themselves the need for such techniques.

In this new edition, the discussion in Chap. 2 of the relation between stress and strain in various materials has been expanded to include fiber-reinforced composite materials. Also, the study of beams under transverse loads is now covered in two separate chapters. Chapter 5 is devoted to the determination of the normal stresses in a beam and to the design of beams based on the allowable normal stress in the material used (Sec. 5.4). The chapter begins with a discussion of the shear and bending-moment diagrams (Secs. 5.2 and 5.3) and includes an optional section on the use of singularity functions for the determination of the shear and bending moment in a beam (Sec. 5.5). The chapter ends with an expanded optional section on nonprismatic beams (Sec. 5.6).

Chapter 6 is devoted to the determination of shearing stresses in beams and thin-walled members under transverse loadings. In this new edition, and at the suggestion of our reviewers, the formula for the shear flow, $q = VQ/I$, is derived in the traditional way. More advanced aspects of the design of beams, such as the determination of the principal stresses at the junction of the flange and web of a W-beam, have been moved to Chap. 8, an optional chapter that may be covered after the transformations of stresses have been discussed in Chap. 7. The design of transmission shafts has been moved to that chapter for the same reason, as well as the determination of stresses under combined loadings which, in this new position, can now include the determination of the principal stresses, principal planes, and maximum shearing stress at a given point.

Statically indeterminate problems are first discussed in Chap. 2 and considered throughout the text for the various loading conditions encountered. Thus, students are presented at an early stage with a method of solution which combines the analysis of deformations with the conventional analysis of forces used in statics. In this way, they will have become thoroughly familiar with this fundamental method by the end of the course. In addition, this approach helps the students realize that stresses themselves are statically indeterminate and can be computed only by considering the corresponding distribution of strains.

Design concepts are discussed throughout the text whenever appropriate. A discussion of the application of the factor of safety to design can be found in Chap. 1, where the concepts of both allowable stress design and load and resistance factor design are presented.

The concept of plastic deformation is introduced in Chap. 2, where it is applied to the analysis of members under axial loading. Problems involving the plastic deformation of circular shafts and of prismatic beams are also considered in optional sections of Chaps. 3, 4, and 6. While some of this material can be omitted at the choice of the instructor, its inclusion in the body of the text will help students realize the limitations of the assumption of a linear stress-strain relation and serve to caution them against the inappropriate use of the eleastic torsion and flexure formulas.

The determination of the deflection of beams is discussed in Chap. 9. The first part of the chapter is devoted to the integration method and to the method of superposition, with an optional section (Sec. 9.6) based on the use of singularity functions. (This section should be used only if Sec. 5.5 was covered earlier.) The second part of Chap. 9 is optional. It presents the moment-area method in two lessons instead off three as in our previous edition.

Chapter 10 is devoted to columns and contains new material on the design of wood columns. Chapter 11 covers energy methods, including Castigliano's theorem.

Additional topics such as residual stresses, torsion of noncircular and thin-walled members, bending of curved beams, shearing stresses in nonsymmetrical members, and failure criteria, have been included in optional sections for use in courses of varying emphases. To preserve the integrity of the subject, these topics are presented in the proper sequence, wherever they logically belong. Thus, even when not covered in the course, they are highly visible and can be easily referred to by the students if needed in a later course or in engineering practice. For convenience all optional sections have been indicated by asterisks.

Each chapter begins with an introductory section setting the purpose and goals of the chapter and describing in simple terms the material to be covered and its application to the solution of engineering problems. The body of the text has been divided into units, each consisting of one or several theory sections followed by sample problems and a large number of problems to be assigned. Each unit corresponds to a well-defined topic and generally can be covered in one lesson. Each chapter ends with a review and summary of the material covered in the chapter. Notes in the margin have been included to help the students organize their review work, and cross references provided to help them find the portions of material requiring their special attention. The theory sections include many examples designed to illustrate the material being presented and facilitate its understanding. The sample problems are intended to show some of the applications of the theory to the solution of engineering problems. Since they have been set up in much the same form that students will use in solving the assigned problems, the sample problems serve the double purpose of amplifying the text and demonstrating the type of neat and orderly work that students should cultivate in their own solutions. Most of the problems are of a practical nature and should appeal to engineering students. They are primarily designed, however, to illustrate the material presented in the text and help the students understand the basic principles used in mechanics of materials. The problems have been grouped according to the portions of material they illustrate and have been arranged in order of increasing difficulty. Problems requiring special attention have been indicated by asterisks. Answers to problems are given at the end of the book, except for those with a number set in italics.

Because it is essential that students be able to handle effectively both SI metric units and U.S. customary units, half the examples, sample problems, and problems to be assigned have been stated in SI units and half in U.S. customary units. Since a large number of problems are available, instructors can assign problems using each system of units in whatever proportion they find most desirable for their class. The availability of personal computers makes it possible for engineering students to solve a great number of challenging problems. In this new edition of *Mechanics of Materials*, a group of six or more problems designed to be solved with a computer can be found at the end of each chapter. Developing the algorithm required to solve a given problem will benefit the students in two different ways: (1) it will help them gain a better understanding of the mechanics principles involved; (2) it will provide them with an opportunity to apply the skills acquired in their computer programming course to the solution of a meaningful engineering problem.

The authors wish to acknowledge with gratitude the helpful comments and suggestions of the following colleagues from various universities: Candace A. Ammerman (Colorado School of Mines), Subhash C. Anand (Clemson University), Christopher M. Foley (Marquette University), Paul Heyliger (Colorado State University), D. A. Jenkins (University of Florida), Vistap M. Karbhari (University of California, San Diego), Ajit D. Kelkar (North Carolina A&T State University), Timothy C. Kennedy (Oregon State University), Robert M. Korol (McMaster University), Jack Lesko (Virginia Polytechnic Institute), Lee L. Lowery, Jr. (Texas A&M University), Ghyslaine McClure (McGill University), Arnhad Pourmovahed (Kettering University), G. E. Ramey (Auburn University), Ting-Leung Sham (Rensselaer Polytechnic Institute), Stephen W. Swanson (The University of Utah), M. D. A. Thomas (University of Toronto), George Z. Voyiadjis (Louisiana State University), Douglas Winslow (Purdue University), and Manoochehr Zoghi (The University of Dayton).

Their special thanks go to two colleagues on the Lehigh University faculty: Dean Updike, of the Department of Mechanical Engineering and Mechanics, who checked the solutions and answers of all the problems in this edition, and Le-Wu Lu, of the Department of Civil and Environmental Engineering, who provided the photograph used on the cover of this book.

Ferdinand P. Beer
E. Russell Johnston, Jr.
John T. DeWolf

List of Symbols

a	Constant; distance
$\mathbf{A}, \mathbf{B}, \mathbf{C}, \ldots$	Forces; reactions
$A, B, C, \ldots$	Points
A, α	Area
b	Distance; width
c	Constant; distance; radius
C	Centroid
$C_1, C_2, \ldots$	Constants of integration
C_P	Column stability factor
d	Distance; diameter; depth
D	Diameter
e	Distance; eccentricity; dilatation
E	Modulus of elasticity
f	Frequency; function
$\mathbf{F}$	Force
$F.S.$	Factor of safety
G	Modulus of rigidity; shear modulus
h	Distance; height
$\mathbf{H}$	Force
H, J, K	Points
$I, I_x, \ldots$	Moment of inertia
$I_{xy}, \ldots$	Product of inertia
J	Polar moment of inertia
k	Spring constant; shape factor; bulk modulus; constant
K	Stress concentration factor; torsional spring constant
l	Length; span
L	Length; span
L_e	Effective length
m	Mass
$\mathbf{M}$	Couple
$M, M_x, \ldots$	Bending moment
M_D	Bending moment, dead load (LRFD)
M_L	Bending moment, live load (LRFD)
M_U	Bending moment, ultimate load (LRFD)
n	Number; ratio of moduli of elasticity; normal direction
p	Pressure
$\mathbf{P}$	Force; concentrated load
P_D	Dead load (LRFD)
P_L	Live load (LRFD)
P_U	Ultimate load (LRFD)
q	Shearing force per unit length; shear flow
$\mathbf{Q}$	Force
Q	First moment of area
r	Radius; radius of gyration
$\mathbf{R}$	Force; reaction
R	Radius; modulus of rupture
s	Length
S	Elastic section modulus
t	Thickness; distance; tangential deviation
$\mathbf{T}$	Torque
T	Temperature
u, v	Rectangular coordinates
u	Strain-energy density
U	Strain energy; work
$\mathbf{v}$	Velocity
$\mathbf{V}$	Shearing force
V	Volume; shear
w	Width; distance; load per unit length
$\mathbf{W}, W$	Weight, load
x, y, z	Rectangular coordinates; distance; displacements; deflections
$\bar{x}, \bar{y}, \bar{z}$	Coordinates of centroid
Z	Plastic section modulus
α, β, γ	Angles
α	Coefficient of thermal expansion; influence coefficient
γ	Shearing strain; specific weight
γ_D	Load factor, dead load (LRFD)
γ_L	Load factor, live load (LRFD)
δ	Deformation; displacement
ϵ	Normal strain
θ	Angle; slope
λ	Direction cosine
ν	Poisson's ratio
ρ	Radius of curvature; distance; density
σ	Normal stress
τ	Shearing stress
ϕ	Angle; angle of twist; resistance factor
ω	Angular velocity

MECHANICS OF MATERIALS

CHAPTER

1

Introduction—Concept of Stress

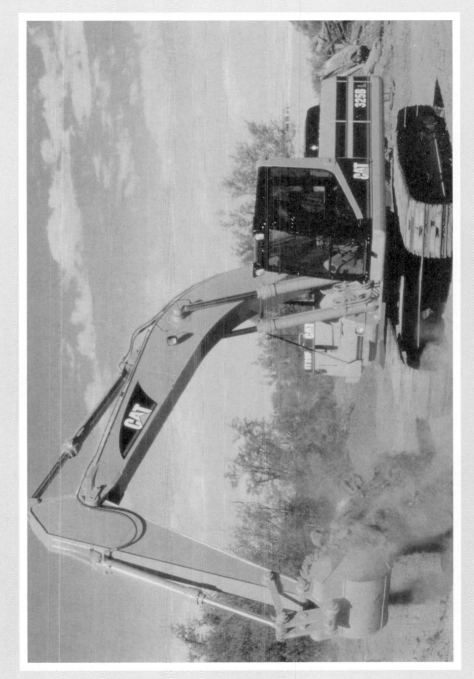

This chapter is devoted to the study of the stresses occurring in many of the elements contained in this excavator, such as two-force members, axles, bolts, and pins.

1.1. INTRODUCTION

The main objective of the study of the mechanics of materials is to provide the future engineer with the means of analyzing and designing various machines and load-bearing structures.

Both the analysis and the design of a given structure involve the determination of *stresses* and *deformations*. This first chapter is devoted to the concept of *stress*.

Section 1.2 is devoted to a short review of the basic methods of statics and to their application to the determination of the forces in the members of a simple structure consisting of pin-connected members. Section 1.3 will introduce you to the concept of *stress* in a member of a structure, and you will be shown how that stress can be determined from the *force* in the member. After a short discussion of engineering analysis and design (Sec. 1.4), you will consider successively the *normal stresses* in a member under axial loading (Sec. 1.5), the *shearing stresses* caused by the application of equal and opposite transverse forces (Sec. 1.6), and the *bearing stresses* created by bolts and pins in the members they connect (Sec. 1.7). These various concepts will be applied in Sec. 1.8 to the determination of the stresses in the members of the simple structure considered earlier in Sec. 1.2.

The first part of the chapter ends with a description of the method you should use in the solution of an assigned problem (Sec. 1.9) and with a discussion of the numerical accuracy appropriate in engineering calculations (Sec. 1.10).

In Sec. 1.11, where a two-force member under axial loading is considered again, it will be observed that the stresses on an *oblique* plane include both *normal* and *shearing* stresses, while in Sec. 1.12 you will note that *six components* are required to describe the state of stress at a point in a body under the most general loading conditions.

Finally, Sec. 1.13 will be devoted to the determination from test specimens of the *ultimate strength* of a given material and to the use of a *factor of safety* in the computation of the *allowable load* for a structural component made of that material.

1.2. A SHORT REVIEW OF THE METHODS OF STATICS

In this section you will review the basic methods of statics while determining the forces in the members of a simple structure.

Consider the structure shown in Fig. 1.1, which was designed to support a 30-kN load. It consists of a boom *AB* with a 30 × 50-mm rectangular cross section and of a rod *BC* with a 20-mm-diameter circular cross section. The boom and the rod are connected by a pin at *B* and are supported by pins and brackets at *A* and *C* respectively. Our first step should be to draw a *free-body diagram* of the structure by detaching it from its supports at *A* and *C*, and showing the reactions that these supports exert on the structure (Fig. 1.2). Note that the sketch of the structure has been simplified by omitting all unnecessary details. Many of you may have recognized at this point that *AB* and *BC* are *two-force members*. For those of you who have not, we will pursue our analysis, ignoring that fact and assuming that the directions of the reactions at *A* and *C* are unknown. Each of these reactions, therefore, will

$d = 20$ mm

600 mm

50 mm

800 mm

30 kN

Fig. 1.1

be represented by two components, A_x and A_y at A, and C_x and C_y at C. We write the following three equilibrium equations:

$$+\circlearrowleft \Sigma M_C = 0: \quad A_x(0.6 \text{ m}) - (30 \text{ kN})(0.8 \text{ m}) = 0$$ (1.1)
$$A_x = +40 \text{ kN}$$

$$\xrightarrow{+} \Sigma F_x = 0: \quad A_x + C_x = 0$$ (1.2)
$$C_x = -A_x \quad C_x = -40 \text{ kN}$$

$$+\uparrow \Sigma F_y = 0: \quad A_y + C_y - 30 \text{ kN} = 0$$ (1.3)
$$A_y + C_y = +30 \text{ kN}$$

We have found two of the four unknowns, but cannot determine the other two from these equations, and no additional independent equation can be obtained from the free-body diagram of the structure. We must now dismember the structure. Considering the free-body diagram of the boom AB (Fig. 1.3), we write the following equilibrium equation:

$$+\circlearrowleft \Sigma M_B = 0: \quad -A_y(0.8 \text{ m}) = 0 \quad A_y = 0$$ (1.4)

Substituting for A_y from (1.4) into (1.3), we obtain $C_y = +30$ kN. Expressing the results obtained for the reactions at A and C in vector form, we have

$$\mathbf{A} = 40 \text{ kN} \rightarrow \quad \mathbf{C}_x = 40 \text{ kN} \leftarrow, \ \mathbf{C}_y = 30 \text{ kN}\uparrow$$

We note that the reaction at A is directed along the axis of the boom AB and causes compression in that member. Observing that the components C_x and C_y of the reaction at C are respectively proportional to the horizontal and vertical components of the distance from B to C, we conclude that the reaction at C is equal to 50 kN, is directed along the axis of the rod BC, and causes tension in that member.

C_y

C

B

C_x

0.6 m

A_y

A

A_x

0.8 m

30 kN

Fig. 1.2

B_y

B_z

A_y

B

A

A_x

0.8 m

30 kN

Fig. 1.3

These results could have been anticipated by recognizing that AB and BC are two-force members, i.e., members that are subjected to forces at only two points, these points being A and B for member AB, and B and C for member BC. Indeed, for a two-force member the lines of action of the resultants of the forces acting at each of the two points are equal and opposite and pass through both points. Using this property, we could have obtained a simpler solution by considering the free-body diagram of pin B. The forces on pin B are the forces $\mathbf{F}_{AB}$ and $\mathbf{F}_{BC}$ exerted, respectively, by members AB and BC, and the 30-kN load (Fig. 1.4a). We can express that pin B is in equilibrium by drawing the corresponding force triangle (Fig. 1.4b).

Since the force $\mathbf{F}_{BC}$ is directed along member BC, its slope is the same as that of BC, namely, 3/4. We can, therefore, write the proportion

$$\frac{F_{AB}}{4} = \frac{F_{BC}}{5} = \frac{30 \text{ kN}}{3}$$

from which we obtain

$$F_{AB} = 40 \text{ kN} \qquad F_{BC} = 50 \text{ kN}$$

The forces $\mathbf{F}'_{AB}$ and $\mathbf{F}'_{BC}$ exerted by pin B respectively on boom AB and rod BC are equal and opposite to $\mathbf{F}_{AB}$ and $\mathbf{F}_{BC}$ (Fig. 1.5).

(a)

(b)

Fig. 1.4

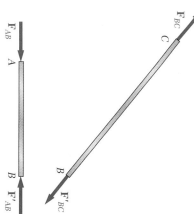

Fig. 1.5

Knowing the forces at the ends of each of the members, we can now determine the internal forces in these members. Passing a section at some arbitrary point D of rod BC, we obtain two portions BD and CD (Fig. 1.6). Since 50-kN forces must be applied at D to both portions of the rod to keep them in equilibrium, we conclude that an internal force of 50 kN is produced in rod BC when a 30-kN load is applied at B. We further check from the directions of the forces $\mathbf{F}_{BC}$ and $\mathbf{F}'_{BC}$ in Fig. 1.6 that the rod is in tension. A similar procedure would enable us to determine that the internal force in boom AB is 40 kN and that the boom is in compression.

Fig. 1.6

1.3. STRESSES IN THE MEMBERS OF A STRUCTURE

While the results obtained in the preceding section represent a first and necessary step in the analysis of the given structure, they do not tell us whether the given load can be safely supported. Whether rod BC, for example, will break or not under this loading depends not only upon the value found for the internal force F_{BC}, but also upon the cross-sectional area of the rod and the material of which the rod is made. Indeed, the internal force F_{BC} actually represents the resultant of elementary forces distributed over the entire area A of the cross section (Fig. 1.7) and the average intensity of these distributed forces is equal to the force per unit area, F_{BC}/A, in the section. Whether or not the rod will break under the given loading clearly depends upon the ability of the material to withstand the corresponding value F_{BC}/A of the intensity of the distributed internal forces. It thus depends upon the force F_{BC}, the cross-sectional area A, and the material of the rod.

The force per unit area, or intensity of the forces distributed over a given section, is called the *stress* on that section and is denoted by the Greek letter σ (sigma). The stress in a member of cross-sectional area A subjected to an axial load $\mathbf{P}$ (Fig. 1.8) is therefore obtained by dividing the magnitude P of the load by the area A:

$$\sigma = \frac{P}{A} \qquad (1.5)$$

A positive sign will be used to indicate a tensile stress (member in tension) and a negative sign to indicate a compressive stress (member in compression).

Since SI metric units are used in this discussion, with P expressed in newtons (N) and A in square meters (m^2). This unit is called a *pascal* (Pa). However, one finds that the pascal is an exceedingly small quantity and that, in practice, multiples of this unit must be used, namely, the kilopascal (kPa), the megapascal (MPa), and the gigapascal (GPa). We have

$$1 \text{ kPa} = 10^3 \text{ Pa} = 10^3 \text{ N/m}^2$$

$$1 \text{ MPa} = 10^6 \text{ Pa} = 10^6 \text{ N/m}^2$$

$$1 \text{ GPa} = 10^9 \text{ Pa} = 10^9 \text{ N/m}^2$$

When U.S. customary units are used, the force P is usually expressed in pounds (lb) or kilopounds (kip), and the cross-sectional area A in square inches (in^2). The stress σ will then be expressed in pounds per square inch (psi) or kilopounds per square inch (ksi).†

---

†The principal SI and U.S. customary units used in mechanics are listed in tables inside the front cover of this book. From the table on the right-hand side, we note that 1 psi is approximately equal to 7 kPa, and 1 ksi approximately equal to 7 MPa.

$$\sigma = \frac{F_{BC}}{A}$$

Fig. 1.7

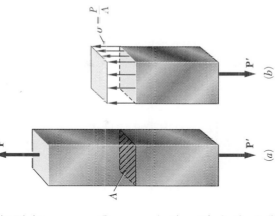

$$\upsilon = \frac{P}{A}$$

Fig. 1.8

1.4. ANALYSIS AND DESIGN

Considering again the structure of Fig. 1.1, let us assume that rod BC is made of a steel with a maximum allowable stress $\sigma_{all} = 165$ MPa. Can rod BC safely support the load to which it will be subjected? The magnitude of the force F_{BC} in the rod was found earlier to be 50 kN. Recalling that the diameter of the rod is 20 mm, we use Eq. (1.5) to determine the stress created in the rod by the given loading. We have

$$P = F_{BC} = +50 \text{ kN} = +50 \times 10^3 \text{ N}$$

$$A = \pi r^2 = \pi \left(\frac{20 \text{ mm}}{2} \right)^2 = \pi (10 \times 10^{-3} \text{ m})^2 = 314 \times 10^{-6} \text{ m}^2$$

$$\sigma = \frac{P}{A} = \frac{+50 \times 10^3 \text{ N}}{314 \times 10^{-6} \text{ m}^2} = +159 \times 10^6 \text{ Pa} = +159 \text{ MPa}$$

Since the value obtained for σ is smaller than the value σ_{all} of the allowable stress in the steel used, we conclude that rod BC can safely support the load to which it will be subjected. To be complete, our analysis of the given structure should also include the determination of the compressive stress in boom AB, as well as an investigation of the stresses produced in the pins and their bearings. This will be discussed later in this chapter. We should also determine whether the deformations produced by the given loading are acceptable. The study of deformations under axial loads will be the subject of Chap. 2. An additional consideration, required for members in compression involves the *stability* of the member, i.e., its ability to support a given load without experiencing a sudden change in configuration. This will be discussed in Chap. 10.

The engineer's role is not limited to the analysis of existing structures and machines subjected to given loading conditions. Of even greater importance to the engineer is the *design* of new structures and machines, that is, the selection of appropriate components to perform a given task. As an example of design, let us return to the structure of Fig. 1.1, and assume that aluminum with an allowable stress $\sigma_{all} = 100$ MPa is to be used. Since the force in rod BC will still be $P = F_{BC} = 50$ kN under the given loading, we must have, from Eq. (1.5),

$$\sigma_{all} = \frac{P}{A} \qquad A = \frac{P}{\sigma_{all}} = \frac{50 \times 10^3 \text{ N}}{100 \times 10^6 \text{ Pa}} = 500 \times 10^{-6} \text{ m}^2$$

and, since $A = \pi r^2$,

$$r = \sqrt{\frac{A}{\pi}} = \sqrt{\frac{500 \times 10^{-6} \text{ m}^2}{\pi}} = 12.62 \times 10^{-3} \text{ m} = 12.62 \text{ mm}$$

$$d = 2r = 25.2 \text{ mm}$$

We conclude that an aluminum rod 26 mm or more in diameter will be adequate.

1.5. AXIAL LOADING; NORMAL STRESS

As we have already indicated, rod *BC* of the example considered in the preceding section is a two-force member and, therefore, the forces $\mathbf{F}_{BC}$ and $\mathbf{F}'_{BC}$ acting on its ends *B* and *C* (Fig. 1.5) are directed along the axis of the rod. We say that the rod is under *axial loading*. An actual example of structural members under axial loading is provided by the members of the bridge truss shown in Fig. 1.9.

Fig. 1.9 This bridge truss consists of two-force members that may be in tension or in compression.

Returning to rod *BC* of Fig. 1.5, we recall that the section we passed through the rod to determine the internal force in the rod and the corresponding stress was perpendicular to the axis of the rod; the internal force was therefore normal to the plane of the section (Fig. 1.7) and the corresponding stress is described as a *normal stress*. Thus, formula (1.5) gives us the *normal stress in a member under axial loading*:

$$\sigma = \frac{P}{A} \tag{1.5}$$

We should also note that, in formula (1.5), σ is obtained by dividing the magnitude *P* of the resultant of the internal forces distributed over the cross section by the area *A* of the cross-section; it represents, therefore, the *average value* of the stress over the cross section, rather than the stress at a specific point of the cross section.

To define the stress at a given point *Q* of the cross section, we should consider a small area ΔA (Fig. 1.10). Dividing the magnitude of $\Delta \mathbf{F}$ by ΔA, we obtain the average value of the stress over ΔA. Letting ΔA approach zero, we obtain the stress at point *Q*:

$$\sigma = \lim_{\Delta A \to 0} \frac{\Delta F}{\Delta A} \tag{1.6}$$

Fig. 1.10

In general, the value obtained for the stress σ at a given point Q of the section is different from the value of the average stress given by formula (1.5), and σ is found to vary across the section. In a slender rod subjected to equal and opposite concentrated loads $\mathbf{P}$ and $\mathbf{P'}$ (Fig. 1.11a), this variation is small in a section away from the points of application of the concentrated loads (Fig. 1.11c), but it is quite noticeable in the neighborhood of these points (Fig. 1.11b and d).

It follows from Eq. (1.6) that the magnitude of the resultant of the distributed internal forces is

Fig. 1.11

(a) $\mathbf{P'}$ $\mathbf{P}$

(b) $\mathbf{P'}$ σ

(c) $\mathbf{P'}$ σ

(d) $\mathbf{P'}$ σ

But the conditions of equilibrium of each of the portions of rod shown in Fig. 1.11 require that this magnitude be equal to the magnitude P of the concentrated loads. We have, therefore,

$$\int dF = \int_A \sigma \, dA$$

$=$

$\mathbf{P}$

C

Fig. 1.12

$$P = \int dF = \int_A \sigma \, dA \qquad (1.7)$$

which means that the volume under each of the stress surfaces in Fig. 1.11 must be equal to the magnitude P of the loads. This, however, is the only information that we can derive from our knowledge of statics, regarding the distribution of normal stresses in the various sections of the rod. The actual distribution of stresses in any given section is *statically indeterminate*. To learn more about this distribution, it is necessary to consider the deformations resulting from the particular mode of application of the loads at the ends of the rod. This will be discussed further in Chap. 2.

In practice, it will be assumed that the distribution of normal stresses in an axially loaded member is uniform, except in the immediate vicinity of the points of application of the loads. The value σ of the stress is then equal to σ_{ave} and can be obtained from formula (1.5). However, we should realize that, when we assume a uniform distribution of stresses in the section, i.e., when we assume that the internal forces are uniformly distributed across the section, it follows from elementary statics[†] that the resultant $\mathbf{P}$ of the internal forces must be applied at the centroid C of the section (Fig. 1.12). This means that *a uniform distribution of stress is possible only if the line of action of the concentrated loads $\mathbf{P}$ and $\mathbf{P'}$ passes through the centroid of the section considered* (Fig. 1.13). This type of loading is called *centric loading* and will be assumed to take place in all straight two-force members found in trusses and pin-connected structures, such as the one considered in Fig. 1.1.

†See Ferdinand P. Beer and E. Russell Johnston, Jr., *Mechanics for Engineers*, 4th ed., McGraw-Hill, New York, 1987, or *Vector Mechanics for Engineers*, 6th ed., McGraw-Hill, New York, 1996, secs. 5.2 and 5.3.

P   **P'**

Fig. 1.13

(a)

(b)

Fig. 1.14

Fig. 1.15

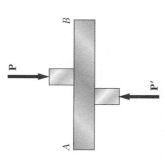

(a)

(b)

Fig. 1.16

However, if a two-force member is loaded axially, but *eccentrically* as shown in Fig. 1.14a, we find from the conditions of equilibrium of the portion of member shown in Fig. 1.14b that the internal forces in a given section must be equivalent to a force **P** applied at the centroid of the section and a couple **M** of moment $M = Pd$. The distribution of forces—and, thus, the corresponding distribution of stresses—*cannot be uniform*. Nor can the distribution of stresses be symmetric as shown in Fig. 1.11. This point will be discussed in detail in Chap. 4.

1.6. SHEARING STRESS

The internal forces and the corresponding stresses discussed in Secs. 1.2 and 1.3 were normal to the section considered. A very different type of stress is obtained when transverse forces **P** and **P'** are applied to a member *AB* (Fig. 1.15). Passing a section at *C* between the points of application of the two forces (Fig. 1.16a), we obtain the diagram of portion *AC* shown in Fig. 1.16b. We conclude that internal forces must exist in the plane of the section, and that their resultant is equal to **P**. These elementary internal forces are called *shearing forces*, and the magnitude *P* of their resultant is the *shear* in the section. Dividing the shear

P by the area A of the cross section, we obtain the *average shearing stress* in the section. Denoting the shearing stress by the Greek letter τ (tau), we write

$$\tau_{ave} = \frac{P}{A} \qquad (1.8)$$

It should be emphasized that the value obtained is an average value of the shearing stress over the entire section. Contrary to what we said earlier for normal stresses, the distribution of shearing stresses across the section *cannot* be assumed uniform. As you will see in Chap. 6, the actual value τ of the shearing stress varies from zero at the surface of the member to a maximum value τ_{max} that may be much larger than the average value τ_{ave}.

Fig. 1.17 Cutaway view of a connection with a bolt in shear.

Shearing stresses are commonly found in bolts, pins, and rivets used to connect various structural members and machine components (Fig. 1.17). Consider the two plates A and B, which are connected by a bolt CD (Fig. 1.18). If the plates are subjected to tension forces of magnitude F, stresses will develop in the section of bolt corresponding to the plane EE'. Drawing the diagrams of the bolt and of the portion located above the plane EE' (Fig. 1.19), we conclude that the shear P in the section is equal to F. The average shearing stress in the section is obtained, according to formula (1.8), by dividing the shear $P = F$ by the area A of the cross section:

$$\tau_{ave} = \frac{P}{A} = \frac{F}{A} \qquad (1.9)$$

Fig. 1.18

Fig. 1.19

(a)

(b)

Fig. 1.20

(a)

(b)

Fig. 1.21

The bolt we have just considered is said to be in *single shear*. Different loading situations may arise, however. For example, if splice plates C and D are used to connect plates A and B (Fig. 1.20), shear will take place in bolt HJ in each of the two planes KK' and LL' (and similarly in bolt EG). The bolts are said to be in *double shear*. To determine the average shearing stress in each plane, we draw free-body diagrams of bolt HJ and of the portion of bolt located between the two planes (Fig. 1.21). Observing that the shear P in each of the sections is $P = F/2$, we conclude that the average shearing stress is

$$\tau_{ave} = \frac{P}{A} = \frac{F/2}{A} = \frac{F}{2A} \qquad (1.10)$$

1.7. BEARING STRESS IN CONNECTIONS

Bolts, pins, and rivets create stresses in the members they connect, along the *bearing surface*, or surface of contact. For example, consider again the two plates A and B connected by a bolt CD that we have discussed in the preceding section (Fig. 1.18). The bolt exerts on plate A a force $\mathbf{P}$ equal and opposite to the force $\mathbf{F}$ exerted by the plate on the bolt (Fig. 1.22). The force $\mathbf{P}$ represents the resultant of elementary forces distributed on the inside surface of a half-cylinder of diameter d and of length t equal to the thickness of the plate. Since the distribution of these forces—and of the corresponding stresses—is quite complicated, one uses in practice an average nominal value σ_b of the stress, called the *bearing stress*, obtained by dividing the load P by the area of the rectangle representing the projection of the bolt on the plate section (Fig. 1.23). Since this area is equal to td, where t is the plate thickness and d the diameter of the bolt, we have

$$\sigma_b = \frac{P}{A} = \frac{P}{td} \qquad (1.11)$$

Fig. 1.22

Fig. 1.23

1.8. APPLICATION TO THE ANALYSIS AND DESIGN OF SIMPLE STRUCTURES

We are now in a position to determine the stresses in the members and connections of various simple two-dimensional structures and, thus, to design such structures.

As an example, let us return to the structure of Fig. 1.1 that we have already considered in Sec. 1.2 and let us specify the supports and connections at A, B, and C. As shown in Fig. 1.24, the 20-mm-diameter rod BC has flat ends of 20×40-mm rectangular cross section, while boom AB has a 30×50-mm rectangular cross section and is fitted with a clevis at end B. Both members are connected at B by a pin from which the 30-kN load is suspended by means of a U-shaped bracket. Boom AB is supported at A by a pin fitted into a double bracket, while rod BC is connected at C to a single bracket. All pins are 25 mm in diameter.

Fig. 1.24

a. Determination of the Normal Stress in Boom *AB* and Rod *BC*.
As we found in Secs. 1.2 and 1.4, the force in rod BC is $F_{BC} = 50$ kN (tension) and the area of its circular cross section is $A = 314 \times 10^{-6}$ m^2; the corresponding average normal stress is $\sigma_{BC} = +159$ MPa. However, the flat parts of the rod are also under

tension and at the narrowest section, where a hole is located, we have

$$A = (20 \text{ mm})(40 \text{ mm} - 25 \text{ mm}) = 300 \times 10^{-6} \text{ m}^2$$

The corresponding average value of the stress, therefore, is

$$(\sigma_{BC})_{\text{end}} = \frac{P}{A} = \frac{50 \times 10^3 \text{ N}}{300 \times 10^{-6} \text{ m}^2} = 167 \text{ MPa}$$

Note that this is an *average value*; close to the hole, the stress will actually reach a much larger value, as you will see in Sec. 2.18. It is clear that, under an increasing load, the rod will fail near one of the holes rather than in its cylindrical portion; its design, therefore, could be improved by increasing the width or the thickness of the flat ends of the rod.

Turning now our attention to boom AB, we recall from Sec. 1.2 that the force in the boom is $F_{AB} = 40$ kN (compression). Since the area of the boom's rectangular cross section is $A = 30$ mm $\times$ 50 mm $= 1.5 \times 10^{-3}$ m^2, the average value of the normal stress in the main part of the rod, between pins A and B, is

$$\sigma_{AB} = -\frac{40 \times 10^3 \text{ N}}{1.5 \times 10^{-3} \text{ m}^2} = -26.7 \times 10^6 \text{Pa} = -26.7 \text{ MPa}$$

Note that the sections of minimum area at A and B are not under stress, since the boom is in compression, and, therefore, *pushes* on the pins (instead of *pulling* on the pins as rod BC does).

b. Determination of the Shearing Stress in Various Connections.

To determine the shearing stress in a connection such as a bolt, pin, or rivet, we first clearly show the forces exerted by the various members it connects. Thus, in the case of pin C of our example (Fig. 1.25a), we draw Fig. 1.25b, showing the 50-kN force exerted by member BC on the pin, and the equal and opposite force exerted by the bracket. Drawing now the diagram of the portion of the pin located below the plane DD' where shearing stresses occur (Fig. 1.25c), we conclude that the shear in that plane is $P = 50$ kN. Since the cross-sectional area of the pin is

$$A = \pi r^2 = \pi \left(\frac{25 \text{ mm}}{2} \right)^2 = \pi (12.5 \times 10^{-3} \text{ m})^2 = 491 \times 10^{-6} \text{ m}^2$$

we find that the average value of the shearing stress in the pin at C is

$$\tau_{\text{ave}} = \frac{P}{A} = \frac{50 \times 10^3 \text{ N}}{491 \times 10^{-6} \text{ m}^2} = 102 \text{ MPa}$$

Considering now the pin at A (Fig. 1.26), we note that it is in double shear. Drawing the free-body diagrams of the pin and of the portion of pin located between the planes DD' and EE' where shearing stresses occur, we conclude that $P = 20$ kN and that

$$\tau_{\text{ave}} = \frac{P}{A} = \frac{20 \text{ kN}}{491 \times 10^{-6} \text{ m}^2} = 40.7 \text{ MPa}$$

(a)

(b)

(c)

50 kN   50 kN $d = 25$ mm $\mathbf{P}$ 50 kN $\mathbf{F}_b$ D D' C

Fig. 1.25

(a)

(b)

(c)

40 kN $d = 25$ mm $\mathbf{F}_b$ D D' E E' $\mathbf{F}_b$ 40 kN A $\mathbf{P}$ $\mathbf{P}$ 40 kN

Fig. 1.26

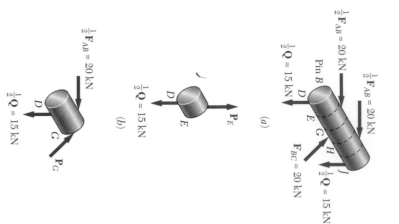

$\frac{1}{2}\mathbf{F}_{AB} = 20$ kN

$\frac{1}{2}\mathbf{F}_{AB} = 20$ kN

$\frac{1}{2}\mathbf{Q} = 15$ kN

Pin B

D E G H

$\frac{1}{2}\mathbf{Q} = 15$ kN

P_E

$\mathbf{F}_{BC} = 20$ kN

J

(a)

D E P_E

$\frac{1}{2}\mathbf{Q} = 15$ kN

(b)

$\frac{1}{2}\mathbf{F}_{AB} = 20$ kN

D G P_G

$\frac{1}{2}\mathbf{Q} = 15$ kN

(c)

Fig. 1.27

Considering the pin at B (Fig. 1.27a), we note that the pin may be divided into five portions which are acted upon by forces exerted by the boom, rod, and bracket. Considering successively the portions DE (Fig. 1.27b) and DG (Fig. 1.27c), we conclude that the shear in section E is $P_E = 15$ kN, while the shear in section G is $P_G = 25$ kN. Since the loading of the pin is symmetric, we conclude that the maximum value of the shear in pin B is $P_G = 25$ kN, and that the largest shearing stresses occur in sections G and H, where

$$\tau_{ave} = \frac{P_G}{A} = \frac{25 \text{ kN}}{491 \times 10^{-6} \text{ m}^2} = 50.9 \text{ MPa}$$

c.   Determination of the Bearing Stresses.   To determine the nominal bearing stress at A in member AB, we use formula (1.11) of Sec. 1.7. From Fig. 1.24, we have $t = 30$ mm and $d = 25$ mm. Recalling that $P = F_{AB} = 40$ kN, we have

$$\sigma_b = \frac{P}{td} = \frac{40 \text{ kN}}{(30 \text{ mm})(25 \text{ mm})} = 53.3 \text{ MPa}$$

To obtain the bearing stress in the bracket at A, we use $t = 2(25 \text{ mm}) = 50$ mm and $d = 25$ mm:

$$\sigma_b = \frac{P}{td} = \frac{40 \text{ kN}}{(50 \text{ mm})(25 \text{ mm})} = 32.0 \text{ MPa}$$

The bearing stresses at B in member AB, at B and C in member BC, and in the bracket at C are found in a similar way.

1.9. METHOD OF PROBLEM SOLUTION

You should approach a problem in mechanics of materials as you would approach an actual engineering situation. By drawing on your own experience and intuition, you will find it easier to understand and formulate the problem. Once the problem has been clearly stated, however, there is no place in its solution for your particular fancy. Your solution must be based on the fundamental principles of statics and on the principles you will learn in this course. Every step you take must be justified on that basis, leaving no room for your "intuition." After an answer has been obtained, it should be checked. Here again, you may call upon your common sense and personal experience. If not completely satisfied with the result obtained, you should carefully check your formulation of the problem, the validity of the methods used in its solution, and the accuracy of your computations.

The *statement* of the problem should be clear and precise. It should contain the given data and indicate what information is required. A simplified drawing showing all essential quantities involved should be included. The solution of most of the problems you will encounter will necessitate that you first determine the *reactions at supports and inter-*

nal forces and couples. This will require the drawing of one or several *free-body diagrams*, as was done in Sec. 1.2, from which you will write *equilibrium equations.* These equations can be solved for the unknown forces, from which the required *stresses* and *deformations* will be computed.

After the answer has been obtained, it should be *carefully checked.* Mistakes in *reasoning* can often be detected by carrying the units through your computations and checking the units obtained for the answer. For example, in the design of the rod discussed in Sec. 1.4, we found, after carrying the units through our computations, that the required diameter of the rod was expressed in millimeters, which is the correct unit for a dimension; if another unit had been found, we would have known that some mistake had been made.

Errors in *computation* will usually be found by substituting the numerical values obtained into an equation which has not yet been used and verifying that the equation is satisfied. The importance of correct computations in engineering cannot be overemphasized.

1.10. NUMERICAL ACCURACY

The accuracy of the solution of a problem depends upon two items: (1) the accuracy of the given data and (2) the accuracy of the computations performed.

The solution cannot be more accurate than the less accurate of these two items. For example, if the loading of a beam is known to be 75,000 lb with a possible error of 100 lb either way, the relative error which measures the degree of accuracy of the data is

$$\frac{100 \text{ lb}}{75,000 \text{ lb}} = 0.0013 = 0.13\%$$

In computing the reaction at one of the beam supports, it would then be meaningless to record it as 14,322 lb. The accuracy of the solution cannot be greater than 0.13%, no matter how accurate the computations are, and the possible error in the answer may be as large as $(0.13/100)(14,322 \text{ lb}) \approx 20 \text{ lb}$. The answer should be properly recorded as $14,320 \pm 20 \text{ lb}$.

In engineering problems, the data are seldom known with an accuracy greater than 0.2%. It is therefore seldom justified to write the answers to such problems with an accuracy greater than 0.2 percent. A practical rule is to use 4 figures to record numbers beginning with a "1" and 3 figures in all other cases. Unless otherwise indicated, the data given in a problem should be assumed known with a comparable degree of accuracy. A force of 40 lb, for example, should be read 40.0 lb, and a force of 15 lb should be read 15.00 lb.

Pocket calculators and computers are widely used by practicing engineers and engineering students. The speed and accuracy of these devices facilitate the numerical computations in the solution of many problems. However, students should not record more significant figures than can be justified merely because they are easily obtained. As noted above, an accuracy greater than 0.2% is seldom necessary or meaningful in the solution of practical engineering problems.

SAMPLE PROBLEM 1.1

In the hanger shown the upper portion of link ABC is $\frac{3}{8}$ in. thick and the lower portions are each $\frac{1}{4}$ in. thick. Epoxy resin is used to bond the upper and lower portions together at B. The pin at A is of $\frac{3}{8}$-in. diameter while a $\frac{1}{4}$-in.-diameter pin is used at C. Determine (a) the shearing stress in pin A, (b) the shearing stress in pin C, (c) the largest normal stress in link ABC, (d) the average shearing stress on the bonded surfaces at B, (e) the bearing stress in the link at C.

SOLUTION

Free Body: Entire Hanger. Since the link ABC is a two-force member, the reaction at A is vertical; the reaction at D is represented by its components $\mathbf{D}_x$ and $\mathbf{D}_y$. We write

$$+\!\uparrow\; \Sigma M_D = 0: \qquad (500\ \text{lb})(15\ \text{in.}) - F_{AC}(10\ \text{in.}) = 0$$
$$F_{AC} = +750\ \text{lb} \qquad F_{AC} = 750\ \text{lb}$$

a. Shearing Stress in Pin A. Since this $\frac{3}{8}$-in.-diameter pin is in single shear, we write

$$\tau_A = \frac{F_{AC}}{A} = \frac{750\ \text{lb}}{\frac{1}{4}\pi(0.375\ \text{in.})^2} \qquad \tau_A = 6790\ \text{psi} \blacktriangleleft$$

b. Shearing Stress in Pin C. Since this $\frac{1}{4}$-in.-diameter pin is in double shear, we write

$$\tau_C = \frac{\frac{1}{2}F_{AC}}{A} = \frac{375\ \text{lb}}{\frac{1}{4}\pi(0.25\ \text{in.})^2} \qquad \tau_C = 7640\ \text{psi} \blacktriangleleft$$

c. Largest Normal Stress in Link ABC. The largest stress is found where the area is smallest; this occurs at the cross section at A where the $\frac{3}{8}$-in. hole is located. We have

$$\sigma_A = \frac{F_{AC}}{A_{\text{net}}} = \frac{750\ \text{lb}}{(\frac{3}{8}\ \text{in.})(1.25\ \text{in.} - 0.375\ \text{in.})} = \frac{750\ \text{lb}}{0.328\ \text{in}^2} \qquad \sigma_A = 2290\ \text{psi} \blacktriangleleft$$

d. Average Shearing Stress at B. We note that bonding exists on both sides of the upper portion of the link and that the shear force on each side is $F_1 = (750\ \text{lb})/2 = 375\ \text{lb}$. The average shearing stress on each surface is thus

$$\tau_B = \frac{F_1}{A} = \frac{375\ \text{lb}}{(1.25\ \text{in.})(1.75\ \text{in.})} \qquad \tau_B = 171.4\ \text{psi} \blacktriangleleft$$

e. Bearing Stress in Link at C. For each portion of the link, $F_1 = 375$ lb and the nominal bearing area is $(0.25\ \text{in.})(0.25\ \text{in.}) = 0.0625\ \text{in}^2$.

$$\sigma_b = \frac{F_1}{A} = \frac{375\ \text{lb}}{0.0625\ \text{in}^2} \qquad \sigma_b = 6000\ \text{psi} \blacktriangleleft$$

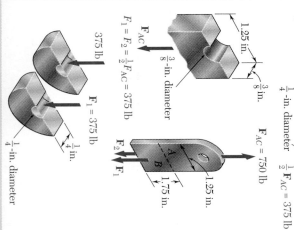

SAMPLE PROBLEM 1.2

The steel tie bar shown is to be designed to carry a tension force of magnitude $P = 120$ kN when bolted between double brackets at A and B. The bar will be fabricated from 20-mm-thick plate stock. For the grade of steel to be used the maximum allowable stresses are: $\sigma = 175$ MPa, $\tau = 100$ MPa, $\sigma_b = 350$ MPa. Design the tie bar by determining the required values of (a) the diameter d of the bolt, (b) the dimension b at each end of the bar, (c) the dimension h of the bar.

SOLUTION

a. Diameter of the Bolt. Since the bolt is in double shear, $F_1 = \frac{1}{2}P = 60$ kN.

$$\tau = \frac{F_1}{A} = \frac{60 \text{ kN}}{\frac{1}{4}\pi d^2} \qquad 100 \text{ MPa} = \frac{60 \text{ kN}}{\frac{1}{4}\pi d^2} \qquad d = 27.6 \text{ mm}$$

We will use $d = 28$ mm ◀

At this point we check the bearing stress between the 20-mm-thick plate and the 28-mm-diameter bolt.

$$\tau_b = \frac{P}{td} = \frac{120 \text{ kN}}{(0.020 \text{ m})(0.028 \text{ m})} = 214 \text{ MPa} < 350 \text{ MPa} \qquad \text{OK}$$

b. Dimension b at Each End of the Bar. We consider one of the end portions of the bar. Recalling that the thickness of the steel plate is $t = 20$ mm and that the average tensile stress must not exceed 175 MPa, we write

$$\sigma = \frac{\frac{1}{2}P}{ta} \qquad 175 \text{ MPa} = \frac{60 \text{ kN}}{(0.02 \text{ m})a} \qquad a = 17.14 \text{ mm}$$

$$b = d + 2a = 28 \text{ mm} + 2(17.14 \text{ mm}) \qquad b = 62.3 \text{ mm} ◀$$

c. Dimension h of the Bar. Recalling that the thickness of the steel plate is $t = 20$ mm, we have

$$\sigma = \frac{P}{th} \qquad 175 \text{ MPa} = \frac{120 \text{ kN}}{(0.020 \text{ m})h} \qquad h = 34.3 \text{ mm}$$

We will use $h = 35$ mm ◀

PROBLEMS

1.1 Two solid cylindrical rods AB and BC are welded together at B and loaded as shown. Knowing that $d_1 = 30$ mm and $d_2 = 50$ mm, find the average normal stress in the midsection of (a) rod AB, (b) rod BC.

Fig. P1.1 and P1.2

1.2 Two solid cylindrical rods AB and BC are welded together at B and loaded as shown. Knowing that the average normal stress must not exceed 150 MPa in either rod, determine the smallest allowable values of the diameters d_1 and d_2.

1.3 Two solid cylindrical rods AB and BC are welded together at B and loaded as shown. Knowing that $d_1 = 1.25$ in. and $d_2 = 0.75$ in., find the normal stress at the midpoint of (a) rod AB, (b) rod BC.

1.4 Two solid cylindrical rods AB and BC are welded together at B and loaded as shown. Knowing that the normal stress must not exceed 25 ksi in either rod, determine the smallest allowable values of the diameters d_1 and d_2.

Fig. P1.3 and P1.4

1.5 A strain gage located at C on the surface of bone AB indicates that the average normal stress in the bone is 3.80 MPa when the bone is subjected to two 1200-N forces as shown. Assuming the cross section of the bone at C to be annular and knowing that its outer diameter is 25 mm, determine the inner diameter of the bone's cross section at C.

Fig. P1.5

1.6 Two steel plates are to be held together by means of $\frac{1}{4}$-in.-diameter high-strength steel bolts fitting snugly inside cylindrical brass spacers. Knowing that the average normal stress must not exceed 30 ksi in the bolts and 18 ksi in the spacers, determine the outer diameter of the spacers that yields the most economical and safe design.

Fig. P1.6

Fig. P1.7

Fig. P1.8

1.7 Link *BD* consists of a single bar 30 mm wide and 12 mm thick. Knowing that each pin has a 10-mm diameter, determine the maximum value of the average normal stress in link *BD* if (*a*) $\theta = 0$, (*b*) $\theta = 90°$.

1.8 Each of the four vertical links has an 8 × 36-mm uniform rectangular cross section and each of the four pins has a 16-mm diameter. Determine the maximum value of the average normal stress in the links connecting (*a*) points *B* and *D*, (*b*) points *C* and *E*.

1.9 Two horizontal 5-kip forces are applied to pin *B* of the assembly shown. Knowing that a pin of 0.8-in. diameter is used at each connection, determine the maximum value of the average normal stress (*a*) in link *AB*, (*b*) in link *BC*.

1.10 The frame shown consists of *four* wooden members, *ABC*, *DEF*, *BE*, and *CF*. Knowing that each member has a 2 × 4-in. rectangular cross section and that each pin has a $\frac{1}{2}$-in. diameter, determine the maximum value of the average normal stress (*a*) in member *BE*, (*b*) in member *CF*.

Fig. P1.9

Fig. P1.10

1.11   For the Pratt bridge truss and loading shown, determine the average normal stress in member BE, knowing that the cross-sectional area of that member is 5.87 in.^{2}.

Fig. P1.11 and P1.12

1.12   Knowing that the average normal stress in member CE of the Pratt bridge truss shown must not exceed 21 ksi for the given loading, determine the cross-sectional area of that member which will yield the most economical and safe design. Assume that both ends of the member will be adequately reinforced.

1.13   A couple M of magnitude 1500 N · m is applied to the crank of an engine. For the position shown, determine (a) the force P required to hold the engine system in equilibrium, (b) the average normal stress in the connecting rod BC, which has a 450-mm^2 uniform cross section.

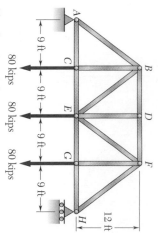

Fig. P1.13

1.14   Two hydraulic cylinders are used to control the position of the robotic arm ABC. Knowing that the control rods attached at A and D each have a 20-mm diameter and happen to be parallel in the position shown, determine the average normal stress in (a) member AE, (b) member DG.

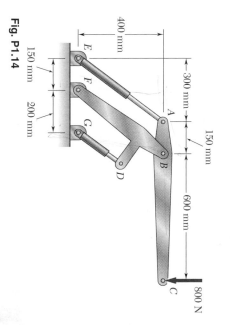

Fig. P1.14

1.15   The wooden members A and B are to be joined by plywood splice plates which will be fully glued on the surfaces in contact. As part of the design of the joint and knowing that the clearance between the ends of the members is to be 8 mm, determine the smallest allowable length L if the average shearing stress in the glue is not to exceed 800 kPa.

Fig. P1.15

1.16 Determine the diameter of the largest circular hole that can be punched into a sheet of polystyrene 6 mm thick, knowing that the force exerted by the punch is 45 kN and that a 55-MPa average shearing stress is required to cause the material to fail.

Fig. P1.17

1.17 Two wooden planks, each $\frac{7}{8}$ in. thick and 6 in. wide, are joined by the glued mortise joint shown. Knowing that the joint will fail when the average shearing stress in the glue reaches 120 psi, determine the smallest allowable length d of the cuts if the joint is to withstand an axial load of magnitude $P = 1200$ lb.

Fig. P1.18

1.18 A load **P** is applied to a steel rod supported as shown by an aluminum plate into which a 0.6-in.-diameter hole has been drilled. Knowing that the shearing stress must not exceed 18 ksi in the steel rod and 10 ksi in the aluminum plate, determine the largest load **P** that may be applied to the rod.

1.19 The axial force in the column supporting the timber beam shown is $P = 75$ kN. Determine the smallest allowable length L of the bearing plate if the bearing stress in the timber is not to exceed 3.0 MPa.

Fig. P1.19

Fig. P1.20

1.20 An axial load **P** is supported by a short W250 × 67 column of cross-sectional area $A = 8580$ mm² and is distributed to a concrete foundation by a square plate as shown. Knowing that the average normal stress in the column must not exceed 150 MPa and that the bearing stress on the concrete foundation must not exceed 12.5 MPa, determine the side a of the plate that will provide the most economical and safe design.

Fig. P1.21

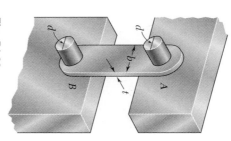

Fig. P1.22

1.21 Three wooden planks are fastened together by a series of bolts to form a column. The diameter of each bolt is $\frac{1}{2}$ in. and the inner diameter of each washer is $\frac{5}{8}$ in., which is slightly larger than the diameter of the holes in the planks. Determine the smallest allowable outer diameter d of the washers, knowing that the average normal stress in the bolts is 5 ksi and that the bearing stress between the washers and the planks must not exceed 1.2 ksi.

1.22 Link AB, of width $b = 2$ in. and thickness $t = \frac{1}{4}$ in., is used to support the end of a horizontal beam. Knowing that the average normal stress in the link is -20 ksi and that the average shearing stress in each of the two pins is 12 ksi, determine (a) the diameter d of the pins, (b) the average bearing stress in the link.

1.23 For the assembly and loading of Prob. 1.8, determine (a) the average shearing stress in the pin at B, (b) the average bearing stress at B in link BD, (c) the average bearing stress at B in member ABC, knowing that this member has a 10×50-mm uniform rectangular cross section.

1.24 For the assembly and loading of Prob. 1.8, determine (a) the average shearing stress in the pin at C, (b) the average bearing stress at C in link CE, (c) the average bearing stress at C in member ABC, knowing that this member has a 10×50-mm uniform rectangular cross section.

1.25 For the assembly and loading of Prob. 1.9, determine (a) the average shearing stress in the pin at A, (b) the average bearing stress at A in member AB.

1.26 For the assembly and loading of Prob. 1.9, determine (a) the average shearing stress in the pin at C, (b) the average bearing stress at C in member BC, (c) the average bearing stress at B in member BC.

1.27 Knowing that $\theta = 40°$ and $P = 9$ kN, determine (a) the smallest allowable diameter of the pin at B if the average shearing stress in the pin is not to exceed 120 MPa, (b) the corresponding average bearing stress in member AB at B, (c) the corresponding average bearing stress in each of the support brackets at B.

1.28 Determine the largest load $\mathbf{P}$ that may be applied at A when $\theta = 60°$, knowing that the average shearing stress in the 10-mm-diameter pin at B must not exceed 120 MPa and that the average bearing stress in member AB and in the bracket at B must not exceed 90 MPa.

Fig. P1.27 and P1.28

1.11. STRESS ON AN OBLIQUE PLANE UNDER AXIAL LOADING

In the preceding sections, axial forces exerted on a two-force member (Fig. 1.28a) were found to cause normal stresses in that member (Fig. 1.28b), while transverse forces exerted on bolts and pins (Fig. 1.29a) were found to cause shearing stresses in those connections (Fig. 1.29b). The reason such a relation was observed between axial forces and normal stresses on one hand, and transverse forces and shearing stresses on the other, was because stresses were being determined only on planes perpendicular to the axis of the member or connection. As you will see in this section, axial forces cause both normal and shearing stresses on planes which are not perpendicular to the axis of the member. Similarly, transverse forces exerted on a bolt or a pin cause both normal and shearing stresses on planes which are not perpendicular to the axis of the bolt or pin.

(a)

(b)

Fig. 1.28

(a)

(b)

Fig. 1.29

Consider the two-force member of Fig. 1.28, which is subjected to axial forces **P** and **P′**. If we pass a section forming an angle θ with a normal plane (Fig. 1.30a) and draw the free-body diagram of the portion of member located to the left of that section (Fig. 1.30b), we find from the equilibrium conditions of the free body that the distributed forces acting on the section must be equivalent to the force **P**.

Resolving **P** into components **F** and **V**, respectively normal and tangential to the section (Fig. 1.30c), we have

$$F = P \cos \theta \qquad V = P \sin \theta \qquad (1.12)$$

The force **F** represents the resultant of normal forces distributed over the section, and the force **V** the resultant of shearing forces (Fig. 1.30d). The average values of the corresponding normal and shearing stresses are obtained by dividing, respectively, F and V by the area A_θ of the section:

$$\sigma = \frac{F}{A_\theta} \qquad \tau = \frac{V}{A_\theta} \qquad (1.13)$$

Substituting for F and V from (1.12) into (1.13), and observing from Fig. 1.30c that $A_0 = A_\theta \cos \theta$, or $A_\theta = A_0/\cos \theta$, where A_0 denotes the

(a)

(b)

(c)

(d)

Fig. 1.30

area of a section perpendicular to the axis of the member, we obtain

(a) Axial loading

$$\sigma = \frac{P\cos\theta}{A_0/\cos\theta} \qquad \tau = \frac{P\sin\theta}{A_0/\cos\theta} \qquad (1.14)$$

or

$$\sigma = \frac{P}{A_0}\cos^2\theta \qquad \tau = \frac{P}{A_0}\sin\theta\cos\theta \qquad (1.15)$$

We note from the first of Eqs. (1.14) that the normal stress σ is maximum when $\theta = 0$, i.e., when the plane of the section is perpendicular to the axis of the member, and that it approaches zero as θ approaches $90°$. We check that the value of σ when $\theta = 0$ is

(b) Stresses for $\theta = 0$

$$\sigma_m = \frac{P}{A_0} \qquad (1.15)$$

as we found earlier in Sec. 1.3. The second of Eqs. (1.14) shows that the shearing stress τ is zero for $\theta = 0$ and $\theta = 90°$, and that for $\theta = 45°$ it reaches its maximum value

(c) Stresses for $\theta = 45°$

$$\tau_m = \frac{P}{A_0}\sin 45°\cos 45° = \frac{P}{2A_0} \qquad (1.16)$$

The first of Eqs. (1.14) indicates that, when $\theta = 45°$, the normal stress σ' is also equal to $P/2A_0$:

$$\sigma' = \frac{P}{A_0}\cos^2 45° = \frac{P}{2A_0} \qquad (1.17)$$

(d) Stresses for $\theta = -45°$

Fig. 1.31

The results obtained in Eqs. (1.15), (1.16), and (1.17) are shown graphically in Fig. 1.31. We note that the same loading may produce either a normal stress $\sigma_m = P/A_0$ and no shearing stress (Fig. 1.31b), or a normal and a shearing stress of the same magnitude $\sigma' = \tau_m = P/2A_0$ (Fig. 1.31c and d), depending upon the orientation of the section.

1.12. STRESS UNDER GENERAL LOADING CONDITIONS; COMPONENTS OF STRESS

The examples of the previous sections were limited to members under axial loading and connections under transverse loading. Most structural members and machine components are under more involved loading conditions.

Consider a body subjected to several loads P_1, P_2, etc. (Fig. 1.32). To understand the stress condition created by these loads at some point Q within the body, we shall first pass a section through Q, using a plane parallel to the yz plane. The portion of the body to the left of the section is subjected to some of the original loads, and to normal and shearing forces distributed over the section. We shall denote by $\Delta\mathbf{F}^x$ and $\Delta\mathbf{V}^x$, respectively, the normal and the shearing forces acting on a small

Fig. 1.32

Fig. 1.33

(a)

(b)

Fig. 1.34

Fig. 1.35

area ΔA surrounding point Q (Fig. 1.33a). Note that the superscript x is used to indicate that the forces $\Delta\mathbf{F}^x$ and $\Delta\mathbf{V}^x$ act on a surface perpendicular to the x axis. While the normal force $\Delta\mathbf{F}^x$ has a well-defined direction, the shearing force $\Delta\mathbf{V}^x$ may have any direction in the plane of the section. We therefore resolve $\Delta\mathbf{V}^x$ into two component forces, $\Delta\mathbf{V}_y^x$ and $\Delta\mathbf{V}_z^x$, in directions parallel to the y and z axes, respectively (Fig. 1.33 b). Dividing now the magnitude of each force by the area ΔA, and letting ΔA approach zero, we define the three stress components shown in Fig. 1.34:

$$\sigma_x = \lim_{\Delta A \to 0} \frac{\Delta F^x}{\Delta A}$$

$$\tau_{xy} = \lim_{\Delta A \to 0} \frac{\Delta V_y^x}{\Delta A} \qquad \tau_{xz} = \lim_{\Delta A \to 0} \frac{\Delta V_z^x}{\Delta A} \qquad (1.18)$$

We note that the first subscript in σ_x, τ_{xy}, and τ_{xz} is used to indicate that the stresses under consideration are exerted *on a surface perpendicular to the x axis.* The second subscript in τ_{xy} and τ_{xz} identifies *the direction of the component.* The normal stress σ_x is positive if the corresponding arrow points in the positive x direction, i.e., if the body is in tension, and negative otherwise. Similarly, the shearing stress components τ_{xy} and τ_{xz} are positive if the corresponding arrows point, respectively, in the positive y and z directions.

The above analysis may also be carried out by considering the portion of body located to the right of the vertical plane through Q (Fig. 1.35). The same magnitudes, but opposite directions, are obtained for the normal and shearing forces $\Delta\mathbf{F}^x$, $\Delta\mathbf{V}_y^x$, and $\Delta\mathbf{V}_z^x$. Therefore, the same values are also obtained for the corresponding stress components, but since the section in Fig. 1.35 now faces the *negative x axis*, a positive sign for σ_x will indicate that the corresponding arrow points *in the negative x direction.* Similarly, positive signs for τ_{xy} and τ_{xz} will indicate that the corresponding arrows point, respectively, in the negative y and z directions, as shown in Fig. 1.35.

Passing a section through Q parallel to the zx plane, we define in the same manner the stress components, σ_y, τ_{yz}, and τ_{yx}. Finally, a section through Q parallel to the xy plane yields the components σ_z, τ_{zx}, and τ_{zy}.

To facilitate the visualization of the stress condition at point Q, we shall consider a small cube of side a centered at Q and the stresses exerted on each of the six faces of the cube (Fig. 1.36). The stress components shown in the figure are σ_x, σ_y, and σ_z, which represent the normal stress on faces respectively perpendicular to the x, y, and z axes, and the six shearing stress components τ_{xy}, τ_{xz}, etc. We recall that, according to the definition of the shearing stress components, τ_{xy} represents the y component of the shearing stress exerted on the face perpendicular to the x axis, while τ_{yx} represents the x component of the shearing stress exerted on the face perpendicular to the y axis. Note that only three shearing stress components act on the hidden faces. While the stresses acting on the faces of the cube differ slightly from the stresses acting at Q, the error involved is small and vanishes as side a of the cube approaches zero.

Important relations among the shearing stress components will now be derived. Let us consider the free-body diagram of the small cube centered at point Q (Fig. 1.37). The normal and shearing forces acting on the various faces of the cube are obtained by multiplying the corresponding stress components by the area ΔA of each face. We first write the following three equilibrium equations:

$$\Sigma F_x = 0 \qquad \Sigma F_y = 0 \qquad \Sigma F_z = 0 \qquad (1.19)$$

Since forces equal and opposite to the forces actually shown in Fig. 1.37 are acting on the hidden faces of the cube, it is clear that Eqs. (1.19) are satisfied. Considering now the moments of the forces about axes Qx', Qy', and Qz' drawn from Q in directions respectively parallel to the x, y, and z axes, we write the three additional equations

$$\Sigma M_x = 0 \qquad \Sigma M_y = 0 \qquad \Sigma M_z = 0 \qquad (1.20)$$

Using a projection on the $x'y'$ plane (Fig. 1.38), we note that the only forces with moments about the z axis different from zero are the shearing forces. These forces form two couples, one of counterclockwise (positive) moment $(\tau_{xy} \, \Delta A)a$, the other of clockwise (negative) moment $-(\tau_{yx} \, \Delta A)a$. The last of the three Eqs. (1.20) yields, therefore,

$$+\circlearrowleft \Sigma M_z = 0: \qquad (\tau_{xy} \, \Delta A)a - (\tau_{yx} \, \Delta A)a = 0$$

from which we conclude that

$$\tau_{xy} = \tau_{yx} \qquad (1.21)$$

The relation obtained shows that the y component of the shearing stress exerted on a face perpendicular to the x axis is equal to the x compo-

Fig. 1.36

Fig. 1.37

Fig. 1.38

nent of the shearing stress exerted on a face perpendicular to the y axis. From the remaining two equations (1.20), we derive in a similar manner the relations

$$\tau_{yz} = \tau_{zy} \qquad \tau_{zx} = \tau_{xz} \tag{1.22}$$

Fig. 1.39

We conclude from Eqs. (1.21) and (1.22) that only six stress components are required to define the condition of stress at a given point Q, instead of nine as originally assumed. These six components are $\sigma_x, \sigma_y, \sigma_z, \tau_{xy}, \tau_{yz}$, and τ_{zx}. We also note that, at a given point, *shear cannot take place in one plane only*; an equal shearing stress must be exerted on another plane perpendicular to the first one. For example, considering again the bolt of Fig. 1.29 and a small cube at the center Q of the bolt (Fig. 1.39a), we find that shearing stresses of equal magnitude must be exerted on the two horizontal faces of the cube and on the two faces which are perpendicular to the forces **P** and **P'** (Fig. 1.39b).

Before concluding our discussion of stress components, let us consider again the case of a member under axial loading. If we consider a small cube with faces respectively parallel to the faces of the member and recall the results obtained in Sec. 1.11, we find that the conditions of stress in the member may be described as shown in Fig. 1.40a; the only stresses are normal stresses σ_x exerted on the faces of the cube which are perpendicular to the x axis. However, if the small cube is rotated by 45° about the z axis so that its new orientation matches the orientation of the sections considered in Fig. 1.31c and d, we conclude that normal and shearing stresses of equal magnitude are exerted on four faces of the cube (Fig. 1.40b). We thus observe that the same loading condition may lead to different interpretations of the stress situation at a given point, depending upon the orientation of the element considered. More will be said about this in Chap 7.

Fig. 1.40

1.13. DESIGN CONSIDERATIONS

In the preceding sections you learned to determine the stresses in rods, bolts, and pins under simple loading conditions. In later chapters you will learn to determine stresses in more complex situations. In engineering applications, however, the determination of stresses is seldom an end in itself. Rather, the knowledge of stresses is used by engineers to assist in their most important task, namely, the design of structures and machines that will safely and economically perform a specified function.

a. Determination of the Ultimate Strength of a Material. An important element to be considered by a designer is how the material that has been selected will behave under a load. For a given material this is determined by performing specific tests on prepared samples of the material. For example, a test specimen of steel may be prepared and placed in a laboratory testing machine to be subjected to a known centric axial tensile force, as described in Sec. 2.3. As the magnitude of the force is increased, various changes in the specimen are measured, for example, changes in its length and its diameter. Eventually the largest force which may be applied to the specimen is reached, and the specimen either breaks or begins to carry less load. This largest force is called the *ultimate load* for the test specimen and is denoted by P_U. Since the applied load is centric, we may divide the ultimate load by the original cross-sectional area of the rod to obtain the *ultimate normal stress* of the material used. This stress, also known as the *ultimate strength in tension* of the material, is

$$\sigma_U = \frac{P_U}{A} \tag{1.23}$$

Several test procedures are available to determine the *ultimate shearing stress*, or *ultimate strength in shear*, of a material. The one most commonly used involves the twisting of a circular tube (Sec. 3.5). A more direct, if less accurate, procedure consists in clamping a rectangular or round bar in a shear tool (Fig. 1.41) and applying an increasing load P until the ultimate load P_U for single shear is obtained. If the free end of the specimen rests on both of the hardened dies (Fig. 1.42), the ultimate load for double shear is obtained. In either case, the ultimate shearing stress τ_U is obtained by dividing the ultimate load by the total area over which shear has taken place. We recall that, in the case of single shear, this area is the cross-sectional area A of the specimen, while in double shear it is equal to twice the cross-sectional area.

b.   Allowable Load and Allowable Stress; Factor of Safety.   The maximum load that a structural member or a machine component will be allowed to carry under normal conditions of utilization is considerably smaller than the ultimate load. This smaller load is referred to as the *allowable load* and, sometimes, as the *working load* or *design load*. Thus, only a fraction of the ultimate-load capacity of the member is utilized when the allowable load is applied. The remaining portion of the load-carrying capacity of the member is kept in reserve to assure its safe performance. The ratio of the ultimate load to the allowable load is used to define the *factor of safety*.[†] We have

$$\text{Factor of safety} = F.S. = \frac{\text{ultimate load}}{\text{allowable load}} \tag{1.24}$$

An alternative definition of the factor of safety is based on the use of stresses:

$$\text{Factor of safety} = F.S. = \frac{\text{ultimate stress}}{\text{allowable stress}} \tag{1.25}$$

The two expressions given for the factor of safety in Eqs. (1.24) and (1.25) are identical when a linear relationship exists between the load and the stress. In most engineering applications, however, this relationship ceases to be linear as the load approaches its ultimate value, and the factor of safety obtained from Eq. (1.25) does not provide a

Fig. 1.41

Fig. 1.42

[†]In some fields of engineering, notably aeronautical engineering, the *margin of safety* is used in place of the factor of safety. The margin of safety is defined as the factor of safety minus one; that is, margin of safety = F.S. − 1.00.

true assessment of the safety of a given design. Nevertheless, the *allowable-stress method* of design, based on the use of Eq. (1.25), is widely used.

c. Selection of an Appropriate Factor of Safety. The selection of the factor of safety to be used for various applications is one of the most important engineering tasks. On the one hand, if a factor of safety is chosen too small, the possibility of failure becomes unacceptably large; on the other hand, if a factor of safety is chosen unnecessarily large, the result is an uneconomical or nonfunctional design. The choice of the factor of safety that is appropriate for a given design application requires engineering judgment based on many considerations, such as the following:

1. *Variations that may occur in the properties of the member under consideration.* The composition, strength, and dimensions of the member are all subject to small variations during manufacture. In addition, material properties may be altered and residual stresses introduced through heating or deformation that may occur during manufacture, storage, transportation, or construction.

2. *The number of loadings that may be expected during the life of the structure or machine.* For most materials the ultimate stress decreases as the number of load applications is increased. This phenomenon is known as *fatigue* and, if ignored, may result in sudden failure (see Sec. 2.7).

3. *The type of loadings that are planned for in the design, or that may occur in the future.* Very few loadings are known with complete accuracy—most design loadings are engineering estimates. In addition, future alterations or changes in usage may introduce changes in the actual loading. Larger factors of safety are also required for dynamic, cyclic, or impulsive loadings.

4. *The type of failure that may occur.* Brittle materials fail suddenly, usually with no prior indication that collapse is imminent. On the other hand, ductile materials, such as structural steel, normally undergo a substantial deformation called *yielding* before failing, thus providing a warning that overloading exists. However, most buckling or stability failures are sudden, whether the material is brittle or not. When the possibility of sudden failure exists, a larger factor of safety should be used than when failure is preceded by obvious warning signs.

5. *Uncertainty due to methods of analysis.* All design methods are based on certain simplifying assumptions which result in calculated stresses being approximations of actual stresses.

6. *Deterioration that may occur in the future because of poor maintenance or because of unpreventable natural causes.* A larger factor of safety is necessary in locations where conditions such as corrosion and decay are difficult to control or even to discover.

7. *The importance of a given member to the integrity of the whole structure.* Bracing and secondary members may in many cases be designed with a factor of safety lower than that used for primary members.

In addition to the above considerations, there is the additional consideration concerning the risk to life and property that a failure would produce. Where a failure would produce no risk to life and only minimal risk to property, the use of a smaller factor of safety can be considered. Finally, there is the practical consideration that, unless a careful design with a nonexcessive factor of safety is used, a structure or machine might not perform its design function. For example, high factors of safety may have an unacceptable effect on the weight of an aircraft.

For the majority of structural and machine applications, factors of safety are specified by design specifications or building codes written by committees of experienced engineers working with professional societies, with industries, or with federal, state, or city agencies. Examples of such design specifications and building codes are

1. *Steel:* American Institute of Steel Construction, Specifications for Structural Steel Buildings

2. *Concrete:* American Concrete Institute, Building Code Requirement for Structural Concrete

3. *Timber:* American Forest and Paper Association, National Design Specification for Wood Construction

4. *Highway bridges:* American Association of State Highway Officials, Standard Specifications for Highway Bridges

*d. Load and Resistance Factor Design. As we saw above, the allowable-stress method requires that all the uncertainties associated with the design of a structure or machine element be grouped into a single factor of safety. An alternative method of design, which is gaining acceptance chiefly among structural engineers, makes it possible through the use of three different factors to distinguish between the uncertainties associated with the structure itself and those associated with the load it is designed to support. This method, referred to as *Load and Resistance Factor Design (LRFD),* further allows the designer to distinguish between uncertainties associated with the *live load, P_L,* that is, with the load to be supported by the structure, and the *dead load, P_D,* that is, with the weight of the portion of structure contributing to the total load.

When this method of design is used, the *ultimate load, P_U,* of the structure, that is, the load at which the structure ceases to be useful, should first be determined. The proposed design is then acceptable if the following inequality is satisfied:

$$\gamma_D P_D + \gamma_L P_L \le \phi P_U \qquad (1.26)$$

The coefficient ϕ is referred to as the *resistance factor;* it accounts for the uncertainties associated with the structure itself and will normally be less than 1. The coefficients γ_D and γ_L are referred to as the *load factors;* they account for the uncertainties associated, respectively, with the dead load and live load and will normally be greater than 1, with γ_L generally larger than γ_D. While a few examples or assigned problems using LRFD are included in this chapter and in Chaps. 5 and 10, the allowable-stress method of design will be used in this text.

SAMPLE PROBLEM 1.3

Two forces are applied to the bracket *BCD* as shown. (*a*) Knowing that the control rod *AB* is to be made of a steel having an ultimate normal stress of 600 MPa, determine the diameter of the rod for which the factor of safety with respect to failure will be 3.3. (*b*) The pin at *C* is to be made of a steel having an ultimate shearing stress of 350 MPa. Determine the diameter of the pin *C* for which the factor of safety with respect to shear will also be 3.3. (*c*) Determine the required thickness of the bracket supports at *C* knowing that the allowable bearing stress of the steel used is 300 MPa.

SOLUTION

Free Body: Entire Bracket. The reaction at *C* is represented by its components $\mathbf{C}_x$ and $\mathbf{C}_y$.

$+\circlearrowleft \Sigma M_C = 0$: $\quad P(0.6\text{ m}) - (50\text{ kN})(0.3\text{ m}) - (15\text{ kN})(0.6\text{ m}) = 0 \qquad P = 40\text{ kN}$

$\Sigma F_x = 0$: $\qquad C_x = 40\text{ k}$

$\Sigma F_y = 0$: $\qquad C_y - 65\text{ kN}$ $\qquad C = \sqrt{C_x^2 + C_y^2} = 76.3\text{ kN}$

***a*. Control Rod *AB*.** Since the factor of safety is to be 3.3, the allowable stress is

$$\sigma_{\text{all}} = \frac{\sigma_U}{F.S.} = \frac{600\text{ MPa}}{3.3} = 181.8\text{ MPa}$$

For $P = 40$ kN the cross-sectional area required is

$$A_{\text{req}} = \frac{P}{\sigma_{\text{all}}} = \frac{40\text{ kN}}{181.8\text{ MPa}} = 220 \times 10^{-6}\text{ m}^2$$

$$A_{\text{req}} = \frac{\pi}{4}d_{AB}^2 = 220 \times 10^{-6}\text{ m}^2 \qquad d_{AB} = 16.74\text{ mm} \;\blacktriangleleft$$

***b*. Shear in Pin *C*.** For a factor of safety of 3.3, we have

$$\tau_{\text{all}} = \frac{\tau_U}{F.S.} = \frac{350\text{ MPa}}{3.3} = 106.1\text{ MPa}$$

Since the pin is in double shear, we write

$$A_{\text{req}} = \frac{C/2}{\tau_{\text{all}}} = \frac{(76.3\text{ kN})/2}{106.1\text{ MPa}} = 360\text{ mm}^2$$

$$A_{\text{req}} = \frac{\pi}{4}d_C^2 = 360\text{ mm}^2 \qquad d_C = 21.4\text{ mm} \qquad \text{Use: } d_C = 22\text{ mm} \;\blacktriangleleft$$

The next larger size pin available is of 22-mm diameter and should be used.

***c*. Bearing at *C*.** Using $d = 22$ mm, the nominal bearing area of each bracket is 22*t*. Since the force carried by each bracket is *C*/2 and the allowable bearing stress is 300 MPa, we write

$$A_{\text{req}} = \frac{C/2}{\sigma_{\text{all}}} = \frac{(76.3\text{ kN})/2}{300\text{ MPa}} = 127.2\text{ mm}^2$$

Thus $22t = 127.2$ $\qquad t = 5.78$ mm $\qquad$ Use: $t = 6$ mm $\;\blacktriangleleft$

SAMPLE PROBLEM 1.4

The rigid beam BCD is attached by bolts to a control rod at B, to a hydraulic cylinder at C, and to a fixed support at D. The diameters of the bolts used are: $d_B = d_D = \frac{3}{8}$ in., $d_C = \frac{1}{2}$ in. Each bolt acts in double shear and is made of a steel for which the ultimate shearing stress is $\tau_U = 40$ ksi. The control rod AB has a diameter $d_A = \frac{7}{16}$ in. and is made of a steel for which the ultimate tensile stress is $\sigma_U = 60$ ksi. If the minimum factor of safety is to be 3.0 for the entire unit, determine the largest upward force which may be applied by the hydraulic cylinder at C.

SOLUTION

The factor of safety with respect to failure must be 3.0 or more in each of the three bolts and in the control rod. These four independent criteria will be considered separately.

Free Body: Beam BCD.   We first determine the force at C in terms of the force at B and in terms of the force at D.

$+\curvearrowleft \Sigma M_D = 0$: $B(14 \text{ in.}) - C(8 \text{ in.}) = 0$ $C = 1.750B$ (1)
$+\curvearrowleft \Sigma M_B = 0$: $-D(14 \text{ in.}) + C(6 \text{ in.}) = 0$ $C = 2.33D$ (2)

Control Rod.   For a factor of safety of 3.0 we have

$$\sigma_{all} = \frac{\sigma_U}{F.S.} = \frac{60 \text{ ksi}}{3.0} = 20 \text{ ksi}$$

The allowable force in the control rod is

$$B = \sigma_{all}(A) = (20 \text{ ksi}) \tfrac{1}{4}\pi (\tfrac{7}{16} \text{ in.})^2 = 3.01 \text{ kips}$$

Using Eq. (1) we find the largest permitted value of C:

$$C = 1.750B = 1.750(3.01 \text{ kips}) \qquad C = 5.27 \text{ kips} \;\triangle$$

Bolt at B. $\tau_{all} = \tau_U/F.S. = (40 \text{ ksi})/3 = 13.33$ ksi. Since the bolt is in double shear, the allowable magnitude of the force $\mathbf{B}$ exerted on the bolt is

$$B = 2F_1 = 2(\tau_{all}A) = 2(13.33 \text{ ksi})(\tfrac{1}{4}\pi)(\tfrac{3}{8} \text{ in.})^2 = 2.94 \text{ kips}$$

From Eq. (1): $C = 1.750B = 1.750(2.94 \text{ kips})$ $C = 5.15$ kips $\triangle$

Bolt at D.   Since this bolt is the same as bolt B, the allowable force is $D = B = 2.94$ kips. From Eq. (2):

$$C = 2.33D = 2.33(2.94 \text{ kips}) \qquad C = 6.85 \text{ kips} \;\triangle$$

Bolt at C.   We again have $\tau_{all} = 13.33$ ksi and write

$$C = 2F_2 = 2(\tau_{all}A) = 2(13.33 \text{ ksi})(\tfrac{1}{4}\pi)(\tfrac{1}{2} \text{ in.})^2 \qquad C = 5.23 \text{ kips} \;\triangle$$

Summary.   We have found separately four maximum allowable values of the force C. In order to satisfy all these criteria we must choose the smallest value, namely:

$$C = 5.15 \text{ kips} \quad \blacktriangle$$

PROBLEMS

1.29 The 6-kN load **P** is supported by two wooden members of 75×125-mm uniform rectangular cross section that are joined by the simple glued scarf splice shown. Determine the normal and shearing stresses in the glued splice.

Fig. P1.29 and P1.30

1.30 Two wooden members of 75×125-mm uniform rectangular cross section are joined by the simple glued scarf splice shown. Knowing that the maximum allowable tensile stress in the glued splice is 500 kPa, determine (*a*) the largest load **P'** which can be safely supported, (*b*) the corresponding shearing stress in the splice.

1.31 Two wooden members of 3×6-in. uniform rectangular cross section are joined by the simple glued scarf splice shown. Knowing that the maximum allowable shearing stress in the glued splice is 90 psi, determine (*a*) the largest load **P** which can be safely applied, (*b*) the corresponding tensile stress in the splice.

Fig. P1.31 and P1.32

1.32 Two wooden members of 3×6-in. uniform rectangular cross section are joined by the simple glued scarf splice shown. Knowing that $P = 2400$ lb, determine the normal and shearing stresses in the glued splice.

1.33 A centric load **P** is applied to the granite block shown. Knowing that the resulting maximum value of the shearing stress in the block is 2.5 ksi, determine (*a*) the magnitude of **P**, (*b*) the orientation of the surface on which the maximum shearing stress occurs, (*c*) the normal stress exerted on that surface, (*d*) the maximum value of the normal stress in the block.

1.34 A 240-kip load **P** is applied to the granite block shown. Determine the resulting maximum value of (*a*) the normal stress, (*b*) the shearing stress. Specify the orientation of the plane on which each of these maximum values occurs.

Fig. *P1.33 and P1.34*

Fig. P1.35 and P1.36

1.35 A steel pipe of 300-mm outer diameter is fabricated from 6-mm-thick plate by welding along a helix which forms an angle of 25° with a plane perpendicular to the axis of the pipe. Knowing that a 250-kN axial force **P** is applied to the pipe, determine the normal and shearing stresses in directions respectively normal and tangential to the weld.

1.36 A steel pipe of 300-mm outer diameter is fabricated from 6-mm-thick plate by welding along a helix which forms an angle of 25° with a plane perpendicular to the axis of the pipe. Knowing that the maximum allowable normal and shearing stresses in directions respectively normal and tangential to the weld are $\sigma = 50$ MPa and $\tau = 30$ MPa, determine the magnitude P of the largest axial force that can be applied to the pipe.

1.37 Link BC is 6 mm thick, has a width $w = 25$ mm, and is made of a steel with a 480-MPa ultimate strength in tension. What was the safety factor used if the structure shown was designed to support a 16-kN load **P?**

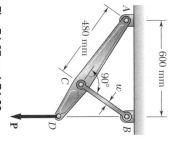

Fig. P1.37 and P1.38

1.38 Link BC is 6 mm thick and is made of a steel with a 450-MPa ultimate strength in tension. What should be its width w if the structure shown is being designed to support a 20-kN load **P** with a factor of safety of 3?

1.39 Member ABC, which is supported by a pin and bracket at C and a cable BD, was designed to support the 4-kip load **P** as shown. Knowing that the ultimate load for cable BD is 25 kips, determine the factor of safety with respect to cable failure.

1.40 Knowing that the ultimate load for cable BD is 25 kips and that a factor of safety of 3.2 with respect to cable failure is required, determine the magnitude of the largest force **P** which can be safely applied as shown to member ABC.

Fig. P1.39 and P1.40

1.41 Members AB and AC of the truss shown consist of bars of square cross section made of the same alloy. It is known that a 20-mm-square bar of the same alloy was tested to failure and that an ultimate load of 120 kN was recorded. If bar AB has a 15-mm-square cross section, determine (a) the factor of safety for bar AB, (b) the dimensions of the cross section of bar AC if it is to have the same factor of safety as bar AB.

1.42 Members AB and AC of the truss shown consist of bars of square cross section made of the same alloy. It is known that a 20-mm-square bar of the same alloy was tested to failure and that an ultimate load of 120 kN was recorded. If a factor of safety of 3.2 is to be achieved for both bars, determine the required dimensions of the cross section of (a) bar AB, (b) bar AC.

Fig. P1.41 and P1.42

1.43 The two wooden members shown, which support a 20-kN load, are joined by plywood splices fully glued on the surfaces in contact. The ultimate shearing stress in the glue is 2.8 MPa and the clearance between the members is 8 mm. Determine the factor of safety, knowing that the length of each splice is $L = 200$ mm.

Fig. P1.43

1.44 For the joint and loading of Prob. 1.43, determine the required length L of each splice if a factor of safety of 3.5 is to be achieved.

1.45 Three $\frac{3}{4}$-in.-diameter steel bolts are to be used to attach the steel plate shown to a wooden beam. Knowing that the plate will support a 24-kip load and that the ultimate shearing stress for the steel used is 52 ksi, determine the factor of safety for this design.

1.46 Three steel bolts are to be used to attach the steel plate shown to a wooden beam. Knowing that the plate will support a 24-kip load, that the ultimate shearing stress for the steel used is 52 ksi, and that a factor of safety of 3.37 is desired, determine the required diameter of the bolts.

Fig. P1.45 and P1.46

1.47 A load **P** is supported as shown by a steel pin that has been inserted in a short wooden member hanging from the ceiling. The ultimate strength of the wood used is 60 MPa in tension and 7.5 MPa in shear, while the ultimate strength of the steel is 150 MPa in shear. Knowing that the diameter of the pin is $d = 16$ mm and that the magnitude of the load is $P = 20$ kN, determine (a) the factor of safety for the pin, (b) the required values of b and c if the factor of safety for the wooden member is to be the same as that found in part a for the pin.

1.48 For the support of Prob. 1.47, knowing that $b = 40$ mm, $c = 55$ mm, and $d = 12$ mm, determine the allowable load **P** if an overall factor of safety of 3.2 is desired.

Fig. P1.47

Fig. P1.53

Fig. P1.51 and P1.52

Fig. P1.49 and P1.50

1.49 In the structure shown, an 8-mm-diameter pin is used at A, and 12-mm-diameter pins are used at B and D. Knowing that the ultimate shearing stress is 100 MPa at all connections and that the ultimate normal stress is 250 MPa in each of the two links joining B and D, determine the allowable load **P** if an overall factor of safety of 3.0 is desired.

1.50 In an alternative design for the structure of Prob. 1.49, a pin of 10-mm-diameter is to be used at A. Assuming that all other specifications remain unchanged, determine the allowable load **P** if an overall factor of safety of 3.0 is desired.

1.51 Each of the steel links AB and CD is connected to a support and to member BCE by $\frac{1}{2}$-in.-diameter steel pins acting in single shear. Knowing that the ultimate shearing stress is 24 ksi for the steel used in the pins and that the ultimate normal stress is 60 ksi for the steel used in the links, determine the allowable load **P** if an overall factor of safety of 3.2 is desired. (Note that the links are not reinforced around the pin holes.)

1.52 An alternative design is being considered to support member BCE of Prob. 1.51, in which link CD will be replaced by two links, each of $\frac{1}{8} \times 1$-in. cross section, causing the pins at C and D to be in double shear. Assuming that all other specifications remain unchanged, determine the allowable load **P** if an overall factor of safety of 3.2 is desired.

1.53 Each of the two vertical links CF connecting the two horizontal members AD and EG has a 10×40-mm uniform rectangular cross section and is made of a steel with an ultimate strength in tension of 400 MPa, while each of the pins at C and F has a 20-mm diameter and is made of a steel with an ultimate strength in shear of 150 MPa. Determine the overall factor of safety for the links CF and the pins connecting them to the horizontal members.

1.54 Solve Prob. 1.53, assuming that the pins at C and F have been replaced by pins with a 30-mm diameter.

Fig. P1.55

1.55 A steel plate $\frac{5}{16}$ in. thick is embedded in a horizontal concrete slab and is used to anchor a high-strength vertical cable as shown. The diameter of the hole in the plate is $\frac{3}{4}$ in., the ultimate strength of the steel used is 36 ksi, and the ultimate bonding stress between plate and concrete is 300 psi. Knowing that a factor of safety of 3.60 is desired when $P = 2.5$ kips, determine (a) the required width a of the plate, (b) the minimum depth b to which a plate of that width should be embedded in the concrete slab. (Neglect the normal stresses between the concrete and the lower end of the plate.)

1.56 Determine the factor of safety for the cable anchor of Prob. 1.55 when $P = 3$ kips, knowing that $a = 2$ in. and $b = 7.5$ in.

Fig. P1.57

*1.57 A 40-kg platform is attached to the end B of a 50-kg wooden beam AB, which is supported as shown by a pin at A and by a slender steel rod BC with a 12-kN ultimate load. (a) Using the Load and Resistance Factor Design method with a resistance factor $\phi = 0.90$ and load factors $\gamma_D = 1.25$ and $\gamma_L = 1.6$, determine the largest load which can be safely placed on the platform. (b) What is the corresponding conventional factor of safety for rod BC?

Fig. P1.58

*1.58 The Load and Resistance Factor Design method is to be used to select the two cables which will raise and lower a platform supporting two window washers. The platform weighs 160 lb and each of the window washers is assumed to weigh 195 lb with equipment. Since these workers are free to move on the platform, 75% of their total weight and of the weight of their equipment will be used as the design live load of each cable. (a) Assuming a resistance factor $\phi = 0.85$ and load factors $\gamma_D = 1.2$ and $\gamma_L = 1.5$, determine the required minimum ultimate load of one cable. (b) What is the conventional factor of safety for the selected cables?

REVIEW AND SUMMARY

FOR CHAPTER 1

This chapter was devoted to the concept of stress and to an introduction to the methods used for the analysis and design of machines and load-bearing structures.

Section 1.2 presented a short review of the methods of statics and of their application to the determination of the reactions exerted by its supports on a simple structure consisting of pin-connected members. Emphasis was placed on the use of a *free-body-diagram* to obtain equilibrium equations which were solved for the unknown reactions. Free-body diagrams were also used to find the internal forces in the various members of the structure.

The concept of *stress* was first introduced in Sec. 1.3 by considering a two-force member under an *axial loading*. The *normal stress* in that member was obtained by dividing the magnitude P of the load by the cross-sectional area A of the member (Fig. 1.8a). We wrote

Fig. 1.8a

Axial loading. Normal stress

$$\sigma = \frac{P}{A} \qquad (1.5)$$

Section 1.4 was devoted to a short discussion of the two principal tasks of an engineer, namely, the *analysis* and the *design* of structures and machines.

As noted in Sec. 1.5, the value of σ obtained from Eq. (1.5) represents the *average stress* over the section rather than the stress at a specific point Q of the section. Considering a small area ΔA surrounding Q and the magnitude ΔF of the force exerted on ΔA, we defined the stress at point Q as

$$\sigma = \lim_{\Delta A \to 0} \frac{\Delta F}{\Delta A} \qquad (1.6)$$

In general, the value obtained for the stress σ at point Q is different from the value of the average stress given by formula (1.5) and is found to vary across the section. However, this variation is small in any section away from the points of application of the loads. In practice, therefore, the distribution of the normal stresses in an axially loaded member is assumed to be *uniform*, except in the immediate vicinity of the points of application of the loads.

However, for the distribution of stresses to be uniform in a given section, it is necessary that the line of action of the loads $\mathbf{P}$ and $\mathbf{P'}$ pass through the centroid C of the section. Such a loading is called a *centric* axial loading. In the case of an *eccentric* axial loading, the distribution of stresses is *not* uniform. Stresses in members subjected to an eccentric axial loading will be discussed in Chap. 4.

Transverse Forces. Shearing stress

Fig. 1.16a

Single and double shear

Fig. 1.18

Fig. 1.20

Bearing stress

Fig. 1.22

Method of solution

When equal and opposite *transverse forces* **P** and **P′** of magnitude P are applied to a member AB (Fig. 1.16*a*), *shearing stresses* τ are created over any section located between the points of application of the two forces [Sec. 1.6]. These stresses vary greatly across the section and their distribution *cannot* be assumed uniform. However dividing the magnitude P—referred to as the *shear* in the section—by the cross-sectional area A, we defined the *average shearing stress* over the section:

$$\tau_{ave} = \frac{P}{A} \qquad (1.8)$$

Shearing stresses are found in bolts, pins, or rivets connecting two structural members or machine components. For example, in the case of bolt CD (Fig. 1.18), which is in *single shear*, we wrote

$$\tau_{ave} = \frac{P}{A} = \frac{F}{A} \qquad (1.9)$$

while, in the case of bolts EG and HJ (Fig. 1.20), which are both in *double shear*, we had

$$\tau_{ave} = \frac{P}{A} = \frac{F/2}{A} = \frac{F}{2A} \qquad (1.10)$$

Bolts, pins, and rivets also create stresses in the members they connect, along the *bearing surface*, or surface of contact [Sec. 1.7]. The bolt CD of Fig. 1.18, for example, creates stresses on the semi-cylindrical surface of plate A with which it is in contact (Fig. 1.22). Since the distribution of these stresses is quite complicated, one uses in practice an average nominal value σ_b of the stress, called *bearing stress*, obtained by dividing the load P by the area of the rectangle representing the projection of the bolt on the plate section. Denoting by t the thickness of the plate and by d the diameter of the bolt, we wrote

$$\sigma_b = \frac{P}{A} = \frac{P}{td} \qquad (1.11)$$

In Sec. 1.8, we applied the concept introduced in the previous sections to the analysis of a simple structure consisting of two pin-connected members supporting a given load. We determined successively the normal stresses in the two members, paying special attention to their narrowest sections, the shearing stresses in the various pins, and the bearing stress at each connection.

The method you should use in solving a problem in mechanics of materials was described in Sec. 1.9. Your solution should begin with a clear and precise *statement* of the problem. You will then *draw* one or several *free-body diagrams* that you will use to write *equilibrium equations*. These equations will be solved for *unknown forces*, from which the required *stresses* and *deformations* can be computed. Once the answer has been obtained, it should be *carefully checked*.

Stresses on an oblique section

Fig. 1.30a

Stress under general loading

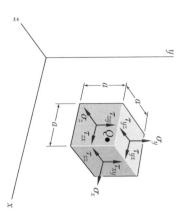

Fig. 1.36

Factor of safety

Load and Resistance Factor Design

The first part of the chapter ended with a discussion of *numerical accuracy* in engineering, which stressed the fact that the accuracy of an answer can never be greater than the accuracy of the given data [Sec. 1.10].

In Sec. 1.11, we considered the stresses created on an *oblique section* in a two-force member under axial loading. We found that both *normal* and *shearing* stresses occurred in such a situation. Denoting by θ the angle formed by the section with a normal plane (Fig. 1.30a) and by A_0 the area of a section perpendicular to the axis of the member, we derived the following expressions for the normal stress σ and the shearing stress τ on the oblique section:

$$\sigma = \frac{P}{A_0} \cos^2 \theta \qquad \tau = \frac{P}{A_0} \sin \theta \cos \theta \qquad (1.14)$$

We observed from these formulas that the normal stress is maximum and equal to $\sigma_m = P/A_0$ for $\theta = 0$, while the shearing stress is maximum and equal to $\tau_m = P/2A_0$ for $\theta = 45°$. We also noted that $\tau = 0$ when $\theta = 0$, while $\sigma = P/2A_0$ when $\theta = 45°$.

Next, we discussed the state of stress at a point Q in a body under the most general loading condition [Sec. 1.12]. Considering a small cube centered at Q (Fig. 1.36), we denoted by σ_x the normal stress exerted on a face of the cube perpendicular to the x axis, and by τ_{xy} and τ_{xz}, respectively, the y and z components of the shearing stress exerted on the same face of the cube. Repeating this procedure for the other two faces of the cube and observing that $\tau_{xy} = \tau_{yx}$, $\tau_{yz} = \tau_{zy}$, and $\tau_{zx} = \tau_{xz}$, we concluded that *six stress components* are required to define the state of stress at a given point Q, namely, σ_x, σ_y, σ_z, τ_{xy}, τ_{yz}, and τ_{zx}.

Section 1.13 was devoted to a discussion of the various concepts used in the design of engineering structures. The *ultimate load* of a given structural member or machine component is the load at which the member or component is expected to fail; it is computed from the *ultimate stress* or *ultimate strength* of the material used, as determined by a laboratory test on a specimen of that material. The *ultimate load* should be considerably larger than the *allowable load*, i.e., the load that the member or component will be allowed to carry under normal conditions. The ratio of the ultimate load to the allowable load is defined as the *factor of safety*:

$$\text{Factor of safety} = F.S. = \frac{\text{ultimate load}}{\text{allowable load}} \qquad (1.26)$$

The determination of the factor of safety which should be used in the design of a given structure depends upon a number of considerations, some of which were listed in this section.

Section 1.13 ended with the discussion of an alternative approach to design, known as *Load and Resistance Factor Design*, which allows the engineer to distinguish between the uncertainties associated with the structure and those associated with the load.

REVIEW PROBLEMS

1.59 For the truss and loading shown, determine the average normal stress in member *DF*, knowing that the cross-sectional area of that member is 2500 mm².

Fig. P1.59

1.60 Link *AC* has a uniform rectangular cross section $\frac{1}{8}$ in. thick and 1 in. wide. Determine the normal stress in the central portion of that link.

Fig. P1.60

1.61 When the force **P** reached 8 kN, the wooden specimen shown failed in shear along the surface indicated by the dashed line. Determine the average shearing stress along that surface at the time of failure.

Fig. P1.61

1.62 Two wooden planks, each 12 mm thick and 225 mm wide, are joined by the dry mortise joint shown. Knowing that the wood used shears off along its grain when the average shearing stress reaches 8 MPa, determine the magnitude *P* of the axial load which will cause the joint to fail.

Fig. P1.62

1.63 The hydraulic cylinder *CF*, which partially controls the position of rod *DE*, has been locked in the position shown. Member *BD* is $\frac{5}{8}$ in. thick and is connected to the vertical rod by a $\frac{3}{8}$-in.-diameter bolt. Determine (*a*) the average shearing stress in the bolt, (*b*) the bearing stress at *C* in member *BD*.

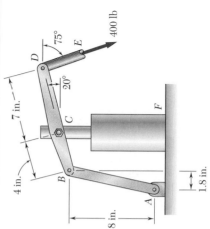

Fig. P1.63

1.64 A $\frac{1}{2}$-in.-diameter steel rod AB is fitted to a round hole near end C of the wooden member CD. For the loading shown, determine (a) the maximum average normal stress in the wood, (b) the distance b for which the average shearing stress is 90 psi on the surfaces indicated by the dashed lines, (c) the average bearing stress on the wood.

Fig. P1.64

1.65 Two plates, each 3 mm thick, are used to splice a plastic strip as shown. Knowing that the ultimate shearing stress of the bonding between the surfaces is 900 kPa, determine the factor of safety with respect to shear when $P = 1500$ N.

Fig. P1.65

1.66 Two wooden members of 3.5×5.5-in. uniform rectangular cross section are joined by the simple glued scarf splice shown. Knowing that the maximum allowable shearing stress in the glued splice is 75 psi, determine the largest axial load P that can be safely applied.

Fig. P1.66

1.67 A steel loop $ABCD$ of length 1.2 m and of 10-mm diameter is placed as shown around a 24-mm-diameter aluminum rod AC. Cables BE and DF, each of 12-mm diameter, are used to apply the load Q. Knowing that the ultimate strength of the aluminum used for the rod is 260 MPa and that the ultimate strength of the steel used for the loop and the cables is 480 MPa, determine the largest load Q that can be applied if an overall factor of safety of 3 is desired.

Fig. P1.67

1.68 Link AC has a uniform $\frac{1}{4} \times \frac{1}{2}$-in. rectangular cross section and is made of a steel with a 60-ksi ultimate normal stress. It is connected to a support at A and to member BCD at C by $\frac{3}{8}$-in.-diameter pins, while member BCD is connected to a support at B by a $\frac{5}{16}$-in.-diameter pin. All of the pins are in single shear and are made of a steel with a 25-ksi ultimate shearing stress. Knowing that an overall factor of safety of 3.25 is desired, determine the largest load P that can be safely applied at D. Note that link AC is not reinforced around the pin holes.

Fig. P1.68

1.69 The two portions of member AB are glued together along a plane forming an angle θ with the horizontal. Knowing that the ultimate stress for the glued joint is 17 MPa in tension and 9 MPa in shear, determine the range of values of θ for which the factor of safety of the member is at least 3.0.

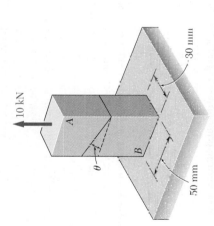

Fig. P1.69 and P1.70

1.70 The two portions of member AB are glued together along a plane forming an angle θ with the horizontal. Knowing that the ultimate stress for the glued joint is 17 MPa in tension and 9 MPa in shear, determine (a) the value of θ for which the factor of safety of the member is maximum, (b) the corresponding value of the factor of safety. (*Hint:* Equate the expressions obtained for the factors of safety with respect to normal stress and shear.)

COMPUTER PROBLEMS

The following problems are designed to be solved with a computer.

1.C1 A solid steel rod consisting of n cylindrical elements welded together is subjected to the loading shown. The diameter of element i is denoted by d_i and the load applied to its lower end by P_i, with the magnitude P_i of this load being assumed positive if P_i is directed downward as shown and negative otherwise. (a) Write a computer program which can be used with either SI or U.S. customary units to determine the average stress in each element of the rod. (b) Use this program to solve Probs. 1.1 and 1.3.

Fig. P1.C1

1.C2 A 20-kN force is applied as shown to the horizontal member ABC. Member ABC has a 10×50-mm uniform rectangular cross section and is supported by four vertical links, each of 8×36-mm uniform rectangular cross section. Each of the four pins at A, B, C, and D has the same diameter d and is in double shear. (a) Write a computer program to calculate for values of d from 10 to 30 mm, using 1-mm increments, (1) the maximum value of the average normal stress in the links connecting pins B and D, (2) the average normal stress in the links connecting pins C and E, (3) the average normal stress in pin B, (4) the average shearing stress in pin C, (5) the average bearing stress at B in member ABC, (6) the average bearing stress at C in member ABC. (b) Check your program by comparing the values obtained for $d = 16$ mm with the answers given for Probs. 1.8, 1.23, and 1.24. (c) Use this program to find the permissible values of the diameter d of the pins, knowing that the allowable values of the normal, shearing, and bearing stresses for the steel used are, respectively, 150 MPa, 90 MPa, and 230 MPa. (d) Solve part c, assuming that the thickness of member ABC has been reduced from 10 to 8 mm.

Fig. P1.C2

1.C3 Two horizontal 5-kip forces are applied to pin B of the assembly shown. Each of the three pins at A, B, and C has the same diameter d and is in double shear. (a) Write a computer program to calculate for values of d from 0.50 to 1.50 in., using 0.05-in. increments, (1) the maximum value of the average normal stress in member AB, (2) the average normal stress in member BC, (3) the average shearing stress in pin A, (4) the average shearing stress in pin C, (5) the average bearing stress at A in member AB, (6) the average bearing stress at B in member AB, (7) the average bearing stress at B in member BC. (b) Check your program by comparing the values obtained for $d = 0.8$ in. with the answers given for Probs. 1.9, 1.25, and 1.26. (c) Use this program to find the permissible values of the diameter d of the pins, knowing that the allowable values of the normal, shearing, and bearing stresses for the steel used are, respectively, 22 ksi, 13 ksi, and 36 ksi. (d) Solve part c, assuming that a new design is being investigated, in which the thickness and width of the two members are changed, respectively, from 0.5 to 0.3 in. and from 1.8 in. to 2.4 in.

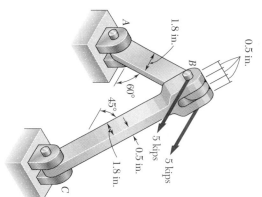

Fig. P1.C3

1.C4 A 4-kip force $\mathbf{P}$ forming an angle α with the vertical is applied as shown to member ABC, which is supported by a pin and bracket at C and by a cable BD forming an angle β with the horizontal. (a) Knowing that the ultimate load of the cable is 25 kips, write a computer program to construct a table of the values of the factor of safety of the cable for values of α and β from 0 to $45°$, using increments in α and β corresponding to 0.1 increments in $\tan \alpha$ and $\tan \beta$. (b) Check that for any given value of α the maximum value of the factor of safety is obtained for $\beta = 38.66°$ and explain why. (c) Determine the smallest possible value of the factor of safety for $\beta = 38.66°$, as well as the corresponding value of α, and explain the result obtained.

Fig. P1.C4

1.C5 A load **P** is supported as shown by two wooden members of uniform rectangular cross section which are joined by a simple glued scarf splice. (*a*) Denoting by σ_U and τ_U, respectively, the ultimate strength of the joint in tension and in shear, write a computer program which, for given values of a, b, P, σ_U, and τ_U, expressed in either SI or U.S. customary units, and for values of α from 5 to 85° at 5° intervals, can be used to calculate (1) the normal stress in the joint, (2) the shearing stress in the joint, (3) the factor of safety relative to failure in tension, (4) the factor of safety relative to failure in shear, (5) the overall factor of safety for the glued joint. (*b*) Apply this program, using the dimensions and loading of the members of Probs. 1.29 and 1.32, knowing that $\sigma_U = 1.26$ MPa and $\tau_U = 1.50$ MPa for the glue used in Prob. 1.29, and that $\sigma_U = 150$ psi and $\tau_U = 214$ psi for the glue used in Prob. 1.32. (*c*) Verify in each of these two cases that the shearing stress is maximum for $\alpha = 45°$.

Fig. P1.C5

1.C6 Member *ABC* is supported by a pin and bracket at *A* and by two links, which are pin-connected to the member at *B* and to a fixed support at *D*. (*a*) Write a computer program to calculate the allowable load P_{all} for any given values of (1) the diameter d_1 of the pin at *A*, (2) the common diameter d_2 of the pins at *B* and *D*, (3) the ultimate normal stress σ_U in each of the two links, (4) the ultimate shearing stress τ_U in each of the three pins, (5) the desired overall factor of safety *F.S.* Your program should also indicate which of the following three stresses is critical: the normal stress in the links, the shearing stress in the pin at *A*, or the shearing stress in the pins at *B* and *D*. (*b* and *c*) Check your program by using the data of Probs. 1.49 and 1.50, respectively, and comparing the answers obtained for P_{all} with those given in the text. (*d*) Use your program to determine the allowable load P_{all}, as well as which of the stresses is critical, when $d_1 = d_2 = 15$ mm, $\sigma_U = 110$ MPa for aluminum links, $\tau_U = 100$ MPa for steel pins, and *F.S.* = 3.2.

Fig. P1.C6

CHAPTER

2

Stress and Strain—Axial Loading

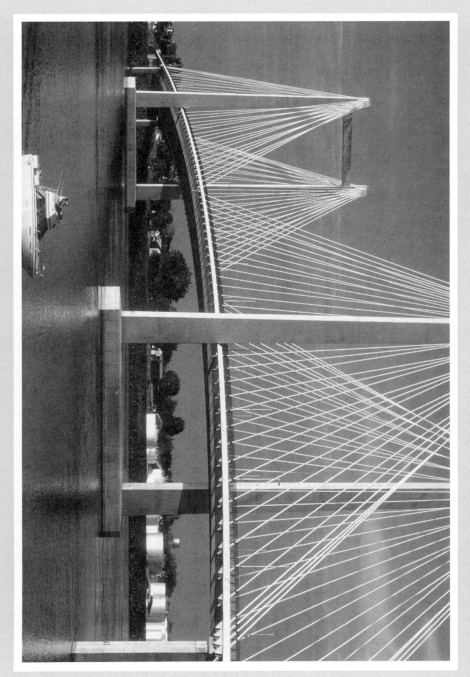

This chapter is devoted to the study of deformations occurring in structural components subjected to axial loading. The change in length of the diagonal stays was carefully accounted for in the design of this cable-stayed bridge that crosses the Columbia River in Eastern Washington.

2.1. INTRODUCTION

In Chap. 1 we analyzed the stresses created in various members and connections by the loads applied to a structure or machine. We also learned to design simple members and connections so that they would not fail under specified loading conditions. Another important aspect of the analysis and design of structures relates to the *deformations* caused by the loads applied to a structure. Clearly, it is important to avoid deformations so large that they may prevent the structure from fulfilling the purpose for which it was intended. But the analysis of deformations may also help us in the determination of stresses. Indeed, it is not always possible to determine the forces in the members of a structure by applying only the principles of statics. This is because statics is based on the assumption of undeformable, rigid structures. By considering engineering structures as *deformable* and analyzing the deformations in their various members, it will be possible for us to compute forces which are *statically indeterminate*, i.e., indeterminate within the framework of statics. Also, as we indicated in Sec. 1.5, the distribution of stresses in a given member is statically indeterminate, even when the force in that member is known. To determine the actual distribution of stresses within a member, it is thus necessary to analyze the deformations which take place in that member. In this chapter, you will consider the deformations of a structural member such as a rod, bar, or plate under *axial loading*.

First, the *normal stress* ϵ in a member will be defined as the *deformation of the member per unit length*. Plotting the stress σ versus the strain ϵ as the load applied to the member is increased will yield a *stress-strain diagram* for the material used. From such a diagram we can determine some important properties of the material, such as its *modulus of elasticity*, and whether the material is *ductile* or *brittle* (Secs. 2.2 to 2.5). You will also see in Sec. 2.5 that, while the behavior of most materials is independent of the direction in which the load is applied, the response of fiber-reinforced composite materials depends upon the direction of the load.

From the stress-strain diagram, we can also determine whether the strains in the specimen will disappear after the load has been removed—in which case the material is said to behave *elastically*—or whether *a permanent set or plastic deformation* will result (Sec. 2.6).

Section 2.7 is devoted to the phenomenon of *fatigue*, which causes structural or machine components to fail after a very large number of repeated loadings, even though the stresses remain in the elastic range.

The first part of the chapter ends with Sec. 2.8, which is devoted to the determination of the deformation of various types of members under various conditions of axial loading.

In Secs. 2.9 and 2.10, *statically indeterminate problems* will be considered, i.e., problems in which the reactions and the internal forces *cannot* be determined from statics alone. The equilibrium equations derived from the free-body diagram of the member under consideration must be complemented by relations involving deformations; these relations will be obtained from the geometry of the problem.

In Secs. 2.11 to 2.15, additional constants associated with isotropic materials—i.e., materials with mechanical characteristics independent of direction—will be introduced. They include *Poisson's ratio*, which relates lateral and axial strain, the *bulk modulus*, which characterizes the change in volume of a material under hydrostatic pressure, and the *modulus of rigidity*, which relates the components of the shearing stress and shearing strain. Stress-strain relationships for an isotropic material under a multi-axial loading will also be derived.

In Sec. 2.16, stress-strain relationships involving several distinct values of the modulus of elasticity, Poisson's ratio, and the modulus of rigidity, will be developed for fiber-reinforced composite materials under a multiaxial loading. While these materials are not isotropic, they usually display special properties, known as *orthotropic* properties, which facilitate their study.

In the text material described so far, stresses are assumed uniformly distributed in any given cross section; they are also assumed to remain within the elastic range. The validity of the first assumption is discussed in Sec. 2.17, while *stress concentrations* near circular holes and fillets in flat bars are considered in Sec. 2.18. Sections 2.19 and 2.20 are devoted to the discussion of stresses and deformations in members made of a ductile material when the *yield point* of the material is exceeded. As you will see, permanent *plastic deformations* and *residual stresses* result from such loading conditions.

2.2. NORMAL STRAIN UNDER AXIAL LOADING

Let us consider a rod *BC*, of length *L* and uniform cross-sectional area *A*, which is suspended from *B* (Fig. 2.1a). If we apply a load **P** to end *C*, the rod elongates (Fig. 2.1b). Plotting the magnitude *P* of the load against the deformation δ (Greek letter delta), we obtain a certain load-deformation diagram (Fig. 2.2). While this diagram contains information useful to the analysis of the rod under consideration, it cannot be used directly to predict the deformation of a rod of the same material but of different dimensions. Indeed, we observe that, if a deformation δ is produced in rod *BC* by a load **P**, a load **2P** is required to cause the same deformation in a rod *B'C'* of the same length *L*, but of cross-sectional area 2*A* (Fig. 2.3). We note that, in both cases, the value of the stress is the same: σ = *P*/A. On the other hand, a load **P** applied

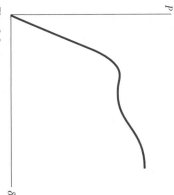

Fig. 2.1

(a)

(b)

Fig. 2.2

Fig. 2.3

Fig. 2.4

to a rod $B''C''$, of the same cross-sectional area A, but of length $2L$, causes a deformation 2δ in that rod (Fig. 2.4), i.e., a deformation twice as large as the deformation δ it produces in rod BC. But in both cases the ratio of the deformation over the length of the rod is the same; it is equal to δ/L. This observation brings us to introduce the concept of *strain*: We define the *normal strain* in a rod under axial loading as the *deformation per unit length* of that rod. Denoting the normal strain by ϵ (Greek letter epsilon), we write

$$\epsilon = \frac{\delta}{L} \tag{2.1}$$

Plotting the stress $\sigma = P/A$ against the strain $\epsilon = \delta/L$, we obtain a curve that is characteristic of the properties of the material and does not depend upon the dimensions of the particular specimen used. This curve is called a *stress strain diagram* and will be discussed in detail in Sec. 2.3.

Since the rod BC considered in the preceding discussion had a uniform cross section of area A, the normal stress σ could be assumed to have a constant value P/A throughout the rod. Thus, it was appropriate to define the strain ϵ as the ratio of the total deformation δ over the total length L of the rod. In the case of a member of variable cross-sectional area A, however, the normal stress $\sigma = P/A$ varies along the member, and it is necessary to define the strain at a given point Q by considering a small element of undeformed length Δx (Fig. 2.5). Denoting by $\Delta\delta$ the deformation of the element under the given loading, we define the *normal strain at point Q* as

$$\epsilon = \lim_{\Delta x \to 0} \frac{\Delta\delta}{\Delta x} = \frac{d\delta}{dx} \tag{2.2}$$

Fig. 2.5

Since deformation and length are expressed in the same units, the normal strain ϵ obtained by dividing δ by L (or $d\delta$ by dx) is a *dimensionless quantity*. Thus, the same numerical value is obtained for the normal strain in a given member, whether SI metric units or U.S. customary units are used. Consider, for instance, a bar of length $L = 0.600$ m and uniform cross section, which undergoes a deformation $\delta = 150 \times 10^{-6}$ m. The corresponding strain is

$$\epsilon = \frac{\delta}{L} = \frac{150 \times 10^{-6}\ \text{m}}{0.600\ \text{m}} = 250 \times 10^{-6}\ \text{m/m} = 250 \times 10^{-6}$$

Note that the deformation could have been expressed in micrometers: $\delta = 150\ \mu\text{m}$. We would then have written

$$\epsilon = \frac{\delta}{L} = \frac{150\ \mu\text{m}}{0.600\ \text{m}} = 250\ \mu\text{m/m} = 250\ \mu$$

and read the answer as "250 micros." If U.S. customary units are used, the length and deformation of the same bar are, respectively, $L = 23.6$ in. and $\delta = 5.91 \times 10^{-3}$ in. The corresponding strain is

$$\epsilon = \frac{\delta}{L} = \frac{5.91 \times 10^{-3}\ \text{in.}}{23.6\ \text{in.}} = 250 \times 10^{-6}\ \text{in./in.}$$

which is the same value that we found using SI units. It is customary, however, when lengths and deformations are expressed in inches or microinches (μin.), to keep the original units in the expression obtained for the strain. Thus, in our example, the strain would be recorded as $\epsilon = 250 \times 10^{-6}$ in./in. or, alternatively, as $\epsilon = 250\ \mu$in./in.

2.3. STRESS-STRAIN DIAGRAM

We saw in Sec. 2.2 that the diagram representing the relation between stress and strain in a given material is an important characteristic of the material. To obtain the stress-strain diagram of a material, one usually conducts a *tensile test* on a specimen of the material. One type of specimen commonly used is shown in Fig. 2.6. The cross-sectional area of the cylindrical central portion of the specimen has been accurately determined and two gage marks have been inscribed on that portion at a distance L_0 from each other. The distance L_0 is known as the *gage length* of the specimen.

Fig. 2.6 Typical tensile-test specimen.

Fig. 2.7 This machine is used to test tensile test specimens, such as those shown in this chapter.

Fig. 2.8 Test specimen with tensile load.

The test specimen is then placed in a testing machine (Fig. 2.7), which is used to apply a centric load **P**. As the load **P** increases, the distance L between the two gage marks also increases (Fig. 2.8). The distance L is measured with a dial gage, and the elongation $\delta = L - L_0$ is recorded for each value of P. A second dial gage is often used simultaneously to measure and record the change in diameter of the specimen. From each pair of readings P and δ, the stress σ is computed by dividing P by the original cross-sectional area A_0 of the specimen, and the strain ϵ by dividing the elongation δ by the original distance L_0 between the two gage marks. The stress-strain diagram may then be obtained by plotting ϵ as an abscissa and σ as an ordinate.

Stress-strain diagrams of various materials vary widely, and different tensile tests conducted on the same material may yield different results, depending upon the temperature of the specimen and the speed of loading. It is possible, however, to distinguish some common characteristics among the stress-strain diagrams of various groups of materials and to divide materials into two broad categories on the basis of these characteristics, namely, the *ductile* materials and the *brittle* materials.

Ductile materials, which comprise structural steel, as well as many alloys of other metals, are characterized by their ability to *yield* at normal temperatures. As the specimen is subjected to an increasing load, its length first increases linearly with the load and at a very slow rate. Thus, the initial portion of the stress-strain diagram is a straight line

Fig. 2.9 Stress-strain diagrams of two typical ductile materials.

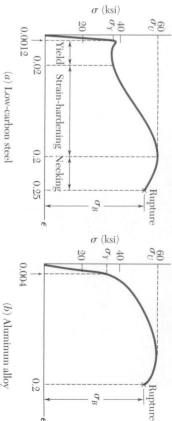

(a) Low-carbon steel

(b) Aluminum alloy

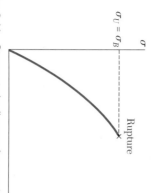

Fig. 2.10 Tested specimen of a ductile material.

Fig. 2.11 Stress-strain diagram for a typical brittle material.

with a steep slope (Fig. 2.9). However, after a critical value σ_Y of the stress has been reached, the specimen undergoes a large deformation with a relatively small increase in the applied load. This deformation is caused by slippage of the material along oblique surfaces and is due, therefore, primarily to shearing stresses. As we can note from the stress-strain diagrams of two typical ductile materials (Fig. 2.9), the elongation of the specimen after it has started to yield can be 200 times as large as its deformation before yield. After a certain maximum value of the load has been reached, the diameter of a portion of the specimen begins to decrease, because of local instability (Fig. 2.10a). This phenomenon is known as *necking*. After necking has begun, somewhat lower loads are sufficient to keep the specimen elongating further, until it finally ruptures (Fig. 2.10b). We note that rupture occurs along a cone-shaped surface which forms an angle of approximately 45° with the original surface of the specimen. This indicates that shear is primarily responsible for the failure of ductile materials, and confirms the fact that, under an axial load, shearing stresses are largest on surfaces forming an angle of 45° with the load (cf. Sec. 1.11). The stress σ_Y at which yield is initiated is called the *yield strength* of the material, the stress σ_U corresponding to the maximum load applied to the specimen is known as the *ultimate strength*, and the stress σ_B corresponding to rupture is called the *breaking strength*.

Brittle materials, which comprise cast iron, glass, and stone, are characterized by the fact that rupture occurs without any noticeable prior change in the rate of elongation (Fig. 2.11). Thus, for brittle materials, there is no difference between the ultimate strength and the breaking strength. Also, the strain at the time of rupture is much smaller for brittle than for ductile materials. From Fig. 2.12, we note the absence of any necking of the specimen in the case of a brittle material, and observe that rupture occurs along a surface perpendicular to the load. We conclude from this observation that normal stresses are primarily responsible for the failure of brittle materials.[†]

[†]The tensile tests described in this section were assumed to be conducted at normal temperatures. However, a material that is ductile at normal temperatures may display the characteristics of a brittle material at very low temperatures, while a normally brittle material may behave in a ductile fashion at very high temperatures. At temperatures other than normal, therefore, one should refer to *a material in a ductile state* or to *a material in a brittle state*, rather than to a ductile or brittle material.

Fig. 2.12 Tested specimen of a brittle material.

The stress-strain diagrams of Fig. 2.9 show that structural steel and aluminum, while both ductile, have different yield characteristics. In the case of structural steel (Fig. 2.9a), the stress remains constant over a large range of values of the strain after the onset of yield. Later the stress must be increased to keep elongating the specimen, until the maximum value σ_U has been reached. This is due to a property of the material known as strain-hardening. The yield strength of structural steel can be determined during the tensile test by watching the load shown on the display of the testing machine. After increasing steadily, the load is observed to suddenly drop to a slightly lower value, which is maintained for a certain period while the specimen keeps elongating. In a very carefully conducted test, one may be able to distinguish between the *upper yield point*, which corresponds to the load reached just before yield starts, and the *lower yield point*, which corresponds to the load required to maintain yield. Since the upper yield point is transient, the lower yield point should be used to determine the yield strength of the material.

In the case of aluminum (Fig. 2.9b) and of many other ductile materials, the onset of yield is not characterized by a horizontal portion of the stress-strain curve. Instead, the stress keeps increasing—although not linearly—until the ultimate strength is reached. Necking then begins, leading eventually to rupture. For such materials, the yield strength σ_Y can be defined by the offset method. The yield strength at 0.2% offset, for example, is obtained by drawing through the point of the horizontal axis of abscissa $\epsilon = 0.2\%$ (or $\epsilon = 0.002$), a line parallel to the initial straight-line portion of the stress-strain diagram (Fig. 2.13). The stress σ_Y corresponding to the point Y obtained in this fashion is defined as the yield strength at 0.2% offset.

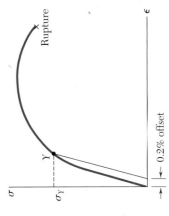

Fig. 2.13 Determination of yield strength by offset method.

A standard measure of the ductility of a material is its *percent elon-gation*, which is defined as

$$\text{Percent elongation} = 100 \frac{L_B - L_0}{L_0}$$

where L_0 and L_B denote, respectively, the initial length of the tensile test specimen and its final length at rupture. The specified minimum elongation for a 2-in. gage length for commonly used steels with yield strengths up to 50 ksi is 21%. We note that this means that the average strain at rupture should be at least 0.21 in./in.

Another measure of ductility which is sometimes used is the *per-cent reduction in area*, defined as

$$\text{Percent reduction in area} = 100 \frac{A_0 - A_B}{A_0}$$

where A_0 and A_B denote, respectively, the initial cross-sectional area of the specimen and its minimum cross-sectional area at rupture. For struc-tural steel, percent reductions in area of 60 to 70 percent are common.

Thus far, we have discussed only tensile tests. If a specimen made of a ductile material were loaded in compression instead of tension, the stress-strain curve obtained would be essentially the same through its initial straight-line portion and through the beginning of the portion cor-responding to yield and strain-hardening. Particularly noteworthy is the fact that for a given steel, the yield strength is the same in both tension and compression. For larger values of the strain, the tension and com-pression stress-strain curves diverge, and it should be noted that neck-ing cannot occur in compression. For most brittle materials, one finds that the ultimate strength in compression is much larger than the ulti-mate strength in tension. This is due to the presence of flaws, such as microscopic cracks or cavities, which tend to weaken the material in tension, while not appreciably affecting its resistance to compressive failure.

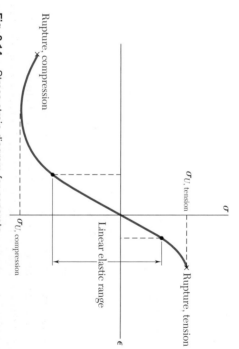

Fig. 2.14   Stress-strain diagram for concrete.

An example of brittle material with different properties in tension and compression is provided by *concrete*, whose stress-strain diagram is shown in Fig. 2.14. On the tension side of the diagram, we first observe a linear elastic range in which the strain is proportional to the stress. After the yield point has been reached, the strain increases faster than the stress until rupture occurs. The behavior of the material in compression is different. First, the linear elastic range is significantly larger. Second, rupture does not occur as the stress reaches its maximum value. Instead, the stress decreases in magnitude while the strain keeps increasing until rupture occurs. Note that the modulus of elasticity, which is represented by the slope of the stress-strain curve in its linear portion, is the same in tension and compression. This is true of most brittle materials.

*2.4. TRUE STRESS AND TRUE STRAIN

We recall that the stress plotted in the diagrams of Figs. 2.9 and 2.11 was obtained by dividing the load P by the cross-sectional area A_0 of the specimen measured before any deformation had taken place. Since the cross-sectional area of the specimen decreases as P increases, the stress plotted in our diagrams does not represent the actual stress in the specimen. The difference between the *engineering stress* $\sigma = P/A_0$ that we have computed and the *true stress* $\sigma_t = P/A$ obtained by dividing P by the cross-sectional area A of the deformed specimen becomes apparent in ductile materials after yield has started. While the engineering stress σ, which is directly proportional to the load P, decreases with P during the necking phase, the true stress σ_t, which is proportional to P but also inversely proportional to A, is observed to keep increasing until rupture of the specimen occurs.

Many scientists also use a definition of strain different from that of the *engineering strain* $\epsilon = \delta/L_0$. Instead of using the total elongation δ and the original value L_0 of the gage length, they use all the successive values of L that they have recorded. Dividing each increment ΔL of the distance between the gage marks, by the corresponding value of L, they obtain the elementary strain $\Delta \epsilon = \Delta L/L$. Adding the successive values of $\Delta \epsilon$, they define the *true strain* ϵ_t:

$$\epsilon_t = \Sigma \Delta \epsilon = \Sigma (\Delta L/L)$$

With the summation replaced by an integral, they can also express the true strain as follows:

$$\epsilon_t = \int_{L_0}^{L} \frac{dL}{L} = \ln \frac{L}{L_0} \qquad (2.3)$$

The diagram obtained by plotting true stress versus true strain (Fig. 2.15) reflects more accurately the behavior of the material. As we have already noted, there is no decrease in true stress during the necking phase. Also, the results obtained from tensile and from compressive

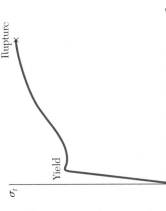

Fig. 2.15 True stress versus true strain for a typical ductile material.

σ_t

Rupture ×

Yield

ϵ_t

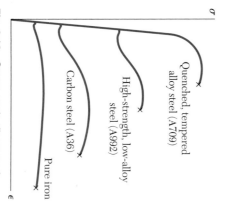

Fig. 2.16 Stress-strain diagrams for iron and different grades of steel.

tests will yield essentially the same plot when true stress and true strain are used. This is not the case for large values of the strain when the engineering stress is plotted versus the engineering strain. However, engineers, whose responsibility is to determine whether a load P will produce an acceptable stress and an acceptable deformation in a given member, will want to use a diagram based on the engineering stress $\sigma = P/A_0$ and the engineering strain $\epsilon = \delta/L_0$, since these expressions involve data that are available to them, namely the cross-sectional area A_0 and the length L_0 of the member in its undeformed state.

2.5. HOOKE'S LAW; MODULUS OF ELASTICITY

Most engineering structures are designed to undergo relatively small deformations, involving only the straight-line portion of the corresponding stress-strain diagram. For that initial portion of the diagram (Fig. 2.9), the stress σ is directly proportional to the strain ϵ, and we can write

$$\sigma = E\epsilon \tag{2.4}$$

This relation is known as *Hooke's law*, after the English mathematician Robert Hooke (1635–1703). The coefficient E is called the *modulus of elasticity* of the material involved, or also *Young's modulus*, after the English scientist Thomas Young (1773–1829). Since the strain ϵ is a dimensionless quantity, the modulus E is expressed in the same units as the stress σ, namely in pascals or one of its multiples if SI units are used, and in psi or ksi if U.S. customary units are used.

The largest value of the stress for which Hooke's law can be used for a given material is known as the *proportional limit* of that material. In the case of ductile materials possessing a well-defined yield point, as in Fig. 2.9a, the proportional limit almost coincides with the yield point. For other materials, the proportional limit cannot be defined as easily, since it is difficult to determine with accuracy the value of the stress σ for which the relation between σ and ϵ ceases to be linear. But from this very difficulty we can conclude for such materials that using Hooke's law for values of the stress slightly larger than the actual proportional limit will not result in any significant error.

Some of the physical properties of structural metals, such as strength, ductility, and corrosion resistance, can be greatly affected by alloying, heat treatment, and the manufacturing process used. For example, we note from the stress-strain diagrams of pure iron and of three different grades of steel (Fig. 2.16) that large variations in the yield strength, ultimate strength, and final strain (ductility) exist among these four metals. All of them, however, possess the same modulus of elasticity; in other words, their "stiffness," or ability to resist a deformation within the linear range, is the same. Therefore, if a high-strength steel is substituted for a lower-strength steel in a given structure, and if all dimensions are kept the same, the structure will have an increased load-carrying capacity, but its stiffness will remain unchanged.

Fig. 2.17 Layer of fiber-reinforced composite material.

For each of the materials considered so far, the relation between normal stress and normal strain, $\sigma = E\epsilon$, is independent of the direction of loading. This is because the mechanical properties of each material, including its modulus of elasticity E, are independent of the direction considered. Such materials are said to be *isotropic*. Materials whose properties depend upon the direction considered are said to be *anisotropic*. An important class of anisotropic materials consists of *fiber-reinforced composite materials*.

These composite materials are obtained by embedding fibers of a strong, stiff material into a weaker, softer material, referred to as a *matrix*. Typical materials used as fibers are graphite, glass, and polymers, while various types of resins are used as a matrix. Figure 2.17 shows a layer, or *lamina*, of a composite material consisting of a large number of parallel fibers embedded in a matrix. An axial load applied to the lamina along the x axis, that is, in a direction parallel to the fibers, will create a normal stress σ_x in the lamina and a corresponding normal strain ϵ_x which will satisfy Hooke's law as the load is increased and as long as the elastic limit of the lamina is not exceeded. Similarly, an axial load applied along the y axis, that is, in a direction perpendicular to the lamina, will create a normal stress σ_y and a normal strain ϵ_y satisfying Hooke's law, and an axial load applied along the z axis will create a normal stress σ_z and a normal strain ϵ_z which again satisfy Hooke's law. However, the moduli of elasticity E_x, E_y, and E_z corresponding, respectively, to each of the above loadings will be different. Because the fibers are parallel to the x axis, the lamina will offer a much stronger resistance to a loading directed along the x axis than to a loading directed along the y or z axis, and E_x will be much larger than either E_y or E_z.

A flat *laminate* is obtained by superposing a number of layers or laminas. If the laminate is to be subjected only to an axial load causing tension, the fibers in all layers should have the same orientation as the load in order to obtain the greatest possible strength. But if the laminate may be in compression, the matrix material may not be sufficiently strong to prevent the fibers from kinking or buckling. The lateral stability of the laminate may then be increased by positioning some of the layers so that their fibers will be perpendicular to the load. Positioning some layers so that their fibers are oriented at 30°, 45°, or 60° to the load may also be used to increase the resistance of the laminate to in-plane shear. Fiber-reinforced composite materials will be further discussed in Sec. 2.16, where their behavior under multiaxial loadings will be considered.

2.6. ELASTIC VERSUS PLASTIC BEHAVIOR OF A MATERIAL

If the strains caused in a test specimen by the application of a given load disappear when the load is removed, the material is said to behave *elastically*. The largest value of the stress for which the material behaves elastically is called the *elastic limit* of the material.

If the material has a well-defined yield point as in Fig. 2.9a, the elastic limit, the proportional limit (Sec. 2.5), and the yield point are essentially equal. In other words, the material behaves elastically and

Fig. 2.18

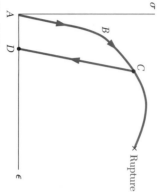

Fig. 2.19

linearly as long as the stress is kept below the yield point. If the yield point is reached, however, yield takes place as described in Sec. 2.3 and, when the load is removed, the stress and strain decrease in a linear fashion, along a line CD parallel to the straight-line portion AB of the loading curve (Fig. 2.18). The fact that ϵ does not return to zero after the load has been removed indicates that a *permanent set* or *plastic deformation* of the material has taken place. For most materials, the plastic deformation depends not only upon the maximum value reached by the stress, but also upon the time elapsed before the load is removed. The stress-dependent part of the plastic deformation is referred to as *slip*, and the time-dependent part—which is also influenced by the temperature—as *creep*.

When a material does not possess a well-defined yield point, the elastic limit cannot be determined with precision. However, assuming the elastic limit equal to the yield strength as defined by the offset method (Sec. 2.3) results in only a small error. Indeed, referring to Fig. 2.13, we note that the straight line used to determine point Y also represents the unloading curve after a maximum stress σ_Y has been reached. While the material does not behave truly elastically, the resulting plastic strain is as small as the selected offset.

If, after being loaded and unloaded (Fig. 2.19), the test specimen is loaded again, the new loading curve will closely follow the earlier unloading curve until it almost reaches point C; it will then bend to the right and connect with the curved portion of the original stress-strain diagram. We note that the straight-line portion of the new loading curve is longer than the corresponding portion of the initial one. Thus, the proportional limit and the elastic limit have increased as a result of the strain-hardening that occurred during the earlier loading of the specimen. However, since the point of rupture R remains unchanged, the ductility of the specimen, which should now be measured from point D, has decreased.

We have assumed in our discussion that the specimen was loaded twice in the same direction, i.e., that both loads were tensile loads. Let us now consider the case when the second load is applied in a direction opposite to that of the first one.

We assume that the material is mild steel, for which the yield strength is the same in tension and in compression. The initial load is tensile and is applied until point C has been reached on the stress-strain diagram (Fig. 2.20). After unloading (point D), a compressive load is applied, causing the material to reach point H, where the stress is equal to $-\sigma_Y$. We note that portion DH of the stress-strain diagram is curved and does not show any clearly defined yield point. This is referred to as the *Bauschinger effect*. As the compressive load is maintained, the material yields along line HJ.

If the load is removed after point J has been reached, the stress returns to zero along line JK, and we note that the slope of JK is equal to the modulus of elasticity E. The resulting permanent set AK may be positive, negative, or zero, depending upon the lengths of the segments BC and HJ. If a tensile load is applied again to the test specimen, the portion of the stress-strain diagram beginning at K (dashed line) will curve up and to the right until the yield stress σ_Y has been reached.

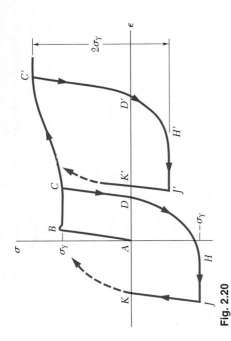

Fig. 2.20

If the initial loading is large enough to cause strain-hardening of the material (point C'), unloading takes place along line $C'D'$. As the reverse load is applied, the stress becomes compressive, reaching its maximum value at H' and maintaining it as the material yields along line $H'J'$. We note that while the maximum value of the compressive stress is less than σ_Y, the total change in stress between C' and H' is still equal to $2\sigma_Y$.

If point K or K' coincides with the origin A of the diagram, the permanent set is equal to zero, and the specimen may appear to have returned to its original condition. However, internal changes will have taken place and, while the same loading sequence may be repeated, the specimen will rupture without any warning after relatively few repetitions. This indicates that the excessive plastic deformations to which the specimen was subjected have caused a radical change in the characteristics of the material. Reverse loadings into the plastic range, therefore, are seldom allowed, and only under carefully controlled conditions. Such situations occur in the straightening of damaged material and in the final alignment of a structure or machine.

2.7. REPEATED LOADINGS; FATIGUE

In the preceding sections we have considered the behavior of a test specimen subjected to an axial loading. We recall that, if the maximum stress in the specimen does not exceed the elastic limit of the material, the specimen returns to its initial condition when the load is removed. You might conclude that a given loading may be repeated many times, provided that the stresses remain in the elastic range. Such a conclusion is correct for loadings repeated a few dozen or even a few hundred times. However, as you will see, it is not correct when loadings are repeated thousands or millions of times. In such cases rupture will occur at a stress much lower than the static breaking strength; this phenomenon is known as *fatigue*. A fatigue failure is of a brittle nature, even for materials that are normally ductile.

Fatigue must be considered in the design of all structural and machine components that are subjected to repeated or to fluctuating loads. The number of loading cycles that may be expected during the useful life of a component varies greatly. For example, a beam supporting an industrial crane may be loaded as many as two million times in 25 years (about 300 loadings per working day), an automobile crankshaft will be loaded about half a billion times if the automobile is driven 200,000 miles, and an individual turbine blade may be loaded several hundred billion times during its lifetime.

Some loadings are of a fluctuating nature. For example, the passage of traffic over a bridge will cause stress levels that will fluctuate about the stress level due to the weight of the bridge. A more severe condition occurs when a complete reversal of the load occurs during the loading cycle. The stresses in the axle of a railroad car, for example, are completely reversed after each half-revolution of the wheel.

The number of loading cycles required to cause the failure of a specimen through repeated successive loadings and reverse loadings may be determined experimentally for any given maximum stress level. If a series of tests is conducted, using different maximum stress levels, the resulting data may be plotted as a σ-n curve. For each test, the maximum stress σ is plotted as an ordinate and the number of cycles n as an abscissa; because of the large number of cycles required for rupture, the cycles n are plotted on a logarithmic scale.

A typical σ-n curve for steel is shown in Fig. 2.21. We note that, if the applied maximum stress is high, relatively few cycles are required to cause rupture. As the magnitude of the maximum stress is reduced, the number of cycles required to cause rupture increases, until a stress, known as the *endurance limit*, is reached. The endurance limit is the stress for which failure does not occur, even for an indefinitely large number of loading cycles. For a low-carbon steel, such as structural steel, the endurance limit is about one-half of the ultimate strength of the steel.

For nonferrous metals, such as aluminum and copper, a typical σ-n curve (Fig. 2.21) shows that the stress at failure continues to decrease as the number of loading cycles is increased. For such metals, one defines the *fatigue limit* as the stress corresponding to failure after a specified number of loading cycles, such as 500 million.

Examination of test specimens, of shafts, of springs, and of other components that have failed in fatigue shows that the failure was initiated at a microscopic crack or at some similar imperfection. At each loading, the crack was very slightly enlarged. During successive loading cycles, the crack propagated through the material until the amount of undamaged material was insufficient to carry the maximum load, and an abrupt, brittle failure occurred. Because fatigue failure may be initiated at any crack or imperfection, the surface condition of a specimen has an important effect on the value of the endurance limit obtained in testing. The endurance limit for machined and polished specimens is higher than for rolled or forged components, or for components that are corroded. In applications in or near seawater, or in other applications where corrosion is expected, a reduction of up to 50% in the endurance limit can be expected.

Fig. 2.21

Stress (ksi)

Number of completely reversed cycles

Steel (1020HR)

Aluminum (2024)

2.8. DEFORMATIONS OF MEMBERS UNDER AXIAL LOADING

Consider a homogeneous rod BC of length L and uniform cross section of area A subjected to a centric axial load $\mathbf{P}$ (Fig. 2.22). If the resulting axial stress $\sigma = P/A$ does not exceed the proportional limit of the material, we may apply Hooke's law and write

$$\sigma = E\epsilon \tag{2.4}$$

from which it follows that

$$\epsilon = \frac{\sigma}{E} = \frac{P}{AE} \tag{2.5}$$

Recalling that the strain ϵ was defined in Sec. 2.2 as $\epsilon = \delta/L$, we have

$$\delta = \epsilon L \tag{2.6}$$

and, substituting for ϵ from (2.5) into (2.6):

$$\delta = \frac{PL}{AE} \tag{2.7}$$

Fig. 2.22

Equation (2.7) may be used only if the rod is homogeneous (constant E), has a uniform cross section of area A, and is loaded at its ends. If the rod is loaded at other points, or if it consists of several portions of various cross sections and possibly of different materials, we must divide it into component parts that satisfy individually the required conditions for the application of formula (2.7). Denoting respectively by P_i, L_i, A_i, and E_i the internal force, length, cross-sectional area, and modulus of elasticity corresponding to part i, we express the deformation of the entire rod as

$$\delta = \sum_i \frac{P_i L_i}{A_i E_i} \tag{2.8}$$

We recall from Sec. 2.2 that, in the case of a rod of variable cross section (Fig. 2.5), the strain ϵ depends upon the position of the point Q where it is computed and is defined as $\epsilon = d\delta/dx$. Solving for $d\delta$ and substituting for ϵ from Eq. (2.5), we express the deformation of an element of length dx as

$$d\delta = \epsilon\, dx = \frac{P\, dx}{AE}$$

The total deformation δ of the rod is obtained by integrating this expression over the length L of the rod:

$$\delta = \int_0^L \frac{P\, dx}{AE} \tag{2.9}$$

Formula (2.9) should be used in place of (2.7), not only when the cross-sectional area A is a function of x, but also when the internal force P depends upon x, as is the case for a rod hanging under its own weight.

EXAMPLE 2.01

Determine the deformation of the steel rod shown in Fig. 2.23a under the given loads ($E = 29 \times 10^{-6}$ psi).

Fig. 2.23

We divide the rod into three component parts shown in Fig. 2.23b and write

$$L_1 = L_2 = 12 \text{ in.} \qquad L_3 = 16 \text{ in.}$$
$$A_1 = A_2 = 0.9 \text{ in}^2 \qquad A_3 = 0.3 \text{ in}^2$$

To find the internal forces P_1, P_2, and P_3, we must pass sections through each of the component parts, drawing each time the free-body diagram of the portion of rod located to the right of the section (Fig. 2.23c). Expressing that each of the free bodies is in equilibrium, we obtain successively

$$P_1 = 60 \text{ kips} = 60 \times 10^3 \text{ lb}$$
$$P_2 = -15 \text{ kips} = -15 \times 10^3 \text{ lb}$$
$$P_3 = 30 \text{ kips} = 30 \times 10^3 \text{ lb}$$

Carrying the values obtained into Eq. (2.8), we have

$$\delta = \sum_i \frac{P_i L_i}{A_i E_i} = \frac{1}{E} \left(\frac{P_1 L_1}{A_1} + \frac{P_2 L_2}{A_2} + \frac{P_3 L_3}{A_3} \right)$$

$$= \frac{1}{29 \times 10^6} \left[\frac{(60 \times 10^3)(12)}{0.9} + \frac{(-15 \times 10^3)(12)}{0.9} + \frac{(30 \times 10^3)(16)}{0.3} \right]$$

$$\delta = \frac{2.20 \times 10^6}{29 \times 10^6} = 75.9 \times 10^{-3} \text{ in.}$$

The rod BC of Fig. 2.22, which was used to derive formula (2.7), and the rod AD of Fig. 2.23, which has just been discussed in Example 2.01, both had one end attached to a fixed support. In each case, therefore, the deformation δ of the rod was equal to the displacement of its free end. When both ends of a rod move, however, the deformation of the rod is measured by the *relative displacement* of one end of the rod with respect to the other. Consider, for instance, the assembly shown in Fig. 2.24a, which consists of three elastic bars of length L connected by a rigid pin at A. If a load $\mathbf{P}$ is applied at B (Fig. 2.24b), each of the three bars will deform. Since the bars AC and AC' are attached to fixed supports at C and C', their common deformation is measured by the displacement δ_A of point A. On the other hand, since both ends of bar AB move, the deformation of AB is measured by the difference between the displacements δ_A and δ_B of points A and B, i.e., by the relative displacement of B with respect to A. Denoting this relative displacement by $\delta_{B/A}$, we write

$$\delta_{B/A} = \delta_B - \delta_A = \frac{PL}{AE} \tag{2.10}$$

where A is the cross-sectional area of AB and E is its modulus of elasticity.

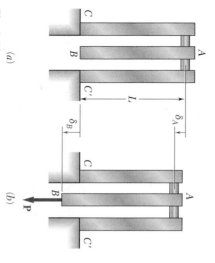

(a)

(b)

Fig. 2.24

SAMPLE PROBLEM 2.1

The rigid bar BDE is supported by two links AB and CD. Link AB is made of aluminum ($E = 70$ GPa) and has a cross-sectional area of 500 mm²; link CD is made of steel ($E = 200$ GPa) and has a cross-sectional area of 600 mm². For the 30-kN force shown, determine the deflection (a) of B, (b) of D, (c) of E.

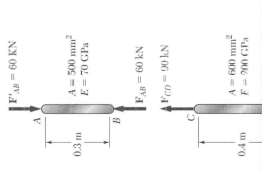

SOLUTION

Free Body: Bar BDE

$+\curvearrowleft \Sigma M_B = 0$:

$$-(30 \text{ kN})(0.6 \text{ m}) + F_{CD}(0.2 \text{ m}) = 0$$
$$F_{CD} = +90 \text{ kN} \qquad F_{CD} = 90 \text{ kN} \quad \text{tension}$$

$+\curvearrowleft \Sigma M_D = 0$:

$$-(30 \text{ kN})(0.4 \text{ m}) - F_{AB}(0.2 \text{ m}) = 0$$
$$F_{AB} = -60 \text{ kN} \qquad F_{AB} = 60 \text{ kN} \quad \text{compression}$$

a. Deflection of B. Since the internal force in link AB is compressive, we have $P = -60$ kN

$$\delta_B = \frac{PL}{AE} = \frac{(-60 \times 10^3 \text{ N})(0.3 \text{ m})}{(500 \times 10^{-6} \text{ m}^2)(70 \times 10^9 \text{ Pa})} = -514 \times 10^{-6} \text{ m}$$

The negative sign indicates a contraction of member AB, and, thus, an upward deflection of end B:

$$\delta_B = 0.514 \text{ mm} \uparrow \qquad \blacktriangledown$$

b. Deflection of D. Since in rod CD, $P = 90$ kN, we write

$$\delta_D = \frac{PL}{AE} = \frac{(90 \times 10^3 \text{ N})(0.4 \text{ m})}{(600 \times 10^{-6} \text{ m}^2)(200 \times 10^9 \text{ Pa})}$$
$$= 300 \times 10^{-6} \text{ m} \qquad \delta_D = 0.300 \text{ mm} \rightarrow \qquad \blacktriangledown$$

c. Deflection of E. We denote by B' and D' the displaced positions of points B and D. Since the bar BDE is rigid, points B', D', and E' lie in a straight line and we write

$$\frac{BB'}{DD'} = \frac{BH}{HD} \qquad \frac{0.514 \text{ mm}}{0.300 \text{ mm}} = \frac{(200 \text{ mm}) - x}{x} \qquad x = 73.7 \text{ mm}$$

$$\frac{EE'}{DD'} = \frac{HE}{HD} \qquad \frac{\delta_E}{0.300 \text{ mm}} = \frac{(400 \text{ mm}) + (73.7 \text{ mm})}{73.7 \text{ mm}}$$

$$\delta_E = 1.928 \text{ mm} \rightarrow \qquad \blacktriangledown$$

SAMPLE PROBLEM 2.2

The rigid castings A and B are connected by two $\frac{3}{4}$-in.-diameter steel bolts CD and GH and are in contact with the ends of a 1.5-in.-diameter aluminum rod EF. Each bolt is single-threaded with a pitch of 0.1 in., and after being snugly fitted, the nuts at D and H are both tightened one-quarter of a turn. Knowing that E is 29×10^6 psi for steel and 10.6×10^6 psi for aluminum, determine the normal stress in the rod.

SOLUTION

Deformations

Bolts CD and GH. Tightening the nuts causes tension in the bolts. Because of symmetry, both are subjected to the same internal force P_b and undergo the same deformation δ_b. We have

$$\delta_b = +\frac{P_b L_b}{A_b E_b} = +\frac{P_b (18 \text{ in.})}{\frac{1}{4}\pi(0.75 \text{ in.})^2(29 \times 10^6 \text{ psi})} = +1.405 \times 10^{-6} P_b \quad (1)$$

Rod EF. The rod is in compression. Denoting by P_r the magnitude of the force in the rod and by δ_r the deformation of the rod, we write

$$\delta_r = -\frac{P_r L_r}{A_r E_r} = -\frac{P_r(12 \text{ in.})}{\frac{1}{4}\pi(1.5 \text{ in.})^2(10.6 \times 10^6 \text{ psi})} = -0.6406 \times 10^{-6} P_r \quad (2)$$

Displacement of D Relative to B. Tightening the nuts one-quarter of a turn causes ends D and H of the bolts to undergo a displacement of $\frac{1}{4}(0.1$ in.) *relative to casting B.* Considering end D, we write

$$\delta_{D/B} = \tfrac{1}{4}(0.1 \text{ in.}) = 0.025 \text{ in.} \quad (3)$$

But $\delta_{D/B} = \delta_D - \delta_B$, where δ_D and δ_B represent the displacements of D and B. If we assume that casting A is held in a fixed position while the nuts at D and H are being tightened, these displacements are equal to the deformations of the bolts and of the rod, respectively. We have, therefore,

$$\delta_{D/B} = \delta_b - \delta_r \quad (4)$$

Substituting from (1), (2), and (3) into (4), we obtain

$$0.025 \text{ in.} = 1.405 \times 10^{-6} P_b + 0.6406 \times 10^{-6} P_r \quad (5)$$

Free Body: Casting B

$$\xrightarrow{+} \Sigma F = 0: \qquad P_r - 2P_b = 0 \qquad P_r = 2P_b \quad (6)$$

Forces in Bolts and Rod

Substituting for P_r from (6) into (5), we have

$$0.025 \text{ in.} = 1.405 \times 10^{-6} P_b + 0.6406 \times 10^{-6}(2P_b)$$
$$P_b = 9.307 \times 10^3 \text{ lb} = 9.307 \text{ kips}$$
$$P_r = 2P_b = 2(9.307 \text{ kips}) = 18.61 \text{ kips}$$

Stress in Rod

$$\sigma_r = \frac{P_r}{A_r} = \frac{18.61 \text{ kips}}{\frac{1}{4}\pi(1.5 \text{ in.})^2} \qquad \sigma_r = 10.53 \text{ ksi} \quad \blacktriangleleft$$

PROBLEMS

2.1 A steel rod is 2.2 m long and must not stretch more than 1.2 mm when a 8.5-kN load is applied to it. Knowing that $E = 200$ GPa, determine (*a*) the smallest diameter rod that should be used, (*b*) the corresponding normal stress caused by the load.

2.2 A 4.8-ft-long steel wire of $\frac{1}{4}$-in.-diameter steel wire is subjected to a 750-lb tensile load. Knowing that $E = 29 \times 10^6$ psi, determine (*a*) the elongation of the wire, (*b*) the corresponding normal stress.

2.3 Two gage marks are placed exactly 10 in. apart on a $\frac{1}{2}$-in.-diameter aluminum rod with $E = 10.1 \times 10^6$ psi and an ultimate strength of 16 ksi. Knowing that the distance between the gage marks is 10.009 in. after a load is applied, determine (*a*) the stress in the rod, (*b*) the factor of safety.

2.4 A control rod made of yellow brass must not stretch more than 3 mm when the tension in the wire is 4 kN. Knowing that $E = 105$ GPa and that the maximum allowable normal stress is 180 MPa, determine (*a*) the smallest diameter that can be selected for the rod, (*b*) the corresponding maximum length of the rod.

2.5 A 9-m length of 6-mm-diameter steel wire is to be used in a hanger. It is noted that the wire stretches 18 mm when a tensile force **P** is applied. Knowing that $E = 200$ GPa, determine (*a*) the magnitude of the force **P**, (*b*) the corresponding normal stress in the wire.

2.6 A 4.5-ft aluminum pipe should not stretch more than 0.05 in. when it is subjected to a tensile load. Knowing that $E = 10.1 \times 10^6$ psi and that the allowable tensile strength is 14 ksi, determine (*a*) the maximum allowable length of the pipe, (*b*) the required area of the pipe if the tensile load is 127.5 kips.

2.7 A nylon thread is subjected to a 8.5-N tension force. Knowing that $E = 3.3$ GPa and that the length of the thread increases by 1.1%, determine (*a*) the diameter of the thread, (*b*) the stress in the thread.

2.8 A cast-iron tube is used to support a compressive load. Knowing that $E = 10 \times 10^6$ psi and that the maximum allowable change in length is 0.025%, determine (*a*) the maximum normal stress in the tube, (*b*) the minimum wall thickness for a load of 1600 lb if the outside diameter of the tube is 2.0 in.

2.9 A block of 10-in. length and 1.8×1.6-in. cross section is to support a centric compressive load **P**. The material to be used is a bronze for which $E = 14 \times 10^6$ psi. Determine the largest load that can be applied, knowing that the normal stress must not exceed 18 ksi and that the decrease in length of the block should be at most 0.12% of its original length.

2.10   A 9-kN tensile load will be applied to a 50-m length of steel wire with $E = 200$ GPa. Determine the smallest-diameter wire that can be used, knowing that the normal stress must not exceed 150 MPa and that the increase in the length of the wire should be at most 25 mm.

2.11   The 4-mm-diameter cable BC is made of a steel with $E = 200$ GPa. Knowing that the maximum stress in the cable must not exceed 190 MPa and that the elongation of the cable must not exceed 6 mm, find the maximum load **P** that can be applied as shown.

Fig. P2.11

2.12   Rod BD is made of steel ($E = 29 \times 10^6$ psi) and is used to brace the axially compressed member ABC. The maximum force that can be developed in member BD is $0.02P$. If the stress must not exceed 18 ksi and the maximum change in length of BD must not exceed 0.001 times the length of ABC, determine the smallest-diameter rod that can be used for member BD.

Fig. P2.12

2.13   A single axial load of magnitude $P = 58$ kN is applied at end C of the brass rod ABC. Knowing that $E = 105$ GPa, determine the diameter d of portion BC for which the deflection of point C will be 3 mm.

2.14   Both portions of the rod ABC are made of an aluminum for which $E = 73$ GPa. Knowing that the diameter of portion BC is $d = 20$ mm, determine the largest force **P** that can be applied if $\sigma_{all} = 160$ MPa and the corresponding deflection at point C is not to exceed 4 mm.

Fig. P2.13 and P2.14

2.15   The specimen shown is made from a 1-in.-diameter cylindrical steel rod with two 1.5-in.-outer-diameter sleeves bonded to the rod as shown. Knowing that $E = 29 \times 10^6$ psi, determine (a) the load **P** so that the total deformation is 0.002 in., (b) the corresponding deformation of the central portion BC.

Fig. P2.15

2.16 Both portions of the rod *ABC* are made of an aluminum for which $E = 70$ GPa. Knowing that the magnitude of **P** is 4 kN, determine (*a*) the value of **Q** so that the deflection at *A* is zero, (*b*) the corresponding deflection of *B*.

2.17 The rod *ABC* is made of an aluminum for which $E = 70$ GPa. Knowing that $P = 6$ kN and $Q = 42$ kN, determine the deflection of (*a*) point *A*, (*b*) point *B*.

20-mm diameter

60-mm diameter

0.4 m

0.5 m

Fig. P2.16 and P2.17

2 m

3 m

2.5 m

Steel: $E = 200$ GPa

Brass: $E = 105$ GPa

50 kN

100 kN

Fig. P2.18

3/64 in.

15.0 in.

Fig. P2.19

2.18 The 36-mm-diameter steel rod *ABC* and a brass rod *CD* of the same diameter are joined at point *C* to form the 7.5-m rod *ABCD*. For the loading shown, and neglecting the weight of the rod, determine the deflection of (*a*) point *C*, (*b*) point *D*.

2.19 The brass tube $AB(E = 15 \times 10^6$ psi) has a cross-sectional area of 0.22 in² and is fitted with a plug at *A*. The tube is attached at *B* to a rigid plate which is itself attached at *C* to the bottom of an aluminum cylinder $(E = 10.4 \times 10^6$ psi) with a cross-sectional area of 0.40 in². The cylinder is then hung from a support at *D*. In order to close the cylinder, the plug must move down through $\frac{3}{64}$ in. Determine the force **P** that must be applied to the cylinder.

2.20 A 1.2-m section of aluminum pipe of cross-sectional area 1100 mm² rests on a fixed support at *A*. The 15-mm-diameter steel rod *BC* hangs from a rigid bar that rests on the top of the pipe at *B*. Knowing that the modulus of elasticity is 200 GPa for steel and 72 GPa for aluminum, determine the deflection of point *C* when a 60-kN force is applied at *C*.

1.2 m

0.9 m

Fig. P2.20

2.21 The steel frame ($E = 200$ GPa) shown has a diagonal brace BD with an area of 1920 mm². Determine the largest allowable load **P** if the change in length of member BD is not to exceed 1.6 mm.

Fig. P2.21

Fig. P2.22

2.22 For the steel truss ($E = 200$ GPa) and loading shown, determine the deformations of members AB and AD, knowing that their cross-sectional areas are 2400 mm² and 1800 mm², respectively.

2.23 Members AB and BC are made of steel ($E = 29 \times 10^6$ psi) with cross-sectional areas of 0.80 in² and 0.64 in², respectively. For the loading shown, determine the elongation of (*a*) member AB, (*b*) member BC.

2.24 Members AB and CD are $1\frac{1}{8}$-in.-diameter steel rods, and members BC and AD are $\frac{7}{8}$-in.-diameter steel rods. When the turnbuckle is tightened, the diagonal member AC is put in tension. Knowing that $E = 29 \times 10^6$ psi and $h = 4$ ft, determine the largest allowable tension in AC so that the deformations in members AB and CD do not exceed 0.04 in.

2.25 For the structure in Prob. 2.24, determine (*a*) the distance h so that the deformations in members AB, BC, CD, and AD are all equal to 0.04 in., (*b*) the corresponding tension in member AC.

Fig. P2.23

Fig. P2.24

Fig. P2.26

Fig. P2.27

Fig. P2.28

Fig. P2.30

2.26 Members *ABC* and *DEF* are joined with steel links ($E = 200$ GPa). Each of the links is made of a pair of a pair of 25×35-mm plates. Determine the change in length of (*a*) member *BE*, (*b*) member *CF*.

2.27 Each of the links *AB* and *CD* is made of aluminum ($E = 75$ GPa) and has a cross-sectional area of 125 mm². Knowing that they support the rigid member *BC*, determine the deflection of point *E*.

2.28 Link *BD* is made of brass ($E = 15 \times 10^6$ psi) and has a cross-sectional area of 0.40 in². Link *CE* is made of aluminum ($E = 10.4 \times 10^6$ psi) and has a cross-sectional area of 0.50 in². Determine the maximum force **P** that can be applied vertically at point *A* if the deflection of *A* is not to exceed 0.014 in.

2.29 A homogeneous cable of length *L* and uniform cross section is suspended from one end. (*a*) Denoting by ρ the density (mass per unit volume) of the cable and by *E* its modulus of elasticity, determine the elongation of the cable due to its own weight. (*b*) Assuming now the cable to be horizontal, determine the force that should be applied to each end of the cable to obtain the same elongation as in part *a*.

2.30 Determine the deflection of the apex *A* of a homogeneous circular cone of height *h*, density ρ, and modulus of elasticity *E*, due to its own weight.

2.31 The volume of a tensile specimen is essentially constant while plastic deformation occurs. If the initial diameter of the specimen is *d*, show that when the diameter is d_1, the true strain is $\epsilon_t = 2 \ln (d_1/d)$.

2.32 Denoting by ϵ the "engineering strain" in a tensile specimen, show that the true strain is $\epsilon_t = \ln (1 + \epsilon)$.

2.9. STATICALLY INDETERMINATE PROBLEMS

In the problems considered in the preceding section, we could always use free-body diagrams and equilibrium equations to determine the internal forces produced in the various portions of a member under given loading conditions. The values obtained for the internal forces were then entered into Eq. (2.8) or (2.9) to obtain the deformation of the member.

There are many problems, however, in which the internal forces cannot be determined from statics alone. In fact, in most of these problems the reactions themselves—which are external forces—cannot be determined by simply drawing a free-body diagram of the member and writing the corresponding equilibrium equations. The equilibrium equations must be complemented by relations involving deformations obtained by considering the geometry of the problem. Because statics is not sufficient to determine either the reactions or the internal forces, problems of this type are said to be *statically indeterminate*. The following examples will show how to handle this type of problem.

EXAMPLE 2.02

A rod of length L, cross-sectional area A_1, and modulus of elasticity E_1, has been placed inside a tube of the same length L, but of cross-sectional area A_2 and modulus of elasticity E_2 (Fig. 2.25a). What is the deformation of the rod and tube when a force **P** is exerted on a rigid end plate as shown?

(a)

(b)

(c)

(d)

Fig. 2.25

Denoting by P_1 and P_2, respectively, the axial forces in the rod and in the tube, we draw free-body diagrams of all three elements (Fig. 2.25b, c, d). Only the last of the diagrams yields any significant information, namely:

$$P_1 + P_2 = P \qquad (2.11)$$

Clearly, one equation is not sufficient to determine the two unknown internal forces P_1 and P_2. The problem is statically indeterminate.

However, the geometry of the problem shows that the deformations δ_1 and δ_2 of the rod and tube must be equal. Recalling Eq. (2.7), we write

$$\delta_1 = \frac{P_1 L}{A_1 E_1} \qquad \delta_2 = \frac{P_2 L}{A_2 E_2} \qquad (2.12)$$

Equating the deformations δ_1 and δ_2, we obtain:

$$\frac{P_1}{A_1 E_1} = \frac{P_2}{A_2 E_2} \qquad (2.13)$$

Equations (2.11) and (2.13) can be solved simultaneously for P_1 and P_2:

$$P_1 = \frac{A_1 E_1 P}{A_1 E_1 + A_2 E_2} \qquad P_2 = \frac{A_2 E_2 P}{A_1 E_1 + A_2 E_2}$$

Either of Eqs. (2.12) can then be used to determine the common deformation of the rod and tube.

EXAMPLE 2.03

A bar AB of length L and uniform cross section is attached to rigid supports at A and B before being loaded. What are the stresses in portions AC and BC due to the application of a load P at point C (Fig. 2.26a)?

Drawing the free-body diagram of the bar (Fig. 2.26b), we obtain the equilibrium equation

$$R_A + R_B = P \qquad (2.14)$$

Since this equation is not sufficient to determine the two unknown reactions R_A and R_B, the problem is statically indeterminate.

However, the reactions may be determined if we observe from the geometry that the total elongation δ of the bar must be zero. Denoting by δ_1 and δ_2, respectively, the elongations of the portions AC and BC, we write

$$\delta = \delta_1 + \delta_2 = 0$$

or, expressing δ_1 and δ_2 in terms of the corresponding internal forces P_1 and P_2:

$$\delta = \frac{P_1 L_1}{AE} + \frac{P_2 L_2}{AE} = 0 \qquad (2.15)$$

But we note from the free-body diagrams shown respectively in parts b and c of Fig. 2.27 that $P_1 = R_A$ and $P_2 = -R_B$. Carrying these values into (2.15), we write

$$R_A L_1 - R_B L_2 = 0 \qquad (2.16)$$

Equations (2.14) and (2.16) can be solved simultaneously for R_A and R_B; we obtain $R_A = PL_2/L$ and $R_B = PL_1/L$. The desired stresses σ_1 in AC and σ_2 in BC are obtained by dividing, respectively, $P_1 = R_A$ and $P_2 = -R_B$ by the cross-sectional area of the bar:

$$\sigma_1 = \frac{PL_2}{AL} \qquad \sigma_2 = -\frac{PL_1}{AL}$$

Fig. 2.26

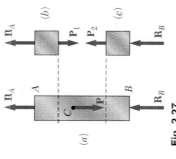

Fig. 2.27

Superposition Method.

We observe that a structure is statically indeterminate whenever it is held by more supports than are required to maintain its equilibrium. This results in more unknown reactions than available equilibrium equations. It is often found convenient to designate one of the reactions as *redundant* and to eliminate the corresponding support. Since the stated conditions of the problem cannot be arbitrarily changed, the redundant reaction must be maintained in the solution. But it will be treated as an *unknown load* which, together with the other loads, must produce deformations that are compatible with the original constraints. The actual solution of the problem is carried out by considering separately the deformations caused by the given loads and by the redundant reaction, and by adding—or *superposing*—the results obtained.†

†The general conditions under which the combined effect of several loads can be obtained in this way are discussed in Sec. 2.12.

EXAMPLE 2.04

Determine the reactions at A and B for the steel bar and loading shown in Fig. 2.28, assuming a close fit at both supports before the loads are applied.

$A = 250 \text{ mm}^2$

$A = 400 \text{ mm}^2$

300 kN

600 kN

150 mm
150 mm
150 mm
150 mm

Fig. 2.28

Following the same procedure as in Example 2.01, we write

$$P_1 = 0 \quad P_2 = P_3 = 600 \times 10^3 \text{ N} \quad P_4 = 900 \times 10^3 \text{ N}$$

$$A_1 = A_2 = 400 \times 10^{-6} \text{ m}^2 \quad A_3 = A_4 = 250 \times 10^{-6} \text{ m}^2$$

$$L_1 = L_2 = L_3 = L_4 = 0.150 \text{ m}$$

We consider the reaction at B as redundant and release the bar from that support. The reaction $\mathbf{R}_B$ is now considered as an unknown load (Fig. 2.29a) and will be determined from the condition that the deformation δ of the rod must be equal to zero. The solution is carried out by considering separately the deformation δ_L caused by the given loads (Fig. 2.29b) and the deformation δ_R due to the redundant reaction $\mathbf{R}_B$ (Fig. 2.29c).

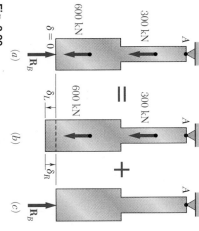

300 kN

600 kN

$\delta = 0$ $\mathbf{R}_B$

δ_L

δ_R

(a)   (b)   (c)

Fig. 2.29

The deformation δ_L is obtained from Eq. (2.8) after the bar has been divided into four portions, as shown in Fig. 2.30.

300 kN

600 kN

150 mm
150 mm
150 mm
150 mm

Fig. 2.30

Substituting these values into Eq. (2.8), we obtain

$$\delta_L = \sum_{i=1}^{4} \frac{P_i L_i}{A_i E} = \left(0 + \frac{600 \times 10^3 \text{ N}}{400 \times 10^{-6} \text{ m}^2} \right.$$

$$\left. + \frac{600 \times 10^3 \text{ N}}{250 \times 10^{-6} \text{ m}^2} + \frac{900 \times 10^3 \text{ N}}{250 \times 10^{-6} \text{ m}^2} \right) \frac{0.150 \text{ m}}{E}$$

$$\delta_L = \frac{1.125 \times 10^9}{E} \qquad (2.17)$$

Considering now the deformation δ_R due to the redundant reaction R_B, we divide the bar into two portions, as shown in Fig. 2.31, and write

$$P_1 = P_2 = -R_B$$

$$A_1 = 400 \times 10^{-6} \text{ m}^2 \quad A_2 = 250 \times 10^{-6} \text{ m}^2$$

$$L_1 = L_2 = 0.300 \text{ m}$$

300 mm
300 mm

Fig. 2.31

Substituting these values into Eq. (2.8), we obtain

$$\delta_R = \frac{P_1 L_1}{A_1 E} + \frac{P_2 L_2}{A_2 E} = -\frac{(1.95 \times 10^3) R_B}{E} \quad (2.18)$$

Expressing that the total deformation δ of the bar must be zero, we write

$$\delta = \delta_L + \delta_R = 0 \quad (2.19)$$

and, substituting for δ_L and δ_R from (2.17) and (2.18) into (2.19),

$$\delta = \frac{1.125 \times 10^9}{E} - \frac{(1.95 \times 10^3) R_B}{E} = 0$$

Solving for R_B, we have

$$R_B = 577 \times 10^3 \, \text{N} = 577 \, \text{kN}$$

The reaction R_A at the upper support is obtained from the free-body diagram of the bar (Fig. 2.32). We write

$$+\uparrow \Sigma F_y = 0: \quad R_A - 300 \, \text{kN} - 600 \, \text{kN} + R_B = 0$$
$$R_A = 900 \, \text{kN} - R_B = 900 \, \text{kN} - 577 \, \text{kN} = 323 \, \text{kN}$$

300 kN

600 kN

Fig. 2.32

Once the reactions have been determined, the stresses and strains in the bar can easily be obtained. It should be noted that, while the total deformation of the bar is zero, each of its component parts *does deform* under the given loading and restraining conditions.

EXAMPLE 2.05

Determine the reactions at A and B for the steel bar and loading of Example 2.04, assuming now that a 4.50-mm clearance exists between the bar and the ground before the loads are applied (Fig. 2.33). Assume $E = 200 \, \text{GPa}$.

We follow the same procedure as in Example 2.04. Considering the reaction at B as redundant, we compute the deformations δ_L and δ_R caused respectively by the given loads and by the redundant reaction $\mathbf{R}_B$. However, in this case the total deformation is not zero, but $\delta - 4.5$ mm. We write therefore

$$\delta = \delta_L + \delta_R = 4.5 \times 10^{-3} \, \text{m} \quad (2.20)$$

Substituting for δ_L and δ_R from (2.17) and (2.18) into (2.20), and recalling that $E = 200 \, \text{GPa} = 200 \times 10^9$ Pa, we have

$$\delta = \frac{1.125 \times 10^9}{200 \times 10^9} - \frac{(1.95 \times 10^3) R_B}{200 \times 10^9} = 4.5 \times 10^{-3} \, \text{m}$$

Solving for R_B, we obtain

$$R_B = 115.4 \times 10^3 \, \text{N} = 115.4 \, \text{kN}$$

The reaction at A is obtained from the free-body diagram of the bar (Fig. 2.32):

$$+\uparrow \Sigma F_y = 0: \quad R_A - 300 \, \text{kN} - 600 \, \text{kN} + R_B = 0$$
$$R_A = 900 \, \text{kN} - R_B = 900 \, \text{kN} - 115.4 \, \text{kN} = 785 \, \text{kN}$$

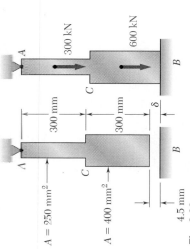

$A = 250 \, \text{mm}^2$

300 mm

$A = 400 \, \text{mm}^2$

300 mm

300 kN

600 kN

δ

4.5 mm

Fig. 2.33

2.10. PROBLEMS INVOLVING TEMPERATURE CHANGES

All of the members and structures that we have considered so far were assumed to remain at the same temperature while they were being loaded. We are now going to consider various situations involving changes in temperature.

Let us first consider a homogeneous rod AB of uniform cross section, which rests freely on a smooth horizontal surface (Fig. 2.34a). If the temperature of the rod is raised by ΔT, we observe that the rod elongates by an amount δ_T which is proportional to both the temperature change ΔT and the length L of the rod (Fig. 2.34b). We have

$$\delta_T = \alpha(\Delta T)L \tag{2.21}$$

where α is a constant characteristic of the material, called the *coefficient of thermal expansion*. Since δ_T and L are both expressed in units of length, α represents a quantity *per degree C*, or *per degree F*, depending whether the temperature change is expressed in degrees Celsius or in degrees Fahrenheit.

With the deformation δ_T must be associated a strain $\epsilon_T = \delta_T/L$. Recalling Eq. (2.21), we conclude that

$$\epsilon_T = \alpha\,\Delta T \tag{2.22}$$

The strain ϵ_T is referred to as a *thermal strain*, since it is caused by the change in temperature of the rod. In the case we are considering here, there is *no stress associated with the strain* ϵ_T.

Let us now assume that the same rod AB of length L is placed between two fixed supports at a distance L from each other (Fig. 2.35a). Again, there is neither stress nor strain in this initial condition. If we raise the temperature by ΔT, the rod cannot elongate because of the restraints imposed on its ends; the elongation δ_T of the rod is thus zero. Since the rod is homogeneous and of uniform cross section, the strain ϵ_T at any point is $\epsilon_T = \delta_T/L$ and, thus, also zero. However, the supports will exert equal and opposite forces $\mathbf{P}$ and $\mathbf{P}'$ on the rod after the temperature has been raised, to keep it from elongating (Fig. 2.35b). It thus follows that a state of stress (with no corresponding strain) is created in the rod.

Fig. 2.34

Fig. 2.35

As we prepare to determine the stress σ created by the temperature change ΔT, we observe that the problem we have to solve is statically indeterminate. Therefore, we should first compute the magnitude P of the reactions at the supports from the condition that the elongation of the rod is zero. Using the superposition method described in Sec. 2.9, we detach the rod from its support B (Fig. 2.36a) and let it elongate freely as it undergoes the temperature change ΔT (Fig. 2.36b). According to formula (2.21), the corresponding elongation is

$$\delta_T = \alpha(\Delta T)L$$

Applying now to end B the force $\mathbf{P}$ representing the redundant reaction, and recalling formula (2.7), we obtain a second deformation (Fig. 2.36c)

$$\delta_P = \frac{PL}{AE}$$

Expressing that the total deformation δ must be zero, we have

$$\delta = \delta_T + \delta_P = \alpha(\Delta T)L + \frac{PL}{AE} = 0$$

from which we conclude that

$$P = -AE\alpha(\Delta T)$$

and that the stress in the rod due to the temperature change ΔT is

$$\sigma = \frac{P}{A} = -E\alpha(\Delta T) \qquad (2.23)$$

It should be kept in mind that the result we have obtained here and our earlier remark regarding the absence of any strain in the rod *apply only in the case of a homogeneous rod of uniform cross section.* Any other problem involving a restrained structure undergoing a change in temperature must be analyzed on its own merits. However, the same general approach can be used; i.e., we can consider separately the deformation due to the temperature change and the deformation due to the redundant reaction and superpose the solutions obtained.

(a)

(b)

(c)

Fig. 2.36

EXAMPLE 2.06

Determine the values of the stress in portions AC and CB of the steel bar shown (Fig. 2.37) when the temperature of the bar is $-50°F$, knowing that a close fit exists at both of the rigid supports when the temperature is $+75°F$. Use the values $E = 29 \times 10^6$ psi and $\alpha = 6.5 \times 10^6/°F$ for steel.

We first determine the reactions at the supports. Since the problem is statically indeterminate, we detach the bar from its support at B and let it undergo the temperature change

$$\Delta T = (-50°F) - (75°F) = -125°F$$

Fig. 2.37

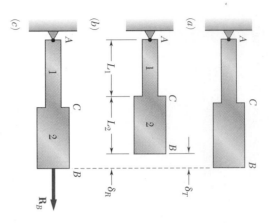

Fig. 2.38

The corresponding deformation (Fig. 2.38b) is

$$\delta_T = \alpha(\Delta T)L = (6.5 \times 10^{-6}/°F)(-125°F)(24 \text{ in.})$$
$$= -19.50 \times 10^{-3} \text{ in.}$$

Applying now the unknown force $\mathbf{R}_B$ at end B (Fig. 2.38c), we use Eq. (2.8) to express the corresponding deformation δ_R. Substituting

$$L_1 = L_2 = 12 \text{ in.}$$
$$A_1 = 0.6 \text{ in}^2 \qquad A_2 = 1.2 \text{ in}^2$$
$$P_1 = P_2 = R_B \qquad E = 29 \times 10^6 \text{ psi}$$

into Eq. (2.8), we write

$$\delta_R = \frac{P_1 L_1}{A_1 E} + \frac{P_2 L_2}{A_2 E}$$
$$= \frac{R_B}{29 \times 10^6 \text{ psi}} \left(\frac{12 \text{ in.}}{0.6 \text{ in}^2} + \frac{12 \text{ in.}}{1.2 \text{ in}^2} \right)$$
$$= (1.0345 \times 10^{-6} \text{ in./lb})R_B$$

Expressing that the total deformation of the bar must be zero as a result of the imposed constraints, we write

$$\delta = \delta_T + \delta_R = 0$$
$$= -19.50 \times 10^{-3} \text{ in.} + (1.0345 \times 10^{-6} \text{ in./lb})R_B = 0$$

from which we obtain

$$R_B = 18.85 \times 10^3 \text{ lb} = 18.85 \text{ kips}$$

The reaction at A is equal and opposite.

Noting that the forces in the two portions of the bar are $P_1 = P_2 = 18.85$ kips, we obtain the following values of the stress in portions AC and CB of the bar:

$$\sigma_1 = \frac{P_1}{A_1} = \frac{18.85 \text{ kips}}{0.6 \text{ in}^2} = +31.42 \text{ ksi}$$
$$\sigma_2 = \frac{P_2}{A_2} = \frac{18.85 \text{ kips}}{1.2 \text{ in}^2} = +15.71 \text{ ksi}$$

We cannot emphasize too strongly the fact that, while the *total deformation* of the bar must be zero, the deformations of the portions *AC and CB are not zero*. A solution of the problem based on the assumption that these deformations are zero would therefore be wrong. Neither can the values of the strain in AC or CB be assumed equal to zero. To amplify this point, let us determine the strain ϵ_{AC} in portion AC of the bar. The strain ϵ_{AC} can be divided into two component parts; one is the thermal strain ϵ_T produced in the unrestrained bar by the temperature change ΔT (Fig. 2.38b). From Eq. (2.22) we write

$$\epsilon_T = \alpha \Delta T = (6.5 \times 10^{-6}/°F)(-125°F)$$
$$= -812.5 \times 10^{-6} \text{ in./in.}$$

The other component of ϵ_{AC} is associated with the stress σ_1 due to the force $\mathbf{R}_B$ applied to the bar (Fig. 2.38c). From Hooke's law, we express this component of the strain as

$$\frac{\sigma_1}{E} = \frac{+31.42 \times 10^3 \text{ psi}}{29 \times 10^6 \text{ psi}} = +1083.4 \times 10^{-6} \text{ in./in.}$$

Adding the two components of the strain in AC, we obtain

$$\epsilon_{AC} = \epsilon_T + \frac{\sigma_1}{E} = -812.5 \times 10^{-6} + 1083.4 \times 10^{-6}$$
$$= +271 \times 10^{-6} \text{ in./in.}$$

A similar computation yields the strain in portion CB of the bar:

$$\epsilon_{CB} = \epsilon_T + \frac{\sigma_2}{E} = -812.5 \times 10^{-6} + 541.7 \times 10^{-6}$$
$$= -271 \times 10^{-6} \text{ in./in.}$$

The deformations δ_{AC} and δ_{CB} of the two portions of the bar are expressed respectively as

$$\delta_{AC} = \epsilon_{AC}(AC) = (+271 \times 10^{-6})(12 \text{ in.})$$
$$= +3.25 \times 10^{-3} \text{ in.}$$
$$\delta_{CB} = \epsilon_{CB}(CB) = (-271 \times 10^{-6})(12 \text{ in.})$$
$$= -3.25 \times 10^{-3} \text{ in.}$$

We thus check that, while the sum $\delta = \delta_{AC} + \delta_{CB}$ of the two deformations is zero, neither of the deformations is zero.

SAMPLE PROBLEM 2.3

The $\frac{1}{2}$-in.-diameter rod CE and the $\frac{3}{4}$-in.-diameter rod DF are attached to the rigid bar $ABCD$ as shown. Knowing that the rods are made of aluminum and using $E = 10.6 \times 10^6$ psi, determine (a) the force in each rod caused by the loading shown, (b) the corresponding deflection of point A.

SOLUTION

Statics. Considering the free body of bar $ABCD$, we note that the reaction at B and the forces exerted by the rods are indeterminate. However, using statics, we may write

$$+\uparrow \Sigma M_B = 0: \quad (10 \text{ kips})(18 \text{ in.}) - F_{CE}(12 \text{ in.}) - F_{DF}(20 \text{ in.}) = 0$$

$$12F_{CE} + 20F_{DF} = 180 \tag{1}$$

Geometry. After application of the 10-kip load, the position of the bar is $A'BC'D'$. From the similar triangles BAA', BCC', and BDD' we have

$$\frac{\delta_C}{12 \text{ in.}} = \frac{\delta_D}{20 \text{ in.}} \qquad \delta_C = 0.6\delta_D \tag{2}$$

$$\frac{\delta_A}{18 \text{ in.}} = \frac{\delta_D}{20 \text{ in.}} \qquad \delta_A = 0.9\delta_D \tag{3}$$

Deformations. Using Eq. (2.7), we have

$$\delta_C = \frac{F_{CE}L_{CE}}{A_{CE}E} \qquad \delta_D = \frac{F_{DF}L_{DF}}{A_{DF}E}$$

Substituting for δ_C and δ_D into (2), we write

$$\delta_C = 0.6\delta_D \qquad \frac{F_{CE}L_{CE}}{A_{CE}E} = 0.6 \frac{F_{DF}L_{DF}}{A_{DF}E}$$

$$F_{CE} = 0.6 \frac{L_{DF}}{L_{CE}} \frac{A_{CE}}{A_{DF}} F_{DF} = 0.6 \left(\frac{30 \text{ in.}}{24 \text{ in.}}\right)\left[\frac{\frac{1}{4}\pi(\frac{1}{2}\text{ in.})^2}{\frac{1}{4}\pi(\frac{3}{4}\text{ in.})^2}\right] F_{DF} \qquad F_{CE} = 0.333F_{DF}$$

Force in Each Rod. Substituting for F_{CE} into (1) and recalling that all forces have been expressed in kips, we have

$$12(0.333F_{DF}) + 20F_{DF} = 180$$

$$F_{DF} = 7.50 \text{ kips} \qquad \blacktriangledown$$

$$F_{CE} = 0.333F_{DF} = 0.333(7.50 \text{ kips})$$

$$F_{CE} = 2.50 \text{ kips} \qquad \blacktriangledown$$

Deflections. The deflection of point D is

$$\delta_D = \frac{F_{DF}L_{DF}}{A_{DF}E} = \frac{(7.50 \times 10^3 \text{ lb})(30 \text{ in.})}{\frac{1}{4}\pi(\frac{3}{4}\text{ in.})^2(10.6 \times 10^6 \text{ psi})}$$

$$\delta_D = 48.0 \times 10^{-3} \text{ in.}$$

Using (3), we write

$$\delta_A = 0.9\delta_D = 0.9(48.0 \times 10^{-3} \text{ in.}) \qquad \delta_A = 43.2 \times 10^{-3} \text{ in.} \qquad \blacktriangledown$$

77

SAMPLE PROBLEM 2.4

The rigid bar CDE is attached to a pin support at E and rests on the 30-mm-diameter brass cylinder BD. A 22-mm-diameter steel rod AC passes through a hole in the bar and is secured by a nut which is snugly fitted when the temperature of the entire assembly is 20°C. The temperature of the brass cylinder is then raised to 50°C while the steel rod remains at 20°C. Assuming that no stresses were present before the temperature change, determine the stress in the cylinder.

Rod AC: Steel	Cylinder BD: Brass
$E = 200$ GPa	$E = 105$ GPa
$\alpha = 11.7 \times 10^{-6}/°C$	$\alpha = 20.9 \times 10^{-6}/°C$

SOLUTION

Statics. Considering the free body of the entire assembly, we write

$$+\circlearrowleft \Sigma M_E = 0: \quad R_A(0.75 \text{ m}) - R_B(0.3 \text{ m}) = 0 \qquad R_A = 0.4R_B \quad (1)$$

Deformations. We use the method of superposition, considering R_B as redundant. With the support at B removed, the temperature rise of the cylinder causes point B to move down through δ_T. The reaction R_B must cause a deflection δ_1 equal to δ_T so that the final deflection of B will be zero (Fig. 3).

Deflection δ_T. Because of a temperature rise of $50° - 20° = 30°C$, the length of the brass cylinder increases by δ_T.

$$\delta_T = L(\Delta T)\alpha = (0.3 \text{ m})(30°C)(20.9 \times 10^{-6}/°C) = 188.1 \times 10^{-6} \text{ m} \downarrow$$

Deflection δ_1. We note that $\delta_D = 0.4\delta_C$ and $\delta_1 = \delta_D + \delta_{B/D}$.

$$\delta_C = \frac{R_A L}{AE} = \frac{R_A(0.9 \text{ m})}{\frac{1}{4}\pi(0.022 \text{ m})^2(200 \text{ GPa})} = 11.84 \times 10^{-9}R_A \uparrow$$

$$\delta_D = 0.40\delta_C = 0.4(11.84 \times 10^{-9}R_A) = 4.74 \times 10^{-9}R_A \uparrow$$

$$\delta_{B/D} = \frac{R_B L}{AE} = \frac{R_B(0.3 \text{ m})}{\frac{1}{4}\pi(0.03 \text{ m})^2(105 \text{ GPa})} = 4.04 \times 10^{-9}R_B \uparrow$$

We recall from (1) that $R_A = 0.4R_B$ and write

$$\delta_1 = \delta_D + \delta_{B/D} = [4.74(0.4R_B) + 4.04R_B]10^{-9} = 5.94 \times 10^{-9}R_B \uparrow$$

But $\delta_T = \delta_1$: $\quad 188.1 \times 10^{-6} \text{ m} = 5.94 \times 10^{-9}R_B \qquad R_B = 31.7 \text{ kN}$

Stress in Cylinder: $\quad \sigma_B = \frac{R_B}{A} = \frac{31.7 \text{ kN}}{\frac{1}{4}\pi(0.03)^2} \qquad \sigma_B = 44.8 \text{ MPa} \blacktriangleleft$

PROBLEMS

2.33 An axial force of 60 kN is applied to the assembly shown by means of rigid end plates. Determine (a) the normal stress in the brass shell, (b) the corresponding deformation of the assembly.

2.34 The length of the assembly decreases by 0.15 mm when an axial force is applied by means of rigid end plates. Determine (a) the magnitude of the applied force, (b) the corresponding stress in the steel core.

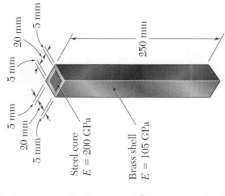

Steel core
$E = 200$ GPa

Brass shell
$E = 105$ GPa

Fig. P2.33 and P2.34

2.35 The 4.5-ft concrete post is reinforced with six steel bars, each with a $1\frac{1}{8}$-in. diameter. Knowing that $E_s = 29 \times 10^6$ psi and $E_c = 4.2 \times 10^6$ psi, determine the normal stresses in the steel and in the concrete when a 350-kip axial centric force **P** is applied to the post.

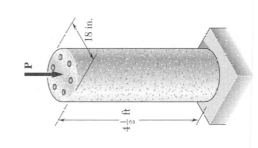

Fig. P2.35

2.36 An axial centric force of magnitude $P = 450$ kN is applied to the composite block shown by means of a rigid end plate. Knowing that $h = 10$ mm, determine the normal stress in (a) the brass core, (b) the aluminum plates.

2.37 For the composite block shown in Prob. 2.36, determine (a) the value of h if the portion of the load carried by the aluminum plates is half the portion of the load carried by the brass core, (b) the total load if the stress in the brass is 80 MPa.

2.38 For the post of Prob. 2.35, determine the maximum centric force that may be applied if the allowable normal stress is 20 ksi in the steel and 2.4 ksi in the concrete.

Rigid end plate

Brass core
($E = 105$ GPa)

Aluminum plates
($E = 70$ GPa)

Fig. P2.36 and P2.37

2.39 Three steel rods ($E = 200$ GPa) support a 36-kN load **P**. Each of the rods AB and CD has a 200-mm² cross-sectional area and rod EF has a 625-mm² cross-sectional area. Determine the (a) the change in length of rod EF, (b) the stress in each rod.

Fig. P2.39

2.40 A brass bolt ($E_b = 15 \times 10^6$ psi) with a $\frac{3}{8}$-in. diameter is fitted inside a steel tube ($E_s = 29 \times 10^6$ psi) with a $\frac{7}{8}$-in. outer diameter and $\frac{1}{8}$-in. wall thickness. After the nut has been fit snugly, it is tightened one-quarter of a full turn. Knowing that the bolt is single-threaded with a 0.1-in. pitch, determine the normal stress (a) in the bolt, (b) in the tube.

Fig. P2.40

2.41 Two cylindrical rods, CD made of steel ($E = 29 \times 10^6$ psi) and AC made of aluminum ($E = 10.4 \times 10^6$ psi), are joined at C and restrained by rigid supports at A and D. Determine (a) the reactions at A and D, (b) the deflection of point C.

Fig. P2.41

2.42 A steel tube ($E = 200$ GPa) with a 32-mm outer diameter and a 4-mm thickness is placed in a vise that is adjusted so that its jaws just touch the ends of the tube without exerting any pressure on them. The two forces shown are then applied to the tube. After these forces are applied, the vise is adjusted to decrease the distance between its jaws by 0.2 mm. Determine (a) the forces exerted by the vise on the tube at A and D, (b) the change in length of the portion BC of the tube.

Fig. P2.42

2.43 Solve Prob. 2.42, assuming that after the forces have been applied, the vise is adjusted to decrease the distance between its jaws by 0.1 mm.

2.44 Three wires are used to suspend the plate shown. Aluminum wires are used at A and B with a diameter of $\frac{1}{8}$ in. and a steel wire is used at C with a diameter of $\frac{1}{12}$ in. Knowing that the allowable stress for aluminum ($E = 10.4 \times 10^6$ psi) is 14 ksi and that the allowable stress for steel ($E = 29 \times 10^6$ psi) is 18 ksi, determine the maximum load **P** that may be applied.

Fig. P2.44

2.45 The rigid bar AD is supported by two steel wires of $\frac{1}{16}$-in. diameter ($E = 29 \times 10^6$ psi) and a pin and bracket at D. Knowing that the wires were initially taught, determine (a) the additional tension in each wire when a 220-lb load P is applied at D, (b) the corresponding deflection of point D.

Fig. P2.45

2.46 The steel rods BE and CD each have a diameter of $\frac{5}{8}$ in. ($E = 29 \times 10^6$ psi). The ends are threaded with a pitch of 0.1 in. Knowing that after being snugly fit, the nut at B is tightened one full turn, determine (a) the tension in rod CD, (b) the deflection of point C of the rigid member ABC.

2.47 The rigid rod ABC is suspended from three wires of the same material. The cross sectional area of the wire at B is equal to half of the cross-sectional area of the wires at A and C. Determine the tension in each wire caused by the load P.

Fig. P2.47

Fig. P2.46

Fig. P2.48

2.48 The rigid bar $ABCD$ is suspended from four identical wires. Determine the tension in each wire caused by the load P.

2.49 A steel railroad track ($E = 200$ GPa, $\alpha = 11.7 \times 10^{-6}/°C$) was laid out at a temperature of 6°C. Determine the normal stress in the rails when the temperature reaches 48°C, assuming that the rails (a) are welded to form a continuous track, (b) are 10 m long with 3-mm gaps between them.

2.50 The aluminum shell is fully bonded to the brass core, and the assembly is unstressed at a temperature of 78°F. Considering only axial deformations, determine the stress when the temperature reaches 180°F (a) in the brass core, (b) in the aluminum shell.

Brass core
$E = 15 \times 10^6$ psi
$\alpha = 11.6 \times 10^{-6}/°F$

Aluminum shell
$E = 10.6 \times 10^6$ psi
$\alpha = 12.9 \times 10^{-6}/°F$

Fig. P2.50

2.51 The brass shell ($\alpha_b = 20.9 \times 10^{-6}$/°C) is fully bonded to the steel core ($\alpha_s = 11.7 \times 10^{-6}$/°C). Determine the largest allowable increase in temperature if the stress in the steel core is not to exceed 55 MPa.

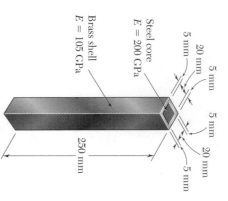

Steel core
$E = 200$ GPa

Brass shell
$E = 105$ GPa

20 mm
5 mm
5 mm
5 mm
20 mm
5 mm
250 mm

Fig. P2.51

2.52 The concrete post ($E_c = 25$ GPa and $\alpha_c = 9.9 \times 10^{-6}$/°C) is reinforced with six steel bars, each of 22-mm diameter ($E_s = 200$ GPa and $\alpha_s = 11.7 \times 10^{-6}$/°C). Determine the normal stresses induced in the steel and in the concrete by a temperature rise of 35°C.

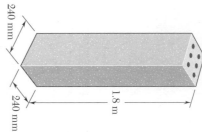

240 mm
240 mm
1.8 m

Fig. P2.52

2.53 A rod consisting of two cylindrical portions AB and BC is restrained at both ends. Portion AB is made of steel ($E_s = 29 \times 10^6$ psi, $\alpha_s = 6.5 \times 10^{-6}$/°F) and portion BC is made of brass ($E_b = 15 \times 10^6$ psi, $\alpha_b = 10.4 \times 10^{-6}$/°F). Knowing that the rod is initially unstressed, determine (a) the normal stresses induced in portions AB and BC by a temperature rise of 65°F, (b) the corresponding deflection of point B.

15 in.
12 in.
C
B
A
$2\frac{1}{4}$-in. diameter
$1\frac{1}{4}$-in. diameter

Fig. P2.53

2.54 A rod consisting of two cylindrical portions AB and BC is restrained at both ends. Portion AB is made of brass ($E_b = 105$ GPa, $\alpha_b = 20.9 \times 10^{-6}$/°C) and portion BC is made of aluminum ($E_a = 72$ GPa, $\alpha_a = 23.9 \times 10^{-6}$/°C). Knowing that the rod is initially unstressed, determine (a) the normal stresses induced in portions AB and BC by a temperature rise of 42°C, (b) the corresponding deflection of point B.

1.1 m
1.3 m
C
B
A
40-mm diameter
60-mm diameter

Fig. P2.54

2.55 The assembly shown consists of an aluminum shell $(E_a = 10.6 \times 10^6 \text{ psi}, \alpha_a = 12.9 \times 10^{-6}/°\text{F})$ fully bonded to a steel core $(E_s = 29 \times 10^6 \text{ psi}, \alpha_s = 6.5 \times 10^{-6}/°\text{F})$ and is unstressed. Determine (a) the largest allowable change in temperature if the stress in the aluminum shell is not to exceed 6 ksi, (b) the corresponding change in length of the assembly.

Fig. P2.55

2.56 For the rod of Prob. 2.53, determine the maximum allowable temperature change if the stress in the steel portion AB is not to exceed 18 ksi and if the stress in the brass portion CB is not to exceed 7 ksi.

2.57 Determine (a) the compressive force in the bars shown after a temperature rise of 96°C, (b) the corresponding change in length of the bronze bar.

2.58 Knowing that a 0.5-mm gap exists when the temperature is 20°C, determine (a) the temperature at which the normal stress in the aluminum bar will be equal to −90 MPa, (b) the corresponding exact length of the aluminum bar.

Bronze
$A = 1500 \text{ mm}^2$
$E = 105 \text{ GPa}$
$\alpha = 21.6 \times 10^{-6}/°\text{C}$

Aluminum
$A = 1800 \text{ mm}^2$
$E = 73 \text{ GPa}$
$\alpha = 23.2 \times 10^{-6}/°\text{C}$

Fig. P2.57 and P2.58

2.59 At room temperature (70°F) a 0.02-in. gap exists between the ends of the rods shown. At a later time when the temperature has reached 320°F, determine (a) the normal stress in the aluminum rod, (b) the change in length of the aluminum rod.

Aluminum
$A = 2.8 \text{ in}^2$
$E = 10.4 \times 10^6 \text{ psi}$
$\alpha = 13.3 \times 10^{-6}/°\text{F}$

Stainless steel
$A = 1.2 \text{ in}^2$
$E = 28.0 \times 10^6 \text{ psi}$
$\alpha = 9.6 \times 10^{-6}/°\text{C}$

Fig. P2.59

2.60 A brass link $(E_b = 15 \times 10^6 \text{ psi}, \alpha_b = 10.4 \times 10^{-6}/°\text{F})$ and a steel rod $(E_s = 29 \times 10^6 \text{ ksi}, \alpha_s = 6.5 \times 10^{-6}/°\text{F})$ have the dimensions shown at a temperature of 65°F. The steel rod is cooled until it fits freely into the link. The temperature of the whole assembly is then raised to 100°F. Determine (a) the final normal stress in the steel rod, (b) the final length of the steel rod.

Fig. P2.60

2.61 Two steel bars $(E_s = 200 \text{ GPa}$ and $\alpha_s = 11.7 \times 10^{-6}/°\text{C})$ are used to reinforce a brass bar $(E_b = 105 \text{ GPa}, \alpha_b = 20.9 \times 10^{-6}/°\text{C})$ which is subjected to a load $P = 25 \text{ kN}$. When the steel bars were fabricated, the distance between the centers of the holes which were to fit on the pins was made 0.5 mm smaller than the 2 m needed. The steel bars were then placed in an oven to increase their length so that they would just fit on the pins. Following fabrication, the temperature in the steel bars dropped back to room temperature. Determine (a) the increase in temperature that was required to fit the steel bars on the pins, (b) the stress in the brass bar after the load is applied to it.

Fig. P2.61

2.62 Determine the maximum load P that may be applied to the brass bar of Prob. 2.61 if the allowable stress in the steel bars is 30 MPa and the allowable stress in the brass bar is 25 MPa.

Fig. 2.39

(a)

(b)

Fig. 2.40

2.11. POISSON'S RATIO

We saw in the earlier part of this chapter that, when a homogeneous slender bar is axially loaded, the resulting stress and strain satisfy Hooke's law, as long as the elastic limit of the material is not exceeded. Assuming that the load **P** is directed along the x axis (Fig. 2.39a), we have $\sigma_x = P/A$, where A is the cross-sectional area of the bar, and, from Hooke's law,

$$\epsilon_x = \sigma_x/E \qquad (2.24)$$

where E is the modulus of elasticity of the material.

We also note that the normal stresses on faces respectively perpendicular to the y and z axes are zero: $\sigma_y = \sigma_z = 0$ (Fig. 2.39b). It would be tempting to conclude that the corresponding strains ϵ_y and ϵ_z are also zero. This, however, *is not the case*. In all engineering materials, the elongation produced by an axial tensile force **P** in the direction of the force is accompanied by a contraction in any transverse direction (Fig. 2.40).† In this section and the following sections (Secs. 2.12 through 2.15), all materials considered will be assumed to be both *homogeneous* and *isotropic*, i.e., their mechanical properties will be assumed independent of both *position* and *direction*. It follows that the strain must have the same value for any transverse direction. Therefore, for the loading shown in Fig. 2.39 we must have $\epsilon_y = \epsilon_z$. This common value is referred to as the *lateral strain*. An important constant for a given material is its *Poisson's ratio*, named after the French mathematician Siméon Denis Poisson (1781–1840) and denoted by the Greek letter ν (nu). It is defined as

$$\nu = -\frac{\text{lateral strain}}{\text{axial strain}} \qquad (2.25)$$

or

$$\nu = -\frac{\epsilon_y}{\epsilon_x} = -\frac{\epsilon_z}{\epsilon_x} \qquad (2.26)$$

for the loading condition represented in Fig. 2.39. Note the use of a minus sign in the above equations to obtain a positive value for ν, the axial and lateral strains having opposite signs for all engineering materials.‡ Solving Eq. (2.26) for ϵ_y and ϵ_z, and recalling (2.24), we write the following relations, which fully describe the condition of strain under an axial load applied in a direction parallel to the x axis:

$$\epsilon_x = \frac{\sigma_x}{E} \qquad \epsilon_y = \epsilon_z = -\frac{\nu\sigma_x}{E} \qquad (2.27)$$

†It would also be tempting, but equally wrong, to assume that the volume of the rod remains unchanged as a result of the combined effect of the axial elongation and transverse contraction (see Sec. 2.13).

‡However, some experimental materials, such as polymer foams, expand laterally when stretched. Since the axial and lateral strains have then the same sign, the Poisson's ratio of these materials is negative. (*See* Roderic Lakes, "Foam Structures with a Negative Poisson's Ratio," *Science*, 27 February 1987, Volume 235, p. 1038–1040.)

EXAMPLE 2.07

A 500-mm-long, 16-mm-diameter rod made of a homogenous, isotropic material is observed to increase in length by 300 μm, and to decrease in diameter by 2.4 μm when subjected to an axial 12-kN load. Determine the modulus of elasticity and Poisson's ratio of the material.

Fig. 2.41

The cross-sectional area of the rod is

$$A = \pi r^2 = \pi (8 \times 10^{-3} \, \text{m})^2 = 201 \times 10^{-6} \, \text{m}^2$$

Choosing the x axis along the axis of the rod (Fig. 2.41), we write

$$\sigma_x = \frac{P}{A} = \frac{12 \times 10^3 \, \text{N}}{201 \times 10^{-6} \, \text{m}^2} = 59.7 \, \text{MPa}$$

$$\epsilon_x = \frac{\delta_x}{L} = \frac{300 \, \mu\text{m}}{500 \, \text{mm}} = 600 \times 10^{-6}$$

$$\epsilon_y = \frac{\delta_y}{d} = \frac{-2.4 \, \mu\text{m}}{16 \, \text{mm}} = -150 \times 10^{-6}$$

From Hooke's law, $\sigma_x = E\epsilon_x$, we obtain

$$E = \frac{\sigma_x}{\epsilon_x} = \frac{59.7 \, \text{MPa}}{600 \times 10^{-6}} = 99.5 \, \text{GPa}$$

and, from Eq. (2.26),

$$v = -\frac{\epsilon_y}{\epsilon_x} = -\frac{-150 \times 10^{-6}}{600 \times 10^{-6}} = 0.25$$

2.12. MULTIAXIAL LOADING; GENERALIZED HOOKE'S LAW

All the examples considered so far in this chapter have dealt with slender members subjected to axial loads, i.e., to forces directed along a single axis. Choosing this axis as the x axis, and denoting by P the internal force at a given location, the corresponding stress components were found to be $\sigma_x = P/A$, $\sigma_y = 0$, and $\sigma_z = 0$.

Let us now consider structural elements subjected to loads acting in the directions of the three coordinate axes and producing normal stresses σ_x, σ_y, and σ_z which are all different from zero (Fig. 2.42). This condition is referred to as a *multiaxial loading*. Note that this is not the general stress condition described in Sec. 1.12, since no shearing stresses are included among the stresses shown in Fig. 2.42.

Fig. 2.42

Consider an element of an isotropic material in the shape of a cube (Fig. 2.43a). We can assume the side of the cube to be equal to unity, since it is always possible to select the side of the cube as a unit of length. Under the given multiaxial loading, the element will deform into a *rectangular parallelepiped* of sides equal, respectively, to $1 + \epsilon_x$, $1 + \epsilon_y$, and $1 + \epsilon_z$, where ϵ_x, ϵ_y, and ϵ_z denote the values of the normal strain in the directions of the three coordinate axes (Fig. 2.43b). You should note that, as a result of the deformations of the other elements of the material, the element under consideration could also undergo a translation, but we are concerned here only with the *actual deformation* of the element, and not with any possible superimposed rigid-body displacement.

In order to express the strain components ϵ_x, ϵ_y, ϵ_z in terms of the stress components σ_x, σ_y, σ_z, we will consider separately the effect of each stress component and combine the results obtained. The approach we propose here will be used repeatedly in this text, and is based on the *principle of superposition*. This principle states that the effect of a given combined loading on a structure can be obtained by *determining separately the effects of the various loads and combining the results obtained*, provided that the following conditions are satisfied:

1. Each effect is linearly related to the load that produces it.
2. The deformation resulting from any given load is small and does not affect the conditions of application of the other loads.

In the case of a multiaxial loading, the first condition will be satisfied if the stresses do not exceed the proportional limit of the material, and the second condition will also be satisfied if the stress on any given face does not cause deformations of the other faces that are large enough to affect the computation of the stresses on those faces.

Considering first the effect of the stress component σ_x, we recall from Sec. 2.11 that σ_x causes a strain equal to σ_x/E in the x direction, and strains equal to $-\nu\sigma_x/E$ in each of the y and z directions. Similarly, the stress component σ_y, if applied separately, will cause a strain σ_y/E in the y direction and strains $-\nu\sigma_y/E$ in the other two directions. Finally, the stress component σ_z causes a strain σ_z/E in the z direction and strains $-\nu\sigma_z/E$ in the x and y directions. Combining the results obtained, we conclude that the components of strain corresponding to the given multiaxial loading are

$$\epsilon_x = +\frac{\sigma_x}{E} - \frac{\nu\sigma_y}{E} - \frac{\nu\sigma_z}{E}$$
$$\epsilon_y = -\frac{\nu\sigma_x}{E} + \frac{\sigma_y}{E} - \frac{\nu\sigma_z}{E}$$
$$\epsilon_z = -\frac{\nu\sigma_x}{E} - \frac{\nu\sigma_y}{E} + \frac{\sigma_z}{E}$$

(2.28)

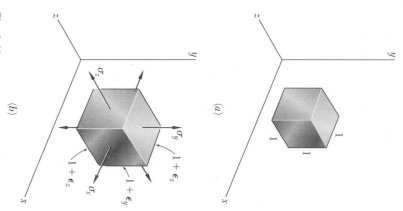

(a)

(b)

Fig. 2.43

The relations (2.28) are referred to as the *generalized Hooke's law for the multiaxial loading of a homogeneous isotropic material.* As we indicated earlier, the results obtained are valid only as long as the stresses do not exceed the proportional limit, and as long as the deformations involved remain small. We also recall that a positive value for a stress component signifies tension, and a negative value compression. Similarly, a positive value for a strain component indicates expansion in the corresponding direction, and a negative value contraction.

EXAMPLE 2.08

The steel block shown (Fig. 2.44) is subjected to a uniform pressure on all its faces. Knowing that the change in length of edge AB is -1.2×10^{-3} in., determine (*a*) the change in length of the other two edges, (*b*) the pressure p applied to the faces of the block. Assume $E = 29 \times 10^6$ psi and $\nu = 0.29$.

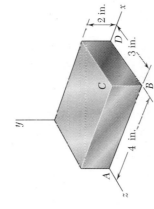

Fig. 2.44

(a) Change in Length of Other Edges. Substituting $\sigma_x = \sigma_y = \sigma_z = -p$ into the relations (2.28), we find that the three strain components have the common value

$$\epsilon_x = \epsilon_y = \epsilon_z = -\frac{p}{E}(1 - 2\nu) \qquad (2.29)$$

Since

$$\epsilon_x = \delta_x/AB = (-1.2 \times 10^{-3} \text{ in.})/(4 \text{ in.})$$
$$= -300 \times 10^{-6} \text{ in./in.}$$

we obtain

$$\epsilon_y = \epsilon_z = \epsilon_x = -300 \times 10^{-6} \text{ in./in.}$$

from which it follows that

$$\delta_y = \epsilon_y(BC) = (-300 \times 10^{-6})(2 \text{ in.}) = -600 \times 10^{-6} \text{ in.}$$
$$\delta_z = \epsilon_z(BD) = (-300 \times 10^{-6})(3 \text{ in.}) = -900 \times 10^{-6} \text{ in.}$$

(b) Pressure. Solving Eq. (2.29) for p, we write

$$p = -\frac{E\epsilon_x}{1 - 2\nu} = -\frac{(29 \times 10^6 \text{ psi})(-300 \times 10^{-6})}{1 - 0.58}$$
$$p = 20.7 \text{ ksi}$$

*2.13. DILATATION; BULK MODULUS

In this section you will examine the effect of the normal stresses σ_x, σ_y, and σ_z on the volume of an element of isotropic material. Consider the element shown in Fig. 2.43. In its unstressed state, it is in the shape of a cube of unit volume; and under the stresses σ_x, σ_y, σ_z, it deforms into a rectangular parallelepiped of volume

$$v = (1 + \epsilon_x)(1 + \epsilon_y)(1 + \epsilon_z)$$

Since the strains ϵ_x, ϵ_y, ϵ_z are much smaller than unity, their products will be even smaller and may be omitted in the expansion of the product. We have, therefore,

$$v = 1 + \epsilon_x + \epsilon_y + \epsilon_z$$

Denoting by e the change in volume of our element, we write

$$e = v - 1 = 1 + \epsilon_x + \epsilon_y + \epsilon_z - 1$$

or

$$e = \epsilon_x + \epsilon_y + \epsilon_z \qquad (2.30)$$

Since the element had originally a unit volume, the quantity e represents *the change in volume per unit volume*; it is referred to as the *dilatation* of the material. Substituting for ϵ_x, ϵ_y, and ϵ_z from Eqs. (2.28) into (2.30), we write

$$e = \frac{\sigma_x + \sigma_y + \sigma_z}{E} - \frac{2v(\sigma_x + \sigma_y + \sigma_z)}{E}$$

$$e = \frac{1 - 2v}{E}(\sigma_x + \sigma_y + \sigma_z) \qquad (2.31)\dagger$$

A case of special interest is that of a body subjected to a uniform hydrostatic pressure p. Each of the stress components is then equal to $-p$ and Eq. (2.31) yields

$$e = -\frac{3(1 - 2v)}{E}p \qquad (2.32)$$

Introducing the constant

$$k = \frac{E}{3(1 - 2v)} \qquad (2.33)$$

we write Eq. (2.32) in the form

$$e = -\frac{p}{k} \qquad (2.34)$$

The constant k is known as the *bulk modulus* or *modulus of compression* of the material. It is expressed in the same units as the modulus of elasticity E, that is, in pascals or in psi.

Observation and common sense indicate that a stable material subjected to a hydrostatic pressure can only *decrease* in volume; thus the dilatation e in Eq. (2.34) is negative, from which it follows that the bulk modulus k is a positive quantity. Referring to Eq. (2.33), we conclude that $1 - 2v > 0$, or $v < \frac{1}{2}$. On the other hand, we recall from Sec. 2.11 that v is positive for all engineering materials. We thus conclude that, for any engineering material,

$$0 < v < \tfrac{1}{2} \qquad (2.35)$$

We note that an ideal material having a value of v equal to zero could be stretched in one direction without any lateral contraction. On the other hand, an ideal material for which $v = \frac{1}{2}$, and thus $k = \infty$, would

†Since the dilatation e represents a change in volume, it must be independent of the orientation of the element considered. It then follows from Eqs. (2.30) and (2.31) that the quantities $\epsilon_x + \epsilon_y + \epsilon_z$ and $\sigma_x + \sigma_y + \sigma_z$ are also independent of the orientation of the element. This property will be verified in Chap. 7.

be perfectly incompressible ($e = 0$). Referring to Eq. (2.31) we also note that, since $\nu < \frac{1}{2}$ in the elastic range, stretching an engineering material in one direction, for example in the x direction ($\sigma_x > 0$, $\sigma_y = \sigma_z = 0$), will result in an increase of its volume ($e > 0$).†

EXAMPLE 2.09

Determine the change in volume ΔV of the steel block shown in Fig. 2.44, when it is subjected to the hydrostatic pressure $p = 180$ MPa. Use $E = 200$ GPa and $\nu = 0.29$.

From Eq. (2.33), we determine the bulk modulus of steel,

$$k = \frac{E}{3(1 - 2\nu)} = \frac{200 \text{ GPa}}{3(1 - 0.58)} = 158.7 \text{ GPa}$$

and, from Eq. (2.31), the dilatation,

$$e = -\frac{p}{k} = -\frac{180 \text{ MPa}}{158.7 \text{ GPa}} = -1.134 \times 10^{-3}$$

Since the volume V of the block in its unstressed state is

$$V = (80 \text{ mm})(40 \text{ mm})(60 \text{ mm}) = 192 \times 10^3 \text{ mm}^3$$

and since e represents the change in volume per unit volume, $e = \Delta V/V$, we have

$$\Delta V = eV = (-1.134 \times 10^{-3})(192 \times 10^3 \text{ mm}^3)$$

$$\Delta V = -218 \text{ mm}^3$$

2.14. SHEARING STRAIN

When we derived in Sec. 2.12 the relations (2.28) between normal stresses and normal strains in a homogeneous isotropic material, we assumed that no shearing stresses were involved. In the more general situation represented in Fig. 2.45, shearing stresses τ_{xy}, τ_{yz}, and τ_{zx} will be present (as well, of course, as the corresponding shearing stresses τ_{yx}, τ_{zy}, and τ_{xz}). These stresses have no direct effect on the normal strains and, as long as all the deformations involved remain small, they will not affect the derivation nor the validity of the relations (2.28). The shearing stresses, however, will tend to deform a cubic element of material into an *oblique* parallelepiped.

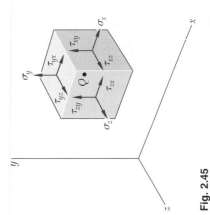

Fig. 2.45

---

†However, in the plastic range, the volume of the material remains nearly constant.

Consider first a cubic element of side one (Fig. 2.46) subjected to no other stresses than the shearing stresses τ_{xy} and τ_{yx} applied to faces of the element respectively perpendicular to the x and y axes. (We recall from Sec. 1.12 that $\tau_{xy} = \tau_{yx}$.) The element is observed to deform into a rhomboid of sides equal to one (Fig. 2.47). Two of the angles formed by the four faces under stress are reduced from $\frac{\pi}{2}$ to $\frac{\pi}{2} - \gamma_{xy}$, while the other two are increased from $\frac{\pi}{2}$ to $\frac{\pi}{2} + \gamma_{xy}$. The small angle γ_{xy} (expressed in radians) defines the *shearing strain* corresponding to the x and y directions. When the deformation involves a *reduction of the angle* formed by the two faces oriented respectively toward the positive x and y axes (as shown in Fig. 2.47), the shearing strain γ_{xy} is said to be *positive*; otherwise, it is said to be negative.

We should note that, as a result of the deformations of the other elements of the material, the element under consideration can also undergo an overall rotation. However, as was the case in our study of normal strains, we are concerned here only with the *actual deformation of* the element, and not with any possible superimposed rigid-body displacement.†

Plotting successive values of τ_{xy} against the corresponding values of γ_{xy}, we obtain the shearing stress-strain diagram for the material under consideration. This can be accomplished by carrying out a torsion test, as you will see in Chap. 3. The diagram obtained is similar to the normal stress-strain diagram obtained for the same material from the tensile test described earlier in this chapter. However, the values obtained for the yield strength, ultimate strength, etc., of a given material are only about half as large in shear as they are in tension. As was the case for normal stresses and strains, the initial portion of the shearing stress-strain diagram is a straight line. For values of the shearing stress

Fig. 2.46

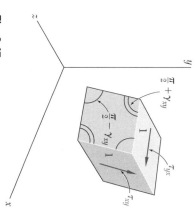

Fig. 2.47

†In defining the strain γ_{xy}, some authors arbitrarily assume that the actual deformation of the element is accompanied by a rigid-body rotation such that the horizontal faces of the element do not rotate. The strain γ_{xy} is then represented by the angle through which the other two faces have rotated (Fig. 2.48). Others assume a rigid-body rotation such that the horizontal faces rotate through $\frac{1}{2}\gamma_{xy}$ counterclockwise and the vertical faces through $\frac{1}{2}\gamma_{xy}$ clockwise (Fig. 2.49). Since both assumptions are unnecessary and may lead to confusion, we prefer in this text to associate the shearing strain γ_{xy} with the *change in the angle* formed by the two faces, rather than with the *rotation of a given face* under restrictive conditions.

Fig. 2.48

Fig. 2.49

which do not exceed the proportional limit in shear, we can therefore write for any homogeneous isotropic material,

$$\tau_{xy} = G\gamma_{xy} \qquad (2.36)$$

This relation is known as *Hooke's law for shearing stress and strain*, and the constant G is called the *modulus of rigidity* or *shear modulus* of the material. Since the strain γ_{xy} was defined as an angle in radians, it is dimensionless, and the modulus G is expressed in the same units as τ_{xy}, that is, in pascals or in psi. The modulus of rigidity G of any given material is less than one-half, but more than one-third of the modulus of elasticity E of that material.†

Considering now a small element of material subjected to shearing stresses τ_{yz} and τ_{zy} (Fig. 2.50*a*), we define the shearing strain γ_{yz} as the change in the angle formed by the faces under stress. The shearing strain γ_{zx} is defined in a similar way by considering an element subjected to shearing stresses τ_{zx} and τ_{xz} (Fig. 2.50*b*). For values of the stress which do not exceed the proportional limit, we can write the two additional relations

$$\tau_{yz} = G\gamma_{yz} \qquad \tau_{zx} = G\gamma_{zx} \qquad (2.37)$$

where the constant G is the same as in Eq. (2.36).

For the general stress condition represented in Fig. 2.45, and as long as none of the stresses involved exceeds the corresponding proportional limit, we can apply the principle of superposition and combine the results obtained in this section and in Sec. 2.12. We obtain the following group of equations representing the generalized Hooke's law for a homogeneous isotropic material under the most general stress condition.

$$\epsilon_x = +\frac{\sigma_x}{E} - \frac{\nu\sigma_y}{E} - \frac{\nu\sigma_z}{E}$$

$$\epsilon_y = -\frac{\nu\sigma_x}{E} + \frac{\sigma_y}{E} - \frac{\nu\sigma_z}{E}$$

$$\epsilon_z = -\frac{\nu\sigma_x}{E} - \frac{\nu\sigma_y}{E} + \frac{\sigma_z}{E} \qquad (2.38)$$

$$\gamma_{xy} = \frac{\tau_{xy}}{G} \qquad \gamma_{yz} = \frac{\tau_{yz}}{G} \qquad \gamma_{zx} = \frac{\tau_{zx}}{G}$$

An examination of Eqs. (2.38) might lead us to believe that three distinct constants, E, ν, and G, must first be determined experimentally, if we are to predict the deformations caused in a given material by an arbitrary combination of stresses. Actually, only two of these constants need be determined experimentally for any given material. As you will see in the next section, the third constant can then be obtained through a very simple computation.

† See Prob. 2.91.

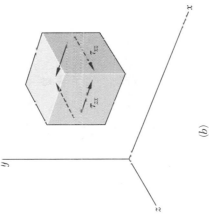

(a)

(b)

Fig. 2.50

EXAMPLE 2.10

A rectangular block of a material with a modulus of rigidity $G = 90$ ksi is bonded to two rigid horizontal plates. The lower plate is fixed, while the upper plate is subjected to a horizontal force $\mathbf{P}$ (Fig. 2.51). Knowing that the upper plate moves through 0.04 in. under the action of the force, determine (a) the average shearing strain in the material, (b) the force $\mathbf{P}$ exerted on the upper plate.

(a) *Shearing Strain.* We select coordinate axes centered at the midpoint C of edge AB and directed as shown (Fig. 2.52). According to its definition, the shearing strain γ_{xy} is equal to the angle formed by the vertical and the line CF joining the midpoints of edges AB and DE. Noting that this is a very small angle and recalling that it should be expressed in radians, we write

not an actual angle as far as degrees go

$$\gamma_{xy} \approx \tan \gamma_{xy} = \frac{0.04 \text{ in.}}{2 \text{ in.}} \qquad \gamma_{xy} = 0.020 \text{ rad}$$

Fig. 2.51

(b) *Force Exerted on Upper Plate.* We first determine the shearing stress τ_{xy} in the material. Using Hooke's law for shearing stress and strain, we have

$$\tau_{xy} = G\gamma_{xy} = (90 \times 10^3 \text{ psi})(0.020 \text{ rad}) = 1800 \text{ psi}$$

The force exerted on the upper plate is thus

$$P = \tau_{xy} A = (1800 \text{ psi})(8 \text{ in.})(2.5 \text{ in.}) = 36.0 \times 10^3 \text{ lb}$$

$$P = 36.0 \text{ kips}$$

Fig. 2.52

2.15. FURTHER DISCUSSION OF DEFORMATIONS UNDER AXIAL LOADING; RELATION AMONG E, ν, AND G

We saw in Sec. 2.11 that a slender bar subjected to an axial tensile load $\mathbf{P}$ directed along the x axis will elongate in the x direction and contract in both of the transverse y and z directions. If ϵ_x denotes the axial strain, the lateral strain is expressed as $\epsilon_y = \epsilon_z = -\nu\epsilon_x$, where ν is Poisson's ratio. Thus, an element in the shape of a cube of side equal to one and oriented as shown in Fig. 2.53a will deform into a rectangular parallelepiped of sides $1 + \epsilon_x$, $1 - \nu\epsilon_x$, and $1 - \nu\epsilon_x$. (Note that only one face of the element is shown in the figure.) On the other hand, if the element is oriented at $45°$ to the axis of the load (Fig. 2.53b), the face shown in the figure is observed to deform into a rhombus. We conclude that the axial load $\mathbf{P}$ causes in this element a shearing strain γ' equal to the amount by which each of the angles shown in Fig. 2.53b increases or decreases.†

†Note that the load $\mathbf{P}$ also produces normal strains in the element shown in Fig. 2.53b (see Prob. 2.74).

Fig. 2.53

The fact that shearing strains, as well as normal strains, result from an axial loading should not come to us as a surprise, since we already observed at the end of Sec. 1.12 that an axial load **P** causes normal and shearing stresses of equal magnitude on four of the faces of an element oriented at 45° to the axis of the member. This was illustrated in Fig. 1.40, which, for convenience, has been repeated here. It was also shown in Sec. 1.11 that the shearing stress is maximum on a plane forming an angle of 45° with the axis of the load. It follows from Hooke's law for shearing stress and strain that the shearing strain γ' associated with the element of Fig. 2.53b is also maximum: $\gamma' = \gamma_m$.

While a more detailed study of the transformations of strain will be postponed until Chap. 7, we will derive in this section a relation between the maximum shearing strain $\gamma' = \gamma_m$ associated with the element of Fig. 2.53b and the normal strain ϵ_x in the direction of the load. Let us consider for this purpose the prismatic element obtained by intersecting the cubic element of Fig. 2.53a by a diagonal plane (Fig. 2.54a and b). Referring to Fig. 2.53a, we conclude that this new element will deform into the element shown in Fig. 2.54c, which has horizontal and vertical sides respectively equal to $1 + \epsilon_x$ and $1 - \nu\epsilon_x$. But the angle formed by the oblique and horizontal faces of the element of Fig. 2.53b is precisely half of one of the right angles of the cubic ele-

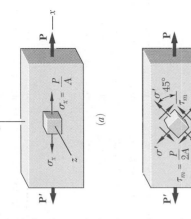

(a)

$\sigma_x = \dfrac{P}{A}$

(b)

$\tau_m = \dfrac{P}{2A}$ $\sigma' = \dfrac{P}{2A}$

Fig. 1.40 (repeated)

(a)   (b)   (c)

$1 - \nu\epsilon_x$ β $1 + \epsilon_x$

Fig. 2.54

ment considered in Fig. 2.53b. The angle β into which this angle deforms must therefore be equal to half of $\pi/2 - \gamma_m$. We write

$$\beta = \frac{\pi}{4} - \frac{\gamma_m}{2}$$

Applying the formula for the tangent of the difference of two angles, we obtain

$$\tan \beta = \frac{\tan\dfrac{\pi}{4} - \tan\dfrac{\gamma_m}{2}}{1 + \tan\dfrac{\pi}{4}\tan\dfrac{\gamma_m}{2}} = \frac{1 - \tan\dfrac{\gamma_m}{2}}{1 + \tan\dfrac{\gamma_m}{2}}$$

or, since $\gamma_m/2$ is a very small angle,

$$\tan \beta = \frac{1 - \dfrac{\gamma_m}{2}}{1 + \dfrac{\gamma_m}{2}} \tag{2.39}$$

But, from Fig. 2.54c, we observe that

$$\tan \beta = \frac{1 - \nu \epsilon_x}{1 + \epsilon_x} \tag{2.40}$$

Equating the right-hand members of (2.39) and (2.40), and solving for γ_m, we write

$$\gamma_m = \frac{(1 + \nu)\epsilon_x}{1 + \dfrac{1 - \nu}{2}\epsilon_x}$$

Since $\epsilon_x \ll 1$, the denominator in the expression obtained can be assumed equal to one; we have, therefore,

$$\gamma_m = (1 + \nu)\epsilon_x \tag{2.41}$$

which is the desired relation between the maximum shearing strain γ_m and the axial strain ϵ_x.

To obtain a relation among the constants E, ν, and G, we recall that, by Hooke's law, $\gamma_m = \tau_m/G$, and that, for an axial loading, $\epsilon_x = \sigma_x/E$. Equation (2.41) can therefore be written as

$$\frac{\tau_m}{G} = (1 + \nu)\frac{\sigma_x}{E}$$

or

$$\frac{E}{G} = (1 + \nu)\frac{\sigma_x}{\tau_m} \tag{2.42}$$

We now recall from Fig. 1.40 that $\sigma_x = P/A$ and $\tau_m = P/2A$, where A is the cross-sectional area of the member. It thus follows that $\sigma_x/\tau_m = 2$. Substituting this value into (2.42) and dividing both members by 2, we obtain the relation

$$\boxed{\frac{E}{2G} = 1 + \nu} \tag{2.43}$$

which can be used to determine one of the constants E, ν, or G from the other two. For example, solving Eq. (2.43) for G, we write

$$G = \frac{E}{2(1 + \nu)} \tag{2.43'}$$

*2.16. STRESS-STRAIN RELATIONSHIPS FOR FIBER-REINFORCED COMPOSITE MATERIALS

Fiber-reinforced composite materials were briefly discussed in Sec. 2.5. It was shown at that time that these materials are obtained by embedding fibers of a strong, stiff material into a weaker, softer material, referred to as a *matrix*. It was also shown that the relationship between the normal stress and the corresponding normal strain created in a lamina, or layer, of a composite material depends upon the direction in which the load is applied. Different moduli of elasticity, E_x, E_y, and E_z, are therefore required to describe the relationship between normal stress and normal strain, according to whether the load is applied in a direction parallel to the fibers, in a direction perpendicular to the layer, or in a transverse direction.

Let us consider again the layer of composite material discussed in Sec. 2.5 and let us subject it to a uniaxial tensile load parallel to its fibers, i.e., in the x direction (Fig. 2.55a). To simplify our analysis, it will be assumed that the properties of the fibers and of the matrix have been combined, or "smeared," into a fictitious equivalent homogeneous

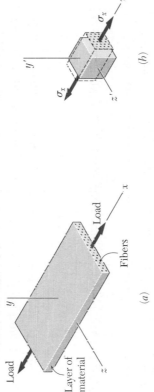

Fig. 2.55

material possessing these combined properties. We now consider a small element of that layer of smeared material (Fig. 2.55b). We denote by σ_x the corresponding normal stress and observe that $\sigma_y = \sigma_z = 0$. As indicated earlier in Sec. 2.5, the corresponding normal strain in the x direction is $\epsilon_x = \sigma_x/E_x$, where E_x is the modulus of elasticity of the composite material in the x direction. As we saw for isotropic materials, the elongation of the material in the x direction is accompanied by contractions in the y and z directions. These contractions depend upon the placement of the fibers in the matrix and will generally be different. It follows that the lateral strains ϵ_y and ϵ_z will also be different, and so will the corresponding Poisson's ratios:

$$\nu_{xy} = -\frac{\epsilon_y}{\epsilon_x} \quad \text{and} \quad \nu_{xz} = -\frac{\epsilon_z}{\epsilon_x} \quad (2.44)$$

Note that the first subscript in each of the Poisson's ratios ν_{xy} and ν_{xz} in Eqs. (2.44) refers to the direction of the load, and the second to the direction of the contraction.

It follows from the above that, in the case of the *multiaxial loading* of a layer of a composite material, equations similar to Eqs. (2.28) of Sec. 2.12 can be used to describe the stress-strain relationship. In the present case, however, three different values of the modulus of elasticity, and six different values of Poisson's ratio will be involved. We write

$$\epsilon_x = \frac{\sigma_x}{E_x} - \frac{\nu_{yx}\sigma_y}{E_y} - \frac{\nu_{zx}\sigma_z}{E_z}$$

$$\epsilon_y = -\frac{\nu_{xy}\sigma_x}{E_x} + \frac{\sigma_y}{E_y} - \frac{\nu_{zy}\sigma_z}{E_z} \qquad (2.45)$$

$$\epsilon_z = -\frac{\nu_{xz}\sigma_x}{E_x} - \frac{\nu_{yz}\sigma_y}{E_y} + \frac{\sigma_z}{E_z}$$

Equations (2.45) may be considered as defining the transformation of stress into strain for the given layer. It follows from a general property of such transformations that the coefficients of the stress components are symmetric, i.e., that

$$\frac{\nu_{xy}}{E_x} = \frac{\nu_{yx}}{E_y} \qquad \frac{\nu_{yz}}{E_y} = \frac{\nu_{zy}}{E_z} \qquad \frac{\nu_{zx}}{E_z} = \frac{\nu_{xz}}{E_x} \qquad (2.46)$$

These equations show that, while different, the Poisson's ratios ν_{xy} and ν_{yx} are not independent; either of them can be obtained from the other if the corresponding values of the modulus of elasticity are known. The same is true of ν_{yz} and ν_{zy}, and of ν_{zx} and ν_{xz}.

Consider now the effect of the presence of shearing stresses on the faces of a small element of smeared layer. As pointed out in Sec. 2.14 in the case of isotropic materials, these stresses come in pairs of equal and opposite vectors applied to opposite sides of the given element and have no effect on the normal strains. Thus, Eqs. (2.45) remain valid. The shearing stresses, however, will create shearing strains which are defined by equations similar to the last three of the equations (2.38) of Sec. 2.14, except that three different values of the modulus of rigidity, G_{xy}, G_{yz}, and G_{zx} must now be used. We have

$$\gamma_{xy} = \frac{\tau_{xy}}{G_{xy}} \qquad \gamma_{yz} = \frac{\tau_{yz}}{G_{yz}} \qquad \gamma_{zx} = \frac{\tau_{zx}}{G_{zx}} \qquad (2.47)$$

The fact that the three components of strain ϵ_x, ϵ_y, and ϵ_z can be expressed in terms of the normal stresses only and do not depend upon any shearing stresses characterizes *orthotropic materials* and distinguishes them from other anisotropic materials.

As we saw in Sec. 2.5, a flat *laminate* is obtained by superposing a number of layers or laminas. If the fibers in all layers are given the same orientation to better withstand an axial tensile load, the laminate itself will be orthotropic. If the lateral stability of the laminate is increased by positioning some of its layers so that their fibers are at a right angle to the fibers of the other layers, the resulting laminate will also be orthotropic. On the other hand, if any of the layers of a laminate are positioned so that their fibers are neither parallel nor perpendicular to the fibers of other layers, the lamina; generally, will not be orthotropic.†

†For more information on fiber-reinforced composite materials, see Hyer, M. W., *Stress Analysis of Fiber-Reinforced Composite Materials*, McGraw-Hill, New York, 1998.

EXAMPLE 2.11

A 60-mm cube is made from layers of graphite epoxy with fibers aligned in the x direction. The cube is subjected to a compressive load of 140 kN in the x direction. The properties of the composite material are: $E_x = 155.0$ GPa, $E_y = 12.10$ GPa, $E_z = 12.10$ GPa, $\nu_{xy} = 0.248$, $\nu_{xz} = 0.248$, and $\nu_{yz} = 0.458$. Determine the changes in the cube dimensions, knowing that (a) the cube is free to expand in the y and z directions (Fig. 2.56); (b) the cube is free to expand in the z direction, but is restrained from expanding in the y direction by two fixed frictionless plates (Fig. 2.57).

Fig. 2.56

Fig. 2.57

(a) Free in y and z Directions. We first determine the stress σ_x in the direction of loading. We have

$$\sigma_x = \frac{\sigma_x}{A} = \frac{-140 \times 10^3 \text{ N}}{(0.060 \text{ m})(0.060 \text{ m})} = -38.89 \text{ MPa}$$

Since the cube is not loaded or restrained in the y and z directions, we have $\sigma_y = \sigma_z = 0$. Thus, the right-hand members of Eqs. (2.45) reduce to their first terms. Substituting the given data into these equations, we write

$$\epsilon_x = \frac{\sigma_x}{E_x} = \frac{-38.89 \text{ MPa}}{155.0 \text{ GPa}} = -250.9 \times 10^{-6}$$

$$\epsilon_y = -\frac{\nu_{xy}\sigma_x}{E_x} = -\frac{(0.248)(-38.89 \text{ MPa})}{155.0 \text{ GPa}} = +62.22 \times 10^{-6}$$

$$\epsilon_z = -\frac{\nu_{xz}\sigma_x}{E_x} = -\frac{(0.248)(-38.89 \text{ MPa})}{155.0 \text{ GPa}} = +62.22 \times 10^{-6}$$

The changes in the cube dimensions are obtained by multiplying the corresponding strains by the length $L = 0.060$ m of the side of the cube:

$$\delta_x = \epsilon_x L = (-250.9 \times 10^{-6})(0.060 \text{ m}) = -15.05 \text{ }\mu\text{m}$$
$$\delta_y = \epsilon_y L = (+62.2 \times 10^{-6})(0.060 \text{ m}) = +3.73 \text{ }\mu\text{m}$$
$$\delta_z = \epsilon_z L = (+62.2 \times 10^{-6})(0.060 \text{ m}) = +3.73 \text{ }\mu\text{m}$$

(b) Free in z Direction, Restrained in y Direction. The stress in the x direction is the same as in part a, namely, $\sigma_x = -38.89$ MPa. Since the cube is free to expand in the z direction as in part a, we again have $\sigma_z = 0$. But since the cube is now restrained in the y direction, we should expect a stress σ_y different from zero. On the other hand, since the cube cannot expand in the y direction, we must have $\delta_y = 0$ and, thus, $\epsilon_y = \delta_y/L = 0$. Making $\sigma_z = 0$ and $\epsilon_y = 0$ in the second of Eqs. (2.45), solving that equation for σ_y, and substituting the given data, we have

$$\sigma_y = \left(\frac{E_y}{E_x}\right)\nu_{xy}\sigma_x = \left(\frac{12.10}{155.0}\right)(0.248)(-38.89 \text{ MPa})$$

$$= -752.9 \text{ kPa}$$

Now that the three components of stress have been determined, we can use the first and last of Eqs. (2.45) to compute the strain components ϵ_x and ϵ_z. But the first of these equations contains

Poisson's ratio ν_{yx} and, as we saw earlier, this ratio is *not equal* to the ratio ν_{xy} which was among the given data. To find ν_{yx} we use the first of Eqs. (2.46) and write

$$\nu_{yx} = \left(\frac{E_y}{E_x}\right)\nu_{xy} = \left(\frac{12.10}{155.0}\right)(0.248) = 0.01936$$

Making $\sigma_z = 0$ in the first and third of Eqs. (2.45) and substituting in these equations the given values of E_x, E_y, ν_{xz}, and ν_{yz}, as well as the values obtained for σ_x, σ_y, and ν_{yx}, we have

$$\epsilon_x = \frac{\sigma_x}{E_x} - \frac{\nu_{yx}\sigma_y}{E_y} = \frac{-38.89 \text{ MPa}}{155.0 \text{ GPa}}$$
$$- \frac{(0.01936)(-752.9 \text{ kPa})}{12.10 \text{ GPa}} = -249.7 \times 10^{-6}$$

$$\epsilon_z = -\frac{\nu_{xz}\sigma_x}{E_x} - \frac{\nu_{yz}\sigma_y}{E_y} = -\frac{(0.248)(-38.89 \text{ MPa})}{155.0 \text{ GPa}}$$
$$- \frac{(0.458)(-752.9 \text{ kPa})}{12.10 \text{ GPa}} = + 90.72 \times 10^{-6}$$

The changes in the cube dimensions are obtained by multiplying the corresponding strains by the length $L = 0.060$ m of the side of the cube:

$$\delta_x = \epsilon_x L = (-249.7 \times 10^{-6})(0.060 \text{ m}) = -14.98 \text{ }\mu\text{m}$$
$$\delta_y = \epsilon_y L = (0)(0.060 \text{ m}) = 0$$
$$\delta_z = \epsilon_z L = (+90.72 \times 10^{-6})(0.060 \text{ m}) = +5.44 \text{ }\mu\text{m}$$

Comparing the results of parts a and b, we note that the difference between the values obtained for the deformation δ_x in the direction of the fibers is negligible. However, the difference between the values obtained for the lateral deformation δ_z is not negligible. This deformation is clearly larger when the cube is restrained from deforming in the y direction.

SAMPLE PROBLEM 2.5

A circle of diameter $d = 9$ in. is scribed on an unstressed aluminum plate of thickness $t = \frac{3}{4}$ in. Forces acting in the plane of the plate later cause normal stresses $\sigma_x = 12$ ksi and $\sigma_z = 20$ ksi. For $E = 10 \times 10^6$ psi and $\nu = \frac{1}{3}$, determine the change in (a) the length of diameter AB, (b) the length of diameter CD, (c) the thickness of the plate, (d) the volume of the plate.

SOLUTION

Hooke's Law. We note that $\sigma_y = 0$. Using Eqs. (2.28) we find the strain in each of the coordinate directions.

$$\epsilon_x = +\frac{\sigma_x}{E} - \frac{\nu\sigma_y}{E} - \frac{\nu\sigma_z}{E}$$

$$= \frac{1}{10 \times 10^6 \text{ psi}}\left[(12 \text{ ksi}) - 0 - \frac{1}{3}(20 \text{ ksi})\right] = +0.533 \times 10^{-3} \text{ in./in.}$$

$$\epsilon_y = -\frac{\nu\sigma_x}{E} + \frac{\sigma_y}{E} - \frac{\nu\sigma_z}{E}$$

$$= \frac{1}{10 \times 10^6 \text{ psi}}\left[-\frac{1}{3}(12 \text{ ksi}) + 0 - \frac{1}{3}(20 \text{ ksi})\right] = -1.067 \times 10^{-3} \text{ in./in.}$$

$$\epsilon_z = -\frac{\nu\sigma_x}{E} - \frac{\nu\sigma_y}{E} + \frac{\sigma_z}{E}$$

$$= \frac{1}{10 \times 10^6 \text{ psi}}\left[-\frac{1}{3}(12 \text{ ksi}) - 0 + (20 \text{ ksi})\right] = +1.600 \times 10^{-3} \text{ in./in.}$$

a. Diameter AB. The change in length is $\delta_{B/A} = \epsilon_x d$.

$$\delta_{B/A} = \epsilon_x d = (+0.533 \times 10^{-3} \text{ in./in.})(9 \text{ in.})$$

$$\delta_{B/A} = +4.8 \times 10^{-3} \text{ in.} \quad \blacktriangle$$

b. Diameter CD.

$$\delta_{C/D} = \epsilon_z d = (+1.600 \times 10^{-3} \text{ in./in.})(9 \text{ in.})$$

$$\delta_{C/D} = +14.4 \times 10^{-3} \text{ in.} \quad \blacktriangle$$

c. Thickness. Recalling that $t = \frac{3}{4}$ in., we have

$$\delta_t = \epsilon_y t = (-1.067 \times 10^{-3} \text{ in./in.})(\tfrac{3}{4} \text{ in.})$$

$$\delta_t = -0.800 \times 10^{-3} \text{ in.} \quad \blacktriangle$$

d. Volume of the Plate. Using Eq. (2.30), we write

$$e = \epsilon_x + \epsilon_y + \epsilon_z = (+0.533 - 1.067 + 1.600)10^{-3} = +1.067 \times 10^{-3}$$

$$\Delta V = eV = +1.067 \times 10^{-3}[(15 \text{ in.})(15 \text{ in.})(\tfrac{3}{4} \text{ in.})] \quad \Delta V = +0.187 \times \text{in}^3 \quad \blacktriangle$$

PROBLEMS

2.63 In a standard tensile test a steel rod of $\frac{7}{8}$-in. diameter is subjected to a tension force of 17 kips. Knowing that $\nu = 0.3$ and $E = 29 \times 10^6$ psi, determine (a) the elongation of the rod in an 8-in. gage length, (b) the change in diameter of the rod.

Fig. P2.63

2.64 A standard tension test is used to determine the properties of an experimental plastic. The test specimen is a 15-mm-diameter rod and it is subjected to a 3.5-kN tensile force. Knowing that an elongation of 11 mm and a decrease in diameter of 0.62 mm are observed in a 120-mm gage length, determine the modulus of elasticity, the modulus of rigidity, and Poisson's ratio of the material.

Fig. P2.64

2.65 A 2-m length of an aluminum pipe of 240-mm outer diameter and 10-mm wall thickness is used as a short column and carries a centric axial load of 640 kN. Knowing that $E = 73$ GPa and $\nu = 0.33$, determine (a) the change in length of the pipe, (b) the change in its outer diameter, (c) the change in its wall thickness.

Fig. P2.65

2.66 The change in diameter of a large steel bolt is carefully measured as the nut is tightened. Knowing that $E = 200$ GPa and $\nu = 0.29$, determine the internal force in the bolt, if the diameter is observed to decrease by 13 μm.

Fig. P2.66

2.67 An aluminum plate ($E = 74$ GPa, $\nu = 0.33$) plate is subjected to a centric axial load that causes a normal stress σ. Knowing that, before loading, a line of slope 2:1 is scribed on the plate, determine the slope of the line when $\sigma = 125$ MPa.

Fig. P2.67

Fig. P2.68

2.68   A 600-lb tensile load is applied to a test coupon made from $\frac{1}{16}$-in. flat steel plate ($E = 29 \times 10^6$ psi, $v = 0.30$). Determine the resulting change (a) in the 2-in. gage length, (b) in the width of portion AB of the test coupon, (c) in the thickness of portion AB, (d) in the cross-sectional area of portion AB.

Fig. P2.69

2.69   A 1-in. square is scribed on the side of a large steel pressure vessel. After pressurization the biaxial stress condition at the square is as shown. Knowing that $E = 29 \times 10^6$ psi and $v = 0.30$, determine the change in length of (a) side AB, (b) side BC, (c) diagonal AC.

2.70   For the square of Prob. 2.69, determine the percent change in the slope of diagonal DB due to the pressurization of the vessel.

2.71   A fabric used in air-inflated structures is subjected to a biaxial loading that results in normal stresses $\sigma_x = 120$ MPa and $\sigma_z = 160$ MPa. Knowing that the properties of the fabric can be approximated as $E = 87$ GPa and $v = 0.34$, determine the change in length of (a) side AB, (b) side BC, (c) diagonal AC.

Fig. P2.71

2.72   The brass rod AD is fitted with a jacket that is used to apply a hydrostatic pressure of 48 MPa to the 240-mm portion BC of the rod. Knowing that $E = 105$ GPa and $v = 0.33$, determine (a) the change in the total length AD, (b) the change in diameter of portion BC of the rod.

Fig. P2.72

2.73 The homogeneous plate $ABCD$ is subjected to a biaxial loading as shown. It is known that $\sigma_z = \sigma_0$ and that the change in length of the plate in the x direction must be zero, that is, $\epsilon_x = 0$. Denoting by E the modulus of elasticity and by ν Poisson's ratio, determine (a) the required magnitude of σ_x, (b) the ratio σ_0/ϵ_z.

Fig. P2.73

2.74 For a member under axial loading, express the normal strain ϵ' in a direction forming an angle of $45°$ with the axis of the load in terms of the axial strain ϵ_x by (a) comparing the hypotenuses of the triangles shown in Fig. 2.54, which represent respectively an element before and after deformation; (b) using the values of the corresponding stresses σ' and σ_x shown in Fig. 1.40, and the generalized Hooke's law.

2.75 In many situations it is known that the normal stress in a given direction is zero, for example, $\sigma_z = 0$ in the case of the thin plate shown. For this case, which is known as *plane stress*, show that if the strains ϵ_x and ϵ_y have been determined experimentally, we can express σ_x, σ_y, and ϵ_z as follows:

$$\sigma_x = E\frac{\epsilon_x + \nu\epsilon_y}{1 - \nu^2}$$

$$\sigma_y = E\frac{\epsilon_y + \nu\epsilon_x}{1 - \nu^2}$$

$$\epsilon_z = -\frac{\nu}{1 - \nu}(\epsilon_x + \epsilon_y)$$

Fig. P2.75

2.76 In many situations physical constraints prevent strain from occurring in a given direction, for example, $\epsilon_z = 0$ in the case shown, where longitudinal movement of the long prism is prevented at every point. Plane sections perpendicular to the longitudinal axis remain plane and the same distance apart. Show that for this situation, which is known as *plane strain*, we can express σ_z, ϵ_x, and ϵ_y as follows:

$$\sigma_z = \nu(\sigma_x + \sigma_y)$$

$$\epsilon_x = \frac{1}{E}[(1 - \nu^2)\sigma_x - \nu(1 + \nu)\sigma_y]$$

$$\epsilon_y = \frac{1}{E}[(1 - \nu^2)\sigma_y - \nu(1 + \nu)\sigma_x]$$

(a)

(b)

Fig. P2.76

2.77 Two blocks of rubber, each of width $w = 60$ mm, are bonded to rigid supports and to the movable plate AB. Knowing that a force of magnitude $P = 19$ kN causes a deflection $\delta = 3$ mm, determine the modulus of rigidity of the rubber used.

Fig. P2.77 and P2.78

2.78 Two blocks of rubber, for which $G = 7.5$ MPa, are bonded to rigid supports and to the movable plate AB. Knowing that the width of each block is $w = 80$ mm, determine the effective spring constant, $k = P/\delta$, of the system.

2.79 The plastic block shown is bonded to a fixed base and to a horizontal rigid plate to which a force P is applied. Knowing that for the plastic used $G = 55$ ksi, determine the deflection of the plate when $P = 9$ kips.

Fig. P2.79

2.80 A vibration isolation unit consists of two blocks of hard rubber bonded to plate AB and to rigid supports as shown. For the type and grade of rubber used $\tau_{\text{all}} = 220$ psi and $G = 1800$ psi. Knowing that a centric vertical force of magnitude $P = 3.2$ kips must cause a 0.1-in. vertical deflection of the plate AB, determine the smallest allowable dimensions a and b of the block.

Fig. P2.80

2.81 An elastomeric bearing ($G = 0.9$ MPa) is used to support a bridge girder as shown to provide flexibility during earthquakes. The beam must not displace more than 10 mm when a 22-kN lateral load is applied as shown. Knowing that the maximum allowable shearing stress is 420 kPa, determine (*a*) the smallest allowable dimension *b*, (*b*) the smallest required thickness *a*.

Fig. P2.81

2.82 For the elastomeric bearing in Prob. 2.81 with $b = 220$ mm and $a = 30$ mm, determine the shearing modulus G and the shear stress τ for a maximum lateral load $P = 19$ kN and a maximum displacement $\delta = 12$ mm.

*2.83 Determine the dilatation e and the change in volume of the 200-mm length of the rod shown if (*a*) the rod is made of steel with $E = 200$ GPa and $\nu = 0.30$, (*b*) the rod is made of aluminum with $E = 70$ GPa and $\nu = 0.35$.

Fig. P2.83

*2.84 Determine the change in volume of the 2-in. gage length segment AB in Prob. 2.68 (*a*) by computing the dilatation of the material, (*b*) by subtracting the original volume of portion AB from its final volume.

*2.85 A 6-in.-diameter solid steel sphere is lowered into the ocean to a point where the pressure is 7.1 ksi (about 3 miles below the surface). Knowing that $E = 29 \times 10^6$ psi and $\nu = 0.30$, determine (*a*) the decrease in diameter of the sphere, (*b*) the decrease in volume of the sphere, (*c*) the percent increase in the density of the sphere.

*2.86 (*a*) For the axial loading shown, determine the change in height and the change in volume of the brass cylinder shown. (*b*) Solve part *a*, assuming that the loading is hydrostatic with $\sigma_x = \sigma_y = \sigma_z = -70$ MPa.

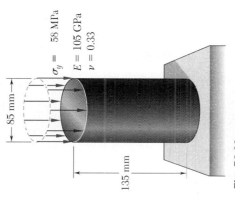

Fig. P2.86

*2.87 A vibration isolation support consists of a rod A of radius $R_1 = \frac{3}{8}$ in. and a tube B of inner radius $R_2 = 1$ in. bonded to a 3-in.-long hollow rubber cylinder with a modulus of rigidity $G = 1.8$ ksi. Determine the largest allowable force P which may be applied to rod A if its deflection is not to exceed 0.1 in.

Fig. P2.87 and P2.88

*2.88 A vibration isolation support consists of a rod A of radius R_1 and a tube B of inner radius R_2 bonded to a 3-in.-long hollow rubber cylinder with a modulus of rigidity $G = 1.6$ ksi. Determine the required value of the ratio R_2/R_1 if a 2-kip force P is to cause a 0.12-in. deflection of rod A.

Fig. P2.89

$E_x = 50$ GPa $\quad \nu_{xz} = 0.254$
$E_y = 15.2$ GPa $\quad \nu_{xy} = 0.254$
$E_z = 15.2$ GPa $\quad \nu_{zy} = 0.428$

*2.89 A composite cube with 40-mm sides and the properties shown is made with glass polymer fibers aligned in the x direction. The cube is constrained against deformations in the y and z directions and is subjected to a tensile load of 65 kN in the x direction. Determine (*a*) the change in the length of the cube in the x direction, (*b*) the stresses σ_x, σ_y, and σ_z.

*2.90 The composite cube of Prob. 2.89 is constrained against deformation in the z direction and elongated in the x direction by 0.035 mm by a tensile load in the x direction. Determine (*a*) the stresses σ_x, σ_y, and σ_z, (*b*) the change in the dimension in the y direction.

*2.91 Show that for any given material, the ratio G/E of the modulus of rigidity over the modulus of elasticity is always less than $\frac{1}{2}$ but more than $\frac{1}{3}$ [*Hint*: Refer to Eq. (2.43) and to Sec. 2.13.]

*2.92 The material constants E, G, k, and ν are related by Eqs. (2.33) and (2.43). Show that any one of these constants may be expressed in terms of any other two constants. For example, show that (*a*) $k = GE/(9G - 3E)$ and (*b*) $\nu = (3k - 2G)/(6k + 2G)$.

2.17. STRESS AND STRAIN DISTRIBUTION UNDER AXIAL LOADING; SAINT-VENANT'S PRINCIPLE

We have assumed so far that, in an axially loaded member, the normal stresses are uniformly distributed in any section perpendicular to the axis of the member. As we saw in Sec. 1.5, such an assumption may be quite in error in the immediate vicinity of the points of application of the loads. However, the determination of the actual stresses in a given section of the member requires the solution of a statically indeterminate problem.

In Sec. 2.9, you saw that statically indeterminate problems involving the determination of *forces* can be solved by considering the *deformations* caused by these forces. It is thus reasonable to conclude that the determination of the *stresses* in a member requires the analysis of the *strains* produced by the stresses in the member. This is essentially the approach found in advanced textbooks, where the mathematical theory of elasticity is used to determine the distribution of stresses corresponding to various modes of application of the loads at the ends of the member. Given the more limited mathematical means at our disposal, our analysis of stresses will be restricted to the particular case when two rigid plates are used to transmit the loads to a member made of a homogeneous isotropic material (Fig. 2.58).

If the loads are applied at the center of each plate,† the plates will move toward each other without rotating, causing the member to get shorter, while increasing in width and thickness. It is reasonable to assume that the member will remain straight, that plane sections will re-

Fig. 2.58

$\mathbf{P}$

$\mathbf{P'}$

†More precisely, the common line of action of the loads should pass through the centroid of the cross section (*cf.* Sec. 1.5).

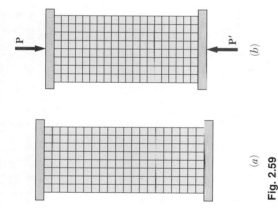

(a)

(b)

Fig. 2.59

main plane, and that all elements of the member will deform in the same way, since such an assumption is clearly compatible with the given end conditions. This is illustrated in Fig. 2.59, which shows a rubber model before and after loading.‡ Now, if all elements deform in the same way, the distribution of strains throughout the member must be uniform. In other words, the axial strain ϵ_y and the lateral strain $\epsilon_x = -\nu\epsilon_y$ are constant. But, if the stresses do not exceed the proportional limit, Hooke's law applies and we may write $\sigma_y = E\epsilon_y$, from which it follows that the normal stress σ_y is also constant. Thus, the distribution of stresses is uniform throughout the member and, at any point,

$$\sigma_y = (\sigma_y)_{\text{ave}} = \frac{P}{A}$$

Fig. 2.60

On the other hand, if the loads are concentrated, as illustrated in Fig. 2.60, the elements in the immediate vicinity of the points of application of the loads are subjected to very large stresses, while other elements near the ends of the member are unaffected by the loading. This may be verified by observing that strong deformations, and thus large strains and large stresses, occur near the points of application of the loads, while no deformation takes place at the corners. As we consider elements farther and farther from the ends, however, we note a progressive equalization of the deformations involved, and thus a more nearly uniform distribution of the strains and stresses across a section of the member. This is further illustrated in Fig. 2.61, which shows the result of the calculation by advanced mathematical methods of the dis-

‡Note that for long, slender members, another configuration is possible, and indeed will prevail, if the load is sufficiently large; the member *buckles* and assumes a curved shape. This will be discussed in Chap. 10.

tribution of stresses across various sections of a thin rectangular plate subjected to concentrated loads. We note that at a distance b from either end, where b is the width of the plate, the stress distribution is nearly uniform across the section, and the value of the stress σ_y at any point of that section can be assumed equal to the average value P/A. Thus, at a distance equal to, or greater than, the width of the member, the distribution of stresses across a given section is the same, whether the member is loaded as shown in Fig. 2.58 or Fig. 2.60. In other words, except in the immediate vicinity of the points of application of the loads, the stress distribution may be assumed independent of the actual mode of application of the loads. This statement, which applies not only to axial loadings, but to practically any type of load, is known as *Saint-Venant's principle*, after the French mathematician and engineer Adhémar Barré de Saint-Venant (1797–1886).

While Saint-Venant's principle makes it possible to replace a given loading by a simpler one for the purpose of computing the stresses in a structural member, you should keep in mind two important points when applying this principle:

1. The actual loading and the loading used to compute the stresses must be *statically equivalent*.

2. Stresses cannot be computed in this manner in the immediate vicinity of the points of application of the loads. Advanced theoretical or experimental methods must be used to determine the distribution of stresses in these areas.

You should also observe that the plates used to obtain a uniform stress distribution in the member of Fig. 2.59 must allow the member to freely expand laterally. Thus, the plates cannot be rigidly attached to the member; you must assume them to be just in contact with the member, and smooth enough not to impede the lateral expansion of the member. While such end conditions can actually be achieved for a member in compression, they cannot be physically realized in the case of a member in tension. It does not matter, however, whether or not an actual fixture can be realized and used to load a member so that the distribution of stresses in the member is uniform. The important thing is to *imagine a model* which will allow such a distribution of stresses, and to keep this model in mind so that you may later compare it with the actual loading conditions.

Fig. 2.61

$\sigma_{min} = 0.973\sigma_{ave}$
$\sigma_{max} = 1.027\sigma_{ave}$

$\sigma_{min} = 0.668\sigma_{ave}$
$\sigma_{max} = 1.387\sigma_{ave}$

$\sigma_{min} = 0.198\sigma_{ave}$
$\sigma_{max} = 2.575\sigma_{ave}$

$\sigma_{ave} = \dfrac{P}{A}$

2.18. STRESS CONCENTRATIONS

As you saw in the preceding section, the stresses near the points of application of concentrated loads can reach values much larger than the average value of the stress in the member. When a structural member contains a discontinuity, such as a hole or a sudden change in cross section, high localized stresses can also occur near the discontinuity. Figures 2.62 and 2.63 show the distribution of stresses in critical sections corresponding to two such situations. Figure 2.62 refers to a flat bar with a *circular hole* and shows the stress distribution in a section passing through the center of the hole. Figure 2.63 refers to a flat bar consisting of two portions of different widths connected by *fillets*; it shows the stress distribution in the narrowest part of the connection, where the highest stresses occur.

Fig. 2.63   Stress distribution near fillets in flat bar under axial loading.

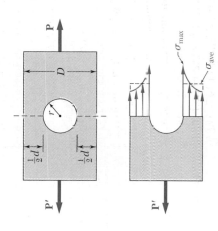

Fig. 2.62   Stress distribution near circular hole in flat bar under axial loading.

These results were obtained experimentally through the use of a photoelastic method. Fortunately for the engineer who has to design a given member and cannot afford to carry out such an analysis, the results obtained are independent of the size of the member and of the material used; they depend only upon the ratios of the geometric parameters involved, i.e., upon the ratio r/d in the case of a circular hole, and upon the ratios r/d and D/d in the case of fillets. Furthermore, the designer is more interested in the *maximum value* of the stress in a given section, than in the actual distribution of stresses in that section, since his main concern is to determine *whether* the allowable stress will be exceeded under a given loading, and not *where* this value will be exceeded. For this reason, one defines the ratio

$$K = \frac{\sigma_{max}}{\sigma_{ave}} \tag{2.48}$$

of the maximum stress over the average stress computed in the critical (narrowest) section of the discontinuity. This ratio is referred to as the *stress-concentration factor* of the given discontinuity. Stress-concentration factors can be computed once and for all in terms of the ratios of the geometric parameters involved, and the results obtained can be expressed in the form of tables or of graphs, as shown in Fig. 2.64. To determine the maximum stress occurring near a discontinuity in a given member subjected to a given axial load P, the designer needs only to compute the average stress $\sigma_{ave} = P/A$ in the critical section, and multiply the result obtained by the appropriate value of the stress-concentration factor K. You should note, however, that this procedure is valid only as long as σ_{max} does not exceed the proportional limit of the material, since the values of K plotted in Fig. 2.64 were obtained by assuming a linear relation between stress and strain.

(a) Flat bars with holes

Fig. 2.64 Stress concentration factors for flat bars under axial loading[†]

Note that the average stress must be computed across the narrowest section: $\sigma_{ave} = P/td$, where t is the thickness of the bar.

(b) Flat bars with fillets

Determine the largest axial load **P** that can be safely supported by a flat steel bar consisting of two portions, both 10 mm thick, and respectively 40 and 60 mm wide, connected by fillets of radius $r = 8$ mm. Assume an allowable normal stress of 165 MPa.

We first compute the ratios

$$\frac{D}{d} = \frac{60 \text{ mm}}{40 \text{ mm}} = 1.50 \qquad \frac{r}{d} = \frac{8 \text{ mm}}{40 \text{ mm}} = 0.20$$

Using the curve in Fig. 2.64b corresponding to $D/d = 1.50$, we find that the value of the stress-concentration factor corresponding to $r/d = 0.20$ is

$$K = 1.82$$

Carrying this value into Eq. (2.48) and solving for σ_{ave}, we have

$$\sigma_{ave} = \frac{\sigma_{max}}{1.82}$$

But σ_{max} cannot exceed the allowable stress $\sigma_{all} = 165$ MPa. Substituting this value for σ_{max}, we find that the average stress in the narrower portion ($d = 40$ mm) of the bar should not exceed the value

$$\sigma_{ave} = \frac{165 \text{ MPa}}{1.82} = 90.7 \text{ MPa}$$

Recalling that $\sigma_{ave} = P/A$, we have

$$P = A\sigma_{ave} = (40 \text{ mm})(10 \text{ mm})(90.7 \text{ MPa}) = 36.3 \times 10^3 \text{ N}$$
$$P = 36.3 \text{ kN}$$

†W. D. Pilkey, *Peterson's Stress Concentration Factors*, 2nd ed., John Wiley & Sons, New York, 1997.

2.19. PLASTIC DEFORMATIONS

The results obtained in the preceding sections were based on the assumption of a linear stress-strain relationship. In other words, we assumed that the proportional limit of the material was never exceeded. This is a reasonable assumption in the case of brittle materials, which rupture without yielding. In the case of ductile materials, however, this assumption implies that the yield strength of the material is not exceeded. The deformations will then remain within the elastic range and the structural member under consideration will regain its original shape after all loads have been removed. If, on the other hand, the stresses in any part of the member exceed the yield strength of the material, plastic deformations occur and most of the results obtained in earlier sections cease to be valid. A more involved analysis, based on a nonlinear stress-strain relationship, must then be carried out.

While an analysis taking into account the actual stress-strain relationship is beyond the scope of this text, we gain considerable insight into plastic behavior by considering an idealized *elastoplastic material* for which the stress-strain diagram consists of the two straight-line segments shown in Fig. 2.65. We may note that the stress-strain diagram for mild steel in the elastic and plastic ranges is similar to this idealization. As long as the stress σ is less than the yield strength σ_Y, the material behaves elastically and obeys Hooke's law, $\sigma = E\epsilon$. When σ reaches the value σ_Y, the material starts yielding and keeps deforming plastically under a constant load. If the load is removed, unloading takes place along a straight-line segment CD parallel to the initial portion AY of the loading curve. The segment AD of the horizontal axis represents the strain corresponding to the permanent set or plastic deformation resulting from the loading and unloading of the specimen. While no actual material behaves exactly as shown in Fig. 2.65, this stress-strain diagram will prove useful in discussing the plastic deformations of ductile materials such as mild steel.

Fig. 2.65

EXAMPLE 2.13

A rod of length $L = 500$ mm and cross-sectional area $A = 60$ mm^2 is made of an elastoplastic material having a modulus of elasticity $E = 200$ GPa in its elastic range and a yield point $\sigma_Y = 300$ MPa. The rod is subjected to an axial load until it is stretched 7 mm and the load is then removed. What is the resulting permanent set?

Referring to the diagram of Fig. 2.65, we find that the maximum strain, represented by the abscissa of point C, is

$$\epsilon_C = \frac{\delta_C}{L} = \frac{7 \text{ mm}}{500 \text{ mm}} = 14 \times 10^{-3}$$

On the other hand, the yield strain, represented by the abscissa of point Y, is

$$\epsilon_Y = \frac{\sigma_Y}{E} = \frac{300 \times 10^6 \text{ Pa}}{200 \times 10^9 \text{ Pa}} = 1.5 \times 10^{-3}$$

The strain after unloading is represented by the abscissa ϵ_D of point D. We note from Fig. 2.65 that

$$\epsilon_D = AD = YC = \epsilon_C - \epsilon_Y$$
$$= 14 \times 10^{-3} - 1.5 \times 10^{-3} = 12.5 \times 10^{-3}$$

The permanent set is the deformation δ_D corresponding to the strain ϵ_D. We have

$$\delta_D = \epsilon_D L = (12.5 \times 10^{-3})(500 \text{ mm}) = 6.25 \text{ mm}$$

EXAMPLE 2.14

A 30-in.-long cylindrical rod of cross-sectional area $A_r = 0.075$ in^2 is placed inside a tube of the same length and of cross-sectional area $A_t = 0.100$ in^2. The ends of the rod and tube are attached to a rigid support on one side, and to a rigid plate on the other, as shown in the longitudinal section of Fig. 2.66. The rod and tube are both assumed to be elastoplastic, with moduli of elasticity $E_r = 30 \times 10^6$ psi and $E_t = 15 \times 10^6$ psi, and yield strengths $(\sigma_r)_y = 36$ ksi and $(\sigma_t)_y = 45$ ksi. Draw the load-deflection diagram of the rod-tube assembly when a load **P** is applied to the plate as shown.

Fig. 2.66

30 in.

Rod

Tube

Plate

P

We first determine the internal force and the elongation of the rod as it begins to yield:

$$(P_r)_y = (\sigma_r)_y A_r = (36 \text{ ksi})(0.075 \text{ in}^2) = 2.7 \text{ kips}$$

$$(\delta_r)_y = (\epsilon_r)_y L = \frac{(\sigma_r)_y}{E_r} L = \frac{36 \times 10^3 \text{ psi}}{30 \times 10^6 \text{ psi}} (30 \text{ in.})$$

$$= 36 \times 10^{-3} \text{ in.}$$

Since the material is elastoplastic, the force-elongation diagram *of the rod alone* consists of an oblique straight line and of a horizontal straight line, as shown in Fig. 2.67a. Following the same procedure for the tube, we have

$$(P_t)_y = (\sigma_t)_y A_t = (45 \text{ ksi})(0.100 \text{ in}^2) = 4.5 \text{ kips}$$

$$(\delta_t)_y = (\epsilon_t)_y L = \frac{(\sigma_t)_y}{E_t} L = \frac{45 \times 10^3 \text{ psi}}{15 \times 10^6 \text{ psi}} (30 \text{ in.})$$

$$= 90 \times 10^{-3} \text{ in.}$$

The load-deflection diagram of the tube alone is shown in Fig. 2.67b. Observing that the load and deflection of the rod-tube combination are, respectively,

$$P = P_r + P_t \qquad \delta = \delta_r = \delta_t$$

we draw the required load-deflection diagram by adding the ordinates of the diagrams obtained for the rod and for the tube (Fig. 2.67c). Points Y_r and Y_t correspond to the onset of yield in the rod and in the tube, respectively.

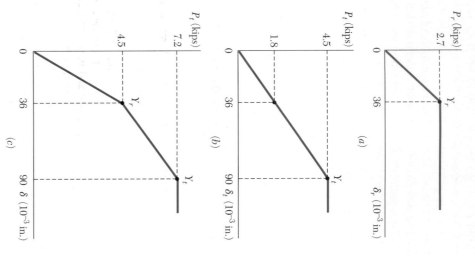

Fig. 2.67

EXAMPLE 2.15

If the load **P** applied to the rod-tube assembly of Example 2.14 is increased from zero to 5.7 kips and decreased back to zero, determine (a) the maximum elongation of the assembly, (b) the permanent set after the load has been removed.

(a) Maximum Elongation.

Referring to Fig. 2.67c, we observe that the load $P_{max} = 5.7$ kips corresponds to a point located on the segment Y_tY_t of the load-deflection diagram of the assembly. Thus, the rod has reached the plastic range, with $P_r = (P_r)_Y = 2.7$ kips and $\sigma_r = (\sigma_r)_Y = 36$ ksi, while the tube is still in the elastic range, with

$$P_t = P - P_r = 5.7 \text{ kips} - 2.7 \text{ kips} = 3.0 \text{ kips}$$

$$\sigma_t = \frac{P_t}{A_t} = \frac{3.0 \text{ kips}}{0.1 \text{ in}^2} = 30 \text{ ksi}$$

$$\delta_t = \epsilon_t L = \frac{\sigma_t}{E_t} L = \frac{30 \times 10^3 \text{ psi}}{15 \times 10^6 \text{ psi}} (30 \text{ in.}) = 60 \times 10^{-3} \text{ in.}$$

The maximum elongation of the assembly, therefore, is

$$\delta_{max} = \delta_t = 60 \times 10^{-3} \text{ in.}$$

(b) Permanent Set.

As the load **P** decreases from 5.7 kips to zero, the internal forces P_r and P_t both decrease along a straight line, as shown in Fig. 2.68a and b, respectively. The force P_r decreases along line CD parallel to the initial portion of the loading curve, while the force P_t decreases along the original loading curve, since the yield stress was not exceeded in the tube. Their sum P, therefore, will decrease along a line CE parallel to the portion $0Y_r$ of the load-deflection curve of the assembly (Fig. 2.68c). Referring to Fig. 2.67c, we find that the slope of $0Y_r$, and thus of CE, is

$$m = \frac{4.5 \text{ kips}}{36 \times 10^{-3} \text{ in.}} = 125 \text{ kips/in.}$$

The segment of line FE in Fig. 2.68c represents the deformation δ' of the assembly during the unloading phase, and the segment $0E$ the permanent set δ_p after the load **P** has been removed. From triangle CEF we have

$$\delta' = -\frac{P_{max}}{m} = -\frac{5.7 \text{ kips}}{125 \text{ kips/in.}} = -45.6 \times 10^{-3} \text{ in.}$$

The permanent set is thus

$$\delta_p = \delta_{max} + \delta' = 60 \times 10^{-3} - 45.6 \times 10^{-3}$$
$$= 14.4 \times 10^{-3} \text{ in.}$$

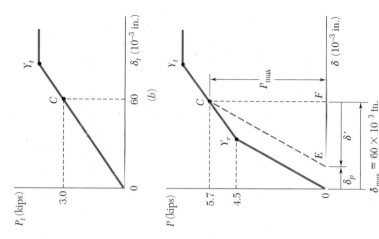

Fig. 2.68

We recall that the discussion of stress concentrations of Sec. 2.18 was carried out under the assumption of a linear stress-strain relationship. The stress distributions shown in Figs. 2.62 and 2.63, and the values of the stress-concentration factors plotted in Fig. 2.64 cannot be used, therefore, when plastic deformations take place, i.e., when the value of σ_{max} obtained from these figures exceeds the yield strength σ_Y.

Let us consider again the flat bar with a circular hole of Fig. 2.62, and let us assume that the material is elastoplastic, i.e., that its stress-strain diagram is as shown in Fig. 2.65. As long as no plastic deformation takes place, the distribution of stresses is as indicated in Sec. 2.18 (Fig. 2.69a). We observe that the area under the stress-distribution curve represents the integral $\int \sigma \, dA$, which is equal to the load P. Thus this area, and the value of σ_{max}, must increase as the load P increases. As long as $\sigma_{max} \leq \sigma_Y$, all the successive stress distributions obtained as P increases will have the shape shown in Fig. 2.62 and repeated in Fig. 2.69a. However, as P is increased beyond the value P_Y corresponding to $\sigma_{max} = \sigma_Y$ (Fig. 2.69b), the stress-distribution curve must flatten in the vicinity of the hole (Fig. 2.69c), since the stress in the material considered cannot exceed the value σ_Y. This indicates that the material is yielding in the vicinity of the hole. As the load P is further increased, the plastic zone where yield takes place keeps expanding, until it reaches the edges of the plate (Fig. 2.69d). At that point, the distribution of stresses across the plate is uniform, $\sigma = \sigma_Y$, and the corresponding value $P = P_U$ of the load is the largest which may be applied to the bar without causing rupture.

It is interesting to compare the maximum value P_Y of the load which can be applied if no permanent deformation is to be produced in the bar, with the value P_U which will cause rupture. Recalling the definition of the average stress, $\sigma_{ave} = P/A$, where A is the net cross-sectional area, and the definition of the stress concentration factor, $K = \sigma_{max}/\sigma_{ave}$, we write

$$P = \sigma_{ave}A = \frac{\sigma_{max}A}{K} \qquad (2.49)$$

for any value of σ_{max} which does not exceed σ_Y. When $\sigma_{max} = \sigma_Y$ (Fig. 2.69b), we have $P = P_Y$, and Eq. (2.49) yields

$$P_Y = \frac{\sigma_Y A}{K} \qquad (2.50)$$

On the other hand, when $P = P_U$ (Fig. 2.69d), we have $\sigma_{ave} = \sigma_Y$ and

$$P_U = \sigma_Y A \qquad (2.51)$$

Comparing Eqs. (2.50) and (2.51), we conclude that

$$P_Y = \frac{P_U}{K} \qquad (2.52)$$

Fig. 2.69 Distribution of stresses in elastoplastic material under increasing load.

(a) σ_{max} σ_Y P

(b) $\sigma_{max} = \sigma_Y$ P_Y

(c) $\sigma_{max} = \sigma_Y$ P

(d) $\sigma_{ave} = \sigma_Y$ P_U

*2.20 RESIDUAL STRESSES

In Example 2.13 of the preceding section, we considered a rod that was stretched beyond the yield point. As the load was removed, the rod did not regain its original length; it had been permanently deformed. However, after the load was removed, all stresses disappeared. You should not assume that this will always be the case. Indeed, when only some of the parts of an indeterminate structure undergo plastic deformations, as in Example 2.15, or when different parts of the structure undergo different plastic deformations, the stresses in the various parts of the structure will not, in general, return to zero after the load has been removed. Stresses, called *residual stresses*, will remain in the various parts of the structure.

While the computation of the residual stresses in an actual structure can be quite involved, the following example will provide you with a general understanding of the method to be used for their determination.

EXAMPLE 2.16

Determine the residual stresses in the rod and tube of Examples 2.14 and 2.15 after the load **P** has been increased from zero to 5.7 kips and decreased back to zero.

We observe from the diagrams of Fig. 2.70 that after the load **P** has returned to zero, the internal forces P_r and P_t are *not* equal to zero. Their values have been indicated by point E in parts a and b, respectively, of Fig. 2.70. It follows that the corresponding stresses are not equal to zero either after the assembly has been unloaded. To determine these residual stresses, we shall determine the reverse stresses σ_r' and σ_t' caused by the unloading and add them to the maximum stresses $\sigma_r = 36$ ksi and $\sigma_t = 30$ ksi found in part a of Example 2.15.

The strain caused by the unloading is the same in the rod and in the tube. It is equal to δ'/L, where δ' is the deformation of the assembly during unloading, which was found in Example 2.15. We have

$$\epsilon' = \frac{\delta'}{L} = \frac{-45.6 \times 10^{-3} \text{ in.}}{30 \text{ in.}} = -1.52 \times 10^{-3} \text{in./in.}$$

The corresponding reverse stresses in the rod and tube are

$$\sigma_r' = \epsilon' E_r = (-1.52 \times 10^{-3})(30 \times 10^6 \text{ psi}) = -45.6 \text{ ksi}$$
$$\sigma_t' = \epsilon' E_t = (-1.52 \times 10^{-3})(15 \times 10^6 \text{ psi}) = -22.8 \text{ ksi}$$

The residual stresses are found by superposing the stresses due to loading and the reverse stresses due to unloading. We have

$$(\sigma_r)_{\text{res}} = \sigma_r + \sigma_r' = 36 \text{ ksi} - 45.6 \text{ ksi} = -9.6 \text{ ksi}$$
$$(\sigma_t)_{\text{res}} = \sigma_t + \sigma_t' = 30 \text{ ksi} - 22.8 \text{ ksi} = +7.2 \text{ ksi}$$

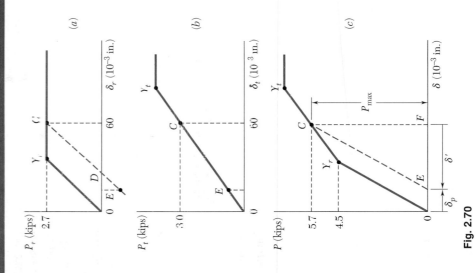

(a)

(b)

(c)

Fig. 2.70

Plastic deformations caused by temperature changes can also result in residual stresses. For example, consider a small plug that is to be welded to a large plate. For discussion purposes consider the plug will be considered as a small rod AB that is to be welded across a small hole in the plate (Fig. 2.71). During the welding process the temperature of the rod will be raised to over 1000°C, at which temperature its modulus of elasticity, and hence its stiffness and stress, will be almost zero. Since the plate is large, its temperature will not be increased significantly above room temperature (20°C). Thus, when the welding is completed, we have rod AB at $T = 1000$°C, with no stress, attached to the plate which is at 20°C.

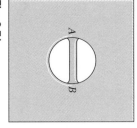

Fig. 2.71

As the rod cools, its modulus of elasticity increases and, at about 500°C, will approach its normal value of about 200 GPa. As the temperature of the rod decreases further, we have a situation similar to that considered in Sec. 2.10 and illustrated in Fig. 2.35. Solving Eq. (2.23) for ΔT and making σ equal to the yield strength, $\sigma_Y = 300$ MPa, of average steel, and $\alpha = 12 \times 10^{-6}$/°C, we find the temperature change which will cause the rod to yield:

$$\Delta T = -\frac{\sigma}{E\alpha} = -\frac{300 \text{ MPa}}{(200 \text{ GPa})(12 \times 10^{-6}/°C)} = -125°C$$

This means that the rod will start yielding at about 375°C and will keep yielding at a fairly constant stress level, as it cools down to room temperature. As a result of the welding operation, a residual stress approximately equal to the yield strength of the steel used is thus created in the plug and in the weld.

Residual stresses also occur as a result of the cooling of metals which have been cast or hot rolled. In these cases, the outer layers cool more rapidly than the inner core. This causes the outer layers to reacquire their stiffness (E returns to its normal value) faster than the inner core. When the entire specimen has returned to room temperature, the inner core will have contracted more than the outer layers. The result is residual longitudinal tensile stresses in the inner core and residual compressive stresses in the outer layers.

Residual stresses due to welding, casting, and hot rolling can be quite large (of the order of magnitude of the yield strength). These stresses can be removed, when necessary, by reheating the entire specimen to about 600°C, and then allowing it to cool slowly over a period of 12 to 24 hours.

SAMPLE PROBLEM 2.6

The rigid beam ABC is suspended from two steel rods as shown and is initially horizontal. The midpoint B of the beam is deflected 10 mm downward by the slow application of the force Q, after which the force is slowly removed. Knowing that the steel used for the rods is elastoplastic with $E = 200$ GPa and $\sigma_Y = 300$ MPa, determine (a) the required maximum value of Q and the corresponding position of the beam, (b) the final position of the beam.

Areas:
$AD = 400$ mm²
$CE = 500$ mm²

SOLUTION

Statics. Since Q is applied at the midpoint of the beam, we have

$$P_{AD} = P_{CE} \qquad \text{and} \qquad Q = 2P_{AD}$$

Elastic Action. The maximum value of Q and the maximum elastic deflection of point A occur when $\sigma = \sigma_Y$ in rod AD.

$$(P_{AD})_{max} = \sigma_Y A = (300 \text{ MPa})(400 \text{ mm}^2) = 120 \text{ kN}$$
$$Q_{max} = 2(P_{AD})_{max} = 2(120 \text{ kN}) \qquad Q_{max} = 240 \text{ kN} \ \blacktriangledown$$
$$\delta_{A_1} = \epsilon L = \frac{\sigma_Y}{E} L = \left(\frac{300 \text{ MPa}}{200 \text{ GPa}}\right)(2 \text{ m}) = 3 \text{ mm}$$

Since $P_{CE} = P_{AD} = 120$ kN, the stress in rod CE is

$$\sigma_{CE} = \frac{P_{CE}}{A} = \frac{120 \text{ kN}}{500 \text{ mm}^2} = 240 \text{ MPa}$$

The corresponding deflection of point C is

$$\delta_{C_1} = \epsilon L = \frac{\sigma_{CE}}{E} L = \left(\frac{240 \text{ MPa}}{200 \text{ GPa}}\right)(5 \text{ m}) = 6 \text{ mm}$$

The corresponding deflection of point B is

$$\delta_{B_1} = \tfrac{1}{2}(\delta_{A_1} + \delta_{C_1}) = \tfrac{1}{2}(3 \text{ mm} + 6 \text{ mm}) = 4.5 \text{ mm}$$

Since we must have $\delta_B = 10$ mm, we conclude that plastic deformation will occur.

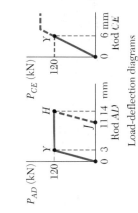

Load-deflection diagrams

Plastic Deformation. For $Q = 240$ kN, plastic deformation occurs in rod AD, where $\sigma_{AD} = \sigma_Y = 300$ MPa. Since the stress in rod CE is within the elastic range, δ_C remains equal to 6 mm. The deflection δ_A for which $\delta_B = 10$ mm is obtained by writing

$$\delta_{B_2} = 10 \text{ mm} = \tfrac{1}{2}(\delta_{A_2} + 6 \text{ mm}) \qquad \delta_{A_2} = 14 \text{ mm}$$

(a) Deflections for $\delta_B = 10$ mm $\ \blacktriangledown$

Unloading. As force Q is slowly removed, the force P_{AD} decreases along line HJ parallel to the initial portion of the load-deflection diagram of rod AD. The final deflection of point A is

$$\delta_{A_3} = 14 \text{ mm} - 3 \text{ mm} = 11 \text{ mm}$$

Since the stress in rod CE remained within the elastic range, we note that the final deflection of point C is zero.

(b) Final deflections $\ \blacktriangledown$

PROBLEMS

2.93 Two holes have been drilled through a long steel bar that is subjected to a centric axial load as shown. For $P = 6.5$ kips, determine the maximum value of the stress (*a*) at *A*, (*b*) at *B*.

Fig. P2.95 and P2.96

Fig. P2.93 and P2.94

2.94 Knowing that $\sigma_{all} = 16$ ksi, determine the maximum allowable value of the centric axial load **P**.

2.95 Knowing that, for the plate shown, the allowable stress is 125 MPa, determine the maximum allowable value of *P* when (*a*) $r = 12$ mm, (*b*) $r = 18$ mm.

2.96 Knowing that $P = 38$ kN, determine the maximum stress when (*a*) $r = 10$ mm, (*b*) $r = 16$ mm, (*c*) $r = 18$ mm.

Fig. P2.98

Fig. P2.97

2.97 Knowing that the hole has a diameter of $\frac{3}{8}$-in., determine (*a*) the radius r_f of the fillets for which the same maximum stress occurs at the hole *A* and at the fillets, (*b*) the corresponding maximum allowable load **P** if the allowable stress is 15 ksi.

2.98 For $P = 8.5$ kips, determine the minimum plate thickness *t* required if the allowable stress is 18 ksi.

2.99 (a) Knowing that the allowable stress is 140 MPa, determine the maximum allowable magnitude of the centric load **P**. (b) Determine the percent change in the maximum allowable magnitude of **P** if the raised portions are removed at the ends of the specimen.

Fig. P2.99

Fig. P2.100

2.100 A centric axial force is applied to the steel bar shown. Knowing that σ_{all} is 135 MPa, determine the maximum allowable load **P**.

2.101 The 30-mm-square bar AB has a length $L = 2.2$ m; it is made of a mild steel that is assumed to be elastoplastic with $E = 200$ GPa and $\sigma_Y = 345$ MPa. A force **P** is applied to the bar until end A has moved down by an amount δ_m. Determine the maximum value of the force **P** and the permanent set of the bar after the force has been removed, knowing that (a) $\delta_m = 4.5$ mm, (b) $\delta_m = 8$ mm.

2.102 The 30-mm-square bar AD has a length $L = 2.5$ m; it is made of mild steel that is assumed to be elastoplastic with $E = 200$ GPa and $\sigma_Y = 345$ MPa. A force **P** is applied to the bar and then removed to give it a permanent set δ_p. Determine the maximum value of the force **P** and the maximum amount δ_m by which the bar should be stretched if the desired value of δ_p is (a) 3.5 mm, (b) 6.5 mm.

Fig. P2.101 and P2.102

2.103 Rod AB is made of a mild steel that is assumed to be elastoplastic with $E = 29 \times 10^6$ ksi and $\sigma_Y = 36$ ksi. After the rod has been attached to a rigid lever CD, it is found that end C is $\frac{3}{8}$ in. too high. A vertical force **Q** is then applied at C until this point has moved to position C'. Determine the required magnitude of **Q** and the deflection δ_1 if the lever is to *snap* back to a horizontal position after **Q** is removed.

Fig. P2.103

2.104 Solve Prob. 2.103, assuming that the yield point of the mild steel is 50 ksi.

2.105 Rods AB and BC are made of a mild steel that is assumed to be elastoplastic with $E = 200$ GPa and $\sigma_Y = 345$ MPa. The rods are stretched until the end has moved down 9 mm. Neglecting stress concentrations, determine (a) the maximum value of the force $\mathbf{P}$, (b) the permanent set measured at points A and B after the force has been removed.

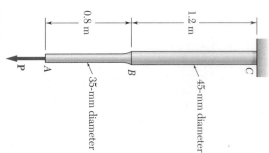

Fig. P2.105

2.106 Solve Prob. 2.105, assuming that the yield point of the mild steel is 250 MPa.

2.107 Each of the three 6-mm-diameter steel cables is made of an elastoplastic material for which $\sigma_Y = 345$ MPa and $E = 200$ GPa. A force $\mathbf{P}$ is applied to the rigid bar ABC until the bar has moved downward a distance $\delta = 2$ mm. Knowing that the cables were initially taut, determine (a) the maximum value of $\mathbf{P}$, (b) the maximum stress that occurs in cable AD, (c) the final displacement of the bar after the load is removed. (*Hint:* In part c, cable BE is not taut.)

Fig. P2.107

2.108 Solve Prob. 2.107, assuming that the cables are replaced by rods of the same cross-sectional area and material. Further assume that the rods are braced so that they can carry compressive forces.

2.109 Rod AB consists of two cylindrical portions AC and BC, each with a cross-sectional area of 1750 mm². Portion AC is made of a mild steel with $E = 200$ GPa and $\sigma_Y = 250$ MPa, and portion BC is made of a high-strength steel with $E = 200$ GPa and $\sigma_Y = 345$ MPa. A load $\mathbf{P}$ is applied at C as shown. Assuming both steels to be elastoplastic, determine (a) the maximum deflection of C if P is gradually increased from zero to 975 kN and then reduced back to zero, (b) the maximum stress in each portion of the rod, (c) the permanent deflection of C.

Fig. P2.109

2.110 For the composite rod of Prob. 2.109, if P is gradually increased from zero until the deflection of point C reaches a maximum value of $\delta_m = 0.3$ mm and then decreased back to zero, determine, (a) the maximum value of P, (b) the maximum stress in each portion of the rod, (c) the permanent deflection of C after the load is removed.

Fig. P2.111

2.111 Two tempered-steel bars, each $\frac{3}{16}$ in. thick, are bonded to a $\frac{1}{2}$-in. mild-steel bar. This composite bar is subjected as shown to a centric axial load of magnitude P. Both steels are elastoplastic with $E = 29 \times 10^6$ psi and with yield strengths equal to 100 ksi and 50 ksi, respectively, for the tempered and mild steel. The load P is gradually increased from zero until the deformation of the bar reaches a maximum value $\delta_m = 0.04$ in. and then decreased back to zero. Determine (a) the maximum value of P, (b) the maximum stress in the tempered-steel bars, (c) the permanent set after the load is removed.

2.112 For the composite bar of Prob. 2.111, if P is gradually increased from zero to 98 kips and then decreased back to zero, determine (a) the maximum deformation of the bar, (b) the maximum stress in the tempered-steel bars, (c) the permanent set after the load is removed.

2.113 Bar AB has a cross-sectional area of 1200 mm^2 and is made of a steel that is assumed to be elastoplastic with $E = 200$ GPa and $\sigma_Y = 250$ MPa. Knowing that the force **F** increases from 0 to 520 kN and then decreases to zero, determine (a) the permanent deflection of point C, (b) the residual stress in the bar.

Fig. P2.113

2.114 Solve Prob. 2.113, assuming that $a = 180$ mm.

*2.115 For the composite bar of Prob. 2.111, determine the residual stresses in the tempered-steel bars if P is gradually increased from zero to 98 kips and then decreased back to zero.

*2.116 For the composite bar in Prob. 2.111, determine the residual stresses in the tempered-steel bars if P is gradually increased from zero until the deformation of the bar reaches a maximum value $\delta_m = 0.4$ in. and is then decreased back to zero.

2.117 A uniform steel rod of cross-sectional area A is attached to rigid supports and is unstressed at a temperature of 8°C. The steel is assumed to be elastoplastic with $\sigma_Y = 250$ MPa and $G = 200$ GPa. Knowing that $\alpha = 11.7 \times 10^{-6}/°C$, determine the stress in the bar (a) when the temperature is raised to 165°C, (b) after the temperature has returned to 8°C.

Fig. P2.117

2.118 A narrow bar of aluminum is bonded to the side of a thick steel plate as shown. Initially, at $T_1 = 20°C$, all stresses are zero. Knowing that the temperature will be slowly raised to T_2 and then reduced to T_1, determine (a) the highest temperature T_2 that does *not* result in residual stresses, (b) the temperature T_2 that will result in a residual stress in the aluminum equal to 100 MPa. Assume $\alpha_a = 23.6 \times 10^{-6}/°C$ for the aluminum and $\alpha_s = 11.7 \times 10^{-6}/°C$ for the steel. Further assume that the aluminum is elastoplastic, with $E = 70$ GPa and $\sigma_Y = 100$ MPa. (*Hint:* Neglect the small stresses in the plate.)

Fig. P2.118

2.119 The steel rod ABC is attached to rigid supports and is unstressed at a temperature of 38°F. The steel is assumed elastoplastic, with $\sigma_Y = 36$ ksi and $E = 29 \times 10^6$ psi. The temperature of both portions of the rod is then raised to 250°F. Knowing that $\alpha = 6.5 \times 10^{-6}/°F$, determine (a) the stress in portion AC, (b) the deflection of point C.

$A = 0.70 \text{ in}^2$ $A = 1.0 \text{ in}^2$

Fig. P2.119

*2.120 Solve Prob. 2.119, assuming that the temperature of the rod is raised to 250°F and then returned to 38°F.

2.121 The rigid bar ABC is supported by two links, AD and BE, of uniform 37.5 × 6-mm rectangular cross section and made of a mild steel that is assumed to be elastoplastic with $E = 200$ GPa and $\sigma_Y = 250$ MPa. The magnitude of the force $\mathbf{Q}$ applied at B is gradually increased from zero to 260 kN. Knowing that $a = 0.640$ m, determine (a) the value of the normal stress in each link, (b) the maximum deflection of point B.

Fig. P2.121

2.122 Solve Prob. 2.121, knowing that $a = 1.76$ m and that the magnitude of the force $\mathbf{Q}$ applied at B is gradually increased from zero to 135 kN.

*2.123 Solve Prob. 2.121, assuming that the magnitude of the force $\mathbf{Q}$ applied at B is gradually increased from zero to 260 kN and then decreased back to zero. Knowing that $a = 0.640$ m, determine (a) the residual stress in each link, (b) the final deflection of point B. Assume that the links are braced so that they can carry compressive forces without buckling.

REVIEW AND SUMMARY
FOR CHAPTER 2

This chapter was devoted to the introduction of the concept of *strain*, to the discussion of the relationship between stress and strain in various types of materials, and to the determination of the deformations of structural components under axial loading.

Considering a rod of length L and uniform cross section and denoting by δ its deformation under an axial load **P** (Fig. 2.1), we defined the *normal strain* ϵ in the rod as the *deformation per unit length* [Sec. 2.2]:

$$\epsilon = \frac{\delta}{L} \qquad (2.1)$$

In the case of a rod of variable cross section, the normal strain was defined at any given point Q by considering a small element of rod at Q. Denoting by Δx the length of the element and by $\Delta \delta$ its deformation under the given load, we wrote

$$\epsilon = \lim_{\Delta x \to 0} \frac{\Delta \delta}{\Delta x} = \frac{d\delta}{dx} \qquad (2.2)$$

Plotting the stress σ versus the strain ϵ as the load increased, we obtained a *stress-strain diagram* for the material used [Sec. 2.3]. From such a diagram, we were able to distinguish between *brittle* and *ductile* materials: A specimen made of a brittle material ruptures without any noticeable prior change in the rate of elongation (Fig. 2.11), while a specimen made of a ductile material *yields* after a critical stress σ_Y, called the *yield strength*, has been reached, i.e., the specimen undergoes a large deformation before rupturing, with a relatively small increase in the applied load (Fig. 2.9). An example of brittle material with different properties in tension and in compression was provided by *concrete*.

Normal strain

(a)

(b)

Fig. 2.1

Stress-strain diagram

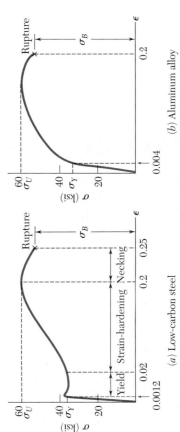

(a) Low-carbon steel

(b) Aluminum alloy

Fig. 2.9

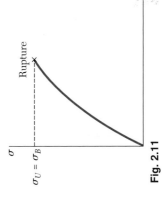

Fig. 2.11

Hooke's law
Modulus of elasticity

We noted in Sec. 2.5 that the initial portion of the stress-strain diagram is a straight line. This means that for small deformations, the stress is directly proportional to the strain:

$$\sigma = Ee \qquad (2.4)$$

This relation is known as *Hooke's law* and the coefficient E as the *modulus of elasticity* of the material. The largest stress for which Eq. (2.4) applies is the *proportional limit* of the material.

Materials considered up to this point were *isotropic*, i.e., their properties were independent of direction. In Sec. 2.5 we also considered a class of *anisotropic* materials, i.e., materials whose properties depend upon direction. They were *fiber-reinforced composite materials*, made of fibers of a strong, stiff material embedded in layers of a weaker, softer material (Fig. 2.17). We saw that different moduli of elasticity had to be used, depending upon the direction of loading.

Fig. 2.17

Elastic limit. Plastic deformation

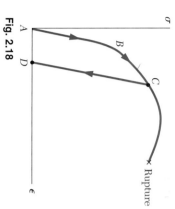

Fig. 2.18

Fatigue. Endurance limit

Elastic deformation under axial loading

Fig. 2.22

If the strains caused in a test specimen by the application of a given load disappear when the load is removed, the material is said to behave *elastically*; and the largest stress for which this occurs is called the *elastic limit* of the material [Sec. 2.6]. If the elastic limit is exceeded, the stress and strain decrease in a linear fashion when the load is removed and the strain does not return to zero (Fig. 2.18), indicating that a *permanent set or plastic deformation* of the material has taken place.

In Sec. 2.7, we discussed the phenomenon of *fatigue*, which causes the failure of structural or machine components after a very large number of repeated loadings, even though the stresses remain in the elastic range. A standard fatigue test consists in determining the number n of successive loading-and-unloading cycles required to cause the failure of a specimen for any given maximum stress level σ, and plotting the resulting σ-n curve. The value of σ for which failure does not occur, even for an indefinitely large number of cycles, is known as the *endurance limit* of the material used in the test.

Section 2.8 was devoted to the determination of the elastic deformations of various types of machine and structural components under various conditions of axial loading. We saw that if a rod of length L and uniform cross section of area A is subjected at its end to a centric axial load **P** (Fig. 2.22), the corresponding deformation is

$$\delta = \frac{PL}{AE} \qquad (2.7)$$

If the rod is loaded at several points or consists of several parts of various cross sections and possibly of different materials, the deformation δ of the rod must be expressed as the sum of the deformations of its component parts [Example 2.01]:

$$\delta = \sum_i \frac{P_i L_i}{A_i E_i} \qquad (2.8)$$

Statically indeterminate problems

Fig. 2.25a

Rod (A_1, E_1)

Tube (A_2, E_2)

End plate

P

L

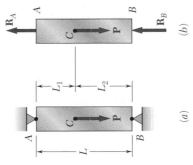

(a)

A

C

P

B

L_1

L_2

L

$\mathbf{R}_A$

A

C

P

B

$\mathbf{R}_B$

(b)

Fig. 2.26

Problems with temperature changes

(a)

A

B

L

B

δ_T

(b)

A

B

(c)

A

B

δ_P

P

L

Fig. 2.36

A

L

B

Fig. 2.35a

Section 2.9 was devoted to the solution of *statically indeterminate problems*, i.e., problems in which the reactions and the internal forces *cannot* be determined from statics alone. The equilibrium equations derived from the free-body diagram of the member under consideration were complemented by relations involving deformations and obtained from the geometry of the problem. The forces in the rod and in the tube of Fig. 2.25a, for instance, were determined by observing, on one hand, that their sum is equal to P, and on the other, that they cause equal deformations in the rod and in the tube [Example 2.02]. Similarly, the reactions at the supports of the bar of Fig. 2.26 could not be obtained from the free-body diagram of the bar alone [Example 2.03]; but they could be determined by expressing that the total elongation of the bar must be equal to zero.

In Sec. 2.10, we considered problems involving *temperature changes*. We first observed that if the temperature of an *unrestrained* rod AB of length L is increased by ΔT, its elongation is

$$\delta_T = \alpha(\Delta T)L \qquad (2.21)$$

where α is the *coefficient of thermal expansion* of the material. We noted that the corresponding strain, called *thermal strain*, is

$$\epsilon_t = \alpha\Delta T \qquad (2.22)$$

and that *no stress* is associated with this strain. However, if the rod AB is *restrained* by fixed supports (Fig. 2.35a), stresses develop in

the rod as the temperature increases, because of the reactions at the supports. To determine the magnitude P of the reactions, we detached the rod from its support at B (Fig. 2.36) and considered separately the deformation δ_T of the rod as it expands freely because of the temperature change, and the deformation δ_P caused by the force P required to bring it back to its original length, so that it may be reattached to the support at B. Writing that the total deformation $\delta = \delta_T + \delta_P$ is equal to zero, we obtained an equation that could be solved for P. While the final strain in rod AB is clearly zero, this will generally not be the case for rods and bars consisting of elements of different cross sections or materials, since the deformations of the various elements will usually *not* be zero [Example 2.06].

Fig. 2.39a

When an axial load **P** is applied to a homogeneous, slender bar (Fig. 2.39a), it causes a strain, not only along the axis of the bar but in any transverse direction as well [Sec. 2.11]. This strain is referred to as the *lateral strain*, and the ratio of the lateral strain over the axial strain is called *Poisson's ratio* and is denoted by ν (Greek letter nu). We wrote

Lateral strain. Poisson's ratio

$$\nu = -\frac{\text{lateral strain}}{\text{axial strain}} \qquad (2.25)$$

Recalling that the axial strain in the bar is $\epsilon_x = \sigma_x/E$, we expressed as follows the condition of strain under an axial loading in the *x* direction:

$$\epsilon_x = \frac{\sigma_x}{E} \qquad \epsilon_y = \epsilon_z = -\frac{\nu \sigma_x}{E} \qquad (2.27)$$

This result was extended in Sec. 2.12 to the case of a *multiaxial loading* causing the state of stress shown in Fig. 2.42. The resulting strain condition was described by the following relations, referred to as the *generalized Hooke's law* for a multiaxial loading.

Fig. 2.42

Multiaxial loading

$$\epsilon_x = +\frac{\sigma_x}{E} - \frac{\nu \sigma_y}{E} - \frac{\nu \sigma_z}{E}$$

$$\epsilon_y = -\frac{\nu \sigma_x}{E} + \frac{\sigma_y}{E} - \frac{\nu \sigma_z}{E}$$

$$\epsilon_z = -\frac{\nu \sigma_x}{E} - \frac{\nu \sigma_y}{E} + \frac{\sigma_z}{E} \qquad (2.28)$$

Dilatation

If an element of material is subjected to the stresses σ_x, σ_y, σ_z, it will deform and a certain change of volume will result [Sec. 2.13]. The *change in volume per unit volume* is referred to as the *dilatation* of the material and is denoted by *e*. We showed that

$$e = \frac{1 - 2\nu}{E}(\sigma_x + \sigma_y + \sigma_z) \qquad (2.31)$$

When a material is subjected to a hydrostatic pressure *p*, we have

$$e = -\frac{p}{k} \qquad (2.34)$$

Bulk modulus

where *k* is known as the *bulk modulus* of the material:

$$k = \frac{E}{3(1 - 2\nu)} \qquad (2.33)$$

Fig. 2.45

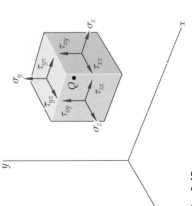

Fig. 2.47

Shearing strain. Modulus of rigidity

As we saw in Chap. 1, the state of stress in a material under the most general loading condition involves shearing stresses, as well as normal stresses (Fig. 2.45). The shearing stresses tend to deform a cubic element of material into an oblique parallelepiped [Sec. 2.14]. Considering, for instance, the stresses τ_{xy} and τ_{yx} shown in Fig. 2.47 (which, we recall, are equal in magnitude), we noted that they cause the angles formed by the faces on which they act to either increase or decrease by a small angle γ_{xy}; this angle, expressed in radians, defines the *shearing strain* corresponding to the x and y directions. Defining in a similar way the shearing strains γ_{yz} and γ_{zx}, we wrote the relations

$$\tau_{xy} = G\gamma_{xy} \qquad \tau_{yz} = G\gamma_{yz} \qquad \tau_{zx} = G\gamma_{zx} \qquad (2.36, 37)$$

which are valid for any homogeneous isotropic material within its proportional limit in shear. The constant G is called the *modulus of rigidity* of the material and the relations obtained express *Hooke's law for shearing stress and strain*. Together with Eqs. (2.28), they form a group of equations representing the generalized Hooke's law for a homogeneous isotropic material under the most general stress condition.

We observed in Sec. 2.15 that while an axial load exerted on a slender bar produces only normal strains—both axial and transverse—on an element of material oriented along the axis of the bar, it will produce both normal and shearing strains on an element rotated through 45° (Fig. 2.53). We also noted that the three constants

Fig. 2.53

Fiber-reinforced composite materials

Saint-Venant's principle

Stress concentrations

Plastic deformations

which may be used to determine any of the three constants in terms of the other two.

$$\frac{E}{2G} = 1 + \nu \qquad (2.43)$$

E, ν, and G are not independent; they satisfy the relation.

Stress-strain relationships for fiber-reinforced composite materials were discussed in an optional section (Sec. 2.16). Equations similar to Eqs. (2.28) and (2.36, 37) were derived for these materials, but we noted that direction-dependent moduli of elasticity, Poisson's ratios, and moduli of rigidity had to be used.

In Sec. 2.17, we discussed *Saint-Venant's principle*, which states that except in the immediate vicinity of the points of application of the loads, the distribution of stresses in a given member is independent of the actual mode of application of the loads. This principle makes it possible to assume a uniform distribution of stresses in a member subjected to concentrated axial loads, except close to the points of application of the loads, where stress concentrations will occur.

Stress concentrations will also occur in structural members near a discontinuity, such as a hole or a sudden change in cross section [Sec. 2.18]. The ratio of the maximum value of the stress occurring near the discontinuity over the average stress computed in the critical section is referred to as the *stress-concentration factor* of the discontinuity and is denoted by K:

$$K = \frac{\sigma_{max}}{\sigma_{ave}} \qquad (2.48)$$

Values of K for circular holes and fillets in flat bars were given in Fig. 2.64 on p. 108.

In Sec. 2.19, we discussed the *plastic deformations* which occur in structural members made of a ductile material when the stresses in some part of the member exceed the yield strength of the material. Our analysis was carried out for an idealized *elastoplastic material* characterized by the stress-strain diagram shown in Fig. 2.65

Fig. 2.65

[Examples 2.13, 2.14, and 2.15]. Finally, in Sec. 2.20, we observed that when an indeterminate structure undergoes plastic deformations, the stresses do not, in general, return to zero after the load has been removed. The stresses remaining in the various parts of the structure are called *residual stresses* and may be determined by adding the maximum stresses reached during the loading phase and the reverse stresses corresponding to the unloading phase [Example 2.16].

REVIEW PROBLEMS

2.124 The aluminum rod ABC ($E = 10.1 \times 10^6$ psi), which consists of two cylindrical portions AB and BC, is to be replaced with a cylindrical steel rod DE ($E = 29 \times 10^6$ psi) of the same overall length. Determine the minimum required diameter d of the steel rod if its vertical deformation is not to exceed the deformation of the aluminum rod under the same load and if the allowable stress in the steel rod is not to exceed 24 ksi.

2.125 A 250-mm-long aluminum tube ($E = 70$ GPa) of 36-mm outer diameter and 28-mm inner diameter may be closed at both ends by means of single-threaded screw-on covers of 1.5-mm pitch. With one cover screwed on tight, a solid brass rod ($E = 105$ GPa) of 25-mm diameter is placed inside the tube and the second cover is screwed on. Since the rod is slightly longer than the tube, it is observed that the cover must be forced against the rod by rotating it one-quarter of a turn before it can be tightly closed. Determine (a) the average normal stress in the tube and in the rod, (b) the deformations of the tube and of the rod.

2.126 In Prob. 2.125, determine the average normal stress in the tube and the rod, assuming that the temperature was 15°C when the nuts were snugly fitted and that the final temperature is 55°C. (For aluminum, $\alpha = 23.6 \times 10^6$/°C; for brass, $\alpha = 20.9 \times 10^6$/°C.)

2.127 The block shown is made of a magnesium alloy for which $E = 6.5 \times 10^6$ psi and $\nu = 0.35$. Knowing that $\sigma_x = -20$ ksi, determine (a) the magnitude of σ_y for which the change in the height of the block will be zero, (b) the corresponding change in the area of the face $ABCD$, (c) the corresponding change in the volume of the block.

2.128 The uniform rods AB and BC are made of steel and are loaded as shown. Knowing that $E = 29 \times 10^6$ psi, determine the magnitude and direction of the deflection of point B when $\theta = 22°$.

2.129 Knowing that $E = 29 \times 10^6$ psi, determine (a) the value of θ for which the deflection of point B is down and to the left along a line forming an angle of 36° with the horizontal, (b) the corresponding magnitude of the deflection of B.

28 kips

D d

E

28 kips

A 1.5 in.

B 2.25 in.

C

12 in.

18 in.

Fig. P2.124

σ_x

σ_y

y

B

E

C

F

D

G

1 in.

$1\tfrac{3}{8}$ in.

4 in.

Fig. P2.127

36 mm

28 mm

250 mm

25 mm

Fig. P2.125

25 in.

C

45 in.

B

A

Area = 0.8 in²

Area = 1.2 in²

25 kips

θ

Fig. P2.128 and P2.129

2.130 The uniform wire ABC, of unstretched length $2l$, is attached to the supports shown and a vertical load $\mathbf{P}$ is applied at the midpoint B. Denoting by A the cross-sectional area of the wire and by E the modulus of elasticity, show that, for $\delta \ll l$, the deflection at the midpoint B is

$$\delta = l\sqrt[3]{\dfrac{P}{AE}}$$

Fig. P2.130

Fig. P2.131

2.131 The steel bars BE and AD each have a 6×18-mm cross section. Knowing that $E = 200$ GPa, determine the deflections of points A, B, and C of the rigid bar ABC.

2.132 In Prob. 2.131, the 3.2-kN force caused point C to deflect to the right. Using $\alpha = 11.7 \times 10^{6}/°C$, determine the ($a$) the overall change in temperature that causes point C to return to its original position, (b) the corresponding total deflection of points A and B.

2.133 A hole is to be drilled in the plate at A. The diameters of the bits available to drill the hole range from 9 to 27 mm in 6-mm increments. (a) Determine the diameter d of the largest bit that can be used if the allowable load at the hole is not to exceed that at the fillets. (b) If the allowable stress in the plate is 145 MPa, what is the corresponding allowable load $\mathbf{P}$?

Fig. P2.133 and P2.134

2.134 (a) For $P = 58$ kN and $d = 12$ mm, determine the maximum stress in the plate shown. (b) Solve part a, assuming that the hole at A is not drilled.

2.135 The steel tensile specimen $ABCD$ ($E = 29 \times 10^{6}$ psi and $\sigma_y = 50$ ksi) is loaded in tension until the maximum strain is $\epsilon = 0.0025$. (a) Neglecting the effect of the fillets on the change in length of the specimen, determine the resulting overall length AD of the specimen after the load is removed. (b) Following the removal of the load in part a, a compressive load is applied until the maximum compressive strain is $\epsilon = 0.0020$. Determine the resulting overall length AD after the compressive load is removed.

Fig. P2.135

COMPUTER PROBLEMS

The following problems are designed to be solved with a computer. Write each program so that it can be used with either SI or U.S. customary units and in such a way that solid cylindrical elements may be defined by either their diameter or their cross-sectional area.

2.C1 A rod consisting of n elements, each of which is homogeneous and of uniform cross section, is subjected to the loading shown. The length of element i is denoted by L_i, its cross-sectional area by A_i, its modulus of elasticity by E_i, and the load applied to its right end by $\mathbf{P}_i$, the magnitude P_i of this load being assumed to be positive if $\mathbf{P}_i$ is directed to the right and negative otherwise. (a) Write a computer program that can be used to determine the average normal stress in each element, the deformation of each element, and the total deformation of the rod. (b) Use this program to solve Probs. 2.17 and 2.18.

Fig. P2.C1

2.C2 Rod AB is horizontal with both ends fixed; it consists of n elements, each of which is homogeneous and of uniform cross section, and is subjected to the loading shown. The length of element i is denoted by L_i, its cross-sectional area by A_i, its modulus of elasticity by E_i, and the load applied to its right end by $\mathbf{P}_i$, the magnitude P_i of this load being assumed to be positive if $\mathbf{P}_i$ is directed to the right and negative otherwise. (Note that $P_1 = 0$.) (a) Write a computer program that can be used to determine the reactions at A and B, the average normal stress in each element, and the deformation of each element. (b) Use this program to solve Prob. 2.41.

Fig. P2.C2

2.C3 Rod AB consists of n elements, each of which is homogeneous and of uniform cross section. End A is fixed, while initially there is a gap δ_0 between end B and the fixed vertical surface on the right. The length of element i is denoted by L_i, its cross-sectional area by A_i, its modulus of elasticity by E_i, and its coefficient of thermal expansion by α_i. After the temperature of the rod has been increased by ΔT, the gap at B is closed and the vertical surfaces exert equal and opposite forces on the rod. (a) Write a computer program that can be used to determine the magnitude of the reactions at A and B, the normal stress in each element, and the deformation of each element. (b) Use this program to solve Probs. 2.53, 2.54, 2.57, and 2.59.

Fig. P2.C3

2.C4 Bar AB has a length L and is made of two different materials of given cross-sectional area, modulus of elasticity, and yield strength. The bar is subjected as shown to a load $\mathbf{P}$ which is gradually increased from zero until the deformation of the bar has reached a maximum value δ_m and then decreased back to zero. (a) Write a computer program that, for each of 25 values of δ_m equally spaced over a range extending from 0 to a value equal to 120% of the deformation causing both materials to yield, can be used to determine the maximum value P_m of the load, the maximum normal stress in each material, the permanent deformation δ_p of the bar, and the residual stress in each material. (b) Use this program to solve Probs. 2.109, 2.111, and 2.112.

Fig. P2.C4

2.C5   The stress concentration factor for a flat bar with a centric hole under axial loading can be expressed as:

$$K = 3.00 - 3.13\left(\frac{2r}{D}\right) + 3.66\left(\frac{2r}{D}\right)^2 - 1.53\left(\frac{2r}{D}\right)^3$$

where r is the radius of the hole and D is the width of the bar. (a) Write a computer program that can be used to determine the allowable load P for given values of r, D, the thickness t of the bar, and the allowable stress σ_{all} of the material. (b) Use this program to solve Prob. 2.94.

Fig. P2.C5

2.C6   A solid truncated cone is subjected to an axial force **P** as shown. Write a computer program that can be used to obtain an approximation of the elongation of the cone by replacing it by n circular cylinders of equal thickness and of radius equal to the mean radius of the portion of cone they replace. Knowing that the exact value of the elongation of the cone is $(PL)/(2\pi c^2 E)$ and using for P, L, c, and E values of your choice, determine the percentage error involved when the program is used with (a) $n = 6$, (b) $n = 12$, (c) $n = 60$.

Fig. P2.C6

C H A P T E R

3

Torsion

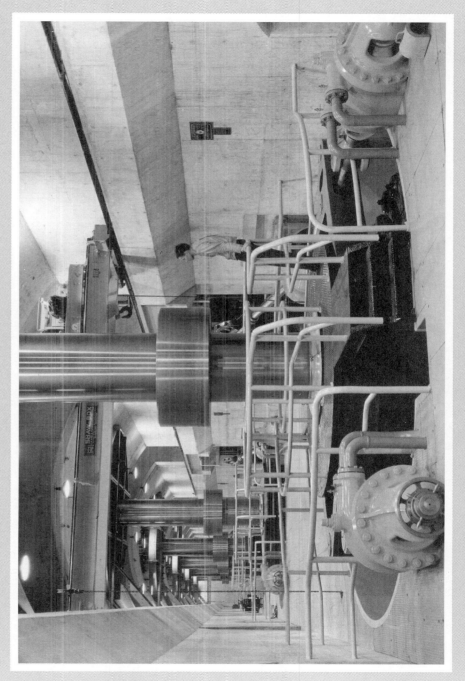

This chapter is devoted to the study of torsion and of the stresses and deformations it causes. In the hydroelectric plant shown here turbines exert torques on the vertical shafts that turn the rotors of electric generators.

3.1. INTRODUCTION

In the two preceding chapters you studied how to calculate the stresses and strains in structural members subjected to axial loads, that is, to forces directed along the axis of the member. In this chapter structural members and machine parts that are in *torsion* will be considered. More specifically, you will analyze the stresses and strains in members of circular cross section subjected to twisting couples, or *torques*, **T** and **T′** (Fig. 3.1). These couples have a common magnitude *T*, and opposite senses. They are vector quantities and can be represented either by curved arrows as in Fig. 3.1*a*, or by couple vectors as in Fig. 3.1*b*.

(a)

(b)

Fig. 3.1

Members in torsion are encountered in many engineering applications. The most common application is provided by *transmission shafts*, which are used to transmit power from one point to another. For example, the shaft shown in Fig. 3.2 is used to transmit power from the engine to the rear wheels of an automobile. These shafts can be either solid, as shown in Fig. 3.1, or hollow.

Fig. 3.2   In the automotive power train shown, the shaft transmits power from the engine to the rear wheels.

Turbine

A

Rotation

B

Generator

(a)

T'

T'

A

T

B

T'

(b)

Fig. 3.3

Consider the system shown in Fig. 3.3*a*, which consists of a steam turbine *A* and an electric generator *B* connected by a transmission shaft *AB*. By breaking the system into its three component parts (Fig. 3.3*b*), you can see that the turbine exerts a twisting couple or torque **T** on the shaft and that the shaft exerts an equal torque on the generator. The generator reacts by exerting the equal and opposite torque **T'** on the shaft, and the shaft by exerting the torque **T'** on the turbine.

You will first analyze the stresses and deformations that take place in circular shafts. In Sec. 3.3, an important property of circular shafts is demonstrated: *When a circular shaft is subjected to torsion, every cross section remains plane and undistorted. In other words, while the various cross sections along the shaft rotate through different angles, each cross section rotates as a solid rigid slab. This property will en-able you to determine the distribution of shearing strains in a circular shaft and to conclude that the shearing strain varies linearly with the distance from the axis of the shaft.*

Considering deformations in the *elastic range* and using Hooke's law for shearing stress and strain, you will determine the *distribution of shearing stresses in a circular shaft* and derive the *elastic torsion formulas* (Sec. 3.4).

In Sec. 3.5, you will learn how to find the *angle of twist* of a circular shaft subjected to a given torque, assuming again elastic deformations. The solution of problems involving *statically indeterminate shafts* is considered in Sec. 3.6.

In Sec. 3.7, you will study the *design of transmission shafts*. In order to accomplish the design, you will learn to determine the required physical characteristics of a shaft in terms of its speed of rotation and the power to be transmitted.

The torsion formulas cannot be used to determine stresses near sections where the loading couples are applied or near a section where an abrupt change in the diameter of the shaft occurs. Moreover, these formulas apply only within the elastic range of the material.

In Sec. 3.8, you will learn how to account for stress concentrations where an abrupt change in diameter of the shaft occurs. In Secs. 3.9 to 3.11, you will consider stresses and deformations in circular shafts made of a ductile material when the yield point of the material is exceeded. You will then learn how to determine the permanent *plastic deformations and residual stresses* that remain in a shaft after it has been loaded beyond the yield point of the material.

In the last sections of this chapter, you will study the torsion of noncircular members (Sec. 3.12) and analyze the distribution of stresses in thin-walled hollow noncircular shafts (Sec. 3.13).

3.2. PRELIMINARY DISCUSSION OF THE STRESSES IN A SHAFT

Considering a shaft AB subjected at A and B to equal and opposite torques $\mathbf{T}$ and $\mathbf{T}'$, we pass a section perpendicular to the axis of the shaft through some arbitrary point C (Fig. 3.4). The free-body diagram of the portion BC of the shaft must include the elementary shearing forces $d\mathbf{F}$, perpendicular to the radius of the shaft, that portion AC ex-

Fig. 3.4

erts on *BC* as the shaft is twisted (Fig. 3.5*a*). But the conditions of equilibrium for *BC* require that the system of these elementary forces be equivalent to an internal torque **T**, equal and opposite to **T′** (Fig. 3.5*b*). Denoting by ρ the perpendicular distance from the force *d***F** to the axis of the shaft, and expressing that the sum of the moments of the shearing forces *d***F** about the axis of the shaft is equal in magnitude to the torque **T**, we write

$$\int \rho \, dF = T$$

or, since $dF = \tau \, dA$, where τ is the shearing stress on the element of area *dA*,

$$\int \rho (\tau \, dA) = T \tag{3.1}$$

While the relation obtained expresses an important condition that must be satisfied by the shearing stresses in any given cross section of the shaft, it does *not* tell us how these stresses are distributed in the cross section. We thus observe, as we already did in Sec. 1.5, that the actual distribution of stresses under a given load is *statically indeterminate*, i.e., this distribution *cannot be determined by the methods of statics*. However, having assumed in Sec. 1.5 that the normal stresses produced by an axial centric load were uniformly distributed, we found later (Sec. 2.17) that this assumption was justified, except in the neighborhood of concentrated loads. A similar assumption with respect to the

Fig. 3.5

distribution of shearing stresses in an elastic shaft *would be wrong*. We must withhold any judgment regarding the distribution of stresses in a shaft until we have analyzed the *deformations* that are produced in the shaft. This will be done in the next section.

One more observation should be made at this point. As was indicated in Sec. 1.12, shear cannot take place in one plane only. Consider the very small element of shaft shown in Fig. 3.6. We know that the torque applied to the shaft produces shearing stresses τ on the faces perpendicular to the axis of the shaft. But the conditions of equilibrium discussed in Sec. 1.12 require the existence of equal stresses on the faces formed by the two planes containing the axis of the shaft. That such shearing stresses actually occur in torsion can be demonstrated.

Axis of shaft

Fig. 3.6

(a)

(b)

Fig. 3.7

(a)

(b)

Fig. 3.8

(a)

(b)

Fig. 3.9

by considering a "shaft" made of separate slats pinned at both ends to disks as shown in Fig. 3.7a. If markings have been painted on two adjoining slats, it is observed that the slats slide with respect to each other when equal and opposite torques are applied to the ends of the "shaft" (Fig. 3.7b). While sliding will not actually take place in a shaft made of a homogeneous and cohesive material, the tendency for sliding will exist, showing that stresses occur on longitudinal planes as well as on planes perpendicular to the axis of the shaft.†

3.3. DEFORMATIONS IN A CIRCULAR SHAFT

Consider a circular shaft that is attached to a fixed support at one end (Fig. 3.8a). If a torque **T** is applied to the other end, the shaft will twist, with its free end rotating through an angle ϕ called *the angle of twist* (Fig. 3.8b). Observation shows that, within a certain range of values of T, the angle of twist ϕ is proportional to T. It also shows that ϕ is proportional to the length L of the shaft. In other words, the angle of twist for a shaft of the same material and same cross section, but twice as long, will be twice as large under the same torque **T**. One purpose of our analysis will be to find the specific relation existing among ϕ, L, and T; another purpose will be to determine the distribution of shearing stresses in the shaft, which we were unable to obtain in the preceding section on the basis of statics alone.

At this point, an important property of circular shafts should be noted: When a circular shaft is subjected to torsion, *every cross section remains plane and undistorted.* In other words, while the various cross sections along the shaft rotate through different amounts, each cross section rotates as a solid rigid slab. This is illustrated in Fig. 3.9a, which shows the deformations in a rubber model subjected to torsion. The property we are discussing is characteristic of circular shafts, whether solid or hollow; it is not enjoyed by members of noncircular cross section. For example, when a bar of square cross section is subjected to torsion, its various cross sections warp and do not remain plane (Fig. 3.9b).

†The twisting of a cardboard tube that has been slit lengthwise provides another demonstration of the existence of shearing stresses on longitudinal planes.

The cross sections of a circular shaft remain plane and undistorted because a circular shaft is *axisymmetric*, i.e., its appearance remains the same when it is viewed from a fixed position and rotated about its axis through an arbitrary angle. (Square bars, on the other hand, retain the same appearance only if they are rotated through 90° or 180°.) As we will see presently, the axisymmetry of circular shafts may be used to prove theoretically that their cross sections remain plane and undistorted.

Consider the points C and D located on the circumference of a given cross section of the shaft, and let C' and D' be the positions they will occupy after the shaft has been twisted (Fig. 3.10a). The axisymmetry of the shaft and of the loading requires that the rotation which would have brought D into C should now bring D' into C'. Thus C' and D' must lie on the circumference of a circle, and the arc $C'D'$ must be equal to the arc CD (Fig. 3.10b). We will now examine whether the circle on which C' and D' lie is different from the original circle. Let us assume that C' and D' do lie on a different circle and that the new circle is located to the left of the original circle, as shown in Fig. 3.10b. The same situation will prevail for any other cross section, since all the cross sections of the shaft are subjected to the same internal torque T, and an observer looking at the shaft from its end A will conclude that the loading causes any given circle drawn on the shaft to move *away*. But an observer located at B, to whom the given loading looks the same (a clockwise couple in the foreground and a counterclockwise couple in the background) will reach the opposite conclusion, i.e., that the circle moves *toward* him. This contradiction proves that our assumption is wrong and that C' and D' lie on the same circle as C and D. Thus, as the shaft is twisted, the original circle just rotates in its own plane. Since the same reasoning may be applied to any smaller, concentric circle located in the cross section under consideration, we conclude that the entire cross section remains plane (Fig. 3.11).

The above argument does not preclude the possibility for the various concentric circles of Fig. 3.11 to rotate by different amounts when the shaft is twisted. But if that were so, a given diameter of the cross section would be distorted into a curve which might look as shown in Fig. 3.12a. An observer looking at this curve from A would conclude that the outer layers of the shaft get more twisted than the inner ones, while an observer looking from B would reach the opposite conclusion (Fig. 3.12b). This inconsistency leads us to conclude that any diameter of a given cross section remains straight (Fig. 3.12c) and, therefore, that any given cross section of a circular shaft remains plane and undistorted.

(a)

(b)

Fig. 3.10

Fig. 3.11

(a)   *(b)*   *(c)*

Fig. 3.12

Our discussion so far has ignored the mode of application of the twisting couples $\mathbf{T}$ and $\mathbf{T}'$. If *all* sections of the shaft, from one end to the other, are to remain plane and undistorted, we must make sure that the couples are applied in such a way that the ends of the shaft themselves remain plane and undistorted. This may be accomplished by applying the couples $\mathbf{T}$ and $\mathbf{T}'$ to rigid plates, which are solidly attached to the ends of the shaft (Fig. 3.13a). We can then be sure that all sections will remain plane and undistorted when the loading is applied, and that the resulting deformations will occur in a uniform fashion throughout the entire length of the shaft (Fig. 3.13a). All of the equally spaced circles shown in Fig. 3.13a will rotate by the same amount relative to their neighbors, and each of the straight lines will be transformed into a curve (helix) intersecting the various circles at the same angle (Fig. 3.13b).

The derivations given in this and the following sections will be based on the assumption of rigid end plates. Loading conditions encountered in practice may differ appreciably from those corresponding to the model of Fig. 3.13. The chief merit of this model is that it helps us define a torsion problem for which we can obtain an exact solution, just as the rigid-end-plates model of Sec. 2.17 made it possible for us to define an axial-load problem which could be easily and accurately solved. By virtue of Saint-Venant's principle, the results obtained for our idealized model may be extended to most engineering applications. However, we should keep these results associated in our mind with the specific model shown in Fig. 3.13.

We will now determine the distribution of *shearing strains* in a circular shaft of length L and radius c which has been twisted through an angle ϕ (Fig. 3.14a). Detaching from the shaft a cylinder of radius ρ, we consider the small square element formed by two adjacent circles and two adjacent straight lines traced on the surface of the cylinder before any load is applied (Fig. 3.14b). As the shaft is subjected to a torsional load, the element deforms into a rhombus (Fig. 3.14c). We now recall from Sec. 2.14 that the element deforms into a rhombus (Fig. 3.14c). We now recall from Sec. 2.14 that the shearing strain γ in a given element is measured by the change in the angles formed by the sides of that element. Since the circles defining two of the sides of the element considered here remain unchanged, the shearing strain γ must be equal to the angle between lines AB and $A'B$. (We recall that γ should be expressed in radians.)

We observe from Fig. 3.14c that, for small values of γ, we can express the arc length AA' as $AA' = L\gamma$. But, on the other hand, we have $AA' = \rho\phi$. It follows that $L\gamma = \rho\phi$, or

$$\gamma = \frac{\rho\phi}{L} \tag{3.2}$$

where γ and ϕ are both expressed in radians. The equation obtained shows, as we could have anticipated, that the shearing strain γ at a given point of a shaft in torsion is proportional to the angle of twist ϕ. It also shows that γ is proportional to the distance ρ from the axis of the shaft to the point under consideration. Thus, *the shearing strain in a circular shaft varies linearly with the distance from the axis of the shaft.*

Fig. 3.13

Fig. 3.14

It follows from Eq. (3.2) that the shearing strain is maximum on the surface of the shaft, where $\rho = c$. We have

$$\gamma_{max} = \frac{c\phi}{L} \qquad (3.3)$$

Eliminating ϕ from Eqs. (3.2) and (3.3), we can express the shearing strain γ at a distance ρ from the axis of the shaft as

$$\gamma = \frac{\rho}{c}\, \gamma_{max} \qquad (3.4)$$

3.4. STRESSES IN THE ELASTIC RANGE

No particular stress-strain relationship has been assumed so far in our discussion of circular shafts in torsion. Let us now consider the case when the torque T is such that all shearing stresses in the shaft remain below the yield strength τ_Y. We know from Chap. 2 that, for all practical purposes, this means that the stresses in the shaft will remain below the proportional limit and below the elastic limit as well. Thus, Hooke's law will apply and there will be no permanent deformation. Recalling Hooke's law for shearing stress and strain from Sec. 2.14, we write

$$\tau = G\gamma \qquad (3.5)$$

where G is the modulus of rigidity or shear modulus of the material. Multiplying both members of Eq. (3.4) by G, we write

$$G\gamma = \frac{\rho}{c}\, G\gamma_{max}$$

or, making use of Eq. (3.5),

$$\tau = \frac{\rho}{c}\, \tau_{max} \qquad (3.6)$$

The equation obtained shows that, as long as the yield strength (or proportional limit) is not exceeded in any part of a circular shaft, *the shearing stress in the shaft varies linearly with the distance ρ from the axis of the shaft*. Figure 3.15a shows the stress distribution in a solid circular shaft of radius c, and Fig. 3.15b in a hollow circular shaft of inner radius c_1 and outer radius c_2. From Eq. (3.6), we find that, in the latter case,

$$\tau_{min} = \frac{c_1}{c_2}\, \tau_{max} \qquad (3.7)$$

(a)

(b)

Fig. 3.15

Fig. 3.15 (repeated)

(a)

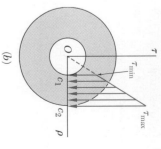

(b)

We now recall from Sec. 3.2 that the sum of the moments of the elementary forces exerted on any cross section of the shaft must be equal to the magnitude T of the torque exerted on the shaft:

$$\int \rho(\tau \, dA) = T \qquad (3.1)$$

Substituting for τ from (3.6) into (3.1), we write

$$T = \int \rho \tau \, dA = \frac{\tau_{max}}{c} \int \rho^2 \, dA$$

But the integral in the last member represents the polar moment of inertia J of the cross section with respect to its center O. We have therefore

$$T = \frac{\tau_{max}}{c} J \qquad (3.8)$$

or, solving for τ_{max},

$$\tau_{max} = \frac{Tc}{J} \qquad (3.9)$$

Substituting for τ_{max} from (3.9) into (3.6), we express the shearing stress at any distance ρ from the axis of the shaft as

$$\boxed{\tau = \frac{T\rho}{J}} \qquad (3.10)$$

Equations (3.9) and (3.10) are known as the *elastic torsion formulas*. We recall from statics that the polar moment of inertia of a circle of radius c is $J = \frac{1}{2}\pi c^4$. In the case of a hollow circular shaft of inner radius c_1 and outer radius c_2, the polar moment of inertia is

$$J = \frac{1}{2}\pi c_2^4 - \frac{1}{2}\pi c_1^4 = \frac{1}{2}\pi(c_2^4 - c_1^4) \qquad (3.11)$$

We note that, if SI metric units are used in Eq. (3.9) or (3.10), T will be expressed in N · m, c or ρ in meters, and J in m⁴; we check that the resulting shearing stress will be expressed in N/m², that is, pascals (Pa). If U.S. customary units are used, T should be expressed in lb · in., c or ρ in inches, and J in in⁴, with the resulting shearing stress expressed in psi.

EXAMPLE 3.01

A hollow cylindrical steel shaft is 1.5 m long and has inner and outer diameters respectively equal to 40 and 60 mm (Fig. 3.16). (*a*) What is the largest torque that can be applied to the shaft if the shearing stress is not to exceed 120 MPa? (*b*) What is the corresponding minimum value of the shearing stress in the shaft?

Fig. 3.16

(a) Largest Permissible Torque. The largest torque **T** that can be applied to the shaft is the torque for which $\tau_{max} = 120$ MPa. Since this value is less than the yield strength for steel, we can use Eq. (3.9). Solving this equation for *T*, we have

$$T = \frac{J\tau_{max}}{c} \qquad (3.12)$$

Recalling that the polar moment of inertia *J* of the cross section is given by Eq. (3.11), where $c_1 = \frac{1}{2}(40 \text{ mm}) = 0.02$ m and $c_2 = \frac{1}{2}(60 \text{ mm}) = 0.03$ m, we write

$$J = \tfrac{1}{2}\pi(c_2^4 - c_1^4) = \tfrac{1}{2}\pi(0.03^4 - 0.02^4) = 1.021 \times 10^{-6} \text{ m}^4$$

Substituting for *J* and τ_{max} into (3.12), and letting $c = c_2 = 0.03$ m, we have

$$T = \frac{J\tau_{max}}{c} = \frac{(1.021 \times 10^{-6} \text{ m}^4)(120 \times 10^6 \text{ Pa})}{0.03 \text{ m}}$$

$$= 4.08 \text{ kN} \cdot \text{m}$$

(b) Minimum Shearing Stress. The minimum value of the shearing stress occurs on the inner surface of the shaft. It is obtained from Eq. (3.7), which expresses that τ_{min} and τ_{max} are respectively proportional to c_1 and c_2:

$$\tau_{min} = \frac{c_1}{c_2}\tau_{max} = \frac{0.02 \text{ m}}{0.03 \text{ m}}(120 \text{ MPa}) = 80 \text{ MPa}$$

The torsion formulas (3.9) and (3.10) were derived for a shaft of uniform circular cross section subjected to torques at its ends. However, they can also be used for a shaft of variable cross section or for a shaft subjected to torques at locations other than its ends (Fig. 3.17*a*). The distribution of shearing stresses in a given cross section *S* of the shaft is obtained from Eq. (3.9), where *J* denotes the polar moment of inertia of that section, and where *T* represents the *internal torque* in that section. The value of *T* is obtained by drawing the free-body diagram of the portion of shaft located on one side of the section (Fig. 3.17*b*) and writing that the sum of the torques applied to that portion, including the internal torque **T**, is zero (see Sample Prob. 3.1).

Up to this point, our analysis of stresses in a shaft has been limited to shearing stresses. This is due to the fact that the element we had selected was oriented in such a way that its faces were either parallel or perpendicular to the axis of the shaft (Fig. 3.6). We know from earlier discussions (Secs. 1.11 and 1.12) that normal stresses, shearing stresses, or a combination of both may be found under the same loading condition, depending upon the orientation of the element which has

Fig. 3.17

been chosen. Consider the two elements a and b located on the surface of a circular shaft subjected to torsion (Fig. 3.18). Since the faces of element a are respectively parallel and perpendicular to the axis of the shaft, the only stresses on the element will be the shearing stresses defined by formula (3.9), namely $\tau_{max} = Tc/J$. On the other hand, the faces of element b, which form arbitrary angles with the axis of the shaft, will be subjected to a combination of normal and shearing stresses.

Let us consider the particular case of an element c (not shown) at $45°$ to the axis of the shaft. In order to determine the stresses on the faces of this element, we consider the two triangular elements shown in Fig. 3.19 and draw their free-body diagrams. In the case of the element BD are the shearing stresses τ_{max} of Fig. 3.19a, we know that the stresses exerted on the faces BC and BD are the shearing stresses $\tau_{max} = Tc/J$. The magnitude of the corresponding shearing forces is thus $\tau_{max}A_0$, where A_0 denotes the area of the face. Observing that the components along DC of the two shearing forces are equal and opposite, we conclude that the force $\mathbf{F}$ exerted on DC must be perpendicular to that face. It is a tensile force, and its magnitude is

$$F = 2(\tau_{max}A_0)\cos 45° = \tau_{max}A_0\sqrt{2} \qquad (3.13)$$

The corresponding stress is obtained by dividing the force F by the area A of face DC. Observing that $A = A_0\sqrt{2}$, we write

$$\sigma = \frac{F}{A} = \frac{\tau_{max}A_0\sqrt{2}}{A_0\sqrt{2}} = \tau_{max} \qquad (3.14)$$

A similar analysis of the element of Fig. 3.19b shows that the stress on the face BE is $\sigma = -\tau_{max}$. We conclude that the stresses exerted on the faces of an element c at $45°$ to the axis of the shaft (Fig. 3.20) are normal stresses equal to $\pm\tau_{max}$. Thus, while the element a in Fig. 3.20 is in pure shear, the element c in the same figure is subjected to a tensile stress on two of its faces, and to a compressive stress on the other two. We also note that all the stresses involved have the same magnitude, Tc/J.†

As you learned in Sec. 2.3, ductile materials generally fail in shear. Therefore, when subjected to torsion, a specimen J made of a ductile material breaks along a plane perpendicular to its longitudinal axis (Fig. 3.21a). On the other hand, brittle materials are weaker in tension than in shear. Thus, when subjected to torsion, a specimen made of a brittle material tends to break along surfaces which are perpendicular to the direction in which tension is maximum, i.e., along surfaces forming a $45°$ angle with the longitudinal axis of the specimen (Fig. 3.21b).

T′ ← → T

Fig. 3.18

(a)

$\tau_{max}A_0$
$\tau_{max}A_0$
D B 45° C $\mathbf{F}$

(b)

B 45° E
C $\tau_{max}A_0$
$\tau_{max}A_0$ $\mathbf{F'}$

Fig. 3.19

T″ ↗
↘ T

$\tau_{max} = \dfrac{Tc}{J}$ $\sigma_{45°} = \pm\dfrac{Tc}{J}$

Fig. 3.20

(a)

T′ — T

(b)

T′ — T

Fig. 3.21

†Stresses on elements of arbitrary orientation, such as element b of Fig. 3.18, will be discussed in Chap. 7.

SAMPLE PROBLEM 3.1

Shaft BC is hollow with inner and outer diameters of 90 mm and 120 mm respectively. Shafts AB and CD are solid and of diameter d. For the loading shown, determine (a) the maximum and minimum shearing stress in shaft BC, (b) the required diameter d of shafts AB and CD if the allowable shearing stress in these shafts is 65 MPa.

0.9 m
0.7 m
0.5 m

$\mathbf{T}_A = 6 \text{ kN} \cdot \text{m}$
$\mathbf{T}_B = 14 \text{ kN} \cdot \text{m}$
$\mathbf{T}_C = 26 \text{ kN} \cdot \text{m}$
$\mathbf{T}_D = 6 \text{ kN} \cdot \text{m}$

120 mm

SOLUTION

Equations of Statics. Denoting by $\mathbf{T}_{AB}$ the torque in shaft AB, we pass a section through shaft AB and, for the free body shown, we write

$$\Sigma M_x = 0: \qquad (6 \text{ kN} \cdot \text{m}) - T_{AB} = 0 \qquad T_{AB} = 6 \text{ kN} \cdot \text{m}$$

We now pass a section through shaft BC and, for the free body shown, we have

$$\Sigma M_x = 0: \quad (6 \text{ kN} \cdot \text{m}) + (14 \text{ kN} \cdot \text{m}) - T_{BC} = 0 \qquad T_{BC} = 20 \text{ kN} \cdot \text{m}$$

$\mathbf{T}_A = 6 \text{ kN} \cdot \text{m}$

$\mathbf{T}_{AB}$

$\mathbf{T}_A = 6 \text{ kN} \cdot \text{m}$
$\mathbf{T}_B = 14 \text{ kN} \cdot \text{m}$
$\mathbf{T}_{BC}$

a. Shaft BC. For this hollow shaft we have

$$J = \frac{\pi}{2}(c_2^4 - c_1^4) = \frac{\pi}{2}[(0.060)^4 - (0.045)^4] = 13.92 \times 10^{-6} \text{ m}^4$$

Maximum Shearing Stress. On the outer surface, we have

$$\tau_{max} = \tau_2 = \frac{T_{BC}c_2}{J} = \frac{(20 \text{ kN} \cdot \text{m})(0.060 \text{ m})}{13.92 \times 10^{-6} \text{ m}^4} \qquad \tau_{min} = 86.2 \text{ MPa} \blacktriangledown$$

Minimum Shearing Stress. We write that the stresses are proportional to the distance from the axis of the shaft.

$$\frac{\tau_{min}}{\tau_{max}} = \frac{c_1}{c_2} \qquad \frac{\tau_{min}}{86.2 \text{ MPa}} = \frac{45 \text{ mm}}{60 \text{ mm}} \qquad \tau_{min} = 64.7 \text{ MPa} \blacktriangledown$$

$c_1 = 45 \text{ mm}$
$c_2 = 60 \text{ mm}$

b. Shafts AB and CD. We note that in both of these shafts the magnitude of the torque is $T = 6 \text{ kN} \cdot \text{m}$ and $\tau_{all} = 65 \text{ MPa}$. Denoting by c the radius of the shafts, we write

$$\tau = \frac{Tc}{J} \qquad 65 \text{ MPa} = \frac{(6 \text{ kN} \cdot \text{m})c}{\frac{\pi}{2}c^4}$$

$$c^3 = 58.8 \times 10^{-6} \text{ m}^3 \qquad c = 38.9 \times 10^{-3} \text{ m} \qquad (38.9 \text{ mm})$$

$$d = 2c = 2(38.9 \text{ mm}) \qquad d = 77.8 \text{ mm} \blacktriangledown$$

$6 \text{ kN} \cdot \text{m}$
$6 \text{ kN} \cdot \text{m}$

SAMPLE PROBLEM 3.2

The preliminary design of a large shaft connecting a motor to a generator calls for the use of a hollow shaft with inner and outer diameters of 4 in. and 6 in., respectively. Knowing that the allowable shearing stress is 12 ksi, determine the maximum torque that can be transmitted (a) by the shaft as designed, (b) by a solid shaft of the same weight, (c) by a hollow shaft of the same weight and of 8-in. outer diameter.

SOLUTION

a. Hollow Shaft as Designed.

For the hollow shaft we have

$$J = \frac{\pi}{2}(c_2^4 - c_1^4) = \frac{\pi}{2}[(3 \text{ in.})^4 - (2 \text{ in.})^4] = 102.1 \text{ in}^4$$

Using Eq. (3.9), we write

$$\tau_{max} = \frac{Tc_2}{J} \qquad 12 \text{ ksi} = \frac{T(3 \text{ in.})}{102.1 \text{ in}^4} \qquad T = 408 \text{ kip} \cdot \text{in.} \quad \blacktriangleleft$$

b. Solid Shaft of Equal Weight.

For the shaft as designed and this solid shaft to have the same weight and length, their cross-sectional areas must be equal.

$$A_{(a)} = A_{(b)}$$
$$\pi[(3 \text{ in.})^2 - (2 \text{ in.})^2] = \pi c_3^2 \qquad c_3 = 2.24 \text{ in.}$$

Since $\tau_{all} = 12$ ksi, we write

$$\tau_{max} = \frac{Tc_3}{J} \qquad 12 \text{ ksi} = \frac{T(2.24 \text{ in.})}{\frac{\pi}{2}(2.24 \text{ in.})^4} \qquad T = 211 \text{ kip} \cdot \text{in.} \quad \blacktriangleleft$$

c. Hollow Shaft of 8-in. Diameter.

For equal weight, the cross-sectional areas again must be equal. We determine the inside diameter of the shaft by writing

$$A_{(a)} = A_{(c)}$$
$$\pi[(3 \text{ in.})^2 - (2 \text{ in.})^2] = \pi[(4 \text{ in.})^2 - c_5^2] \qquad c_5 = 3.317 \text{ in.}$$

For $c_5 = 3.317$ in. and $c_4 = 4$ in.,

$$J = \frac{\pi}{2}[(4 \text{ in.})^4 - (3.317 \text{ in.})^4] = 212 \text{ in}^4$$

With $\tau_{all} = 12$ ksi and $c_4 = 4$ in.,

$$\tau_{max} = \frac{Tc_4}{J} \qquad 12 \text{ ksi} = \frac{T(4 \text{ in.})}{212 \text{ in}^4} \qquad T = 636 \text{ kip} \cdot \text{in.} \quad \blacktriangleleft$$

PROBLEMS

3.1 Determine the torque **T** that causes a maximum shearing stress of 70 MPa in the steel cylindrical shaft shown.

3.2 Determine the maximum shearing stress caused by a torque of magnitude $T = 800$ N · m.

3.3 Knowing that the internal diameter of the hollow shaft shown is $d = 0.9$ in., determine the maximum shearing stress caused by a torque of magnitude $T = 9$ kip · in.

Fig. P3.1 and P3.2

Fig. P3.3 and P3.4

3.4 Knowing that $d = 1.2$ in., determine the torque **T** that causes a maximum shearing stress of 7.5 ksi in the hollow shaft shown.

3.5 (*a*) For the hollow shaft and loading shown, determine the maximum shearing stress. (*b*) Determine the diameter of a solid shaft for which the maximum shearing stress under the loading shown is the same as in part *a*.

Fig. P3.5

3.6 (*a*) Determine the torque that may be applied to a solid shaft of 90-mm outer diameter without exceeding an allowable shearing stress of 75 MPa. (*b*) Solve part *a*, assuming that the solid shaft is replaced by a hollow shaft of the same mass and of 90-mm inner diameter.

3.7 (*a*) For the 3-in.-diameter solid cylinder and loading shown, determine the maximum shearing stress. (*b*) Determine the inner diameter of the hollow cylinder, of 4-in. outer diameter, for which the maximum stress is the same as in part *a*.

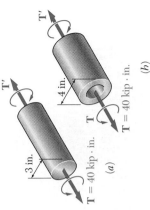

Fig. P3.7

3.8 (*a*) Determine the torque that may be applied to a solid shaft of 0.75-in. diameter without exceeding an allowable shearing stress of 10 ksi. (*b*) Solve part *a*, assuming that the solid shaft has been replaced by a hollow shaft of the same cross-sectional area and with an inner diameter equal to half its own outer diameter.

3.9   The torques shown are exerted on pulleys A and B. Knowing that each shaft is solid, determine the maximum shearing stress (a) in shaft AB, (b) in shaft BC.

3.10   In order to reduce the total mass of the assembly of Prob. 3.9, a new design is being considered in which the diameter of shaft BC will be smaller. Determine the smallest diameter of shaft BC for which the maximum value of the shearing stress in the assembly will not increase.

$T_A = 300$ N · m

30 mm

$T_B = 400$ N · m

46 mm

A

B

C

Fig. P3.9 and P3.10

3.11   Knowing that each of the shafts AB, BC, and CD consists of a solid circular rod, determine (a) the shaft in which the maximum shearing stress oc-curs, (b) the magnitude of that stress.

3.12   Knowing that a 0.30-in.-diameter hole has been drilled through each of the shafts AB, BC, and CD, determine (a) the shaft in which the max-imum shearing stress occurs, (b) the magnitude of that stress.

400 lb · in.

1200 lb · in.

500 lb · in.

A

B

C

D

$d_{AB} = 0.6$ in.

$d_{BC} = 0.75$ in.

$d_{CD} = 0.9$ in.

Fig. P3.11 and P3.12

3.13   Under normal operating conditions, the electric motor exerts a torque of 2.4 kN · m on shaft AB. Knowing that each shaft is solid, determine the maximum shearing stress (a) in shaft AB, (b) in shaft BC, (c) in shaft CD.

3.14   In order to reduce the total mass of the assembly of Prob. 3.13, a new design is being considered in which the diameter of shaft BC will be smaller. Determine the smallest diameter of shaft BC for which the maximum value of the shearing stress in the assembly will not be increased.

A

B

C

D

E

54 mm

$T_B = 1.2$ kN · m

46 mm

$T_C = 0.8$ kN · m

46 mm

$T_D = 0.4$ kN · m

40 mm

Fig. P3.13 and P3.14

3.15 The allowable stress is 15 ksi in the 1.5-in.-diameter steel rod *AB* and 8 ksi in the 1.8-in.-diameter brass rod *BC*. Neglecting the effect of stress concentrations, determine the largest torque that may be applied at *A*.

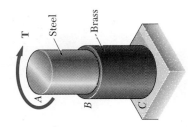

Fig. P3.15 and P3.16

3.16 The allowable stress is 15 ksi in the steel rod *AB* and 8 ksi in the brass rod *BC*. Knowing that a torque of magnitude $T = 10$ kip · in. is applied at *A*, determine the required diameter of (*a*) rod *AB*, (*b*) rod *BC*.

3.17 The solid rod *AB* has a diameter $d_{AB} = 60$ mm. The pipe *CD* has an outer diameter of 90 mm and a wall thickness of 6 mm. Knowing that both the rod and the pipe are made of a steel for which the allowable shearing stress is 75 MPa, determine the largest torque **T** that may be applied at *A*.

3.18 The solid rod *AB* has a diameter $d_{AB} = 60$ mm and is made of a steel for which the allowable shearing stress is 85 MPa. The pipe *CD*, which has an outer diameter of 90 mm and a wall thickness of 6 mm, is made of an aluminum for which the allowable shearing stress is 54 MPa. Determine the largest torque **T** that may be applied at *A*.

Fig. P3.17 and *P3.18*

3.19 The allowable stress is 50 MPa in the brass rod *AB* and 25 MPa in the aluminum rod *BC*. Knowing that a torque of magnitude $T = 1250$ N · m is applied at *A*, determine the required diameter of (*a*) rod *AB*, (*b*) rod *BC*.

Fig. P3.19 and *P3.20*

3.20 The *solid* rod *BC* has a diameter of 30 mm and is made of an aluminum for which the allowable shearing stress is 25 MPa. Rod *AB* is *hollow* and has an outer diameter of 25 mm; it is made of a brass for which the allowable shearing stress is 50 MPa. Determine (*a*) the largest inner diameter of rod *AB* for which the factor of safety is the same for each rod, (*b*) the largest torque that may be applied at *A*.

3.21   A torque of magnitude $T = 1000$ N · m is applied at D as shown. Knowing that the diameter of shaft AB is 56 mm and the diameter of shaft CD is 42 mm, determine the maximum shearing stress in (a) shaft AB, (b) shaft CD.

Fig. P3.21 and P3.22

3.22   A torque of magnitude $T = 1000$ N · m is applied at D as shown. Knowing that the allowable shearing stress is 60 MPa in each shaft, determine the required diameter of (a) shaft AB, (b) shaft CD.

3.23 and 3.24   Under normal operating conditions a motor exerts a torque of magnitude $T_F = 1200$ lb · in. at F. Knowing that the allowable shearing stress is 10.5 ksi in each shaft, determine the required diameter of (a) shaft CDE, (b) shaft FGH.

3.23 $r_D = 8$ in., $r_G = 3$ in.
3.24 $r_D = 3$ in., $r_G = 8$ in.

Fig. P3.23, P3.24, and P3.25

3.25   Under normal operating conditions a motor exerts a torque of magnitude T_F at F. The shafts are made of a steel for which the allowable shearing stress is 12 ksi and have diameters $d_{CDE} = 0.900$ in. and $d_{FGH} = 0.800$ in. Knowing that $r_D = 6.5$ in. and $r_G = 4.5$ in, determine the largest allowable value of T_F.

3.26 The two solid shafts are connected by gears as shown and are made of a steel for which the allowable shearing stress in 60 MPa. Knowing that a torque of magnitude $T_C = 600$ N · m is applied at C and that the assembly is in equilibrium, determine the required diameter of (a) shaft BC, (b) shaft EF.

3.27 The two solid shafts are connected by gears as shown and are made of a steel for which the allowable shearing stress is 50 MPa. Knowing that the diameters of the two shafts are, respectively, $d_{BC} = 40$ mm and $d_{EF} = 32$ mm and that the assembly is in equilibrium, determine the largest torque T_C that may be applied at C.

Fig. P3.26 and P3.27

3.28 In the bevel-gear system shown, $\alpha = 18.43°$. Knowing that the allowable shearing stress is 8 ksi in each shaft and that the system is in equilibrium, determine the largest torque T_A that may be applied at A.

Fig. P3.28

3.29 (a) For a given allowable stress, determine the ratio T/w of the maximum allowable torque T and the weight per unit length w for the hollow shaft shown. (b) Denoting by $(T/w)_0$ the value of this ratio computed for a solid shaft of the same radius c_2, express the ratio T/w for the hollow shaft in terms of $(T/w)_0$ and c_1/c_2.

Fig. P3.29

3.30 While the exact distribution of the shearing stresses in a hollow-cylinder shaft is as shown in Fig. P3.30a, an approximate value may be obtained for τ_{max} by assuming the stresses to be uniformly distributed over the area A of the cross section, as shown in Fig. P3.30b, and then further assuming that all the elementary shearing forces act at a distance from O equal to the mean radius $r_m = \frac{1}{2}(c_1 + c_2)$ of the cross section. This approximate value is $\tau_0 = T/Ar_m$, where T is the applied torque. Determine the ratio τ_{max}/τ_0 of the true value of the maximum shearing stress and its approximate value τ_0 for values of c_1/c_2 respectively equal to 1.00, 0.95, 0.75, 0.50, and 0.

(a) *(b)*

Fig. P3.30

Fig. 3.22

3.5. ANGLE OF TWIST IN THE ELASTIC RANGE

In this section, a relation will be derived between the angle of twist ϕ of a circular shaft and the torque $\mathbf{T}$ exerted on the shaft. The entire shaft will be assumed to remain elastic. Considering first the case of a shaft of length L and of uniform cross section of radius c subjected to a torque $\mathbf{T}$ at its free end (Fig. 3.22), we recall from Sec. 3.3 that the angle of twist ϕ and the maximum shearing strain γ_{max} are related as follows:

$$\gamma_{max} = \frac{c\phi}{L} \qquad (3.3)$$

But, in the elastic range, the yield stress is not exceeded anywhere in the shaft, Hooke's law applies, and we have $\gamma_{max} = \tau_{max}/G$ or, recalling Eq. (3.9),

$$\gamma_{max} = \frac{\tau_{max}}{G} = \frac{Tc}{JG} \qquad (3.15)$$

Equating the right-hand members of Eqs. (3.3) and (3.15), and solving for ϕ, we write

$$\phi = \frac{TL}{JG} \qquad (3.16)$$

where ϕ is expressed in radians. The relation obtained shows that, within the elastic range, *the angle of twist ϕ is proportional to the torque T applied to the shaft.* This is in accordance with the experimental evidence cited at the beginning of Sec. 3.3.

Equation (3.16) provides us with a convenient method for determining the modulus of rigidity of a given material. A specimen of the material, in the form of a cylindrical rod of known diameter and length, is placed in a *torsion testing machine* (Fig. 3.23). Torques of increasing magnitude T are applied to the specimen, and the corresponding values of the angle of twist ϕ in a length L of the specimen are recorded. As long as the yield stress of the material is not exceeded, the points obtained by plotting ϕ against T will fall on a straight line. The slope of this line represents the quantity JG/L, from which the modulus of rigidity G may be computed.

Fig. 3.23 Torsion testing machine.

EXAMPLE 3.02

What torque should be applied to the end of the shaft of Example 3.01 to produce a twist of 2°? Use the value $G = 77$ GPa for the modulus of rigidity of steel.

Solving Eq. (3.16) for T, we write

$$T = \frac{JG}{L}\phi$$

Substituting the given values

$$G = 77 \times 10^9 \text{ Pa} \qquad L = 1.5 \text{ m}$$

$$\phi = 2°\left(\frac{2\pi \text{ rad}}{360°}\right) = 34.9 \times 10^{-3} \text{ rad}$$

and recalling from Example 3.01 that, for the given cross section,

$$J = 1.021 \times 10^{-6} \text{ m}^4$$

we have

$$T = \frac{JG}{L}\phi =$$

$$\frac{(1.021 \times 10^{-6} \text{ m}^4)(77 \times 10^9 \text{ Pa})}{1.5 \text{ m}}(34.9 \times 10^{-3} \text{ rad})$$

$$T = 1.829 \times 10^3 \text{ N} \cdot \text{m} = 1.829 \text{ kN} \cdot \text{m}$$

EXAMPLE 3.03

What angle of twist will create a shearing stress of 70 MPa on the inner surface of the hollow steel shaft of Examples 3.01 and 3.02?

The method of attack for solving this problem that first comes to mind is to use Eq. (3.10) to find the torque T corresponding to the given value of τ, and Eq. (3.16) to determine the angle of twist ϕ corresponding to the value of T just found. A more direct solution, however, may be used. From Hooke's law, we first compute the shearing strain on the inner surface of the shaft:

$$\gamma_{\min} = \frac{\tau_{\min}}{G} = \frac{70 \times 10^6 \text{ Pa}}{77 \times 10^9 \text{ Pa}} = 909 \times 10^{-6}$$

Recalling Eq. (3.2), which was obtained by expressing the length of arc AA' in Fig. 3.14c in terms of both γ and ϕ, we have

$$\phi = \frac{L\gamma_{\min}}{c_1} = \frac{1500 \text{ mm}}{20 \text{ mm}}(909 \times 10^{-6}) = 68.2 \times 10^{-3} \text{ rad}$$

To obtain the angle of twist in degrees, we write

$$\phi = (68.2 \times 10^{-3} \text{ rad})\left(\frac{360°}{2\pi \text{ rad}}\right) = 3.91°$$

Formula (3.16) for the angle of twist can be used only if the shaft is homogeneous (constant G), has a uniform cross section, and is loaded only at its ends. If the shaft is subjected to torques at locations other than its ends, or if it consists of several portions with various cross sections and possibly of different materials, we must divide it into component parts that satisfy individually the required conditions for the application of formula (3.16). In the case of the shaft AB shown in Fig. 3.24, for example, four different parts should be considered: AC, CD, DE, and EB. The total angle of twist of the shaft, i.e., the angle through which end A rotates with respect to end B, is obtained by adding *algebraically* the angles of twist of each component part. Denoting respectively by T_i, L_i, J_i, and G_i the internal torque, length, cross-sectional polar moment of inertia, and modulus of rigidity corresponding to part i, the total angle of twist of the shaft is expressed as

$$\phi = \sum_i \frac{T_i L_i}{J_i G_i} \tag{3.17}$$

The internal torque T_i in any given part of the shaft is obtained by passing a section through that part and drawing the free-body diagram of

Fig. 3.24

the portion of shaft located on one side of the section. This procedure, which has already been explained in Sec. 3.4 and illustrated in Fig. 3.17, is applied in Sample Prob. 3.3.

In the case of a shaft with a variable circular cross section, as shown in Fig. 3.25, formula (3.16) may be applied to a disk of thickness dx. The angle by which one face of the disk rotates with respect to the other is thus

$$d\phi = \frac{T\,dx}{JG}$$

Fig. 3.25

where J is a function of x which may be determined. Integrating in x from 0 to L, we obtain the total angle of twist of the shaft:

$$\phi = \int_0^L \frac{T\,dx}{JG} \qquad (3.18)$$

The shaft shown in Fig. 3.22, which was used to derive formula (3.16), and the shaft of Fig. 3.16, which was discussed in Examples 3.02 and 3.03, both had one end attached to a fixed support. In each case, therefore, the angle of twist ϕ of the shaft was equal to the angle of rotation of its free end. When both ends of a shaft rotate, however, the angle of twist of the shaft is equal to the angle through which one end of the shaft rotates *with respect to the other*. Consider, for instance, the assembly shown in Fig. 3.26a, consisting of two elastic shafts AD and BE, each of length L, radius c, and modulus of rigidity G, which are attached to gears meshed at C. If a torque $\mathbf{T}$ is applied at E (Fig. 3.26b), both shafts will be twisted. Since the end D of shaft AD is fixed, the angle of twist of AD is measured by the angle of rotation ϕ_A of end A. On the other hand, since both ends of shaft BE rotate, the angle of twist of BE is equal to the difference between the angles of rotation ϕ_B and ϕ_E, i.e., the angle of twist is equal to the angle through which end E rotates with respect to end B. Denoting this relative angle of rotation by $\phi_{E/B}$, we write

$$\phi_{E/B} = \phi_E - \phi_B = \frac{TL}{JG}$$

(a)

(b)

Fig. 3.26

EXAMPLE 3.04

For the assembly of Fig. 3.26, knowing that $r_A = 2r_B$, determine the angle of rotation of end E of shaft BE when the torque **T** is applied at E.

We first determine the torque $\mathbf{T}_{AD}$ exerted on shaft AD. Observing that equal and opposite forces **F** and **F**′ are applied on the two gears at C (Fig. 3.27), and recalling that $r_A = 2r_B$, we conclude that the torque exerted on shaft AD is twice as large as the torque exerted on shaft BE; thus, $T_{AD} = 2T$.

Since the end D of shaft AD is fixed, the angle of rotation ϕ_A of gear A is equal to the angle of twist of the shaft and is obtained by writing

$$\phi_A = \frac{T_{AD}L}{JG} = \frac{2TL}{JG}$$

Fig. 3.27

Observing that the arcs CC' and CC'' in Fig. 3.26b must be equal, we write $r_A\phi_A = r_B\phi_B$ and obtain

$$\phi_B = (r_A/r_B)\phi_A = 2\phi_A$$

We have, therefore,

$$\phi_B = 2\phi_A = \frac{4TL}{JG}$$

Considering now shaft BE, we recall that the angle of twist of the shaft is equal to the angle $\phi_{E/B}$ through which end E rotates with respect to end B. We have

$$\phi_{E/B} = \frac{T_{BE}L}{JG} = \frac{TL}{JG}$$

The angle of rotation of end E is obtained by writing

$$\phi_E = \phi_B + \phi_{E/B}$$
$$= \frac{4TL}{JG} + \frac{TL}{JG} = \frac{5TL}{JG}$$

3.6. STATICALLY INDETERMINATE SHAFTS

You saw in Sec. 3.4 that, in order to determine the stresses in a shaft, it was necessary to first calculate the internal torques in the various parts of the shaft. These torques were obtained from statics by drawing the free-body diagram of the portion of shaft located on one side of a given section and writing that the sum of the torques exerted on that portion was zero.

There are situations, however, where the internal torques cannot be determined from statics alone. In fact, in such cases the external torques themselves, i.e., the torques exerted on the shaft by the supports and connections, cannot be determined from the free-body diagram of the entire shaft. The equilibrium equations must be complemented by relations involving the deformations of the shaft and obtained by considering the geometry of the problem. Because statics is not sufficient to determine the external and internal torques, the shafts are said to be *statically indeterminate*. The following example, as well as Sample Prob. 3.5, will show how to analyze statically indeterminate shafts.

EXAMPLE 3.05

A circular shaft AB consists of a 10-in.-long, $\frac{7}{8}$-in.-diameter steel cylinder, in which a 5-in.-long, $\frac{5}{8}$-in.-diameter cavity has been drilled from end B. The shaft is attached to fixed supports at both ends, and a 90 lb · ft torque is applied at its midsection (Fig. 3.28). Determine the torque exerted on the shaft by each of the supports.

5 in.

5 in.

90 lb · ft

A

B

Fig. 3.28

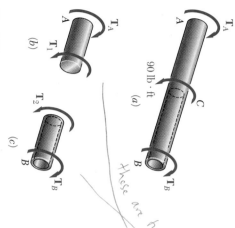

T_A

A

90 lb · ft

C

B

T_B

(a)

T_A

A

T_1

(b)

T_2

B

T_B

(c)

Fig. 3.29

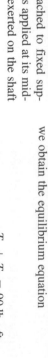

these are twisted in opposite directions, so angle of one is negative

Drawing the free-body diagram of the shaft and denoting by $\mathbf{T}_A$ and $\mathbf{T}_B$ the torques exerted by the supports (Fig. 3.29a), we obtain the equilibrium equation

$$T_A + T_B = 90 \text{ lb} \cdot \text{ft}$$

Since this equation is not sufficient to determine the two unknown torques $\mathbf{T}_A$ and $\mathbf{T}_B$, the shaft is statically indeterminate.

However, $\mathbf{T}_A$ and $\mathbf{T}_B$ can be determined if we observe that the total angle of twist of shaft AB must be zero, since both of its ends are restrained. Denoting by ϕ_1 and ϕ_2, respectively, the angles of twist of portions AC and CB, we write

$$\phi = \phi_1 + \phi_2 = 0$$

From the free-body diagram of a small portion of shaft including end A (Fig. 3.29b), we note that the internal torque T_1 in AC is equal to T_A; from the free-body diagram of a small portion of shaft including end B (Fig. 3.29c), we note that the internal torque T_2 in CB is equal to T_B. Recalling Eq. (3.16) and observing that portions AC and CB of the shaft are twisted in opposite senses, we write

$$\phi = \phi_1 + \phi_2 = \frac{T_A L_1}{J_1 G} - \frac{T_B L_2}{J_2 G} = 0$$

Solving for T_B, we have

$$T_B = \frac{L_1 J_2}{L_2 J_1} T_A$$

Substituting the numerical data

$$L_1 = L_2 = 5 \text{ in.}$$
$$J_1 = \tfrac{1}{2}\pi (\tfrac{7}{16} \text{ in.})^4 = 57.6 \times 10^{-3} \text{ in}^4$$
$$J_2 = \tfrac{1}{2}\pi [(\tfrac{7}{16} \text{ in.})^4 - (\tfrac{5}{16} \text{ in.})^4] = 42.6 \times 10^{-3} \text{ in}^4$$

we obtain

$$T_B = 0.740 \, T_A$$

Substituting this expression into the original equilibrium equation, we write

$$1.740 \, T_A = 90 \text{ lb} \cdot \text{ft}$$

$$T_A = 51.7 \text{ lb} \cdot \text{ft} \qquad T_B = 38.3 \text{ lb} \cdot \text{ft}$$

SAMPLE PROBLEM 3.3

The horizontal shaft AD is attached to a fixed base at D and is subjected to the torques shown. A 44-mm-diameter hole has been drilled into portion CD of the shaft. Knowing that the entire shaft is made of steel for which $G = 77$ GPa, determine the angle of twist at end A.

SOLUTION

Since the shaft consists of three portions AB, BC, and CD, each of uniform cross section and each with a constant internal torque, Eq. (3.17) may be used.

Statics. Passing a section through the shaft between A and B and using the free body shown, we find

$$\Sigma M_x = 0: \quad (250 \text{ N} \cdot \text{m}) - T_{AB} = 0 \qquad T_{AB} = 250 \text{ N} \cdot \text{m}$$

Passing now a section between B and C, we have

$$\Sigma M_x = 0: (250 \text{ N} \cdot \text{m}) + (2000 \text{ N} \cdot \text{m}) - T_{BC} = 0 \qquad T_{BC} = 2250 \text{ N} \cdot \text{m}$$

Since no torque is applied at C,

$$T_{CD} = T_{BC} = 2250 \text{ N} \cdot \text{m}$$

Polar Moments of Inertia

$$J_{AB} = \frac{\pi}{2} c^4 = \frac{\pi}{2} (0.015 \text{ m})^4 = 0.0795 \times 10^{-6} \text{ m}^4$$

$$J_{BC} = \frac{\pi}{2} c^4 = \frac{\pi}{2} (0.030 \text{ m})^4 = 1.272 \times 10^{-6} \text{ m}^4$$

$$J_{CD} = \frac{\pi}{2} (c_2^4 - c_1^4) = \frac{\pi}{2} [(0.030 \text{ m})^4 - (0.022 \text{ m})^4] = 0.904 \times 10^{-6} \text{ m}^4$$

Angle of Twist. Using Eq. (3.17) and recalling that $G = 77$ GPa for the entire shaft, we have

$$\phi_A = \sum_i \frac{T_i L_i}{J_i G} = \frac{1}{G} \left(\frac{T_{AB} L_{AB}}{J_{AB}} + \frac{T_{BC} L_{BC}}{J_{BC}} + \frac{T_{CD} L_{CD}}{J_{CD}} \right)$$

$$\phi_A = \frac{1}{77 \text{ GPa}} \left[\frac{(250 \text{ N} \cdot \text{m})(0.4 \text{ m})}{0.0795 \times 10^{-6} \text{ m}^4} + \frac{(2250)(0.2)}{1.272 \times 10^{-6}} + \frac{(2250)(0.6)}{0.904 \times 10^{-6}} \right]$$

$$= 0.01634 + 0.00459 + 0.01939 = 0.0403 \text{ rad}$$

$$\phi_A = (0.0403 \text{ rad}) \frac{360°}{2\pi \text{ rad}}$$

$$\phi_A = 2.31° \qquad \blacktriangleleft$$

SAMPLE PROBLEM 3.4

Two solid steel shafts are connected by the gears shown. Knowing that for each shaft $G = 11.2 \times 10^6$ psi and that the allowable shearing stress is 8 ksi, determine (a) the largest torque $\mathbf{T}_0$ that may be applied to end A of shaft AB, (b) the corresponding angle through which end A of shaft AB rotates.

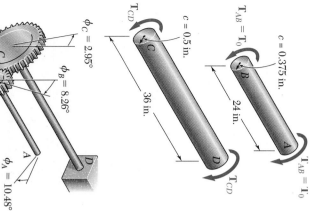

SOLUTION

Statics. Denoting by F the magnitude of the tangential force between gear teeth, we have

Gear B. $\Sigma M_B = 0$: $F(0.875 \text{ in.}) - T_0 = 0$ $T_{CD} = 2.8T_0$ (1)

Gear C. $\Sigma M_C = 0$: $F(2.45 \text{ in.}) - T_{CD} = 0$

Kinematics. Noting that the peripheral motions of the gears are equal, we write

$$r_B \phi_B = r_C \phi_C \qquad \phi_B = \phi_C \frac{r_C}{r_B} = \phi_C \frac{2.45 \text{ in.}}{0.875 \text{ in.}} = 2.8\phi_C \qquad (2)$$

a. Torque T_0

Shaft AB. With $T_{AB} = T_0$ and $c = 0.375$ in., together with a maximum permissible shearing stress of 8000 psi, we write

$$\tau = \frac{T_{AB}c}{J} \qquad 8000 \text{ psi} = \frac{T_0(0.375 \text{ in.})}{\frac{1}{2}\pi(0.375 \text{ in.})^4} \qquad T_0 = 663 \text{ lb} \cdot \text{in} \quad \triangle$$

Shaft CD. From (1) we have $T_{CD} = 2.8T_0$. With $c = 0.5$ in. and $\tau_{all} = 8000$ psi, we write

$$\tau = \frac{T_{CD}c}{J} \qquad 8000 \text{ psi} = \frac{2.8T_0(0.5 \text{ in.})}{\frac{1}{2}\pi(0.5 \text{ in.})^4} \qquad T_0 = 561 \text{ lb} \cdot \text{in} \quad \triangle$$

Maximum Permissible Torque. We choose the smaller value obtained for T_0

$$T_0 = 561 \text{ lb} \cdot \text{in.} \quad \blacktriangleleft$$

b. Angle of Rotation at End A.

We first compute the angle of twist for each shaft.

Shaft AB. For $T_{AB} = T_0 = 561$ lb $\cdot$ in., we have

$$\phi_{A/B} = \frac{T_{AB}L}{JG} = \frac{(561 \text{ lb} \cdot \text{in.})(24 \text{ in.})}{\frac{1}{2}\pi(0.375 \text{ in.})^4(11.2 \times 10^6 \text{ psi})} = 0.0387 \text{ rad} = 2.22°$$

Shaft CD. $T_{CD} = 2.8T_0 = 2.8(561 \text{ lb} \cdot \text{in.})$.

$$\phi_{C/D} = \frac{T_{CD}L}{JG} = \frac{2.8(561 \text{ lb} \cdot \text{in.})(36 \text{ in.})}{\frac{1}{2}\pi(0.5 \text{ in.})^4(11.2 \times 10^6 \text{ psi})} = 0.514 \text{ rad} = 2.95°$$

Since end D of shaft CD is fixed, we have $\phi_C = \phi_{C/D} = 2.95°$. Using (2), we find the angle of rotation of gear B to be

$$\phi_B = 2.8\phi_C = 2.8(2.95°) = 8.26°$$

For end A of shaft AB, we have

$$\phi_A = \phi_B + \phi_{A/B} = 8.26° + 2.22° \qquad \phi_A = 10.48° \quad \blacktriangleleft$$

SAMPLE PROBLEM 3.5

A steel shaft and an aluminum tube are connected to a fixed support and to a rigid disk as shown in the cross section. Knowing that the initial stresses are zero, determine the maximum torque $\mathbf{T}_0$ that can be applied to the disk if the allowable stresses are 120 MPa in the steel shaft and 70 MPa in the aluminum tube. Use $G = 77$ GPa for steel and $G = 27$ GPa for aluminum.

8 mm

50 mm

76 mm

500 mm

SOLUTION

Statics. *Free Body of Disk.* Denoting by $\mathbf{T}_1$ the torque exerted by the tube on the disk and by $\mathbf{T}_2$ the torque exerted by the shaft, we find

$$T_0 = T_1 + T_2 \qquad (1)$$

$\mathbf{T}_1$

$\mathbf{T}_2$

$\mathbf{T}_0$

Deformations. Since both the tube and the shaft are connected to the rigid disk, we have

$$\phi_1 = \phi_2: \qquad \frac{T_1 L_1}{J_1 G_1} = \frac{T_2 L_2}{J_2 G_2}$$

$$\frac{T_1(0.5 \text{ m})}{(2.003 \times 10^{-6} \text{ m}^4)(27 \text{ GPa})} = \frac{T_2(0.5 \text{ m})}{(0.614 \times 10^{-6} \text{ m}^4)(77 \text{ GPa})}$$

$$T_2 = 0.874 T_1 \qquad (2)$$

0.5 m

ϕ_1

38 mm
30 mm

$\mathbf{T}_1$

Aluminum
$G_1 = 27$ GPa
$J_1 = \frac{\pi}{2}[(38 \text{ mm})^4 - (30 \text{ mm})^4]$
$= 2.003 \times 10^{-6} \text{ m}^4$

Shearing Stresses. We *assume* that the requirement $\tau_{\text{alum}} \leq 70$ MPa is critical. For the aluminum tube, we have

$$T_1 = \frac{\tau_{\text{alum}} J_1}{c_1} = \frac{(70 \text{ MPa})(2.003 \times 10^{-6} \text{ m}^4)}{0.038 \text{ m}} = 3690 \text{ N} \cdot \text{m}$$

Using Eq. (2), we compute the corresponding value T_2 and then find the maximum shearing stress in the steel shaft.

$$T_2 = 0.874 T_1 = 0.874(3690) = 3225 \text{ N} \cdot \text{m}$$

$$\tau_{\text{steel}} = \frac{T_2 c_2}{J_2} = \frac{(3225 \text{ N} \cdot \text{m})(0.025 \text{ m})}{0.614 \times 10^{-6} \text{ m}^4} = 131.3 \text{ MPa}$$

We note that the allowable steel stress of 120 MPa is exceeded; our assumption was *wrong*. Thus the maximum torque $\mathbf{T}_0$ will be obtained by making $\tau_{\text{steel}} = 120$ MPa. We first determine the torque $\mathbf{T}_2$.

$$T_2 = \frac{\tau_{\text{steel}} J_2}{c_2} = \frac{(120 \text{ MPa})(0.614 \times 10^{-6} \text{ m}^4)}{0.025 \text{ m}} = 2950 \text{ N} \cdot \text{m}$$

From Eq. (2), we have

$$2950 \text{ N} \cdot \text{m} = 0.874 T_1 \qquad T_1 = 3375 \text{ N} \cdot \text{m}$$

Using Eq. (1), we obtain the maximum permissible torque

$$T_0 = T_1 + T_2 = 3375 \text{ N} \cdot \text{m} + 2950 \text{ N} \cdot \text{m}$$

$$T_0 = 6.325 \text{ kN} \cdot \text{m} \qquad \blacktriangleleft$$

0.5 m

25 mm

$\mathbf{T}_2$

ϕ_2

Steel
$G_1 = 77$ GPa
$J_1 = \frac{\pi}{2}[(25 \text{ mm})^4$
$= 0.614 \times 10^{-6} \text{ m}^4$

PROBLEMS

3.31 (*a*) For the solid steel shaft shown (*G* = 77 GPa), determine the angle of twist at *A*. (*b*) Solve part *a*, assuming that the steel shaft is hollow with a 30-mm outer diameter and a 20-mm inner diameter.

30 mm

1.8 m

A

250 N · m

Fig. P3.31

3.32 For the aluminum shaft shown (*G* = 3.9 × 10⁶ psi), determine (*a*) the torque **T** that causes an angle of twist of 5°, (*b*) the angle of twist caused by the same torque **T** in a solid cylindrical shaft of the same length and cross-sectional area.

0.5 in.

T

0.75 in.

4 ft

Fig. P3.32

3.33 The ship at *A* has just started to drill for oil on the ocean floor at a depth of 5000 ft. Knowing that the top of the 8-in.-diameter steel drill pipe (*G* = 11.2 × 10⁶ psi) rotates through two complete revolutions before the drill bit at *B* starts to operate, determine the maximum shearing stress caused in the pipe by torsion.

A

B

5000 ft

Fig. P3.33

3.34 Determine the largest allowable diameter of a 3-m-long steel rod (*G* = 77 GPa) if the rod is to be twisted through 30° without exceeding a shearing stress of 80 MPa.

3.35 The torques shown are exerted on pulleys *A* and *B*. Knowing that the shafts are solid and made of aluminum (*G* = 77 GPa), determine the angle of twist between (*a*) *A* and *B*, (*b*) *A* and *C*.

T_A = 300 N · m

30 mm

T_B = 400 N · m

46 mm

0.9 m

A

B

C

0.75 m

Fig. P3.35

Fig. P3.36

3.36 The torques shown are exerted on pulleys B, C, and D. Knowing that the entire shaft is made of steel ($G = 27$ GPa), determine the angle of twist between (a) C and B, (b) D and B.

3.37 The solid brass rod AB ($G = 39$ GPa) is bonded to the solid aluminum rod BC ($G = 27$ GPa). Determine the angle of twist (a) at B, (b) at A.

Fig. P3.37

Fig. P3.38

3.38 The aluminum rod BC ($G = 3.9 \times 10^6$ psi) is bonded to the brass rod AB ($G = 5.6 \times 10^6$). Knowing that each rod is solid and has a diameter of 0.5 in., determine the angle of twist (a) at B, (b) at C.

3.39 Two solid steel shafts ($G = 77$ GPa) are connected by the gears shown. Knowing that the radius of gear B is $r_B = 20$ mm, determine the angle through which end A rotates when $T_A = 75$ N · m.

3.40 Solve Prob. 3.39, assuming that a change in design of the assembly resulted in the radius of gear B being increased to 30 mm.

Fig. P3.39

Fig. P3.41

3.41 Two shafts, each of $\frac{3}{4}$-in. diameter, are connected by the gears shown. Knowing that $G = 11.2 \times 10^6$ psi and that the shaft at F is fixed, determine the angle through which end A rotates when a 750-lb · in. torque is applied at A.

3.42 Solve Prob. 3.41, assuming that after a design change the radius of gear B is 4 in. and the radius of gear E is 3 in.

3.43 A coder F, used to record in digital form the rotation of shaft A, is connected to the shaft by means of the gear train shown, which consists of four gears and three solid steel shafts each of diameter d. Two of the gears have a radius r and the other two a radius nr. If the rotation of the coder F is prevented, determine in terms of T, l, G, J, and n the angle through which end A rotates.

Fig. P3.43

3.44 For the gear train described in Prob. 3.43, determine the angle through which end A rotates when $T = 0.75$ N · m, $l = 60$ mm, $d = 4$ mm, $G = 77$ GPa, and $n = 2$.

3.45 The design specifications of a 2-m-long solid circular transmission shaft require that the angle of twist of the shaft not exceed 3° when a torque of 9 kN · m is applied. Determine the required diameter of the shaft, knowing that the shaft is made of (a) a steel with an allowable shearing stress of 90 MPa and a modulus of rigidity of 77 GPa, (b) a bronze with an allowable shearing stress of 35 MPa and a modulus of rigidity of 42 GPa.

3.46 The design specifications of a 4-ft-long solid transmission shaft require that the angle of twist of the shaft not exceed 4° when a torque of 6 kip · in. is applied. Determine the required diameter of the shaft, knowing that the shaft is made of a steel with an allowable shearing stress of 12 ksi and a modulus of rigidity of 11.2×10^6 psi.

Fig. P3.47

3.47 The design of the gear-and-shaft system shown requires that steel shafts of the same diameter be used for both AB and CD. It is further required that $\tau_{max} \leq 60$ MPa and that the angle ϕ_D through which end D of shaft CD rotates not exceed 1.5°. Knowing that $G = 77$ GPa, determine the required diameter of the shafts.

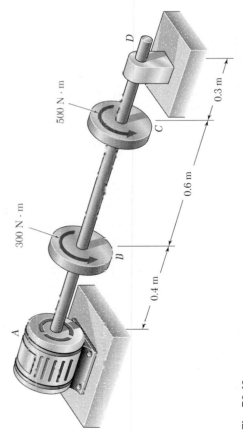

Fig. P3.48

3.48 The electric motor exerts a torque of 800 N · m on the steel shaft $ABCD$ when it is rotating at constant speed. Design specifications require that the diameter of the shaft be uniform from A to D and that the angle of twist between A and D not exceed 1.5°. Knowing that $\tau_{max} \leq 60$ MPa and $G = 77$ GPa, determine the minimum diameter shaft that may be used.

3.49 The solid cylindrical rod BC is attached to the rigid lever AB and to the fixed support at C. The vertical force $\mathbf{P}$ applied at A causes a small displacement Δ at point A. Show that the corresponding maximum shearing stress in the rod is

$$\tau = \frac{Gd}{2La}\,\Delta$$

where d is the diameter of the rod and G its modulus of rigidity.

Fig. P3.49, P3.50, and P3.51

3.50 and 3.51 The solid cylindrical rod BC of length $L = 24$ in. is attached to the rigid lever AB of length $a = 15$ in. and to the support at C. Design specifications require that the displacement of A not exceed 1 in. when a 100-lb force $\mathbf{P}$ is applied at A. For the material indicated determine the required diameter of the rod.

–**3.50** *Steel:* $\tau_{\text{all}} = 15$ ksi, $G = 11.2 \times 10^6$ psi.
3.51 *Aluminum:* $\tau_{\text{all}} = 10$ ksi, $G = 3.9 \times 10^6$ psi.

3.52 A torque of magnitude $T = 4$ kN · m is applied at end A of the composite shaft shown. Knowing that the modulus of rigidity is 77 GPa for the steel and 27 GPa for the aluminum, determine (a) the maximum shearing stress in the steel core, (b) the maximum shearing stress in the aluminum jacket, (c) the angle of twist at A.

Fig. P3.52 and P3.53

3.53 The composite shaft shown is to be twisted by applying a torque $\mathbf{T}$ at end A. Knowing that the modulus of rigidity is 77 GPa for the steel and 27 GPa for the aluminum, determine the largest angle through which end A may be rotated if the following allowable stresses are not to be exceeded: $\tau_{\text{steel}} = 60$ MPa and $\tau_{\text{aluminum}} = 45$ MPa.

3.54 The composite shaft shown consists of a 0.2-in.-thick brass jacket ($G = 5.6 \times 10^6$ psi) bonded to a 1.2-in.-diameter steel core ($G = 11.2 \times 10^6$ psi). Knowing that the shaft is subjected to 5 kip · in. torques, determine (a) the maximum shearing stress in the brass jacket, (b) the maximum shearing stress in the steel core, (c) the angle of twist of end B relative to end A.

3.55 The composite shaft shown consists of a 0.2-in.-thick brass jacket ($G = 5.6 \times 10^6$ psi) bonded to a 1.2-in.-diameter steel core ($G = 11.2 \times 10^6$ psi). Knowing that the shaft is being subjected to the torques shown, determine the largest angle through which it can be twisted if the following allowable stresses are not to be exceeded: $\tau_{\text{steel}} = 15$ ksi and $\tau_{\text{brass}} = 8$ ksi.

Fig. P3.54 and P3.55

3.56 Two solid steel shafts are fitted with flanges which are then connected by fitted bolts so that there is no relative rotation between the flanges. Knowing that $G = 77$ GPa, determine the maximum shearing stress in each shaft when a torque of magnitude $T = 500$ N · m applied to flange B.

Fig. P3.56, P3.57, and P3.58

3.57 and 3.58 Two solid steel shafts are fitted with flanges that are then connected by bolts as shown. The bolts are slightly undersized and permit a 1.5° rotation of one flange with respect to the other before the flanges begin to rotate as a single unit. Knowing that $G = 77$ GPa, determine the maximum shearing stress in each shaft when a torque **T** of magnitude 500 N · m is applied to the flange indicated.

3.57 The torque **T** is applied to flange B.
3.58 The torque **T** is applied to flange C.

3.59 At a time when rotation is prevented at the lower end of each shaft, a 50 N · m torque is applied to end A of shaft AB. Knowing that $G = 77$ GPa for both shafts, determine (*a*) the maximum shearing stress in shaft CD, (*b*) the angle of rotation at A.

Fig. P3.59

3.60 Solve Prob. 3.59, assuming that the 50 N · m torque is applied to end C of shaft CD.

3.61 The steel jacket CD has been attached to the 40-mm-diameter steel shaft AE by means of *rigid* flanges welded to the jacket and to the rod. The outer diameter of the jacket is 80 mm and its wall thickness is 4 mm. If 500 N · m torques are applied as shown, determine the maximum shearing stress in the jacket.

Fig. P3.61

3.62 The mass moment of inertia of a gear is to be determined experimentally by using a torsional pendulum consisting of a 6-ft steel wire. Knowing that $G = 11.2 \times 10^6$ psi, determine the diameter of the wire for which the torsional spring constant will be 4.27 lb $\cdot$ ft/rad.

Fig. P3.62

3.63 A solid shaft and a hollow shaft are made of the same material and are of the same weight and length. Denoting by n the ratio c_1/c_2, show that the ratio T_s/T_h of the torque T_s in the solid shaft to the torque T_h in the hollow shaft is (a) $\sqrt{1 - n^2}/(1 + n^2)$ if the maximum shearing stress is the same in each shaft, (b) $(1 - n^2)/(1 + n^2)$ if the angle of twist is the same for each shaft.

Fig. P3.63

3.64 A torque $\mathbf{T}$ is applied as shown to a solid tapered shaft AB. Show by integration that the angle of twist at A is

$$\phi = \frac{7TL}{12\pi Gc^4}$$

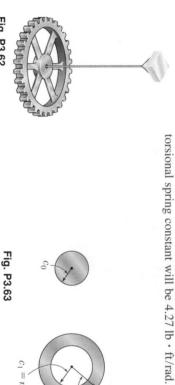

Fig. P3.64

3.65 An annular plate of thickness t and modulus of rigidity G is used to connect shaft AB of radius r_1 to tube CD of inner radius r_2. Knowing that a torque $\mathbf{T}$ is applied to end A of shaft AB and that end D of tube CD is fixed, (a) determine the magnitude and location of the maximum shearing stress in the annular plate, (b) show that the angle through which end B of the shaft rotates with respect to end C of the tube is

$$\phi_{B/C} = \frac{T}{4\pi G t}\left(\frac{1}{r_1^2} - \frac{1}{r_2^2}\right)$$

3.66 An annular aluminum plate ($G = 27$ GPa), of thickness $t = 6$ mm, is used to connect the aluminum shaft AB, of length $L_1 = 90$ mm and radius $r_1 = 30$ mm, to the aluminum tube CD, of length $L_2 = 150$ mm, inner radius $r_2 = 75$ mm and 4 mm thickness. Knowing that a torque of magnitude $T = 2500$ N $\cdot$ m is applied to end A of shaft AB and that end D of tube CD is fixed, determine (a) the maximum shearing stress in the shaft-plate-tube system, (b) the angle through which end A rotates. (*Hint:* Use the formula derived in Prob. 3.65 to solve part b.)

Fig. P3.65 and P3.66

3.7. DESIGN OF TRANSMISSION SHAFTS

The principal specifications to be met in the design of a transmission shaft are the *power* to be transmitted and the *speed of rotation* of the shaft. The role of the designer is to select the material and the dimensions of the cross section of the shaft, so that the maximum shearing stress allowable in the material will not be exceeded when the shaft is transmitting the required power at the specified speed.

To determine the torque exerted on the shaft, we recall from elementary dynamics that the power P associated with the rotation of a rigid body subjected to a torque $\mathbf{T}$ is

$$P = T\omega \qquad (3.19)$$

where ω is the angular velocity of the body expressed in radians per second. But $\omega = 2\pi f$, where f is the frequency of the rotation, i.e., the number of revolutions per second. The unit of frequency is thus 1 s^{-1} and is called a *hertz* (Hz). Substituting for ω into Eq. (3.19), we write

$$P = 2\pi f T \qquad (3.20)$$

If SI units are used we verify that, with f expressed in Hz and T in $\text{N} \cdot \text{m}$, the power will be expressed in $\text{N} \cdot \text{m/s}$, that is, in *watts* (W). Solving Eq. (3.20) for T, we obtain the torque exerted on a shaft transmitting the power P at a frequency of rotation f,

$$T = \frac{P}{2\pi f} \qquad (3.21)$$

where P, f, and T are expressed in the units indicated above.

After having determined the torque $\mathbf{T}$ that will be applied to the shaft and having selected the material to be used, the designer will carry the values of T and of the maximum allowable stress into the elastic torsion formula (3.9). Solving for J/c, we have

$$\frac{J}{c} = \frac{T}{\tau_{\max}} \qquad (3.22)$$

and obtain in this way the minimum value allowable for the parameter J/c. We check that, if SI units are used, T will be expressed in $\text{N} \cdot \text{m}$, $\tau_{\max}$ in Pa (or N/m^2), and J/c will be obtained in m^3. In the case of a solid circular shaft, $J = \frac{1}{2}\pi c^4$, and $J/c = \frac{1}{2}\pi c^3$; substituting this value for J/c into Eq. (3.22) and solving for c yields the minimum allowable value for the radius of the shaft. In the case of a hollow circular shaft, the critical parameter is J/c_2, where c_2 is the outer radius of the shaft; the value of this parameter may be computed from Eq. (3.11) of Sec. 3.4 to determine whether a given cross section will be acceptable.

When U.S. customary units are used, the frequency is usually expressed in rpm and the power in horsepower (hp). It is then necessary, before applying formula (3.21), to convert the frequency into revolutions per second (i.e., hertzes) and the power into ft · lb/s or in · lb/s through the use of the following relations:

$$1 \text{ rpm} = \frac{1}{60} \text{ s}^{-1} = \frac{1}{60} \text{ Hz}$$

$$1 \text{ hp} = 550 \text{ ft} \cdot \text{lb/s} = 6600 \text{ in} \cdot \text{lb/s}$$

If we express the power in in · lb/s, formula (3.21) will yield the value of the torque T in lb · in. Carrying this value of T into Eq. (3.22), and expressing $\tau_{\max}$ in psi, we obtain the value of the parameter J/c in in.^{3}.

Substituting for T and $\tau_{\max}$ into Eq. (3.22), we write

$$\frac{J}{c} = \frac{T}{\tau_{\max}} = \frac{87.54 \text{ lb} \cdot \text{in.}}{8500 \text{ psi}} = 10.30 \times 10^{-3} \text{ in}^3$$

But $J/c = \frac{1}{2}\pi c^3$ for a solid shaft. We have therefore

$$\tfrac{1}{2}\pi c^3 = 10.30 \times 10^{-3} \text{ in}^3$$

$$c = 0.1872 \text{ in.}$$

$$d = 2c = 0.374 \text{ in.}$$

A $\frac{3}{8}$-in. shaft should be used.

EXAMPLE 3.06

What size of shaft should be used for the rotor of a 5-hp motor operating at 3600 rpm if the shearing stress is not to exceed 8500 psi in the shaft?

We first express the power of the motor in in · lb/s and its frequency in cycles per second (or hertzes).

$$P = (5 \text{ hp}) \left(\frac{6600 \text{ in} \cdot \text{lb/s}}{1 \text{ hp}} \right) = 33{,}000 \text{ in} \cdot \text{lb/s}$$

$$f = (3600 \text{ rpm}) \frac{1 \text{ Hz}}{60 \text{ rpm}} = 60 \text{ Hz} = 60 \text{ s}^{-1}$$

The torque exerted on the shaft is given by Eq. (3.21):

$$T = \frac{P}{2\pi f} = \frac{33{,}000 \text{ in} \cdot \text{lb/s}}{2\pi (60 \text{ s}^{-1})} = 87.54 \text{ lb} \cdot \text{in.}$$

EXAMPLE 3.07

A shaft consisting of a steel tube of 50-mm outer diameter is to transmit 100 kW of power while rotating at a frequency of 20 Hz. Determine the tube thickness which should be used if the shearing stress is not to exceed 60 MPa.

The torque exerted on the shaft is given by Eq. (3.21):

$$T = \frac{P}{2\pi f} = \frac{100 \times 10^3 \text{ W}}{2\pi (20 \text{ Hz})} = 795.8 \text{ N} \cdot \text{m}$$

From Eq. (3.22) we conclude that the parameter J/c_2 must be at least equal to

$$\frac{J}{c_2} = \frac{T}{\tau_{\max}} = \frac{795.8 \text{ N} \cdot \text{m}}{60 \times 10^6 \text{ N/m}^2} = 13.26 \times 10^{-6} \text{ m}^3 \quad (3.23)$$

But, from Eq. (3.10) we have

$$\frac{J}{c_2} = \frac{\pi}{2c_2} (c_2^4 - c_1^4) = \frac{\pi}{0.050} [(0.025)^4 - c_1^4] \quad (3.24)$$

Equating the right-hand members of Eqs. (3.23) and (3.24):

$$(0.025)^4 - c_1^4 = \frac{0.050}{\pi} (13.26 \times 10^{-6})$$

$$c_1^4 = 390.6 \times 10^{-9} - 211.0 \times 10^{-9} = 179.6 \times 10^{-9} \text{ m}^4$$

$$c_1 = 20.6 \times 10^{-3} \text{ m} = 20.6 \text{ mm}$$

The corresponding tube thickness is

$$c_2 - c_1 = 25 \text{ mm} - 20.6 \text{ mm} = 4.4 \text{ mm}$$

A tube thickness of 5 mm should be used.

3.8. STRESS CONCENTRATIONS IN CIRCULAR SHAFTS

The torsion formula $\tau_{max} = Tc/J$ was derived in Sec. 3.4 for a circular shaft of uniform cross section. Moreover, we had assumed earlier in Sec. 3.3 that the shaft was loaded at its ends through rigid end plates solidly attached to it. In practice, however, the torques are usually applied to the shaft through flange couplings (Fig. 3.30a) or through gears connected to the shaft by keys fitted into keyways (Fig. 3.30b). In both cases one should expect the distribution of stresses, in and near the section where the torques are applied, to be different from that given by the torsion formula. High concentrations of stresses, for example, will occur in the neighborhood of the keyway shown in Fig. 3.30b. The determination of these localized stresses may be carried out by experimental stress analysis methods or, in some cases, through the use of the mathematical theory of elasticity.

As we indicated in Sec. 3.4, the torsion formula can also be used for a shaft of variable circular cross section. In the case of a shaft with an abrupt change in the diameter of its cross section, however, stress concentrations will occur near the discontinuity, with the highest stresses occurring at A (Fig. 3.31). These stresses may be reduced

(a)

(b)

Fig. 3.30

Fig. 3.31

Fig. 3.32 Stress-concentration factors for fillets in circular shafts.[†]

through the use of a fillet, and the maximum value of the shearing stress at the fillet can be expressed as

$$\tau_{max} = K\frac{Tc}{J} \tag{3.25}$$

where the stress Tc/J is the stress computed for the smaller-diameter shaft, and where K is a stress-concentration factor. Since the factor K depends only upon the ratio of the two diameters and the ratio of the radius of the fillet to the diameter of the smaller shaft, it may be computed once and for all and recorded in the form of a table or a graph, as shown in Fig. 3.32. We should note, however, that this procedure for determining localized shearing stresses is valid only as long as the value of τ_{max} given by Eq. (3.25) does not exceed the proportional limit of the material, since the values of K plotted in Fig. 3.32 were obtained under the assumption of a linear relation between shearing stress and shearing strain. If plastic deformations occur, they will result in values of the maximum stress lower than those indicated by Eq. (3.25).

[†]W. D. Pilkey, *Peterson's Stress Concentration Factors*, 2nd ed., John Wiley & Sons, New York, 1997.

3.75 in.

$r = \frac{9}{16}$ in.

7.50 in.

SAMPLE PROBLEM 3.6

The stepped shaft shown is to rotate at 900 rpm as it transmits power from a turbine to a generator. The grade of steel specified in the design has an allowable shearing stress of 8 ksi. (*a*) For the preliminary design shown, determine the maximum power that can be transmitted. (*b*) If in the final design the radius of the fillet is increased so that $r = \frac{15}{16}$ in., what will be the percent change, relative to the preliminary design, in the power which may be transmitted?

SOLUTION

a. **Preliminary Design.** Using the notation of Fig. 3.32, we have:

$$D = 7.50 \text{ in.}, d = 3.75 \text{ in.}, r = \frac{9}{16} \text{ in.} = 0.5625 \text{ in.}$$

$$\frac{D}{d} = \frac{7.50 \text{ in.}}{3.75 \text{ in.}} = 2 \qquad \frac{r}{d} = \frac{0.5625 \text{ in.}}{3.75 \text{ in.}} = 0.15$$

A stress concentration factor $K = 1.33$ is found from Fig. 3.32.

Torque. Recalling Eq. (3.25), we write

$$\tau_{\max} = K \frac{Tc}{J} \qquad T = \frac{J}{c} \frac{\tau_{\max}}{K} \tag{1}$$

$\tau_m = \dfrac{\tau_{\max}}{K} = 6.02$ ksi

where J/c refers to the smaller-diameter shaft:

$$J/c = \tfrac{1}{2}\pi c^3 = \tfrac{1}{2}\pi (1.875 \text{ in.})^3 = 10.35 \text{ in}^3$$

and where

$$\frac{\tau_{\max}}{K} = \frac{8 \text{ ksi}}{1.33} = 6.02 \text{ ksi}$$

Substituting into Eq. (1), we find $T = (10.35 \text{ in}^3)(6.02 \text{ ksi}) = 62.3 \text{ kip} \cdot \text{in.}$

$\mathbf{T}_a = 62.3$ kip · in. $r = \frac{9}{16}$ in.

Power. Since $f = (900 \text{ rpm})\dfrac{1 \text{ Hz}}{60 \text{ rpm}} = 15 \text{ Hz} = 15 \text{ s}^{-1}$, we write

$$P_a = 2\pi fT = 2\pi(15 \text{ s}^{-1})(62.3 \text{ kip} \cdot \text{in.}) = 5.87 \times 10^6 \text{ in} \cdot \text{lb/s}$$

$$P_a = (5.87 \times 10^6 \text{ in} \cdot \text{lb/s})(1 \text{ hp}/6600 \text{ in} \cdot \text{lb/s}) \qquad P_a = 890 \text{ hp} \blacktriangleleft$$

b. **Final Design.** For $r = \frac{15}{16}$ in. = 0.9375 in.,

$$\frac{D}{d} = 2 \qquad \frac{r}{d} = \frac{0.9375 \text{ in.}}{3.75 \text{ in.}} = 0.250 \qquad K = 1.20$$

Following the procedure used above, we write

$$\frac{\tau_{\max}}{K} = \frac{8 \text{ ksi}}{1.20} = 6.67 \text{ ksi}$$

$\tau_m = \dfrac{\tau_{\max}}{K} = 6.67$ ksi

$$T = \frac{J}{c}\frac{\tau_{\max}}{K} = (10.35 \text{ in}^3)(6.67 \text{ ksi}) = 69.0 \text{ kip} \cdot \text{in.}$$

$$P_b = 2\pi fT = 2\pi (15 \text{ s}^{-1})(69.0 \text{ kip} \cdot \text{in.}) = 6.50 \times 10^6 \text{ in} \cdot \text{lb/s}$$

$$P_b = (6.50 \times 10^6 \text{ in} \cdot \text{lb/s})(1 \text{ hp}/6600 \text{ in} \cdot \text{lb/s}) = 985 \text{ hp}$$

$\mathbf{T}_b = 69.0$ kip · in. $r = \frac{15}{16}$ in.

Percent Change in Power

$$\text{Percent change} = 100 \frac{P_b - P_a}{P_a} = 100 \frac{985 - 890}{890} = +11\% \blacktriangleleft$$

PROBLEMS

3.67 Using an allowable stress of 55 MPa, design a solid steel shaft to transmit 10 kW at a frequency of 15 Hz.

3.68 Using an allowable stress of 5 ksi, design a solid steel shaft to transmit $\frac{1}{2}$ hp at a speed of 1725 rpm.

3.69 Design a solid steel shaft to transmit 100 hp at a speed of 1200 rpm, if the maximum shearing stress is not to exceed 7500 psi.

3.70 Design a solid steel shaft to transmit 0.375 kW at a frequency of 29 Hz, if the shearing stress in the shaft is not to exceed 35 MPa.

3.71 A hollow shaft is to transmit 250 kW at a frequency of 30 Hz. Knowing that the shearing stress must not exceed 50 MPa, design a shaft for which the ratio of the inner diameter to the outer diameter is 0.75.

3.72 One of two hollow drive shafts of an ocean liner is 125 ft long, and its outer and inner diameters are 16 in. and 8 in., respectively. The shaft is made of a steel for which $\tau_{\text{all}} = 8500$ psi and $G = 11.2 \times 10^6$ psi. Knowing that the maximum speed of rotation of the shaft is 165 rpm, determine (*a*) the maximum power that can be transmitted by the one shaft to its propeller, (*b*) the corresponding angle of twist of the shaft.

3.73 While a steel shaft of the cross section shown rotates at 120 rpm, a stroboscopic measurement indicates that the angle of twist is 2° in a 4-m length. Using $G = 77$ GPa, determine the power being transmitted.

Fig. P3.73

3.74 Determine the required thickness of the 50-mm tubular shaft of Example 3.07, if it is to transmit the same power while rotating at a frequency of 40 Hz.

3.75 The design of a machine element calls for a 1.5-in.-outer-diameter shaft to transmit 60 hp. (*a*) If the speed of rotation is 720 rpm, determine the maximum shearing stress in shaft *a*. (*b*) If the speed of rotation can be increased 50% to 1080 rpm, determine the largest inner diameter of shaft *b* for which the maximum shearing stress will be the same in each shaft.

Fig. P3.75

3.76   A steel pipe of 60-mm outer diameter is to be used to transmit a torque of 350 N · m without exceeding an allowable shearing stress of 12 MPa. A series of 60-mm-outer-diameter pipes is available for use. Knowing that the wall thickness of the available pipes varies from 4 mm to 10 mm in 2-mm increments, choose the lightest pipe that can be used.

3.77   A steel drive shaft is 6 ft long and its outer and inner diameters are respectively equal to 2.25 in. and 1.75 in. (a) Knowing that the shaft transmits 240 hp while rotating at 1800 rpm, determine the maximum shearing stress. (b) Using $G = 11.2 \times 10^6$ psi, determine the corresponding angle of twist of the shaft.

3.78   Knowing that the allowable shearing stress of the steel to be used is 7500 psi, determine (a) the smallest permissible diameter of a shaft which must transmit 15 hp while rotating at 2000 rpm, (b) the corresponding angle of twist in a 4-ft length of the shaft ($G = 11.2 \times 10^6$ psi).

3.79   Three shafts and four gears are used to form a gear train that will transmit 7.5 kW from the motor at A to a machine tool at F. (Bearings for the shafts are omitted in the sketch.) Knowing that the frequency of the motor is 30 Hz and that the allowable stress for each shaft is 60 MPa, determine the required diameter of each shaft.

Fig. P3.79 and P3.80

3.80   Three shafts and four gears are used to form a gear train that will transmit power from the motor at A to a machine tool at F. (Bearings for the shafts are omitted in the sketch.) The diameter of each shaft is as follows: $d_{AB} = 16$ mm, $d_{CD} = 20$ mm, $d_{EF} = 28$ mm. Knowing that the frequency of the motor is 24 Hz and that the allowable shearing stress for each shaft is 75 MPa, determine the maximum power that can be transmitted.

3.81   The shaft-disk-belt arrangement shown is used to transmit 3 hp from point A to point D. (a) Using an allowable shearing stress of 9500 psi, determine the required speed of shaft AB. (b) Solve part a, assuming that the diameters of shafts AB and CD are respectively 0.75 in. and 0.625 in.

Fig. P3.81

3.82 A steel shaft must transmit 150 kW at a speed of 360 rpm. Knowing that $G = 77$ GPa, design a solid shaft so that the maximum stress will not exceed 50 MPa and the angle of twist in a 2.5-m length will not exceed 3°.

3.83 A steel shaft of 1.5-m length and 48-mm diameter is to be used to transmit 36 kW between a motor and a machine tool. Knowing that $G = 77$ GPa, determine the lowest speed of rotation of the shaft at which the maximum stress will not exceed 60 MPa and the angle of twist will not exceed 2.5°.

3.84 A 1.5-in.-diameter steel shaft of length 4 ft will be used to transmit 60 hp between a motor and a pump. Knowing that $G = 11.2 \times 10^6$ psi, determine the lowest speed of rotation at which the shearing stress will not exceed 8500 psi and the angle of twist will not exceed 2°.

3.85 A 1.6-m-long tubular shaft of 42-mm outer diameter d_1 is to be made of a steel for which $\tau_{all} = 75$ MPa and $G = 77$ GPa. Knowing that the angle of twist must not exceed 4° when the shaft is subjected to a torque of 900 N · m, determine the largest inner diameter d_2 which can be specified in the design.

d_2

$d_1 = 42$ mm

Fig. P3.85 and P3.86

3.86 A 1.6-m-long tubular steel shaft ($G = 77$ GPa) of 42-mm outer diameter d_1 and 30-mm inner diameter d_2 is to transmit 120 kW between a turbine and a generator. Knowing that the allowable shearing stress is 65 MPa and that the angle of twist must not exceed 3°, determine the minimum frequency at which the shaft may rotate.

3.87 The stepped shaft shown rotates at 450 rpm. Knowing that $r = 10$ mm, determine the maximum power that can be transmitted without exceeding an allowable shearing stress of 45 MPa.

3.88 The stepped shaft shown rotates at 450 rpm. Knowing that $r = 4$ mm, determine the maximum power that can be transmitted without exceeding an allowable shearing stress of 45 MPa.

120 mm

100 mm

r

Fig. P3.87 and P3.88

3.89 Knowing that the stepped shaft shown must transmit 60 hp at a speed of 2100 rpm, determine the minimum radius r of the fillet if an allowable stress of 6000 psi is not to be exceeded.

3.90 The stepped shaft shown must transmit 60 hp. Knowing that the allowable shearing stress in the shaft is 6000 psi and that the radius of the fillet is $r = \frac{1}{4}$ in., determine the smallest permissible speed of the shaft.

T'

2.5 in.

1.25 in.

T

Fig. P3.89 and P3.90

3.91   A torque of magnitude $T = 25$ N · m is applied to the stepped shaft shown, which has a full quarter-circular fillet. Knowing that $D = 24$ mm, determine the maximum shearing stress in the shaft when (a) $d = 20$ mm, (b) $d = 21.6$ mm.

Full quarter-circular fillet extends to edge of larger shaft

Fig. P3.91, P3.92, and P3.93

3.92   In the stepped shaft shown, which has a full quarter-circular fillet, $D = 1.5$ in. and $d = 1.2$ in. Knowing that the speed of the shaft is 1800 rpm and that the allowable shearing stress is 8000 psi, determine the maximum power that may be transmitted by the shaft.

3.93   In the stepped shaft shown, which has a full quarter-circular fillet, the allowable shearing stress is 12 ksi. Knowing that $D = 1.25$ in., determine the largest allowable torque that may be applied to the shaft if (a) $d = 1.1$ in., (b) $d = 1.0$ in.

*3.9. PLASTIC DEFORMATIONS IN CIRCULAR SHAFTS

When we derived Eqs. (3.10) and (3.16), which define respectively the stress distribution and the angle of twist for a circular shaft subjected to a torque **T**, we assumed that Hooke's law applied throughout the shaft. If the yield strength is exceeded in some portion of the shaft, or if the material involved is a brittle material with a nonlinear shearing-stress–strain diagram, these relations cease to be valid. The purpose of this section is to develop a more general method—which may be used when Hooke's law does not apply—for determining the distribution of stresses in a solid circular shaft, and for computing the torque required to produce a given angle of twist.

We first recall that no specific stress-strain relationship was assumed in Sec. 3.3, when we proved that the shearing strain γ varies linearly with the distance ρ from the axis of the shaft (Fig. 3.33). Thus, we may still use this property in our present analysis and write

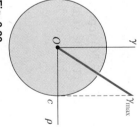

Fig. 3.33

$$\gamma = \frac{\rho}{c}\, \gamma_{max} \tag{3.4}$$

where c is the radius of the shaft.

Assuming that the maximum value τ_{max} of the shearing stress τ has been specified, the plot of τ versus ρ may be obtained as follows. We first determine from the shearing-stress-strain diagram the value of γ_{max} corresponding to τ_{max} (Fig. 3.34), and carry this value into Eq. (3.4).

Fig. 3.34

Then, for each value of ρ, we determine the corresponding value of γ from Eq. (3.4) or Fig. 3.33 and obtain from the stress-strain diagram of Fig. 3.34 the shearing stress τ corresponding to this value of γ. Plotting τ against ρ yields the desired distribution of stresses (Fig. 3.35).

We now recall that, when we derived Eq. (3.1) in Sec. 3.2, we assumed no particular relation between shearing stress and strain. We may therefore use Eq. (3.1) to determine the torque $\mathbf{T}$ corresponding to the shearing-stress distribution obtained in Fig. 3.35. Considering an annular element of radius ρ and thickness $d\rho$, we express the element of area in Eq. (3.1) as $dA = 2\pi\rho\, d\rho$ and write

$$T = \int_0^c \rho\tau(2\pi\rho\, d\rho)$$

or

$$T = 2\pi \int_0^c \rho^2\tau\, d\rho \qquad (3.26)$$

where τ is the function of ρ plotted in Fig. 3.35.

If τ is a known analytical function of γ, Eq. (3.4) may be used to express τ as a function of ρ, and the integral in (3.26) may be determined analytically. Otherwise, the torque $\mathbf{T}$ may be obtained through a numerical integration. This computation becomes more meaningful if we note that the integral in Eq. (3.26) represents the second moment, or moment of inertia, with respect to the vertical axis of the area in Fig. 3.35 located above the horizontal axis and bounded by the stress-distribution curve.

An important value of the torque is the ultimate torque T_U which causes failure of the shaft. This value may be determined from the ultimate shearing stress τ_U of the material by choosing $\tau_{max} = \tau_U$ and carrying out the computations indicated earlier. However, it is found more convenient in practice to determine T_U experimentally by twisting a specimen of a given material until it breaks. Assuming a fictitious linear distribution of stresses, Eq. (3.9) is then used to determine the corresponding maximum shearing stress R_T:

$$R_T = \frac{T_U c}{J} \qquad (3.27)$$

Fig. 3.35

The fictitious stress R_T is called the *modulus of rupture in torsion* of the given material. It may be used to determine the ultimate torque T_U of a shaft made of the same material, but of different dimensions, by solving Eq. (3.27) for T_U. Since the actual and the fictitious linear stress distributions shown in Fig. 3.36 must yield the same value T_U for the

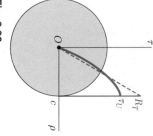

Fig. 3.36

ultimate torque, the areas they define must have the same moment of inertia with respect to the vertical axis. It is thus clear that the modulus of rupture R_T will always be larger than the actual ultimate shearing stress τ_U.

In some cases, we may wish to determine the stress distribution and the torque **T** corresponding to a given angle of twist ϕ. This may be done by recalling the expression obtained in Sec. 3.3 for the shearing strain γ in terms of ϕ, ρ, and the length L of the shaft:

$$\gamma = \frac{\rho\phi}{L} \qquad (3.2)$$

With ϕ and L given, we may determine from Eq. (3.2) the value of γ corresponding to any given value of ρ. Using the stress-strain diagram of the material, we may then obtain the corresponding value of the shearing stress τ and plot τ against ρ. Once the shearing-stress distribution has been obtained, the torque **T** may be determined analytically or numerically as explained earlier.

*3.10. CIRCULAR SHAFTS MADE OF AN ELASTOPLASTIC MATERIAL

Further insight into the plastic behavior of a shaft in torsion is obtained by considering the idealized case of a *solid circular shaft made of an elastoplastic material*. The shearing-stress-strain diagram of such a material is shown in Fig. 3.37. Using this diagram, we can proceed as indicated earlier and find the stress distribution across a section of the shaft for any value of the torque **T**.

As long as the shearing stress τ does not exceed the yield strength τ_Y, Hooke's law applies, and the stress distribution across the section is linear (Fig. 3.38*a*), with τ_{max} given by Eq. (3.9):

$$\tau_{max} = \frac{Tc}{J} \qquad (3.9)$$

Fig. 3.37

As the torque increases, τ_{max} eventually reaches the value τ_Y (Fig. 3.38b). Substituting this value into Eq. (3.9), and solving for the corresponding value of T, we obtain the value T_Y of the torque at the onset of yield:

$$T_Y = \frac{J}{c}\tau_Y \qquad (3.28)$$

The value obtained is referred to as the *maximum elastic torque*, since it is the largest torque for which the deformation remains fully elastic. Recalling that for a solid circular shaft $J/c = \frac{1}{2}\pi c^3$, we have

$$T_Y = \frac{1}{2}\pi c^3 \tau_Y \qquad (3.29)$$

As the torque is further increased, a plastic region develops in the shaft, around an elastic core of radius ρ_Y (Fig. 3.38c). In the plastic region the stress is uniformly equal to τ_Y, while in the elastic core the stress varies linearly with ρ and may be expressed as

$$\tau = \frac{\tau_Y}{\rho_Y}\rho \qquad (3.30)$$

As T is increased, the plastic region expands until, at the limit, the deformation is fully plastic (Fig. 3.38d).

Equation (3.26) will be used to determine the value of the torque T corresponding to a given radius ρ_Y of the elastic core. Recalling that τ is given by Eq. (3.30) for $0 \leq \rho \leq \rho_Y$, and is equal to τ_Y for $\rho_Y \leq \rho \leq c$, we write

$$T = 2\pi \int_0^{\rho_Y} \rho^2 \left(\frac{\tau_Y}{\rho_Y}\rho\right) d\rho + 2\pi \int_{\rho_Y}^c \rho^2 \tau_Y d\rho$$

$$= \frac{1}{2}\pi\rho_Y^3\tau_Y + \frac{2}{3}\pi c^3\tau_Y - \frac{2}{3}\pi\rho_Y^3\tau_Y$$

$$T = \frac{2}{3}\pi c^3\tau_Y\left(1 - \frac{1}{4}\frac{\rho_Y^3}{c^3}\right) \qquad (3.31)$$

or, in view of Eq. (3.29),

$$T = \frac{4}{3}T_Y\left(1 - \frac{1}{4}\frac{\rho_Y^3}{c^3}\right) \qquad (3.32)$$

where T_Y is the maximum elastic torque. We note that, as ρ_Y approaches zero, the torque approaches the limiting value

$$T_p = \frac{4}{3}T_Y \qquad (3.33)$$

This value of the torque, which corresponds to a fully plastic deformation (Fig. 3.38d), is called the *plastic torque* of the shaft considered. We note that Eq. (3.33) is valid only for a *solid circular shaft made of an elastoplastic material*.

Since the distribution of *strain* across the section remains linear after the onset of yield, Eq. (3.2) remains valid and can be used to express the radius ρ_Y of the elastic core in terms of the angle of twist ϕ. If ϕ is large enough to cause a plastic deformation, the radius ρ_Y of the

(a)

(b)

(c)

(d)

Fig. 3.38

elastic core is obtained by making γ equal to the yield strain γ_Y in Eq. (3.2) and solving for the corresponding value ρ_Y of the distance ρ. We have

$$\rho_Y = \frac{L\gamma_Y}{\phi} \qquad (3.34)$$

Let us denote by ϕ_Y the angle of twist at the onset of yield, i.e., when $\rho_Y = c$. Making $\phi = \phi_Y$ and $\rho_Y = c$ in Eq. (3.34), we have

$$c = \frac{L\gamma_Y}{\phi_Y} \qquad (3.35)$$

Dividing (3.34) by (3.35), member by member, we obtain the following relation:†

$$\frac{\rho_Y}{c} = \frac{\phi_Y}{\phi} \qquad (3.36)$$

If we carry into Eq. (3.32) the expression obtained for ρ_Y/c, we express the torque T as a function of the angle of twist ϕ,

$$T = \frac{4}{3}T_Y\left(1 - \frac{1}{4}\frac{\phi_Y^3}{\phi^3}\right) \qquad (3.37)$$

where T_Y and ϕ_Y represent, respectively, the torque and the angle of twist at the onset of yield. Note that Eq. (3.37) may be used only for values of ϕ larger than ϕ_Y. For $\phi < \phi_Y$, the relation between T and ϕ is linear and given by Eq. (3.16). Combining both equations, we obtain the plot of T against ϕ represented in Fig. 3.39. We check that, as ϕ increases indefinitely, T approaches the limiting value $T_p = \frac{4}{3}T_Y$ corresponding to the case of a fully developed plastic zone (Fig. 3.38d). While the value T_p cannot actually be reached, we note from Eq. (3.37) that it is rapidly approached as ϕ increases. For $\phi = 2\phi_Y$, T is within about 3% of T_p, and for $\phi = 3\phi_Y$, within about 1 percent.

Since the plot of T against ϕ that we have obtained for an idealized elastoplastic material (Fig. 3.39) differs greatly from the shearing-stress-strain diagram of that material (Fig. 3.37), it is clear that the shearing-stress-strain diagram of an actual material cannot be obtained directly from a torsion test carried out on a solid circular rod made of that material. However, a fairly accurate diagram may be obtained from a torsion test if the specimen used incorporates a portion consisting of a thin circular tube.‡ Indeed, we may assume that the shearing stress will have a constant value τ in that portion. Equation (3.1) thus reduces to

$$T = \rho A\tau$$

where ρ denotes the average radius of the tube and A its cross-sectional area. The shearing stress is thus proportional to the torque, and successive values of τ can be easily computed from the corresponding values of T. On the other hand, the values of the shearing strain γ may be obtained from Eq. (3.2) and from the values of ϕ and L measured on the tubular portion of the specimen.

---

†Equation (3.36) applies to any ductile material with a well-defined yield point, since its derivation is independent of the shape of the stress-strain diagram beyond the yield point.

‡In order to minimize the possibility of failure by buckling, the specimen should be made so that the length of the tubular portion is no longer than its diameter.

Fig. 3.39

EXAMPLE 3.08

A solid circular shaft, 1.2 m long and 50 mm in diameter, is subjected to a 4.60 kN · m torque at each end (Fig. 3.40). Assuming the shaft to be made of an elastoplastic material with a yield strength in shear of 150 MPa and a modulus of rigidity of 77 GPa, determine (a) the radius of the elastic core, (b) the angle of twist of the shaft.

Fig. 3.40

(a) Radius of Elastic Core. We first determine the torque T_Y at the onset of yield. Using Eq. (3.28) with $\tau_Y = 150$ MPa, $c = 25$ mm, and

$$J = \tfrac{1}{2}\pi c^4 = \tfrac{1}{2}\pi(25 \times 10^{-3} \text{ m})^4 = 614 \times 10^{-9} \text{ m}^4$$

we write

$$T_Y = \frac{J\tau_Y}{c} = \frac{(614 \times 10^{-9} \text{ m}^4)(150 \times 10^6 \text{ Pa})}{25 \times 10^{-3} \text{ m}} = 3.68 \text{ kN} \cdot \text{m}$$

Solving Eq. (3.32) for $(\rho_Y/c)^3$ and substituting the values of T and T_Y, we have

$$\left(\frac{\rho_Y}{c}\right)^3 = 4 - \frac{3T}{T_Y} = 4 - \frac{3(4.60 \text{ kN} \cdot \text{m})}{3.68 \text{ kN} \cdot \text{m}} = 0.250$$

$$\frac{\rho_Y}{c} = 0.630 \qquad \rho_Y = 0.630(25 \text{ mm}) = 15.8 \text{ mm}$$

(b) Angle of Twist. We first determine the angle of twist ϕ_Y at the onset of yield from Eq. (3.16):

$$\phi_Y = \frac{T_Y L}{JG} = \frac{(3.68 \times 10^3 \text{ N} \cdot \text{m})(1.2 \text{ m})}{(614 \times 10^{-9} \text{ m}^4)(77 \times 10^9 \text{ Pa})}$$

$$= 93.4 \times 10^{-3} \text{ rad}$$

Solving Eq. (3.36) for ϕ and substituting the values obtained for ϕ_Y and ρ_Y/c, we write

$$\phi = \frac{\phi_Y}{\rho_Y/c} = \frac{93.4 \times 10^{-3} \text{ rad}}{0.630} = 148.3 \times 10^{-3} \text{ rad}$$

or

$$\phi = (148.3 \times 10^{-3} \text{ rad})\left(\frac{360°}{2\pi \text{ rad}}\right) = 8.50°$$

*3.11. RESIDUAL STRESSES IN CIRCULAR SHAFTS

In the two preceding sections we saw that a plastic region will develop in a shaft subjected to a large enough torque, and that the shearing stress τ at any given point in the plastic region may be obtained from the shearing-stress-strain diagram of Fig. 3.34. If the torque is removed, the resulting reduction of stress and strain at the point considered will take place along a straight line (Fig. 3.41). As you will see further in this section, the final value of the stress will not, in general, be zero. There will be a residual stress at most points, and that stress may be either positive or negative. We note that, as was the case for the normal stress, the shearing stress will keep decreasing until it has reached a value equal to its maximum value at C minus twice the yield strength of the material.

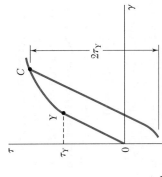

Fig. 3.41

Consider again the idealized case of the elastoplastic material characterized by the shearing-stress-strain diagram of Fig. 3.37. Assuming that the relation between τ and γ at any point of the shaft remains linear as long as the stress does not decrease by more than $2\tau_Y$, we can use Eq. (3.16) to obtain the angle through which the shaft untwists as the torque decreases back to zero. As a result, the unloading of the shaft will be represented by a straight line on the T-ϕ diagram (Fig. 3.42). We note that the angle of twist does not return to zero after the torque has been removed. Indeed, the loading and unloading of the shaft result in a permanent deformation characterized by the angle

$$\phi_p = \phi - \phi' \tag{3.38}$$

where ϕ corresponds to the loading phase and may be obtained from T by solving Eq. (3.38), and where ϕ' corresponds to the unloading phase and may be obtained from Eq. (3.16).

The residual stresses in an elastoplastic material are obtained by applying the principle of superposition in a manner similar to that described in Sec. 2.20 for an axial loading. We consider, on one hand, the stresses due to the application of the given torque $\mathbf{T}$ and, on the other, the stresses due to the equal and opposite torque which is applied to unload the shaft. The first group of stresses reflects the elastoplastic behavior of the material during the loading phase (Fig. 3.43a), and the

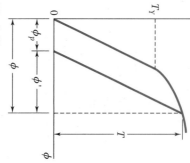

Fig. 3.42

second group the linear behavior of the same material during the unloading phase (Fig. 3.43b). Adding the two groups of stresses, we obtain the distribution of the residual stresses in the shaft (Fig. 3.43c).

We note from Fig. 3.43c that some residual stresses have the same sense as the original stresses, while others have the opposite sense. This was to be expected since, according to Eq. (3.1), the relation

$$\int \rho(\tau\,dA) = 0 \tag{3.39}$$

must be verified after the torque has been removed.

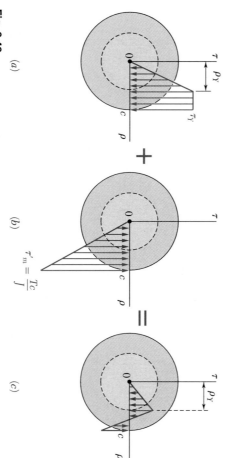

$$(a) \qquad\qquad (b)\qquad\qquad (c)$$

$$\tau'_m = \frac{Tc}{J}$$

Fig. 3.43

EXAMPLE 3.09

For the shaft of Example 3.08 determine (a) the permanent twist, (b) the distribution of residual stresses, after the 4.60 kN · m torque has been removed.

(a) Permanent Twist. We recall from Example 3.08 that the angle of twist corresponding to the given torque is $\phi = 8.50°$. The angle ϕ' through which the shaft untwists as the torque is removed is obtained from Eq. (3.16). Substituting the given data,

$$T = 4.60 \times 10^3 \text{ N} \cdot \text{m}$$

$$L = 1.2 \text{ m}$$

$$G = 77 \times 10^9 \text{ Pa}$$

and the value $J = 614 \times 10^{-9} \text{ m}^4$ obtained in the solution of Example 3.08, we have

$$\phi' = \frac{TL}{JG} = \frac{(4.60 \times 10^3 \text{ N} \cdot \text{m})(1.2 \text{ m})}{(614 \times 10^{-9} \text{ m}^4)(77 \times 10^9 \text{ Pa})}$$

$$= 116.8 \times 10^{-3} \text{ rad}$$

or

$$\phi' = (116.8 \times 10^{-3} \text{ rad}) \frac{360°}{2\pi \text{ rad}} = 6.69°$$

The permanent twist is therefore

$$\phi_p = \phi - \phi' = 8.50° - 6.69° = 1.81°$$

(b) Residual Stresses. We recall from Example 3.08 that the yield strength is $\tau_Y = 150$ MPa and that the radius of the elastic core corresponding to the given torque is $\rho_Y = 15.8$ mm. The distribution of the stresses in the loaded shaft is thus as shown in Fig. 3.44a.

The distribution of stresses due to the opposite 4.60 kN · m torque required to unload the shaft is linear and as shown in Fig. 3.44b. The maximum stress in the distribution of the reverse stresses is obtained from Eq. (3.9):

$$\tau'_{\max} = \frac{Tc}{J} = \frac{(4.60 \times 10^3 \text{ N} \cdot \text{m})(25 \times 10^{-3} \text{ m})}{614 \times 10^{-9} \text{ m}^4}$$

$$= 187.3 \text{ MPa}$$

Superposing the two distributions of stresses, we obtain the residual stresses shown in Fig. 3.44c. We check that, even though the reverse stresses exceed the yield strength τ_Y, the assumption of a linear distribution of these stresses is valid, since they do not exceed $2\tau_Y$.

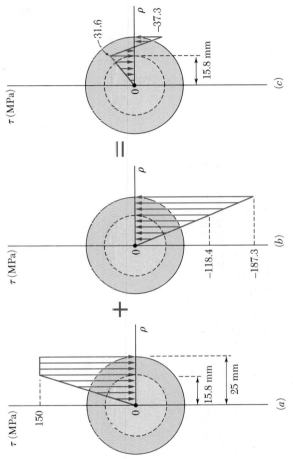

Fig. 3.44

179

SAMPLE PROBLEM 3.7

Shaft AB is made of a mild steel which is assumed to be elastoplastic with $G = 11.2 \times 10^6$ psi and $\tau_Y = 21$ ksi. A torque $\mathbf{T}$ is applied and gradually increased in magnitude. Determine the magnitude of $\mathbf{T}$ and the corresponding angle of twist (a) when yield first occurs, (b) when the deformation has become fully plastic.

SOLUTION

Geometric Properties

The geometric properties of the cross section are

$$c_1 = \tfrac{1}{2}(1.5 \text{ in.}) = 0.75 \text{ in.} \qquad c_2 = \tfrac{1}{2}(2.25 \text{ in.}) = 1.125 \text{ in.}$$
$$J = \tfrac{1}{2}\pi(c_2^4 - c_1^4) = \tfrac{1}{2}\pi[(1.125 \text{ in.})^4 - (0.75 \text{ in.})^4] = 2.02 \text{ in}^4$$

a. **Onset of Yield.** For $\tau_{\max} = \tau_Y = 21$ ksi, we find

$$T_Y = \frac{\tau_Y J}{c_2} = \frac{(21 \text{ ksi})(2.02 \text{ in}^4)}{1.125 \text{ in.}}$$

$$T_Y = 37.7 \text{ kip} \cdot \text{in.} \quad \blacktriangle$$

Making $\rho = c_2$ and $\gamma = \gamma_Y$ in Eq. (3.2) and solving for ϕ, we obtain the value of ϕ_Y:

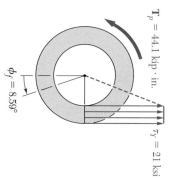

$$\phi_Y = \frac{\gamma_Y L}{c_2} = \frac{\tau_Y L}{c_2 G} = \frac{(21 \times 10^3 \text{ psi})(60 \text{ in.})}{(1.125 \text{ in.})(11.2 \times 10^6 \text{ psi})} = 0.100 \text{ rad}$$

$$\phi_Y = 5.73° \quad \blacktriangle$$

b. **Fully Plastic Deformation.** When the plastic zone reaches the inner surface, the stresses are uniformly distributed as shown. Using Eq. (3.26), we write

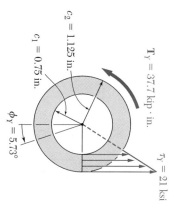

$$T_p = 2\pi\tau_Y \int_{c_1}^{c_2} \rho^2 \, d\rho = \tfrac{2}{3}\pi\tau_Y(c_2^3 - c_1^3)$$
$$= \tfrac{2}{3}\pi(21 \text{ ksi})[(1.125 \text{ in.})^3 - (0.75 \text{ in.})^3]$$

$$T_p = 44.1 \text{ kip} \cdot \text{in.} \quad \blacktriangle$$

When yield first occurs on the inner surface, the deformation is fully plastic; we have from Eq. (3.2):

$$\phi_f = \frac{\gamma_Y L}{c_1} = \frac{\tau_Y L}{c_1 G} = \frac{(21 \times 10^3 \text{ psi})(60 \text{ in.})}{(0.75 \text{ in.})(11.2 \times 10^6 \text{ psi})} = 0.150 \text{ rad}$$

$$\phi_f = 8.59° \quad \blacktriangle$$

For larger angles of twist the torque remains constant; the T-ϕ diagram of the shaft is as shown.

SAMPLE PROBLEM 3.8

For the shaft of Sample Prob. 3.7, determine the residual stresses and the permanent angle of twist after the torque $T_p = 44.1$ kip $\cdot$ in. has been removed.

SOLUTION

Referring to Sample Prob. 3.7, we recall that when the plastic zone first reached the inner surface, the applied torque was $T_p = 44.1$ kip $\cdot$ in. and the corresponding angle of twist was $\phi_f = 8.59°$. These values are shown in Fig. 1.

Elastic Unloading. We unload the shaft by applying a 44.1 kip $\cdot$ in. torque in the sense shown in Fig. 2. During this unloading, the behavior of the material is linear. Recalling from Sample Prob. 3.7 the values found for c_1, c_2, and J, we obtain the following stresses and angle of twist:

$$\tau_{max} = \frac{Tc_2}{J} = \frac{(44.1 \text{ kip} \cdot \text{in.})(1.125 \text{ in.})}{2.02 \text{ in}^4} = 24.56 \text{ ksi}$$

$$\tau_{min} = \tau_{max} \frac{c_1}{c_2} = (24.56 \text{ ksi}) \frac{0.75 \text{ in.}}{1.125 \text{ in.}} = 16.37 \text{ ksi}$$

$$\phi' = \frac{TL}{JG} = \frac{(44.1 \times 10^3 \text{ psi})(60 \text{ in.})}{(2.02 \text{ in}^4)(11.2 \times 10^6 \text{ psi})} = 0.1170 \text{ rad} = 6.70°$$

Residual Stresses and Permanent Twist. The results of the loading (Fig. 1) and the unloading (Fig. 2) are superposed (Fig. 3) to obtain the residual stresses and the permanent angle of twist ϕ_p.

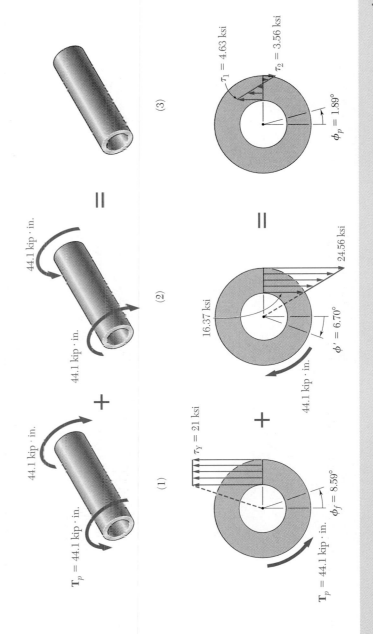

PROBLEMS

3.94 A 54-mm-diameter solid shaft is made of mild steel that is assumed to be elastoplastic with $\tau_Y = 145$ MPa. Determine the maximum shearing stress and the radius of the elastic core caused by the application of a torque of magnitude (*a*) 4 kN · m, (*b*) 5 kN · m.

3.95 A 1.5-in.-diameter solid shaft is made of mild steel that is assumed to be elastoplastic with $\tau_Y = 21$ ksi. Determine the maximum shearing stress and the radius of the elastic core caused by the application of a torque of magnitude (*a*) 12 kip · in., (*b*) 18 kip · in.

3.96 A 30-mm-diameter solid rod is made of an elastoplastic material with $\tau_Y = 3.5$ MPa. Knowing that the elastic core of the rod is 25 mm in diameter, determine the magnitude of the torque applied to the rod.

3.97 It is observed that a straightened steel paper clip can be twisted through several revolutions by the application of a torque of approximately 0.8 lb · in. Knowing that the diameter of the wire in the paper clip is 0.04 in., determine the approximate value of the yield stress of the steel.

3.98 The solid circular shaft shown is made of a steel that is assumed to be elastoplastic with $\tau_Y = 21$ ksi. Determine the magnitude T of the applied torque when the plastic zone is (*a*) 0.6 in. deep, (*b*) 1 in. deep.

$c = 1.25$ in.

Fig. P3.98

3.99 For the shaft and loading of Prob. 3.98, determine the angle of twist in a 4-ft length of shaft.

3.100 A torque **T** is applied to the 20-mm-diameter steel rod AB. Assuming the steel to be elastoplastic with $G = 77$ GPa and $\tau_Y = 145$ MPa, determine (*a*) the torque **T** when the angle of twist at A is 25°, (*b*) the corresponding diameter of the elastic core of the shaft.

20 mm

1.5 m

A

B

Fig. P3.100

3.101 An 18-mm-diameter solid circular shaft is made of a material that is assumed to be elastoplastic with $G = 77$ GPa and $\tau_Y = 145$ MPa. For a 1.2-m length, determine the maximum shearing stress and the angle of twist caused by a 200 N · m torque.

3.102 A solid circular rod is made of a material that is assumed to be elastoplastic. Denoting by τ_Y and ϕ_Y, respectively, the torque and angle of twist at the onset of yield, determine the angle of twist if the torque is increased to (a) $T = 1.1T_Y$, (b) $T = 1.25T_Y$, (c) $T = 1.3T_Y$.

3.103 A 0.75-in.-diameter solid circular shaft is made of a material that is assumed to be elastoplastic with $G = 11.2 \times 10^6$ psi and $\tau_Y = 21$ ksi. For a 5-ft length of the shaft, determine the maximum shearing stress and the angle of twist caused by a 2 kip · in. torque.

3.104 A hollow steel shaft is 0.9 m long and has the cross section shown. The steel is assumed to be elastoplastic with $\tau_Y = 180$ MPa and $G = 77$ GPa. Determine the applied torque and the corresponding angle of twist (a) at the onset of yield, (b) when the plastic zone is 10 mm deep.

3.105 A hollow steel shaft is 0.9 m long and has the cross section shown. The steel is assumed to be elastoplastic with $\tau_Y = 180$ MPa and $G = 77$ GPa. Determine (a) the angle of twist at which the section first becomes fully plastic, (b) the corresponding magnitude of the applied torque.

Fig. P3.104 and P3.105

3.106 A shaft of mild steel is machined to the shape shown and then twisted by torques of magnitude 40 kip · in. Assuming the steel to be elastoplastic with $\tau_Y = 21$ ksi, determine (a) the thickness of the plastic zone in portion CD of the shaft, (b) the length of the portion BE which remains fully elastic.

Fig. P3.106

3.107 The magnitude of the torque **T** applied to the tapered shaft of Prob. 3.106 is slowly increased. Determine (a) the largest torque that may be applied to the shaft, (b) the length of portion BE that remains fully elastic.

3.108 Considering the partially plastic shaft of Fig. 3.38c, derive Eq. (3.32) by recalling that the integral in Eq. (3.26) represents the second moment about the τ axis of the area under the τ-ρ curve.

3.109 Using the stress-strain diagram shown, determine (a) the torque that causes a maximum shearing stress of 15 ksi in a 0.8-diameter solid rod, (b) the corresponding angle of twist in a 20-in. length of the rod.

3.110 A hollow shaft of outer and inner diameters respectively equal to 0.6 in. and 0.2 in. is fabricated from an aluminum alloy for which the stress-strain diagram is given in the sketch. Determine the torque required to twist a 9-in. length of the shaft through 10°.

Fig. P3.109 and P3.110

3.111   A 50-mm-diameter cylinder is made of a brass for which the stress-strain diagram is as shown. Knowing that the angle of twist is 5° in a 725-mm length, determine by approximate means the magnitude T of the torque applied to the shaft.

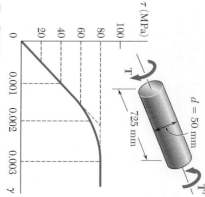

Fig. P3.111

3.112   Three points on the nonlinear stress-strain diagram used in Prob. 3.111 are (0, 0), (0.0015, 55 MPa), and (0.003, 80 MPa). By fitting the polynomial $\tau = A + B\gamma + C\gamma^2$ through these points, the following approximate relation has been obtained.

$$\tau = 46.7 \times 10^9 \gamma - 6.67 \times 10^{12}\gamma^2$$

Solve Prob. 3.111 using this relation, Eq. (3.2), and Eq. (3.26).

Fig. P3.113

3.113   The solid circular drill rod AB is made of a steel that is assumed to be elastoplastic with $\tau_Y = 160$ MPa and $G = 77$ GPa. Knowing that a torque $T = 5$ kN · m is applied to the rod and then removed, determine the maximum residual shearing stress in the rod.

3.114   The solid circular shaft AB is made of a steel that is assumed to be elastoplastic with $G = 11.2 \times 10^6$ psi and $\tau_Y = 21$ ksi. The torque $\mathbf{T}$ is increased until the radius of the elastic core is 0.25 in. Determine the maximum residual shearing stress in the shaft after the torque $\mathbf{T}$ has been removed.

Fig. P3.114

3.115 The hollow shaft AB is made of a steel that is assumed to be elastoplastic with $\tau_Y = 145$ MPa and $G = 77$ GPa. The magnitude T of the torque is slowly increased until the plastic zone first reaches the inner surface; the torque is then removed. Determine (a) the maximum residual shearing stress, (b) the permanent angle of twist of the shaft.

Fig. P3.115

3.116 The solid shaft shown is made of a steel that is assumed to be elastoplastic with $\tau_Y = 145$ MPa and $G = 77$ GPa. The torque **T** is increased in magnitude until the shaft has been twisted through 6°, and the torque is then removed. Determine (a) the magnitude and location of the maximum residual shearing stress, (b) the permanent angle of twist of the shaft.

3.117 In Prob. 3.113, determine the permanent angle of twist of the rod.

3.118 In Prob. 3.114, determine the permanent angle of twist of the shaft.

3.119 A torque **T** applied to a solid rod made of an elastoplastic material is increased until the rod becomes fully plastic, and then the torque is removed. (a) Show that the distribution of residual stresses is as represented in the figure. (b) Determine the magnitude of the torque due to the stresses acting on the portion of the rod located within a circle of radius c_0.

Fig. P3.116

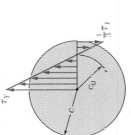

Fig. P3.119

3.120 After the solid shaft of Prob. 3.116 has been loaded and unloaded as described in that problem, a torque $\mathbf{T}_1$ of sense opposite to the original torque **T** is applied to the shaft. Assuming no change in the value of ϕ_Y, determine the angle of twist ϕ_1 for which yield is initiated in this second loading and compare it with the angle ϕ_Y for which the shaft started to yield in the original loading.

3.121 After the solid shaft of Prob. 3.114 has been loaded and unloaded as described in that problem, a torque $\mathbf{T}_1$ of sense opposite to the original torque **T** is applied to the shaft. Assuming no change in the value of τ_Y, determine the magnitude T_1 of the torque $\mathbf{T}_1$ required to initiate yield in this second loading and compare it with the magnitude T_Y of the torque **T** that caused the shaft to yield in the original loading.

Fig. 3.47

Fig. 3.45

(a)

(b)

Fig. 3.46

*3.12. TORSION OF NONCIRCULAR MEMBERS

The formulas obtained in Secs. 3.3 and 3.4 for the distributions of strain and stress under a torsional loading apply only to members with a circular cross section. Indeed, their derivation was based on the assumption that the cross section of the member remained plane and undistorted, and we saw in Sec. 3.3 that the validity of this assumption depends upon the *axisymmetry* of the member, i.e., upon the fact that its appearance remains the same when it is viewed from a fixed position and rotated about its axis through an arbitrary angle.

A square bar, on the other hand, retains the same appearance only when it is rotated through 90° or 180°. Following a line of reasoning similar to that used in Sec. 3.3, one could show that the diagonals of the square cross section of the bar and the lines joining the midpoints of the sides of that section remain straight (Fig. 3.45). However, because of the lack of axisymmetry of the bar, any other line drawn in its cross section will deform when the bar is twisted, and the cross section itself will be warped out of its original plane.

It follows that Eqs. (3.4) and (3.6), which define respectively the distributions of strain and stress in an elastic circular shaft, cannot be used for noncircular members. For example, it would be wrong to assume that the shearing stress in the cross section of a square bar varies linearly with the distance from the axis of the bar and is therefore largest at the corners of the cross section. As you will see presently, the shearing stress is actually zero at these points.

Consider a small cubic element located at a corner of the cross section of a square bar in torsion and select coordinate axes parallel to the edges of the element (Fig. 3.46a). Since the face of the element perpendicular to the y axis is part of the free surface of the bar, all stresses on this face must be zero. Referring to Fig. 3.46b, we write

$$\tau_{yx} = 0 \qquad \tau_{yz} = 0 \qquad (3.40)$$

For the same reason all stresses on the face of the element perpendicular to the z axis must be zero, and we write

$$\tau_{zx} = 0 \qquad \tau_{zy} = 0 \qquad (3.41)$$

It follows from the first of Eqs. (3.4) and the first of Eqs. (3.41) that

$$\tau_{xy} = 0 \qquad \tau_{xz} = 0 \qquad (3.42)$$

Thus, both components of the shearing stress on the face of the element perpendicular to the axis of the bar are zero. We conclude that there is no shearing stress at the corners of the cross section of the bar.

By twisting a rubber model of a square bar, one easily verifies that no deformations—and, thus, no stresses—occur along the edges of the bar, while the largest deformations—and, thus, the largest stresses—occur along the center line of each of the faces of the bar (Fig. 3.47).

The determination of the stresses in noncircular members subjected to a torsional loading is beyond the scope of this text. However, results obtained from the mathematical theory of elasticity for straight bars with a *uniform rectangular cross section* will be indicated here for convenience.†

†See S. P. Timoshenko and J. N. Goodier, *Theory of Elasticity*, 3d ed., McGraw-Hill, New York, 1969, sec. 109.

Denoting by L the length of the bar, by a and b, respectively, the wider and narrower side of its cross section, and by T the magnitude of the torques applied to the bar (Fig. 3.48), we find that the maximum shearing stress

Fig. 3.48

occurs along the center line of the *wider* face of the bar and is equal to

$$\tau_{max} = \frac{T}{c_1 ab^2} \qquad (3.43)$$

The angle of twist, on the other hand, may be expressed as

$$\phi = \frac{TL}{c_2 ab^3 G} \qquad (3.44)$$

The coefficients c_1 and c_2 depend only upon the ratio a/b and are given in Table 3.1 for a number of values of that ratio. Note that Eqs. (3.43) and (3.44) are valid only within the elastic range.

TABLE 3.1. Coefficients for Rectangular Bars in Torsion

a/b	c_1	c_2
1.0	0.208	0.1406
1.2	0.219	0.1661
1.5	0.231	0.1958
2.0	0.246	0.229
2.5	0.258	0.249
3.0	0.267	0.263
4.0	0.282	0.281
5.0	0.291	0.291
10.0	0.312	0.312
∞	0.333	0.333

We note from Table 3.1 that for $a/b \geq 5$, the coefficients c_1 and c_2 are equal. It may be shown that for such values of a/b, we have

$$c_1 = c_2 = \tfrac{1}{3}(1 - 0.630b/a) \qquad \textbf{\textit{(for a/b} $\geq$ \textbf{\textit{5 only)}}} \qquad (3.45)$$

The distribution of shearing stresses in a noncircular member may be visualized more easily by using the *membrane analogy*. A homogeneous elastic membrane attached to a fixed frame and subjected to a uniform pressure on one of its sides happens to constitute an *analog* of the bar in torsion, i.e., the determination of the deformation of the mem-

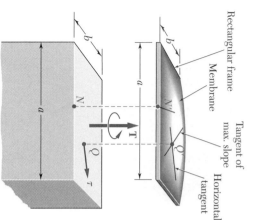

Rectangular frame

Membrane

Tangent of
max. slope

Horizontal
tangent

Fig. 3.49

brane depends upon the solution of the same partial differential equation as the determination of the shearing stresses in the bar.† More specifically, if Q is a point of the cross section of the bar and Q' the corresponding point of the membrane (Fig. 3.49), the shearing stress τ at Q will have the same direction as the horizontal tangent to the membrane at Q', and its magnitude will be proportional to the maximum slope of the membrane at Q'.‡ Furthermore, the applied torque will be proportional to the volume between the membrane and the plane of the fixed frame. In the case of the membrane of Fig. 3.49, which is attached to a rectangular frame, the steepest slope occurs at the midpoint N' of the larger side of the frame. Thus, we verify that the maximum shearing stress in a bar of rectangular cross section will occur at the midpoint N of the larger side of that section.

The membrane analogy may be used just as effectively to visualize the shearing stresses in any straight bar of uniform, noncircular cross section. In particular, let us consider several thin-walled members with the cross sections shown in Fig. 3.50, which are subjected to the same torque. Using the membrane analogy to help us visualize the shearing stresses, we note that, since the same torque is applied to each member, the same volume will be located under each membrane, and the maximum slope will be about the same in each case. Thus, for a thin-walled member of uniform thickness and arbitrary shape, the maximum shearing stress is the same as for a rectangular bar with a very large value of a/b and may be determined from Eq. (3.43) with $c_1 = 0.333.\S$

Fig. 3.50

† See ibid. sec. 107.
‡ This is the slope measured in a direction perpendicular to the horizontal tangent at Q'.
§ It could also be shown that the angle of twist may be determined from Eq. (3.44) with $c_2 = 0.333$.

*3.13. THIN-WALLED HOLLOW SHAFTS

In the preceding section we saw that the determination of stresses in noncircular members generally requires the use of advanced mathematical methods. In the case of thin-walled hollow noncircular shafts, however, a good approximation of the distribution of stresses in the shaft can be obtained by a simple computation.

Consider a hollow cylindrical member of *noncircular* section subjected to a torsional loading (Fig. 3.51).† While the thickness t of the wall may vary within a transverse section, it will be assumed that it remains small compared to the other dimensions of the member. We now detach from the member the colored portion of wall AB bounded by two transverse planes at a distance Δx from each other, and by two longitudinal planes perpendicular to the wall. Since the portion AB is in equilibrium, the sum of the forces exerted on it in the longitudinal x direction must be zero (Fig. 3.52). But the only forces involved are the shearing forces $\mathbf{F}_A$ and $\mathbf{F}_B$ exerted on the ends of portion AB. We have therefore

Fig. 3.51

$$\Sigma F_x = 0: \qquad F_A - F_B = 0 \qquad (3.46)$$

We now express F_A as the product of the longitudinal shearing stress τ_A on the small face at A and of the area $t_A \, \Delta x$ of that face:

$$F_A = \tau_A(t_A \, \Delta x)$$

Fig. 3.52

We note that, while the shearing stress is independent of the x coordinate of the point considered, it may vary across the wall; thus, τ_A represents the average value of the stress computed across the wall. Expressing F_B in a similar way and substituting for F_A and F_B into (3.46), we write

$$\tau_A(t_A \, \Delta x) - \tau_B(t_B \, \Delta x) = 0$$

or

$$\tau_A t_A = \tau_B t_B \qquad (3.47)$$

Since A and B were chosen arbitrarily, Eq. (3.47) expresses that the product τt of the longitudinal shearing stress τ and of the wall thickness t is constant throughout the member. Denoting this product by q, we have

$$q = \tau t = \text{constant} \qquad (3.48)$$

Fig. 3.53

We now detach a small element from the wall portion AB (Fig. 3.53). Since the upper and lower faces of this element are part of the free surface of the hollow member, the stresses on these faces are equal to zero. Recalling relations (1.21) and (1.22) of Sec. 1.12, it follows that the stress components indicated on the other faces by dashed arrows are also zero, while those represented by solid arrows are equal. Thus the shearing stress at any point of a transverse section of the hollow member is parallel to the wall surface (Fig. 3.54) and its average value computed across the wall satisfies Eq. (3.48).

Fig. 3.54

---

†The wall of the member must enclose a single cavity and must not be slit open. In other words, the member should be topologically equivalent to a hollow circular shaft.

At this point we can note an analogy between the distribution of the shearing stresses τ in the transverse section of a thin-walled hollow shaft and the distribution of the velocities v in water flowing through a closed channel of unit depth and variable width. While the velocity v of the water varies from point to point on account of the variation in the width t of the channel, the rate of flow, $q = vt$, remains constant throughout the channel, just as τt in Eq. (3.48). Because of this analogy, the product $q = \tau t$ is referred to as the *shear flow* in the wall of the hollow shaft.

We will now derive a relation between the torque T applied to a hollow member and the shear flow q in its wall. We consider a small element of the wall section, of length ds (Fig. 3.55). The area of the element is $dA = t\, ds$, and the magnitude of the shearing force $d\mathbf{F}$ exerted on the element is

$$dF = \tau\, dA = \tau(t\, ds) = (\tau t)\, ds = q\, ds \qquad (3.49)$$

Fig. 3.55

The moment dM_O of this force about an arbitrary point O within the cavity of the member may be obtained by multiplying dF by the perpendicular distance p from O to the line of action of $d\mathbf{F}$. We have

$$dM_O = p\, dF = p(q\, ds) = q(p\, ds) \qquad (3.50)$$

But the product $p\, ds$ is equal to twice the area $d\mathcal{Q}$ of the colored triangle in Fig. 3.56. We thus have

$$dM_O = q(2d\mathcal{Q}) \qquad (3.51)$$

Fig. 3.56

Since the integral around the wall section of the left-hand member of Eq. (3.51) represents the sum of the moments of all the elementary shearing forces exerted on the wall section, and since this sum is equal to the torque T applied to the hollow member, we have

$$T = \oint dM_O = \oint q(2d\mathcal{Q})$$

The shear flow q being a constant, we write

$$T = 2q\mathcal{Q} \qquad (3.52)$$

where $\mathcal{Q}$ is the area bounded by the center line of the wall cross section (Fig. 3.57).

Fig. 3.57

The shearing stress τ at any given point of the wall may be expressed in terms of the torque T if we substitute for q from (3.48) into (3.52) and solve for τ the equation obtained. We have

$$\tau = \frac{T}{2t\mathcal{Q}} \qquad (3.53)$$

where t is the wall thickness at the point considered and $\mathcal{Q}$ the area bounded by the center line. We recall that τ represents the average value of the shearing stress across the wall. However, for elastic deformations the distribution of stresses across the wall may be assumed uniform, and Eq. (3.53) will yield the actual value of the shearing stress at a given point of the wall.

The angle of twist of a thin-walled hollow shaft may be obtained by using the method of energy (Chap. 11). Assuming an elastic deformation, it may be shown† that the angle of twist of a thin-walled shaft of length L and modulus of rigidity G is

$$\phi = \frac{TL}{4\mathcal{C}^2 G} \oint \frac{ds}{t} \qquad (3.54)$$

where the integral is computed along the center line of the wall section.

EXAMPLE 3.10

Structural aluminum tubing of 2.5×4-in. rectangular cross section was fabricated by extrusion. Determine the shearing stress in each of the four walls of a portion of such tubing when it is subjected to a torque of 24 kip · in., assuming (a) a uniform 0.160-in. wall thickness (Fig. 3.58a), (b) that, as a result of defective fabrication, walls AB and AC are 0.120 in. thick, and walls BD and CD are 0.200-in. thick (Fig. 3.58b).

Fig. 3.58

(a) Tubing of Uniform Wall Thickness. The area $\mathcal{C}$ bounded by the center line (Fig. 3.59) is

$$\mathcal{C} = (3.84 \text{ in.})(2.34 \text{ in.}) = 8.986 \text{ in}^2$$

Since the thickness of each of the four walls is $t = 0.160$ in., we find from Eq. (3.53) that the shearing stress in each wall is

$$\tau = \frac{T}{2t\mathcal{C}} = \frac{24 \text{ kip} \cdot \text{in.}}{2(0.160 \text{ in.})(8.986 \text{ in}^2)} = 8.35 \text{ ksi}$$

Fig. 3.59

(b) Tubing with Variable Wall Thickness. Observing that the area $\mathcal{C}$ bounded by the center line is the same as in part a, and substituting successively $t = 0.120$ in. and $t = 0.200$ in. into Eq. (3.53), we have

$$\tau_{AB} = \tau_{AC} = \frac{24 \text{ kip} \cdot \text{in.}}{2(0.120 \text{ in.})(8.986 \text{ in}^2)} = 11.13 \text{ ksi}$$

and

$$\tau_{BD} = \tau_{CD} = \frac{24 \text{ kip} \cdot \text{in.}}{2(0.200 \text{ in.})(8.986 \text{ in}^2)} = 6.68 \text{ ksi}$$

We note that the stress in a given wall depends only upon its thickness.

†See Prob. 11.83.

SAMPLE PROBLEM 3.9

Using $\tau_{all} = 40$ MPa, determine the largest torque that may be applied to each of the brass bars and to the brass tube shown. Note that the two solid bars have the same cross-sectional area, and that the square bar and square tube have the same outside dimensions.

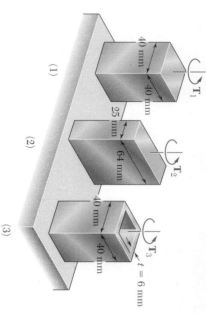

(1)

(2)

(3)

SOLUTION

1. Bar with Square Cross Section. For a solid bar of rectangular cross section the maximum shearing stress is given by Eq. (3.43)

$$\tau_{max} = \frac{T}{c_1 ab^2}$$

where the coefficient c_1 is obtained from Table 3.1 in Sec. 3.12. We have

$$a = b = 0.040 \text{ m} \qquad \frac{a}{b} = 1.00 \qquad c_1 = 0.208$$

For $\tau_{max} = \tau_{all} = 40$ MPa, we have

$$\tau_{max} = \frac{T_1}{c_1 ab^2} \qquad 40 \text{ MPa} = \frac{T_1}{0.208(0.040 \text{ m})^3} \qquad T_1 = 532 \text{ N} \cdot \text{m} \blacktriangle$$

2. Bar with Rectangular Cross Section. We now have

$$a = 0.064 \text{ m} \qquad b = 0.025 \text{ m} \qquad \frac{a}{b} = 2.56$$

Interpolating in Table 3.1: $c_1 = 0.259$

$$\tau_{max} = \frac{T_2}{c_1 ab^2} \qquad 40 \text{ MPa} = \frac{T_2}{0.259(0.064 \text{ m})(0.025 \text{ m})^2} \qquad T_2 = 414 \text{ N} \cdot \text{m} \blacktriangle$$

3. Square Tube. For a tube of thickness t, the shearing stress is given by Eq. (3.53)

$$\tau = \frac{T}{2t\mathcal{Q}}$$

where $\mathcal{Q}$ is the area bounded by the center line of the cross section. We have

$$\mathcal{Q} = (0.034 \text{ m})(0.034 \text{ m}) = 1.156 \times 10^{-3} \text{ m}^2$$

We substitute $\tau = \tau_{all} = 40$ MPa and $t = 0.006$ m and solve for the allowable torque:

$$\tau = \frac{T}{2t\mathcal{Q}}$$

$$40 \text{ MPa} = \frac{T_3}{2(0.006 \text{ m})(1.156 \times 10^{-3} \text{ m}^2)} \qquad T_3 = 555 \text{ N} \cdot \text{m} \blacktriangle$$

PROBLEMS

3.122 Knowing that the magnitude of the torque **T** is 1800 lb · in., determine for each of the aluminum bars shown the maximum shearing stress and the angle of twist at end B. Use $G = 3.9 \times 10^6$ psi.

3.123 Using $\tau_{all} = 10$ ksi, determine for each of the aluminum bars shown the largest torque **T** that may be applied and the corresponding angle of twist. Use $G = 3.9 \times 10^6$ psi.

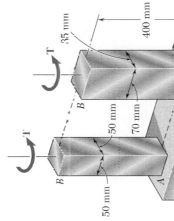

Fig. P3.122 and P3.123

3.124 Knowing that $T = 800$ N · m, determine for each of the cold-rolled yellow brass bars shown the maximum shearing stress and the angle of twist at end B. Use $G = 39$ GPa.

3.125 Using $\tau_{all} = 50$ MPa, determine for each of the cold-rolled yellow brass bars shown the largest torque **T** that may be applied and the corresponding angle of twist. Use $G = 39$ GPa.

Fig. P3.124 and P3.125

3.126 A torque of magnitude $T = 2$ kip · in. is applied to each of the steel bars shown. Knowing that $\tau_{all} = 6$ ksi, determine the required dimension b for each bar.

Fig. P3.126 and P3.127

3.127 A torque of magnitude $T = 300$ N · m is applied to each of the aluminum bars shown. Knowing that $\tau_{all} = 60$ MPa, determine the required dimension b for each bar.

3.128 The torque **T** causes a rotation of 2° at end B of the stainless steel bar shown. Knowing that G = 77 GPa, determine the maximum shearing stress in the bar.

Fig. P3.128

3.129 Two shafts are made of the same material. The cross section of shaft A is a square of side b and that of shaft B is a circle of diameter b. Knowing that the shafts are subjected to the same torque, determine the ratio of the maximum shearing stresses occurring in the shafts.

Fig. P3.129

3.130 Determine the largest allowable square cross section of a steel shaft of length 4 m if the maximum shearing stress is not to exceed 100 MPa when the shaft is twisted through one complete revolution. Use G = 77 GPa.

3.131 Shafts A and B are made of the same material and have the same cross-sectional area, but A has a circular cross section and B has a square cross section. Determine the ratio of the maximum torques T_A and T_B that may be safely applied to A and B, respectively.

3.132 Shafts A and B are made of the same material and have the same length and cross-sectional area, but A has a circular cross section and B has a square cross section. Determine the ratio of the maximum values of the angles ϕ_A and ϕ_B through which shafts A and B, respectively, may be twisted.

Fig. P3.131 and P3.132

3.133 A 1.25-m-long steel angle has an L127 × 76 × 6.4 cross section. From Appendix C we find that the thickness of the section is 6.4 mm and that its area is 1252 mm². Knowing that τ_{all} = 60 MPa and G = 77 GPa, and ignoring the effect of stress concentrations, determine (a) the largest torque **T** that may be applied, (b) the corresponding angle of twist.

Fig. P3.133

3.134 A 3000 lb · in. torque is applied to a 6-ft-long steel angle with an L4 × 4 × $\frac{3}{8}$ cross section. From Appendix C we find that the thickness of the section is $\frac{3}{8}$ in. and that its area is 2.86 in.² Knowing that G = 11.2 × 10⁶ psi, determine (a) the maximum shearing stress along line a-a, (b) the angle of twist.

Fig. P3.134

3.135 An 8-ft-long steel member with a W8 × 31 cross section is subjected to a 5-kip · in torque. From Appendix C we find that the thickness of the section is $\frac{3}{8}$ in. and that its area is 2.86 in². Knowing that $G = 11.2 \times 10^6$ psi, determine (*a*) the maximum shearing stress along line *a-a*, (*b*) the maximum shearing stress along line *b-b*, (*c*) the angle of twist. (*Hint:* Consider the web and flanges separately and obtain a relation between the torques exerted on the web and a flange, respectively, by expressing that the resulting angles of twist are equal.)

W8 × 31

Fig. P3.135

3.136 A 3-m-long steel member has a W250 × 58 cross section. Knowing that $G = 77$ GPa and that the allowable shearing stress is 35 MPa, determine (*a*) the largest torque **T** that may be applied, (*b*) the corresponding angle of twist. Refer to Appendix C for the dimensions of the cross section and neglect the effect of stress concentrations. (See hint of Prob. 3.135.)

W250 × 58

Fig. P3.136

Fig. P3.137

3.137 A torque $T = 750$ N · m is applied to a hollow shaft shown that has a uniform 6-mm wall thickness. Neglecting the effect of stress concentrations, determine the shearing stress at points *a* and *b*.

3.138 A torque $T = 750$ N · m is applied to the hollow shaft shown that has a uniform 8-mm wall thickness. Neglecting the effect of stress concentrations, determine the shearing stress at points *a* and *b*.

Fig. P3.138

3.139 and 3.140 A 50 kip · in. torque is applied to a hollow shaft having the cross section shown. Neglecting the effect of stress concentration, determine the shearing stress at points a and b.

Fig. P3.139

Fig. P3.140

3.141 and 3.142 A hollow member having the cross section shown is formed from sheet metal of 2-mm thickness. Knowing that the shearing stress must not exceed 3 MPa, determine the largest torque that may be applied to the member.

Fig. P3.141

Fig. P3.142

3.143 and 3.144 A hollow member having the cross section shown is to be formed from sheet metal of 0.06 in. thickness. Knowing that a 1250 lb · in. torque will be applied to the member, determine the smallest dimension d that may be used if the shearing stress is not to exceed 750 psi.

Fig. P3.143

Fig. P3.144

3.145 A hollow cylindrical shaft was designed to have a uniform wall thickness of 0.1 in. Defective fabrication, however, resulted in the shaft having the cross section shown. Knowing that a 15 kip · in. torque is applied to the shaft, determine the shearing stress at points a and b.

0.08 in.

a

2.2 in.

b

0.12 in.

2.4 in.

Fig. P3.145

3.146 A cooling tube having the cross section shown is formed from a sheet of stainless steel of 3-mm thickness. The radii $c_1 = 150$ mm and $c_2 = 100$ mm are measured to the center line of the sheet metal. Knowing that a torque of magnitude $T = 3$ kN · m is applied to the tube, determine (a) the maximum shearing stress in the tube, (b) the magnitude of the torque carried by the outer circular shell. Neglect the dimension of the small opening where the outer and inner shells are connected.

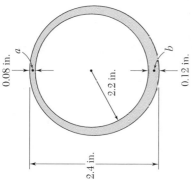

c_1

O

c_2

Fig. P3.146 and P3.147

3.147 A cooling tube having the cross section shown is formed from a sheet of stainless steel of thickness t. The radii c_1 and c_2 are measured to the center line of the sheet metal. Knowing that a torque T is applied to the tube, determine in terms of T, c_1, c_2, and t the maximum shearing stress in the tube.

3.148 Equal torques are applied to thin-walled tubes of the same length L, same thickness t, and same radius c. One of the tubes has been slit lengthwise as shown. Determine (a) the ratio τ_b/τ_a of the maximum shearing stresses in the tubes, (b) the ratio ϕ_b/ϕ_a of the angles of twist of the shafts.

(a)

(b)

Fig. P3.148

3.149 A hollow cylindrical shaft of length L, mean radius c_m, and uniform thickness t is subjected to torques of magnitude T. Consider, on the one hand, the values of the average shearing stress τ_{ave} and the angle of twist ϕ obtained from the elastic torsion formulas developed in Secs. 3.4 and 3.5 and, on the other hand, the corresponding values obtained from the formulas developed in Sec. 3.13 for thin-walled hollow shafts. (a) Show that the relative error introduced by using the thin-wall-shaft formulas rather than the elastic torsion formulas is the same for τ_{ave} and ϕ and that the relative error is positive and proportional to the square of the ratio t/c_m. (b) Compare the percent error corresponding to values of the ratio t/c_m of 0.1, 0.2, and 0.4.

c_m

t

L

T

T'

Fig. P3.149

REVIEW AND SUMMARY
FOR CHAPTER 3

This chapter was devoted to the analysis and design of *shafts* subjected to twisting couples, or *torques*. Except for the last two sections of the chapter our discussion was limited to *circular shafts*.

In a preliminary discussion [Sec. 3.2], it was pointed out that the distribution of stresses in the cross section of a circular shaft is *statically indeterminate*. The determination of these stresses, therefore, requires a prior analysis of the *deformations* occurring in the shaft [Sec. 3.3]. Having demonstrated that in a circular shaft subjected to torsion, *every cross section remains plane and undistorted*, we derived the following expression for the *shearing strain* in a small element with sides parallel and perpendicular to the axis of the shaft and at a distance ρ from that axis:

$$\gamma = \frac{\rho\phi}{L} \qquad (3.2)$$

where φ is the angle of twist for a length L of the shaft (Fig. 3.14). Equation (3.2) shows that the *shearing strain in a circular shaft varies linearly with the distance from the axis of the shaft.* It follows that the strain is maximum at the surface of the shaft, where ρ is equal to the radius c of the shaft. We wrote

$$\gamma_{max} = \frac{c\phi}{L} \qquad \gamma = \frac{\rho}{c}\gamma_{max} \qquad (3.3, 4)$$

Considering *shearing stresses* in a circular shaft within the elastic range [Sec. 3.4] and recalling Hooke's law for shearing stress and strain, τ = Gγ, we derived the relation

$$\tau = \frac{\rho}{c}\tau_{max} \qquad (3.6)$$

which shows that within the elastic range, the *shearing stress τ in a circular shaft also varies linearly with the distance from the axis of the shaft.* Equating the sum of the moments of the elementary forces exerted on any section of the shaft to the magnitude T of the torque applied to the shaft, we derived the *elastic torsion formulas*

$$\tau_{max} = \frac{Tc}{J} \qquad \tau = \frac{T\rho}{J} \qquad (3.9, 10)$$

where c is the radius of the cross section and J its centroidal polar moment of inertia. We noted that $J = \frac{1}{2}\pi c^4$ for a solid shaft and $J = \frac{1}{2}\pi(c_2^4 - c_1^4)$ for a hollow shaft of inner radius c_1 and outer radius c_2.

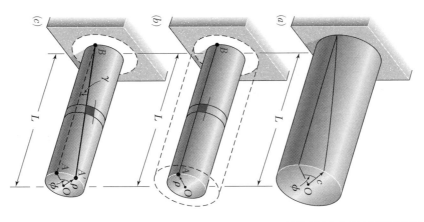

Fig. 3.14

Deformations in circular shafts

Shearing stresses in elastic range

Fig. 3.20

$$\tau_{max} = \frac{Tc}{J} \qquad \sigma_{45°} = \pm \frac{Tc}{J}$$

Angle of twist

Fig. 3.22

(b)

Fig. 3.26b

Statically indeterminate shafts

We noted that while the element a in Fig. 3.20 is in pure shear, the element c in the same figure is subjected to normal stresses of the same magnitude, Tc/J, two of the normal stresses being tensile and two compressive. This explains why in a torsion test ductile materials, which generally fail in shear, will break along a plane perpendicular to the axis of the specimen, while brittle materials, which are weaker in tension than in shear, will break along surfaces forming a 45° angle with that axis.

In Sec. 3.5, we found that within the elastic range, the angle of twist ϕ of a circular shaft is proportional to the torque T applied to it (Fig. 3.22). Expressing ϕ in *radians*, we wrote

$$\phi = \frac{TL}{JG} \qquad (3.16)$$

where L = length of shaft
  J = polar moment of inertia of cross section
  G = modulus of rigidity of material

If the shaft is subjected to torques at locations other than its ends or consists of several parts of various cross sections and possibly of different materials, the angle of twist of the shaft must be expressed as the *algebraic sum* of the angles of twist of its component parts [Sample Prob. 3.3]:

$$\phi = \sum_i \frac{T_i L_i}{J_i G_i} \qquad (3.17)$$

We observed that when both ends of a shaft BE rotate (Fig. 3.26b), the angle of twist of the shaft is equal to the *difference* between the angles of rotation ϕ_B and ϕ_E of its ends. We also noted that when two shafts AD and BE are connected by gears A and B, the torques applied, respectively, by gear A on shaft AD and by gear B on shaft BE are *directly proportional* to the radii r_A and r_B of the two gears—since the forces applied on each other by the gear teeth at C are equal and opposite. On the other hand, the angles ϕ_A and ϕ_B through which the two gears rotate are *inversely proportional* to r_A and r_B—since the arcs CC' and CC'' described by the gear teeth are equal [Example 3.04 and Sample Prob. 3.4].

If the reactions at the supports of a shaft or the internal torques cannot be determined from statics alone, the shaft is said to be *statically indeterminate* [Sec. 3.6]. The equilibrium equations obtained from free-body diagrams must then be complemented by relations involving the deformations of the shaft and obtained from the geometry of the problem [Example 3.05, Sample Prob. 3.5].

Transmission shafts

In Sec. 3.7, we discussed the *design of transmission shafts*. We first observed that the power P transmitted by a shaft is

$$P = 2\pi fT \qquad (3.20)$$

where T is the torque exerted at each end of the shaft and f the *frequency* or speed of rotation of the shaft. The unit of frequency is the revolution per second (s^{-1}) or *hertz* (Hz). If SI units are used, T is expressed in newton-meters $(N \cdot m)$ and P in *watts* (W). If U.S. customary units are used, T is expressed in lb $\cdot$ ft or lb $\cdot$ in., and P in ft $\cdot$ lb/s or in $\cdot$ lb/s; the power may then be converted into *horsepower* (hp) through the use of the relation

$$1\ hp = 550\ ft \cdot lb/s = 6600\ in \cdot lb/s$$

To design a shaft to transmit a given power P at a frequency f, you should first solve Eq. (3.20) for T. Carrying this value and the maximum allowable value of τ for the material used into the elastic formula (3.9), you will obtain the corresponding value of the parameter J/c, from which the required diameter of the shaft may be calculated [Examples 3.06 and 3.07].

Stress concentrations

In Sec. 3.8, we discussed *stress concentrations* in circular shafts. We saw that the stress concentrations resulting from an abrupt change in the diameter of a shaft can be reduced through the use of a *fillet* (Fig. 3.31). The maximum value of the shearing stress at the fillet is

$$\tau_{max} = K\frac{Tc}{J} \qquad (3.25)$$

Fig. 3.31

where the stress Tc/J is computed for the smaller-diameter shaft, and where K is a stress-concentration factor. Values of K were plotted in Fig. 3.32 on p. 167 against the ratio r/d, where r is the radius of the fillet, for various values of D/d.

Plastic deformations

Sections 3.9 through 3.11 were devoted to the discussion of *plastic deformations* and *residual stresses* in circular shafts. We first recalled that even when Hooke's law does not apply, the distribution of *strains* in a circular shaft is always linear [Sec. 3.9]. If the shearing-stress-strain diagram for the material is known, it is then possible to plot the shearing stress τ against the distance ρ from the axis of the shaft for any given value of τ_{max} (Fig. 3.35). Summing the

Fig. 3.35

contributions to the torque of annular elements of radius ρ and thickness $d\rho$, we expressed the torque T as

$$T = \int_0^c \rho\tau(2\pi\rho\,d\rho) = 2\pi\int_0^c \rho^2\tau\,d\rho \qquad (3.26)$$

where τ is the function of ρ plotted in Fig. 3.35.

Modulus of rupture

Fig. 3.36

An important value of the torque is the ultimate torque T_U which causes failure of the shaft. This value can be determined, either experimentally, or by carrying out the computations indicated above with τ_{max} chosen equal to the ultimate shearing stress τ_U of the material. From T_U, and assuming a linear stress distribution (Fig 3.36), we determined the corresponding fictitious stress $R_T = T_U c/J$, known as the *modulus of rupture in torsion* of the given material.

Considering the idealized case of a *solid circular shaft* made of an *elastoplastic material* [Sec. 3.10], we first noted that, as long as τ_{max} does not exceed the yield strength τ_Y of the material, the stress distribution across a section of the shaft is linear (Fig. 3.38a). The torque T_Y corresponding to $\tau_{max} = \tau_Y$ (Fig. 3.38b) is known as the *maximum elastic torque*; for a solid circular shaft of radius c, we have

$$T_Y = \tfrac{1}{2}\pi c^3 \tau_Y \qquad (3.29)$$

As the torque increases, a plastic region develops in the shaft around an elastic core of radius ρ_Y. The torque T corresponding to a given value of ρ_Y was found to be

$$T = \frac{4}{3}T_Y\left(1 - \frac{1}{4}\frac{\rho_Y^3}{c^3}\right) \qquad (3.32)$$

Solid shaft of elastoplastic material

   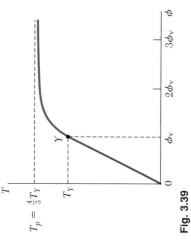

(a) (b) (c) (d)

Fig. 3.38

We noted that as ρ_Y approaches zero, the torque approaches a limiting value T_p, called the *plastic torque* of the shaft considered:

$$T_p = \frac{4}{3}T_Y \qquad (3.33)$$

Plotting the torque T against the angle of twist ϕ of a solid circular shaft (Fig. 3.39), we obtained the segment of straight line $0Y$ defined by Eq. (3.16), followed by a curve approaching the straight line $T = T_p$ and defined by the equation

$$T = \frac{4}{3}T_Y\left(1 - \frac{1}{4}\frac{\phi_Y^3}{\phi^3}\right) \qquad (3.37)$$

Fig. 3.39

Permanent deformation. Residual stresses

Loading a circular shaft beyond the onset of yield and unloading it [Sec. 3.11] results in a *permanent deformation* characterized by the angle of twist $\phi_p = \phi - \phi'$, where ϕ corresponds to the loading phase described in the previous paragraph, and ϕ' to the unloading phase represented by a straight line in Fig. 3.42. There will

Fig. 3.42

also be *residual stresses* in the shaft, which can be determined by adding the maximum stresses reached during the loading phase and the reverse stresses corresponding to the unloading phase [Example 3.09].

The last two sections of the chapter dealt with the torsion of *noncircular members*. We first recalled that the derivation of the formulas for the distribution of strain and stress in circular shafts was based on the fact that due to the axisymmetry of these members, cross sections remain plane and undistorted. Since this property does not hold for noncircular members, none of the formulas derived earlier can be used in their analysis [Sec. 3.12].

It was indicated in Sec. 3.12 that in the case of straight bars with a *uniform rectangular cross section* (Fig. 3.48), the maximum shearing stress occurs along the center line of the *wider* face of the bar. Formulas for the maximum shearing stress and the angle of twist were given without proof. The *membrane analogy* for visualizing the distribution of stresses in a noncircular member was also discussed.

We next analyzed the distribution of stresses in noncircular *thin-walled hollow shafts* [Sec. 3.13]. We saw that the shearing stress is parallel to the wall surface and varies both across the wall and along the wall cross section. Denoting by τ the average value of the shearing stress computed across the wall at a given point of the cross section, and by t the thickness of the wall at that point (Fig. 3.57), we showed that the product $q = \tau t$, called the *shear flow*, is constant along the cross section.

Furthermore, denoting by T the torque applied to the hollow shaft and by $\textcircled{a}$ the area bounded by the center line of the wall cross section, we expressed as follows the average shearing stress τ at any given point of the cross section:

$$\tau = \frac{T}{2t\textcircled{a}} \qquad (3.53)$$

Torsion of noncircular members

Fig. 3.45

Bars of rectangular cross section

Fig. 3.48

Thin-walled hollow shafts

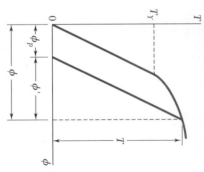

Fig. 3.57

REVIEW PROBLEMS

3.150 For the solid brass shaft shown, determine the maximum shearing stress in (a) portion AB, (b) portion BC.

$T_B = 12.5$ kip · in.

$T_C = 3.5$ kip · in.

$d = 1.75$ in.

$d = 1.25$ in.

C

B

A

30 in.

24 in.

Fig. P3.150 and P3.151

3.151 Knowing that $G = 5.6 \times 10^6$ psi for the solid brass shaft shown, determine the angle of twist at point C.

3.152 The stepped shaft shown rotates at 900 rpm. Knowing that $\tau_{all} = 42$ MPa, determine the maximum power that can be transmitted if the radius r of the fillet is (a) 12 mm, (b) 20 mm.

80 mm

160 mm

r

Fig. P3.152

3.153 The long, hollow, tapered shaft AB has a uniform thickness t. Denoting by G the modulus of rigidity, show that the angle of twist at end A is

$$\phi_A = \frac{TL}{4\pi Gt} \frac{c_A + c_B}{c_A^2 c_B^2}$$

$\mathbf{T}$

c_A

t

A

L

c_B

B

Fig. P3.153

3.154 Two solid steel shafts, each of 30-mm diameter, are connected by the gears shown. Knowing that $G = 77$ GPa, determine the angle through which end A rotates when a torque of magnitude $T = 200$ N · m is applied at A.

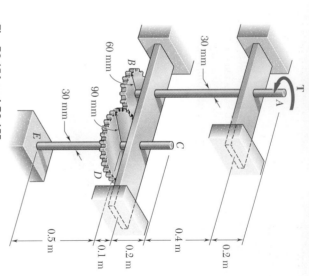

3.155 The angle of rotation of end A of the gear-and-shaft system shown must not exceed 4°. Knowing that the shafts are made of a steel for which $\tau_{all} = 65$ MPa and $G = 77$ GPa, determine the largest torque **T** that can be safely applied at end A.

Fig. P3.154 and P3.155

(3 in. − w)

w

Fig. P3.156

3.156 A sheet-metal strip of width 6 in. and 0.12 in. thickness is to be formed into a tube of rectangular cross section. Knowing that $\tau_{all} = 4$ ksi, determine the largest torque that may be applied to the tube when (a) $w = 1.5$ in., (b) $w = 1.2$ in., (c) $w = 1$ in.

3.157 Two solid brass rods AB and CD are brazed to a brass sleeve EF. Determine the ratio d_2/d_1 for which the same maximum shearing stress occurs in the rods and in the sleeve.

Fig. P3.157

3.158 One of the two hollow steel drive shafts of an ocean liner is 75 m long and has the cross section shown. Knowing that $G = 77$ GPa and that the shaft transmits 44 MW to its propeller when rotating at 144 rpm, determine (a) the maximum shearing stress in the shaft, (b) the angle of twist of the shaft.

320 mm

580 mm

Fig. P3.158

3.159 The shaft *AB* is made of a material that is elastoplastic with $\tau_Y = 12.5$ ksi and $G = 4 \times 10^6$ psi. For the loading shown, determine (*a*) the radius of the elastic core of the shaft, (*b*) the angle of twist at end *B*.

3.160 If the 3 kip · in. torque applied to the shaft of Prob. 3.159 is removed, determine (*a*) the magnitude and location of the maximum residual shearing stress in the shaft, (*b*) the permanent angle of twist of the shaft.

Fig. P3.159

3.161 The composite shaft shown is twisted by applying a torque **T** at end *A*. Knowing that the maximum shearing stress in the steel shell is 150 MPa, determine the corresponding maximum shearing stress in the aluminum core. Use $G = 77$ GPa for steel and $G = 27$ GPa for aluminum.

Fig. P3.161

COMPUTER PROBLEMS

The following problems are designed to be solved with a computer. Write each program so that it can be used with either SI or U.S. customary units.

3.C1 Shaft *AB* consists of *n* homogeneous cylindrical elements, which can be solid or hollow. Its end *A* is fixed, while its end *B* is free, and it is subjected to the loading shown. The length of element *i* is denoted by L_i, its outer diameter by OD_i, its inner diameter by ID_i, its modulus of rigidity by G_i, and the torque applied to its right end by $\mathbf{T}_i$, the magnitude T_i of this torque being assumed to be positive if $\mathbf{T}_i$ is observed as counterclockwise from end *B* and negative otherwise. (Note that $ID_i = 0$ if the element is solid.) (*a*) Write a computer program that can be used to determine the maximum shearing stress in each element, the angle of twist of each element, and the angle of twist of the entire shaft. (*b*) Use this program to solve Probs. 3.9, 3.35, 3.37, 3.150, and 3.151.

Fig. P3.C1

3.C2 The assembly shown consists of n cylindrical shafts, which can be solid or hollow, connected by gears and supported by brackets (not shown). End A_1 of the first shaft is free and is subjected to a torque $\mathbf{T}_0$, while end B_n of the last shaft is fixed. The length of shaft A_iB_i is denoted by L_i, its outer diameter by OD_i, its inner diameter by ID_i, and its modulus of rigidity by G_i. (Note that $ID_i = 0$ if the element is solid.) The radius of gear B_i is denoted by a_i, and the radius of gear B_i by b_i. (a) Write a computer program that can be used to determine the maximum shearing stress in each shaft, the angle of twist of each shaft, and the angle through which end A_i rotates. (b) Use this program to solve Probs. 3.21, 3.39, 3.41, 3.42, and 3.154.

Fig. P3.C2

3.C3 Shaft AB consists of n homogeneous cylindrical elements, which can be solid or hollow. Both of its ends are fixed, and it is subjected to the loading shown. The length of element i is denoted by L_i, its outer diameter by OD_i, its inner diameter by ID_i, its modulus of rigidity by G_i, and the torque applied to its right end by $\mathbf{T}_i$, the magnitude T_i of this torque being assumed to be positive if $\mathbf{T}_i$ is observed as counterclockwise from end B and negative otherwise. Note that $ID_i = 0$ if the element is solid and also that $T_1 = 0$. Write a computer program that can be used to determine the reactions at A and B, the maximum shearing stress in each element, and the angle of twist of each element. Use this program (a) to solve Prob. 3.56, (b) to determine the maximum shearing stress in the shaft of Example 3.05.

Fig. P3.C3

3.C4 The homogeneous, solid cylindrical shaft AB has a length L, a diameter d, a modulus of rigidity G, and a yield strength τ_Y. It is subjected to a torque $\mathbf{T}$ that is gradually increased from zero until the angle of twist of the shaft has reached a maximum value ϕ_m and then decreased back to zero. (a) Write a computer program that, for each of 16 values of ϕ_m equally spaced over a range extending from 0 to a value 3 times as large as the angle of twist at the onset of yield, can be used to determine the maximum value T_m of the torque, the radius of the elastic core, the maximum shearing stress, the permanent twist, and the residual shearing stress both at the surface of the shaft and at the interface of the elastic core and the plastic region. (b) Use this program to obtain approximate answers to Probs. 3.95, 3.113, 3.159, and 3.160.

Fig. P3.C4

3.C5 The exact expression is given in Prob. 3.64 for the angle of twist of the solid tapered shaft AB when a torque **T** is applied as shown. Derive an approximate expression for the angle of twist of the tapered shaft by n cylindrical shafts of equal length and of radius $r_i = (n + i - \frac{1}{2})(c/n)$, where $i = 1, 2, \ldots, n$. Using for T, L, G, and c values of your choice, determine the percentage error in the approximate expression when (a) $n = 4$, (b) $n = 8$, (c) $n = 20$, (d) $n = 100$.

Fig. P3.C5

3.C6 A torque **T** is applied as shown to the long, hollow, tapered shaft AB of uniform thickness t. The exact expression for the angle of twist of the shaft can be obtained from the expression given in Prob. 3.153. Derive an approximate expression for the angle of twist by replacing the tapered shaft by n cylindrical rings of equal length and of radius $r_i = (n + i - \frac{1}{2})(c/n)$, where $i = 1, 2, \ldots, n$. Using for T, L, G, c and t values of your choice, determine the percentage error in the approximate expression when (a) $n = 4$, (b) $n = 8$, (c) $n = 20$, (d) $n = 100$.

Fig. P3.C6

4

Pure Bending

The athlete shown holds the barbell with his hands placed at equal distances from the weights. This results in pure bending in the center portion of the bar. The normal stresses and the curvature resulting from pure bending will be determined in this chapter.

4.1. INTRODUCTION

In the preceding chapters you studied how to determine the stresses in prismatic members subjected to axial loads or to twisting couples. In this chapter and in the following two you will analyze the stresses and strains in prismatic members subjected to *bending*. Bending is a major concept used in the design of many machine and structural components, such as beams and girders.

This chapter will be devoted to the analysis of prismatic members subjected to equal and opposite couples **M** and **M′** acting in the same longitudinal plane. Such members are said to be in *pure bending*. In most of the chapter, the members will be assumed to possess a plane of symmetry and the couples **M** and **M′** to be acting in that plane (Fig. 4.1).

Fig. 4.1

An example of pure bending is provided by the bar of a typical barbell as it is held overhead by a weight lifter as shown on the opposite page. The bar carries equal weights at equal distances from the hands of the weight lifter. Because of the symmetry of the free-body diagram of the bar (Fig. 4.2a), the reactions at the hands must be equal and opposite to the weights. Therefore, as far as the middle portion *CD* of the bar is concerned, the weights and the reactions can be replaced by two equal and opposite 960-lb · in. couples (Fig. 4.2b), showing that the middle portion of the bar is in pure bending. A similar analysis of the axle of a small trailer (Fig. 4.3) would show that, between the two points where it is attached to the trailer, the axle is in pure bending.

As interesting as the direct applications of pure bending may be, devoting an entire chapter to its study would not be justified if it were not for the fact that the results obtained will be used in the analysis of other types of loadings as well, such as *eccentric axial loadings* and *transverse loadings*.

Fig. 4.2

Fig. 4.3 In the small trailer shown the center portion of the axle is in pure bending.

Figure 4.4 shows a 12-in. steel bar clamp used to exert 150-lb forces on two pieces of lumber as they are being glued together. Figure 4.5a shows the equal and opposite forces exerted by the lumber on the clamp. These forces result in an *eccentric loading* of the straight portion of the clamp. In Fig. 4.5b a section CC′ has been passed through the clamp and a free-

Fig. 4.4

(a)

(b)

Fig. 4.5

body diagram has been drawn of the upper half of the clamp, from which we conclude that the internal forces in the section are equivalent to a 150-lb axial tensile force **P** and a 750 lb · in. couple **M**. We can thus combine our knowledge of the stresses under a *centric* load and the results of our forthcoming analysis of stresses in pure bending to obtain the distribution of stresses under an *eccentric* load. This will be further discussed on Sec. 4.12.

The study of pure bending will also play an essential role in the study of beams, i.e., the study of prismatic members subjected to various types of *transverse loads*. Consider, for instance, a cantilever beam *AB* supporting a concentrated load **P** at its free end (Fig. 4.6a). If we pass a section through *C* at a distance *x* from *A*, we observe from the free-body diagram of *AC* (Fig. 4.6b) that the internal forces in the section consist of a force **P′** equal and opposite to **P** and a couple **M** of magnitude *M* = *Px*. The distribution of normal stresses in the section can be obtained from the couple **M** as if the beam were in pure bending. On the other hand, the shearing stresses in the section depend on the force **P′**, and you will learn in Chap. 6 how to determine their distribution over a given section.

The first part of the chapter is devoted to the analysis of the stresses and deformations caused by pure bending in a homogeneous member possessing a plane of symmetry and made of a material following Hooke's law. In a preliminary discussion of the stresses due to bending (Sec. 4.2), the methods of statics will be used to derive three fundamental equations which must be satisfied by the normal stresses in any given cross section of the member. In Sec. 4.3, it will be proved that *transverse sections re-*

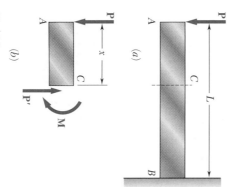

Fig. 4.6

A   **P**

L

C

B

(a)

A   **P**

x

C   **P′**   **M**

(b)

main plane in a member subjected to pure bending, while in Sec. 4.4 formulas will be developed that can be used to determine the *normal stresses*, as well as the *radius of curvature* for that member within the elastic range.

In Sec. 4.6, you will study the stresses and deformations in *composite members* made of more than one material, such as reinforced-concrete beams, which utilize the best features of steel and concrete and are extensively used in the construction of buildings and bridges. You will learn to draw a *transformed section* representing the section of a member made of a homogeneous material that undergoes the same deformations as the composite member under the same loading. The transformed section will be used to find the stresses and deformations in the original composite member. Section 4.7 is devoted to the determination of the *stress concentrations* occurring at locations where the cross section of a member undergoes a sudden change.

In the next part of the chapter you will study *plastic deformations* in bending, i.e., the deformations of members which are made of a material which does not follow Hooke's law and are subjected to bending. After a general discussion of the deformations of such members (Sec. 4.8), you will investigate the stresses and deformations in members made of an *elasto-plastic material* (Sec. 4.9). Starting with the *maximum elastic moment* M_Y, which corresponds to the onset of yield, you will consider the effects of increasingly larger moments until the *plastic moment* M_p is reached, at which time the member has yielded fully. You will also learn to determine the *permanent deformations* and *residual stresses* that result from such loadings (Sec. 4.11). It should be noted that during the past half-century the elastoplastic property of steel has been widely used to produce designs resulting in both improved safety and economy.

In Sec. 4.12, you will learn to analyze an *eccentric axial loading* in a plane of symmetry, such as the one shown in Fig. 4.4, by superposing the stresses due to pure bending and the stresses due to a centric axial loading.

Your study of the bending of prismatic members will conclude with the analysis of *unsymmetric bending* (Sec. 4.13), and the study of the general case of *eccentric axial loading* (Sec. 4.14). The final section of the chapter will be devoted to the determination of the stresses in *curved members* (Sec. 4.15).

4.2. SYMMETRIC MEMBER IN PURE BENDING

Consider a prismatic member AB possessing a plane of symmetry and subjected to equal and opposite couples $\mathbf{M}$ and $\mathbf{M}'$ acting in that plane (Fig. 4.7a). We observe that if a section is passed through the member AB at some arbitrary point C, the conditions of equilibrium of the portion AC of the member require that the internal forces in the section be equivalent to the couple $\mathbf{M}$ (Fig. 4.7b). Thus, the internal forces in any cross section of a symmetric member in pure bending are equivalent to a couple. The moment M of that couple is referred to as the *bending moment* in the section. Following the usual convention, a positive sign will be assigned to M when the member is bent as shown in Fig. 4.7a, i.e., when the concavity of the beam faces upward, and a negative sign otherwise.

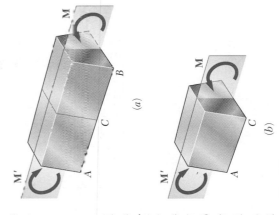

Fig. 4.7

Denoting by σ_x the normal stress at a given point of the cross section and by τ_{xy} and τ_{xz} the components of the shearing stress, we express that the system of the elementary internal forces exerted on the section is equivalent to the couple $\mathbf{M}$ (Fig. 4.8).

Fig. 4.8

We recall from statics that a couple $\mathbf{M}$ actually consists of two equal and opposite forces. The sum of the components of these forces in any direction is therefore equal to zero. Moreover, the moment of the couple is the same about *any* axis perpendicular to its plane, and is zero about any axis contained in that plane. Selecting arbitrarily the z axis as shown in Fig. 4.8, we express the equivalence of the elementary internal forces and of the couple $\mathbf{M}$ by writing that the sums of the components and of the moments of the elementary forces are equal to the corresponding components and moments of the couple $\mathbf{M}$:

$$x \text{ components:} \qquad \int \sigma_x \, dA = 0 \qquad (4.1)$$

$$\text{moments about } y \text{ axis:} \qquad \int z \sigma_x \, dA = 0 \qquad (4.2)$$

$$\text{moments about } z \text{ axis:} \qquad \int (-y \sigma_x \, dA) = M \qquad (4.3)$$

Three additional equations could be obtained by setting equal to zero the sums of the y components, z components, and moments about the x axis, but these equations would involve only the components of the shearing stress and, as you will see in the next section, the components of the shearing stress are both equal to zero.

Two remarks should be made at this point: (1) The minus sign in Eq. (4.3) is due to the fact that a tensile stress ($\sigma_x > 0$) leads to a negative moment (clockwise) of the normal force $\sigma_x \, dA$ about the z axis. (2) Equation (4.2) could have been anticipated, since the application of couples in the plane of symmetry of member AB will result in a distribution of normal stresses that is symmetric about the y axis.

Once more, we note that the actual distribution of stresses in a given cross section cannot be determined from statics alone. It is *statically indeterminate* and may be obtained only by analyzing the *deformations* produced in the member.

4.3. DEFORMATIONS IN A SYMMETRIC MEMBER IN PURE BENDING

Let us now analyze the deformations of a prismatic member possessing a plane of symmetry and subjected at its ends to equal and opposite couples **M** and **M'** acting in the plane of symmetry. The member will bend under the action of the couples, but will remain symmetric with respect to that plane (Fig. 4.9). Moreover, since the bending mo-

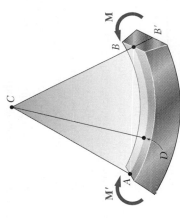

Fig. 4.9

ment M is the same in any cross section, the member will bend uniformly. Thus, the line AB along which the upper face of the member intersects the plane of the couples will have a constant curvature. In other words, the line AB, which was originally a straight line, will be transformed into a circle of center C, and so will the line $A'B'$ (not shown in the figure) along which the lower face of the member intersects the plane of symmetry. We also note that the line AB will decrease in length when the member is bent as shown in the figure, i.e., when $M > 0$, while $A'B'$ will become longer.

Next we will prove that any cross section perpendicular to the axis of the member remains plane, and that the plane of the section passes through C. If this were not the case, we could find a point E of the original section through D (Fig. 4.10a) which, after the member has been bent, would *not* lie in the plane perpendicular to the plane of symmetry that contains line CD (Fig. 4.10b). But, because of the symmetry of the member, there would be another point F that would be transformed exactly in the same way. Let us assume that, after the beam has been bent, both points would be located to the left of the plane defined by CD, as shown in Fig. 4.10b. Since the bending moment M is the same throughout the member, a similar situation would prevail in any other cross section, and the points corresponding to E and E' would also move to the left. Thus, an observer at A would conclude that the loading causes the points E and E' in the various cross sections to move forward (toward the observer). But an observer at B, to whom the loading looks the same, and who observes the points E and E' in the same positions (except that they are now inverted) would reach the opposite conclusion. This inconsistency leads us to conclude that E and E' will lie in the plane defined by CD and, therefore, that the section remains plane and passes through C. We should note, however, that this discussion does not rule out the possibility of deformations *within* the plane of the section (see Sec. 4.5).

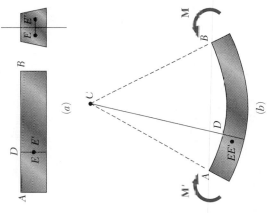

Fig. 4.10

Suppose that the member is divided into a large number of small cubic elements with faces respectively parallel to the three coordinate planes. The property we have established requires that these elements be transformed as shown in Fig. 4.11 when the member is subjected to the couples **M** and **M**′. Since all the faces represented in the two projections of Fig. 4.11 are at 90° to each other, we conclude that $\gamma_{xy} = \gamma_{zx} = 0$ and, thus, that $\tau_{xy} = \tau_{xz} = 0$. Regarding the three stress components that we have not yet discussed, namely, σ_y, σ_z, and τ_{yz}, we note that they must be zero on the surface of the member. Since, on the other hand, the deformations involved do not require any interaction between the elements of a given transverse cross section, we can assume that these three stress components are equal to zero throughout the member. This assumption is verified, both from experimental evidence and from the theory of elasticity, for slender members undergoing small deformations.† We conclude that the only nonzero stress component exerted on any of the small cubic elements considered here is the normal component σ_x. Thus, at any point of a slender member in pure bending, we have a state of *uniaxial stress*. Recalling that, for $M > 0$, lines AB and $A'B'$ are observed, respectively, to decrease and increase in length, we note that the strain ϵ_x and the stress σ_x are negative in the upper portion of the member (*compression*) and positive in the lower portion (*tension*).

It follows from the above that there must exist a surface parallel to the upper and lower faces of the member, where ϵ_x and σ_x are zero. This surface is called the *neutral surface*. The neutral surface intersects the plane of symmetry along an arc of circle DE (Fig. 4.12a), and it intersects a transverse section along a straight line called the *neutral axis* of the section (Fig. 4.12b). The origin of coordinates will now be se-

lected on the neutral surface, rather than on the lower face of the member as done earlier, so that the distance from any point to the neutral surface will be measured by its coordinate y.

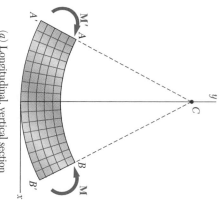

M′ **M**

(a) Longitudinal, vertical section (plane of symmetry)

(b) Longitudinal, horizontal section

Fig. 4.11

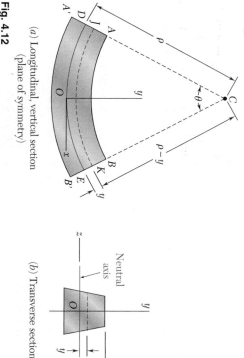

(a) Longitudinal, vertical section (plane of symmetry)

(b) Transverse section

Fig. 4.12

Denoting by ρ the radius of arc DE (Fig. 4.12a), by θ the central angle corresponding to DE, and observing that the length of DE is equal to the length L of the undeformed member, we write

$$L = \rho\theta \qquad (4.4)$$

Considering now the arc JK located at a distance y above the neutral surface, we note that its length L' is

$$L' = (\rho - y)\theta \qquad (4.5)$$

Since the original length of arc JK was equal to L, the deformation of JK is

$$\delta = L' - L \qquad (4.6)$$

or, if we substitute from (4.4) and (4.5) into (4.6),

$$\delta = (\rho - y)\theta - \rho\theta = -y\theta \qquad (4.7)$$

The longitudinal strain ϵ_x in the elements of JK is obtained by dividing δ by the original length L of JK. We write

$$\epsilon_x = \frac{\delta}{L} = \frac{-y\theta}{\rho\theta}$$

or

$$\epsilon_x = -\frac{y}{\rho} \qquad (4.8)$$

The minus sign is due to the fact that we have assumed the bending moment to be positive and, thus, the beam to be concave upward.

Because of the requirement that transverse sections remain plane, identical deformations will occur in all planes parallel to the plane of symmetry. Thus the value of the strain given by Eq. (4.8) is valid anywhere, and we conclude that the *longitudinal normal strain* ϵ_x *varies linearly with the distance* y *from the neutral surface.*

The strain ϵ_x reaches its maximum absolute value when y itself is largest. Denoting by c the largest distance from the neutral surface (which corresponds to either the upper or the lower surface of the member), and by ϵ_m the *maximum absolute value* of the strain, we have

$$\epsilon_m = \frac{c}{\rho} \qquad (4.9)$$

Solving (4.9) for ρ and substituting the value obtained into (4.8), we can also write

$$\epsilon_x = -\frac{y}{c}\epsilon_m \qquad (4.10)$$

We conclude our analysis of the deformations of a member in pure bending by observing that we are still unable to compute the strain or stress at a given point of the member, since we have not yet located the neutral surface in the member. In order to locate this surface, we must first specify the stress-strain relation of the material used.†

---

†Let us note, however, that if the member possesses both a vertical and a horizontal plane of symmetry (e.g., a member with a rectangular cross section), and if the stress-strain curve is the same in tension and compression, the neutral surface will coincide with the plane of symmetry (cf. Sec. 4.8).

4.4. STRESSES AND DEFORMATIONS IN THE ELASTIC RANGE

We now consider the case when the bending moment M is such that the normal stresses in the member remain below the yield strength σ_Y. This means that, for all practical purposes, the stresses in the member will remain below the proportional limit and the elastic limit as well. There will be no permanent deformation, and Hooke's law for uniaxial stress applies. Assuming the material to be homogeneous, and denoting by E its modulus of elasticity, we have in the longitudinal x direction

$$\sigma_x = E\epsilon_x \qquad (4.11)$$

Recalling Eq. (4.10), and multiplying both members of that equation by E, we write

$$E\epsilon_x = -\frac{y}{c}(E\epsilon_m)$$

or, using (4.11),

$$\sigma_x = -\frac{y}{c}\sigma_m \qquad (4.12)$$

where σ_m denotes the *maximum absolute value* of the stress. This result shows that, *in the elastic range, the normal stress varies linearly with the distance from the neutral surface* (Fig. 4.13).

It should be noted that, at this point, we do not know the location of the neutral surface, nor the maximum value σ_m of the stress. Both can be found if we recall the relations (4.1) and (4.3) which were obtained earlier from statics. Substituting first for σ_x from (4.12) into (4.1), we write

$$\int \sigma_x \, dA = \int \left(-\frac{y}{c}\sigma_m \right) dA = -\frac{\sigma_m}{c} \int y \, dA = 0$$

from which it follows that

$$\int y \, dA = 0 \qquad (4.13)$$

This equation shows that the first moment of the cross section about its neutral axis must be zero.[†] In other words, *for a member subjected to pure bending, and as long as the stresses remain in the elastic range, the neutral axis passes through the centroid of the section.*

We now recall Eq. (4.3), which was derived in Sec. 4.2 with respect to an *arbitrary* horizontal z axis,

$$\int (-y\sigma_x \, dA) = M \qquad (4.3)$$

Specifying that the z axis should coincide with the neutral axis of the cross section, we substitute for σ_x from (4.12) into (4.3) and write

Fig. 4.13

Neutral surface

σ_m

c

y

σ_x

[†]See Appendix A for a discussion of the moments of areas.

or

$$\int (-y)\left(-\frac{y}{c}\sigma_m\right) dA = M \qquad (4.14)$$

$$\frac{\sigma_m}{c}\int y^2 dA = M \qquad (4.14)$$

Recalling that in the case of pure bending the neutral axis passes through the centroid of the cross section, we note that I is the moment of inertia, or second moment, of the cross section with respect to a centroidal axis perpendicular to the plane of the couple $\mathbf{M}$. Solving (4.14) for σ_m, we write therefore[†]

$$\sigma_m = \frac{Mc}{I} \qquad (4.15)$$

Substituting for σ_m from (4.15) into (4.12), we obtain the normal stress σ_x at any distance y from the neutral axis:

$$\sigma_x = -\frac{My}{I} \qquad (4.16)$$

Equations (4.15) and (4.16) are called the *elastic flexure formulas*, and the normal stress σ_x caused by the bending or "flexing" of the member is often referred to as the *flexural stress*. We verify that the stress is compressive ($\sigma_x < 0$) above the neutral axis ($y > 0$) when the bending moment M is positive, and tensile ($\sigma_x > 0$) when M is negative.

Returning to Eq. (4.15), we note that the ratio I/c depends only upon the geometry of the cross section. This ratio is called the *elastic section modulus* and is denoted by S. We have

$$\text{Elastic section modulus} = S = \frac{I}{c} \qquad (4.17)$$

Substituting S for I/c into Eq. (4.15), we write this equation in the alternative form

$$\sigma_m = \frac{M}{S} \qquad (4.18)$$

Since the maximum stress σ_m is inversely proportional to the elastic section modulus S, it is clear that beams should be designed with as large a value of S as practicable. For example, in the case of a wooden beam with a rectangular cross section of width b and depth h, we have

$$S = \frac{I}{c} = \frac{\frac{1}{12}bh^3}{h/2} = \frac{1}{6}bh^2 = \frac{1}{6}Ah \qquad (4.19)$$

where A is the cross-sectional area of the beam. This shows that, of two beams with the same cross-sectional area A (Fig. 4.14), the beam with the larger depth h will have the larger section modulus and, thus, will be the more effective in resisting bending.[‡]

$A = 24\text{ in}^2$

$h = 6$ in.

$b = 4$ in.

$h = 8$ in.

$b = 3$ in.

Fig. 4.14

[†] We recall that the bending moment was assumed to be positive. If the bending moment is negative, M should be replaced in Eq. (4.15) by its absolute value $|M|$.

[‡] However, large values of the ratio h/b could result in lateral instability of the beam.

In the case of structural steel, American standard beams (S-beams) and wide-flange beams (W-beams), Fig. 4.15, are preferred to other

Fig. 4.15 Wide-flange steel beams form the frame of many buildings.

shapes because a large portion of their cross section is located far from the neutral axis (Fig. 4.16). Thus, for a given cross-sectional area and a given depth, their design provides large values of I and, consequently, of S. Values of the elastic section modulus of commonly manufactured beams can be obtained from tables listing the various geometric properties of such beams. To determine the maximum stress σ_m in a given section of a standard beam, the engineer needs only to read the value of the elastic section modulus S in a table, and divide the bending moment M in the section by S.

The deformation of the member caused by the bending moment M is measured by the *curvature* of the neutral surface. The curvature is defined as the reciprocal of the radius of curvature ρ, and can be obtained by solving Eq. (4.9) for $1/\rho$:

$$\frac{1}{\rho} = \frac{\epsilon_m}{c} \tag{4.20}$$

But, in the elastic range, we have $\epsilon_m = \sigma_m/E$. Substituting for ϵ_m into (4.20), and recalling (4.15), we write

$$\frac{1}{\rho} = \frac{\sigma_m}{Ec} = \frac{1}{Ec}\frac{Mc}{I}$$

or

$$\boxed{\frac{1}{\rho} = \frac{M}{EI}} \tag{4.21}$$

(a) S-beam   (b) W-beam

Fig. 4.16

EXAMPLE 4.01

A steel bar of 0.8×2.5-in. rectangular cross section is subjected to two equal and opposite couples acting in the vertical plane of symmetry of the bar (Fig. 4.17). Determine the value of the bending moment M that causes the bar to yield. Assume $\sigma_Y = 36$ ksi.

Fig. 4.17

Since the neutral axis must pass through the centroid C of the cross section, we have $c = 1.25$ in. (Fig. 4.18). On the other hand, the centroidal moment of inertia of the rectangular cross section is

$$I = \tfrac{1}{12}bh^3 = \tfrac{1}{12}(0.8 \text{ in.})(2.5 \text{ in.})^3 = 1.042 \text{ in}^4$$

Solving Eq. (4.15) for M, and substituting the above data, we have

$$M = \frac{I}{c}\sigma_m = \frac{1.042 \text{ in}^4}{1.25 \text{ in.}} (36 \text{ ksi})$$

$$M = 30 \text{ kip} \cdot \text{in.}$$

Fig. 4.18

EXAMPLE 4.02

An aluminum rod with a semicircular cross section of radius $r = 12$ mm (Fig. 4.19) is bent into the shape of a circular arc of mean radius $\rho = 2.5$ m. Knowing that the flat face of the rod is turned toward the center of curvature of the arc, determine the maximum tensile and compressive stress in the rod. Use $E = 70$ GPa.

Fig. 4.19

We could use Eq. (4.21) to determine the bending moment M corresponding to the given radius of curvature ρ, and then Eq. (4.15) to determine σ_m. However, it is simpler to use Eq. (4.9) to determine ϵ_m, and Hooke's law to obtain σ_m.

The ordinate $\bar{y}$ of the centroid C of the semicircular cross section is

$$\bar{y} = \frac{4r}{3\pi} = \frac{4(12 \text{ mm})}{3\pi} = 5.093 \text{ mm}$$

The neutral axis passes through C (Fig. 4.20) and the distance c to the point of the cross section farthest away from the neutral axis is

$$c = r - \bar{y} = 12 \text{ mm} - 5.093 \text{ mm} = 6.907 \text{ mm}$$

Using Eq. (4.9), we write

$$\epsilon_m = \frac{c}{\rho} = \frac{6.907 \times 10^{-3} \text{ m}}{2.5 \text{ m}} = 2.763 \times 10^{-3}$$

and, applying Hooke's law,

$$\sigma_m = E\epsilon_m = (70 \times 10^9 \text{ Pa})(2.763 \times 10^{-3}) = 193.4 \text{ MPa}$$

Since this side of the rod faces away from the center of curvature, the stress obtained is a tensile stress. The maximum compressive stress occurs on the flat side of the rod. Using the fact that the stress is proportional to the distance from the neutral axis, we write

$$\sigma_{comp} = -\frac{\bar{y}}{c}\sigma_m = -\frac{5.093 \text{ mm}}{6.907 \text{ mm}}(193.4 \text{ MPa})$$

$$= -142.6 \text{ MPa}$$

Fig. 4.20

4.5. DEFORMATIONS IN A TRANSVERSE CROSS SECTION

When we proved in Sec. 4.3 that the transverse cross section of a member in pure bending remains plane, we did not rule out the possibility of deformations within the plane of the section. That such deformations will exist is evident, if we recall from Sec. 2.11 that elements in a state of uniaxial stress, $\sigma_x \neq 0$, $\sigma_y = \sigma_z = 0$, are deformed in the transverse y and z directions, as well as in the axial x direction. The normal strains ϵ_y and ϵ_z depend upon Poisson's ratio ν for the material used and are expressed as

$$\epsilon_y = -\nu\epsilon_x \qquad \epsilon_z = -\nu\epsilon_x$$

or, recalling Eq. (4.8),

$$\epsilon_y = \frac{\nu y}{\rho} \qquad \epsilon_z = \frac{\nu y}{\rho} \qquad (4.22)$$

The relations we have obtained show that the elements located above the neutral surface ($y > 0$) will expand in both the y and z directions, while the elements located below the neutral surface ($y < 0$) will contract. In the case of a member of rectangular cross section, the expansion and contraction of the various elements in the vertical direction will compensate, and no change in the vertical dimension of the cross section will be observed. As far as the deformations in the horizontal transverse z direction are concerned, however, the expansion of the elements located above the neutral surface and the corresponding contraction of the elements located below that surface will result in the various horizontal lines in the section being bent into arcs of circle (Fig. 4.21). The situation observed here is similar to that observed earlier in a longitudinal cross section. Comparing the second of Eqs. (4.22) with Eq. (4.8), we conclude that the neutral axis of the transverse section will be bent into a circle of radius $\rho' = \rho/\nu$. The center C' of this circle is located below the neutral surface (assuming $M > 0$), i.e., on the side opposite to the center of curvature C of the member. The reciprocal of the radius of curvature ρ' represents the curvature of the transverse cross section and is called the *anticlastic curvature*. We have

$$\text{Anticlastic curvature} = \frac{1}{\rho'} = \frac{\nu}{\rho} \qquad (4.23)$$

In our discussion of the deformations of a symmetric member in pure bending, in this section and in the preceding ones, we have ignored the manner in which the couples $\mathbf{M}$ and $\mathbf{M}'$ were actually applied to the member. If *all* transverse sections of the member, from one end to the other, are to remain plane and free of shearing stresses, we

Fig. 4.21

Fig. 4.22

must make sure that the couples are applied in such a way that the ends of the member themselves remain plane and free of shearing stresses. This can be accomplished by applying the couples **M** and **M′** to the member through the use of rigid and smooth plates (Fig. 4.22). The elementary forces exerted by the plates on the member will be normal to the end sections, and these sections, while remaining plane, will be free to deform as described earlier in this section.

We should note that these loading conditions cannot be actually realized, since they require each plate to exert tensile forces on the corresponding end section below its neutral axis, while allowing the section to freely deform in its own plane. The fact that the rigid-end-plates model of Fig. 4.22 cannot be physically realized, however, does not detract from its importance, which is to allow us to *visualize* the loading conditions corresponding to the relations derived in the preceding sections. Actual loading conditions may differ appreciably from this idealized model. By virtue of Saint-Venant's principle, however, the relations obtained can be used to compute stresses in engineering situations, as long as the section considered is not too close to the points where the couples are applied.

SAMPLE PROBLEM 4.1

The rectangular tube shown is extruded from an aluminum alloy for which $\sigma_Y = 40$ ksi, $\sigma_U = 60$ ksi, and $E = 10.6 \times 10^6$ psi. Neglecting the effect of fillets, determine (a) the bending moment M for which the factor of safety will be 3.00, (b) the corresponding radius of curvature of the tube.

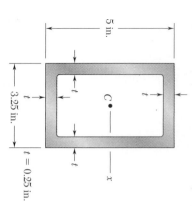

$t = 0.25$ in.

SOLUTION

Moment of Inertia. Considering the cross-sectional area of the tube as the difference between the two rectangles shown and recalling the formula for the centroidal moment of inertia of a rectangle, we write

$$I = \tfrac{1}{12}(3.25)(5)^3 - \tfrac{1}{12}(2.75)(4.5)^3 \qquad I = 12.97 \text{ in}^4$$

Allowable Stress. For a factor of safety of 3.00 and an ultimate stress of 60 ksi, we have

$$\sigma_{all} = \frac{\sigma_U}{F.S.} = \frac{60 \text{ ksi}}{3.00} = 20 \text{ ksi}$$

Since $\sigma_{all} < \sigma_Y$, the tube remains in the elastic range and we can apply the results of Sec. 4.4.

a. Bending Moment. With $c = \tfrac{1}{2}(5 \text{ in.}) = 2.5$ in., we write

$$\sigma_{all} = \frac{Mc}{I} \qquad M = \frac{I}{c}\sigma_{all} = \frac{12.97 \text{ in}^4}{2.5 \text{ in.}}(20 \text{ ksi}) \qquad M = 103.8 \text{ kip} \cdot \text{in.} \quad \blacktriangleleft$$

b. Radius of Curvature. Recalling that $E = 10.6 \times 10^6$ psi, we substitute this value and the values obtained for I and M into Eq. (4.21) and find

$$\frac{1}{\rho} = \frac{M}{EI} = \frac{103.8 \times 10^3 \text{ lb} \cdot \text{in.}}{(10.6 \times 10^6 \text{ psi})(12.97 \text{ in}^4)} = 0.755 \times 10^{-3} \text{ in}^{-1}$$

$$\rho = 1325 \text{ in.} \qquad\qquad \rho = 110.4 \text{ ft} \quad \blacktriangleleft$$

Alternative Solution. Since we know that the maximum stress is $\sigma_{all} = 20$ ksi, we can determine the maximum strain ϵ_m and then use Eq. (4.9),

$$\epsilon_m = \frac{\sigma_{all}}{E} = \frac{20 \text{ ksi}}{10.6 \times 10^6 \text{ psi}} = 1.887 \times 10^{-3} \text{ in./in.}$$

$$\epsilon_m = \frac{c}{\rho} \qquad \rho = \frac{c}{\epsilon_m} = \frac{2.5 \text{ in.}}{1.887 \times 10^{-3} \text{ in./in.}}$$

$$\rho = 1325 \text{ in.} \qquad\qquad \rho = 110.4 \text{ ft} \quad \blacktriangleleft$$

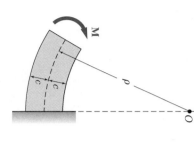

SAMPLE PROBLEM 4.2

A cast-iron machine part is acted upon by the 3 kN · m couple shown. Knowing that $E = 165$ GPa and neglecting the effect of fillets, determine (a) the maximum tensile and compressive stresses in the casting, (b) the radius of curvature of the casting.

$M = 3$ kN · m

SOLUTION

Centroid. We divide the T-shaped cross section into the two rectangles shown and write

	Area, mm²	$\bar{y}$, mm	$\bar{y}A$, mm³
1	$(20)(90) = 1800$	50	90×10^3
2	$(40)(30) = 1200$	20	24×10^3
	$\Sigma A = 3000$		$\Sigma \bar{y}A = 114 \times 10^3$

$$\bar{Y}\Sigma A = \Sigma \bar{y}A$$
$$\bar{Y}(3000) = 114 \times 10^6$$
$$\bar{Y} - 38 \text{ mm}$$

$\bar{y}_1 = 50$ mm
$\bar{y}_2 = 20$ mm
$\bar{Y} = 38$ mm

Centroidal Moment of Inertia. The parallel-axis theorem is used to determine the moment of inertia of each rectangle with respect to the axis x' that passes through the centroid of the composite section. Adding the moments of inertia of the rectangles, we write

$$I_{x'} = \Sigma(\bar{I} + Ad^2) = \Sigma(\tfrac{1}{12}bh^3 + Ad^2)$$
$$= \tfrac{1}{12}(90)(20)^3 + (90 \times 20)(12)^2 + \tfrac{1}{12}(30)(40)^3 + (30 \times 40)(18)^2$$
$$= 868 \times 10^3 \text{ mm}^4$$
$$I = 868 \times 10^{-9} \text{ m}^4$$

a. Maximum Tensile Stress. Since the applied couple bends the casting downward, the center of curvature is located below the cross section. The maximum tensile stress occurs at point A, which is farthest from the center of curvature.

$$\sigma_A = \frac{Mc_A}{I} = \frac{(3 \text{ kN} \cdot \text{m})(0.022 \text{ m})}{868 \times 10^{-9} \text{ m}^4} \qquad \sigma_A = +76.0 \text{ MPa} \blacktriangledown$$

Maximum Compressive Stress. This occurs at point B; we have

$$\sigma_B = -\frac{Mc_B}{I} = -\frac{(3 \text{ kN} \cdot \text{m})(0.038 \text{ m})}{868 \times 10^{-9} \text{ m}^4} \qquad \sigma_B = -131.3 \text{ MPa} \blacktriangledown$$

b. Radius of Curvature. From Eq. (4.21), we have

$$\frac{1}{\rho} = \frac{M}{EI} = \frac{3 \text{ kN} \cdot \text{m}}{(165 \text{ GPa})(868 \times 10^{-9} \text{ m}^4)}$$
$$= 20.95 \times 10^{-3} \text{ m}^{-1} \qquad\qquad \sigma = 47.7 \text{ m} \blacktriangledown$$

$c_A = 0.022$ m
$c_B = 0.038$ m

Center of curvature

PROBLEMS

4.1 and 4.2 Knowing that the couple shown acts in a vertical plane, determine the stress at (*a*) point *A*, (*b*) point *B*.

Fig. P4.1

Fig. P4.2

4.3 The wide-flange beam shown is made of a high-strength, low-alloy steel for which $\sigma_Y = 345$ MPa and $\sigma_U = 450$ MPa. Using a factor of safety of 3.0, determine the largest couple that can be applied to the beam when it is bent about the *z* axis. Neglect the effect of fillets.

4.4 Solve Prob. 4.3, assuming that the beam is bent about the *y* axis.

4.5 Using an allowable stress of 16 ksi, determine the largest couple that can be applied to each pipe.

Fig. P4.3

(*a*)   (*b*)

Fig. P4.5

4.6 A nylon spacing bar has the cross section shown. Knowing that the allowable stress for the grade of nylon used is 24 MPa, determine the largest couple M_z that can be applied to the bar.

Fig. P4.6

224

4.7 and 4.8 Two W4 × 13 rolled sections are welded together as shown. Knowing that for the steel alloy used σ_Y = 36 ksi and σ_U = 58 ksi and using a factor of safety of 3.0, determine the largest couple that can be applied when the assembly is bent about the z axis.

Fig. P4.7

Fig. P4.8

4.9 through 4.11 Two vertical forces are applied to a beam of the cross section shown. Determine the maximum tensile and compressive stresses in portion BC of the beam.

Fig. P4.9

Fig. P4.10

Fig. P4.11

4.12 Two equal and opposite couples of magnitude M = 15 kN · m are applied to the channel-shaped beam AB. Observing that the couples cause the beam to bend in a horizontal plane, determine the stress (a) at point C, (b) at point D, (c) at point E.

Fig. P4.12

4.13 Knowing that a beam of the cross section shown is bent about a horizontal axis and that the bending moment is 8 kN · m, determine the total force acting on the top flange.

4.14 Knowing that a beam of the cross section shown is bent about a vertical axis and that the bending moment is 4 kN · m, determine the total force acting on the shaded portion of the lower flange.

Fig. P4.13 and P4.14

4.15 Knowing that a beam of the cross section shown is bent about a horizontal axis and that the bending moment is 3.5 kip · in., determine the total force acting on the shaded portion of the beam.

4.16 Solve Prob. 4.15, assuming that the beam is bent about a vertical axis and that the bending moment is 6 kip · in.

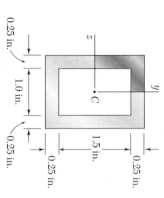

Fig. P4.15

4.17 Knowing that a beam of the cross section shown is bent about a horizontal axis and that the bending moment is 6 kip · in., determine the total force acting on the shaded portion of the beam.

4.18 Solve Prob. 4.17, assuming that the beam is bent about a vertical axis and that the bending moment is 6 kip · in.

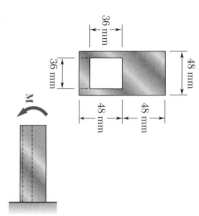

Fig. P4.17

4.19 and 4.20 Knowing that for the extruded beam shown the allowable stress is 120 MPa in tension and 150 MPa in compression, determine the largest couple **M** that can be applied.

Fig. P4.19

Fig. P4.20

4.21 Knowing that for the extruded beam shown the allowable stress is 12 ksi in tension and 16 ksi in compression, determine the largest couple **M** that can be applied.

Fig. P4.21

4.22 The beam shown is made of a nylon for which the allowable stress is 24 MPa in tension and 30 MPa in compression. Determine the largest couple **M** that can be applied to the beam.

Fig. *P4.22*

4.23 Solve Prob. 4.22, assuming that $d = 80$ mm.

4.24 Knowing that for the beam shown the allowable stress is 12 ksi in tension and 16 ksi in compression, determine the largest couple **M** that can be applied.

Fig. **P4.24**

4.25 Knowing that $\sigma_{all} = 24$ ksi for the steel strip AB, determine (a) the largest couple **M** that can be applied, (b) the corresponding radius of curvature. Use $E = 29 \times 10^6$ psi.

Fig. **P4.25**

4.26 Straight rods of 0.30 in. diameter and 200 ft length are sometimes used to clear underground conduits of obstructions or to thread wires through a new conduit. The rods are made of high-strength steel and, for storage and transportation, are wrapped on spools of 5-ft diameter. Assuming that the yield strength is not exceeded, determine (a) the maximum stress in a rod, when the rod, which was initially straight, is wrapped on a spool, (b) the corresponding bending moment in the rod. Use $E = 29 \times 10^6$ psi.

Fig. **P4.26**

4.27 It is observed that a thin steel strip of 0.06-in. width can be bent into a circle of $\frac{3}{4}$-in. diameter without any resulting permanent deformation. Knowing that $E = 29 \times 10^6$ psi, determine (a) the maximum stress in the bent strip, (b) the magnitude of the couples required to bend the strip.

Fig. P4.27

4.28 A 3 kip · in. couple is applied to the steel bar shown. (a) Assuming that the couple is applied about the z axis as shown, determine the maximum stress and the radius of curvature of the bar. (b) Solve part a, assuming that the couple is applied about the y axis. Use $E = 29 \times 10^6$ psi.

Fig. P4.28

4.29 A couple of magnitude M is applied to a square bar of side a. For each of the orientations shown, determine the maximum stress and the curvature of the bar.

(a)

(b)

Fig. P4.29

4.30 A 24 kN · m couple is applied to the W200 × 46.1 rolled-steel beam shown. (a) Assuming that the couple is applied about the z axis as shown, determine the maximum stress and the radius of curvature of the beam. (b) Solve part a, assuming that the couple is applied about the y axis. Use $E = 200$ GPa.

Fig. P4.30

4.31 (a) Using an allowable stress of 120 MPa, determine the largest couple **M** that can be applied to a beam of the cross section shown. (b) Solve part a, assuming that the cross section of the beam is an 80-mm square.

Fig. P4.31

4.32 A portion of a square bar is removed by milling, so that its cross section is as shown. The bar is then bent about its horizontal diagonal by a couple **M.** Considering the case where $h = 0.9h_0$, express the maximum stress in the bar in the form $\sigma_m = k\sigma_0$, where σ_0 is the maximum stress that would have occurred if the original square bar had been bent by the same couple **M,** and determine the value of k.

Fig. P4.32

4.33 In Prob. 4.32, determine (a) the value of h for which the maximum stress σ_m is as small as possible, (b) the corresponding value of k.

4.34 A couple **M** will be applied to a beam of rectangular cross section that is to be sawed from a log of circular cross section. Determine the ratio d/b for which (a) the maximum stress σ_m will be as small as possible, (b) the radius of curvature of the beam will be maximum.

Fig. P4.34

4.35 For the bar and loading of Example 4.01, determine (a) the radius of curvature ρ, (b) the radius of curvature ρ' of a transverse cross section, (c) the angle between the sides of the bar that were originally vertical. Use $E = 29 \times 10^6$ psi and $\nu = 0.29$.

4.36 For the aluminum bar and loading of Sample Prob. 4.1, determine (a) the radius of curvature ρ' of a transverse cross section, (b) the angle between the sides of the bar that were originally vertical. Use $E = 10.6 \times 10^6$ psi and $\nu = 0.33$.

Fig. P4.37

4.37 A W200 × 31.3 rolled-steel beam is subjected to a couple **M** of moment 45 kN · m. Knowing that $E = 200$ GPa and $\nu = 0.29$. determine (a) the radius of curvature ρ, (b) the radius of curvature ρ' of a transverse cross section.

4.38 It was assumed in Sec. 4.3 that the normal stresses σ_y in a member in pure bending are negligible. For an initially straight elastic member of rectangular cross section, (a) derive an approximate expression for σ_y as a function of y, (b) show that $(\sigma_y)_{max} \approx -(c/2\rho)(\sigma_x)_{max}$ and, thus, that σ_y can be neglected in all practical situations. (*Hint*: Consider the free-body diagram of the portion of beam located below the surface of ordinate y and assume the distribution of the stress σ_x is still linear.)

Fig. P4.38

4.6. BENDING OF MEMBERS MADE OF SEVERAL MATERIALS

The derivations given in Sec. 4.4 were based on the assumption of a homogeneous material with a given modulus of elasticity E. If the member subjected to pure bending is made of two or more materials with different moduli of elasticity, our approach to the determination of the stresses in the member must be modified.

Consider, for instance, a bar consisting of two portions of different materials bonded together as shown in cross section in Fig. 4.23. This composite bar will deform as described in Sec. 4.3, since its cross section remains the same throughout its entire length, and since no assumption was made in Sec. 4.3 regarding the stress-strain relationship of the material or materials involved. Thus, the normal strain ϵ_x still varies linearly with the distance y from the neutral axis of the section (Fig. 4.24a and b), and formula (4.8) holds:

$$\epsilon_x = -\frac{y}{\rho} \tag{4.8}$$

Fig. 4.23

However, we cannot assume that the neutral axis passes through the centroid of the composite section, and one of the goals of the present analysis will be to determine the location of this axis.

Since the moduli of elasticity E_1 and E_2 of the two materials are different, the expressions obtained for the normal stress in each material will also be different. We write

$$\sigma_1 = E_1\epsilon_x = -\frac{E_1 y}{\rho}$$

$$\sigma_2 = E_2\epsilon_x = -\frac{E_2 y}{\rho} \tag{4.24}$$

and obtain a stress-distribution curve consisting of two segments of straight line (Fig. 4.24c). It follows from Eqs. (4.24) that the force dF_1 exerted on an element of area dA of the upper portion of the cross section is

$$dF_1 = \sigma_1 \, dA = -\frac{E_1 y}{\rho} \, dA \tag{4.25}$$

while the force dF_2 exerted on an element of the same area dA of the

(a)

(b)

(c)

Fig. 4.24   Strain and stress distribution in bar made of two materials.

lower portion is

$$dF_2 = \sigma_2 dA = -\frac{E_2 y}{\rho} dA \qquad (4.26)$$

But, denoting by n the ratio E_2/E_1 of the two moduli of elasticity, we can express dF_2 as

$$dF_2 = -\frac{(nE_1)y}{\rho} dA = -\frac{E_1 y}{\rho}(n\,dA) \qquad (4.27)$$

Comparing Eqs. (4.25) and (4.27), we note that the same force dF_2 would be exerted on an element of area $n\,dA$ of the first material. In other words, the resistance to bending of the bar would remain the same if both portions were made of the first material, provided that the width of each element of the lower portion were multiplied by the factor n. Note that this widening (if $n > 1$), or narrowing (if $n < 1$), must be effected *in a direction parallel to the neutral axis of the section*, since it is essential that the distance y of each element from the neutral axis remain the same. The new cross section obtained in this way is called the *transformed section* of the member (Fig. 4.25).

Since the transformed section represents the cross section of a member made of a *homogeneous material* with a modulus of elasticity E_1, the method described in Sec. 4.4 can be used to determine the neutral axis of the section and the normal stress at various points of the section. The neutral axis will be drawn *through the centroid of the transformed section* (Fig. 4.26), and the stress σ_x at any point of the corresponding

Fig. 4.25   Transformed section for composite bar.

fictitious homogeneous member will be obtained from Eq. (4.16)

$$\sigma_x = -\frac{My}{I} \qquad (4.16)$$

where y is the distance from the neutral surface, and I *the moment of inertia of the transformed section* with respect to its centroidal axis.

To obtain the stress σ_1 at a point located in the upper portion of the cross section of the original composite bar, we simply compute the stress σ_x at the corresponding point of the transformed section. However, to obtain the stress σ_2 at a point in the lower portion of the cross section, we must *multiply by n* the stress σ_x computed at the corresponding point of the transformed section. Indeed, as we saw earlier,

Fig. 4.26   Distribution of stresses in transformed section.

the same elementary force dF_2 is applied to an element of area $n\ dA$ of the transformed section and to an element of area dA of the original section. Thus, the stress σ_2 at a point of the original section must be n times larger than the stress at the corresponding point of the transformed section.

The deformations of a composite member can also be determined by using the transformed section. We recall that the transformed section represents the cross section of a member, made of a homogeneous material of modulus E_1, which deforms in the same manner as the composite member. Therefore, using Eq. (4.21), we write that the curvature of the composite member is

$$\frac{1}{\rho} = \frac{M}{E_1 I}$$

where I is the moment of inertia of the transformed section with respect to its neutral axis.

EXAMPLE 4.03

A bar obtained by bonding together pieces of steel ($E_s = 29 \times 10^6$ psi) and brass ($E_b = 15 \times 10^6$ psi) has the cross section shown (Fig. 4.27). Determine the maximum stress in the steel and in the brass when the bar is in pure bending with a bending moment $M = 40$ kip · in.

The transformed section corresponding to an equivalent bar made entirely of brass is shown in Fig. 4.28. Since

$$n = \frac{E_s}{E_b} = \frac{29 \times 10^6 \text{ psi}}{15 \times 10^6 \text{ psi}} = 1.933$$

the width of the central portion of brass, which replaces the original steel portion, is obtained by multiplying the original width by 1.933, we have

$$(0.75 \text{ in.})(1.933) = 1.45 \text{ in.}$$

Note that this change in dimension occurs in a direction parallel to the neutral axis. The moment of inertia of the trans-

formed section about its centroidal axis is

$$I = \tfrac{1}{12} bh^3 = \tfrac{1}{12}(2.25 \text{ in.})(3 \text{ in.})^3 = 5.063 \text{ in}^4$$

and the maximum distance from the neutral axis is $c = 1.5$ in. Using Eq. (4.15), we find the maximum stress in the transformed section:

$$\sigma_m = \frac{Mc}{I} = \frac{(40 \text{ kip} \cdot \text{in.})(1.5 \text{ in.})}{5.063 \text{ in.}} = 11.85 \text{ ksi}$$

The value obtained also represents the maximum stress in the brass portion of the original composite bar. The maximum stress in the steel portion, however, will be larger than the value obtained for the transformed section, since the area of the central portion must be reduced by the factor $n = 1.933$ when we return from the transformed section to the original one. We thus conclude that

$$(\sigma_{\text{brass}})_{\text{max}} = 11.85 \text{ ksi}$$
$$(\sigma_{\text{steel}})_{\text{max}} = (1.933)(11.85 \text{ ksi}) = 22.9 \text{ ksi}$$

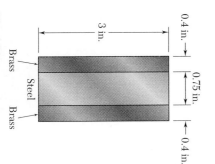

3 in.

0.4 in. → ← 0.4 in.

0.75 in.

Brass
Steel
Brass

Fig. 4.27

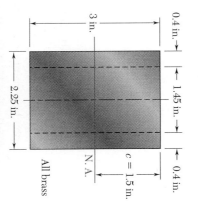

3 in.

0.4 in. → ← 1.45 in. → ← 0.4 in.

2.25 in.

$c = 1.5$ in.

N.A.

All brass

Fig. 4.28

Fig. 4.29

An important example of structural members made of two different materials is furnished by *reinforced concrete beams* (Fig. 4.29). These beams, when subjected to positive bending moments, are reinforced by steel rods placed a short distance above their lower face (Fig. 4.30a). Since concrete is very weak in tension, it will crack below the neutral surface and the steel rods will carry the entire tensile load, while the upper part of the concrete beam will carry the compressive load.

To obtain the transformed section of a reinforced concrete beam, we replace the total cross-sectional area A_s of the steel bars by an equivalent area nA_s, where n is the ratio E_s/E_c of the moduli of elasticity of steel and concrete (Fig. 4.30b). On the other hand, since the concrete in the beam acts effectively only in compression, only the portion of the cross section located above the neutral axis should be used in the transformed section.

The position of the neutral axis is obtained by determining the distance x from the upper face of the beam to the centroid C of the transformed section. Denoting by b the width of the beam, and by d the distance from the upper face to the center line of the steel rods, we write that the first moment of the transformed section with respect to the neutral axis must be zero. Since the first moment of each of the two portions of the transformed section is obtained by multiplying its area by the distance of its own centroid from the neutral axis, we have

$$(bx)\frac{x}{2} - nA_s(d - x) = 0$$

or

$$\frac{1}{2}bx^2 + nA_sx - nA_sd = 0 \qquad (4.28)$$

Solving this quadratic equation for x, we obtain both the position of the neutral axis in the beam, and the portion of the cross section of the concrete beam which is effectively used.

The determination of the stresses in the transformed section is carried out as explained earlier in this section (see Sample Prob. 4.4). The distribution of the compressive stresses in the concrete and the resultant $\mathbf{F}_s$ of the tensile forces in the steel rods are shown in Fig. 4.30c.

Fig. 4.30

4.7. STRESS CONCENTRATIONS

The formula $\sigma_m = Mc/I$ was derived in Sec. 4.4 for a member with a plane of symmetry and a uniform cross section, and we saw in Sec. 4.5 that it was accurate throughout the entire length of the member only if the couples $\mathbf{M}$ and $\mathbf{M'}$ were applied through the use of rigid and smooth plates. Under other conditions of application of the loads, stress concentrations will exist near the points where the loads are applied.

Higher stresses will also occur if the cross section of the member undergoes a sudden change. Two particular cases of interest have been studied,† the case of a flat bar with a sudden change in width, and the case of a flat bar with grooves. Since the distribution of stresses in the critical cross sections depends only upon the geometry of the members, stress-concentration factors can be determined for various ratios of the parameters involved and recorded as shown in Figs. 4.31 and 4.32. The

Fig. 4.31 Stress-concentration factors for *flat bars* with fillets under pure bending.†

Fig. 4.32 Stress-concentration factors for *flat bars* with grooves under pure bending.†

†W. D. Pilkey, *Peterson's Stress Concentration Factors*, 2d ed., John Wiley & Sons, New York, 1997.

value of the maximum stress in the critical cross section can then be expressed as

$$\sigma_m = K \frac{Mc}{I} \qquad (4.29)$$

where K is the stress-concentration factor, and where c and I refer to the critical section, i.e., to the section of width d in both of the cases considered here. An examination of Figs. 4.31 and 4.32 clearly shows the importance of using fillets and grooves of radius r as large as practical.

Finally, we should point out that, as was the case for axial loading and torsion, the values of the factors K have been computed under the assumption of a linear relation between stress and strain. In many applications, plastic deformations will occur and result in values of the maximum stress lower than those indicated by Eq. (4.29).

EXAMPLE 4.04

Grooves 10 mm deep are to be cut in a steel bar which is 60 mm wide and 9 mm thick (Fig. 4.33). Determine the smallest allowable width of the grooves if the stress in the bar is not to exceed 150 MPa when the bending moment is equal to 180 N · m.

Fig. 4.33

We note from Fig. 4.33a that

$$d = 60 \text{ mm} - 2(10 \text{ mm}) = 40 \text{ mm}$$
$$c = \tfrac{1}{2}d = 20 \text{ mm} \qquad b = 9 \text{ mm}$$

The moment of inertia of the critical cross section about its neutral axis is

$$I = \tfrac{1}{12}bd^3 = \tfrac{1}{12}(9 \times 10^{-3} \text{ m})(40 \times 10^{-3} \text{ m})^3$$
$$= 48 \times 10^{-9} \text{ m}^4$$

The value of the stress Mc/I is thus

$$\frac{Mc}{I} = \frac{(180 \text{ N} \cdot \text{m})(20 \times 10^{-3} \text{ m})}{48 \times 10^{-9} \text{ m}^4} = 75 \text{ MPa}$$

Substituting this value for Mc/I into Eq. (4.29) and making $\sigma_m = 150$ MPa, we write

$$150 \text{ MPa} = K(75 \text{ MPa})$$
$$K = 2$$

We have, on the other hand,

$$\frac{D}{d} = \frac{60 \text{ mm}}{40 \text{ mm}} = 1.5$$

Using the curve of Fig. 4.32 corresponding to $D/d = 1.5$, we find that the value $K = 2$ corresponds to a value of r/d equal to 0.13. We have, therefore,

$$\frac{r}{d} = 0.13$$

$$r = 0.13d = 0.13(40 \text{ mm}) = 5.2 \text{ mm}$$

The smallest allowable width of the grooves is thus

$$2r = 2(5.2 \text{ mm}) = 10.4 \text{ mm}$$

SAMPLE PROBLEM 4.3

Two steel plates have been welded together to form a beam in the shape of a T that has been strengthened by securely bolting to it the two oak timbers shown. The modulus of elasticity is 12.5 GPa for the wood and 200 GPa for the steel. Knowing that a bending moment $M = 50$ kN · m is applied to the composite beam, determine (a) the maximum stress in the wood, (b) the stress in the steel along the top edge.

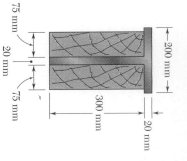

SOLUTION

Transformed Section. We first compute the ratio

$$n = \frac{E_s}{E_w} = \frac{200 \text{ GPa}}{12.5 \text{ GPa}} = 16$$

Multiplying the horizontal dimensions of the steel portion of the section by $n = 16$, we obtain a transformed section made entirely of wood.

Neutral Axis. The neutral axis passes through the centroid of the transformed section. Since the section consists of two rectangles, we have

$$\bar{Y} = \frac{\Sigma \bar{y}A}{\Sigma A} = \frac{(0.160 \text{ m})(3.2 \text{ m} \times 0.020 \text{ m}) + 0}{3.2 \text{ m} \times 0.020 \text{ m} + 0.470 \text{ m} \times 0.300 \text{ m}} = 0.050 \text{ m}$$

Centroidal Moment of Inertia. Using the parallel-axis theorem:

$$I = \frac{1}{12}(0.470)(0.300)^3 + (0.470 \times 0.300)(0.050)^2$$
$$+ \frac{1}{12}(3.2)(0.020)^3 + (3.2 \times 0.020)(0.160 - 0.050)^2$$
$$I = 2.19 \times 10^{-3} \text{ m}^4$$

a. Maximum Stress in Wood. The wood farthest from the neutral axis is located along the bottom edge, where $c_2 = 0.200$ m.

$$\sigma_w = \frac{Mc_2}{I} = \frac{(50 \times 10^3 \text{ N} \cdot \text{m})(0.200 \text{ m})}{2.19 \times 10^{-3} \text{ m}^4} \qquad \sigma_w = 4.57 \text{ MPa} \blacktriangleleft$$

b. Stress in Steel. Along the top edge $c_1 = 0.120$ m. From the transformed section we obtain an equivalent stress in wood, which must be multiplied by n to obtain the stress in steel.

$$\sigma_s = n\frac{Mc_1}{I} = (16)\frac{(50 \times 10^3 \text{ N} \cdot \text{m})(0.120 \text{ m})}{2.19 \times 10^{-3} \text{ m}^4} \qquad \sigma_s = 43.8 \text{ MPa} \blacktriangleleft$$

SAMPLE PROBLEM 4.4

A concrete floor slab is reinforced by $\frac{5}{8}$-in.-diameter steel rods placed 1 in. above the lower face of the slab and spaced 6 in. on centers. The modulus of elasticity is 3.6×10^6 psi for the concrete used and 29×10^6 psi for the steel. Knowing that a bending moment of 40 kip · in. is applied to each 1-ft width of the slab, determine (a) the maximum stress in the concrete, (b) the stress in the steel.

SOLUTION

Transformed Section. We consider a portion of the slab 12 in. wide, in which there are two $\frac{5}{8}$-in.-diameter rods having a total cross-sectional area

$$A_s = 2\left[\frac{\pi}{4}\left(\frac{5}{8}\text{ in.}\right)^2\right] = 0.614 \text{ in}^2$$

Since concrete acts only in compression, all tensile forces are carried by the steel rods, and the transformed section consists of the two areas shown. One is the portion of concrete in compression (located above the neutral axis), and the other is the transformed steel area nA_s. We have

$$n = \frac{E_s}{E_c} = \frac{29 \times 10^6 \text{ psi}}{3.6 \times 10^6 \text{ psi}} = 8.06$$

$$nA_s = 8.06(0.614 \text{ in}^2) = 4.95 \text{ in}^2$$

Neutral Axis. The neutral axis of the slab passes through the centroid of the transformed section. Summing moments of the transformed area about the neutral axis, we write

$$12x\left(\frac{x}{2}\right) - 4.95(4 - x) = 0 \qquad x = 1.450 \text{ in.}$$

Moment of Inertia. The centroidal moment of inertia of the transformed area is

$$I = \tfrac{1}{3}(12)(1.450)^3 + 4.95(4 - 1.450)^2 = 44.4 \text{ in}^4$$

a. Maximum Stress in Concrete. At the top of the slab, we have $c_1 = 1.450$ in. and

$$\sigma_c = \frac{Mc_1}{I} = \frac{(40 \text{ kip} \cdot \text{in.})(1.450 \text{ in.})}{44.4 \text{ in}^4} \qquad \sigma_c = 1.306 \text{ ksi} \quad \blacktriangledown$$

b. Stress in Steel. For the steel, we have $c_2 = 2.55$ in., $n = 8.06$ and

$$\sigma_s = n\frac{Mc_2}{I} = 8.06\frac{(40 \text{ kip} \cdot \text{in.})(2.55 \text{ in.})}{44.4 \text{ in}^4} \qquad \sigma_s = 18.52 \text{ ksi} \quad \blacktriangledown$$

PROBLEMS

4.39 and 4.40 Two metal strips are securely bonded to a metal bar of 30 × 30-mm square cross section. Using the data given below, determine the largest permissible bending moment when the composite bar is bent about a horizontal axis.

	Aluminum	Brass
Modulus of elasticity	70 GPa	105 GPa
Allowable stress	100 MPa	160 MPa

Fig. P4.39

Fig. P4.40

4.41 and 4.42 For the composite bar indicated, determine the permissible bending moment when the bar is bent about a vertical axis.

   4.41 Bar of Prob. 4.39.

   4.42 Bar of Prob. 4.40.

4.43 and 4.44 Wooden beams and steel plates are securely bolted together to form the composite members shown. Using the data given below, determine the largest permissible bending moment when the composite member is bent about a horizontal axis.

	Wood	Steel
Modulus of elasticity	2×10^6 psi	30×10^6 psi
Allowable stress	2000 psi	22 ksi

Fig. P4.43

Fig. P4.44

4.45 and 4.46 A copper strip ($E_c = 105$ GPa) and an aluminum strip ($E_a = 75$ GPa) are bonded together to form the composite bar shown. Knowing that the bar is bent about a horizontal axis by a couple of moment 35 N · m, determine the maximum stress in (a) the aluminum strip, (b) the copper strip.

Aluminum

Copper

24 mm

6 mm

6 mm

Fig. P4.45

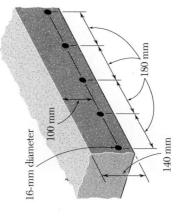

Aluminum

Copper

24 mm

9 mm

3 mm

Fig. P4.46

4.47 and 4.48 A 6 × 10-in. timber beam has been strengthened by bolting to it the steel straps shown. The modulus of elasticity is 1.5×10^6 psi for the wood and 30×10^6 psi for the steel. Knowing that the beam is bent about a horizontal axis by a couple of moment 200 kip · in., determine the maximum stress in (a) the wood, (b) the steel.

10 in.

6 in.

$5 \times \frac{1}{2}$ in.

Fig. P4.47

10 in.

6 in.

$2 \times \frac{3}{8}$ in.

$2 \times \frac{3}{8}$ in.

Fig. P4.48

4.49 and 4.50 For the composite bar indicated, determine the radius of curvature caused by the couple of moment 35 N · m.

4.49 Bar of Prob. 4.45.

4.50 Bar of Prob. 4.46.

4.51 and 4.52 For the composite beam indicated, determine the radius of curvature caused by the couple of moment 200 kip · in.

4.51 Beam of Prob. 4.47.

4.52 Beam of Prob. 4.48.

4.53 A concrete slab is reinforced by 16-mm-diameter steel rods placed on 180-mm centers as shown. The modulus of elasticity is 20 GPa for concrete and 200 GPa for steel. Using an allowable stress of 9 MPa for the concrete and of 120 MPa for the steel, determine the largest allowable positive bending moment in a portion of slab 1 m wide.

4.54 Solve Prob. 4.53, assuming that the spacing of the 16-mm-diameter rods is increased to 225 mm on centers.

16-mm diameter

100 mm

180 mm

140 mm

Fig. P4.53

4.55 The reinforced concrete beam shown is subjected to a positive bending moment of 175 kN · m. Knowing that the modulus of elasticity is 25 GPa for the concrete and 200 GPa for the steel, determine (*a*) the stress in the steel, (*b*) the maximum stress in the concrete.

Fig. P4.55

4.56 Solve Prob. 4.55 assuming that the 450-mm depth of the beam is increased to 500 mm.

4.57 Knowing that the bending moment in the reinforced concrete beam shown is +150 kip · ft and that the modulus of elasticity is 3.75 × 10⁶ psi for the concrete and 30 × 10⁶ psi for the steel, determine (*a*) the stress in the steel, (*b*) the maximum stress in the concrete.

Fig. P4.57

4.58 A concrete beam is reinforced by three steel rods placed as shown. The modulus of elasticity is 3 × 10⁶ psi for the concrete and 30 × 10⁶ psi for the steel. Using an allowable stress of 1350 psi for the concrete and 20 ksi for the steel, determine the largest allowable positive bending moment in the beam.

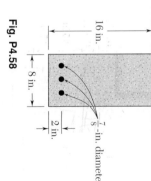

Fig. P4.58

4.59 Solve Prob. 4.58, assuming that the width of the concrete beam is increased to 10 in.

4.60 The design of a reinforced concrete beam is said to be *balanced* if the maximum stresses in the steel and concrete are equal, respectively, to the allowable stresses σ_s and σ_c. Show that to achieve a balanced design the distance *x* from the top of the beam to the neutral axis must be

$$x = \frac{d}{1 + \dfrac{\sigma_s E_c}{\sigma_c E_s}}$$

where E_c and E_s are the moduli of elasticity of concrete and steel, respectively, and *d* is the distance from the top of the beam to the reinforcing steel.

Fig. P4.60

4.61 For the concrete beam shown, the modulus of elasticity is 3.5×10^6 psi for the concrete and 29×10^6 psi for the steel. Knowing that $b = 8$ in. and $d = 22$ in., and using an allowable stress of 1800 psi for the concrete and 20 ksi for the steel, determine (*a*) the required area A_s of the steel reinforcement if the design of the beam is to be balanced, (*b*) the largest allowable bending moment. (See Prob. 4.60 for definition of a balanced beam.)

Fig. P4.61

4.62 and 4.63 Five metal strips, each of 0.5×1.5-in. cross section, are bonded together to form the composite beam shown. The modulus of elasticity is 30×10^6 psi for the steel, 15×10^6 psi for the brass, and 10×10^6 psi for the aluminum. Knowing that the beam is bent about a horizontal axis by couples of moment 12 kip · in., determine (*a*) the maximum stress in each of the three metals, (*b*) the radius of curvature of the composite beam.

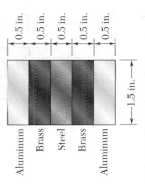

Fig. P4.62

Fig. P4.63

4.64 The composite beam shown is formed by bonding together a brass rod and an aluminum rod of semicircular cross sections. The modulus of elasticity is 15×10^6 psi for the brass and 10×10^6 psi for the aluminum. Knowing that the composite beam is bent about a horizontal axis by couples of moment 8 kip · in., determine the maximum stress (*a*) in the brass, (*b*) in the aluminum.

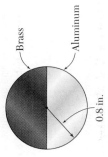

Fig. P4.64

4.65   A steel pipe and an aluminum pipe are securely bonded together to form the composite beam shown. The modulus of elasticity is 210 GPa for the steel and 70 GPa for the aluminum. Knowing that the composite beam is bent by couples of moment 500 N · m, determine the maximum stress (a) in the aluminum, (b) in the steel.

Fig. P4.65

4.66   Solve Prob. 4.65, assuming that the 6-mm-thick inner pipe is made of aluminum and that the 3-mm-thick outer pipe is made of steel.

4.67   The rectangular beam shown is made of a plastic for which the value of the modulus of elasticity in tension is one-half its value in compression. For a bending moment $M = 600$ N · m, determine the maximum (a) tensile stress, (b) compressive stress.

Fig. P4.67

*4.68   A rectangular beam is made of a material for which the modulus of elasticity is E_t in tension and E_c in compression. Show that the curvature of the beam in pure bending is

$$\frac{1}{\rho} = \frac{M}{E_r I}$$

where

$$E_r = \frac{4 E_t E_c}{(\sqrt{E_t} + \sqrt{E_c})^2}$$

4.69   Knowing that $M = 250$ N · m, determine the maximum stress in the beam shown when the radius r of the fillets is (a) 4 mm, (b) 8 mm.

4.70   Knowing that the allowable stress for the beam shown is 90 MPa, determine the allowable bending moment M when the radius r of the fillets is (a) 8 mm, (b) 12 mm.

Fig. P4.69 and P4.70

4.71   Semicircular grooves of radius r must be milled as shown in the sides of a steel member. Using an allowable stress of 8 ksi, determine the largest bending moment that can be applied to the member when the radius r of the semicircular grooves is (a) $\frac{3}{8}$ in., (b) $\frac{3}{4}$ in.

4.72   Semicircular grooves of radius r must be milled as shown in the sides of a steel member. Knowing that $M = 4$ kip · in., determine the maximum stress in the member when (a) $r = \frac{3}{8}$ in., (b) $r = \frac{3}{4}$ in.

Fig. P4.71 and P4.72

4.73 The allowable stress used in the design of a steel bar is 80 MPa. Determine the largest couple **M** that can be applied to the bar (a) if the bar is designed with grooves having semicircular portions of radius $r = 15$ mm, as shown in Fig. P4.73a, (b) if the bar is redesigned by removing the material above the grooves as shown in Fig. P4.73b.

4.74 A couple of moment $M = 2$ kN · m is to be applied to the end of a steel bar. Determine the maximum stress in the bar (a) if the bar is designed with grooves having semicircular portions of radius $r = 10$ mm, as shown in Fig. P4.74a, (b) if the bar is redesigned by removing the material above the grooves as shown in Fig. P4.74b.

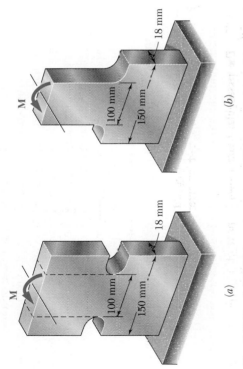

Fig. P4.73 and P4.74

*4.8. PLASTIC DEFORMATIONS

When we derived the fundamental relation $\sigma_x = -My/I$ in Sec. 4.4, we assumed that Hooke's law applied throughout the member. If the yield strength is exceeded in some portion of the member, or if the material involved is a brittle material with a nonlinear stress-strain diagram, this relation ceases to be valid. The purpose of this section is to develop a more general method for the determination of the distribution of stresses in a member in pure bending, which can be used when Hooke's law does not apply.

We first recall that no specific stress-strain relationship was assumed in Sec. 4.3, when we proved that the normal strain ϵ_x varies linearly with the distance y from the neutral surface. Thus, we can still use this property in our present analysis and write

$$\epsilon_x = -\frac{y}{c}\epsilon_m \tag{4.10}$$

where y represents the distance of the point considered from the neutral surface, and c the maximum value of y.

However, we cannot assume any more that, in a given section, the neutral axis passes through the centroid of that section, since this property was derived in Sec. 4.4 under the assumption of elastic deformations. In general, the neutral axis must be located by trial and error, until a distribution of stresses has been found, which satisfies Eqs. (4.1) and (4.3) of Sec. 4.2. However, in the particular case of a member possessing both a vertical and a horizontal plane of symmetry, and made of a material characterized by the same stress-strain relation in tension and in compression, the neutral axis will coincide with the horizontal axis of symmetry of the section. Indeed, the properties of the material require that the stresses be symmetric with respect to the neutral axis, i.e., with respect to *some* horizontal axis, and it is clear that this condition will be met, and Eq. (4.1) satisfied at the same time, only if that axis is the horizontal axis of symmetry itself.

Our analysis will first be limited to the special case we have just described. The distance y in Eq. (4.10) is thus measured from the horizontal axis of symmetry z of the cross section, and the distribution of strain ϵ_x is linear and symmetric with respect to that axis (Fig. 4.34). On the other hand, the stress-strain curve is symmetric with respect to the origin of coordinates (Fig. 4.35).

Fig. 4.34

Fig. 4.35

The distribution of stresses in the cross section of the member, i.e., the plot of σ_x versus y, is obtained as follows. Assuming that σ_{max} has been specified, we first determine the corresponding value of ϵ_m from the stress-strain diagram and carry this value into Eq. (4.10). Then, for each value of y, we determine the corresponding value of ϵ_x from Eq. (4.10) or Fig. 4.34, and obtain from the stress-strain diagram of Fig. 4.35 the stress σ_x corresponding to this value of ϵ_x. Plotting σ_x against y yields the desired distribution of stresses (Fig. 4.36).

We now recall that, when we derived Eq. (4.3) in Sec. 4.2, we assumed no particular relation between stress and strain. We can therefore use Eq. (4.3) to determine the bending moment M corresponding to the stress distribution obtained in Fig. 4.36. Considering the particular case of a member with a rectangular cross section of width b, we express the element of area in Eq. (4.3) as $dA = b\,dy$ and write

$$M = -b \int_{-c}^{c} y\sigma_x \, dy \tag{4.30}$$

where σ_x is the function of y plotted in Fig. 4.36. Since σ_x is an odd

Fig. 4.36

function of y, we can write Eq. (4.30) in the alternative form

$$M = -2b \int_0^c y\sigma_x \, dy \qquad (4.31)$$

If σ_x is a known analytical function of ϵ_x, Eq. (4.10) can be used to express σ_x as a function of y, and the integral in (4.31) can be determined analytically. Otherwise, the bending moment M can be obtained through a numerical integration. This computation becomes more meaningful if we note that the integral in Eq. (4.31) represents the first moment with respect to the horizontal axis of the area in Fig. 4.36 that is located above the horizontal axis and is bounded by the stress-distribution curve and the vertical axis.

An important value of the bending moment is the ultimate bending moment M_U that causes failure of the member. This value can be determined from the ultimate strength σ_U of the material by choosing $\sigma_{\max} = \sigma_U$ and carrying out the computations indicated earlier. However, it is found more convenient in practice to determine M_U experimentally for a specimen of a given material. Assuming a fictitious linear distribution of stresses, Eq. (4.15) is then used to determine the corresponding maximum stress R_B:

$$R_B = \frac{M_U c}{I} \qquad (4.32)$$

The fictitious stress R_B is called the *modulus of rupture in bending* of the given material. It can be used to determine the ultimate bending moment M_U of a member made of the same material and having a cross

Fig. 4.37

section of the same shape, but of different dimensions, by solving Eq. (4.32) for M_U. Since, in the case of a member with a rectangular cross section, the actual and the fictitious linear stress distributions shown in Fig. 4.37 must yield the same value M_U for the ultimate bending moment, the areas they define must have the same first moment with respect to the horizontal axis. It is thus clear that the modulus of rupture R_B will always be larger than the actual ultimate strength σ_U.

Fig. 4.38

*4.9. MEMBERS MADE OF AN ELASTOPLASTIC MATERIAL

In order to gain a better insight into the plastic behavior of a member in bending, let us consider the case of a member made of an *elastoplastic material* and first assume the member to have a *rectangular cross section* of width b and depth $2c$ (Fig. 4.38). We recall from Sec. 2.17 that the stress-strain diagram for an idealized elastoplastic material is as shown in Fig. 4.39.

Fig. 4.39

As long as the normal stress σ_x does not exceed the yield strength σ_Y, Hooke's law applies, and the stress distribution across the section is linear (Fig. 4.40a). The maximum value of the stress is

$$\sigma_m = \frac{Mc}{I} \tag{4.15}$$

As the bending moment increases, σ_m eventually reaches the value σ_Y (Fig. 4.40b). Substituting this value into Eq. (4.15), and solving for the corresponding value of M, we obtain the value M_Y of the bending moment at the onset of yield:

$$M_Y = \frac{I}{c}\sigma_Y \tag{4.33}$$

The moment M_Y is referred to as the *maximum elastic moment*, since it is the largest moment for which the deformation remains fully elastic. Recalling that, for the rectangular cross section considered here,

$$\frac{I}{c} = \frac{b(2c)^3}{12c} = \frac{2}{3}bc^2 \tag{4.34}$$

we write

$$M_Y = \frac{2}{3}bc^2\sigma_Y \tag{4.35}$$

As the bending moment further increases, plastic zones develop in the member, with the stress uniformly equal to $-\sigma_Y$ in the upper zone, and to $+\sigma_Y$ in the lower zone (Fig. 4.40c). Between the plastic zones, an elastic core subsists, in which the stress σ_x varies linearly with y,

$$\sigma_x = -\frac{\sigma_Y}{y_Y}y \tag{4.36}$$

where y_Y represents half the thickness of the elastic core. As M increases, the plastic zones expand until, at the limit, the deformation is fully plastic (Fig. 4.40d).

Equation (4.31) will be used to determine the value of the bending moment M corresponding to a given thickness $2y_Y$ of the elastic core.

Recalling that σ_x is given by Eq. (4.36) for $0 \leq y \leq y_Y$, and is equal to $-\sigma_Y$ for $y_Y \leq y \leq c$, we write

$$M = -2b \int_0^{y_Y} y \left(-\frac{\sigma_Y}{y_Y} y\right) dy - 2b \int_{y_Y}^c y(-\sigma_Y) \, dy$$

$$= \frac{2}{3} b y_Y^2 \sigma_Y + bc^2 \sigma_Y - b y_Y^2 \sigma_Y$$

or, in view of Eq. (4.35),

$$M = bc^2 \sigma_Y \left(1 - \frac{1}{3}\frac{y_Y^2}{c^2}\right) \tag{4.37}$$

$$M = \frac{3}{2}M_Y\left(1 - \frac{1}{3}\frac{y_Y^2}{c^2}\right) \tag{4.38}$$

where M_Y is the maximum elastic moment. Note that as y_Y approaches zero, the bending moment approaches the limiting value

$$M_p = \frac{3}{2}M_Y \tag{4.39}$$

This value of the bending moment, which corresponds to a fully plastic deformation (Fig. 4.40d), is called the *plastic moment* of the member considered. Note that Eq. (4.39) is valid only for a *rectangular member made of an elastoplastic material.*

You should keep in mind that the distribution of *strain* across the section remains linear after the onset of yield. Therefore, Eq. (4.8) of Sec. 4.3 remains valid and can be used to determine the half-thickness y_Y of the elastic core. We have

$$y_Y = \epsilon_Y \rho \tag{4.40}$$

where ϵ_Y is the yield strain and ρ the radius of curvature corresponding to a bending moment $M \geq M_Y$. When the bending moment is equal to M_Y, we have $y_Y = c$ and Eq. (4.40) yields

$$c = \epsilon_Y \rho_Y \tag{4.41}$$

where ρ_Y is the radius of curvature corresponding to the maximum elastic moment M_Y. Dividing (4.40) by (4.41) member by member, we obtain the relation[†]

$$\frac{y_Y}{c} = \frac{\rho}{\rho_Y} \tag{4.42}$$

Substituting for y_Y/c from (4.42) into Eq. (4.38), we express the bending moment M as a function of the radius of curvature ρ of the neutral surface:

$$M = \frac{3}{2}M_Y\left(1 - \frac{1}{3}\frac{\rho^2}{\rho_Y^2}\right) \tag{4.43}$$

Note that Eq. (4.43) is valid only after the onset of yield, i.e., for values of M larger than M_Y. For $M < M_Y$, Eq. (4.21) of Sec. 4.4 should be used.

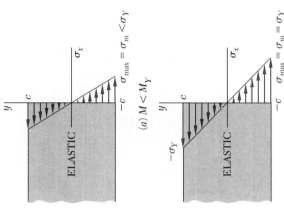
(a) $M < M_Y$
$\sigma_{max} = \sigma_m < \sigma_Y$

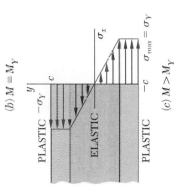
(b) $M = M_Y$
$\sigma_{max} = \sigma_m = \sigma_Y$

(c) $M > M_Y$
$\sigma_{max} = \sigma_Y$

(d) $M = M_p$

Fig. 4.40

[†]Equation (4.42) applies to any member made of any ductile material with a well-defined yield point, since its derivation is independent of the shape of the cross section and of the shape of the stress-strain diagram beyond the yield point.

We observe from Eq. (4.43) that the bending moment reaches the value $M_p = \frac{3}{2}M_Y$ only when $\rho = 0$. Since we clearly cannot have a zero radius of curvature at every point of the neutral surface, we conclude that a fully plastic deformation cannot develop in pure bending. As you will see in Chap. 5, however, such a situation may occur at one point in the case of a beam under a transverse loading.

The stress distributions in a rectangular member corresponding respectively to the maximum elastic moment M_Y and to the limiting case of the plastic moment M_p have been represented in three dimensions in Fig. 4.41. Since, in both cases, the resultants of the elementary tensile and compressive forces must pass through the centroids of the volumes representing the stress distributions and be equal in magnitude to these volumes, we check that

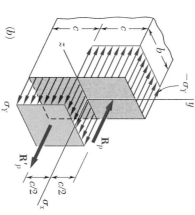

(a)

$$R_Y = \tfrac{1}{2}bc\sigma_Y$$

and

$$R_p = bc\sigma_Y$$

and that the moments of the corresponding couples are, respectively,

$$M_Y = (\tfrac{4}{3}c)R_Y = \tfrac{2}{3}bc^2\sigma_Y \qquad (4.44)$$

and

$$M_p = cR_p = bc^2\sigma_Y \qquad (4.45)$$

We thus verify that, for a rectangular member, $M_p = \frac{3}{2}M_Y$ as required by Eq. (4.39).

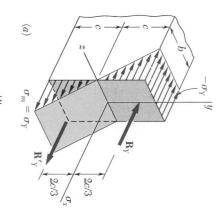

(b)

Fig. 4.41

For beams of *nonrectangular cross section*, the computation of the maximum elastic moment M_Y and of the plastic moment M_p will usually be simplified if a graphical method of analysis is used, as shown in Sample Prob. 4.5. It will be found in this more general case that the ratio $k = M_p/M_Y$ is generally not equal to $\frac{3}{2}$. For structural shapes such as wide-flange beams, for example, this ratio varies approximately from 1.08 to 1.14. Because it depends only upon the shape of the cross section, the ratio $k = M_p/M_Y$ is referred to as the *shape factor* of the cross section. We note that, if the shape factor k and the maximum elastic moment M_Y of a beam are known, the plastic moment M_p of the beam can be obtained by multiplying M_Y by k:

$$M_p = kM_Y \qquad (4.46)$$

The ratio M_p/σ_Y obtained by dividing the plastic moment M_p of a member by the yield strength σ_Y of its material is called the *plastic section modulus* of the member and is denoted by Z. When the plastic section modulus Z and the yield strength σ_Y of a beam are known, the plastic moment M_p of the beam can be obtained by multiplying σ_Y by Z:

$$M_p = Z\sigma_Y$$

Recalling from Eq. (4.18) that $M_Y = S\sigma_Y$, and comparing this relation with Eq. (4.47), we note that the shape factor $k = M_p/M_Y$ of a given

(4.47)

cross section can be expressed as the ratio of the plastic and elastic section moduli:

$$k = \frac{M_p}{M_Y} = \frac{Z\sigma_Y}{S\sigma_Y} = \frac{Z}{S} \qquad (4.48)$$

Considering the particular case of a rectangular beam of width b and depth h, we note from Eqs. (4.45) and (4.47) that the *plastic section modulus* of a rectangular beam is

$$Z = \frac{M_p}{\sigma_Y} = \frac{bc^2\sigma_Y}{\sigma_Y} = bc^2 = \tfrac{1}{4}bh^2$$

On the other hand, we recall from Eq. (4.19) of Sec. 4.4 that the *elastic section modulus* of the same beam is

$$S = \tfrac{1}{6}bh^2$$

Substituting into Eq. (4.48) the values obtained for Z and S, we verify that the shape factor of a rectangular beam is

$$k = \frac{Z}{S} = \frac{\tfrac{1}{4}bh^2}{\tfrac{1}{6}bh^2} = \frac{3}{2}$$

EXAMPLE 4.05

A member of uniform rectangular cross section 50 by 120 mm (Fig. 4.42) is subjected to a bending moment $M = 36.8$ kN · m. Assuming that the member is made of an elastoplastic material with a yield strength of 240 MPa and a modulus of elasticity of 200 GPa, determine (*a*) the thickness of the elastic core, (*b*) the radius of curvature of the neutral surface.

Fig. 4.42

(a) Thickness of Elastic Core. We first determine the maximum elastic moment M_Y. Substituting the given data into Eq. (4.34), we have

$$\frac{I}{c} = \frac{2}{3}bc^2 = \frac{2}{3}(50 \times 10^{-3}\text{ m})(60 \times 10^{-3}\text{ m})^2$$
$$= 120 \times 10^{-6}\text{ m}^3$$

and carrying this value, as well as $\sigma_Y = 240$ MPa, into Eq. (4.33),

$$M_Y = \frac{I}{c}\sigma_Y = (120 \times 10^{-6}\text{ m}^3)(240\text{ MPa}) = 28.8\text{ kN} \cdot \text{m}$$

Substituting the values of M and M_Y into Eq. (4.38), we have

$$36.8\text{ kN} \cdot \text{m} = \frac{3}{2}(28.8\text{ kN} \cdot \text{m})\left(1 - \frac{1}{3}\frac{y_Y^2}{c^2}\right)$$

$$\left(\frac{y_Y}{c}\right)^2 = 0.444 \qquad \frac{y_Y}{c} = 0.666$$

and, since $c = 60$ mm,

$$y_Y = 0.666(60\text{ mm}) = 40\text{ mm}$$

The thickness $2y_Y$ of the elastic core is thus 80 mm.

(b) Radius of Curvature. We note that the yield strain is

$$\epsilon_Y = \frac{\sigma_Y}{E} = \frac{240 \times 10^6\text{ Pa}}{200 \times 10^9\text{ Pa}} = 1.2 \times 10^{-3}$$

Solving Eq. (4.40) for ρ and substituting the values obtained for y_Y and ϵ_Y, we write

$$\rho = \frac{y_Y}{\epsilon_Y} = \frac{40 \times 10^{-3}\text{ m}}{1.2 \times 10^{-3}} = 33.3\text{ m}$$

*4.10. PLASTIC DEFORMATIONS OF MEMBERS WITH A SINGLE PLANE OF SYMMETRY

In our discussion of plastic deformations, we have assumed so far that the member in bending had two planes of symmetry, one containing the couples **M** and **M'**, and the other perpendicular to that plane. Let us now consider the more general case when the member possesses only one plane of symmetry containing the couples **M** and **M'**. However, our analysis will be limited to the situation where the deformation is fully plastic, with the normal stress uniformly equal to $-\sigma_Y$ above the neutral surface, and to $+\sigma_Y$ below that surface (Fig. 4.43a).

As indicated in Sec. 4.8, the neutral axis cannot be assumed to coincide with the centroidal axis of the cross section when the cross section is not symmetric with respect to that axis. To locate the neutral axis, we consider the resultant $\mathbf{R}_1$ of the elementary compressive forces exerted on the portion A_1 of the cross section located above the neutral axis, and the resultant $\mathbf{R}_2$ of the tensile forces exerted on the portion A_2 located below the neutral axis (Fig. 4.43b). Since the forces $\mathbf{R}_1$ and $\mathbf{R}_2$ form a couple equivalent to the couple applied to the member, they must have the same magnitude. We have therefore $R_1 = R_2$, or $A_1\sigma_Y = A_2\sigma_Y$, from which we conclude that $A_1 = A_2$. In other words, *the neutral axis divides the cross section into portions of equal areas.* Note that the axis obtained in this fashion will *not*, in general, be a centroidal axis of the section.

We also observe that the lines of action of the resultants $\mathbf{R}_1$ and $\mathbf{R}_2$ pass through the centroids C_1 and C_2 of the two portions we have just defined. Denoting by d the distance between C_1 and C_2, and by A the total area of the cross section, we express the plastic moment of the member as

$$M_p = \left(\tfrac{1}{2}A\sigma_Y\right)d$$

An example of the actual computation of the plastic moment of a member with only one plane of symmetry is given in Sample Prob. 4.6.

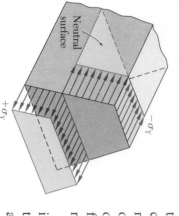

(a)

=

(b)

Fig. 4.43

*4.11. RESIDUAL STRESSES

We saw in the preceding sections that plastic zones will develop in a member made of an elastoplastic material if the bending moment is large enough. When the bending moment is decreased back to zero, the corresponding reduction in stress and strain at any given point can be represented by a straight line on the stress-strain diagram, as shown in Fig. 4.44. As you will see presently, the final value of the stress at a point will not, in general, be zero. There will be a residual stress at most points, and that stress may or may not have the same sign as the maximum stress reached at the end of the loading phase.

Since the linear relation between σ_x and ϵ_x applies at all points of the member during the unloading phase, Eq. (4.16) can be used to obtain the change in stress at any given point. In other words, the unloading phase can be handled by assuming the member to be fully elastic.

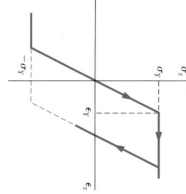

Fig. 4.44

The residual stresses are obtained by applying the principle of superposition in a manner similar to that described in Sec. 2.20 for an axial centric loading and used again in Sec. 3.11 for torsion. We consider, on one hand, the stresses due to the application of the given bending moment M and, on the other, the reverse stresses due to the equal and opposite bending moment $-M$ which is applied to unload the member. The first group of stresses reflect the *elastoplastic* behavior of the material during the loading phase, and the second group the *linear* behavior of the same material during the unloading phase. Adding the two groups of stresses, we obtain the distribution of residual stresses in the member.

EXAMPLE 4.06

For the member of Example 4.05, determine (*a*) the distribution of the residual stresses, (*b*) the radius of curvature, after the bending moment has been decreased from its maximum value of 36.8 kN · m back to zero.

(*a*) *Distribution of Residual Stresses.*   We recall from Example 4.05 that the yield strength is $\sigma_Y = 240$ MPa and that the thickness of the elastic core is $2y_Y = 80$ mm. The distribution of the stresses in the loaded member is thus as shown in Fig. 4.45*a*.

The distribution of the reverse stresses due to the opposite 36.8 kN · m bending moment required to unload the member is linear and as shown in Fig. 4.45*b*. The maximum stress σ'_m in that distribution is obtained from Eq. (4.15). Recalling from Example 4.05 that $I/c = 120 \times 10^{-6}$ m³, we write

$$\sigma'_m = \frac{Mc}{I} = \frac{36.8 \text{ kN} \cdot \text{m}}{120 \times 10^{-6} \text{ m}^3} = 306.7 \text{ MPa}$$

Superposing the two distributions of stresses, we obtain the residual stresses shown in Fig. 4.45*c*. We check that, even

though the reverse stresses exceed the yield strength σ_Y, the assumption of a linear distribution of the reverse stresses is valid, since they do not exceed $2\sigma_Y$.

(*b*) *Radius of Curvature After Unloading.*   We can apply Hooke's law at any point of the core $|y| < 40$ mm, since no plastic deformation has occurred in that portion of the member. Thus, the residual strain at the distance $y = 40$ mm is

$$\epsilon_x = \frac{\sigma_x}{E} = \frac{-35.5 \times 10^6 \text{ Pa}}{200 \times 10^9 \text{ Pa}} = -177.5 \times 10^{-6}$$

Solving Eq. (4.8) for ρ and substituting the appropriate values of y and ϵ_x, we write

$$\rho = -\frac{y}{\epsilon_x} = \frac{40 \times 10^{-3} \text{ m}}{177.5 \times 10^{-6}} = 225 \text{ m}$$

The value obtained for ρ after the load has been removed represents a permanent deformation of the member.

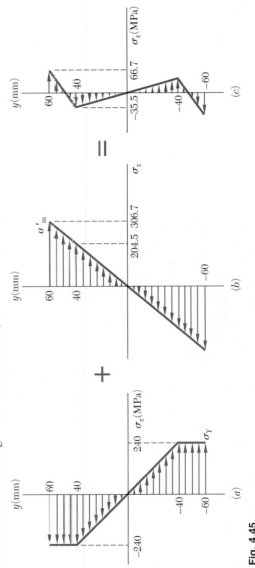

Fig. 4.45

SAMPLE PROBLEM 4.5

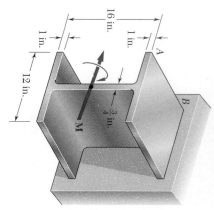

Beam AB has been fabricated from a high-strength low-alloy steel that is assumed to be elastoplastic with $E = 29 \times 10^6$ psi and $\sigma_Y = 50$ ksi. Neglecting the effect of fillets, determine the bending moment M and the corresponding radius of curvature (a) when yield first occurs, (b) when the flanges have just become fully plastic.

SOLUTION

a. Onset of Yield.
The centroidal moment of inertia of the section is

$$I = \frac{1}{12}(12 \text{ in.})(16 \text{ in.})^3 - \frac{1}{12}(12 \text{ in.})(14 \text{ in.})^3 = 1524 \text{ in}^4$$

Bending Moment. For $\sigma_{max} = \sigma_Y = 50$ ksi and $c = 8$ in., we have

$$M_Y = \frac{\sigma_Y I}{c} = \frac{(50 \text{ ksi})(1524 \text{ in}^4)}{8 \text{ in.}} \qquad M_Y = 9525 \text{ kip} \cdot \text{in.} \;\blacktriangleleft$$

Radius of Curvature. Noting that, at $c = 8$ in., the strain is $\epsilon_Y = \sigma_Y/E = (50 \text{ ksi})/(29 \times 10^6 \text{ psi}) = 0.001724$, we have from Eq. (4.41)

$$c = \epsilon_Y \rho_Y \qquad 8 \text{ in.} = 0.001724 \rho_Y \qquad \rho_Y = 4640 \text{ in.} \;\blacktriangleleft$$

$\sigma_Y = 50$ ksi

Strain distribution / Stress distribution

$\epsilon_Y = 0.001724$

b. Flanges Fully Plastic.
When the flanges have just become fully plastic, the strains and stresses in the section are as shown in the figure below.
We replace the elementary compressive forces exerted on the top flange and on the top half of the web by their resultants R_1 and R_2, and similarly replace the tensile forces by R_3 and R_4.

$$R_1 = R_4 = (50 \text{ ksi})(12 \text{ in.})(1 \text{ in.}) = 600 \text{ kips}$$
$$R_2 = R_3 = \tfrac{1}{2}(50 \text{ ksi})(7 \text{ in.})(0.75 \text{ in.}) = 131.3 \text{ kips}$$

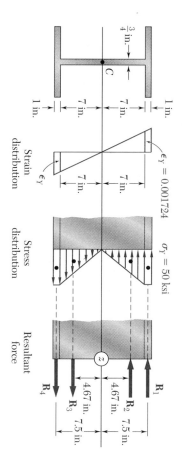

Strain distribution / Stress distribution / Resultant force

$\epsilon_Y = 0.001724$ $\sigma_Y = 50$ ksi

Bending Moment. Summing the moments of R_1, R_2, R_3, and R_4 about the z axis, we write

$$M = 2[R_1(7.5 \text{ in.}) + R_2(4.67 \text{ in.})]$$
$$= 2[(600)(7.5) + (131.3)(4.67)] \qquad M = 10{,}230 \text{ kip} \cdot \text{in.} \;\blacktriangleleft$$

Radius of Curvature. Since $y_Y = 7$ in. for this loading, we have from Eq. (4.40)

$$y_Y = \epsilon_Y \rho \qquad 7 \text{ in.} = (0.001724)\rho \qquad \rho = 4060 \text{ in.} = 338 \text{ ft.} \;\blacktriangleleft$$

SAMPLE PROBLEM 4.6

Determine the plastic moment M_p of a beam with the cross section shown when the beam is bent about a horizontal axis. Assume that the material is elasto-plastic with a yield strength of 240 MPa.

SOLUTION

Neutral Axis. When the deformation is fully plastic, the neutral axis divides the cross section into two portions of equal areas. Since the total area is

$$A = (100)(20) + (80)(20) + (60)(20) = 4800 \text{ mm}^2$$

the area located above the neutral axis must be 2400 mm². We write

$$(20)(100) + 20y = 2400 \qquad y = 20 \text{ mm}$$

Note that the neutral axis does *not* pass through the centroid of the cross section.

Plastic Moment. The resultant $\mathbf{R}_i$ of the elementary forces exerted on the partial area A_i is equal to

$$R_i = A_i \sigma_Y$$

and passes through the centroid of that area. We have

$$R_1 = A_1 \sigma_Y = [(0.100 \text{ m})(0.020 \text{ m})]240 \text{ MPa} = 480 \text{ kN}$$
$$R_2 = A_2 \sigma_Y = [(0.020 \text{ m})(0.020 \text{ m})]240 \text{ MPa} = 96 \text{ kN}$$
$$R_3 = A_3 \sigma_Y = [(0.020 \text{ m})(0.060 \text{ m})]240 \text{ MPa} = 288 \text{ kN}$$
$$R_4 = A_4 \sigma_Y = [(0.060 \text{ m})(0.020 \text{ m})]240 \text{ MPa} = 288 \text{ kN}$$

The plastic moment M_p is obtained by summing the moments of the forces about the z axis.

$$
\begin{aligned}
M_p &= (0.030 \text{ m})R_1 + (0.010 \text{ m})R_2 + (0.030 \text{ m})R_3 + (0.070 \text{ m})R_4 \\
&= (0.030 \text{ m})(480 \text{ kN}) + (0.010 \text{ m})(96 \text{ kN}) \\
&\quad + (0.030 \text{ m})(288 \text{ kN}) + (0.070 \text{ m})(288 \text{ kN}) \\
&= 44.16 \text{ kN} \cdot \text{m}
\end{aligned}
$$

$$M_p = 44.2 \text{ kN} \cdot \text{m} \quad \blacktriangledown$$

Note: Since the cross section is *not* symmetric about the z axis, the sum of the moments of $\mathbf{R}_1$ and $\mathbf{R}_2$ is *not* equal to the sum of the moments of $\mathbf{R}_3$ and $\mathbf{R}_4$.

SAMPLE PROBLEM 4.7

For the beam of Sample Prob. 4.5, determine the residual stresses and the permanent radius of curvature after the 10,230 kip · in. couple M has been removed.

SOLUTION

Loading. In Sample Prob. 4.5 a couple of moment $M = 10,230$ kip · in. was applied and the stresses shown in Fig. 1 were obtained.

Elastic Unloading. The beam is unloaded by the application of a couple of moment $M = -10,230$ kip · in. (which is equal and opposite to the couple originally applied). During this unloading, the action of the beam is fully elastic; recalling from Sample Prob. 4.5 that $I = 1524$ in⁴, we compute the maximum stress

$$\sigma'_m = \frac{Mc}{I} = \frac{(10,230 \text{ kip} \cdot \text{in.})(8 \text{ in.})}{1524 \text{ in}^4} = 53.70 \text{ ksi}$$

The stresses caused by the unloading are shown in Fig. 2.

Residual Stresses. We superpose the stresses due to the loading (Fig. 1) and to the unloading (Fig. 2) and obtain the residual stresses in the beam (Fig. 3).

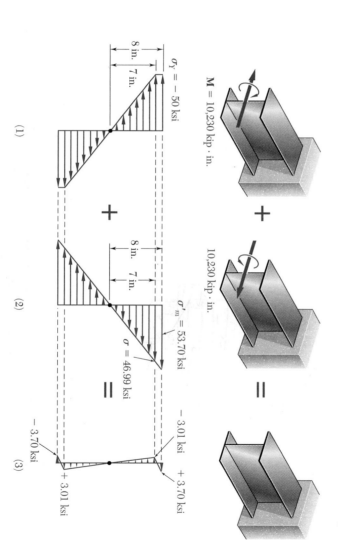

$M = 10,230$ kip · in.

$\sigma_Y = -50$ ksi

8 in.
7 in.

(1)

$+$

10,230 kip · in.

$\sigma'_m = 53.70$ ksi

8 in.
7 in.

$\sigma = 46.99$ ksi

(2)

$=$

-3.01 ksi -3.70 ksi

$+3.01$ ksi $+3.70$ ksi

-3.70 ksi

(3)

Permanent Radius of Curvature. At $y = 7$ in. the residual stress is $\sigma = -3.01$ ksi. Since no plastic deformation occurred at this point, Hooke's law can be used and we have $\epsilon_x = \sigma/E$. Recalling Eq. (4.8), we write

$$\rho = -\frac{y}{\epsilon_x} = -\frac{yE}{\sigma} = -\frac{(7 \text{ in.})(29 \times 10^6 \text{ psi})}{-3.01 \text{ ksi}} = +67,400 \text{ in.} \quad \rho = 5620 \text{ ft} \blacktriangleleft$$

We note that the residual stress is tensile on the upper face of the beam and compressive on the lower face, even though the beam is concave upward.

ρ

$\sigma = +3.70$ ksi (tension)

$\sigma = -3.70$ ksi (compression)

PROBLEMS

4.75 A bar of rectangular cross section, made of a steel assumed to be elastoplastic with $\sigma_Y = 320$ MPa, is subjected to a couple **M** parallel to the z axis. Determine the moment M of the couple for which (a) yield first occurs, (b) the plastic zones at the top and bottom of the bar are 5 mm thick.

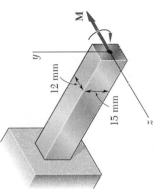

Fig. P4.75

4.76 Solve Prob. 4.75, assuming that the couple **M** is parallel to the y axis.

4.77 The prismatic bar shown, made of a steel assumed to be elastoplastic with $\sigma_Y = 42$ ksi, is subjected to a couple **M** parallel to the x axis. Determine the moment M of the couple for which (a) yield first occurs, (b) the elastic core of the bar is 0.1 in. thick.

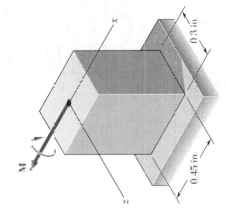

Fig. P4.77

4.78 Solve Prob. 4.77, assuming that the couple **M** is parallel to the z axis.

4.79 A solid square rod of side 0.75 in. is made of a steel that is assumed to be elastoplastic with $E = 29 \times 10^6$ psi and $\sigma_Y = 40$ ksi. Determine the maximum stress and the radius of curvature caused by a 4 kip · in. couple applied and maintained about an axis parallel to a side of the cross section.

4.80 The prismatic rod shown is made of a steel that is assumed to be elastoplastic with $E = 200$ GPa and $\sigma_Y = 280$ MPa. Knowing that couples **M** and **M′** of moment 525 N · m are applied and maintained about axes parallel to the y axis, determine (a) the thickness of the elastic core, (b) the radius of curvature of the bar.

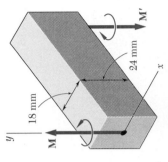

Fig. P4.80

4.81 Solve Prob. 4.80, assuming that the couples **M** and **M′** are applied and maintained about axes parallel to the x axis.

4.82 A solid square rod of side 0.5 in. is made of a steel that is assumed to be elastoplastic with $E = 29 \times 10^6$ psi and $\sigma_Y = 42$ ksi. Knowing that a couple **M** is applied and maintained about an axis parallel to a side of the cross section, determine the moment M of the couple for which the radius of curvature is (*a*) 5 ft, (*b*) 2 ft.

4.83 and 4.84 A bar of the cross section shown is made of a steel that is assumed to be elastoplastic with $E = 200$ GPa and $\sigma_Y = 240$ MPa. For bending about the *z* axis, determine the bending moment at which (*a*) yield first occurs, (*b*) the plastic zones at the top and bottom of the bar are 20 mm thick.

Fig. P4.83

Fig. P4.84

4.85 and 4.86 A bar of the cross section shown is made of a steel that is assumed to be elastoplastic with $E = 29 \times 10^6$ psi and $\sigma_Y = 42$ ksi. For bending about the *z* axis, determine the bending moment at which (*a*) yield first occurs, (*b*) the plastic zones at the top and bottom of the bar are 3 in. thick.

Fig. P4.85

Fig. P4.86

4.87 through 4.90 For the bar indicated, determine (*a*) the fully plastic moment M_p, (*b*) the shape factor of the cross section.

> **4.87** Bar of Prob. 4.83.
> **4.88** Bar of Prob. 4.84.
> **4.89** Bar of Prob. 4.85.
> **4.90** Bar of Prob. 4.86.

4.91 and 4.92 Determine the plastic moment M_p of a steel beam of the cross section shown, assuming the steel to be elastoplastic with a yield strength of 240 MPa.

Fig. P4.92

Fig. P4.91

4.93 and 4.94 Determine the plastic moment M_p of a steel beam of the cross section shown, assuming the steel to be elastoplastic with a yield strength of 36 ksi.

Fig. P4.94

Fig. P4.93

4.95 A thick-walled pipe of the cross section shown is made of a steel that is assumed to be elastoplastic with a yield strength σ_Y. Derive an expression for the plastic moment M_p of the pipe in terms of c_1, c_2, and σ_Y.

4.96 Determine the plastic moment M_p of a thick-walled pipe of the cross section shown, knowing that $c_1 = 60$ mm, $c_2 = 40$ mm, and $\sigma_Y = 240$ MPa.

Fig. P4.95 and P4.96

4.97 and 4.98 For the beam indicated a couple of moment equal to the fully plastic moment M_p is applied and then removed. Using a yield strength of 240 MPa, determine the residual stress at $y = 30$ mm.

    4.97 Beam of Prob. 4.83.
    4.98 Beam of Prob. 4.84.

4.99 and 4.100 For the beam indicated, a couple of moment equal to the fully plastic moment M_p is applied and then removed. Using a yield strength of 42 ksi, determine the residual stress at $y = 4.5$ in.

    4.99 Beam of Prob. 4.85.
    4.100 Beam of Prob. 4.86.

4.101 and 4.102 A bending couple is applied to the bar indicated, causing plastic zones 20 mm thick to develop at the top and bottom of the bar. After the couple has been removed, determine (a) the residual stress at $y = 30$ mm, (b) the points where the residual stress is zero, (c) the radius of curvature corresponding to the permanent deformation of the bar.

4.101 Bar of Prob. 4.83.
4.102 Bar of Prob. 4.84.

4.103 and 4.104 A bending couple is applied to the bar indicated, causing plastic zones 3 in. thick to develop at the top and bottom of the bar. After the couple has been removed, determine (a) the residual stress at $y = 4.5$ in., (b) the points where the residual stress is zero, (c) the radius of curvature corresponding to the permanent deformation of the bar.

4.103 Bar of Prob. 4.85.
4.104 Bar of Prob. 4.86.

***4.105** A rectangular bar that is straight and unstressed is bent into an arc of circle of radius ρ by two couples of moment M. After the couples are removed, it is observed that the radius of curvature of the bar is ρ_R. Denoting by ρ_Y the radius of curvature of the bar at the onset of yield, show that the radii of curvature satisfy the following relation:

$$\frac{1}{\rho_R} = \frac{1}{\rho}\left\{1 - \frac{3}{2}\frac{\rho}{\rho_Y}\left[1 - \frac{1}{3}\left(\frac{\rho}{\rho_Y}\right)^2\right]\right\}$$

4.106 A solid bar of rectangular cross section is made of a material that is assumed to be elastoplastic. Denoting by M_Y and ρ_Y, respectively, the bending moment and radius of curvature at the onset of yield, determine (a) the radius of curvature when a couple of moment $M = 1.25M_Y$ is applied to the bar, (b) the radius of curvature after the couple is removed. Check the results obtained by using the relation derived in Prob. 4.105.

4.107 Solve Prob. 4.106, assuming that the moment of the couple applied to the bar is $1.40M_Y$.

4.108 The prismatic bar shown is made of a steel that is assumed to be elastoplastic and for which $E = 200$ GPa. Knowing that the radius of curvature of the bar is 2.4 m when a couple of moment $M = 420$ N · m is applied as shown, determine (a) the yield strength σ_Y of the steel, (b) the thickness of the elastic core of the bar.

16 mm
24 mm
M

Fig. P4.108

4.109 The prismatic bar AB is made of an aluminum alloy for which the tensile stress-strain diagram is as shown. Assuming that the σ-ϵ diagram is the same in compression as in tension, determine (a) the radius of curvature of the bar when the maximum stress is 250 MPa, (b) the corresponding value of the bending moment. (*Hint:* For part b, plot σ versus y and use an approximate method of integration.)

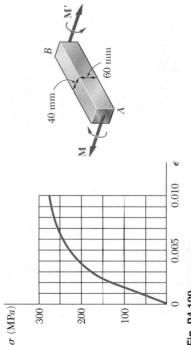

Fig. P4.109

4.110 For the bar of Prob. 4.109, determine (a) the maximum stress when the radius of curvature of the bar is 3 m, (b) the corresponding value of the bending moment. (See hint given in Prob. 4.109.)

4.111 The prismatic bar AB is made of a bronze alloy for which the tensile stress-strain diagram is as shown. Assuming that the σ-ϵ diagram is the same in compression as in tension, determine (a) the maximum stress in the bar when the radius of curvature of the bar is 100 in., (b) the corresponding value of the bending moment. (See hint given in Prob. 4.109.)

4.112 For the bar of Prob. 4.111, determine (a) the radius of curvature of the bar when the maximum stress is 45 ksi, (b) the corresponding value of the bending moment. (See hint given in Prob. 4.109.)

4.113 A prismatic bar of rectangular cross section is made of an alloy for which the stress-strain diagram can be represented by the relation $\epsilon = k\sigma^n$ for $\sigma > 0$ and $\epsilon = -|k\sigma^n|$ for $\sigma < 0$. If a couple M is applied to the bar, show that the maximum stress is

$$\sigma_m = \frac{1+2n}{3n}\frac{Mc}{I}$$

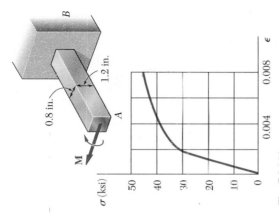

Fig. P4.111

Fig. P4.113 and P4.114

4.114 A prismatic bar of rectangular cross section is made of an alloy for which the stress-strain diagram can be represented by the relation $\epsilon = k\sigma^3$. If a couple M is applied to the bar, show that the maximum stress is

$$\sigma_m = \frac{7}{9}\frac{Mc}{I}$$

4.12. ECCENTRIC AXIAL LOADING IN A PLANE OF SYMMETRY

We saw in Sec. 1.5 that the distribution of stresses in the cross section of a member under axial loading can be assumed uniform only if the line of action of the loads $\mathbf{P}$ and $\mathbf{P'}$ passes through the centroid of the cross section. Such a loading is said to be *centric*. Let us now analyze the distribution of stresses when the line of action of the loads does *not* pass through the centroid of the cross section, i.e., when the loading is *eccentric*.

Two examples of an eccentric loading are shown in Figs. 4.46 and 4.47. In the case of the highway light, the weight of the lamp causes an eccentric loading on the post. Likewise, the vertical forces exerted on the press cause an eccentric loading on the back column of the press.

Fig. 4.46   **Fig. 4.47**

In this section, our analysis will be limited to members which possess a plane of symmetry, and it will be assumed that the loads are applied in the plane of symmetry of the member (Fig. 4.48a). The internal forces acting on a given cross section may then be represented by a force $\mathbf{F}$ applied at the centroid C of the section and a couple $\mathbf{M}$ acting in the plane of symmetry of the member (Fig. 4.48b). The conditions of equilibrium of the free body AC require that the force $\mathbf{F}$ be equal and opposite to $\mathbf{P'}$ and that the moment of the couple $\mathbf{M}$ be equal and opposite to the moment of $\mathbf{P'}$ about C. Denoting by d the distance from the centroid C to the line of action AB of the forces $\mathbf{P}$ and $\mathbf{P'}$, we have

$$F = P \qquad \text{and} \qquad M = Pd \qquad (4.49)$$

We now observe that the internal forces in the section would have been represented by the same force and couple if the straight portion DE of member AB had been detached from AB and subjected simultaneously to the centric loads $\mathbf{P}$ and $\mathbf{P'}$ and to the bending couples $\mathbf{M}$ and $\mathbf{M'}$ (Fig. 4.49). Thus, the stress distribution due to the original eccentric loading can be obtained by superposing the uniform stress distri-

Fig. 4.48

Fig. 4.49

bution corresponding to the centric loads **P** and **P'** and the linear distribution corresponding to the bending couples **M** and **M'** (Fig. 4.50). We write

$$\sigma_x = (\sigma_x)_{\text{centric}} + (\sigma_x)_{\text{bending}}$$

Fig. 4.50

or, recalling Eqs. (1.5) and (4.16):

$$\sigma_x = \frac{P}{A} - \frac{My}{I} \qquad (4.50)$$

where A is the area of the cross section and I its centroidal moment of inertia, and where y is measured from the centroidal axis of the cross section. The relation obtained shows that the distribution of stresses across the section is *linear but not uniform*. Depending upon the geometry of the cross section and the eccentricity of the load, the combined stresses may all have the same sign, as shown in Fig. 4.50, or some may be positive and others negative, as shown in Fig. 4.51. In the latter case, there will be a line in the section, along which $\sigma_x = 0$. This line represents the *neutral axis of the section*. We note that the neutral axis does *not* coincide with the centroidal axis of the section, since $\sigma_x \neq 0$ for $y = 0$.

Fig. 4.51

The results obtained are valid only to the extent that the conditions of applicability of the superposition principle (Sec. 2.12) and of Saint-Venant's principle (Sec. 2.17) are met. This means that the stresses involved must not exceed the proportional limit of the material, that the deformations due to bending must not appreciably affect the distance d in Fig. 4.48a, and that the cross section where the stresses are computed must not be too close to points D or E in the same figure. The first of these requirements clearly shows that the superposition method cannot be applied to plastic deformations.

EXAMPLE 4.07

An open-link chain is obtained by bending low-carbon steel rods of 0.5-in. diameter into the shape shown (Fig. 4.52). Knowing that the chain carries a load of 160 lb, determine (a) the largest tensile and compressive stresses in the straight portion of a link, (b) the distance between the centroidal and the neutral axis of a cross section.

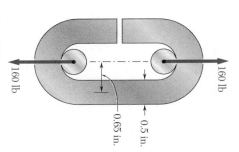

Fig. 4.52

(a) Largest Tensile and Compressive Stresses.
The internal forces in the cross section are equivalent to a centric force **P** and a bending couple **M** (Fig. 4.53) of magnitudes

$$P = 160 \text{ lb}$$
$$M = Pd = (160 \text{ lb})(0.65 \text{ in.}) = 104 \text{ lb} \cdot \text{in.}$$

$d = 0.65$ in.

Fig. 4.53

The corresponding stress distributions are shown in parts a and b of Fig. 4.54. The distribution due to the centric force **P** is uniform and equal to $\sigma_0 = P/A$. We have

$$A = \pi c^2 = \pi(0.25 \text{ in.})^2 = 0.1963 \text{ in}^2$$
$$\sigma_0 = \frac{P}{A} = \frac{160 \text{ lb}}{0.1963 \text{ in}^2} = 815 \text{ psi}$$

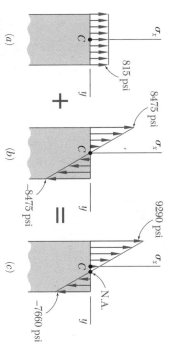

(a)　　(b)　　(c)

Fig. 4.54

The distribution due to the bending couple **M** is linear with a maximum stress $\sigma_m = Mc/I$. We write

$$I = \tfrac{1}{4}\pi c^4 = \tfrac{1}{4}\pi(0.25 \text{ in.})^4 = 3.068 \times 10^{-3} \text{ in}^4$$
$$\sigma_m = \frac{Mc}{I} = \frac{(104 \text{ lb} \cdot \text{in.})(0.25 \text{ in.})}{3.068 \times 10^{-3} \text{ in}^4} = 8475 \text{ psi}$$

Superposing the two distributions, we obtain the stress distribution corresponding to the given eccentric loading (Fig. 4.54c). The largest tensile and compressive stresses in the section are found to be respectively

$$\sigma_t = \sigma_0 + \sigma_m = 815 + 8475 = 9290 \text{ psi}$$
$$\sigma_c = \sigma_0 - \sigma_m = 815 - 8475 = -7660 \text{ psi}$$

(b) Distance Between Centroidal and Neutral Axes.
The distance y_0 from the centroidal to the neutral axis of the section is obtained by setting $\sigma_x = 0$ in Eq. (4.50) and solving for y_0:

$$0 = \frac{P}{A} - \frac{My_0}{I}$$
$$y_0 = \left(\frac{P}{A}\right)\left(\frac{I}{M}\right) = (815 \text{ psi})\frac{3.068 \times 10^{-3} \text{ in}^4}{104 \text{ lb} \cdot \text{in.}}$$
$$y_0 = 0.0240 \text{ in.}$$

SAMPLE PROBLEM 4.8

Knowing that for the cast iron link shown the allowable stresses are 30 MPa in tension and 120 MPa in compression, determine the largest force **P** which can be applied to the link. (*Note:* The T-shaped cross section of the link has previously been considered in Sample Prob. 4.2.)

SOLUTION

Properties of Cross Section. From Sample Prob. 4.2, we have

$$A = 3000 \text{ mm}^2 = 3 \times 10^{-3} \text{ m}^2 \qquad \bar{Y} = 38 \text{ mm} = 0.038 \text{ m}$$
$$I = 868 \times 10^{-9} \text{ m}^4$$

We now write: $\qquad d = (0.038 \text{ m}) - (0.010 \text{ m}) = 0.028 \text{ m}$

Section *a–a*

Force and Couple at C. We replace **P** by an equivalent force-couple system at the centroid C.

$$P = P \qquad M = P(d) = P(0.028 \text{ m}) = 0.028\,P$$

The force **P** acting at the centroid causes a uniform stress distribution (Fig. 1). The bending couple **M** causes a linear stress distribution (Fig. 2).

$$\sigma_0 = \frac{P}{A} = \frac{P}{3 \times 10^{-3}} = 333P \qquad \text{(Compression)}$$

$$\sigma_1 = \frac{Mc_A}{I} = \frac{(0.028P)(0.022)}{868 \times 10^{-9}} = 710P \qquad \text{(Tension)}$$

$$\sigma_2 = \frac{Mc_B}{I} = \frac{(0.028P)(0.038)}{868 \times 10^{-9}} = 1226P \qquad \text{(Compression)}$$

Superposition. The total stress distribution (Fig. 3) is found by superposing the stress distributions caused by the centric force **P** and by the couple **M**. Since tension is positive, and compression negative, we have

$$\sigma_A = -\frac{P}{A} + \frac{Mc_A}{I} = -333P + 710P = +377P \qquad \text{(Tension)}$$

$$\sigma_B = -\frac{P}{A} - \frac{Mc_B}{I} = -333P - 1226P = -1559P \qquad \text{(Compression)}$$

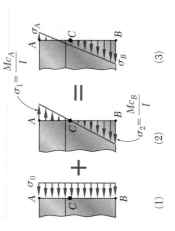

Largest Allowable Force. The magnitude of **P** for which the tensile stress at point A is equal to the allowable tensile stress of 30 MPa is found by writing

$$\sigma_A = 377P = 30 \text{ MPa} \qquad P = 79.6 \text{ kN} \quad \triangledown$$

We also determine the magnitude of **P** for which the stress at B is equal to the allowable compressive stress of 120 MPa.

$$\sigma_B = -1559P = -120 \text{ MPa} \qquad P = 77.0 \text{ kN} \quad \triangledown$$

The magnitude of the largest force **P** that can be applied without exceeding either of the allowable stresses is the smaller of the two values we have found.

$$P = 77.0 \text{ kN} \quad \blacktriangledown$$

PROBLEMS

4.115 Determine the stress at points A and B, (a) for the loading shown, (b) if the 60-kN loads are applied at points 1 and 2 only.

Fig. P4.115 and P4.116

4.116 Determine the stress at points A and B, (a) for the loading shown, (b) if the 60-kN loads applied at points 2 and 3 are removed.

4.117 Knowing that the magnitude of the horizontal force **P** is 8 kN, determine the stress at (a) point A, (b) point B.

Fig. P4.117

4.118 The vertical portion of the press shown consists of a rectangular tube having a wall thickness $t = \frac{1}{2}$ in. Knowing that the press has been tightened on wooden planks being glued together until $P = 6$ kips, determine the stress (a) at point A, (b) at point B.

Fig. P4.118

4.119 Solve Prob. 4.118, assuming that $t = \frac{3}{8}$ in.

4.120 As many as three axial loads, each of magnitude $P = 50$ kN, can be applied to the end of a W200 × 31.1 rolled-steel shape. Determine the stress at point A, (a) for the loading shown, (b) if loads are applied at points 1 and 2 only.

Fig. P4.120 and P4.121

4.121 As many as three axial loads, each of magnitude $P = 50$ kN, can be applied to the end of a W200 × 31.1 rolled-steel shape. Determine the stress at point A, (a) for the loading shown, (b) if loads are applied at points 2 and 3 only.

4.122 An offset h must be introduced into a solid circular rod of diameter d. Knowing that the maximum stress after the offset is introduced must not exceed 4 times the stress in the rod when it was straight, determine the largest offset that can be used.

Fig. P4.122

4.123 An offset h must be introduced into a metal tube of 18-mm outer diameter and 2-mm wall thickness. Knowing that the maximum stress after the offset is introduced must not exceed 4 times the stress in the tube when it was straight, determine the largest offset that can be used.

4.124 A short column is made by nailing two 1 × 4-in. planks to a 2 × 4-in. timber. Determine the largest compressive stress created in the column by a 12-kip load applied as shown at the center of the top section of the timber if (a) the column is as described, (b) plank 1 is removed, (c) both planks are removed.

Fig. P4.124

4.125 The two forces shown are applied to a rigid plate supported by a steel pipe of 140-mm outer diameter and 120-mm inner diameter. Knowing that the allowable compressive stress is 100 MPa, determine the range of allowable values of *P*.

Fig. P4.125 and P4.126

4.126 The two forces shown are applied to a rigid plate supported by a steel pipe of 140-mm outer diameter and 120-mm inner diameter. Determine the range of allowable values of *P* for which all stresses in the pipe are compressive and less than 100 MPa.

4.127 A milling operation was used to remove a portion of a solid bar of square cross section. Knowing that *a* = 1.2 in., *d* = 0.8 in., and σ_{all} = 8 ksi, determine the magnitude *P* of the largest forces that can be safely applied at the centers of the ends of the bar.

4.128 A milling operation was used to remove a portion of a solid bar of square cross section. Forces of magnitude *P* = 4 kips are applied at the centers of the ends of the bar. Knowing that *a* = 1.2 in. and σ_{all} = 8 ksi, determine the smallest allowable depth *d* of the milled portion of the rod.

Fig. P4.127 and *P4.128*

4.129 Three steel plates, each of 1 × 6-in. cross section, are welded together to form a short H-shaped column. Later, for architectural reasons, a 1-in. strip is removed from each side of one of the flanges. Knowing that the load remains centric with respect to the original cross section, and that the allowable stress is 15 ksi, determine the largest force **P**, (*a*) that could be applied to the original column, (*b*) that can be applied to the modified column.

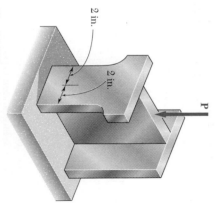

Fig. P4.129

4.130 A steel rod is welded to a steel plate to form the machine element shown. Knowing that the allowable stress is 135 MPa, determine (*a*) the largest force **P** that can be applied to the element, (*b*) the corresponding location of the neutral axis. *Given:* Centroid of the cross section is at *C* and $I_z = 4195$ mm^4.

Fig. P4.130

4.131 Knowing that the allowable stress is 150 MPa in section *a-a* of the hanger shown, determine (*a*) the largest vertical force **P** that can be applied at point *A*, (*b*) the corresponding location of the neutral axis of section *a-a*.

Fig. P4.131

4.132 Solve Prob. 4.131, assuming that the vertical force **P** is applied at point *B*.

4.133 Knowing that the clamp shown has been tightened on wooden planks being glued together until *P* = 75 lb, determine in section *a-a* (*a*) the stress at point *A*, (*b*) the stress at point *D*, (*c*) the location of the neutral axis.

Fig. P4.133

4.134   A vertical force **P** of magnitude 20 kips is applied at a point C located on the axis of symmetry of the cross section of a short column. Knowing that $y = 5$ in., determine (a) the stress at point A, (b) the stress at point B, (c) the location of the neutral axis.

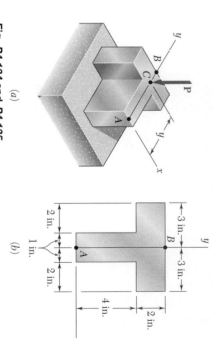

(a)

(b)

Fig. P4.134 and P4.135

4.135   A vertical force **P** is applied at a point C located on the axis of symmetry of the cross section of a short column. Determine the range of values of y for which tensile stresses do not occur in the column.

4.136   The four bars shown have the same cross-sectional area. For the given loadings, show that (a) the maximum compressive stresses are in the ratio 4:5:7:9, (b) the maximum tensile stresses are in the ratio 2:3:5:3. (*Note:* The cross section of the triangular bar is an equilateral triangle.)

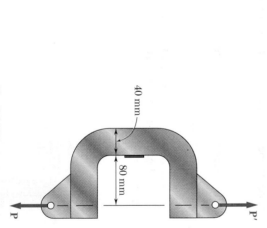

Fig. P4.136

4.137   The C-shaped steel bar is used as a dynamometer to determine the magnitude P of the forces shown. Knowing that the cross section of the bar is a square of side 40 mm and that strain on the inner edge was measured and found to be 450 μ, determine the magnitude P of the forces. Use $E = 200$ GPa.

Fig. P4.137

4.138 A short length of a rolled-steel column supports a rigid plate on which two loads **P** and **Q** are applied as shown. The strains at two points *A* and *B* on the center lines of the outer faces of the flanges have been measured and found to be

$$\epsilon_A = -400 \times 10^{-6} \text{ in./in.} \qquad \epsilon_B = -300 \times 10^{-6} \text{ in./in.}$$

Knowing that $E = 29 \times 10^6$ psi, determine the magnitude of each load.

$A = 10.0 \text{ in}^2$
$I_z = 273 \text{ in}^4$

Fig. P4.138

4.139 Solve Prob. 4.138, assuming that the measured strains are

$$\epsilon_A = 350 \times 10^{-6} \text{ in./in.} \qquad \boldsymbol{\epsilon}_B = 50 \times 10^{-6} \text{ in./in.}$$

4.140 An eccentric axial force **P** is applied as shown to a steel bar of 25×90 mm cross section. The strains at *A* and *B* have been measured and found to be

$$\epsilon_A = +350 \, \mu \qquad \epsilon_B = -70 \, \mu$$

Knowing that $E = 200$ GPa, determine (*a*) the distance *d*, (*b*) the magnitude of the force **P**.

Fig. P4.140

4.141 Solve Prob. 4.140, assuming that the measured strains are

$$\epsilon_A = +600 \, \mu \qquad \epsilon_B = +420 \, \mu$$

4.142 The shape shown was formed by bending a thin steel plate. Assuming that the thickness *t* is small compared to the length *a* of a side of the shape, determine the stress (*a*) at *A*, (*b*) at *B*, (*c*) at *C*.

Fig. P4.142

4.143   The eccentric axial force **P** acts at point D, which must be located 25 mm below the top surface of the steel bar shown. For $P = 60$ kN, determine (a) the depth d of the bar for which the tensile stress at point A is maximum, (b) the corresponding stress at point A.

Fig. P4.143

4.144   For the bar and loading of Prob. 4.143, determine (a) the depth d of the bar for which the compressive stress at point B is maximum, (b) the corresponding stress at point B.

4.13. UNSYMMETRIC BENDING

Our analysis of pure bending has been limited so far to members possessing at least one plane of symmetry and subjected to couples acting in that plane. Because of the symmetry of such members and of their loadings, we concluded that the members would remain symmetric with respect to the plane of the couples and thus bend in that plane (Sec. 4.3). This is illustrated in Fig. 4.55; part a shows the cross section of a member possessing two planes of symmetry, one vertical and one horizontal, and part b the cross section of a member with a single, vertical plane of symmetry. In both cases the couple exerted on the section acts in the vertical plane of symmetry of the member and is represented by the horizontal couple vector **M**, and in both cases the neutral axis of the cross section is found to coincide with the axis of the couple.

Let us now consider situations where the bending couples do *not* act in a plane of symmetry of the member, either because they act in a different plane, or because the member does not possess any plane of symmetry. In such situations, we cannot assume that the member will bend in the plane of the couples. This is illustrated in Fig. 4.56. In each part of the figure, the couple exerted on the section has again been as-

(a)

(b)

Fig. 4.55

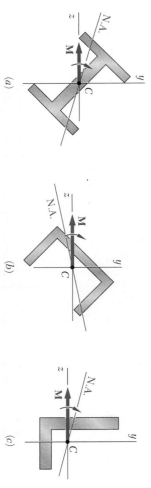

(a)

(b)

(c)

Fig. 4.56

sumed to act in a vertical plane and has been represented by a horizontal couple vector **M**. However, since the vertical plane is not a plane of symmetry, *we cannot expect the member to bend in that plane, or the neutral axis of the section to coincide with the axis of the couple.*

We propose to determine the precise conditions under which the neutral axis of a cross section of arbitrary shape coincides with the axis of the couple **M** representing the forces acting on that section. Such a section is shown in Fig. 4.57, and both the couple vector **M** and the neutral axis have been assumed to be directed along the z axis. We re-

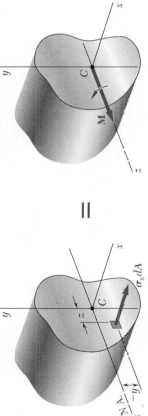

Fig. 4.57

call from Sec. 4.2 that, if we then express that the elementary internal forces $\sigma_x \, dA$ form a system equivalent to the couple **M**, we obtain

x components: $\qquad \int \sigma_x \, dA = 0 \qquad$ (4.1)

moments about y axis: $\qquad \int z \sigma_x \, dA = 0 \qquad$ (4.2)

moments about z axis: $\qquad \int (-y \sigma_x \, dA) = M \qquad$ (4.3)

As we saw earlier, when all the stresses are within the proportional limit, the first of these equations leads to the requirement that the neutral axis be a centroidal axis, and the last to the fundamental relation $\sigma_x = -My/I$. Since we had assumed in Sec. 4.2 that the cross section was symmetric with respect to the y axis, Eq. (4.2) was dismissed as trivial at that time. Now that we are considering a cross section of arbitrary shape, Eq. (4.2) becomes highly significant. Assuming the stresses to remain within the proportional limit of the material, we can substitute $\sigma_x = -\sigma_m y/c$ into Eq. (4.2) and write

$$\int z \left(-\frac{\sigma_m y}{c} \right) dA = 0 \qquad \text{or} \qquad \int yz \, dA = 0 \qquad (4.51)$$

The integral $\int yz \, dA$ represents the product of inertia I_{yz} of the cross section with respect to the y and z axes, and will be zero if these axes are the *principal centroidal axes of the cross section.*[†] We thus conclude that the neutral axis of the cross section will coincide with the axis of the couple **M** representing the forces acting on that section *if, and only if, the couple vector **M** is directed along one of the principal centroidal axes of the cross section.*

[†]See Ferdinand P. Beer and E. Russell Johnston, Jr., *Mechanics for Engineers*, 4th ed., McGraw-Hill, New York, 1987, or *Vector Mechanics for Engineers*, 6th ed., McGraw-Hill, New York, 1996, secs. 9.8–9.10.

We note that the cross sections shown in Fig. 4.55 are symmetric with respect to at least one of the coordinate axes. It follows that, in each case, the y and z axes are the principal centroidal axes of the section. Since the couple vector $\mathbf{M}$ is directed along one of the principal centroidal axes, we verify that the neutral axis will coincide with the axis of the couple. We also note that, if the cross sections are rotated through $90°$ (Fig. 4.58), the couple vector $\mathbf{M}$ will still be directed along a principal centroidal axis, and the neutral axis will again coincide with

Fig. 4.58

the axis of the couple, even though in case b the couple does *not* act in a plane of symmetry of the member.

In Fig. 4.56, on the other hand, neither of the coordinate axes is an axis of symmetry for the sections shown, and the coordinate axes are not principal axes. Thus, the couple vector $\mathbf{M}$ is not directed along a principal centroidal axis, and the neutral axis does not coincide with the axis of the couple. However, any given section possesses principal centroidal axes, even if it is unsymmetric, as the section shown in Fig. 4.56c, and these axes may be determined analytically or by using Mohr's circle.† If the couple vector $\mathbf{M}$ is directed along one of the principal centroidal axes of the section, the neutral axis will coincide with the axis of the couple (Fig. 4.59) and the equations derived in Secs. 4.3 and 4.4 for symmetric members can be used to determine the stresses in this case as well.

Fig. 4.59

As you will see presently, the principle of superposition can be used to determine stresses in the most general case of unsymmetric bending.

†See Ferdinand P. Beer and E. Russell Johnston, Jr., *Mechanics for Engineers*, 4th ed., McGraw-Hill, New York, 1987, or *Vector Mechanics for Engineers*, 6th ed., McGraw-Hill, New York, 1996, secs. 9.8–9.10.

Consider first a member with a vertical plane of symmetry, which is subjected to bending couples **M** and **M**' acting in a plane forming an angle θ with the vertical plane (Fig. 4.60). The couple vector **M** representing the forces acting on a given cross section will form the same

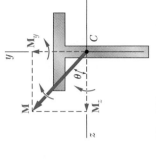

Fig. 4.60

angle θ with the horizontal z axis (Fig. 4.61). Resolving the vector **M** into component vectors $\mathbf{M}_z$ and $\mathbf{M}_y$ along the z and y axes, respectively, we write

$$M_z = M \cos \theta \qquad M_y = M \sin \theta \qquad (4.52)$$

Since the y and z axes are the principal centroidal axes of the cross section, we can use Eq. (4.16) to determine the stresses resulting from the application of either of the couples represented by $\mathbf{M}_z$ and $\mathbf{M}_y$. The couple $\mathbf{M}_z$ acts in a vertical plane and bends the member in that plane (Fig. 4.62). The resulting stresses are

$$\sigma_x = -\frac{M_z y}{I_z} \qquad (4.53)$$

where I_z is the moment of inertia of the section about the principal centroidal z axis. The negative sign is due to the fact that we have compression above the xz plane ($y > 0$) and tension below ($y < 0$). On the other hand, the couple $\mathbf{M}_y$ acts in a horizontal plane and bends the member in that plane (Fig. 4.63). The resulting stresses are found to be

$$\sigma_x = +\frac{M_y z}{I_y} \qquad (4.54)$$

where I_y is the moment of inertia of the section about the principal centroidal y axis, and where the positive sign is due to the fact that we have tension to the left of the vertical xy plane ($z > 0$) and compression to its right ($z < 0$). The distribution of the stresses caused by the original couple **M** is obtained by superposing the stress distributions defined by Eqs. (4.53) and (4.54), respectively. We have

$$\sigma_x = -\frac{M_z y}{I_z} + \frac{M_y z}{I_y} \qquad (4.55)$$

Fig. 4.61

Fig. 4.62

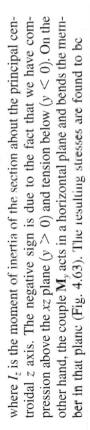

Fig. 4.63

We note that the expression obtained can also be used to compute the stresses in an unsymmetric section, such as the one shown in Fig. 4.64, once the principal centroidal y and z axes have been determined.

Fig. 4.64

On the other hand, Eq. (4.55) is valid only if the conditions of applicability of the principle of superposition are met. In other words, it should not be used if the combined stresses exceed the proportional limit of the material, or if the deformations caused by one of the component couples appreciably affect the distribution of the stresses due to the other.

Equation (4.55) shows that the distribution of stresses caused by unsymmetric bending is linear. However, as we have indicated earlier in this section, the neutral axis of the cross section will not, in general, coincide with the axis of the bending couple. Since the normal stress is zero at any point of the neutral axis, the equation defining that axis can be obtained by setting $\sigma_x = 0$ in Eq. (4.55). We write

$$-\frac{M_z y}{I_z} + \frac{M_y z}{I_y} = 0$$

or, solving for y and substituting for M_z and M_y from Eqs. (4.52),

$$y = \left(\frac{I_z}{I_y}\tan\theta\right)z \qquad (4.56)$$

The equation obtained is that of a straight line of slope $m = (I_z/I_y)\tan\theta$. Thus, the angle ϕ that the neutral axis forms with the z axis (Fig. 4.65) is defined by the relation

$$\boxed{\tan\phi = \frac{I_z}{I_y}\tan\theta} \qquad (4.57)$$

where θ is the angle that the couple vector $\mathbf{M}$ forms with the same axis. Since I_z and I_y are both positive, ϕ and θ have the same sign. Furthermore, we note that $\phi > \theta$ when $I_z > I_y$, and $\phi < \theta$ when $I_z < I_y$. Thus, the neutral axis is always located between the couple vector $\mathbf{M}$ and the principal axis corresponding to the minimum moment of inertia.

Fig. 4.65

EXAMPLE 4.08

A 1600 lb · in. couple is applied to a wooden beam, of rectangular cross section 1.5 by 3.5 in., in a plane forming an angle of 30° with the vertical (Fig. 4.66). Determine (a) the maximum stress in the beam, (b) the angle that the neutral surface forms with the horizontal plane.

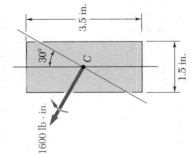

Fig. 4.66

(a) Maximum Stress. The components M_z and M_y of the couple vector are first determined (Fig. 4.67):

$$M_z = (1600 \text{ lb} \cdot \text{in.}) \cos 30° = 1386 \text{ lb} \cdot \text{in.}$$
$$M_y = (1600 \text{ lb} \cdot \text{in.}) \sin 30° = 800 \text{ lb} \cdot \text{in.}$$

Fig. 4.67

We also compute the moments of inertia of the cross section with respect to the z and y axes:

$$I_z = \tfrac{1}{12}(1.5 \text{ in.})(3.5 \text{ in.})^3 = 5.359 \text{ in}^4$$
$$I_y = \tfrac{1}{12}(3.5 \text{ in.})(1.5 \text{ in.})^3 = 0.9844 \text{ in}^4$$

The largest tensile stress due to $\mathbf{M}_z$ occurs along AB and is

$$\sigma_1 = \frac{M_z y}{I_z} = \frac{(1386 \text{ lb} \cdot \text{in.})(1.75 \text{ in.})}{5.359 \text{ in}^4} = 452.6 \text{ psi}$$

The largest tensile stress due to $\mathbf{M}_y$ occurs along AD and is

$$\sigma_2 = \frac{M_y z}{I_y} = \frac{(800 \text{ lb} \cdot \text{in.})(0.75 \text{ in.})}{0.9844 \text{ in}^4} = 609.5 \text{ psi}$$

The largest tensile stress due to the combined loading, therefore, occurs at A and is

$$\sigma_{\max} = \sigma_1 + \sigma_2 = 452.6 + 609.5 = 1062 \text{ psi}$$

The largest compressive stress has the same magnitude and occurs at E.

(b) Angle of Neutral Surface with Horizontal Plane. The angle ϕ that the neutral surface forms with the horizontal plane (Fig. 4.68) is obtained from Eq. (4.57):

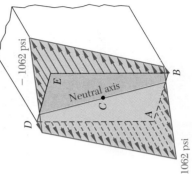

Fig. 4.68

$$\tan \phi = \frac{I_z}{I_y} \tan \theta = \frac{5.359 \text{ in}^4}{0.9844 \text{ in}^4} \tan 30° = 3.143$$
$$\phi = 72.4°$$

The distribution of the stresses across the section is shown in Fig. 4.69.

Fig. 4.69

4.14. GENERAL CASE OF ECCENTRIC AXIAL LOADING

In Sec. 4.12 you analyzed the stresses produced in a member by an eccentric axial load applied in a plane of symmetry of the member. You will now study the more general case when the axial load is not applied in a plane of symmetry.

Consider a straight member AB subjected to equal and opposite eccentric axial forces $\mathbf{P}$ and $\mathbf{P}'$ (Fig. 4.70a), and let a and b denote the distances from the line of action of the forces to the principal centroidal axes of the cross section of the member. The eccentric force $\mathbf{P}$ is statically equivalent to the system consisting of a centric force $\mathbf{P}$ and of the two couples $\mathbf{M}_y$ and $\mathbf{M}_z$ of moments $M_y = Pa$ and $M_z = Pb$ represented in Fig. 4.70b. Similarly, the eccentric force $\mathbf{P}'$ is equivalent to the centric force $\mathbf{P}'$ and the couples $\mathbf{M}'_y$ and $\mathbf{M}'_z$.

By virtue of Saint-Venant's principle (Sec. 2.17), we can replace the original loading of Fig. 4.70a by the statically equivalent loading of Fig. 4.70b in order to determine the distribution of stresses in a section S of the member, as long as that section is not too close to either end of the member. Furthermore, the stresses due to the loading of Fig. 4.70b can be obtained by superposing the stresses corresponding to the centric axial load $\mathbf{P}$ and to the bending couples $\mathbf{M}_y$ and $\mathbf{M}_z$, as long as the conditions of applicability of the principle of superposition are satisfied (Sec. 2.12). The stresses due to the centric load $\mathbf{P}$ are given by Eq. (1.5), and the stresses due to the bending couples by Eq. (4.55), since the corresponding couple vectors are directed along the principal centroidal axes of the section. We write, therefore,

$$\sigma_x = \frac{P}{A} - \frac{M_z y}{I_z} + \frac{M_y z}{I_y} \qquad (4.58)$$

where y and z are measured from the principal centroidal axes of the section. The relation obtained shows that the distribution of stresses across the section is *linear*.

In computing the combined stress σ_x from Eq. (4.58), care should be taken to correctly determine the sign of each of the three terms in the right-hand member, since each of these terms can be positive or negative, depending upon the sense of the loads $\mathbf{P}$ and $\mathbf{P}'$ and the location of their line of action with respect to the principal centroidal axes of the cross section. Depending upon the geometry of the cross section and the location of the line of action of $\mathbf{P}$ and $\mathbf{P}'$, the combined stresses σ_x obtained from Eq. (4.58) at various points of the section may all have the same sign, or some may be positive and others negative. In the latter case, there will be a line in the section, along which the stresses are zero. Setting $\sigma_x = 0$ in Eq. (4.58), we obtain the equation of a straight line, which represents the *neutral axis* of the section:

$$\frac{M_z}{I_z} y - \frac{M_y}{I_y} z = \frac{P}{A}$$

(a)

(b)

Fig. 4.70

EXAMPLE 4.09

A vertical 4.80-kN load is applied as shown on a wooden post of rectangular cross section, 80 by 120 mm (Fig. 4.71). (a) Determine the stress at points A, B, C, and D. (b) Locate the neutral axis of the cross section.

Fig. 4.71

Fig. 4.72

(a) Stresses. The given eccentric load is replaced by an equivalent system consisting of a centric load $\mathbf{P}$ and two couples $\mathbf{M}_x$ and $\mathbf{M}_z$ represented by vectors directed along the principal centroidal axes of the section (Fig. 4.72). We have

$$M_x = (4.80 \text{ kN})(40 \text{ mm}) = 192 \text{ N} \cdot \text{m}$$
$$M_z = (4.80 \text{ kN})(60 \text{ mm} - 35 \text{ mm}) = 120 \text{ N} \cdot \text{m}$$

We also compute the area and the centroidal moments of inertia of the cross section:

$$A = (0.080 \text{ m})(0.120 \text{ m}) = 9.60 \times 10^{-3} \text{ m}^2$$
$$I_x = \tfrac{1}{12}(0.120 \text{ m})(0.080 \text{ m})^3 = 5.12 \times 10^{-6} \text{ m}^4$$
$$I_z = \tfrac{1}{12}(0.080 \text{ m})(0.120 \text{ m})^3 = 11.52 \times 10^{-6} \text{ m}^4$$

The stress σ_0 due to the centric load $\mathbf{P}$ is negative and uniform across the section. We have

$$\sigma_0 = \frac{P}{A} = \frac{-4.80 \text{ kN}}{9.60 \times 10^{-3} \text{ m}^2} = -0.5 \text{ MPa}$$

The stresses due to the bending couples $\mathbf{M}_x$ and $\mathbf{M}_z$ are linearly distributed across the section, with maximum values equal, respectively, to

$$\sigma_1 = \frac{M_x z_{\max}}{I_x} = \frac{(192 \text{ N} \cdot \text{m})(40 \text{ mm})}{5.12 \times 10^{-6} \text{ m}^4} = 1.5 \text{ MPa}$$
$$\sigma_2 = \frac{M_z x_{\max}}{I_z} = \frac{(120 \text{ N} \cdot \text{m})(60 \text{ mm})}{11.52 \times 10^{-6} \text{ m}^4} = 0.625 \text{ MPa}$$

The stresses at the corners of the section are

$$\sigma_y = \sigma_0 \pm \sigma_1 \pm \sigma_2$$

where the signs must be determined from Fig. 4.72. Noting that the stresses due to $\mathbf{M}_x$ are positive at C and D, and negative at A and B, and that the stresses due to $\mathbf{M}_z$ are positive at B and C, and negative at A and D, we obtain

$$\sigma_A = -0.5 - 1.5 - 0.625 = -2.625 \text{ MPa}$$
$$\sigma_B = -0.5 - 1.5 + 0.625 = -1.375 \text{ MPa}$$
$$\sigma_C = -0.5 + 1.5 + 0.625 = +1.625 \text{ MPa}$$
$$\sigma_D = -0.5 + 1.5 - 0.625 = +0.375 \text{ MPa}$$

Fig. 4.73

(b) Neutral Axis. We note that the stress will be zero at a point G between B and C, and at a point H between D and A (Fig. 4.73). Since the stress distribution is linear, we write

$$\frac{BG}{80 \text{ mm}} = \frac{1.375}{1.625 + 1.375} \qquad BG = 36.7 \text{ mm}$$
$$\frac{HA}{80 \text{ mm}} = \frac{2.625}{2.625 + 0.375} \qquad HA = 70 \text{ mm}$$

The neutral axis can be drawn through points G and H (Fig. 4.74).

Fig. 4.74

The distribution of the stresses across the section is shown in Fig. 4.75.

Fig. 4.75

SAMPLE PROBLEM 4.9

A horizontal load **P** is applied as shown to a short section of an S10 × 25.4 rolled-steel member. Knowing that the compressive stress in the member is not to exceed 12 ksi, determine the largest permissible load **P**.

SOLUTION

Properties of Cross Section. The following data are taken from Appendix C.

$$Area: A = 7.46 \text{ in}^2$$
$$Section\ moduli: S_x = 24.7 \text{ in}^3 \qquad S_y = 2.91 \text{ in}^3$$

Force and Couple at C. We replace **P** by an equivalent force-couple system at the centroid C of the cross section.

$$M_x = (4.75 \text{ in.})P \qquad M_y = (1.5 \text{ in.})P$$

Note that the couple vectors $\mathbf{M}_x$ and $\mathbf{M}_y$ are directed along the principal axes of the cross section.

Normal Stresses. The absolute values of the stresses at points A, B, D, and E due, respectively, to the centric load **P** and to the couples $\mathbf{M}_x$ and $\mathbf{M}_y$ are

$$\sigma_1 = \frac{P}{A} = \frac{P}{7.46 \text{ in}^2} = 0.1340P$$

$$\sigma_2 = \frac{M_x}{S_x} = \frac{4.75P}{24.7 \text{ in}^3} = 0.1923P$$

$$\sigma_3 = \frac{M_y}{S_y} = \frac{1.5P}{2.91 \text{ in}^3} = 0.5155P$$

Superposition. The total stress at each point is found by superposing the stresses due to **P**, $\mathbf{M}_y$, and $\mathbf{M}_y$. We determine the sign of each stress by carefully examining the sketch of the force-couple system.

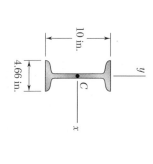

$$\sigma_A = -\sigma_1 + \sigma_2 + \sigma_3 = -0.1340P + 0.1923P + 0.5155P = +0.574P$$
$$\sigma_B = -\sigma_1 + \sigma_2 - \sigma_3 = -0.1340P + 0.1923P - 0.5155P = -0.457P$$
$$\sigma_D = -\sigma_1 - \sigma_2 + \sigma_3 = -0.1340P - 0.1923P + 0.5155P = +0.189P$$
$$\sigma_E = -\sigma_1 - \sigma_2 - \sigma_3 = -0.1340P - 0.1923P - 0.5155P = -0.842P$$

Largest Permissible Load. The maximum compressive stress occurs at point E. Recalling that $\sigma_{all} = -12$ ksi, we write

$$\sigma_{all} = \sigma_E \qquad -12 \text{ ksi} = -0.842P \qquad P = 14.3 \text{ kips} \quad \blacktriangleleft$$

*SAMPLE PROBLEM 4.10

A couple of magnitude $M_0 = 1.5$ kN · m acting in a vertical plane is applied to a beam having the Z-shaped cross section shown. Determine (a) the stress at point A, (b) the angle that the neutral axis forms with the horizontal plane. The moments and product of inertia of the section with respect to the y and z axes have been computed and are as follows:

$$I_y = 3.25 \times 10^{-6} \text{ m}^4$$
$$I_z = 4.18 \times 10^{-6} \text{ m}^4$$
$$I_{yz} = 2.87 \times 10^{-6} \text{ m}^4.$$

SOLUTION

Principal Axes. We draw Mohr's circle and determine the orientation of the principal axes and the corresponding principal moments of inertia.

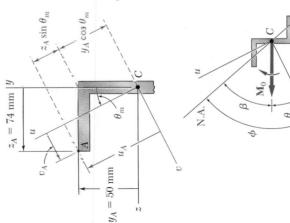

$$\tan 2\theta_m = \frac{FZ}{EF} = \frac{2.87}{0.465} \qquad 2\theta_m = 80.8° \qquad \theta_m = 40.4°$$

$$R^2 = (EF)^2 + (FZ)^2 = (0.465)^2 + (2.87)^2 \qquad R = 2.91 \times 10^{-6} \text{ m}^4$$

$$I_u = I_{\min} = OU = I_{ave} - R = 3.72 - 2.91 = 0.810 \times 10^{-6} \text{ m}^4$$
$$I_v = I_{\max} = OV = I_{ave} + R = 3.72 + 2.91 = 6.63 \times 10^{-6} \text{ m}^4$$

Loading. The applied couple M_0 is resolved into components parallel to the principal axes.

$$M_u = M_0 \sin \theta_m = 1500 \sin 40.4° = 972 \text{ N} \cdot \text{m}$$
$$M_v = M_0 \cos \theta_m = 1500 \cos 40.4° = 1142 \text{ N} \cdot \text{m}$$

a. Stress at A. The perpendicular distances from each principal axis to point A are

$$u_A = y_A \cos \theta_m + z_A \sin \theta_m = 50 \cos 40.4° + 74 \sin 40.4° = 86.0 \text{ mm}$$
$$v_A = -y_A \sin \theta_m + z_A \cos \theta_m = -50 \sin 40.4° + 74 \cos 40.4° = 23.9 \text{ mm}$$

Considering separately the bending about each principal axis, we note that M_u produces a tensile stress at point A while M_v produces a compressive stress at the same point

$$\sigma_A = +\frac{M_u v_A}{I_u} - \frac{M_v u_A}{I_v} = +\frac{(972 \text{ N} \cdot \text{m})(0.0239 \text{ m})}{0.810 \times 10^{-6} \text{ m}^4} - \frac{(1142 \text{ N} \cdot \text{m})(0.0860 \text{ m})}{6.63 \times 10^{-6} \text{ m}^4}$$

$$= +(28.68 \text{ MPa}) - (14.81 \text{ MPa})$$

$$\sigma_A = +13.87 \text{ MPa} \blacktriangleleft$$

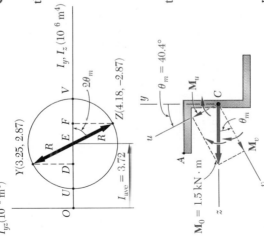

b. Neutral Axis. Using Eq. (4.57), we find the angle ϕ that the neutral axis forms with the v axis.

$$\tan \phi = \frac{I_v}{I_u} \tan \theta_m = \frac{6.63}{0.810} \tan 40.4° \qquad \phi = 81.8°$$

The angle β formed by the neutral axis and the horizontal is

$$\beta = \phi - \theta_m = 81.8° - 40.4° = 41.4° \qquad \beta = 41.4° \blacktriangleleft$$

PROBLEMS

4.145 through 4.147 The couple **M** is applied to a beam of the cross section shown in a plane forming an angle β with the vertical. Determine the stress at (a) point A, (b) point B, (c) point D.

Fig. P4.145

Fig. P4.148

Fig. P4.146

Fig. P4.147

4.148 through 4.150 The couple **M** is applied to a beam of the cross section shown in a plane forming an angle β with the vertical. Determine the stress at (a) point A, (b) point B, (c) point D.

Fig. P4.149

Fig. P4.150

4.151 The couple **M** acts in a vertical plane and is applied to a beam oriented as shown. Determine (a) the angle that the neutral axis forms with the horizontal plane, (b) the maximum tensile stress in the beam.

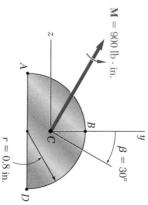

Fig. P4.151

4.152 and 4.153 The couple **M** acts in a vertical plane and is applied to a beam oriented as shown. Determine (*a*) the angle that the neutral axis forms with the horizontal plane, (*b*) the maximum tensile stress in the beam.

S150 × 18.6

M = 1.5 kN · m

Fig. P4.152

C8 × 11.5

M = 25 kip · in.

Fig. P4.153

4.154 through 4.156 The couple **M** acts in a vertical plane and is applied to a beam oriented as shown. Determine (*a*) the angle that the neutral axis forms with the horizontal plane, (*b*) the maximum tensile stress in the beam.

M = 400 N · m

$I_{y'} = 281 \times 10^3 \text{ mm}^4$
$I_{z'} = 176.9 \times 10^3 \text{ mm}^4$

Fig. P4.154

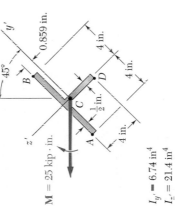

M = 25 kip · in.

$I_{y'} = 6.74 \text{ in}^4$
$I_{z'} = 21.4 \text{ in}^4$

Fig. P4.155

M = 4 kN · m

23.33 mm

Fig. P4.156

***4.157 and 4.158** The couple **M** acts in a vertical plane and is applied to a beam of the cross section shown. Determine the stress at point *A*.

M = 60 kip · in.

$I_y = 8.7 \text{ in}^4$
$I_z = 24.5 \text{ in}^4$
$I_{yz} = +8.3 \text{ in}^4$

Fig. P4.157

M = 1.2 kN · in.

$I_y = 1.894 \times 10^6 \text{ mm}^4$
$I_z = 0.614 \times 10^6 \text{ mm}^4$
$I_{yz} = +0.800 \times 10^6 \text{ mm}^4$

Fig. P4.158

*4.159 A 4 × 10-in. timber has been trimmed to form a beam of the cross section shown. Knowing that the couple **M** acts in a vertical plane, determine the stress at point A.

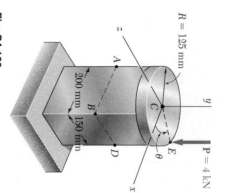

$I_y = 291$ in⁴

$I_z = 39.3$ in⁴

$I_{yz} = -22.5$ in⁴

Fig. P4.159

4.160 A rigid circular plate of 125-mm radius is attached to a solid 150 × 200-mm rectangular post, with the center of the plate directly above the center of the post. If a 4-kN force **P** is applied at E with $\theta = 30°$, determine (a) the stress at point A, (b) the stress at point B, (c) the point where the neutral axis intersects line ABD.

4.161 In Prob. 4.160, determine (a) the value of θ for which the stress at D reaches its largest value, (b) the corresponding values of the stress at A, B, C, and D.

Fig. P4.160

4.162 The tube shown has a uniform wall thickness of 0.5 in. For the given loading, determine (a) the stress at points A and B, (b) the point where the neutral axis intersects line ABD.

Fig. P4.162

4.163 Solve Prob. 4.162, assuming that the 6-kip force at point E is removed.

4.164 An axial load **P** of magnitude 50 kN is applied as shown to a short section of a W150 × 24 rolled-steel member. Determine the largest distance a for which the maximum compressive stress does not exceed 90 MPa.

Fig. P4.164

4.165 An axial load **P** of magnitude 30 kN is applied as shown to a short section of a C150 × 12.2 rolled-steel channel. Determine the distance a for which the maximum compressive stress in the channel is 60 MPa.

P = 30 kN

Fig. P4.165

4.166 A horizontal load **P** is applied to the beam shown. Knowing that $a = 20$ mm and that the tensile stress in the beam is not to exceed 75 MPa, determine the largest permissible load **P.**

4.167 A horizontal load **P** of magnitude 100 kN is applied to the beam shown. Determine the largest distance a for which the maximum tensile stress in the beam does not exceed 75 MPa.

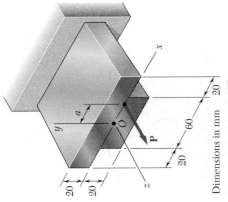

Dimensions in mm

Fig. P4.166 and P4.167

4.168 A beam having the cross section shown is subjected to a couple M_0 that acts in a vertical plane. Determine the largest permissible value of the moment M_0 of the couple if the maximum stress in the beam is not to exceed 12 ksi. *Given:* $I_y = I_z = 11.3$ in^4, $A = 4.75$ in^2, $k_{min} = 0.983$ in. (*Hint:* By reason of symmetry, the principal axes form an angle of 45° with the coordinate axes. Use the relations $I_{min} = Ak_{min}^2$ and $I_{min} + I_{max} = I_y + I_z$)

4.169 Solve Prob. 4.168, assuming that the couple **M**$_0$ acts in a horizontal plane.

Fig. P4.168

4.170 The Z section shown is subjected to a couple **M**$_0$ acting in a vertical plane. Determine the largest permissible value of the moment M_0 of the couple if the maximum stress is not to exceed 80 MPa. *Given:* $I_{max} = 2.28 \times 10^6$ mm^4, $I_{min} = 0.23 \times 10^6$ mm^4, principal axes 25.7° ⦨ and 64.3° ⦨.

4.171 Solve Prob. 4.170, assuming that the couple **M**$_0$ acts in a horizontal plane.

Fig. P4.170

4.172 An extruded aluminum member having the cross section shown is subjected to a couple **M**$_0$ acting in a vertical plane. Determine the largest permissible value of the moment M_0 of the couple if the maximum stress is not to exceed 12 ksi. *Given:* $I_{max} = 0.957$ in^6, $I_{min} = 0.427$ in^4, principal axes 29.4° ⦨ and 60.6° ⦩.

Fig. P4.172

4.173 A beam having the cross section shown is subjected to a couple $\mathbf{M}_0$ acting in a vertical plane. Determine the largest permissible value of the moment M_0 of the couple if the maximum stress is not to exceed 100 MPa.
Given: $I_y = I_z = b^4/36$ and $I_{yz} = b^4/72$.

Fig. P4.173

4.174 A couple $\mathbf{M}_0$ acting in a vertical plane is applied to a W12 × 16 rolled-steel beam, whose web forms an angle θ with the vertical. Denoting by σ_0 the maximum stress in the beam when $\theta = 0$, determine the angle of inclination θ of the beam for which the maximum stress is $2\,\sigma_0$.

Fig. P4.174

4.175 Show that, if a solid rectangular beam is bent by a couple applied in a plane containing one diagonal of the rectangular cross section, the neutral axis will lie along the other diagonal.

Fig. P4.175

4.176 A beam of unsymmetric cross section is subjected to a couple $\mathbf{M}_0$ acting in the vertical xy plane. Show that the stress at point A, of coordinates y and z, is

$$\sigma_A = -\frac{yI_y - zI_{yz}}{I_yI_z - I_{yz}^2}M_z$$

where I_y, I_z and I_{yz} denote the moments and product of inertia of the cross section with respect to centroidal axes, and M_z the moment of the couple.

4.177 A beam of unsymmetric cross section is subjected to a couple $\mathbf{M}_0$ acting in the horizontal xz plane. Show that the stress at point A is

$$\sigma_A = -\frac{zI_z - yI_{yz}}{I_yI_z - I_{yz}^2}M_y$$

where I_y, I_z and I_{yz} denote the moments and product of inertia of the cross section with respect to centroidal axes, and M_y the moment of the couple.

Fig. P4.176 and P4.177

4.178 (*a*) Show that, if a vertical force **P** is applied at point *A* of the section shown, the equation of the neutral axis *BD* is

$$\left(\frac{x_A}{k_z^2}\right)x + \left(\frac{z_A}{k_x^2}\right)z = -1$$

where k_z and k_x denote the radius of gyration of the cross section with respect to the z axis and the x axis, respectively. (*b*) Further show that, if a vertical force **Q** is applied at any point located on line *BD*, the stress at point *A* will be zero.

Fig. P4.178

4.179 (*a*) Show that the stress at corner *A* of the prismatic member shown in Fig. P4.179*a* will be zero if the vertical force **P** is applied at a point located on the line

$$\frac{x}{b/6} + \frac{z}{h/6} = 1$$

(*b*) Further show that, if no tensile stress is to occur in the member, the force **P** must be applied at a point located within the area bounded by the line found in part *a* and the three similar lines corresponding to the condition of zero stress at *B*, *C*, and *D*, respectively. This area, shown in Fig. P4.179*b*, is known as the *kern* of the cross section.

(a)

Fig. P4.179

---

*4.15. BENDING OF CURVED MEMBERS

Our analysis of stresses has been restricted so far to bending members. In this section we will consider the stresses caused by the application of equal and opposite couples to members that are initially curved. Our discussion will be limited to curved members of uniform cross section possessing a plane of symmetry in which the bending couples are applied, and it will be assumed that all stresses remain below the proportional limit.

If the initial curvature of the member is small, i.e., if its radius of curvature is large compared to the depth of its cross section, a good approximation can be obtained for the distribution of stresses by assuming the member to be straight and using the formulas derived in Secs. 4.3 and 4.4[†] However, when the radius of curvature and the dimensions of the cross section of the member are of the same order of magnitude, we must use a different method of analysis, which was first introduced by the German engineer E. Winkler (1835–1888).

[†]See Prob. 4.185.

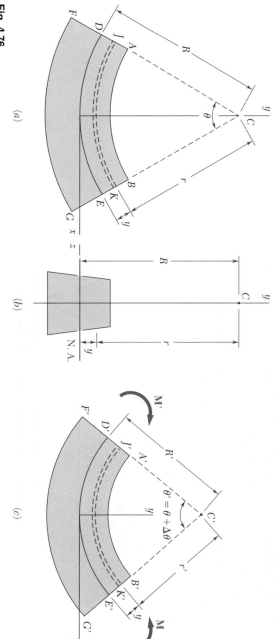

Fig. 4.76

Consider the curved member of uniform cross section shown in Fig. 4.76. Its transverse section is symmetric with respect to the y axis (Fig. 4.76b) and, in its unstressed state, its upper and lower surfaces intersect the vertical xy plane along arcs of circle AB and FG centered at C (Fig. 4.76a).

We now apply two equal and opposite couples $\mathbf{M}$ and $\mathbf{M}'$ in the plane of symmetry of the member (Fig. 4.76c). A reasoning similar to that of Sec. 4.3 would show that any transverse plane section containing C will remain plane, and that the various arcs of circle indicated in Fig. 4.76a will be transformed into circular and concentric arcs with a center C' different from C. More specifically, if the couples $\mathbf{M}$ and $\mathbf{M}'$ are directed as shown, the curvature of the various arcs of circle will increase; that is $A'C' < AC$. We also note that the couples $\mathbf{M}$ and $\mathbf{M}'$ will cause the length of the upper surface of the member to decrease $(A'B' < AB)$ and the length of the lower surface to increase $(F'G' > FG)$. We conclude that a *neutral surface* must exist in the member, the length of which remains constant. The intersection of the neutral surface with the xy plane has been represented in Fig. 4.76a by the arc DE of radius R, and in Fig. 4.76c by the arc $D'E'$ of radius R'. Denoting by θ and θ' the central angles corresponding respectively to DE and $D'E'$, we express the fact that the length of the neutral surface remains constant by writing

$$R\theta = R'\theta' \qquad (4.59)$$

Considering now the arc of circle JK located at a distance y above the neutral surface, and denoting respectively by r and r' the radius of this arc before and after the bending couples have been applied, we express the deformation of JK as

$$\delta = r'\theta' - r\theta \qquad (4.60)$$

Observing from Fig. 4.76 that

$$r = R - y \qquad r' = R' - y \qquad (4.61)$$

and substituting these expressions into Eq. (4.60), we write

$$\delta = (R' - y)\theta' - (R - y)\theta$$

or, recalling Eq. (4.59) and setting $\theta' - \theta = \Delta\theta$,

$$\delta = -y\,\Delta\theta \qquad (4.62)$$

The normal strain ϵ_x in the elements of JK is obtained by dividing the deformation δ by the original length $r\theta$ of arc JK. We write

$$\epsilon_x = \frac{\delta}{r\theta} = -\frac{y\,\Delta\theta}{r\theta}$$

or, recalling the first of the relations (4.61),

$$\epsilon_x = -\frac{\Delta\theta}{\theta}\,\frac{y}{R - y} \qquad (4.63)$$

The relation obtained shows that, while each transverse section remains plane, the normal strain ϵ_x *does not vary linearly* with the distance y from the neutral surface.

The normal stress σ_x can now be obtained from Hooke's law, $\sigma_x = E\epsilon_x$, by substituting for ϵ_x from Eq. (4.63). We have

$$\sigma_x = -\frac{E\,\Delta\theta}{\theta}\,\frac{y}{R - y} \qquad (4.64)$$

or, alternatively, recalling the first of Eqs. (4.61),

$$\sigma_x = -\frac{E\,\Delta\theta}{\theta}\,\frac{R - r}{r} \qquad (4.65)$$

Equation (4.64) shows that, like ϵ_x, the normal stress σ_x *does not vary linearly* with the distance y from the neutral surface. Plotting σ_x versus y, we obtain an arc of hyperbola (Fig. 4.77).

Fig. 4.77

In order to determine the location of the neutral surface in the member and the value of the coefficient $E \Delta\theta/\theta$ used in Eqs. (4.64) and (4.65), we now recall that the elementary forces acting on any transverse section must be statically equivalent to the bending couple **M**. Expressing, as we did in Sec. 4.2 for a straight member, that the sum of the elementary forces acting on the section must be zero, and that the sum of their moments about the transverse z axis must be equal to the bending moment M, we write the equations

$$\int \sigma_x \, dA = 0 \qquad (4.1)$$

and

$$\int (-y \sigma_x \, dA) = M \qquad (4.3)$$

Substituting for σ_x from (4.65) into Eq. (4.1), we write

$$-\int \frac{E \, \Delta\theta}{\theta} \frac{R - r}{r} \, dA = 0$$

$$\int \frac{R - r}{r} \, dA = 0$$

$$R \int \frac{dA}{r} - \int dA = 0$$

from which it follows that the distance R from the center of curvature C to the neutral surface is defined by the relation

$$R = \frac{A}{\displaystyle\int \frac{dA}{r}} \qquad (4.66)$$

We note that the value obtained for R is not equal to the distance $\bar{r}$ from C to the centroid of the cross section, since $\bar{r}$ is defined by a different relation, namely,

$$\bar{r} = \frac{1}{A} \int r \, dA \qquad (4.67)$$

We thus conclude that, *in a curved member, the neutral axis of a transverse section does not pass through the centroid of that section* (Fig. 4.78).† Expressions for the radius R of the neutral surface will be derived for some specific cross-sectional shapes in Example 4.10 and in Probs. 4.207 through 4.209. For convenience, these expressions are shown in Fig. 4.79.

† However, an interesting property of the neutral surface can be noted if we write Eq. (4.66) in the alternative form

$$\frac{1}{R} = \frac{1}{A} \int \frac{1}{r} \, dA \qquad (4.66')$$

Equation (4.66') shows that, if the member is divided into a large number of fibers of cross-sectional area dA, the curvature $1/R$ of the neutral surface will be equal to the average value of the curvature $1/r$ of the various fibers.

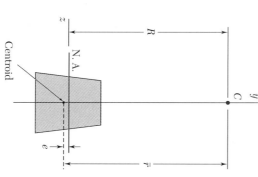

Fig. 4.78

Centroid

N.A.

C

R

$\bar{r}$

e

y

z

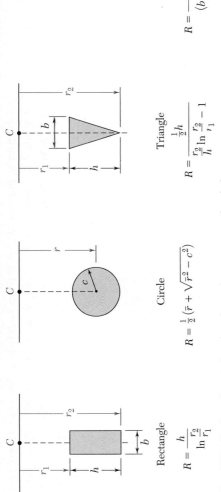

Rectangle

$$R = \frac{h}{\ln \dfrac{r_2}{r_1}}$$

Circle

$$R = \frac{1}{2}(\bar{r} + \sqrt{\bar{r}^2 - c^2})$$

Triangle

$$R = \frac{\frac{1}{2}h}{\dfrac{r_2}{h}\ln \dfrac{r_2}{r_1} - 1}$$

Trapizoid

$$R = \frac{\frac{1}{2}h^2(b_1 + b_2)}{(b_1 r_2 - b_2 r_1)\ln \dfrac{r_2}{r_1} - h(b_1 - b_2)}$$

Fig. 4.79 Radius of neutral surface for various cross-sectional shapes.

Substituting now for σ_x from (4.65) into Eq. (4.3), we write

$$\int \frac{E\,\Delta\theta}{\theta}\frac{R - r}{r}\,y\,dA = M$$

or, since $y = R - r$,

$$\frac{E\,\Delta\theta}{\theta}\int \frac{(R - r)^2}{r}\,dA = M$$

Expanding the square in the integrand, we obtain after reductions

$$\frac{E\,\Delta\theta}{\theta}\left[R^2 \int \frac{dA}{r} - 2RA + \int r\,dA\right] = M$$

Recalling Eqs. (4.66) and (4.67), we note that the first term in the brackets is equal to RA, while the last term is equal to $\bar{r}A$. We have, therefore,

$$\frac{E\,\Delta\theta}{\theta}(RA - 2RA + \bar{r}A) = M$$

and, solving for $E\,\Delta\theta/\theta$,

$$\frac{E\,\Delta\theta}{\theta} = \frac{M}{A(\bar{r} - R)} \qquad (4.68)$$

Referring to Fig. 4.76, we note that $\Delta\theta > 0$ for $M > 0$. It follows that $\bar{r} - R > 0$, or $R < \bar{r}$, regardless of the shape of the section. Thus, the neutral axis of a transverse section is always located between the centroid of the section and the center of curvature of the member (Fig. 4.78). Setting $\bar{r} - R = e$, we write Eq. (4.68) in the form

$$\frac{E\,\Delta\theta}{\theta} = \frac{M}{Ae} \qquad (4.69)$$

Substituting now for $E \Delta\theta/\theta$ from (4.69) into Eqs. (4.64) and (4.65), we obtain the following alternative expressions for the normal stress σ_x in a curved beam:

$$\sigma_x = -\frac{My}{Ae(R-y)}$$

(4.70)

and

$$\sigma_x = \frac{M(r-R)}{Aer}$$

(4.71)

We should note that the parameter e in the above equations is a small quantity obtained by subtracting two lengths of comparable size, R and $\bar{r}$. In order to determine σ_x with a reasonable degree of accuracy, it is therefore necessary to compute R and $\bar{r}$ very accurately, particularly when both of these quantities are large, i.e., when the curvature of the member is small. However, as we indicated earlier, it is possible in such a case to obtain a good approximation for σ_x by using the formula $\sigma_x = -My/I$ developed for straight members.

Let us now determine the change in curvature of the neutral surface caused by the bending moment M. Solving Eq. (4.59) for the curvature $1/R'$ of the neutral surface in the deformed member, we write

$$\frac{1}{R'} = \frac{1}{R}\frac{\theta'}{\theta}$$

or, setting $\theta' = \theta + \Delta\theta$ and recalling Eq. (4.69),

$$\frac{1}{R'} = \frac{1}{R}\left(1 + \frac{\Delta\theta}{\theta}\right) = \frac{1}{R}\left(1 + \frac{M}{EAe}\right)$$

from which it follows that the change in curvature of the neutral surface is

$$\frac{1}{R'} - \frac{1}{R} = \frac{M}{EAeR}$$

(4.72)

EXAMPLE 4.10

A curved rectangular bar has a mean radius $\bar{r} = 6$ in. and a cross section of width $b = 2.5$ in. and depth $h = 1.5$ in. (Fig. 4.80). Determine the distance e between the centroid and the neutral axis of the cross section.

We first derive the expression for the radius R of the neutral surface. Denoting by r_1 and r_2, respectively, the inner and

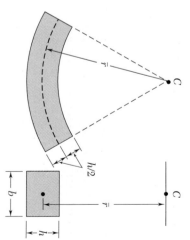

Fig. 4.80

outer radius of the bar (Fig. 4.81), we use Eq. (4.66) and write

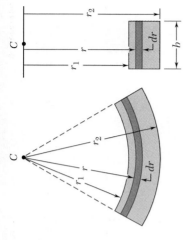

$$R = \frac{A}{\int_{r_1}^{r_2} \frac{dA}{r}} = \frac{bh}{\int_{r_1}^{r_2} \frac{b\, dr}{r}} = \frac{h}{\int_{r_1}^{r_2} \frac{dr}{r}}$$

$$R = \frac{h}{\ln \dfrac{r_2}{r_1}} \qquad (4.73)$$

Fig. 4.81

For the given data, we have

$$r_1 = \bar{r} - \tfrac{1}{2}h = 6 - 0.75 = 5.25 \text{ in.}$$
$$r_2 = \bar{r} + \tfrac{1}{2}h = 6 + 0.75 = 6.75 \text{ in.}$$

Substituting for h, r_1, and r_2 into Eq. (4.73), we have

$$R = \frac{h}{\ln \dfrac{r_2}{r_1}} = \frac{1.5 \text{ in.}}{\ln \dfrac{6.75}{5.25}} = 5.9686 \text{ in.}$$

The distance between the centroid and the neutral axis of the cross section (Fig. 4.82) is thus

$$e = \bar{r} - R = 6 - 5.9686 = 0.0314 \text{ in.}$$

We note that it was necessary to calculate R with five significant figures in order to obtain e with the usual degree of accuracy.

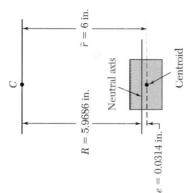

Fig. 4.82

For the bar of Example 4.10, determine the largest tensile and compressive stresses, knowing that the bending moment in the bar is $M = 8$ kip $\cdot$ in.

We use Eq. (4.71) with the given data,

$$M = 8 \text{ kip} \cdot \text{in.} \qquad A = bh = (2.5 \text{ in.})(1.5 \text{ in.}) = 3.75 \text{ in}^2$$

and the values obtained in Example 4.10 for R and e,

$$R = 5.969 \qquad e = 0.0314 \text{ in.}$$

Making first $r = r_2 = 6.75$ in. in Eq. (4.71), we write

$$\sigma_{\max} = \frac{M(r_2 - R)}{Aer_2}$$
$$= \frac{(8 \text{ kip} \cdot \text{in.})(6.75 \text{ in.} - 5.969 \text{ in.})}{(3.75 \text{ in}^2)(0.0314 \text{ in.})(6.75 \text{ in.})}$$
$$\sigma_{\max} = 7.86 \text{ ksi}$$

Making now $r = r_1 = 5.25$ in. in Eq. (4.71), we have

$$\sigma_{\min} = \frac{M(r_1 - R)}{Aer_1}$$
$$= \frac{(8 \text{ kip} \cdot \text{in.})(5.25 \text{ in.} - 5.969 \text{ in.})}{(3.75 \text{ in}^2)(0.0314 \text{ in.})(5.25 \text{ in.})}$$
$$\sigma_{\min} = -9.30 \text{ ksi}$$

Remark. Let us compare the values obtained for $\sigma_{\max}$ and $\sigma_{\min}$ with the result we would get for a straight bar. Using Eq. (4.15) of Sec. 4.4, we write

$$\sigma_{\max,\min} = \pm \frac{Mc}{I}$$
$$= \pm \frac{(8 \text{ kip} \cdot \text{in.})(0.75 \text{ in.})}{\frac{1}{12}(2.5 \text{ in.})(1.5 \text{ in.})^3} = \pm 8.53 \text{ ksi}$$

SAMPLE PROBLEM 4.11

A machine component has a T-shaped cross section and is loaded as shown. Knowing that the allowable compressive stress is 50 MPa, determine the largest force **P** that can be applied to the component.

Section *a-a*

SOLUTION

Centroid of the Cross Section. We locate the centroid *D* of the cross section

	A_i, mm²	$\bar{r}_i$, mm	$\bar{r}_i A_i$, mm³
1	$(20)(80) = 1600$	40	64×10^3
2	$(40)(20) = 800$	70	56×10^3
	$\Sigma A_i = 2400$		$\Sigma \bar{r}_i A_i = 120 \times 10^3$

$$\bar{r}\Sigma A_i = \Sigma \bar{r}_i A_i$$
$$\bar{r}(2400) = 120 \times 10^3$$
$$\bar{r} = 50 \text{ mm} = 0.050 \text{ m}$$

Force and Couple at D. The internal forces in section *a-a* are equivalent to a force **P** acting at *D* and a couple **M** of moment

$$M = P(50 \text{ mm} + 60 \text{ mm}) = (0.110 \text{ m})P$$

Superposition. The centric force **P** causes a uniform compressive stress on section *a-a*. The bending couple **M** causes a varying stress distribution [Eq. (4.71)]. We note that the couple **M** tends to increase the curvature of the member and is therefore positive (cf. Fig. 4.76). The total stress at a point of section *a-a* located at distance *r* from the center of curvature *C* is

$$\sigma = -\frac{P}{A} + \frac{M(r - R)}{Aer} \qquad (1)$$

Radius of Neutral Surface. We now determine the radius *R* of the neutral surface by using Eq. (4.66).

$$R = \frac{A}{\int \frac{dA}{r}} = \frac{A}{\int_{r_1}^{r_2} \frac{(80 \text{ mm})}{r} dr + \int_{r_2}^{r_3} \frac{(20 \text{ mm})}{r} dr}$$

$$= \frac{2400 \text{ mm}^2}{80 \ln \frac{50}{30} + 20 \ln \frac{90}{50}} = \frac{2400}{40.866 + 11.756} = 45.61 \text{ mm}$$

$$= 0.04561 \text{ m}$$

We also compute: $e = \bar{r} - R = 0.05000 \text{ m} - 0.04561 \text{ m} = 0.00439 \text{ m}$

Allowable Load. We observe that the largest compressive stress will occur at point *A* where $r = 0.030$ m. Recalling that $\sigma_{\text{all}} = 50$ MPa and using Eq. (1), we write

$$-50 \times 10^6 \text{ Pa} = -\frac{P}{2.4 \times 10^{-3} \text{ m}^2} + \frac{(0.110 \text{ P})(0.030 \text{ m} - 0.04561 \text{ m})}{(2.4 \times 10^{-3} \text{ m}^2)(0.00439 \text{ m})(0.030 \text{ m})}$$

$$-50 \times 10^6 = -417P - 5432P$$

$$P = 8.55 \text{ kN} \blacktriangleleft$$

PROBLEMS

4.180 For the curved bar and loading shown, determine the stress at point A when (a) $r_1 = 30$ mm, (b) $r_1 = 50$ mm.

4.181 For the curved bar and loading shown, determine the stress at points A and B when $r_1 = 40$ mm.

Fig. P4.180 and P4.181

4.182 For the curved bar and loading shown, determine the stress at point A when (a) $h = 2.5$ in., (b) $h = 3$ in.

4.183 For the curved bar and loading shown, determine the stress at points A and B when $h = 2.75$ in.

Fig. P4.182 and P4.183

4.184 The curved bar shown has a cross section of 40×60 mm and an inner radius $r_1 = 15$ mm. For the loading shown determine the largest tensile and compressive stresses.

4.185 For the curved bar and loading shown, determine the percent error introduced in the computation of the maximum stress by assuming that the bar is straight. Consider the case when (a) $r_1 = 20$ mm, (b) $r_1 = 200$ mm, (c) $r_1 = 2$ m.

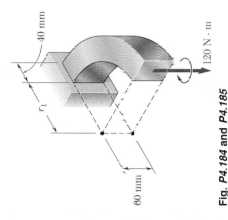

Fig. P4.184 and P4.185

4.186 Steel links having the cross section shown are available with different central angles β. Knowing that the allowable stress is 15 ksi, determine the largest force **P** that can be applied to a link for which $\beta = 90°$.

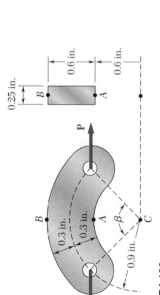

Fig. P4.186

4.187 Solve Prob. 4.186, assuming that $\beta = 60°$.

4.188 The curved portion of the bar shown has an inner radius of 20 mm. Knowing that the line of action of the 3-kN force is located at a distance $a = 60$ mm from the vertical plane containing the center of curvature of the bar, determine the largest compressive stress in the bar.

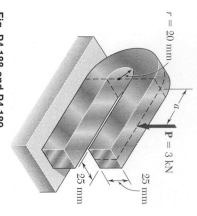

Fig. P4.188 and P4.189

4.189 Knowing that the allowable stress in the bar is 150 MPa, determine the largest permissible distance a from the line of action of the 3-kN force to the vertical plane containing the center of curvature of the bar.

4.190 Three plates are welded together to form the curved beam shown. For the given loading, determine the distance e between the neutral axis and the centroid of the cross section.

Fig. P4.190 and P4.191

0.5 in.

3 in.

3 in.

2 in.

2 in.

0.5 in.

0.5 in.

3 in.

4.191 Three plates are welded together to form the curved beam shown. For $M = 8$ kip · in., determine the stress at (a) point A, (b) point B, (c) the centroid of the cross section.

4.192 and 4.193 Knowing that $M = 20$ kN · m, determine the stress at (a) point A, (b) point B.

Fig. P4.192

M

M'

B   C

A

150 mm

45 mm

135 mm

36 mm

Fig. P4.193

M

M'

B   C

A

150 mm

45 mm

135 mm

36 mm

4.194 The curved bar shown has a circular cross section of 32-mm diameter. Determine the largest couple **M** that can be applied to the bar about a horizontal axis if the maximum stress is not to exceed 60 MPa.

Fig. P4.194

4.195 The bar shown has a circular cross section of 0.6-in. diameter. Knowing that $a = 1.2$ in., determine the stress at (*a*) point *A*, (*b*) point *B*.

4.196 The bar shown has a circular cross section of 0.6-in. diameter. Knowing that the allowable tensile stress is 8 ksi, determine the largest permissible distance *a* from the line of action of the 50-lb forces to the plane containing the center of curvature of the bar.

Fig. P4.195 and P4.196

4.197 The split ring shown has an inner radius $r_1 = 20$ mm and a circular cross section of diameter $d = 32$ mm. For the loading shown, determine the stress at (*a*) point *A*, (*b*) point *B*.

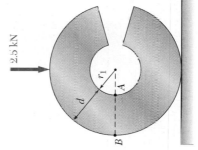

Fig. P4.197 and P4.198

4.198 The split ring shown has an inner radius $r_1 = 16$ mm and a circular cross section of diameter $d = 32$ mm. For the loading shown, determine the stress at (*a*) point *A*, (*b*) point *B*.

4.199 Knowing that the machine component shown has a trapezoidal cross section with $a = 3.5$ in. and $b = 2.5$ in., determine the stress at (*a*) point *A*, (*b*) point *B*.

4.200 Knowing that the machine component shown has a trapezoidal cross section with $a = 2.5$ in. and $b = 3.5$ in., determine the stress at (*a*) point *A*, (*b*) point *B*.

Fig. P4.199 and P4.200

4.201   For the curved beam and loading shown, determine the stress at (*a*) point A, (*b*) point B.

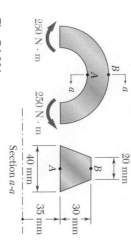

Fig. P4.201

4.202   For the crane hook shown, determine the largest tensile stress in section *a-a*.

Fig. P4.202

4.203 and 4.204   Knowing that *M* = 5 kip · in., determine the stress at (*a*) point A, (*b*) point B.

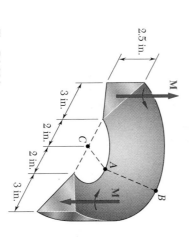

FIG. P4.202

Fig. P4.204

4.205   Knowing that *P* = 3.5 kips, determine the stress at (*a*) point A, (*b*) point B.

Fig. P4.205

Fig. P4.203

4.206 Show that if the cross section of a curved beam consists of two or more rectangles, the radius R of the neutral surface can be expressed as

$$R = \cfrac{A}{\ln\left[\left(\dfrac{r_2}{r_1}\right)^{b_1}\left(\dfrac{r_3}{r_2}\right)^{b_2}\left(\dfrac{r_4}{r_3}\right)^{b_3}\right]}$$

where A is the total area of the cross section.

Fig. P4.206

4.207 through 4.209 Using Eq. (4.66), derive the expression for R given in Fig. 4.79 for

*4.207   A circular cross section

4.208   A trapezoidal section

4.209   A triangular cross section

*4.210 For a curved bar of rectangular cross section subjected to a bending couple $\mathbf{M}$, show that the radial stress at the neutral surface is

$$\sigma_r = \frac{M}{Ae}\left(1 - \frac{r_1}{R} - \ln\frac{R}{r_1}\right)$$

and compute the value of σ_r for the curved bar of Examples 4.10 and 4.11. (*Hint:* consider the free-body diagram of the portion of the beam located above the neutral surface.)

Fig. P4.210

Fig. 4.1

Normal strain in bending

This chapter was devoted to the analysis of members in *pure bending*. That is, we considered the stresses and deformation in members subjected to equal and opposite couples $\mathbf{M}$ and $\mathbf{M}'$ acting in the same longitudinal plane (Fig. 4.1).

We first studied members possessing a plane of symmetry and subjected to couples acting in that plane. Considering possible deformations of the member, we proved that *transverse sections remain plane* as a member is deformed [Sec. 4.3]. We then noted that a member in pure bending has a *neutral surface* along which normal strains and stresses are zero and that the longitudinal *normal strain* ϵ_x *varies linearly* with the distance y from the neutral surface:

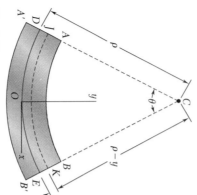

Fig. 4.12a

Normal stress in elastic range

$$\epsilon_x = -\frac{y}{\rho} \qquad (4.8)$$

where ρ is the *radius of curvature* of the neutral surface (Fig. 4.12a). The intersection of the neutral surface with a transverse section is known as the *neutral axis* of the section.

For members made of a material that follows Hooke's law [Sec. 4.4], we found that the *normal stress* σ_x *varies linearly* with the distance from the neutral axis (Fig. 4.13). Denoting by σ_m the maximum stress we wrote

$$\sigma_x = -\frac{y}{c}\sigma_m \qquad (4.12)$$

where c is the largest distance from the neutral axis to a point in the section.

By setting the sum of the elementary forces, $\sigma_x\,dA$, equal to zero, we proved that the *neutral axis passes through the centroid of the cross section* of a member in pure bending. Then by setting the sum of the moments of the elementary forces equal to the bending moment, we derived the *elastic flexure formula* for the maximum normal stress

$$\sigma_m = \frac{Mc}{I} \qquad (4.15)$$

where I is the moment of inertia of the cross section with respect to the neutral axis. We also obtained the normal stress at any distance y from the neutral axis:

$$\sigma_x = -\frac{My}{I} \qquad (4.16)$$

Fig. 4.13

Elastic flexure formula

Noting that I and c depend only on the geometry of the cross section, we introduced the *elastic section modulus*

Elastic section modulus

$$S = \frac{I}{c} \qquad (4.17)$$

and then used the section modulus to write an alternative expression for the maximum normal stress:

$$\sigma_m = \frac{M}{S} \qquad (4.18)$$

Recalling that the curvature of a member is the reciprocal of its radius of curvature, we expressed the *curvature* of the member as

Curvature of member

$$\frac{1}{\rho} = \frac{M}{EI} \qquad (4.21)$$

In Sec. 4.5, we completed our study of the bending of homogeneous members possessing a plane of symmetry by noting that deformations occur in the plane of a transverse cross section and result in *anticlastic curvature* of the members.

Anticlastic curvature

Next we considered the bending of members made of several materials with *different moduli of elasticity* [Sec. 4.6]. While transverse sections remain plane, we found that, in general, the *neutral axis does not pass through the centroid* of the composite cross section (Fig. 4.24). Using the ratio of the moduli of elasticity of the ma-

Members made of several materials

$\epsilon_x = -\dfrac{y}{\rho}$

$\sigma_1 = -\dfrac{E_1 y}{\rho}$

$\sigma_2 = -\dfrac{E_2 y}{\rho}$

(a) (b) (c)

Fig. 4.24

$\sigma_x = -\dfrac{My}{I}$

N.A.

C

Fig. 4.26

terials, we obtained a *transformed section* corresponding to an equivalent member made entirely of one material. We then used the methods previously developed to determine the stresses in this equivalent homogeneous member (Fig. 4.26) and then again used the ratio of the moduli of elasticity to determine the stresses in the composite beam [Sample Probs. 4.3 and 4.4].

In Sec. 4.7, *stress concentrations* that occur in members in pure bending were discussed and charts giving stress concentration factors for flat bars with fillets and grooves were presented in Figs. 4.31 and 4.32.

Stress concentrations

We next investigated members made of materials that do not follow Hooke's law [Sec. 4.8]. A rectangular beam made of an *elasto-plastic material* (Fig. 4.39) was analyzed as the magnitude of the bending moment was increased. The *maximum elastic moment* M_Y occurred when yielding was initiated in the beam (Fig. 4.40). As the bending moment was further increased, plastic zones developed and the size of the elastic core of the member decreased [Sec. 4.9]. Finally the beam became fully plastic and we obtained the maximum or *plastic moment* M_p. In Sec. 4.11, we found that *permanent deformations* and *residual stresses* remain in a member after the loads that caused yielding have been removed.

Fig. 4.39

Plastic deformations

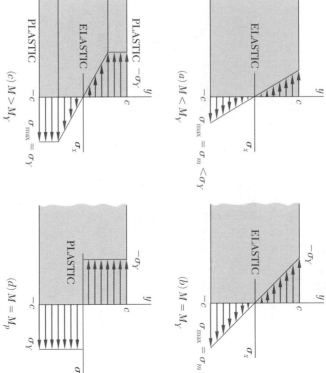

$(a)\ M < M_Y$

$(b)\ M = M_Y$

$(c)\ M > M_Y$

$(d)\ M = M_p$

Fig. 4.40

Eccentric axial loading

Fig. 4.48b

In Sec. 4.12, we studied the stresses in members loaded *eccentrically in a plane of symmetry*. Our analysis made use of methods developed earlier. We replaced the *eccentric load* by a force-couple system located at the centroid of the cross section (Fig. 4.48b) and then superposed stresses due to the centric load and the bending couple (Fig. 4.51):

$$\sigma_x = \frac{P}{A} - \frac{My}{I} \qquad (4.50)$$

Fig. 4.51

Unsymmetric bending

The bending of members of *unsymmetric cross section* was considered next [Sec. 4.13]. We found that the flexure formula may be used, provided that the couple vector **M** is directed along one of the principal centroidal axes of the cross section. When necessary we re-

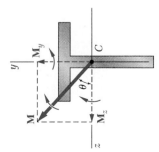

Fig. 4.60

solved **M** into components along the principal axes and superposed the stresses due to the component couples (Figs. 4.60 and 4.61).

Fig. 4.61

$$\sigma_x = -\frac{M_z y}{I_z} + \frac{M_y z}{I_y} \qquad (4.55)$$

For the couple **M** shown in Fig. 4.65, we determined the orientation of the neutral axis by writing

$$\tan \phi = \frac{I_z}{I_y} \tan \theta \qquad (4.57)$$

Fig. 4.65

General eccentric axial loading

The general case of *eccentric axial loading* was considered in Sec. 4.14, where we again replaced the load by a force-couple system located at the centroid. We then superposed the stresses due to the centric load and two component couples directed along the principal axes:

$$\sigma_x = \frac{P}{A} - \frac{M_z y}{I_z} + \frac{M_y z}{I_y} \qquad (4.58)$$

Curved members

The chapter concluded with the analysis of stresses in *curved members* (Fig. 4.76a). While transverse sections remain plane when the member is subjected to bending, we found that the *stresses do not vary linearly* and the neutral surface does not pass through the centroid of the section. The distance R from the center of curvature of the member to the neutral surface was found to be

$$R = \frac{A}{\int \frac{dA}{r}} \qquad (4.66)$$

where A is the area of the cross section. The normal stress at a distance y from the neutral surface was expressed as

$$\sigma_x = -\frac{My}{Ae(R - y)} \qquad (4.70)$$

where M is the bending moment and e the distance from the centroid of the section to the neutral surface.

Fig. 4.76a

4.211 A single vertical force **P** is applied to a short steel post as shown. Gages located at A, B, and C indicate the following strains:

$$\epsilon_A = -500\,\mu \qquad \epsilon_B = -1000\,\mu \qquad \epsilon_C = -200\,\mu$$

Knowing that $E = 29 \times 10^6$ psi, determine (a) the magnitude of **P**, (b) the line of action of **P**, (c) the corresponding strain at the hidden edge of the post, where $x = -2.5$ in. and $z = -1.5$ in.

Fig. P4.211

4.212 The couple **M**, which acts in a vertical plane ($\beta = 0$), is applied to an aluminum beam of the cross section shown. Determine (a) the stress at point A, (b) the stress at point B, (c) the radius of curvature of the beam. Use $E = 72$ GPa.

4.213 The couple **M** is applied to a beam of the cross section shown in a plane forming an angle $\beta = 15°$ with the vertical. Determine (a) the stress at point A, (b) the stress at point B, (c) the angle that the neutral axis forms with the horizontal.

Fig. P4.212 and P4.213

4.214 Determine the maximum stress in each of the two machine elements shown.

4.215 The four forces shown are applied to a rigid plate supported by a solid steel post of radius a. Determine the maximum stress in the post when (a) all four forces are applied, (b) the force at D is removed, (c) the forces at C and D are removed.

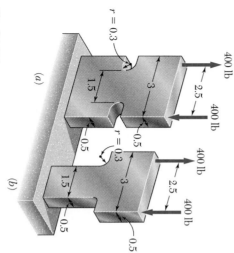

$r = 0.3$
1.5
3
2.5
400 lb
400 lb
0.5
0.5
$r = 0.3$
(a)

400 lb
1.5
3
2.5
400 lb
0.5
0.5
(b)

Fig. P4.214

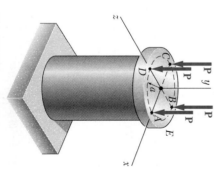

Fig. P4.215

4.216 In order to increase corrosion resistance, a 0.08-in.-thick cladding of aluminum has been added to a steel bar as shown. The modulus of elasticity is 29×10^6 psi for steel and 10.4×10^6 psi for aluminum. For a bending moment of 12 kip · in., determine (*a*) the maximum stress in the steel, (*b*) the maximum stress in the aluminum, (*c*) the radius of curvature of the bar.

1.34 in.

1.5 in.

1.84 in.

2 in.

M

Fig. *P4.216*

4.217 A couple **M** of moment 8 kN · m acting in a vertical plane is applied to a W200 × 19.3 rolled-steel beam as shown. Determine (*a*) the angle that the neutral axis forms with the horizontal plane, (*b*) the maximum stress in the beam.

W200 × 19.3

5°

y'

A

B

C

z'

E

D

M = 8 kN · m

Fig. P4.217

4.218 Three 120 × 10-mm steel plates have been welded together to form the beam shown. Assuming that the steel is elastoplastic with $E = 200$ GPa and $\sigma_Y = 300$ MPa, determine (*a*) the bending moment for which the plastic zones at the top and bottom of the beam are 40 mm thick, (*b*) the corresponding radius of curvature of the beam.

10 mm

120 mm

10 mm

120 mm

120 mm

10 mm

M

Fig. P4.218

4.219 Two thin strips of the same material and same cross section are bent by couples of the same magnitude and glued together. After the two surfaces of contact have been securely bonded, the couples are removed. Denoting by σ_1 the maximum stress and by ρ_1 the radius of curvature of each strip while the couples were applied, determine (a) the final radius of curvature, (b) the final stresses at points A, B, C, and D, (b) the final radius of curvature.

Fig. P4.219

4.220 Knowing that the hollow beam shown has a uniform wall thickness of 0.25 in. determine (a) the largest couple that can be applied without exceeding the allowable stress of 20 ksi, (b) the corresponding radius of curvature of the beam. Use $E = 10.6 \times 10^6$ psi.

Fig. P4.220

4.221 A beam of the cross section shown is extruded from an aluminum alloy for which $E = 72$ GPa. Knowing that the couple shown acts in a vertical plane, determine (a) the maximum stress in the beam, (b) the corresponding radius of curvature.

$M = 3$ kN · m

10 40 10 40 10

10 20 10

Dimensions in mm

Fig. P4.221

4.222 For the machine element and loading shown, determine the stress at point A, knowing that (a) $h = 0.9$ in., (b) $h = 1.5$ in.

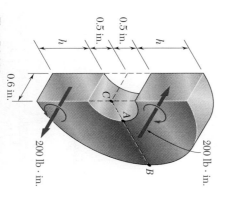

0.5 in.

0.5 in.

h

h

0.6 in.

200 lb · in.

200 lb · in.

Fig. P4.222

COMPUTER PROBLEMS

The following problems are designed to be solved with a computer.

4.C1 Two aluminum strips and a steel strip are to be bonded together to form a composite member of width $b = 60$ mm and depth $h = 40$ mm. The modulus of elasticity is 200 GPa for the steel and 75 GPa for the aluminum. Knowing that $M = 1500$ N · m, write a computer program to calculate the maximum stress in the aluminum and in the steel for values of a from 0 to 20 mm using 2-mm increments. Using appropriate smaller increments, determine (a) the largest stress that can occur in the steel, (b) the corresponding value of a.

Fig. P4.C1

4.C2 A beam of the cross section shown, made of a steel that is assumed to be elastoplastic with a yield strength σ_Y and a modulus of elasticity E, is bent about the x axis. (a) Denoting by y_Y the half thickness of the elastic core, write a computer program to calculate the bending moment M and the radius of curvature ρ for values of y_Y from $\frac{1}{2}d$ to $\frac{1}{6}d$ using decrements equal to $\frac{1}{2}t_f$. Neglect the effect of fillets. (b) Use this program to solve Prob. 4.218.

Fig. P4.C2

4.C3 An 8 kip · in. couple **M** is applied to a beam of the cross section shown in a plane forming an angle β with the vertical. Noting that the centroid of the cross section is located at C and that the y and z axes are principal axes, write a computer program to calculate the stress at A, B, C, and D for values of β from 0 to 180° using 10° increments. (Given: $I_y = 6.23$ in⁴ and $I_z = 1.481$ in⁴.)

Fig. P4.C3

Dimensions in inches

4.C4 Couples of moment $M = 2$ kN · m are applied to a curved bar having a rectangular cross section with $h = 100$ mm and $b = 25$ mm. Write a computer program and use it to calculate the stresses at points A and B for values of the ratio r_1/h from 10 to 1 using decrements of 1, and from 1 to 0.1 using decrements of 0.1. Using appropriate smaller increments, determine the ratio r_1/h for which the maximum stress in the curved bar is 50 percent larger than the maximum stress in a strainght bar of the same cross section.

Fig. P4.C4

4.C5 The couple $\mathbf{M}$ is applied to a beam of the cross section shown. (a) Write a computer program that, for loads expressed in either SI or U.S. customary units, can be used to calculate the maximum tensile and compressive stresses in the beam. (b) Use this program to solve Probs. 4.1, 4.10, and 4.11.

Fig. P4.C5

4.C6 A solid rod of radius $c = 1.2$ in. is made of a steel that is assumed to be elastoplastic with $E = 29,000$ ksi and $\sigma_Y = 42$ ksi. The rod is subjected to a couple of moment M that increases from zero to the maximum elastic moment M_Y and then to the plastic moment M_p. Denoting by y_Y the half thickness of the elastic core, write a computer program and use it to calculate the bending moment M and the radius of curvature ρ for values of y_Y from 1.2 in. to 0 using 0.2-in. decrements. (*Hint*: Divide the cross section into 80 horizontal elements of 0.03-in. height.)

Fig. P4.C6

4.C7 The machine element of Prob. 4.204 is to be redesigned by removing part of the triangular cross section. It is believed that the removal of a small triangular area of width a will lower the maximum stress in the element. In order to verify this design concept, write a computer program to calculate the maximum stress in the element for values of a from 0 to 1 in. using 0.1-in. increments. Using appropriate smaller increments, determine the distance a for which the maximum stress is as small as possible and the corresponding value of the maximum stress.

Fig. P4.C7

5

Analysis and Design of Beams for Bending

The beams supporting the multiple overhead cranes system shown in this picture are subjected to transverse loads causing the beams to bend. The normal stresses resulting from such loadings will be determined in this chapter.

5.1. INTRODUCTION

This chapter and most of the next one will be devoted to the analysis and the design of *beams*, i.e., structural members supporting loads applied at various points along the member. Beams are usually long, straight prismatic members, as shown in the photo on the previous page. Steel and aluminum beams play an important part in both structural and mechanical engineering. Timber beams are widely used in home construction (Fig. 5.1). In most cases, the loads are perpendicular to the axis of the beam. Such a *transverse loading* causes only bending and shear in the beam. When the loads are not at a right angle to the beam, they also produce axial forces in the beam.

Fig. 5.1

The transverse loading of a beam may consist of *concentrated loads* $\mathbf{P}_1, \mathbf{P}_2, \ldots$, expressed in newtons, pounds, or their multiples, kilonewtons and kips (Fig. 5.2*a*), of a *distributed load* w, expressed in N/m, kN/m, lb/ft, or kips/ft (Fig. 5.2*b*), or of a combination of both. When the load w per unit length has a constant value over part of the beam (as between A and B in Fig. 5.2*b*), the load is said to be *uniformly distributed* over that part of the beam.

Beams are classified according to the way in which they are supported. Several types of beams frequently used are shown in Fig. 5.3. The distance L shown in the various parts of the figure is called the *span*. Note that the reactions at the supports of the beams in parts *a*, *b*, and *c* of the figure involve a total of only three unknowns and, therefore, can be determined by

(a) Concentrated loads

(b) Distributed load

Fig. 5.2

Statically Determinate Beams

(a) Simply supported beam

(b) Overhanging beam

(c) Cantilever beam

Statically Indeterminate Beams

(d) Continuous beam

(e) Beam fixed at one end and simply supported at the other end

(f) Fixed beam

Fig. 5.3

the methods of statics. Such beams are said to be *statically determinate* and will be discussed in this chapter and the next. On the other hand, the re- actions at the supports of the beams in parts *d*, *e*, and *f* of Fig. 5.3 involve more than three unknowns and cannot be determined by the methods of statics alone. The properties of the beams with regard to their resistance to deformations must be taken into consideration. Such beams are said to be *statically indeterminate* and their analysis will be postponed until Chap. 9, where deformations of beams will be discussed.

Sometimes two or more beams are connected by hinges to form a sin- gle continuous structure. Two examples of beams hinged at a point *H* are shown in Fig. 5.4. It will be noted that the reactions at the supports involve four unknowns and cannot be determined from the free-body diagram of the two-beam system. They can be determined, however, by considering the free-body diagram of each beam separately; six unknowns are involved (including two force components at the hinge), and six equations are avail- able.

Fig. 5.4

It was shown in Sec. 4.1 that if we pass a section through a point *C* of a cantilever beam supporting a concentrated load **P** at its end (Fig. 4.6), the internal forces in the section are found to consist of a shear force **P′** equal and opposite to the load **P** and a bending couple **M** of moment equal to the moment of **P** about *C*. A similar situation prevails for other types of supports and loadings. Consider, for example, a simply supported beam *AB* carrying two concentrated loads and a uniformly distributed load (Fig. 5.5*a*). To determine the internal forces in a section through point *C* we first draw the free-body diagram of the entire beam to obtain the reactions at the supports (Fig. 5.5*b*). Passing a section through *C*, we then draw the free-body diagram of *AC* (Fig. 5.5*c*), from which we determine the shear force **V** and the bending couple **M**.

The bending couple **M** creates *normal stresses* in the cross section, while the shear force **V** creates *shearing stresses* in that section. In most cases the dominant criterion in the design of a beam for strength is the maximum value of the normal stress in the beam. The determination of the normal stresses in a beam will be the subject of this chapter, while shear- ing stresses will be discussed in Chap. 6.

Fig. 5.5

Since the distribution of the normal stresses in a given section depends only upon the value of the bending moment *M* in that section and the geo- metry of the section,† the elastic flexure formulas derived in Sec. 4.4 can be used to determine the maximum stress, as well as the stress at any given point, in the section. We write‡

$$\sigma_m = \frac{|M|c}{I} \qquad \sigma_x = -\frac{My}{I} \qquad (5.1, 5.2)$$

where *I* is the moment of inertia of the cross section with respect to a centroidal axis perpendicular to the plane of the couple, *y* is the dis- tance from the neutral surface, and *c* is the maximum value of that dis- tance (Fig. 4.13). We also recall from Sec. 4.4 that, introducing the elas-

---

† It is assumed that the distribution of the normal stresses in a given cross section is not affected by the deformations caused by the shearing stresses. This assumption will be veri- fied in Sec. 6.5.

‡ We recall from Sec. 4.2 that *M* can be positive or negative, depending upon whether the concavity of the beam at the point considered faces upward or downward. Thus, in the case considered here of a transverse loading, the sign of *M* can vary along the beam. since, on the other hand, σ_m is a positive quantity, the absolute value of *M* is used in Eq. (5.1).

tic section modulus $S = I/c$ of the beam, the maximum value σ_m of the normal stress in the section can be expressed as

$$\sigma_m = \frac{|M|}{S} \tag{5.3}$$

The fact that σ_m is inversely proportional to S underlines the importance of selecting beams with a large section modulus. Section moduli of various rolled-steel shapes are given in Appendix C, while the section modulus of a rectangular shape can be expressed, as shown in Sec. 4.4, as

$$S = \tfrac{1}{6}bh^2 \tag{5.4}$$

where b and h are, respectively, the width and the depth of the cross section.

Equation (5.3) also shows that, for a beam of uniform cross section, σ_m is proportional to $|M|$: Thus, the maximum value of the normal stress in the beam occurs in the section where $|M|$ is largest. It follows that one of the most important parts of the design of a beam for a given loading condition is the determination of the location and magnitude of the largest bending moment.

This task is made easier if a *bending-moment diagram* is drawn, i.e., if the value of the bending moment M is determined at various points of the beam and plotted against the distance x measured from one end of the beam. It is further facilitated if a *shear diagram* is drawn at the same time by plotting the shear V against x.

The sign convention to be used to record the values of the shear and bending moment will be discussed in Sec. 5.2. The values of V and M will then be obtained at various points of the beam by drawing free-body diagrams of successive portions of the beam. In Sec. 5.3 relations among load, shear, and bending moment will be derived and used to obtain the shear and bending-moment diagrams. This approach facilitates the determination of the largest absolute value of the bending moment and, thus, the determination of the maximum normal stress in the beam.

In Sec. 5.4 you will learn to design a beam for bending, i.e., so that the maximum normal stress in the beam will not exceed its allowable value. As indicated earlier, this is the dominant criterion in the design of a beam.

Another method for the determination of the maximum values of the shear and bending moment, based on expressing V and M in terms of *singularity functions*, will be discussed in Sec. 5.5. This approach lends itself well to the use of computers and will be expanded in Chap. 9 to facilitate the determination of the slope and deflection of beams.

Finally, the design of *nonprismatic beams*, i.e., beams with a variable cross section, will be discussed in Sec. 5.6. By selecting the shape and size of the variable cross section so that its elastic section modulus $S = I/c$ varies along the length of the beam in the same way as $|M|$, it is possible to design beams for which the maximum normal stress in each section is equal to the allowable stress of the material. Such beams are said to be of *constant strength*.

5.2. SHEAR AND BENDING-MOMENT DIAGRAMS

As indicated in Sec. 5.1, the determination of the maximum absolute values of the shear and of the bending moment in a beam are greatly facilitated if V and M are plotted against the distance x measured from one end of the beam. Besides, as you will see in Chap. 9, the knowledge of M as a function of x is essential to the determination of the deflection of a beam.

In the examples and sample problems of this section, the shear and bending-moment diagrams will be obtained by determining the values of V and M at selected points of the beam. These values will be found in the usual way, i.e., by passing a section through the point where they are to be determined (Fig. 5.6a) and considering the equilibrium of the portion of beam located on either side of the section (Fig. 5.6b). Since the shear forces V and V' have opposite senses, recording the shear at point C with an up or down arrow would be meaningless, unless we indicated at the same time which of the free bodies AC and CB we are considering. For this reason, the shear V will be recorded with a sign: a *plus sign* if the shearing forces are directed as shown in Fig. 5.6b, and a *minus sign* otherwise. A similar convention will apply for the bending moment M. It will be considered as positive if the bending couples are directed as shown in that figure, and negative otherwise.[†] Summarizing the sign conventions we have presented, we state:

The shear V and the bending moment M at a given point of a beam are said to be positive when the internal forces and couples acting on each portion of the beam are directed as shown in Fig. 5.7a.

These conventions can be more easily remembered if we note that

1. *The shear at any given point of a beam is positive when the* **external** *forces (loads and reactions) acting on the beam tend to shear off the beam at that point as indicated in Fig. 5.7b.*
2. *The bending moment at any given point of a beam is positive when the* **external** *forces acting on the beam tend to bend the beam at that point as indicated in Fig. 5.7c.*

It is also of help to note that the situation described in Fig. 5.7, in which the values of the shear and of the bending moment are positive, is precisely the situation that occurs in the left half of a simply supported beam carrying a single concentrated load at its midpoint. This particular case is fully discussed in the next example.

(a)

(b)

Fig. 5.6

(a) Internal forces
(positive shear and positive bending moment)

Fig. 5.7

(b) Effect of external forces
(positive shear)

(c) Effect of external forces
(positive bending moment)

[†]Note that this convention is the same that we used earlier in Sec. 4.2

EXAMPLE 5.01

Draw the shear and bending-moment diagrams for a simply supported beam AB of span L subjected to a single concentrated load P at it midpoint C (Fig. 5.8).

We first determine the reactions at the supports from the free-body diagram of the entire beam (Fig. 5.9a); we find that the magnitude of each reaction is equal to $P/2$.

Fig. 5.8

Next we cut the beam at a point D between A and C and draw the free-body diagrams of AD and DB (Fig. 5.9b). Assuming that shear and bending moment are positive, we direct the internal forces V and V' and the internal couples M and M' as indicated in Fig. 5.7a. Considering the free body AD and writing that the sum of the vertical components and the sum of the moments about D of the forces acting on the free body are zero, we find $V = +P/2$ and $M = +Px/2$. Both the shear and the bending moment are therefore positive; this may be checked by observing that the reaction at A tends to shear off and to bend the beam at D as indicated in Figs. 5.7b and c. We now plot V and M between A and C (Figs. 5.9d and e); the shear has a constant value $V = P/2$, while the bending moment increases linearly from $M = 0$ at $x = 0$ to $M = PL/4$ at $x = L/2$.

Cutting, now, the beam at a point E between C and B and considering the free body EB (Fig. 5.9c), we write that the sum of the vertical components and the sum of the moments about E of the forces acting on the free body are zero. We obtain $V = -P/2$ and $M = P(L - x)/2$. The shear is therefore negative and the bending moment positive; this can be checked by observing that the reaction at B bends the beam at E as indicated in Fig. 5.7c but tends to shear it off in a manner opposite to that shown in Fig. 5.7b. We can complete, now, the shear and bending-moment diagrams of Figs. 5.9d and e; the shear has a constant value $V = -P/2$ between C and B, while the bending moment decreases linearly from $M = PL/4$ at $x = L/2$ to $M = 0$ at $x = L$.

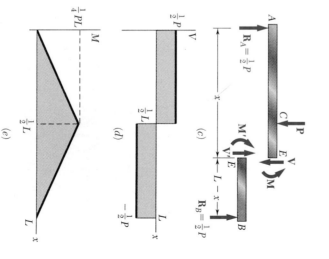

Fig. 5.9

We note from the foregoing example that, when a beam is subjected only to concentrated loads, the shear is constant between loads and the bending moment varies linearly between loads. In such situations, there- fore, the shear and bending-moment diagrams can easily be drawn, once the values of V and M have been obtained at sections selected just to the left and just to the right of the points where the loads and reactions are applied (see Sample Prob. 5.1).

EXAMPLE 5.02

Draw the shear and bending-moment diagrams for a cantilever beam AB of span L supporting a uniformly distributed load w (Fig. 5.10).

(handwritten: For a V to exist, there must be shearing (ie there must be a ran force) If no reaction at endpoint, the $V = 0$ here.)

Fig. 5.10

(a)

We cut the beam at a point C between A and B and draw the free-body diagram of AC (Fig. 5.11a), directing $\mathbf{V}$ and $\mathbf{M}$ as indicated in Fig. 5.7a. Denoting by x the distance from A to C and replacing the distributed load over AC by its result- ant wx applied at the midpoint of AC, we write

$$+\uparrow\Sigma F_y = 0: \qquad -wx - V = 0 \qquad V = -wx$$

$$+\curvearrowleft\Sigma M_C = 0: \qquad wx\left(\frac{x}{2}\right) + M = 0 \qquad M = -\frac{1}{2}wx^2$$

(handwritten: $V = \dfrac{dM}{dx}$)

We note that the shear diagram is represented by an oblique straight line (Fig. 5.11b) and the bending-moment diagram by a parabola (Fig. 5.11c). The maximum values of V and M both occur at B, where we have

$$V_B = -wL \qquad M_B = -\frac{1}{2}wL^2$$

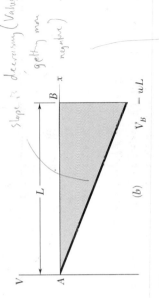

(handwritten: Slope is decreasing (Values getting more negative))

(b)

$V_B = -wL$

(handwritten: Slope is getting more negative)

(c)

$M_B = -\frac{1}{2}wL^2$

Fig. 5.11

SAMPLE PROBLEM 5.1

For the timber beam and loading shown, draw the shear and bending-moment diagrams and determine the maximum normal stress due to bending.

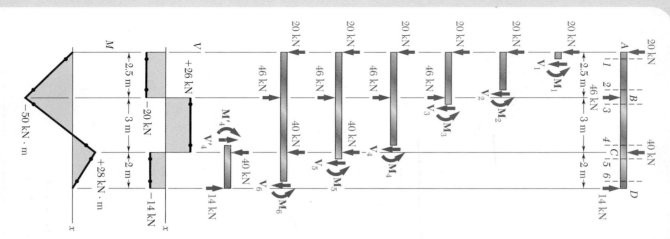

SOLUTION

Reactions. Considering the entire beam as a free body, we find

$$\mathbf{R}_B = 40 \text{ kN} \uparrow \qquad \mathbf{R}_D = 14 \text{ kN} \uparrow$$

Shear and Bending-Moment Diagrams. We first determine the internal forces just to the right of the 20-kN load at A. Considering the stub of beam to the left of section I as a free body and assuming V and M to be positive (according to the standard convention), we write

$$+\uparrow\Sigma F_y = 0: \qquad -20 \text{ kN} - V_1 = 0 \qquad V_1 = -20 \text{ kN}$$
$$+\curvearrowleft\Sigma M_1 = 0: \qquad (20 \text{ kN})(0 \text{ m}) + M_1 = 0 \qquad M_1 = 0$$

We next consider as a free body the portion of beam to the left of section 2 and write

$$+\uparrow\Sigma F_y = 0: \qquad -20 \text{ kN} - V_2 = 0 \qquad V_2 = -20 \text{ kN}$$
$$+\curvearrowleft\Sigma M_2 = 0: \qquad (20 \text{ kN})(2.5 \text{ m}) + M_2 = 0 \qquad M_2 = -50 \text{ kN} \cdot \text{m}$$

The shear and bending moment at sections 3, 4, 5, and 6 are determined in a similar way from the free-body diagrams shown. We obtain

$$V_3 = +26 \text{ kN} \qquad M_3 = -50 \text{ kN} \cdot \text{m}$$
$$V_4 = +26 \text{ kN} \qquad M_4 = +28 \text{ kN} \cdot \text{m}$$
$$V_5 = -14 \text{ kN} \qquad M_5 = +28 \text{ kN} \cdot \text{m}$$
$$V_6 = -14 \text{ kN} \qquad M_6 = 0$$

For several of the latter sections, the results may be more easily obtained by considering as a free body the portion of the beam to the right of the section. For example, for the portion of the beam to the right of section 4, we have

$$+\uparrow\Sigma F_y = 0: \qquad V_4 - 40 \text{ kN} + 14 \text{ kN} = 0 \qquad V_4 = +26 \text{ kN}$$
$$+\curvearrowleft\Sigma M_4 = 0: \qquad -M_4 + (14 \text{ kN})(2 \text{ m}) = 0 \qquad M_4 = +28 \text{ kN} \cdot \text{m}$$

We can now plot the six points shown on the shear and bending-moment diagrams. As indicated earlier in this section, the shear is of constant value between concentrated loads, and the bending moment varies linearly; we obtain therefore the shear and bending-moment diagrams shown.

Maximum Normal Stress. It occurs at B, where $|M|$ is largest. We use Eq. (5.4) to determine the section modulus of the beam:

$$S = \tfrac{1}{6}bh^2 = \tfrac{1}{6}(0.080 \text{ m})(0.250 \text{ m})^2 = 833.33 \times 10^{-6} \text{ m}^3$$

Substituting this value and $|M| = |M_B| = 50 \times 10^3 \text{ N} \cdot \text{m}$ into Eq. (5.3):

$$\sigma_m = \frac{|M_B|}{S} = \frac{(50 \times 10^3 \text{ N} \cdot \text{m})}{833.33 \times 10^{-6}} = 60.00 \times 10^6 \text{ Pa}$$

Maximum normal stress in the beam = 60.0 MPa ◄

SAMPLE PROBLEM 5.2

The structure shown consists of a W10 × 112 rolled-steel beam AB and of two short members welded together and to the beam. (a) Draw the shear and bending-moment diagrams for the beam and the given loading. (b) Determine the maximum normal stress in sections just to the left and just to the right of point D.

SOLUTION

Equivalent Loading of Beam. The 10-kip load is replaced by an equivalent force-couple system at D. The reaction at B is determined by considering the beam as a free body.

a. Shear and Bending-Moment Diagrams

From A to C. We determine the internal forces at a distance x from point A by considering the portion of beam to the left of section 1. That part of the distributed load acting on the free body is replaced by its resultant, and we write

$$+\uparrow\Sigma F_y = 0: \qquad -3x - V = 0 \qquad V = -3x \text{ kips}$$
$$+\curvearrowleft\Sigma M_1 = 0: \qquad 3(\tfrac{1}{2}x) + M = 0 \qquad M = -1.5x^2 \text{ kip} \cdot \text{ft}$$

Since the free-body diagram shown can be used for all values of x smaller than 8 ft, the expressions obtained for V and M are valid in the region $0 < x < 8$ ft.

From C to D. Considering the portion of beam to the left of section 2 and again replacing the distributed load by its resultant, we obtain

$$+\uparrow\Sigma F_y = 0: \qquad -24 - V = 0 \qquad V = -24 \text{ kips}$$
$$+\curvearrowleft\Sigma M_2 = 0: \qquad 24(x-4) + M = 0 \qquad M = 96 - 24x \qquad \text{kip} \cdot \text{ft}$$

These expressions are valid in the region 8 ft < x < 11 ft.

From D to B. Using the position of beam to the left of section 3, we obtain for the region 11 ft < x < 16 ft

$$V = -34 \text{ kips} \qquad M = 226 - 34x \qquad \text{kip} \cdot \text{ft}$$

The shear and bending-moment diagrams for the entire beam can now be plotted. We note that the couple of moment 20 kip · ft applied at point D introduces a discontinuity into the bending-moment diagram.

b. Maximum Normal Stress to the Left and Right of Point D.

From Appendix C we find that for the W10 × 112 rolled-steel shape, $S = 126$ in³ about the X-X axis.

To the left of D: We have $|M| = 168$ kip · ft $= 2016$ kip · in. Substituting for $|M|$ and S into Eq. (5.3), we write

$$\sigma_m = \frac{|M|}{S} = \frac{2016 \text{ kip} \cdot \text{in.}}{126 \text{ in}^3} = 16.00 \text{ ksi} \qquad \sigma_m = 16.00 \text{ ksi} \blacktriangledown$$

To the right of D: We have $|M| = 148$ kip · ft $= 1776$ kip · in. Substituting for $|M|$ and S into Eq. (5.3), we write

$$\sigma_m = \frac{|M|}{S} = \frac{1776 \text{ kip} \cdot \text{in.}}{126 \text{ in}^3} = 14.10 \text{ ksi} \qquad \sigma_m = 14.10 \text{ ksi} \blacktriangledown$$

PROBLEMS

5.1 through 5.6 Draw the shear and bending-moment diagrams for the beam and loading shown.

Fig. P5.1 and P5.7

Fig. P5.5 and P5.11

Fig. P5.3 and P5.9

Fig. P5.2 and P5.8

Fig. P5.4 and P5.10

Fig. P5.6 and P5.12

5.7 through 5.12 Determine the equations of the shear and bending-moment curves for the beam and loading shown. (Place the origin at point A.)

5.13 and 5.14   Draw the shear and bending-moment diagrams for the beam and loading shown, and determine the maximum absolute value (*a*) of the shear, (*b*) of the bending moment.

Fig. P5.13

Dimensions in mm

Fig. P5.14

Dimensions in mm

5.15 and 5.16   Draw the shear and bending-moment diagrams for the beam and loading shown, and determine the maximum absolute value (*a*) of the shear, (*b*) of the bending moment.

Fig. P5.15

Fig. P5.16

5.17 and 5.18   Draw the shear and bending-moment diagrams for the beam and loading shown, and determine the maximum absolute value (*a*) of the shear, (*b*) of the bending moment.

Dimensions in mm

Fig. P5.17

Fig. P5.18

5.19 and 5.20 Assuming the upward reaction of the ground to be uniformly distributed, draw the shear and bending-moment diagrams for the beam AB and determine the maximum value (a) of the shear, (b) of the bending moment.

10 kN/m

36 kN

10 kN/m

A   C   D   E   B

0.9 m 0.9 m 0.9 m 0.9 m 0.9 m

Fig. P5.19

3 kips   3 kips

A   C   D   B

1.5 ft   4.5 ft   1.5 ft

Fig. P5.20

5.21 For the beam and loading shown, determine the maximum normal stress on a transverse section at C.

10 kN

3 kN/m

A   C   B

1.5 m   1.5 m   2.2 m

100 mm

200 mm

Fig. P5.21

5.22 For the beam and loading shown, determine the maximum normal stress on a transverse section at the center of the beam.

750 lb   750 lb

A   C   D   B

150 lb/ft

4 ft   4 ft   4 ft   4 ft

3 in.

12 in.

Fig. P5.22

5.23 For the beam and loading shown, determine the maximum normal stress on section a-a.

30 kN 50 kN 50 kN 50 kN 30 kN

A

a   a

b   b

B

W250 × 67

2 m

5 @ 0.8 m = 4 m

Fig. P5.23

5.24 For the beam and loading shown, determine the maximum normal stress due to bending on a transverse section at C.

30 kips   30 kips

A   C   D   E   B

6 kips/ft

W18 × 76

2.5 ft   2.5 ft   7.5 ft   2.5 ft

Fig. P5.24

25 25 10 10 10 10
kN kN kN kN kN kN

C D E F G

A B

S200 × 27.4

6 @ 0.375 m = 2.25 m

Fig. P5.25

8 kN

3 kN/m

C B

A

1.5 m 2.1 m

W310 × 60

Fig. P5.26

5.25 and 5.26 For the beam and loading shown, determine the maximum normal stress on a transverse section at C.

5.27 Draw the shear and bending-moment diagrams for the beam and loading shown and determine the maximum normal stress due to bending.

5.28 and 5.29 Draw the shear and bending moment diagrams for the beam and loading shown and determine the maximum normal stress due to bending.

20 kN   40 kN

A C D B

1.6 m 2.4 m 1.6 m

W360 × 64

Fig. P5.27

2 kN

4 kN/m

A C B

600 mm 100 mm

S100 × 11.5

Fig. P5.28

25 kips 25 kips 25 kips 25 kips

A C D E B

1 ft 2 ft 2 ft 6 ft 2 ft

S12 × 35

Fig. P5.29

5.30 Knowing that W = 3 kips, draw the shear and bending-moment diagrams for beam AD and determine the maximum normal stress due to bending.

2 kips   W   2 kips

C D E B

A

3 ft 3 ft 3 ft 3 ft 3 ft

W12 × 16

Fig. P5.30

5.31 and 5.32   Draw the shear and bending-moment diagrams for the beam and loading shown and determine the maximum normal stress due to bending.

Fig. P5.31

Fig. P5.32

5.33   Determine (*a*) the distance *a* for which the maximum absolute value of the bending moment in the beam is as small as possible, (*b*) the corresponding maximum normal stress due to bending. (*Hint:* Draw the bending-moment diagram and then equate the absolute values of the largest positive and negative bending moments obtained.)

Fig. P5.33

5.34   For the beam and loading shown, determine (*a*) the distance *a* for which the maximum absolute value of the bending moment in the beam is as small as possible, (*b*) the corresponding maximum normal stress due to bending. (See hint of Prob. 5.33.)

Fig. P5.34

5.35   Determine (*a*) the magnitude of the counterweight *W* for which the maximum absolute value of the bending moment in the beam is as small as possible, (*b*) the corresponding maximum normal stress due to bending. (See hint of Prob. 5.33.)

Fig. P5.35

5.36 For the beam and loading shown, determine (*a*) the distance *a* for which the maximum absolute value of the bending moment in the beam is as small as possible, (*b*) the corresponding maximum normal stress due to bending. (See hint of Prob. 5.33).

Fig. P5.36

5.37 and 5.38 Draw the shear and bending-moment diagrams for the beam and loading shown and determine the maximum normal stress due to bending.

Fig. P5.37

5.39 A solid steel bar has a square cross section of side *b* and is supported as shown. Knowing that for steel $\rho = 7860$ kg/m^3, determine the dimension *b* of the bar for which the maximum normal stress due to bending is (*a*) 10 MPa, (*b*) 50 MPa.

Fig. P5.39

Fig. P5.38

5.40 A solid steel rod of diameter *d* is supported as shown. Knowing that for steel $\gamma = 490$ lb/ft^3, determine the smallest diameter *d* that can be used if the normal stress due to bending is not to exceed 4 ksi.

Fig. P5.40

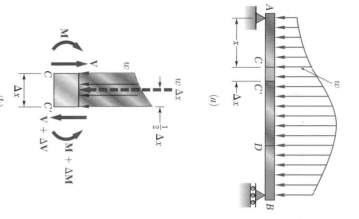

Fig. P5.12

5.3. RELATIONS AMONG LOAD, SHEAR, AND BENDING MOMENT

When a beam carries more than two or three concentrated loads, or when it carries distributed loads, the method outlined in Sec. 5.2 for plotting shear and bending moment can prove quite cumbersome. The construction of the shear diagram and, especially, of the bending-moment diagram will be greatly facilitated if certain relations existing among load, shear, and bending moment are taken into consideration.

Let us consider a simply supported beam AB carrying a distributed load w per unit length (Fig. 5.12a), and let C and C' be two points of the beam at a distance Δx from each other. The shear and bending moment at C will be denoted by V and M, respectively, and will be assumed positive; the shear and bending moment at C' will be denoted by $V + \Delta V$ and $M + \Delta M$.

We now detach the portion of beam CC' and draw its free-body diagram (Fig. 5.12b). The forces exerted on the free body include a load of magnitude $w\,\Delta x$ and internal forces and couples at C and C'. Since shear and bending moment have been assumed positive, the forces and couples will be directed as shown in the figure.

Relations between Load and Shear. Writing that the sum of the vertical components of the forces acting on the free body CC' is zero, we have

$$+\uparrow\Sigma F_y = 0: \qquad V - (V + \Delta V) - w\,\Delta x = 0$$
$$\Delta V = -w\,\Delta x$$

Dividing both members of the equation by Δx and then letting Δx approach zero, we obtain

$$\frac{dV}{dx} = -w \tag{5.5}$$

Equation (5.5) indicates that, for a beam loaded as shown in Fig. 5.12a, the slope dV/dx of the shear curve is negative; the numerical value of the slope at any point is equal to the load per unit length at that point.

Integrating (5.5) between points C and D, we write

$$V_D - V_C = -\int_{x_C}^{x_D} w\,dx \tag{5.6}$$

$$V_D - V_C = -(\text{area under load curve between } C \text{ and } D) \tag{5.6'}$$

Note that this result could also have been obtained by considering the equilibrium of the portion of beam CD, since the area under the load curve represents the total load applied between C and D.

It should be observed that Eq. (5.5) is not valid at a point where a concentrated load is applied; the shear curve is discontinuous at such a point, as seen in Sec. 5.2. Similarly, Eqs. (5.6) and (5.6') cease to be valid when concentrated loads are applied between C and D, since they do not take into account the sudden change in shear caused by a concentrated load. Equations (5.6) and (5.6'), therefore, should be applied only between successive concentrated loads.

Relations between Shear and Bending Moment. Returning to the free-body diagram of Fig. 5.12b, and writing now that the sum of the moments about C' is zero, we have

$+\curvearrowleft\Sigma M_{C'} = 0:$ $(M + \Delta M) - M - V\,\Delta x + w\,\Delta x \dfrac{\Delta x}{2} = 0$

$$\Delta M = V\,\Delta x - \frac{1}{2}w\,(\Delta x)^2$$

Dividing both members of the equation by Δx and then letting Δx approach zero, we obtain

$$\boxed{\dfrac{dM}{dx} = V} \qquad (5.7)$$

Equation (5.7) indicates that the slope dM/dx of the bending-moment curve is equal to the value of the shear. This is true at any point where the shear has a well-defined value, i.e., at any point where no concentrated load is applied. Equation (5.7) also shows that $V = 0$ at points where M is maximum. This property facilitates the determination of the points where the beam is likely to fail under bending.

Integrating (5.7) between points C and D, we write

$$M_D - M_C = \int_{x_C}^{x_D} V\,dx \qquad (5.8)$$

$$\boxed{M_D - M_C = \text{area under shear curve between } C \text{ and } D} \qquad (5.8')$$

Note that the area under the shear curve should be considered positive where the shear is positive and negative where the shear is negative. Equations (5.8) and (5.8') are valid even when concentrated loads are applied between C and D, as long as the shear curve has been correctly drawn. The equations cease to be valid, however, if a couple is applied at a point between C and D, since they do not take into account the sudden change in bending moment caused by a couple (see Sample Prob. 5.6).

Draw the shear and bending-moment diagrams for the simply supported beam shown in Fig. 5.13 and determine the maximum value of the bending moment.

From the free-body diagram of the entire beam, we determine the magnitude of the reactions at the supports.

$$R_A = R_B = \tfrac{1}{2}wL$$

Next, we draw the shear diagram. Close to the end A of the beam, the shear is equal to R_A, that is, to $\tfrac{1}{2}wL$, as we can check by considering as a free body a very small portion of the beam.

Fig. 5.13

Using Eq. (5.6), we then determine the shear V at any distance x from A; we write

$$V - V_A = -\int_0^x w\,dx = -wx$$

$$V = V_A - wx = \tfrac{1}{2}wL - wx = w(\tfrac{1}{2}L - x)$$

The shear curve is thus an oblique straight line which crosses the x axis at $x = L/2$ (Fig. 5.14a). Considering, now, the bending moment, we first observe that $M_A = 0$. The value M of the bending moment at any distance x from A may then be obtained from Eq. (5.8); we have

$$M - M_A = \int_0^x V\,dx$$

$$M = \int_0^x w(\tfrac{1}{2}L - x)\,dx = \tfrac{1}{2}w(Lx - x^2)$$

The bending-moment curve is a parabola. The maximum value of the bending moment occurs when $x = L/2$, since V (and thus dM/dx) is zero for that value of x. Substituting $x = L/2$ in the last equation, we obtain $M_{max} = wL^2/8$ (Fig. 5.14b).

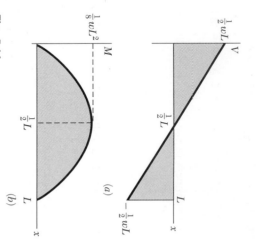

Fig. 5.14

In most engineering applications, one needs to know the value of the bending moment only at a few specific points. Once the shear diagram has been drawn, and after M has been determined at one of the ends of the beam, the value of the bending moment can then be obtained at any given point by computing the area under the shear curve and using Eq. (5.8′). For instance, since $M_A = 0$ for the beam of Example 5.03, the maximum value of the bending moment for that beam can be obtained simply by measuring the area of the shaded triangle in the shear diagram of Fig. 5.14a. We have

$$M_{max} = \frac{1}{2}\frac{L}{2}\frac{wL}{2} = \frac{wL^2}{8}$$

We note that, in this example, the load curve is a horizontal straight line, the shear curve an oblique straight line, and the bending-moment curve a parabola. If the load curve had been an oblique straight line (first degree), the shear curve would have been a parabola (second degree) and the bending-moment curve a cubic (third degree). The shear and bending-moment curves will always be, respectively, one and two degrees higher than the load curve. With this in mind, we should be able to sketch the shear and bending-moment diagrams without actually determining the functions $V(x)$ and $M(x)$, once a few values of the shear and bending moment have been computed. The sketches obtained will be more accurate if we make use of the fact that, at any point where the curves are continuous, the slope of the shear curve is equal to $-w$ and the slope of the bending-moment curve is equal to V.

SAMPLE PROBLEM 5.3

Draw the shear and bending-moment diagrams for the beam and loading shown.

SOLUTION

Reactions. Considering the entire beam as a free body, we write

$+\curvearrowleft \Sigma M_A = 0$:

$$D(24\text{ ft}) - (20\text{ kips})(6\text{ ft}) - (12\text{ kips})(14\text{ ft}) - (12\text{ kips})(28\text{ ft}) = 0$$
$$D = +26\text{ kips} \qquad \mathbf{D} = 26\text{ kips} \uparrow$$

$+\uparrow \Sigma F_y = 0$: $\quad A_y - 20\text{ kips} - 12\text{ kips} + 26\text{ kips} - 12\text{ kips} = 0$
$$A_y = +18\text{ kips} \qquad \mathbf{A}_y = 18\text{ kips} \uparrow$$

$\xrightarrow{+} \Sigma F_x = 0$: $\quad A_x - 0 \qquad\qquad\qquad \mathbf{A}_x - 0$

We also note that at both A and E the bending moment is zero; thus two points (indicated by dots) are obtained on the bending-moment diagram.

Shear Diagram. Since $dV/dx = -w$, we find that between concentrated loads and reactions the slope of the shear diagram is zero (i.e., the shear is constant). The shear at any point is determined by dividing the beam into two parts and considering either part as a free body. For example, using the portion of beam to the left of section 1, we obtain the shear between B and C:

$+\uparrow \Sigma F_y = 0$: $\quad +18\text{ kips} - 20\text{ kips} - V = 0 \qquad V = -2\text{ kips}$

We also find that the shear is $+12$ kips just to the right of D and zero at end E. Since the slope $dV/dx = -w$ is constant between D and E, the shear diagram between these two points is a straight line.

Bending-Moment Diagram. We recall that the area under the shear curve between two points is equal to the change in bending moment between the same two points. For convenience, the area of each portion of the shear diagram is computed and is indicated in parentheses on the diagram. Since the bending moment M_A at the left end is known to be zero, we write

$$M_B - M_A = +108 \qquad M_B = +108\text{ kip} \cdot \text{ft}$$
$$M_C - M_B = -16 \qquad M_C = +92\text{ kip} \cdot \text{ft}$$
$$M_D - M_C = -140 \qquad M_D = -48\text{ kip} \cdot \text{ft}$$
$$M_E - M_D = +48 \qquad M_E = 0$$

Since M_E is known to be zero, a check of the computations is obtained.

Between the concentrated loads and reactions the shear is constant; thus, the slope dM/dx is constant and the bending-moment diagram is drawn by connecting the known points with straight lines. Between D and E where the shear diagram is an oblique straight line, the bending-moment diagram is a parabola.

From the V and M diagrams we note that $V_{max} = 18$ kips and $M_{max} = 108$ kip $\cdot$ ft.

SAMPLE PROBLEM 5.4

The W360 × 79 rolled-steel beam AC is simply supported and carries the uniformly distributed load shown. Draw the shear and bending-moment diagrams for the beam and determine the location and magnitude of the maximum normal stress due to bending.

SOLUTION

Reactions. Considering the entire beam as a free body, we find

$$R_A = 80\text{ kN} \uparrow \qquad R_C = 40\text{ kN} \uparrow$$

Shear Diagram. The shear just to the right of A is $V_A = +80$ kN. Since the change in shear between two points is equal to *minus* the area under the load curve between the same two points, we obtain V_B by writing

$$V_B - V_A = -(20\text{ kN/m})(6\text{ m}) = -120\text{ kN}$$
$$V_B = -120 + V_A = -120 + 80 = -40\text{ kN}$$

The slope $dV/dx = -w$ being constant between A and B, the shear diagram between these two points is represented by a straight line. Between B and C, the area under the load curve is zero; therefore,

$$V_C - V_B = 0 \qquad V_C = V_B = -40\text{ kN}$$

and the shear is constant between B and C.

Bending-Moment Diagram. We note that the bending moment at each end of the beam is zero. In order to determine the maximum bending moment, we locate the section D of the beam where $V = 0$. We write

$$V_D - V_A = -wx$$
$$0 - 80\text{ kN} = -(20\text{ kN/m})x$$

and, solving for x:

$$x = 4\text{ m} \qquad \blacktriangleleft$$

The maximum bending moment occurs at point D, where we have $dM/dx = V = 0$. The areas of the various portions of the shear diagram are computed and are given (in parentheses) on the diagram. Since the area of the shear diagram between two points is equal to the change in bending moment between the same two points, we write

$$M_D - M_A = +160\text{ kN} \cdot \text{m} \qquad M_D = +160\text{ kN} \cdot \text{m}$$
$$M_B - M_D = -40\text{ kN} \cdot \text{m} \qquad M_B = +120\text{ kN} \cdot \text{m}$$
$$M_C - M_B = -120\text{ kN} \cdot \text{m} \qquad M_C = 0$$

The bending-moment diagram consists of an arc of parabola followed by a segment of straight line; the slope of the parabola at A is equal to the value of V at that point.

Maximum Normal Stress. It occurs at D, where $|M|$ is largest. From Appendix C we find that for a W360 × 79 rolled-steel shape, $S = 1280$ mm³ about a horizontal axis. Substituting this value and $|M_D| = 160 \times 10^3$ N · m into Eq. (5.3), we write

$$\sigma_m = \frac{|M_D|}{S} = \frac{160 \times 10^3\text{ N} \cdot \text{m}}{1280 \times 10^{-6}\text{ m}^3} = 125.0 \times 10^6\text{ Pa}$$

Maximum normal stress in the beam = 125.0 MPa $\qquad \blacktriangleleft$

SAMPLE PROBLEM 5.5

to get from dm/dx to shear

Start at a
Support, realizing
most get back to zero

Sketch the shear and bending-moment diagrams for the cantilever beam shown.

SOLUTION

Shear Diagram. At the free end of the beam, we find $V_A = 0$. Between A and B, the area under the load curve is $\frac{1}{2} w_0 a$; we find V_B by writing

$$V_B - V_A = -\frac{1}{2} w_0 a \qquad V_B = -\frac{1}{2} w_0 a$$

Between B and C, the beam is not loaded; thus $V_C = V_B$. At A, we have $w = w_0$ and, according to Eq. (5.5), the slope of the shear curve is $dV/dx = -w_0$, while at B the slope is $dV/dx = 0$. Between A and B, the loading decreases linearly, and the shear diagram is parabolic. Between B and C, $w = 0$, and the shear diagram is a horizontal line.

Bending-Moment Diagram. The bending moment M_A at the free end of the beam is zero. We compute the area under the shear curve and write

$$M_B - M_A = -\frac{1}{3} w_0 a^2 \qquad M_B = -\frac{1}{3} w_0 a^2$$
$$M_C - M_B = -\frac{1}{2} w_0 a(L - a)$$
$$M_C = -\frac{1}{6} w_0 a(3L - a)$$

The sketch of the bending-moment diagram is completed by recalling that $dM/dx = V$. We find that between A and B the diagram is represented by a cubic curve with zero slope at A, and between B and C by a straight line.

of no forces, it's straight line
a reduction of area getting bigger

SAMPLE PROBLEM 5.6

The simple beam AC is loaded by a couple of moment T applied at point B. Draw the shear and bending-moment diagrams of the beam.

SOLUTION

The entire beam is taken as a free body, and we obtain

$$\mathbf{R}_A = \frac{T}{L} \uparrow \qquad \mathbf{R}_C = \frac{T}{L} \downarrow$$

The shear at any section is constant and equal to T/L. Since a couple is applied at B, the bending-moment diagram is discontinuous at B; it is represented by two oblique straight lines and decreases suddenly at B by an amount equal to T.

PROBLEMS

5.41 Using the method of Sec. 5.3, solve Prob. 5.1.

5.42 Using the method of Sec. 5.3, solve Prob. 5.2.

5.43 Using the method of Sec. 5.3, solve Prob. 5.3.

5.44 Using the method of Sec. 5.3, solve Prob. 5.4.

5.45 Using the method of Sec. 5.3, solve Prob. 5.5.

5.46 Using the method of Sec. 5.3, solve Prob. 5.6.

5.47 Using the method of Sec. 5.3, solve Prob. 5.13.

5.48 Using the method of Sec. 5.3, solve Prob. 5.14.

5.49 Using the method of Sec. 5.3, solve Prob. 5.15.

5.50 Using the method of Sec. 5.3, solve Prob. 5.16.

5.51 and 5.52 Draw the shear and bending-moment diagrams for the beam and loading shown, and determine the maximum absolute value (*a*) of the shear, (*b*) of the bending moment.

Fig. P5.51

Fig. P5.52

5.53 Using the method of Sec. 5.3, solve Prob. 5.21.

5.54 Using the method of Sec. 5.3, solve Prob. 5.22.

5.55 Using the method of Sec. 5.3, solve Prob. 5.23.

5.56 Using the method of Sec. 5.3, solve Prob. 5.24.

5.57 and 5.58 Determine (*a*) the equations of the shear and bending-moment curves for the given beam and loading, (*b*) the maximum absolute value of the bending moment in the beam.

$$w = w_0\left(1 - \frac{x}{L}\right)$$

Fig. P5.57

$$w = w_0 \sin \frac{\pi x}{L}$$

Fig. P5.58

5.59 Determine (*a*) the equations of the shear and bending-moment curves for the given beam and loading, (*b*) the maximum absolute value of the bending moment in the beam.

$$w = w_0\,(x/L)^{1/2}$$

Fig. P5.59

Fig. P5.60

5.60 For the beam and loading shown, determine the equations of the shear and bending-moment curves and the maximum absolute value of the bending moment in the beam, knowing that (*a*) $k = 1$, (*b*) $k = 0.5$.

5.61 and 5.62 Draw the shear and bending-moment diagrams for the beam and loading shown and determine the maximum normal stress due to bending.

9 kN

12 kN/m

W200 × 19.3

0.9 m

3 m

Fig. P5.61

16 kN/m

S150 × 18.6

1.5 m

1 m

Fig. P5.62

5.63 and 5.64 Draw the shear and bending-moment diagrams for the beam and loading shown and determine the maximum normal stress due to bending.

Fig. P5.63

Fig. P5.64

5.65 and 5.66 Draw the shear and bending-moment diagrams for the beam and loading shown and determine the maximum normal stress due to bending.

Fig. P5.65

Fig. P5.66

5.67 and 5.68 Draw the shear and bending-moment diagrams for the beam and loading shown and determine the maximum normal stress due to bending.

Fig. P5.67

Fig. P5.68

5.69 Beam AB, of length L and square cross section of side a, is supported by a pivot at C and loaded as shown. (*a*) Check that the beam is in equilibrium. (*b*) Show that the maximum normal stress due to bending occurs at C and is equal to $w_0 L^2/(1.5a)^3$.

Fig. P5.69

5.70 Knowing that rod AB is in equilibrium under the loading shown, draw the shear and bending-moment diagrams and determine the maximum normal stress due to bending.

Fig. P5.70

*5.71 Beam AB supports a uniformly distributed load of 2 kN/m and two concentrated loads $\mathbf{P}$ and $\mathbf{Q}$. It has been experimentally determined that the normal stress due to bending on the bottom edge of the beam is −56.9 MPa at A and −29.9 MPa at C. Draw the shear and bending-moment diagrams for the beam and determine the magnitudes of the loads $\mathbf{P}$ and $\mathbf{Q}$.

Fig. P5.71

*5.72 Beam AB supports a uniformly distributed load of 1000 lb/ft and two concentrated loads $\mathbf{P}$ and $\mathbf{Q}$. It has been experimentally determined that the normal stress due to bending on the bottom edge of the lower flange of the W10 × 22 rolled-steel beam is +2.07 ksi at D and +0.776 ksi at E. (*a*) Draw the shear and bending-moment diagrams for the beam. (*b*) Determine the maximum normal stress due to bending that occurs in the beam.

Fig. P5.72

*5.73 Beam AB supports a uniformly distributed load of 8 kN/m and two concentrated loads $\mathbf{P}$ and $\mathbf{Q}$. It has been experimentally determined that the normal stress due to bending on the bottom edge of the lower flange of the W200 × 52 rolled-steel beam is 100 MPa at D and 70 MPa at E. (*a*) Draw the shear and bending-moment diagrams for the beam. (*b*) Determine the maximum normal stress due to bending that occurs in the beam.

Fig. P5.73

*5.74 Beam AB supports two concentrated loads $\mathbf{P}$ and $\mathbf{Q}$. It has been experimentally determined that the normal stress due to bending on the bottom edge of the beam is $+15$ MPa at C and $+22$ MPa at E. (a) Draw the shear and bending-moment diagrams for the beam. (b) Determine the maximum normal stress due to bending that occurs in the beam.

Fig. P5.74

5.4. DESIGN OF PRISMATIC BEAMS FOR BENDING

As indicated in Sec. 5.1, the design of a beam is usually controlled by the maximum absolute value $|M|_{\max}$ of the bending moment that will occur in the beam. The largest normal stress σ_m in the beam is found at the surface of the beam in the critical section where $|M|_{\max}$ occurs and can be obtained by substituting $|M|_{\max}$ for $|M|$ in Eq. (5.1) or Eq. (5.3).† We write

$$\sigma_m = \frac{|M|_{\max} c}{I} \qquad \sigma_m = \frac{|M|_{\max}}{S} \qquad (5.1', 5.3')$$

A safe design requires that $\sigma_m \leq \sigma_{\text{all}}$, where σ_{all} is the allowable stress for the material used. Substituting σ_{all} for σ_m in (5.3') and solving for S yields the minimum allowable value of the section modulus for the beam being designed:

$$S_{\min} = \frac{|M|_{\max}}{\sigma_{\text{all}}} \qquad (5.9)$$

The design of common types of beams, such as timber beams of rectangular cross section and rolled-steel beams of various cross-sectional shapes, will be considered in this section. A proper procedure should lead to the most economical design. This means that, among beams of the same type and the same material, and other things being equal, the beam with the smallest weight per unit length—and, thus, the smallest cross-sectional area—should be selected, since this beam will be the least expensive.

---

†For beams that are not symmetrical with respect to their neutral surface, the largest of the distances from the neutral surface to the surfaces of the beam should be used for c in Eq. (5.1) and in the computation of the section modulus $S = I/c$.

The design procedure will include the following steps†:

1. First determine the value of σ_{all} for the material selected from a table of properties of materials or from design specifications. You can also compute this value by dividing the ultimate strength σ_U of the material by an appropriate factor of safety (Sec. 1.13). Assuming for the time being that the value of σ_{all} is the same in tension and in compression, proceed as follows.

2. Draw the shear and bending-moment diagrams corresponding to the specified loading conditions, and determine the maximum absolute value $|M|_{max}$ of the bending moment in the beam.

3. Determine from Eq. (5.9) the minimum allowable value S_{min} of the section modulus of the beam.

4. For a timber beam, the depth h of the beam, its width b, or the ratio h/b characterizing the shape of its cross section will probably have been specified. The unknown dimensions may then be selected by recalling from Eq. (4.19) of Sec. 4.4 that b and h must satisfy the relation $\frac{1}{6} bh^2 = S \geq S_{min}$.

5. For a rolled-steel beam, consult the appropriate table in Appendix C. Of the available beam sections, consider only those with a section modulus $S \geq S_{min}$ and select from this group the section with the smallest weight per unit length. This is the most economical of the sections for which $S \geq S_{min}$. Note that this is not necessarily the section with the smallest value of S (see Example 5.04). In some cases, the selection of a section may be limited by other considerations, such as the allowable depth of the cross section, or the allowable deflection of the beam (cf. Chap. 9).

The foregoing discussion was limited to materials for which σ_{all} is the same in tension and in compression. If σ_{all} is different in tension and in compression, you should make sure to select the beam section in such a way that $\sigma_m \leq \sigma_{all}$ for both tensile and compressive stresses. If the cross section is not symmetric about its neutral axis, the largest tensile and the largest compressive stresses will not necessarily occur in the section where $|M|$ is maximum. One may occur where M is maximum and the other where M is minimum. Thus, step 2 should include the determination of both M_{max} and M_{min}, and step 3 should be modified to take into account both tensile and compressive stresses.

Finally, keep in mind that the design procedure described in this section takes into account only the normal stresses occurring on the surface of the beam. Short beams, especially those made of timber, may fail in shear under a transverse loading. The determination of shearing stresses in beams will be discussed in Chap. 6. Also, in the case of rolled-steel beams, normal stresses larger than those considered here may occur at the junction of the web with the flanges. This will be discussed in Chap. 8.

---

†We assume that all beams considered in this chapter are adequately braced to prevent lateral buckling, and that bearing plates are provided under concentrated loads applied to rolled-steel beams to prevent local buckling (crippling) of the web.

EXAMPLE 5.04

Select a wide-flange beam to support the 15-kip load as shown in Fig. 5.15. The allowable normal stress for the steel used is 24 ksi.

Fig. 5.15

1. The allowable normal stress is given: $\sigma_{all} = 24$ ksi.

2. The shear is constant and equal to 15 kips. The bending moment is maximum at B. We have

$$|M|_{max} = (15 \text{ kips})(8 \text{ ft}) = 120 \text{ kip} \cdot \text{ft} = 1440 \text{ kip} \cdot \text{in.}$$

3. The minimum allowable section modulus is

$$S_{min} = \frac{|M|_{max}}{\sigma_{all}} = \frac{1440 \text{ kip} \cdot \text{in.}}{24 \text{ ksi}} = 60.0 \text{ in}^3$$

4. Referring to the table of *Properties of Rolled-Steel Shapes* in Appendix C, we note that the shapes are arranged in groups of the same depth and that in each group they are listed in order of decreasing weight. We choose in each group the lightest beam having a section modulus $S = I/c$ at least as large as S_{min} and record the results in the following table.

Shape	S, in³
W21 × 44	81.6
W18 × 50	88.9
W16 × 40	64.7
W14 × 43	62.7
W12 × 50	64.7
W10 × 54	60.0

The most economical is the W16 × 40 shape since it weighs only 40 lb/ft, even though it has a larger section modulus than two of the other shapes. We also note that the total weight of the beam will be (8 ft) × (40 lb) = 320 lb. This weight is small compared to the 15,000-lb load and can be neglected in our analysis.

Load and Resistance Factor Design. This alternative method of design was briefly described in Sec. 1.13 and applied to members under axial loading. It can readily be applied to the design of beams in bending. Replacing in Eq. (1.26) the loads P_D, P_L, and P_U, respectively, by the bending moments M_D, M_L, and M_U, we write

$$\gamma_D M_D + \gamma_L M_L \le \phi M_U \qquad (5.10)$$

The coefficients γ_D and γ_L are referred to as the *load factors* and the coefficient ϕ as the *resistance factor*. The moments M_D and M_L are the bending moments due, respectively, to the dead and the live loads, while M_U is equal to the product of the ultimate strength σ_U of the material and the section modulus S of the beam: $M_U = S\sigma_U$.

SAMPLE PROBLEM 5.7

A 12-ft-long overhanging timber beam AC with an 8-ft span AB is to be designed to support the distributed and concentrated loads shown. Knowing that timber of 4-in. nominal width (3.5-in. actual width) with a 1.75-ksi allowable stress is to be used, determine the minimum required depth h of the beam.

SOLUTION

Reactions. Considering the entire beam as a free body, we write

$+\circlearrowleft\Sigma M_A = 0$: $B(8\text{ ft}) - (3.2\text{ kips})(4\text{ ft}) - (4.5\text{ kips})(12\text{ ft}) = 0$
$B = 8.35\text{ kips}$ $\mathbf{B} = 8.35\text{ kips}\uparrow$

$\xrightarrow{+}\Sigma F_x = 0$: $A_x = 0$

$+\uparrow\Sigma F_y = 0$: $A_y + 8.35\text{ kips} - 3.2\text{ kips} - 4.5\text{ kips} = 0$
$A_y = -0.65\text{ kips}$ $\mathbf{A} = 0.65\text{ kips}\downarrow$

Shear Diagram. The shear just to the right of A is $V_A = A_y = -0.65$ kips. Since the change in shear between A and B is equal to *minus* the area under the load curve between these two points, we obtain V_B by writing

$V_B - V_A = -(400\text{ lb/ft})(8\text{ ft}) = -3200\text{ lb} = -3.20\text{ kips}$
$V_B = V_A - 3.20\text{ kips} = -0.65\text{ kips} - 3.20\text{ kips} = -3.85\text{ kips}.$

The reaction at B produces a sudden increase of 8.35 kips in V, resulting in a value of the shear equal to 4.50 kips to the right of B. Since no load is applied between B and C, the shear remains constant between these two points.

Determination of $|M|_{max}$. We first observe that the bending moment is equal to zero at both ends of the beam: $M_A = M_C = 0$. Between A and B the bending moment decreases by an amount equal to the area under the shear curve, and between B and C it increases by a corresponding amount. Thus, the maximum absolute value of the bending moment is $|M|_{max} = 18.00$ kip · ft.

Minimum Allowable Section Modulus. Substituting into Eq. (5.9) the given value of σ_{all} and the value of $|M|_{max}$ that we have found, we write

$$S_{min} = \frac{|M|_{max}}{\sigma_{all}} = \frac{(18\text{ kip}\cdot\text{ft})(12\text{ in./ft})}{1.75\text{ ksi}} = 123.43\text{ in}^3$$

Minimum Required Depth of Beam. Recalling the formula developed in part 4 of the design procedure described in Sec. 5.4 and substituting the values of b and S_{min}, we have

$$\tfrac{1}{6}bh^2 \geq S_{min} \quad \tfrac{1}{6}(3.5\text{ in.})h^2 \geq 123.43\text{ in}^3 \quad h \geq 14.546\text{ in.}$$

The minimum required depth of the beam is $\qquad h = 14.55\text{ in.}$ ▼

SAMPLE PROBLEM 5.8

A 5-m-long, simply supported steel beam AD is to carry the distributed and concentrated loads shown. Knowing that the allowable normal stress for the grade of steel to be used is 160 MPa, select the wide-flange shape that should be used.

SOLUTION

Reactions. Considering the entire beam as a free body, we write

$$+\!\uparrow\!\!\Sigma M_A = 0: \quad D(5\text{ m}) - (60\text{ kN})(1.5\text{ m}) - (50\text{ kN})(4\text{ m}) = 0$$
$$D = 58.0\text{ kN} \qquad \mathbf{D} = 58.0\text{ kN}\!\uparrow$$
$$\xrightarrow{+}\Sigma F_x = 0: \qquad A_x = 0$$
$$+\!\uparrow\!\!\Sigma F_y = 0: A_y + 58.0\text{ kN} - 60\text{ kN} - 50\text{ kN} = 0$$
$$A_y = 52.0\text{ kN} \qquad \mathbf{A} = 52.0\text{ kN}\!\uparrow$$

Shear Diagram. The shear just to the right of A is $V_A = A_y = +52.0$ kN. Since the change in shear between A and B is equal to *minus* the area under the load curve between these two points, we have

$$V_B = 52.0\text{ kN} - 60\text{ kN} = -8\text{ kN}$$

The shear remains constant between B and C, where it drops to -58 kN, and keeps this value between C and D. We locate the section E of the beam where $V = 0$ by writing

$$V_E - V_A = -wx$$
$$0 - 52.0\text{ kN} = -(20\text{ kN/m})x$$

Solving for x we find $x = 2.60$ m.

Determination of $|M|_{\max}$. The bending moment is maximum at E, where $V = 0$. Since M is zero at the support A, its maximum value at E is equal to the area under the shear curve between A and E. We have, therefore, $|M|_{\max} = M_E = 67.6$ kN · m

Minimum Allowable Section Modulus. Substituting into Eq. (5.9) the given value of σ_{all} and the value of $|M|_{\max}$ that we have found, we write

$$S_{\min} = \frac{|M|_{\max}}{\sigma_{\text{all}}} = \frac{67.6\text{ kN} \cdot \text{m}}{160\text{ MPa}} = 422.5 \times 10^{-6}\text{ m}^3 = 422.5 \times 10^3\text{ mm}^3$$

Selection of Wide-Flange Shape. From Appendix C we compile a list of shapes that have a section modulus larger than $S_{\min}$ and are also the lightest shape in a given depth group.

Shape	S, mm^3
W410 × 38.8	637
W360 × 32.9	474
W310 × 38.7	549
W250 × 44.8	535
W200 × 46.1	448

We select the lightest shape available, namely

$$\text{W360} \times 32.9 \quad \blacktriangleleft$$

PROBLEMS

5.75 and 5.76 For the beam and loading shown, design the cross section of the beam, knowing that the grade of timber used has an allowable normal stress of 12 MPa.

Fig. P5.75

5.77 and 5.78 For the beam and loading shown, design the cross section of the beam, knowing that the grade of timber used has an allowable normal stress of 1.75 ksi.

Fig. P5.77

5.79 and 5.80 For the beam and loading shown, design the cross section of the beam, knowing that the grade of timber used has an allowable normal stress of 12 MPa.

Fig. P5.79

Fig. P5.76

Fig. P5.78

Fig. P5.80

5.81 and 5.82 Knowing that the allowable normal stress for the steel used is 160 MPa, select the most economical metric wide-flange beam to support the loading shown.

Fig. P5.81

Fig. P5.82

5.83 and 5.84 Knowing that the allowable normal stress for the steel used is 24 ksi, select the most economical wide-flange beam to support the loading shown.

Fig. P5.83

Fig. P5.84

5.85 and 5.86 Knowing that the allowable normal stress for the steel used is 160 MPa, select the most economical metric S-shape beam to support the loading shown.

Fig. P5.85

Fig. P5.86

5.87 and 5.88 Knowing that the allowable normal stress for the steel used is 24 ksi, select the most economical S-shape beam to support the loading shown.

Fig. P5.87

Fig. P5.88

5.89 Two metric rolled-steel channels are to be welded along their edges and used to support the loading shown. Knowing that the allowable normal stress for the steel used is 150 MPa, determine the most economical channels that can be used.

Fig. P5.89

5.90 Two metric rolled-steel channels are to be welded back to back and used to support the loading shown. Knowing that the allowable normal stress for the steel used is 190 MPa, determine the most economical channels that can be used.

Fig. P5.90

5.91 Two L4 × 3 rolled-steel angles are bolted together to support the loading shown. Knowing that the allowable normal stress for the steel used is 24 ksi, determine the minimum angle thickness that can be used.

Fig. P5.91

5.92 A steel pipe of 4-in. diameter is to support the loading shown. Knowing that the stock of pipes available has thicknesses varying from $\frac{1}{4}$ in. to 1 in. in $\frac{1}{8}$-in. increments, and that the allowable normal stress for the steel used is 24 ksi, determine the minimum wall thickness t that can be used.

Fig. P5.92

5.93 Assuming the upward reaction of the ground to be uniformly distributed and knowing that the allowable normal stress for the steel used is 170 MPa, select the most economical metric wide-flange beam to support the loading shown.

Fig. P5.93

5.94 Assuming the upward reaction of the ground to be uniformly distributed and knowing that the allowable normal stress for the steel used is 24 ksi, select the most economical S-shape beam to support the loading shown.

Fig. P5.94

5.95 and 5.96 Determine the largest permissible value of **P** for the beam and loading shown, knowing that the allowable normal stress is $+80$ MPa in tension and -140 MPa in compression.

Fig. P5.95

Fig. P5.96

5.97 Determine the largest permissible uniformly distributed load w for the beam shown, knowing that the allowable normal stress is $+12$ ksi in tension and -19.5 ksi in compression.

Fig. P5.97

5.98 Solve Prob. 5.97, assuming that the cross section of the beam is reversed, with the flange of the beam resting on the supports at B and C.

5.99 Beams AB, BC, and CD have the cross section shown and are pin-connected at B and C. Knowing that the allowable normal stress is $+110$ MPa in tension and -150 MPa in compression, determine (a) the largest permissible value of w if beam BC is not to be overstressed, (b) the corresponding maximum distance a for which the cantilever beams AB and CD are not overstressed.

Fig. P5.99

5.100 Beams AB, BC, and CD have the cross section shown and are pin-connected at B and C. Knowing that the allowable normal stress is $+110$ MPa in tension and -150 MPa in compression, determine (a) the largest permissible value of **P** if beam BC is not to be overstressed, (b) the corresponding maximum distance a for which the cantilever beams AB and CD are not overstressed.

Fig. P5.100

5.101 Each of the three rolled-steel beams shown (numbered 1, 2, and 3) is to carry a 64-kip load uniformly distributed over the beam. Each of these beams has a 12-ft span and is to be supported by the two 24-ft rolled-steel girders *AC* and *BD*. Knowing that the allowable normal stress for the steel used is 24 ksi, select (*a*) the most economical S shape for the three beams, (*b*) the most economical W shape for the two girders.

Fig. P5.101

5.102 Solve Prob. 5.101, assuming that the 64-kip distributed loads are replaced by 64-kip concentrated loads applied at the midpoints of the three beams.

5.103 A 240 kN load is to be supported at the center of the 5-m span shown. Knowing that the allowable normal stress for the steel used is 165 MPa, determine (*a*) the smallest allowable length *l* of beam *CD* if the W310 × 74 beam *AB* is not to be overstressed, (*b*) the W shape which should be used for beam *CD*. Neglect the weight of both beams.

Fig. P5.103

5.104 A uniformly distributed load of 84 kN/m is to be supported over the 5-m span shown. Knowing that the allowable normal stress for the steel used is 165 MPa, determine (a) the smallest allowable length l of beam CD if the W310 × 74 beam AB is not to be overstressed, (b) the W shape which should be used for beam CD. Neglect the weight of both beams.

84 kN/m

84 kN/m

W310 × 74

L = 5 m

Fig. P5.104

*5.105 A bridge of length L = 48 ft is to be built on a secondary road whose access to trucks is limited to two-axle vehicles of medium weight. It will consist of a concrete slab and of simply supported steel beams with an ultimate strength σ_U = 60 ksi. The combined weight of the slab and beams can be approximated by a uniformly distributed load w = 0.75 kip/ft on each beam. For the purpose of the design, it is assumed that a truck with axles located at a distance a = 14 ft from each other will be driven across the bridge and that the resulting concentrated loads P₁ and P₂ exerted on each beam could be as large as 24 kips and 6 kips, respectively. Determine the most economical wide-flange shape for the beams, using LRFD with the load factors γ_D = 1.25, γ_L = 1.75 and the resistance factor φ = 0.9 [*Hint.* It can be shown that the maximum value of |M_L| occurs under the larger load when that load is located to the left of the center of the beam at a distance equal to aP₂/2(P₁ + P₂).]

A

x

a

P₁

P₂

L

B

Fig. P5.105

*5.106 Assuming that the front and rear axle loads remain in the same ratio as for the truck of Prob. 5.105, determine how much heavier a truck could safely cross the bridge designed in that problem.

*5.107 A roof structure consisting of plywood and roofing material is supported by several timber beams of length L = 16 m. The dead load carried by each beam, including the estimated weight of the beam, can be represented by a uniformly distributed load w_D = 350 N/m. The live loads consist of the snow load, represented by a uniformly distributed load w_L = 600 N/m, and a 6-kN concentrated load P applied at the midpoint C of each beam. Knowing that the ultimate strength for the timber used is σ_U = 50 MPa and that the width of the beams is b = 75 mm, determine the minimum allowable depth h of the beams, using LRFD with the load factors γ_D = 1.2, γ_L = 1.6 and the resistance factor φ = 0.9.

A

w_D + w_L

C

P

½ L

½ L

B

b

h

Fig. P5.107

*5.108 Solve Prob. 5.107, assuming that the 6-kN concentrated load P applied to each beam is replaced by 3-kN concentrated loads P₁ and P₂ applied at a distance of 4 m from each end of the beams.

*5.5. USING SINGULARITY FUNCTIONS TO DETERMINE SHEAR AND BENDING MOMENT IN A BEAM

Reviewing the work done in the preceding sections, we note that the shear and bending moment could only rarely be described by single analytical functions. In the case of the cantilever beam of Example 5.02 (Fig. 5.10), which supported a uniformly distributed load w, the shear and bending moment *could* be represented by single analytical functions, namely, $V = -wx$ and $M = -\frac{1}{2}wx^2$; this was due to the fact that *no discontinuity* existed in the loading of the beam. On the other hand, in the case of the simply supported beam of Example 5.01, which was loaded only at its midpoint C, the load P applied at C represented a *singularity* in the beam loading. This singularity resulted in discontinuities in the shear and bending moment and required the use of different analytical functions to represent V and M in the portions of beam located, respectively, to the left and to the right of point C. In Sample Prob. 5.2, the beam had to be divided into three portions, in each of which different functions were used to represent the shear and the bending moment. This situation led us to rely on the graphical representation of the functions V and M provided by the shear and bending-moment diagrams and, later in Sec. 5.3, on a graphical method of integration to determine V and M from the distributed load w.

The purpose of this section is to show how the use of *singularity functions* makes it possible to represent the shear V and the bending moment M by single mathematical expressions.

Consider the simply supported beam AB, of length $2a$, which carries a uniformly distributed load w_0 extending from its midpoint C to its right-hand support B (Fig. 5.16). We first draw the free-body diagram of the entire beam (Fig. 5.17a); replacing the distributed load by an equivalent concentrated load and, summing moments about B, we write

$+\circlearrowleft\Sigma M_B = 0$: $(w_0a)(\frac{1}{2}a) - R_A(2a) = 0$ $R_A = \frac{1}{4}w_0a$

Next we cut the beam at a point D between A and C. From the free-body diagram of AD (Fig. 5.17b) we conclude that, over the interval $0 < x < a$, the shear and bending moment are expressed, respectively, by the functions

$$V_1(x) = \frac{1}{4}w_0a \quad \text{and} \quad M_1(x) = \frac{1}{4}w_0ax$$

Cutting, now, the beam at a point E between C and B, we draw the free-body diagram of portion AE (Fig. 5.17c). Replacing the distributed load by an equivalent concentrated load, we write

$+\uparrow\Sigma F_y = 0$: $\frac{1}{4}w_0a - w_0(x - a) - V_2 = 0$

$+\circlearrowleft\Sigma M_E = 0$: $-\frac{1}{4}w_0ax + w_0(x - a)[\frac{1}{2}(x - a)] + M_2 = 0$

and conclude that, over the interval $a < x < 2a$, the shear and bending moment are expressed, respectively, by the functions

$$V_2(x) = \frac{1}{4}w_0a - w_0(x - a) \quad \text{and} \quad M_2(x) = \frac{1}{4}w_0ax - \frac{1}{2}w_0(x - a)^2$$

Fig. 5.16

(a)

(b)

(c)

$R_A = \frac{1}{4}w_0a$

Fig. 5.17

As we pointed out earlier in this section, the fact that the shear and bending moment are represented by different functions of x, depending upon whether x is smaller or larger than a, is due to the discontinuity in the loading of the beam. However, the functions $V_1(x)$ and $V_2(x)$ can be represented by the single expression

$$V(x) = \tfrac{1}{4} w_0 a - w_0 \langle x - a \rangle \qquad (5.11)$$

if we specify that the second term should be included in our computations when $x \geq a$ and ignored when $x < a$. In other words, *the brackets* $\langle \ \rangle$ *should be replaced by ordinary parentheses* $(\ )$ *when* $x \geq a$ *and by zero when* $x < a$. With the same convention, the bending moment can be represented at any point of the beam by the single expression

$$M(x) = \tfrac{1}{4} w_0 a x - \tfrac{1}{2} w_0 \langle x - a \rangle^2 \qquad (5.12)$$

From the convention we have adopted, it follows that brackets $\langle \ \rangle$ can be differentiated or integrated as ordinary parentheses. Instead of calculating the bending moment from free-body diagrams, we could have used the method indicated in Sec. 5.3 and integrated the expression obtained for $V(x)$:

$$M(x) - M(0) = \int_0^x V(x)\,dx = \int_0^x \tfrac{1}{4} w_0 a\,dx - \int_0^x w_0 \langle x - a \rangle\,dx$$

After integration, and observing that $M(0) = 0$, we obtain as before

$$M(x) = \tfrac{1}{4} w_0 a x - \tfrac{1}{2} w_0 \langle x - a \rangle^2$$

Furthermore, using the same convention again, we note that the distributed load at any point of the beam can be expressed as

$$w(x) = w_0 \langle x - a \rangle^0 \qquad (5.13)$$

Indeed, the brackets should be replaced by zero for $x < a$ and by parentheses for $x \geq a$; we thus check that $w(x) = 0$ for $x < a$ and, defining the zero power of any number as unity, that $\langle x - a \rangle^0 = (x - a)^0 = 1$ and $w(x) = w_0$ for $x \geq a$. From Sec. 5.3 we recall that the shear could have been obtained by integrating the function $-w(x)$. Observing that $V = \tfrac{1}{4} w_0 a$ for $x = 0$, we write

$$V(x) - V(0) = -\int_0^x w(x)\,dx = -\int_0^x w_0 \langle x - a \rangle^0\,dx$$

$$V(x) - \tfrac{1}{4} w_0 a = w_0 \langle x - a \rangle^1$$

Solving for $V(x)$ and dropping the exponent 1, we obtain again

$$V(x) = \tfrac{1}{4} w_0 a - w_0 \langle x - a \rangle$$

The expressions $\langle x - a \rangle^0$, $\langle x - a \rangle$, $\langle x - a \rangle^2$ are called *singularity functions*. By definition, we have, for $n \geq 0$,

$$\langle x - a \rangle^n = \begin{cases} (x - a)^n & \text{when } x > a \\ 0 & \text{when } x < a \end{cases} \qquad (5.14)$$

We also note that whenever the quantity between brackets is positive or zero, the brackets should be replaced by ordinary parentheses, and whenever that quantity is negative, the bracket itself is equal to zero.

$\langle x - a \rangle^0$ $\langle x - a \rangle^1$ $\langle x - a \rangle^2$

(a) n = 0   (b) n = 1   (c) n = 2

Fig. 5.18

The three singularity functions corresponding respectively to $n = 0$, $n = 1$, and $n = 2$ have been plotted in Fig. 5.18. We note that the function $\langle x - a \rangle^0$ is discontinuous at $x = a$ and is in the shape of a "step." For that reason it is referred to as the *step function*. According to (5.14), and with the zero power of any number defined as unity, we have[†]

$$\langle x - a \rangle^0 = \begin{cases} 1 & \text{when } x \geq a \\ 0 & \text{when } x < a \end{cases} \qquad (5.15)$$

It follows from the definition of singularity functions that

$$\int \langle x - a \rangle^n \, dx = \frac{1}{n+1} \langle x - a \rangle^{n+1} \qquad \text{for } n \geq 0 \qquad (5.16)$$

and

$$\frac{d}{dx} \langle x - a \rangle^n = n \langle x - a \rangle^{n-1} \qquad \text{for } n \geq 1 \qquad (5.17)$$

Most of the beam loadings encountered in engineering practice can be broken down into the basic loadings shown in Fig. 5.19. Whenever applicable, the corresponding functions $w(x)$, $V(x)$, and $M(x)$ have been expressed in terms of singularity functions and plotted against a color background. A heavier color background was used to indicate for each loading the expression that is most easily derived or remembered and from which the other functions can be obtained by integration.

[†]Since $(x - a)^0$ is discontinuous at $x - a$, it can be argued that this function should be left undefined for $x = a$ or that it should be assigned both of the values 0 and 1 for $x = a$. However, defining $(x - a)^0$ as equal to 1 when $x = a$, as stated in (5.15), has the advantage of being unambiguous and, thus, readily applicable to computer programming (cf. page 348).

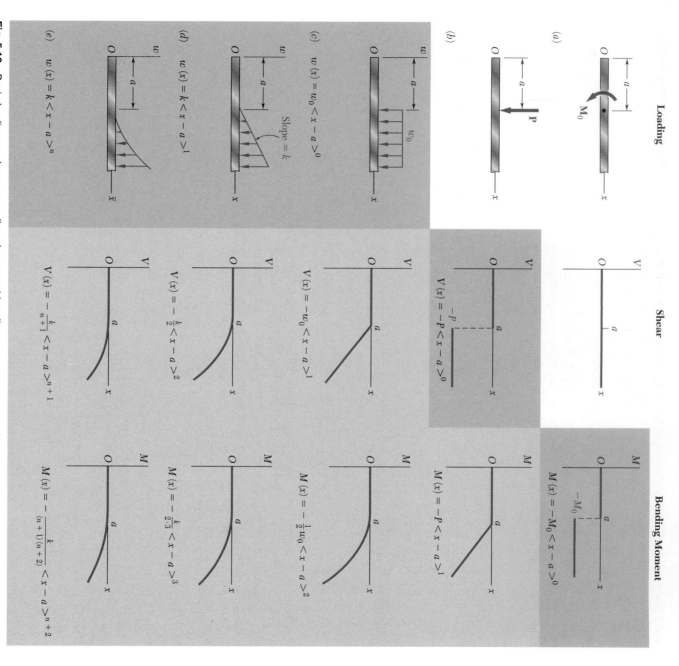

Fig. 5.19 Basic loadings and corresponding shears and bending moments expressed in terms of singularity functions.

After a given beam loading has been broken down into the basic loadings of Fig. 5.19, the functions $V(x)$ and $M(x)$ representing the shear and bending moment at any point of the beam can be obtained by adding the corresponding functions associated with each of the basic loadings and reactions. Since all the distributed loadings shown in Fig. 5.19 are open-ended to the right, a distributed loading that does not extend to the right end of the beam or that is discontinuous should be replaced as shown in Fig. 5.20 by an equivalent combination of open-ended loadings. (See also Example 5.05 and Sample Prob. 5.9.)

As you will see in Sec. 9.6, the use of singularity functions also greatly simplifies the determination of beam deflections. It was in connection with that problem that the approach used in this section was first suggested in 1862 by the German mathematician A. Clebsch (1833–1872). However, the British mathematician and engineer W. H. Macaulay (1853–1936) is usually given credit for introducing the singularity functions in the form used here, and the brackets $\langle\ \rangle$ are generally referred to as *Macaulay's brackets*.†

†W. H. Macaulay, "Note on the Deflection of Beams," *Messenger of Mathematics*, vol. 48, pp. 129–130, 1919.

$$w(x) = w_0\langle x - a\rangle^0 - w_0\langle x - b\rangle^0$$

Fig. 5.20

EXAMPLE 5.05

For the beam and loading shown (Fig. 5.21a) and using singularity functions, express the shear and bending moment as functions of the distance x from the support at A.

We first determine the reaction at A by drawing the free-body diagram of the beam (Fig. 5.21b) and writing

$$\xrightarrow{+}\Sigma F_x = 0: \qquad A_x = 0$$

$$+\;\gamma\Sigma M_B = 0: \qquad -A_y(3.6\text{ m}) + (1.2\text{ kN})(3\text{ m})$$
$$+ (1.8\text{ kN})(2.4\text{ m}) + 1.44\text{ kN}\cdot\text{m} = 0$$
$$A_y = 2.60\text{ kN}$$

Next, we replace the given distributed loading by two equivalent open-ended loadings (Fig. 5.21c) and express the distributed load $w(x)$ as the sum of the corresponding step functions:

$$w(x) = +w_0\langle x - 0.6\rangle^0 - w_0\langle x - 1.8\rangle^0$$

The function $V(x)$ is obtained by integrating $w(x)$, reversing the $+$ and $-$ signs, and adding to the result the constants A_y and $-P\langle x - 0.6\rangle^0$ representing the respective contributions to the shear of the reaction at A and of the concentrated load. (No other constant of integration is required.) Since the concentrated couple does not directly affect the shear, it should be ignored in this computation. We write

$$V(x) = -w_0\langle x - 0.6\rangle^1 + w_0\langle x - 1.8\rangle^1 + A_y - P\langle x - 0.6\rangle^0$$

(a)

(b)

(c)

Fig. 5.21

$A_y = 2.6$ kN

Fig. 5.21c (repeated)

In a similar way, the function $M(x)$ is obtained by integrating $V(x)$ and adding to the result the constant $-M_0\langle x - 2.6\rangle^0$ representing the contribution of the concentrated couple to the bending moment. We have

$$M(x) = -\tfrac{1}{2}w_0\langle x - 0.6\rangle^2 + \tfrac{1}{2}w_0\langle x - 1.8\rangle^2$$
$$+ A_y x - P\langle x - 0.6\rangle^1 - M_0\langle x - 2.6\rangle^0$$

Substituting the numerical values of the reaction and loads into the expressions obtained for $V(x)$ and $M(x)$ and being careful *not* to compute any product or expand any square involving a bracket, we obtain the following expressions for the shear and bending moment at any point of the beam:

$$V(x) = -1.5\langle x - 0.6\rangle^1 + 1.5\langle x - 1.8\rangle^1$$
$$+ 2.6 - 1.2\langle x - 0.6\rangle^0$$

$$M(x) = -0.75\langle x - 0.6\rangle^2 + 0.75\langle x - 1.8\rangle^2$$
$$+ 2.6x - 1.2\langle x - 0.6\rangle^1 - 1.44\langle x - 2.6\rangle^0$$

EXAMPLE 5.06

For the beam and loading of Example 5.05, determine the numerical values of the shear and bending moment at the midpoint D.

Making $x = 1.8$ m in the expressions found for $V(x)$ and $M(x)$ in Example 5.05, we obtain

$$V(1.8) = -1.5\langle 1.2\rangle^1 + 1.5\langle 0\rangle^1 + 2.6 - 1.2\langle 1.2\rangle^0$$

$$V(1.8) = -1.5\langle 1.2\rangle^1 + 1.5\langle 0\rangle^1 + 2.6 - 1.2\langle 1.2\rangle^0$$
$$= -1.5(1.2) + 1.5(0) + 2.6 - 1.2(1)$$
$$= -1.8 + 0 + 2.6 - 1.2$$

$$V(1.8) = -0.4 \text{ kN}$$

and

$$M(1.8) = -0.75\langle 1.2\rangle^2 + 0.75\langle 0\rangle^2$$
$$+ 2.6(1.8) - 1.2\langle 1.2\rangle^1 - 1.44\langle -0.8\rangle^0$$

$$M(1.8) = -0.75(1.2)^2 + 0.75(0)^2$$
$$+ 2.6(1.8) - 1.2(1.2)^1 - 1.44(0)$$
$$= -1.08 + 0 + 4.68 - 1.44 - 0$$

$$M(1.8) = +2.16 \text{ kN} \cdot \text{m}$$

Recalling that whenever a quantity between brackets is positive or zero, the brackets should be replaced by ordinary parentheses, and whenever the quantity is negative, the bracket itself is equal to zero, we write

Application to Computer Programming. Singularity functions are particularly well suited to the use of computers. First we note that the step function $\langle x - a\rangle^0$, which will be represented by the symbol STP, can be defined by an IF/THEN/ELSE statement as being equal to 1 for $X \geq A$ and to 0 otherwise. Any other singularity function $\langle x - a\rangle^n$, with $n \geq 1$, can then be expressed as the product of the ordinary algebraic function $(x - a)^n$ and the step function $\langle x - a\rangle^0$.

When k different singularity functions are involved, such as $\langle x - a_i\rangle^n$, where $i = 1, 2, \ldots, k$, then the corresponding step functions STP(I), where I $= 1, 2, \ldots$, K, can be defined by a loop containing a single IF/THEN/ELSE statement.

SAMPLE PROBLEM 5.9

For the beam and loading shown, determine (a) the equations defining the shear and bending moment at any point, (b) the shear and bending moment at points C, D, and E.

SOLUTION

Reactions. The total load is $\frac{1}{2}w_0 L$; because of symmetry, each reaction is equal to half that value, namely, $\frac{1}{4}w_0 L$.

Distributed Load. The given distributed loading is replaced by two equivalent open-ended loadings as shown. Using a singularity function to express the second loading, we write

$$w(x) = k_1 x + k_2\langle x - \tfrac{1}{2}L\rangle = \frac{2w_0}{L}x - \frac{4w_0}{L}\langle x - \tfrac{1}{2}L\rangle \qquad (1)$$

a. Equations for Shear and Bending Moment. We obtain $V(x)$ by integrating (1), changing the signs, and adding a constant equal to R_A:

$$V(x) = -\frac{w_0}{L}x^2 + \frac{2w_0}{L}\langle x - \tfrac{1}{2}L\rangle^2 + \tfrac{1}{4}w_0 L \qquad (2) \blacktriangledown$$

We obtain $M(x)$ by integrating (2); since there is no concentrated couple, no constant of integration is needed:

$$M(x) = -\frac{w_0}{3L}x^3 + \frac{2w_0}{3L}\langle x - \tfrac{1}{2}L\rangle^3 + \tfrac{1}{4}w_0 Lx \qquad (3) \blacktriangledown$$

b. Shear and Bending Moment at C, D, and E

At Point C: Making $x = \frac{1}{2}L$ in Eqs. (2) and (3) and recalling that whenever a quantity between brackets is positive or zero, the brackets may be replaced by parentheses, we have

$$V_C = -\frac{w_0(\tfrac{1}{2}L)^2}{L} + \frac{2w_0}{L}\langle 0\rangle^2 + \tfrac{1}{4}w_0 L \qquad V_C = 0 \blacktriangledown$$

$$M_C = -\frac{w_0(\tfrac{1}{2}L)^3}{3L} + \frac{2w_0}{3L}\langle 0\rangle^3 + \tfrac{1}{4}w_0 L(\tfrac{1}{2}L) \qquad M_C = \frac{1}{12}w_0 L^2 \blacktriangledown$$

At Point D: Making $x = \frac{1}{4}L$ in Eqs. (2) and (3) and recalling that a bracket containing a negative quantity is equal to zero, we write

$$V_D = -\frac{w_0(\tfrac{1}{4}L)^2}{L} + \frac{2w_0}{L}\langle -\tfrac{1}{4}L\rangle^2 + \tfrac{1}{4}w_0 L \qquad V_D = \frac{3}{16}w_0 L \blacktriangledown$$

$$M_D = -\frac{w_0(\tfrac{1}{4}L)^3}{3L} + \frac{2w_0}{3L}\langle -\tfrac{1}{4}L\rangle^3 + \tfrac{1}{4}w_0 L(\tfrac{1}{4}L) \qquad M_D = \frac{11}{192}w_0 L^2 \blacktriangledown$$

At Point E: Making $x = \frac{3}{4}L$ in Eqs. (2) and (3), we have

$$V_E = -\frac{w_0(\tfrac{3}{4}L)^2}{L} + \frac{2w_0(\tfrac{1}{4}L)^2}{L} + \tfrac{1}{4}w_0 L \qquad V_E = -\frac{3}{16}w_0 L \blacktriangledown$$

$$M_E = -\frac{w_0(\tfrac{3}{4}L)^3}{3L} + \frac{2w_0(\tfrac{1}{4}L)^3}{3L} + \tfrac{1}{4}w_0 L(\tfrac{3}{4}L) \qquad M_E = \frac{11}{192}w_0 L^2 \blacktriangledown$$

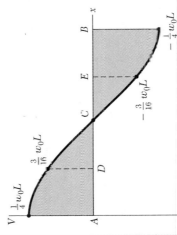

SAMPLE PROBLEM 5.10

The rigid bar *DEF* is welded at point *D* to the steel beam *AB*. For the loading shown, determine (*a*) the equations defining the shear and bending moment at any point of the beam, (*b*) the location and magnitude of the largest bending moment.

SOLUTION

Reactions. We consider the beam and bar as a free body and observe that the total load is 960 lb. Because of symmetry, each reaction is equal to 480 lb.

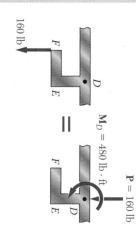

Modified Loading Diagram. We replace the 160-lb load applied at *F* by an equivalent force-couple system at *D*. We thus obtain a loading diagram consisting of a concentrated couple, three concentrated loads (including the two reactions), and a uniformly distributed load

$$w(x) = 50 \text{ lb/ft} \tag{1}$$

a. **Equations for Shear and Bending Moment.** We obtain $V(x)$ by integrating (1), changing the sign, and adding constants representing the respective contributions of $\mathbf{R}_A$ and $\mathbf{P}$ to the shear. Since $\mathbf{P}$ affects $V(x)$ only for values of x larger than 11 ft, we use a step function to express its contribution.

$$V(x) = -50x + 480 - 160\langle x - 11\rangle^0 \tag{2} \blacktriangle$$

We obtain $M(x)$ by integrating (2) and using a step function to represent the contribution of the concentrated couple $\mathbf{M}_D$:

$$M(x) = -25x^2 + 480x - 160\langle x - 11\rangle^1 - 480\langle x - 11\rangle^0 \tag{3} \blacktriangle$$

b. **Largest Bending Moment.** Since *M* is maximum or minimum when $V = 0$, we set $V = 0$ in (2) and solve that equation for *x* to find the location of the largest bending moment. Considering first values of *x* less than 11 ft and noting that for such values the bracket is equal to zero, we write

$$-50x + 480 = 0 \qquad x = 9.60 \text{ ft}$$

Considering now values of *x* larger than 11 ft, for which the bracket is equal to 1, we have

$$-50x + 480 - 160 = 0 \qquad x = 6.40 \text{ ft}$$

Since this value is *not* larger than 11 ft, it must be rejected. Thus, the value of *x* corresponding to the largest bending moment is

$$x_m = 9.60 \text{ ft} \blacktriangle$$

Substituting this value for *x* into Eq. (3), we obtain

$$M_{\max} = -25(9.60)^2 + 480(9.60) - 160\langle -1.40\rangle^1 - 480\langle -1.40\rangle^0$$

and, recalling that brackets containing a negative quantity are equal to zero,

$$M_{\max} = -25(9.60)^2 + 480(9.60) \qquad M_{\max} = 2304 \text{ lb} \cdot \text{ft} \blacktriangle$$

The bending-moment diagram has been plotted. Note the discontinuity at point *D* due to the concentrated couple applied at that point. The values of *M* just to the left and just to the right of *D* were obtained by making *x* = 11 in Eq. (3) and replacing the step function $\langle x - 11\rangle^0$ by 0 and 1, respectively.

PROBLEMS

5.109 through 5.111 (a) Using singularity functions, write the equations defining the shear and bending moment for the beam and loading shown. (b) Use the equation obtained for M to determine the bending moment at point E and check your answer by drawing the free-body diagram of the portion of beam to the right of E.

Fig. P5.109

Fig. P5.110

Fig. P5.111

5.112 through 5.114 (a) Using singularity functions, write the equations defining the shear and bending moment for the beam and loading shown. (b) Use the equation obtained for M to determine the bending moment at point C and check your answer by drawing the free-body diagram of the entire beam.

Fig P5.112

Fig. P5.113

Fig. P5.114

5.115 and 5.116 (a) Using singularity functions, write the equations defining the shear and bending moment for beam ABC under the loading shown. (b) Use the equation obtained for M to determine the bending moment just to the right of point B.

Fig. P5.115

Fig. P5.116

351

5.117 through 5.120 (*a*) Using singularity functions, write the equations defining the shear and bending moment for the beam and loading shown. (*b*) Determine the maximum value of the bending moment in the beam.

Fig. P5.117

Fig. P5.118

Fig. P5.119

Fig. P5.120

5.121 and 5.122 (*a*) Using singularity functions, write the equations defining the shear and bending moment for the beam and loading shown. (*b*) Determine the maximum normal stress due to bending.

Fig. P5.121

Fig. P5.122

5.123 and 5.124 (*a*) Using singularity functions, find the magnitude and location of the maximum bending moment for the beam and loading shown. (*b*) Determine the maximum normal stress due to bending.

Fig. P5.123

Fig. P5.124

5.125 and 5.126 A beam is being designed to be supported and loaded as shown. (*a*) Using singularity functions, determine the magnitude and location of the maximum bending moment in the beam. (*b*) Knowing that the allowable stress for the steel to be used is 24 ksi, find the most economical wide-flange shape that should be selected.

Fig. P5.125

Fig. P5.126

5.127 and 5.128 A timber beam is being designed to be supported and loaded as shown. (*a*) Using singularity functions, determine the magnitude and location of the maximum bending moment in the beam. (*b*) Knowing that the available stock consists of beams with a 12-MPa allowable stress and a rectangular cross section of 30-mm width and depth *h* varying from 80 to 160 mm in 10-mm increments, determine the most economical cross section that can be used.

Fig. *P5.127*

Fig. *P5.128*

5.129 through 5.132 Using a computer and step functions, calculate the shear and bending moment for the beam and loading shown. Use the specified increments ΔL, starting at point *A* and ending at the right-hand support.

Fig. P5.129

Fig. P5.130

Fig. P5.131

Fig. P5.132

5.133 and 5.134

For the beam and loading shown, and using a computer and step functions, (a) tabulate the shear, bending moment, and maximum normal stress in sections of the beam from $x = 0$ to $x = L$, using the increments ΔL indicated, (b) using smaller increments if necessary, determine with a 2 percent accuracy the maximum normal stress in the beam. Place the origin of the x axis at end A of the beam.

Fig. P5.133

Fig. P5.134

5.135 and 5.136

For the beam and loading shown, and using a computer and step functions, (a) tabulate the shear, bending moment, and maximum normal stress in sections of the beam from $x = 0$ to $x = L$, using the increments ΔL indicated, (b) using smaller increments if necessary, determine with a 2 percent accuracy the maximum normal stress in the beam. Place the origin of the x axis at end A of the beam.

Fig. P5.135

Fig. P5.136

*5.6. NONPRISMATIC BEAMS

Our analysis has been limited so far to prismatic beams, i.e., to beams of uniform cross section. As we saw in Sec. 5.4, prismatic beams are designed so that the normal stresses in their critical sections are at most equal to the allowable value of the normal stress for the material being used. It follows that, in all other sections, the normal stresses will be smaller, possibly much smaller, than their allowable value. A prismatic beam, therefore, is almost always overdesigned, and considerable savings of material can be realized by using nonprismatic beams, i.e., beams of variable cross section. The cantilever cast beam used in the testing machine for soils shown in Fig. 5.22 is such a beam.

Since the maximum normal stresses σ_m usually control the design of a beam, the design of a nonprismatic beam will be optimum if the section modulus $S = I/c$ of every cross section satisfies Eq. (5.3) of Sec. 5.1. Solving that equation for S, we write

$$S = \frac{|M|}{\sigma_{\text{all}}} \tag{5.18}$$

A beam designed in this manner is referred to as a *beam of constant strength*.

Fig. 5.22

For a forged or cast structural or machine component, it is possible to vary the cross section of the component along its length and to climinate most of the unnecessary material (see Example 5.07). For a timber beam or a rolled-steel beam, however, it is not possible to vary the cross section of the beam. But considerable savings of material can be achieved by gluing wooden planks of appropriate lengths to a timber beam (see Sample Prob. 5.11) and using cover plates in portions of a rolled-steel beam where the bending moment is large (see Sample Prob. 5.12).

EXAMPLE 5.07

A cast-aluminum plate of uniform thickness b is to support a uniformly distributed load w as shown in Fig. 5.23. (a) Determine the shape of the plate that will yield the most economical design. (b) Knowing that the allowable normal stress for the aluminum used is 72 MPa and that $b = 40$ mm, $L = 800$ mm, and $w = 135$ kN/m, determine the maximum depth h_0 of the plate.

Fig. 5.23

Bending Moment. Measuring the distance x from A and observing that $V_A = M_A = 0$, we use Eqs. (5.6) and (5.8) of Sec. 5.3 and write

$$V(x) = -\int_0^x w\,dx = -wx$$

$$M(x) = \int_0^x V(x)\,dx = -\int_0^x wx\,dx = -\tfrac{1}{2}wx^2$$

(a) Shape of Plate. We recall from Sec. 5.4 that the modulus S of a rectangular cross section of width b and depth h is $S = \tfrac{1}{6}bh^2$. Carrying this value into Eq. (5.18) and solving for h^2, we have

$$h^2 = \frac{6|M|}{b\sigma_{\text{all}}} \qquad (5.19)$$

and, after substituting $|M| = \tfrac{1}{2}wx^2$,

$$h^2 = \frac{3wx^2}{b\sigma_{\text{all}}} \qquad \text{or} \qquad h = \left(\frac{3w}{b\sigma_{\text{all}}}\right)^{1/2} x \qquad (5.20)$$

Since the relation between h and x is linear, the lower edge of the plate is a straight line. Thus, the plate providing the most economical design is of *triangular shape.*

(b) Maximum Depth h_0. Making $x = L$ in Eq. (5.20) and substituting the given data, we obtain

$$h_0 = \left[\frac{3(135 \text{ kN/m})}{(0.040 \text{ m})(72 \text{ MPa})}\right]^{1/2} (800 \text{ mm}) = 300 \text{ mm}$$

SAMPLE PROBLEM 5.11

A 12-ft-long beam made of a timber with an allowable normal stress of 2.40 ksi and an allowable shearing stress of 0.40 ksi is to carry two 4.8-kip loads located at its third points. As shown in Chap. 6, a beam of uniform rectangular cross section, 4 in. wide and 4.5 in. deep, would satisfy the allowable shearing stress requirement. Since such a beam would not satisfy the allowable normal stress requirement, it will be reinforced by gluing planks of the same timber, 4 in. wide and 1.25 in. thick, to the top and bottom of the beam in a symmetric manner. Determine (a) the required number of pairs of planks, (b) the length of the planks in each pair that will yield the most economical design.

SOLUTION

Bending Moment. We draw the free-body diagram of the beam and find the following expressions for the bending moment:

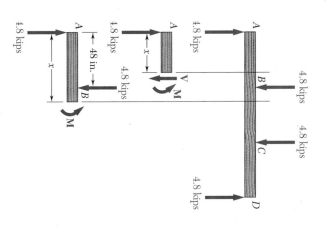

From A to B $(0 \le x \le 48 \text{ in.})$: $M = (4.80 \text{ kips})x$

From B to C $(48 \text{ in.} \le x \le 96 \text{ in.})$:

$$M = (4.80 \text{ kips})x - (4.80 \text{ kips})(x - 48 \text{ in.}) = 230.4 \text{ kip} \cdot \text{in.}$$

a. Number of Pairs of Planks. We first determine the required total depth of the reinforced beam between B and C. We recall from Sec. 5.4 that $S = \frac{1}{6}bh^2$ for a beam with a rectangular cross section of width b and depth h. Substituting this value into Eq. (5.17) and solving for h^2, we have

$$h^2 = \frac{6|M|}{b\sigma_{\text{all}}} \qquad (1)$$

Substituting the value obtained for M from B to C and the given values of b and σ_{all}, we write

$$h^2 = \frac{6(230.4 \text{ kip} \cdot \text{in.})}{(4 \text{ in.})(2.40 \text{ ksi})} = 144 \text{ in.}^2 \qquad h = 12.00 \text{ in.}$$

Since the original beam has a depth of 4.50 in., the planks must provide an additional depth of 7.50 in. Recalling that each pair of planks is 2.50 in. thick:

Required number of pairs of planks = 3 ▲

b. Length of Planks. The bending moment was found to be $M = (4.80 \text{ kips})x$ in the portion AB of the beam. Substituting this expression and the given values of b and σ_{all} into Eq. (1) and solving for x, we have

$$x = \frac{(4 \text{ in.})(2.40 \text{ ksi})}{6(4.80 \text{ kips})}h^2 \qquad x = \frac{h^2}{3 \text{ in.}} \qquad (2)$$

Equation (2) defines the maximum distance x from end A at which a given depth h of the cross section is acceptable. Making $h = 4.50$ in., we find the distance x_1 from A at which the original prismatic beam is safe: $x_1 = 6.75$ in. From that point on, the original beam should be reinforced by the first pair of planks. Making $h = 4.50$ in. + 2.50 in. = 7.00 in. yields the distance $x_2 = 16.33$ in. from which the second pair of planks should be used, and making $h = 9.50$ in. yields the distance $x_3 = 30.08$ in. from which the third pair of planks should be used. The length l_i of the planks of the pair i, where $i = 1, 2, 3$, is obtained by subtracting $2x_i$ from the 144-in. length of the beam. We find

$$l_1 = 130.5 \text{ in.,} \quad l_2 = 111.3 \text{ in.,} \quad l_3 = 83.8 \text{ in.} \quad ▲$$

The corners of the various planks lie on the parabola defined by Eq. (2).

SAMPLE PROBLEM 5.12

Two steel plates, each 16 mm thick, are welded as shown to a W690 × 125 beam to reinforce it. Knowing that σ_{all} = 160 MPa for both the beam and the plates, determine the required value of (*a*) the length of the plates, (*b*) the width of the plates.

SOLUTION

Bending Moment. We first find the reactions. From the free body of a portion of beam of length $x \leq 4$ m, we obtain M between A and C:

$$M = (250 \text{ kN})x \qquad (1)$$

a. **Required Length of Plates.** We first determine the maximum allowable length x_m of the portion AD of the unreinforced beam. From Appendix C we find that the section modulus of a W690 × 125 beam is $S = 3510 \times 10^6$ mm³, or $S = 3.51 \times 10^{-3}$ m³. Substituting for S and σ_{all} into Eq. (5.17) and solving for M, we write

$$M = S\sigma_{all} = (3.51 \times 10^{-3} \text{ m}^3)(160 \times 10^3 \text{ kN/m}^2) = 561.6 \text{ kN} \cdot \text{m}$$

Substituting for M in Eq. (1), we have

$$561.6 \text{ kN} \cdot \text{m} = (250 \text{ kN})x_m \qquad x_m = 2.246 \text{ m}$$

The required length l of the plates is obtained by subtracting $2x_m$ from the length of the beam:

$$l = 8 \text{ m} - 2(2.246 \text{ m}) = 3.508 \text{ m} \qquad l = 3.51 \text{ m} \blacktriangledown$$

b. **Required Width of Plates.** The maximum bending moment occurs in the midsection C of the beam. Making $x = 4$ m in Eq. (1), we obtain the bending moment in that section:

$$M = (250 \text{ kN})(4 \text{ m}) = 1000 \text{ kN} \cdot \text{m}$$

In order to use Eq. (5.1) of Sec. 5.1, we now determine the moment of inertia of the cross section of the reinforced beam with respect to a centroidal axis and the distance c from that axis to the outer surfaces of the plates. From Appendix C we find that the moment of inertia of a W690 × 125 beam is $I_b = 1190 \times 10^6$ mm⁴ and its depth is $d = 678$ mm. On the other hand, denoting by t the thickness of one plate, by b its width, and by $\bar{y}$ the distance of its centroid from the neutral axis, we express the moment of inertia I_p of the two plates with respect to the neutral axis:

$$I_p = 2\left(\tfrac{1}{12}bt^3 + A\bar{y}^2\right) = \left(\tfrac{1}{6}t^3\right)b + 2bt\left(\tfrac{1}{2}d + \tfrac{1}{2}t\right)^2$$

Substituting $t = 16$ mm and $d = 678$ mm, we obtain $I_p = (3.854 \times 10^6 \text{ mm}^3)b$. The moment of inertia I of the beam and plates is

$$I = I_b + I_p = 1190 \times 10^6 \text{ mm}^4 + (3.854 \times 10^6 \text{ mm}^3)b \qquad (2)$$

and the distance from the neutral axis to the surface is $c = \tfrac{1}{2}d + t = 355$ mm. Solving Eq. (5.1) for I and substituting the values of M, σ_{all}, and c, we write

$$I = \frac{|M|c}{\sigma_{all}} = \frac{(1000 \text{ kN} \cdot \text{m})(355 \text{ mm})}{160 \text{ MPa}} = 2.219 \times 10^{-3} \text{ m}^4 = 2219 \times 10^6 \text{ mm}^4$$

Replacing I by this value in Eq. (2) and solving for b, we have

$$2219 \times 10^6 \text{ mm}^4 = 1190 \times 10^6 \text{ mm}^4 + (3.854 \times 10^6 \text{ mm}^3)b$$

$$b = 267 \text{ mm} \blacktriangledown$$

PROBLEMS

5.137 and 5.138   The cantilever beam AB, consisting of a cast-iron plate of uniform thickness b and length L, is to support the distributed load $w(x)$ shown. (*a*) Knowing that the beam is to be of constant strength, express h in terms of x, L, and h_0. (*b*) Determine the smallest allowable value of h_0 if $L = 750$ mm, $b = 30$ mm, $w_0 = 300$ kN/m, and $\sigma_{all} = 250$ MPa.

Fig. P5.137

$w = w_0 \dfrac{x}{L}$

Fig. P5.138

$w = w_0 \sin \dfrac{\pi x}{2L}$

5.139 and 5.140   The beam AB, consisting of a cast-iron plate of uniform thickness b and length L, is to support the load shown. (*a*) Knowing that the beam is to be of constant strength, express h in terms of x, L, and h_0. (*b*) Determine the maximum allowable load if $L = 36$ in., $h_0 = 12$ in., $b = 1.25$ in., and $\sigma_{all} = 36$ ksi.

Fig. P5.139

Fig. P5.141

5.140

Fig. P5.140

w

5.141 and 5.142   The beam AB, consisting of a cast-aluminum plate of uniform thickness b and length L, is to support the load shown. (*a*) Knowing that the beam is to be of constant strength, express h in terms of x, L, and h_0 for portion AC of the beam. (*b*) Determine the maximum allowable load if $L = 800$ mm, $h_0 = 200$ mm, $b = 25$ mm, and $\sigma_{all} = 72$ MPa.

Fig. P5.142

5.143 and 5.144 A preliminary design based on the use of a simply supported prismatic timber beam indicated that a beam with a rectangular cross section 50 mm wide and 200 mm deep would be required to safely support the load shown in part *a* of the figure. It was then decided to replace that beam with a built-up beam obtained by gluing together, as shown in part *b* of the figure, four pieces of the same timber as the original beam and of 50 × 50-mm cross section. Determine the length *l* of the two outer pieces of timber that will yield the same factor of safety as the original design.

(a)

(b)

Fig. P5.143

5.145 and 5.146 A preliminary design based on the use of a cantilever prismatic beam indicated that a beam with a rectangular cross section 2 in. wide and 10 in. deep would be required to safely support the load shown in part *a* of the figure. It was then decided to replace that beam with a built-up beam obtained by gluing together, as shown in part *b* of the figure, five pieces of the same timber as the original beam and of 2 × 2-in. cross section. Determine the respective lengths l_1 and l_2 of the two inner and two outer pieces of timber that will yield the same factor of safety as the original design.

(a)

(b)

Fig. P5.145

(a)

(b)

Fig. P5.144

(a)

(b)

Fig. P5.146

5.147 A cantilevered machine element of cast aluminum and in the shape of a solid of revolution of variable diameter d is being designed to support a horizontal concentrated load **P** as shown. (a) Knowing that the machine element is to be of constant strength, express d in terms of y, L, and d_0. (b) Determine the maximum allowable value of **P** if $L = 300$ mm, $d_0 = 60$ mm, and $\sigma_{all} = 72$ MPa.

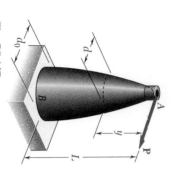

Fig. P5.147

5.148 A cantilevered machine element of cast aluminum and in the shape of a solid of revolution of variable diameter d is being designed to support a horizontal distributed load w as shown. (a) Knowing that the machine element is to be of constant strength, express d in terms of y, L, and d_0. (b) Determine the smallest allowable value of d_0 if $L = 300$ mm, $w = 20$ kN/m, and $\sigma_{all} = 72$ MPa.

Fig. P5.148

5.149 A cantilever beam AB consisting of a steel plate of uniform depth h and variable width b is to support a concentrated load **P** at point A. (a) Knowing that the beam is to be of constant strength, express b in terms of x, L, and b_0. (b) Determine the smallest allowable value of h if $L = 12$ in., $b_0 = 15$ in., $P = 3.2$ kips, and $\sigma_{all} = 24$ ksi.

Fig. P5.149

5.150 A cantilever beam AB consisting of a steel plate of uniform depth h and variable width b is to support a distributed load w along its center line AB. (a) Knowing that the beam is to be of constant strength, express b in terms of x, L, and b_0. (b) Determine the maximum allowable value of w if $L = 15$ in., $b_0 = 18$ in., $h = 0.75$ in., and $\sigma_{all} = 24$ ksi.

Fig. P5.150

Fig. P5.151 and P5.155

Fig. P5.154 and P5.157

Fig. P5.153 and P5.156

Fig. P5.158

5.151 Two cover plates, each 7.5 mm thick, are welded to a W460 × 74 beam as shown. Knowing that $l = 5$ m and $b = 200$ mm, determine the maximum normal stress on a transverse section (a) through the center of the beam, (b) just to the left of D.

5.152 Assuming that the length and width of the cover plates used with the beam of Sample Prob. 5.12 are, respectively, $l = 4$ m and $b = 285$ mm, and recalling that the thickness of each plate is 16 mm, determine the maximum normal stress on a transverse section (a) through the center of the beam, (b) just to the left of D.

5.153 Two cover plates, each $\frac{1}{2}$ in. thick, are welded to a W27 × 84 beam as shown. Knowing that $l = 10$ ft and $b = 10.5$ in., determine the maximum normal stress on a transverse section (a) through the center of the beam, (b) just to the left of D.

5.154 Two cover plates, each $\frac{5}{8}$ in. thick, are welded to a W30 × 99 beam as shown. Knowing that $l = 9$ ft and $b = 12$ in., determine the maximum normal stress on a transverse section (a) through the center of the beam, (b) just to the left of D.

5.155 Two cover plates, each 7.5 mm thick, are welded to a W460 × 74 beam as shown. Knowing that $\sigma_{\text{all}} = 150$ MPa for both the beam and the plates, determine the required value of (a) the length of the plates, (b) the width of the plates.

5.156 Two cover plates, each $\frac{1}{2}$ in. thick, are welded to a W27 × 84 beam as shown. Knowing that $\sigma_{\text{all}} = 24$ ksi for both the beam and the plates, determine the required value of (a) the length of the plates, (b) the width of the plates.

5.157 Two cover plates, each $\frac{5}{8}$ in. thick, are welded to a W30 × 99 beam as shown. Knowing that $\sigma_{\text{all}} = 22$ ksi for both the beam and the plates, determine the required value of (a) the length of the plates, (b) the width of the plates.

5.158 Knowing that $\sigma_{\text{all}} = 165$ MPa, determine the largest concentrated load **P** that can be applied at end E of the beam shown. Neglect the weights of the beam and of the plates.

5.159 For the tapered beam shown, and knowing that $P = 150$ kN, determine (a) the transverse section in which the maximum normal stress occurs, (b) the corresponding value of the normal stress.

5.160 For the tapered beam shown, and knowing that $w = 160$ kN/m, determine (a) the transverse section in which the maximum normal stress occurs, (b) the corresponding value of the normal stress.

Fig. P5.159 and P5.161

Fig. P5.160 and P5.162

5.161 For the tapered beam shown, determine (a) the transverse section in which the maximum normal stress occurs, (b) the largest concentrated load **P** that can be applied, knowing that $\sigma_{all} = 140$ MPa.

5.162 For the tapered beam shown, determine (a) the transverse section in which the maximum normal stress occurs, (b) the largest distributed load w that can be applied, knowing that $\sigma_{all} = 140$ MPa.

5.163 For the tapered beam shown, determine (a) the transverse section in which the maximum normal stress occurs, (b) the largest concentrated load **P** that can be applied, knowing that $\sigma_{all} = 24$ ksi.

Fig. P5.163

Fig. P5.164

5.164 For the tapered beam shown, determine (a) the transverse section in which the maximum normal stress occurs, (b) the largest distributed load w that can be applied, knowing that $\sigma_{all} = 24$ ksi.

REVIEW AND SUMMARY
FOR CHAPTER 5

This chapter was devoted to the analysis and design of beams under transverse loadings. Such loadings can consist of concentrated loads or distributed loads and the beams themselves are classified according to the way they are supported (Fig. 5.3). Only *statically determinate* beams were considered in this chapter, the analysis of statically indeterminate beams being postponed until Chap. 9.

Considerations for the design of prismatic beams

Statically
Determinate
Beams

(*a*) Simply supported beam

(*b*) Overhanging beam

(*c*) Cantilever beam

Statically
Indeterminate
Beams

(*d*) Continuous beam

(*e*) Beam fixed at one end
and simply supported
at the other end

(*f*) Fixed beam

Fig. 5.3

While transverse loadings cause both bending and shear in a beam, the normal stresses caused by bending are the dominant criterion in the design of a beam for strength [Sec. 5.1]. Therefore, this chapter dealt only with the determination of the normal stresses in a beam, the effect of shearing stresses being examined in the next one.

We recalled from Sec. 4.4 the flexure formula for the determination of the maximum value σ_m of the normal stress in a given section of the beam,

$$\sigma_m = \frac{|M|c}{I} \qquad (5.1)$$

where I is the moment of inertia of the cross section with respect to a centroidal axis perpendicular to the plane of the bending couple $\mathbf{M}$ and c is the maximum distance from the neutral surface (Fig. 4.13).

Normal stresses due to bending

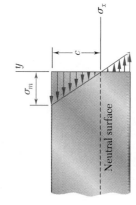

Fig. 4.13

We also recalled from Sec. 4.4 that, introducing the elastic section modulus $S = I/c$ of the beam, the maximum value σ_m of the normal stress in the section can be expressed as

$$\sigma_m = \frac{|M|}{S} \qquad (5.3)$$

It follows from Eq. (5.1) that the maximum normal stress occurs in the section where $|M|$ is largest, at the point farthest from the neutral axis. The determination of the maximum value of $|M|$ and of the critical section of the beam in which it occurs is greatly simplified if we draw a *shear diagram* and a *bending-moment diagram*. These diagrams represent, respectively, the variation of the shear and of the bending moment along the beam and were obtained by determining the values of V and M at selected points of the beam [Sec. 5.2]. These values were found by passing a section through the point where they were to be determined and drawing the free-body diagram of either of the portions of beam obtained in this fashion. To avoid any confusion regarding the sense of the shearing force V and of the bending couple $\mathbf{M}$ (which act in opposite sense on the two portions of the beam), we followed the sign convention adopted earlier in the text and illustrated in Fig. 5.7a [Examples 5.01 and 5.02, Sample Probs. 5.1 and 5.2].

The construction of the shear and bending-moment diagrams is facilitated if the following relations are taken into account [Sec. 5.3]. Denoting by w the distributed load per unit length (assumed positive if directed downward), we wrote

$$\frac{dV}{dx} = -w \qquad \frac{dM}{dx} = V \qquad (5.5, 5.7)$$

or, in integrated form,

$$V_D - V_C = -(\text{area under load curve between } C \text{ and } D) \qquad (5.6')$$
$$M_D - M_C = \text{area under shear curve between } C \text{ and } D \qquad (5.8')$$

Equation (5.6') makes it possible to draw the shear diagram of a beam from the curve representing the distributed load on that beam and the value of V at one end of the beam. Similarly, Eq. (5.8') makes it possible to draw the bending-moment diagram from the shear diagram and the value of M at one end of the beam. However, concentrated loads and concentrated couples introduce discontinuities in the shear diagram and bending-moment diagram, none of which is accounted for in these equations [Sample Probs. 5.3 and 5.6]. Finally, we noted from Eq. (5.7) that the points of the beam where the bending moment is maximum or minimum are also the points where the shear is zero [Sample Prob. 5.4].

(a) Internal forces
(positive shear and positive bending moment)

Fig. 5.7a

Shear and bending-moment
diagrams

Relations among load, shear,
and bending moment

A proper procedure for the design of a prismatic beam was described in Sec. 5.4 and is summarized here:

Having determined σ_{all} for the material used and assuming that the design of the beam is controlled by the maximum normal stress in the beam, compute the minimum allowable value of the section modulus:

$$S_{min} = \frac{|M|_{max}}{\sigma_{all}} \qquad (5.9)$$

For a timber beam of rectangular cross section, $S = \frac{1}{6}bh^2$, where b is the width of the beam and h its depth. The dimensions of the section, therefore, must be selected so that $\frac{1}{6}bh^2 \geq S_{min}$.

For a rolled-steel beam, consult the appropriate table in Appendix C. Of the available beam sections, consider only those with a section modulus $S \geq S_{min}$ and select from this group the section with the smallest weight per unit length. This is the most economical of the sections for which $S \geq S_{min}$.

In Sec. 5.5, we discussed an alternative method for the determination of the maximum values of the shear and bending moment based on the use of the *singularity functions* $\langle x - a \rangle^n$. By definition, and for $n \geq 0$, we had

$$\langle x - a \rangle^n = \begin{cases} (x-a)^n & \text{when } x \geq a \\ 0 & \text{when } x < a \end{cases} \qquad (5.14)$$

We noted that whenever the quantity between brackets is positive or zero, the brackets should be replaced by ordinary parentheses, and whenever that quantity is negative, the bracket itself is equal to zero. We also noted that singularity functions can be integrated and differentiated as ordinary binomials. Finally, we observed that the singularity function corresponding to $n = 0$ is discontinuous at $x = a$ (Fig. 5.18a). This function is called the *step function*. We wrote.

$$\langle x - a \rangle^0 = \begin{cases} 1 & \text{when } x \geq a \\ 0 & \text{when } x < a \end{cases} \qquad (5.15)$$

$\langle x - a \rangle^0$

(a) $n = 0$

Fig. 5.18a

The use of singularity functions makes it possible to represent the shear or the bending moment in a beam by a single expression, valid at any point of the beam. For example, the contribution to the shear of the concentrated load $\mathbf{P}$ applied at the midpoint C of a simply supported beam (Fig. 5.8) can be represented by $-P\langle x - \frac{1}{2}L\rangle^0$, since this expression is equal to zero to the left of C, and to $-P$ to the right of C. Adding the contribution of the reaction $R_A = \frac{1}{2}P$ at A, we express the shear at any point of the beam as

$$V(x) = \frac{1}{2}P - P\langle x - \frac{1}{2}L\rangle^0$$

The bending moment is obtained by integrating this expression:

$$M(x) = \frac{1}{2}Px - P\langle x - \frac{1}{2}L\rangle^1$$

The singularity functions representing, respectively, the load, shear, and bending moment corresponding to various basic loadings were given in Fig. 5.19 on page 346. We noted that a distributed loading which does not extend to the right end of the beam, or which is discontinuous, should be replaced by an equivalent combination of open-ended loadings. For instance, a uniformly distributed load extending from $x = a$ to $x = b$ (Fig. 5.20) should be expressed as

$$w(x) = w_0\langle x - a\rangle^0 - w_0\langle x - b\rangle^0$$

Fig. 5.8

Equivalent open-ended loadings

Fig. 5.20

The contribution of this load to the shear and to the bending moment can be obtained through two successive integrations. Care should be taken, however, to also include in the expression for $V(x)$ the contribution of concentrated loads and reactions, and to include in the expression for $M(x)$ the contribution of concentrated couples [Examples 5.05 and 5.06, Sample Probs. 5.9 and 5.10]. We also observed that singularity functions are particularly well suited to the use of computers.

Using singularity functions to express shear and bending moment

We were concerned so far only with prismatic beams, i.e., beams of uniform cross section. Considering in Sec. 5.6 the design of nonprismatic beams, i.e., beams of variable cross section, we saw that by selecting the shape and size of the cross section so that its elastic section modulus $S = I/c$ varied along the beam in the same way as the bending moment M, we were able to design beams for which σ_m at each section was equal to σ_{all}. Such beams, called *beams of constant strength*, clearly provide a more effective use of the material than prismatic beams. Their section modulus at any section along the beam was defined by the relation

Nonprismatic beams

$$S = \frac{M}{\sigma_{\text{all}}} \tag{5.18}$$

Beams of constant strength

REVIEW PROBLEMS

5.165 Draw the shear and bending-moment diagrams for the beam loading shown, and determine the maximum absolute value (*a*) of the shear, (*b*) of the bending moment.

Fig. P5.165

5.166 For the beam and loading shown, determine the maximum normal stress due to bending on a transverse section at *D*.

Fig. P5.166

5.167 Determine (*a*) the distance *a* for which the maximum absolute value of the bending moment in the beam is as small as possible, (*b*) the corresponding maximum normal stress due to bending. (*Hint:* Draw the bending-moment diagram and then equate the absolute values of the largest positive and negative bending moments obtained.)

Fig. P5.167

5.168 Draw the shear and bending-moment diagrams for the beam and loading shown, and determine the maximum absolute value (*a*) of the shear, (*b*) of the bending moment.

Fig. P5.168

5.169 For the beam and loading shown, determine (*a*) the maximum value of the bending moment, (*b*) the maximum normal stress due to bending.

Fig. P5.169

5.170 (*a*) Using singularity functions, write the equations defining the shear and bending moment for the beam and loading shown. (*b*) Determine the maximum value of the bending moment in the beam.

Fig. P5.170

5.171 For the beam and loading shown, design the cross section of the beam, knowing that the grade of timber used has an allowable normal stress of 1.75 ksi.

Fig. P5.171

5.172 For the beam and loading shown, design the cross section of the beam, knowing that the grade of timber used has an allowable normal stress of 12 MPa.

Fig. P5.172

5.173 Knowing that the allowable normal stress for the steel used is 160 MPa, select the most economical metric wide-flange beam to support the loading shown.

Fig. P5.173

5.174 Knowing that the allowable normal stress for the steel used is 24 ksi, select the most economical wide-flange beam to support the loading shown.

Fig. P5.174

5.175 A machine element of cast aluminum and in the shape of a solid of revolution of variable diameter d is being designed to support a distributed load w as shown. (*a*) Knowing that the machine element is to be of constant strength, express d in terms of y, L, and d_0. (*b*) Determine the smallest allowable value of d_0 if $L = 250$ mm, $w = 30$ kN/m, and $\sigma_{all} = 72$ MPa.

Fig. P5.175

5.176 Two cover plates, each 0.5 in. thick and 10 in. wide, are welded to a W14 × 53 beam as shown. Knowing that the allowable normal stress for the steel used is 22 ksi, determine the largest vertical force **P** that can be safely applied at point *D*.

Fig. P5.176

COMPUTER PROBLEMS

The following problems are designed to be solved with a computer.

5.C1 Several concentrated loads P_i ($i = 1, 2, \ldots, n$) can be applied to a beam as shown. Write a computer program that can be used to calculate the shear, bending moment, and normal stress at any point of the beam for a given loading of the beam and a given value of its section modulus. Use this program to solve Probs. 5.23, 5.27, and 5.29. (*Hint:* Maximum values will occur at a support or under a load.)

Fig. P5.C1

5.C2 A timber beam is to be designed to support a distributed load and up to two concentrated loads as shown. One of the dimensions of its uniform rectangular cross section has been specified and the other is to be determined so that the maximum normal stress in the beam will not exceed a given allowable value σ_{all}. Write a computer program that can be used to calculate at given intervals ΔL the shear, the bending moment, and the smallest acceptable value of the unknown dimension. Apply this program to solve the following problems, using the intervals ΔL indicated: (*a*) Prob. 5.75 ($\Delta L = 0.1$ m), (*b*) Prob. 5.79 ($\Delta L = 0.2$ m), (*c*) Prob. 5.80 ($\Delta L = 0.3$ m),

Fig. P5.C2

5.C3 Two cover plates, each of thickness t, are to be welded to a wide-flange beam of length L, which is to support a uniformly distributed load w. Denoting by σ_{all} the allowable normal stress in the beam and in the plates, by d the depth of the beam, and by I_b and S_b, respectively, the moment of inertia and the section modulus of the cross section of the unreinforced beam about a horizontal centroidal axis, write a computer program that can be used to calculate the required value of (*a*) the length a of the plates, (*b*) the width b of the plates. Use this program to solve Probs. 5.155 and 5.157.

Fig. P5.C3

Fig. P5.C4

5.C4 Two 25-kip loads are maintained 6 ft apart as they are moved slowly across the 18-ft beam AB. Write a computer program and use it to calculate the bending moment under each load and at the midpoint C of the beam for values of x from 0 to 24 ft at intervals $\Delta x = 1.5$ ft.

5.C5 Write a computer program that can be used to plot the shear and bending-moment diagrams for the beam and loading shown. Apply this program with a plotting interval $\Delta L = 0.2$ ft to the beam and loading of (a) Prob. 5.83, (b) Prob. 5.125.

Fig. P5.C5

5.C6 Write a computer program that can be used to plot the shear and bending-moment diagrams for the beam and loading shown. Apply this program with a plotting interval $\Delta L = 0.025$ m to the beam and loading of Prob. 5.124.

Fig. P5.C6

6

Shearing Stresses in Beams and Thin-Walled Members

A reinforced concrete deck will be attached to each of the steel sections shown to form a composite box girder bridge. In this chapter the shearing stresses will be determined in various types of beams and girders.

6.1. INTRODUCTION

You saw in Sec. 5.1 that a transverse loading applied to a beam will result in normal and shearing stresses in any given transverse section of
the beam. The normal stresses are created by the bending couple M in
that section and the shearing stresses by the shear V. Since the dominant criterion in the design of a beam for strength is the maximum value
of the normal stress in the beam, our analysis was limited in Chap. 5
to the determination of the normal stresses. Shearing stresses, however,
can be important, particularly in the design of short, stubby beams, and
their analysis will be the subject of the first part of this chapter.

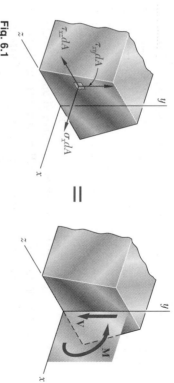

Fig. 6.1

Figure 6.1 expresses graphically that the elementary normal and
shearing forces exerted on a given transverse section of a prismatic beam
with a vertical plane of symmetry are equivalent to the bending couple
M and the shearing force V. Six equations can be written to express
that fact. Three of these equations involve only the normal forces $\sigma_x\,dA$
and have already been discussed in Sec. 4.2; they are Eqs. (4.1), (4.2),
and (4.3), which express that the sum of the normal forces is zero and
that the sums of their moments about the y and z axes are equal to zero
and M, respectively. Three more equations involving the shearing forces
$\tau_{xy}\,dA$ and $\tau_{xz}\,dA$ can now be written. One of them expresses that the
sum of the moments of the shearing forces about the x axis is zero and
can be dismissed as trivial in view of the symmetry of the beam with
respect to the xy plane. The other two involve the y and z components
of the elementary forces and are

$$y \text{ components:} \qquad \int \tau_{xy}\,dA = -V \qquad (6.1)$$
$$z \text{ components:} \qquad \int \tau_{xz}\,dA = 0 \qquad (6.2)$$

The first of these equations shows that vertical shearing stresses must
exist in a transverse section of a beam under transverse loading. The
second equation indicates that the average horizontal shearing stress in
any section is zero. However, this does not mean that the shearing stress
τ_{xz} is zero everywhere.

Let us now consider a small cubic element located in the vertical
plane of symmetry of the beam (where we know that τ_{xz} must be zero)
and examine the stresses exerted on its faces (Fig. 6.2). As we have just
seen, a normal stress σ_x and a shearing stress τ_{xy} are exerted on each
of the two faces perpendicular to the x axis. But we know from Chap.
1 that, when shearing stresses τ_{xy} are exerted on the vertical faces of an

Fig. 6.2

element, equal stresses must be exerted on the horizontal faces of the same element. We thus conclude that longitudinal shearing stresses must exist in any member subjected to a transverse loading. This can be verified by considering a cantilever beam made of separate planks clamped together at one end (Fig. 6.3a). When a transverse load **P** is applied to the free end of this composite beam, the planks are observed to slide with respect to each other (Fig. 6.3b). In contrast, if a couple **M** is applied to the free end of the same composite beam (Fig. 6.3c), the various planks will bend into concentric arcs of circle and will not slide with respect to each other, thus verifying the fact that shear does not occur in a beam subjected to pure bending (cf. Sec. 4.3).

While sliding does not actually take place when a transverse load **P** is applied to a beam made of a homogeneous and cohesive material such as steel, the tendency to slide does exist, showing that stresses occur on horizontal longitudinal planes as well as on vertical transverse planes. In the case of timber beams, whose resistance to shear is weaker between fibers, failure due to shear will occur along a longitudinal plane rather than a transverse plane (Fig. 6.4).

In Sec. 6.2, a beam element of length Δx bounded by two transverse planes and a horizontal one will be considered and the shearing force $\Delta \mathbf{H}$ exerted on its horizontal face will be determined, as well as the shear per unit length, q, also known as *shear flow*. A formula for the shearing stress in a beam with a vertical plane of symmetry will be derived in Sec. 6.3 and used in Sec. 6.4 to determine the shearing stresses in common types of beams. The distribution of stresses in a narrow rectangular beam will be further discussed in Sec. 6.5.

The derivation given in Sec. 6.2 will be extended in Sec. 6.6 to cover the case of a beam element bounded by two transverse planes and a curved surface. This will allow us in Sec. 6.7 to determine the shearing stresses at any point of a symmetric thin-walled member, such as the flanges of wide-flange beams and box beams. The effect of plastic deformations on the magnitude and distribution of shearing stresses will be discussed in Sec. 6.8.

In the last section of the chapter (Sec. 6.9), the unsymmetric loading of thin-walled members will be considered and the concept of *shear center* will be introduced. You will then learn to determine the distribution of shearing stresses in such members.

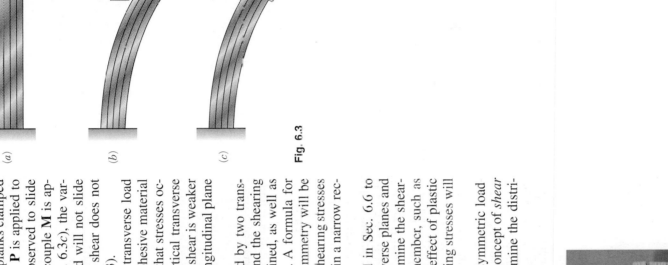

(a)

(b)

(c)

Fig. 6.3

Fig. 6.4

6.2. SHEAR ON THE HORIZONTAL FACE
OF A BEAM ELEMENT

Fig. 6.5

Consider a prismatic beam AB with a vertical plane of symmetry that supports various concentrated and distributed loads (Fig. 6.5). At a distance x from end A we detach from the beam an element $CDD'C'$ of length Δx extending across the width of the beam from the upper surface of the beam to a horizontal plane located at a distance y_1 from the neutral axis (Fig. 6.6). The forces exerted on this element consist of

Fig. 6.6

vertical shearing forces $\mathbf{V}'_C$ and $\mathbf{V}'_D$, a horizontal shearing force $\mathbf{\Delta H}$ exerted on the lower face of the element, elementary horizontal normal forces $\sigma_C\, dA$ and $\sigma_D\, dA$, and possibly a load $w\, \Delta x$ (Fig. 6.7). We write the equilibrium equation

$$\xrightarrow{+}\ \Sigma F_x = 0: \qquad \Delta H + \int_a (\sigma_D - \sigma_C)\, dA = 0$$

where the integral extends over the shaded area a of the section located above the line $y = y_1$. Solving this equation for ΔH and using Eq. (5.2) of Sec. 5.1, $\sigma = My/I$, to express the normal stresses in terms of the bending moments at C and D, we have

$$\Delta H = \frac{M_D - M_C}{I} \int_a y\, dA \qquad (6.3)$$

Fig. 6.7

The integral in (6.3) represents the *first moment* with respect to the neutral axis of the portion $\mathcal{C}$ of the cross section of the beam that is located above the line $y = y_1$ and will be denoted by Q. On the other hand, recalling Eq. (5.7) of Sec. 5.5, we can express the increment $M_D - M_C$ of the bending moment as

$$M_D - M_C = \Delta M = (dM/dx)\,\Delta x = V\,\Delta x$$

Substituting into (6.3), we obtain the following expression for the horizontal shear exerted on the beam element

$$\Delta H = \frac{VQ}{I}\,\Delta x \qquad (6.4)$$

The same result would have been obtained if we had used as a free body the lower element $C'D'D''C''$, rather than the upper element

Fig. 6.8

$CDD'C'$ (Fig. 6.8), since the shearing forces $\Delta\mathbf{H}$ and $\Delta\mathbf{H}'$ exerted by the two elements on each other are equal and opposite. This leads us to observe that the first moment Q of the portion $\mathcal{C}'$ of the cross section located below the line $y = y_1$ (Fig. 6.8) is equal in magnitude and opposite in sign to the first moment of the portion $\mathcal{C}$ located above that line (Fig. 6.6). Indeed, the sum of these two moments is equal to the moment of the area of the entire cross section with respect to its centroidal axis and, thus, must be zero. This property can sometimes be used to simplify the computation of Q. We also note that Q is maximum for $y_1 = 0$, since the elements of the cross section located above the neutral axis contribute positively to the integral (5.5) that defines Q, while the elements located below that axis contribute negatively.

The *horizontal shear per unit length*, which will be denoted by the letter q, is obtained by dividing both members of Eq. (6.4) by Δx:

$$q = \frac{\Delta H}{\Delta x} = \frac{VQ}{I} \qquad (6.5)$$

We recall that Q is the first moment with respect to the neutral axis of the portion of the cross section located either above or below the point at which q is being computed, and that I is the centroidal moment of inertia of the *entire* cross-sectional area. For a reason that will become apparent later (Sec. 6.7), the horizontal shear per unit length q is also referred to as the *shear flow*.

EXAMPLE 6.01

A beam is made of three planks, 20 by 100 mm in cross section, nailed together (Fig. 6.9), Knowing that the spacing between nails is 25 mm and that the vertical shear in the beam is $V = 500$ N, determine the shearing force in each nail.

We first determine the horizontal force per unit length, q, exerted on the lower face of the upper plank. We use Eq. (6.5), where Q represents the first moment with respect to the neutral axis of the shaded area A shown in Fig. 6.10a, and where I is the moment of inertia about the same axis of the entire cross-sectional area (Fig. 6.10b). Recalling that the first moment of an area with respect to a given axis is equal to the product of the area and of the distance from its centroid to the axis,† we have

$$Q = A\bar{y} = (0.020 \text{ m} \times 0.100 \text{ m})(0.060 \text{ m})$$
$$= 120 \times 10^{-6} \text{ m}^3$$

$$I = \tfrac{1}{12}(0.020 \text{ m})(0.100 \text{ m})^3$$
$$+2[\tfrac{1}{12}(0.100 \text{ m})(0.020 \text{ m})^3$$
$$+(0.020 \text{ m} \times 0.100 \text{ m})(0.060 \text{ m})^2]$$
$$= 1.667 \times 10^{-6} + 2(0.0667 + 7.2)10^{-6}$$
$$= 16.20 \times 10^{-6} \text{ m}^4$$

Substituting into Eq. (6.5), we write

$$q = \frac{VQ}{I} = \frac{(500 \text{ N})(120 \times 10^{-6} \text{ m}^3)}{16.20 \times 10^{-6} \text{ m}^4} = 3704 \text{ N/m}$$

only calculate for parts containing nails

† where q is being calculated for parts containing nails

Since the spacing between the nails is 25 mm, the shearing force in each nail is

$$F = (0.025 \text{ m})q = (0.025 \text{ m})(3704 \text{ N/m}) = 92.6 \text{ N}$$

Fig. 6.9

Fig. 6.10

(a)

(b)

6.3. DETERMINATION OF THE SHEARING STRESSES IN A BEAM

Consider again a beam with a vertical plane of symmetry, subjected to various concentrated or distributed loads applied in that plane. We saw in the preceding section that if, through two vertical cuts and one horizontal cut, we detach from the beam an element of length Δx (Fig. 6.11), the magnitude ΔH of the shearing force exerted on the horizontal face of the element can be obtained from Eq. (6.4). The *average shearing stress* τ_{ave} on that face of the element is obtained by dividing ΔH by the area ΔA of the face. Observing that $\Delta A = t \Delta x$, where t is the width of the element at the cut, we write

$$\tau_{\text{ave}} = \frac{\Delta H}{\Delta A} = \frac{\Delta H}{t \Delta x} = \frac{VQ}{I} \frac{\Delta x}{t \Delta x}$$

or

$$\tau_{\text{ave}} = \frac{VQ}{It} \tag{6.6}$$

Fig. 6.11

We note that, since the shearing stresses τ_{xy} and τ_{yx} exerted respectively on a transverse and a horizontal plane through D' are equal, the expression obtained also represents the average value of τ_{xy} along the line $D'_1 D'_2$ (Fig. 6.12).

Fig. 6.12

We observe that $\tau_{yx} = 0$ on the upper and lower faces of the beam, since no forces are exerted on these faces. It follows that $\tau_{xy} = 0$ along the upper and lower edges of the transverse section (Fig. 6.13). We also note that, while Q is maximum for $y = 0$ (see Sec. 6.2), we cannot conclude that τ_{ave} will be maximum along the neutral axis, since τ_{ave} depends upon the width t of the section as well as upon Q.

As long as the width of the beam cross section remains small compared to its depth, the shearing stress varies only slightly along the line $D'_1 D'_2$ (Fig. 6.12) and Eq. (6.6) can be used to compute τ_{xy} at any point along $D'_1 D'_2$. Actually, τ_{xy} is larger at points D'_1 and D'_2 than at D', but the theory of elasticity shows[†] that, for a beam of rectangular section of width b and depth h, and as long as $b \leq h/4$, the value of the shearing stress at points C_1 and C_2 (Fig. 6.14) does not exceed by more than 0.8% the average value of the stress computed along the neutral axis.[‡]

Fig. 6.13

Fig. 6.14

6.4. SHEARING STRESSES τ_{xy} IN COMMON TYPES OF BEAMS

We saw in the preceding section that, for a *narrow rectangular beam*, i.e., for a beam of rectangular section of width b and depth h with $b \leq \frac{1}{4}h$, the variation of the shearing stress τ_{xy} across the width of the beam is less than 0.8% of τ_{ave}. We can, therefore, use Eq. (6.6) in practical applications to determine the shearing stress at any point of the cross section of a narrow rectangular beam and write

$$\tau_{xy} = \frac{VQ}{It} \tag{6.7}$$

where t is equal to the width b of the beam, and where Q is the first

[†] See S. P. Timoshenko and J. N. Goodier, *Theory of Elasticity*, McGraw Hill, New York, 3d ed., 1970, sec. 124.

[‡] On the other hand, for large values of b/h, the value τ_{max} of the stress at C_1 and C_2 may be many times larger than the average value τ_{ave} computed along the neutral axis, as we may see from the following table:

b/h	0.25	0.5	1	2	4	6	10	20	50
τ_{max}/τ_{ave}	1.008	1.033	1.126	1.396	1.988	2.582	3.770	6.740	15.65
τ_{min}/τ_{ave}	0.996	0.983	0.940	0.856	0.805	0.800	0.800	0.800	0.800

moment with respect to the neutral axis of the shaded area A (Fig. 6.15).
Observing that the distance from the neutral axis to the centroid C'
of A is $\bar{y} = \frac{1}{2}(c + y)$, and recalling that $Q = A\bar{y}$; we write

$$Q = A\bar{y} = b(c - y)\tfrac{1}{2}(c + y) = \tfrac{1}{2}b(c^2 - y^2) \qquad (6.8)$$

Recalling, on the other hand, that $I = bh^3/12 = \tfrac{2}{3}bc^3$, we have

$$\tau_{xy} = \frac{VQ}{Ib} = \frac{3}{4}\frac{c^2 - y^2}{bc^3}V$$

or, noting that the cross-sectional area of the beam is $A = 2bc$,

$$\tau_{xy} = \frac{3}{4}\frac{V}{A}\left(1 - \frac{y^2}{c^2}\right) \qquad (6.9)$$

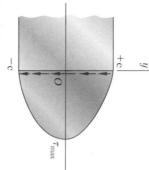

Fig. 6.15

Equation (6.9) shows that the distribution of shearing stresses in a
transverse section of a rectangular beam is *parabolic* (Fig. 6.16). As we
have already observed in the preceding section, the shearing stresses
are zero at the top and bottom of the cross section $(y = \pm c)$. Making
$y = 0$ in Eq. (6.9), we obtain the value of the maximum shearing stress
in a given section of a *narrow rectangular beam*:

$$\tau_{max} = \frac{3}{2}\frac{V}{A} \qquad (6.10)$$

Fig. 6.16

The relation obtained shows that the maximum value of the shearing
stress in a beam of rectangular cross section is 50% larger than the value
V/A which would be obtained by wrongly assuming a uniform stress
distribution across the entire cross section.

In the case of an *American standard beam* (S-beam) or a *wide-
flange beam* (W-beam), Eq. (6.6) can be used to determine the average
value of the shearing stress τ_{xy} over a section aa' or bb' of the transverse cross section of the beam (Figs. 6.17a and b). We write

$$\tau_{ave} = \frac{VQ}{It} \qquad (6.6)$$

where V is the vertical shear, t the width of the section at the elevation
considered, Q the first moment of the shaded area with respect to the
neutral axis cc', and I the moment of inertia of the entire cross-sectional
area about cc'. Plotting τ_{ave} against the vertical distance y, we obtain
the curve shown in Fig. 6.17c. We note the discontinuities existing in
this curve, which reflect the difference between the values of t corresponding respectively to the flanges $ABGD$ and $A'B'G'D'$ and to the
web $EFF'E'$.

Fig. 6.17

In the case of the web, the shearing stress τ_{xy} varies only very slightly across the section bb', and can be assumed equal to its average value τ_{ave}. This is not true, however, for the flanges. For example, considering the horizontal line $DEFG$, we note that τ_{xy} is zero between D and E and between F and G, since these two segments are part of the free surface of the beam. On the other hand the value of τ_{xy} between E and F can be obtained by making $t = EF$ in Eq. (6.6). In practice, one usually assumes that the entire shear load is carried by the web, and that a good approximation of the maximum value of the shearing stress in the cross section can be obtained by dividing V by the cross-sectional area of the web.

$$\tau_{max} = \frac{V}{A_{web}} \qquad (6.11)$$

We should note, however, that while the vertical component τ_{xy} of the shearing stress in the flanges can be neglected, its horizontal component τ_{xz} has a significant value that will be determined in Sec. 6.7.

[Handwritten margin notes and figure:]

analyzing @ a point Q ↷ all that changes

$Q = A_1 \bar{y}_1 + A_2 \bar{y}_2$

EXAMPLE 6.02

Knowing that the allowable shearing stress for the timber beam of Sample Prob. 5.7 is $\tau_{all} = 0.250$ ksi, check that the design obtained in that sample problem is acceptable from the point of view of the shearing stresses.

We recall from the shear diagram of Sample Prob. 5.7 that $V_{max} = 4.50$ kips. The actual width of the beam was given as $b = 3.5$ in. and the value obtained for its depth was

$h = 14.55$ in. Using Eq. (6.10) for the maximum shearing stress in a narrow rectangular beam, we write

$$\tau_{max} = \frac{3}{2}\frac{V}{A} = \frac{3}{2}\frac{V}{bh} = \frac{3(4.50 \text{ kips})}{2(3.5 \text{ in.})(14.55 \text{ in.})} = 0.1325 \text{ ksi}$$

Since $\tau_{max} < \tau_{all}$, the design obtained in Sample Prob. 5.7 is acceptable.

EXAMPLE 6.03

Knowing that the allowable shearing stress for the steel beam of Sample Prob. 5.8 is $\tau_{all} = 90$ MPa, check that the W360 × 32.9 shape obtained in that sample problem is acceptable from the point of view of the shearing stresses.

We recall from the shear diagram of Sample Prob. 5.8 that the maximum absolute value of the shear in the beam is $|V|_{max} = 58$ kN. As we saw in Sec. 6.4, it may be assumed in practice that the entire shear load is carried by the web and that the maximum value of the shearing stress in the beam can be obtained from Eq. (6.11). From Appendix C we find that for a W360 × 32.9 shape the depth of the beam and the

thickness of its web are, respectively, $d = 349$ mm and $t_w = 5.8$ mm. We thus have

$$A_{web} = d\,t_w = (349 \text{ mm})(5.8 \text{ mm}) = 2024 \text{ mm}^2$$

Substituting the values of $|V|_{max}$ and A_{web} into Eq. (6.11), we obtain

$$\tau_{max} = \frac{|V|_{max}}{A_{web}} = \frac{58 \text{ kN}}{2024 \text{ mm}^2} = 28.7 \text{ MPa}$$

Since $\tau_{max} < \tau_{all}$, the design obtained in Sample Prob. 5.8 is acceptable.

*6.5. FURTHER DISCUSSION OF THE DISTRIBUTION OF STRESSES IN A NARROW RECTANGULAR BEAM

Fig. 6.18

Consider a narrow cantilever beam of rectangular cross section of width b and depth h subjected to a load $\mathbf{P}$ at its free end (Fig. 6.18). Since the shear V in the beam is constant and equal in magnitude to the load $\mathbf{P}$, Eq. (6.9) yields

$$\tau_{xy} = \frac{3}{2}\frac{P}{A}\left(1 - \frac{y^2}{c^2}\right) \tag{6.12}$$

We note from Eq. (6.12) that the shearing stresses depend only upon the distance y from the neutral surface. They are independent, therefore, of the distance from the point of application of the load; it follows that all elements located at the same distance from the neutral surface undergo the same shear deformation (Fig. 6.19). While plane sections do *not* remain plane, the distance between two corresponding points D and D' located in different sections remains the same. This indicates that the normal strains ϵ_x, and thus the normal stresses σ_x, are unaffected by the shearing stresses, and that the assumption made in Sec. 5.1 is justified for the loading condition of Fig. 6.18.

We conclude that our analysis of the stresses in a cantilever beam of rectangular cross section, subjected to a concentrated load $\mathbf{P}$ at its free end, is valid. The correct values of the shearing stresses in the beam are given by Eq. (6.12), and the normal stresses at a distance x from the free end are obtained by making $M = -Px$ in Eq. (5.2) of Sec. 5.1. We have

$$\sigma_x = +\frac{Pxy}{I} \tag{6.13}$$

Fig. 6.19

The validity of the above statement, however, depends upon the end conditions. If Eq. (6.12) is to apply everywhere, then the load P must be distributed parabolically over the free-end section. Moreover, the fixed-end support must be of such a nature that it will allow the type of shear deformation indicated in Fig. 6.19. The resulting model (Fig. 6.20) is highly unlikely to be encountered in practice. However, it follows from Saint-Venant's principle that, for other modes of application of the load and for other types of fixed-end supports, Eqs. (6.12) and (6.13) still provide us with the correct distribution of stresses, except close to either end of the beam.

Fig. 6.20

When a beam of rectangular cross section is subjected to several concentrated loads (Fig. 6.21), the principle of superposition can be used to determine the normal and shearing stresses in sections located between the points of application of the loads. However, since the loads P_2, P_3, etc., are applied on the surface of the beam and cannot be assumed to be distributed parabolically throughout the cross section, the results obtained cease to be valid in the immediate vicinity of the points of application of the loads.

Fig. 6.21

When the beam is subjected to a distributed load (Fig. 6.22), the shear varies with the distance from the end of the beam, and so does the shearing stress at a given elevation y. The resulting shear deformations are such that the distance between two corresponding points of different cross sections, such as D_1 and D_1', or D_2 and D_2', will depend upon their elevation. This indicates that the assumption that plane sections remain plane, under which Eqs. (6.12) and (6.13) were derived, must be rejected for the loading condition of Fig. 6.22. The error involved, however, is small for the values of the span-depth ratio encountered in practice.

We should also note that, in portions of the beam located under a concentrated or distributed load, normal stresses σ_y will be exerted on the horizontal faces of a cubic element of material, in addition to the stresses τ_{xy} shown in Fig. 6.2.

Fig. 6.22

SAMPLE PROBLEM 6.1

Beam AB is made of three planks glued together and is subjected, in its plane of symmetry, to the loading shown. Knowing that the width of each glued joint is 20 mm, determine the average shearing stress in each glued joint at section n-n of the beam. The location of the centroid of the section is given in the sketch and the centroidal moment of inertia is known to be $I = 8.63 \times 10^{-6}$ m⁴.

SOLUTION

Vertical Shear at Section n-n. Since the beam and loading are both symmetric with respect to the center of the beam, we have $\mathbf{A} = \mathbf{B} = 1.5$ kN ↑.

Considering the portion of the beam to the left of section n-n as a free body, we write

$$+\uparrow \Sigma F_y = 0: \qquad 1.5\text{ kN} - V = 0 \qquad V = 1.5\text{ kN}$$

Shearing Stress in Joint a. We pass the section a-a through the glued joint and separate the cross-sectional area into two parts. We choose to determine Q by computing the first moment with respect to the neutral axis of the area above section a-a.

$$Q = A\bar{y}_1 = [(0.100\text{ m})(0.020\text{ m})](0.0417\text{ m}) = 83.4 \times 10^{-6}\text{ m}^3$$

Recalling that the width of the glued joint is $t = 0.020$ m, we use Eq. (6.7) to determine the average shearing stress in the joint.

$$\tau_{ave} = \frac{VQ}{It} = \frac{(1500\text{ N})(83.4 \times 10^{-6}\text{ m}^3)}{(8.63 \times 10^{-6}\text{ m}^4)(0.020\text{ m})} \qquad \tau_{ave} = 725\text{ kPa} \blacktriangleleft$$

Shearing Stress in Joint b. We now pass section b-b and compute Q by using the area below the section.

$$Q = A\bar{y}_2 = [(0.060\text{ m})(0.020\text{ m})](0.0583\text{ m}) = 70.0 \times 10^{-6}\text{ m}^3$$

$$\tau_{ave} = \frac{VQ}{It} = \frac{(1500\text{ N})(70.0 \times 10^{-6}\text{ m}^3)}{(8.63 \times 10^{-6}\text{ m}^4)(0.020\text{ m})} \qquad \tau_{ave} = 608\text{ kPa} \blacktriangleleft$$

SAMPLE PROBLEM 6.2

A timber beam AB of span 10 ft and nominal width 4 in. (actual width = 3.5 in.) is to support the three concentrated loads shown. Knowing that for the grade of timber used $\sigma_{all} = 1800$ psi and $\tau_{all} = 120$ psi, determine the minimum required depth d of the beam.

SOLUTION

Maximum Shear and Bending Moment. After drawing the shear and bending-moment diagrams, we note that

$$M_{max} = 7.5 \text{ kip} \cdot \text{ft} = 90 \text{ kip} \cdot \text{in.}$$
$$V_{max} = 3 \text{ kips}$$

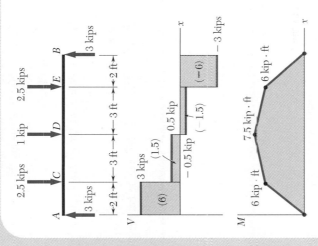

Design Based on Allowable Normal Stress. We first express the elastic section modulus S in terms of the depth d. We have

$$I = \frac{1}{12}bd^3 \qquad S = \frac{I}{c} = \frac{1}{6}bd^2 = \frac{1}{6}(3.5)d^2 = 0.5833d^2$$

For $M_{max} = 90$ kip $\cdot$ in. and $\sigma_{all} = 1800$ psi, we write

$$S = \frac{M_{max}}{\sigma_{all}} \qquad 0.5833d^2 = \frac{90 \times 10^3 \text{ lb} \cdot \text{in.}}{1800 \text{ psi}}$$
$$d^2 = 85.7 \qquad\qquad d = 9.26 \text{ in.}$$

We have satisfied the requirement that $\sigma_m \leq 1800$ psi.

Check Shearing Stress. For $V_{max} = 3$ kips and $d = 9.26$ in., we find

$$\tau_m = \frac{3}{2}\frac{V_{max}}{A} = \frac{3}{2}\frac{3000 \text{ lb}}{(3.5 \text{ in.})(9.26 \text{ in.})} \qquad \tau_m = 138.8 \text{ psi}$$

Since $\tau_{all} = 120$ psi, the depth $d = 9.26$ in. is *not* acceptable and we must redesign the beam on the basis of the requirement that $\tau_m \leq 120$ psi.

Design Based on Allowable Shearing Stress. Since we now know that the allowable shearing stress controls the design, we write

$$\tau_m = \tau_{all} = \frac{3}{2}\frac{V_{max}}{A}$$
$$120 \text{ psi} = \frac{3}{2}\frac{3000 \text{ lb}}{(3.5 \text{ in.})d}$$
$$d = 10.71 \text{ in.} \qquad \blacktriangleleft$$

The normal stress is, of course, less than $\sigma_{all} = 1800$ psi, and the depth of 10.71 in. is fully acceptable.

Comment. Since timber is normally available in depth increments of 2 in., a 4×12-in. nominal size timber should be used. The actual cross section would then be 3.5×11.25 in.

PROBLEMS

6.1   Three full-size 50 × 100-mm boards are nailed together to form a beam that is subjected to a vertical shear of 1500 N. Knowing that the allowable shearing force in each nail is 400 N, determine the largest longitudinal spacing s that can be used between each pair of nails.

Fig. P6.1

50 mm
50 mm
50 mm

s
s
s
s

100 mm

6.2   For the built-up beam of Prob. 6.1, determine the allowable shear if the spacing between each pair of nails is $s = 45$ mm.

6.3   A square box beam is made of two $4\frac{3}{4} \times 5$-in. planks and two $\frac{3}{4} \times 5$-in. planks nailed together as shown. Knowing that the spacing between nails is $s = 1.25$ in. and that the vertical shear in the beam is $V = 250$ lb, determine (a) the shearing force in each nail, (b) the maximum shearing stress in the beam.

Fig. P6.3 and P6.4

$\frac{3}{4}$ in.
$\frac{3}{4}$ in.
3.5 in.

$\frac{3}{4}$ in.

5 in.

s
s
s

6.4   A square box beam is made of two $4\frac{3}{4} \times 3.5$-in. planks and two $\frac{3}{4} \times 5$-in. planks nailed together as shown. Knowing that the spacing between nails is $s = 2$ in. and that the allowable shearing force in each nail is 75 lb, determine (a) the largest allowable vertical shear in the beam, (b) the corresponding maximum shearing stress in the beam.

6.5   The beam shown has been reinforced by attaching to it two 12 × 175-mm plates, using bolts of 18-mm diameter spaced longitudinally every 125 mm. Knowing that the average allowable shearing stress in the bolts is 85 MPa, determine the largest permissible vertical shearing force.

Fig. P6.5

W250 × 44.8

6.6   Solve Prob. 6.5, assuming that the reinforcing plates are only 9 mm thick.

6.7 and 6.8   A column is fabricated by connecting the rolled-steel members shown by bolts of $\frac{3}{4}$-in. diameter spaced longitudinally every 5 in. Determine the average shearing stress in the bolts caused by a shearing force of 30 kips parallel to the y axis.

Fig. P6.7

C8 × 13.75

S10 × 25.4

z

C

y

Fig. P6.8

z

C

y

14 in. × $\frac{3}{8}$ in.

C10 × 25

6.9 through 6.12 For the beam and loading shown, consider section *n-n* and determine (*a*) the largest shearing stress in that section, (*b*) the shearing stress at point *a*.

15 kips 20 kips 15 kips

2 ft 2 ft 2 ft 2 ft

1 ft

n

n

10 in.

10 in.

0.6 in.

0.6 in.

0.375 in.

1 in.

a

Fig. P6.9

10 kips 10 kps

8 in.

16 in. 12 in. 16 in.

n

n

4 in.

4 in.

½ in.

½ in.

a

Fig. P6.10

10 kN

0.3 m

1.5 m

n

n

150 mm

12 mm 12 mm

40 mm

100 mm

200 mm

a

Fig. P6.11

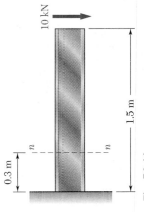

125 kN

450 mm

600 mm

n

n

72 mm 72 mm 72 mm

192 mm

t = 6 mm

t

t

t

a

Fig. P6.12

6.13 Two steel plates of 15×220-mm rectangular cross section are welded to the W250 × 67 beam as shown. Determine the largest allowable vertical shear if the shearing stress in the beam is not to exceed 100 MPa.

— 220 mm —

W250 × 67

257 mm

15 mm

15 mm

Fig. P6.13

6.14 Solve Prob. 6.13, assuming that the two steel plates are (*a*) replaced by steel plates of 10×220-mm rectangular cross section, (*b*) removed.

6.15 Knowing that the allowable shearing stress for the timber used is 825 kPa, check whether the design obtained for the beam indicated is acceptable and, if not, redesign the cross section of the beam. Consider the beam of (*a*) Prob. 5.75, (*b*) Prob. 5.76.

6.16 Knowing that the allowable shearing stress for the timber used is 130 psi, check whether the design obtained for the beam indicated is acceptable and, if not, redesign the cross section of the beam. Consider the beam of (*a*) Prob. 5.77, (*b*) Prob. 5.78.

6.17 Determine the average shearing stress in the web of the beam indicated and check whether the design obtained earlier for that beam is acceptable, knowing that the allowable shearing stress for the timber used is acceptable. Consider the beam of (*a*) Prob. 5.81, (*b*) Prob. 5.82.

6.18 Determine the average shearing stress in the web of the beam indicated and check whether the design obtained earlier for that beam is acceptable, knowing that the allowable shearing stress for the steel used is 14.5 ksi. Consider the beam of (*a*) Prob. 5.83, (*b*) Prob. 5.84.

6.19 A simply supported timber beam AB of rectangular cross section carries a single concentrated load **P** at its midpoint C. (*a*) Show that the ratio τ_m / σ_m of the maximum values of the shearing and normal stresses in the beam is equal to $h/2L$, where h and L are, respectively, the depth and the length of the beam. (*b*) Determine the depth h and width b of the beam, knowing that $L = 2$ m, $P = 40$ kN, $\tau_m = 960$ kPa, and $\sigma_m = 12$ MPa.

A
C
P
L/2 ⟶ ⟵ L/2
B

|← b →|

h

Fig. P6.19

6.20 A timber beam AB of length L and rectangular cross section carries a uniformly distributed load w and is supported as shown. (*a*) Show that the ratio τ_m / σ_m of the maximum values of the shearing and normal stresses in the beam is equal to $2h/L$, where h and L are, respectively, the depth and the length of the beam. (*b*) Determine the depth h and width b of the beam, knowing that $L = 5$ m, $w = 8$ kN/m, $\tau_m = 1.08$ MPa, and $\sigma_m = 12$ MPa.

A
C
w
D
B
L/4 ⟶ ⟵ L/2 ⟶ ⟵ L/4

|← b →|

h

Fig. P6.20

6.21 and 6.22 For the beam and loading shown, consider section n-n and determine the shearing stress at (a) point a, (b) point b.

200 kN   200 kN

n

A

B

0.75 m

1.2 m

0.75 m

0.75 m

Fig. P6.21 and P6.23

50 mm

50 mm

150 mm

50 mm

• a

• b

75 mm 75 mm 75 mm

25 kips   25 kips

n

n

A

B

20 in.

10 in.

20 in.

20 in.

Fig. P6.22 and P6.24

$\frac{3}{4}$ in.

7.25 in.

b
a

1.5 in.
1.5 in.

$\frac{3}{4}$ in.

8 in.

$\frac{3}{4}$ in.

6.23 and 6.24 For the beam and loading shown, determine the largest shearing stress in section n-n.

6.25 Two W200 × 46.1 rolled steel sections are to be welded at A and B in either of the two ways shown to form a composite beam. Knowing that for each weld the allowable horizontal shearing force is 500 kN per meter of weld, determine the maximum allowable shear in the composite beam for each of the two arrangements shown.

A   B

(a)

A   B

(b)

Fig. P6.25

6.26 through 6.28 A beam having the cross section shown is subjected to a vertical shear **V**. Determine (a) the horizontal line along which the shearing stress is maximum, (b) the constant k in the following expression for the maximum shearing stress

$$\tau_{max} = k \frac{V}{A}$$

where A is the cross-sectional area of the beam.

c

Fig. P6.26

r_m

t_m

Fig. P6.27

b

h

Fig. P6.28

Fig. 6.23

(a)

(b)

6.6. LONGITUDINAL SHEAR ON A BEAM ELEMENT OF ARBITRARY SHAPE

Consider a box beam obtained by nailing together four planks, as shown in Fig. 6.23a. You learned in Sec. 6.2 how to determine the shear per unit length, q, on the horizontal surfaces along which the planks are joined. But could you determine q if the planks had been joined along *vertical* surfaces, as shown in Fig. 6.23b? We examined in Sec. 6.4 the distribution of the vertical components τ_{xy} of the stresses on a transverse section of a W-beam or an S-beam and found that these stresses had a fairly constant value in the web of the beam and were negligible in its flanges. But what about the *horizontal* components τ_{xz} of the stresses in the flanges?

To answer these questions we must extend the procedure developed in Sec. 6.2 for the determination of the shear per unit length, q, so that it will apply to the cases just described.

Fig. 6.5 (repeated)

Consider the prismatic beam AB of Fig. 6.5, which has a vertical plane of symmetry and supports the loads shown. At a distance x from end A we detach again an element $CDD'C'$ of length Δx. This element, however, will now extend from two sides of the beam to an arbitrary curved surface (Fig. 6.24). The forces exerted on the element include

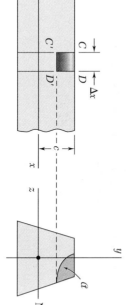

Fig. 6.24

vertical shearing forces $\mathbf{V}'_C$ and $\mathbf{V}'_D$, elementary horizontal normal forces $\sigma_C\,dA$ and $\sigma_D\,dA$, possibly a load $w\,\Delta x$, and a longitudinal shearing force $\Delta\mathbf{H}$ representing the resultant of the elementary longitudinal shearing forces exerted on the curved surface (Fig. 6.25). We write the equilibrium equation

$$\xrightarrow{+} \sum F_x = 0: \qquad \Delta H + \int_{\mathcal{C}} (\sigma_D - \sigma_C)\,dA = 0$$

where the integral is to be computed over the shaded area $\mathcal{C}$ of the section. We observe that the equation obtained is the same as the one we

Fig. 6.25

obtained in Sec. 6.2, but that the shaded area $\mathfrak{C}$ over which the integral is to be computed now extends to the curved surface.

The remainder of the derivation is the same as in Sec. 6.2. We find that the longitudinal shear exerted on the beam element is

$$\Delta H = \frac{VQ}{I} \Delta x \qquad (6.4)$$

where I is the centroidal moment of inertia of the entire section, Q the first moment of the shaded area $\mathfrak{C}$ with respect to the neutral axis, and V the vertical shear in the section. Dividing both members of Eq. (6.4) by Δx, we obtain the horizontal shear per unit length, or shear flow:

$$q = \frac{\Delta H}{\Delta x} = \frac{VQ}{I} \qquad (6.5)$$

EXAMPLE 6.04

A square box beam is made of two 0.75×3-in. planks and two 0.75×4.5-in. planks, nailed together as shown (Fig. 6.26). Knowing that the spacing between nails is 1.75 in. and that the beam is subjected to a vertical shear of magnitude $V = 600$ lb, determine the shearing force in each nail.

We isolate the upper plank and consider the total force per unit length, q, exerted on its two edges. We use Eq. (6.5), where Q represents the first moment with respect to the neutral axis of the shaded area A' shown in Fig. 6.27a, and where I is the moment of inertia about the same axis of the entire cross-sectional area of the box beam (Fig. 6.27b). We have

$$Q = A'\bar{y} = (0.75 \text{ in.})(3 \text{ in.})(1.875 \text{ in.}) = 4.22 \text{ in}^3$$

Recalling that the moment of inertia of a square of side a about a centroidal axis is $I = \frac{1}{12}a^4$, we write

$$I = \tfrac{1}{12}(4.5 \text{ in.})^4 - \tfrac{1}{12}(3 \text{ in.})^4 = 27.42 \text{ in}^4$$

Substituting into Eq. (6.5), we obtain

$$q = \frac{VQ}{I} = \frac{(600 \text{ lb})(4.22 \text{ in}^3)}{27.42 \text{ in}^4} = 92.3 \text{ lb/in.}$$

Because both the beam and the upper plank are symmetric with respect to the vertical plane of loading, equal forces are exerted on both edges of the plank. The force per unit length on each of these edges is thus $\frac{1}{2}q = \frac{1}{2}(92.3) = 46.15$ lb/in. Since the spacing between nails is 1.75 in., the shearing force in each nail is

$$F = (1.75 \text{ in.})(46.15 \text{ lb/in.}) = 80.8 \text{ lb}$$

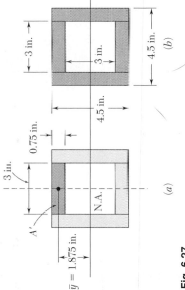

0.75 in. 3 in. 0.75 in.

0.75 in.

0.75 in.

4.5 in.

Fig. 6.26

3 in.

3 in.

3 in.

4.5 in.

0.75 in.

4.5 in.

$\bar{y} = 1.875$ in.

A'

3 in.

0.75 in.

N.A.

(a)

(b)

Fig. 6.27

6.7. SHEARING STRESSES IN THIN-WALLED MEMBERS

We saw in the preceding section that Eq. (6.4) may be used to determine the longitudinal shear ΔH exerted on the walls of a beam element of arbitrary shape and Eq. (6.5) to determine the corresponding shear flow q. These equations will be used in this section to calculate both the shear flow and the average shearing stress in thin-walled members such as the flanges of wide-flange beams (Fig. 6.28) and box beams, or the walls of structural tubes (Fig. 6.29).

Fig. 6.28

Fig. 6.29

Consider, for instance, a segment of length Δx of a wide-flange beam (Fig. 6.30a) and let $\mathbf{V}$ be the vertical shear in the transverse section shown. Let us detach an element $ABB'A'$ of the upper flange (Fig. 6.30b). The longitudinal shear $\Delta \mathbf{H}$ exerted on that element can be obtained from Eq. (6.4):

$$\Delta H = \frac{VQ}{I} \Delta x \qquad (6.4)$$

Dividing ΔH by the area $\Delta A = t\, \Delta x$ of the cut, we obtain for the average shearing stress exerted on the element the same expression that we had obtained in Sec. 6.3 in the case of a horizontal cut:

$$\tau_{ave} = \frac{VQ}{It} \qquad (6.6)$$

Note that τ_{ave} now represents the average value of the shearing stress τ_{zx} over a vertical cut, but since the thickness t of the flange is small, there is very little variation of τ_{zx} across the cut. Recalling that $\tau_{xz} = \tau_{zx}$ (Fig. 6.31), we conclude that the horizontal component τ_{xz} of the shearing stress at any point of a transverse section of the flange can be obtained from Eq. (6.6), where Q is the first moment of the shaded area about the neutral axis (Fig. 6.32a). We recall that a similar result was obtained in Sec. 6.4 for the vertical component τ_{xy} of the shearing stress in the web (Fig. 6.32b). Equation (6.6) can be used to determine shear-

Fig. 6.30

Fig. 6.31

Fig. 6.32

(a)

(b)

Fig. 6.33

Fig. 6.34

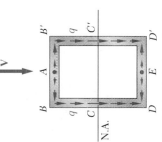

Fig. 6.35 Variation of q in box-beam section.

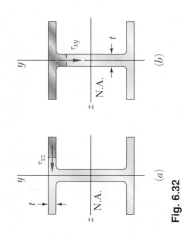

Fig. 6.36 Variation of q in wide-flange beam section.

ing stresses in box beams (Fig. 6.33), half pipes (Fig. 6.34), and other thin-walled members, as long as the loads are applied in a plane of symmetry of the member. In each case, the cut must be perpendicular to the surface of the member, and Eq. (6.6) will yield the component of the shearing stress in the direction of the tangent to that surface. (The other component may be assumed equal to zero, in view of the proximity of the two free surfaces.)

Comparing Eqs. (6.5) and (6.6), we note that the product of the shearing stress τ at a given point of the section and of the thickness t of the section at that point is equal to q. Since V and I are constant in any given section, q depends only upon the first moment Q and, thus, can easily be sketched on the section. In the case of a box beam, for example (Fig. 6.35), we note that q grows smoothly from zero at A to a maximum value at C and C' on the neutral axis, and then decreases back to zero as E is reached. We also note that there is no sudden variation in the magnitude of q as we pass a corner at B, D, B', or D', and that the sense of q in the horizontal portions of the section may be easily obtained from its sense in the vertical portions (which is the same as the sense of the shear **V**). In the case of a wide-flange section (Fig. 6.36), the values of q in portions AB and $A'B$ of the upper flange are distributed symmetrically. As we turn at B into the web, the values of q corresponding to the two halves of the flange must be combined to obtain the value of q at the top of the web. After reaching a maximum value at C on the neutral axis, q decreases, and at D splits into two equal parts corresponding to the two halves of the lower flange. The name of *shear flow* commonly used to refer to the shear per unit length, q, reflects the similarity between the properties of q that we have just described and some of the characteristics of a fluid flow through an open channel or pipe.[†]

So far we have assumed that all the loads were applied in a plane of symmetry of the member. In the case of members possessing two planes of symmetry, such as the wide-flange beam of Fig. 6.32 or the box beam of Fig. 6.33, any load applied through the centroid of a given

[†]We recall that the concept of shear flow was used to analyze the distribution of shearing stresses in thin-walled hollow shafts (Sec. 3.13). However, while the shear flow in a hollow shaft is constant, the shear flow in a member under a transverse loading is not.

cross section can be resolved into components along the two axes of
symmetry of the section. Each component will cause the member to
bend in a plane of symmetry, and the corresponding shearing stresses
can be obtained from Eq. (6.6). The principle of superposition can then
be used to determine the resulting stresses.

However, if the member considered possesses no plane of symmetry, of if it possesses a single plane of symmetry and is subjected to a
load that is not contained in that plane, the member is observed to *bend
and twist* at the same time, except when the load is applied at a specific point, called the *shear center*. Note that the shear center generally
does *not* coincide with the centroid of the cross section. The determination of the shear center of various thin-walled shapes is discussed in
Sec. 6.9.

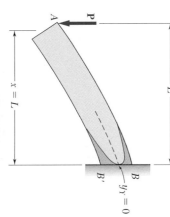

Fig. 6.37 ($PL \leq M_Y$)

Fig. 6.38 ($PL > M_Y$)

*6.8. PLASTIC DEFORMATIONS

Consider a cantilever beam AB of length L and rectangular cross section, subjected at its free end A to a concentrated load $\mathbf{P}$ (Fig. 6.37).
The largest value of the bending moment occurs at the fixed end B and
is equal to $M = PL$. As long as this value does not exceed the maximum elastic moment M_Y, that is, as long as $PL \leq M_Y$, the normal stress
σ_x will not exceed the yield strength σ_Y anywhere in the beam. However, as P is increased beyond the value M_Y/L, yield is initiated at points
B and B' and spreads toward the free end of the beam. Assuming the
material to be elastoplastic, and considering a cross section CC' located
at a distance x from the free end A of the beam (Fig. 6.38), we obtain
the half-thickness y_Y of the elastic core in that section by making
$M = Px$ in Eq. (4.38) of Sec. 4.9. We have

$$Px = \frac{3}{2} M_Y \left(1 - \frac{1}{3} \frac{y_Y^2}{c^2} \right) \qquad (6.14)$$

where c is the half-depth of the beam. Plotting y_Y against x, we obtain
the boundary between the elastic and plastic zones.

As long as $PL < \frac{3}{2} M_Y$, the parabola defined by Eq. (6.14) intersects the line BB', as shown in Fig. 6.38. However, when PL reaches
the value $\frac{3}{2} M_Y$, that is, when $PL = M_p$, where M_p is the plastic moment
defined in Sec. 4.9, Eq. (6.14) yields $y_Y = 0$ for $x = L$, which shows
that the vertex of the parabola is now located in section BB', and that
this section has become fully plastic (Fig. 6.39). Recalling Eq. (4.40)
of Sec. 4.9, we also note that the radius of curvature ρ of the neutral
surface at that point is equal to zero, indicating the presence of a sharp
bend in the beam at its fixed end. We say that a *plastic hinge* has developed at that point. The load $P = M_p/L$ is the largest load that can
be supported by the beam.

Fig. 6.39 ($PL = M_P = \frac{3}{2} M_Y$)

The above discussion was based only on the analysis of the normal stresses in the beam. Let us now examine the distribution of the shearing stresses in a section that has become partly plastic. Consider the portion of beam $CC''D''D$ located between the transverse sections CC' and DD', and above the horizontal plane $D''C''$ (Fig. 6.40a). If this portion is located entirely in the plastic zone, the normal stresses exerted on the faces CC'' and DD'' will be uniformly distributed and equal to the yield strength σ_Y (Fig. 6.40b). The equilibrium of the free

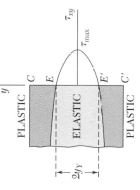

Fig. 6.40

body $CC''D''D$ thus requires that the horizontal shearing force $\Delta \mathbf{H}$ exerted on its lower face be equal to zero. It follows that the average value of the horizontal shearing stress τ_{yx} across the beam at C'' is zero, as well as the average value of the vertical shearing stress τ_{xy}. We thus conclude that the vertical shear $V = P$ in section CC' must be distributed entirely over the portion EE' of that section that is located within the elastic zone (Fig. 6.41). It can be shown† that the distribution of the shearing stresses over EE' is the same as in an elastic rectangular beam of the same width b as beam AB, and of depth equal to the thickness $2y_Y$ of the elastic zone. Denoting by A' the area $2by_Y$ of the elastic portion of the cross section, we have

$$\tau_{xy} = \frac{3}{2}\frac{P}{A'}\left(1 - \frac{y^2}{y_Y^2}\right) \tag{6.15}$$

The maximum value of the shearing stress occurs for $y = 0$ and is

$$\tau_{max} = \frac{3}{2}\frac{P}{A'} \tag{6.16}$$

As the area A' of the elastic portion of the section decreases, τ_{max} increases and eventually reaches the yield strength in shear τ_Y. Thus, shear contributes to the ultimate failure of the beam. A more exact analysis of this mode of failure should take into account the combined effect of the normal and shearing stresses.

Fig. 6.41

†See Prob. 6.55.

SAMPLE PROBLEM 6.3

Knowing that the vertical shear is 50 kips in a W10 × 68 rolled-steel beam, determine the horizontal shearing stress in the top flange at a point a located 4.31 in. from the edge of the beam. The dimensions and other geometric data of the rolled-steel section are given in Appendix C.

$t_f = 0.770$ in.

$I_x = 394$ in.^{4}

5.2 in.

10.4 in.

C

a

$\dfrac{5.2 - 0.770}{2} = 4.815$ in.

4.31 in.

SOLUTION

We isolate the shaded portion of the flange by cutting along the dashed line which passes through point a.

$$Q = (4.31 \text{ in.})(0.770 \text{ in.})(4.815 \text{ in.}) = 15.98 \text{ in}^3$$

$$\tau = \frac{VQ}{It} = \frac{(50 \text{ kips})(15.98 \text{ in}^3)}{(394 \text{ in}^4)(0.770 \text{ in.})} \qquad \tau = 2.63 \text{ ksi} \ \blacktriangle$$

For Q, the area is the area (area) that shear would separate away (shaded area)

SAMPLE PROBLEM 6.4

Solve Sample Prob. 6.3, assuming that 0.75 × 12-in. plates have been attached to the flanges of the W10 × 68 beam by continuous fillet welds as shown.

Welds

0.75 in. × 12 in.

a

4.31 in.

SOLUTION

For the composite beam the centroidal moment of inertia is

$$I = 394 \text{ in}^4 + 2\left[\tfrac{1}{12}(12 \text{ in.})(0.75 \text{ in.})^3 + (12 \text{ in.})(0.75 \text{ in.})(5.575 \text{ in.})^2\right]$$

$$I = 954 \text{ in}^4$$

Since the top plate and the flange are connected only at the welds, we find the shearing stress at a by passing a section through the flange at a, *between* the plate and the flange, and again through the flange at the symmetric point a'.

0.75 in.

10.4 in.

C

12 in.

0.375 in.

5.575 in.

5.2 in.

0.75 in.

5.2 in.

0.770 in.

a' a

4.31 in.

C

4.31 in.

4.815 in.

5.575 in.

12 in.

For the shaded area that we have isolated, we have

$$t = 2t_f = 2(0.770 \text{ in.}) = 1.540 \text{ in.}$$

$$Q = 2[(4.31 \text{ in.})(0.770 \text{ in.})(4.815 \text{ in.})] + (12 \text{ in.})(0.75 \text{ in.})(5.575 \text{ in.})$$

$$Q = 82.1 \text{ in}^3$$

$$\tau = \frac{VQ}{It} = \frac{(50 \text{ kips})(82.1 \text{ in}^3)}{(954 \text{ in}^4)(1.540 \text{ in.})} \qquad \tau = 2.79 \text{ ksi} \ \blacktriangle$$

SAMPLE PROBLEM 6.5

The thin-walled extruded beam shown is made of aluminum and has a uniform 3-mm wall thickness. Knowing that the shear in the beam is 5 kN, determine (*a*) the shearing stress at point *A*, (*b*) the maximum shearing stress in the beam. *Note:* The dimensions given are to lines midway between the outer and inner surfaces of the beam.

SOLUTION

Centroid. We note that $AB = AD = 65$ mm.

$$\bar{Y} = \frac{\Sigma \bar{y}A}{\Sigma A} = \frac{2[(65 \text{ mm})(3 \text{ mm})(30 \text{ mm})]}{2[(65 \text{ mm})(3 \text{ mm})] + (50 \text{ mm})(3 \text{ mm})}$$

$$\bar{Y} = 21.67 \text{ mm}$$

Centroidal Moment of Inertia. Each side of the thin-walled beam can be considered as a parallelogram and we recall that for the case shown $I_{nn} = bh^3/12$ where *b* is measured parallel to the axis *nn*.

$$b = (3 \text{ mm})/\cos \beta = (3 \text{ mm})/(12/13) = 3.25 \text{ mm}$$

$$I = \Sigma(\bar{I} + Ad^2) = 2[\tfrac{1}{12}(3.25 \text{ mm})(60 \text{ mm})^3$$
$$+ (3.25 \text{ mm})(60 \text{ mm})(8.33 \text{ mm})^2] + [\tfrac{1}{12}(50 \text{ mm})(3 \text{ mm})^3$$
$$+ (50 \text{ mm})(3 \text{ mm})(21.67 \text{ mm})^2]$$

$$I = 214.6 \times 10^3 \text{ mm}^4 \qquad I = 0.2146 \times 10^{-6} \text{ m}^4$$

a. **Shearing Stress at *A*.** If a shearing stress τ_A occurs at *A*, the shear flow will be $q_A = \tau_A t$ and must be directed in one of the two ways shown. But the cross section and the loading are symmetric about a vertical line through *A*, and thus the shear flow must also be symmetric. Since neither of the possible shear flows is symmetric, we conclude that

$$\tau_A = 0 \qquad \blacktriangledown$$

b. **Maximum Shearing Stress.** Since the wall thickness is constant, the maximum shearing stress occurs at the neutral axis, where *Q* is maximum. Since we know that the shearing stress at *A* is zero, we cut the section along the dashed line shown and isolate the shaded portion of the beam. In order to obtain the largest shearing stress, the cut at the neutral axis is made perpendicular to the sides, and is of length $t = 3$ mm.

$$Q = [(3.25 \text{ mm})(38.33 \text{ mm})] \left(\frac{38.33 \text{ mm}}{2} \right) = 2387 \text{ mm}^3$$

$$Q = 2.387 \times 10^{-6} \text{ m}^3$$

$$\tau_E = \frac{VQ}{It} = \frac{(5 \text{ kN})(2.387 \times 10^{-6} \text{ m}^3)}{(0.2146 \times 10^{-6} \text{ m}^4)(0.003 \text{ m})}$$

$$\tau_{max} = \tau_E = 18.54 \text{ MPa} \qquad \blacktriangledown$$

6.29 The built-up wooden beam shown is subjected to a vertical shear of 5 kN. Knowing that the longitudinal spacing of the nails is $s = 45$ mm and that each nail is 90 mm long, determine the shearing force in each nail.

Fig. P6.29

6.30 The built-up wooden beam shown is subjected to a vertical shear of 8 kN. Knowing that the nails are spaced longitudinally every 60 mm at A and every 25 mm at B, determine the shearing force in the nails (*a*) at A, (*b*) at B. (*Given:* $I_x = 1.504 \times 10^9$ mm^4.)

Dimensions in mm

Fig. P6.30

6.31 The built-up beam shown is made up by gluing together five planks. Knowing that the allowable average shearing stress in the glued joints is 60 psi, determine the largest permissible vertical shear in the beam.

Fig. P6.31

6.32 The built-up beam shown is made up by gluing together two $\frac{3}{4} \times 10$-in. plywood strips and two 2×4-in. planks. Knowing that the allowable average shearing stress in the glued joints is 50 psi, determine the largest permissible vertical shear in the beam.

Fig. P6.32

6.33 Two 20 × 100-mm and two 20 × 180-mm boards are glued together as shown to form a 120 × 200-mm box beam. Knowing that the beam is subjected to a vertical shear of 3.5 kN, determine the average shearing stress in the glued joint (*a*) at *A*, (*b*) at *R*.

Fig. P6.33

6.34 Knowing that a W360 × 122 rolled-steel beam is subjected to a 250-kN vertical shear, determine the shearing stress (*a*) at point *A*, (*b*) at the centroid *C* of the section.

Fig. P6.34

6.35 and 6.36 An extruded aluminum beam has the cross section shown. Knowing that the vertical shear in the beam is 150 kN, determine the shearing stress at (*a*) point *a*, (*b*) point *b*.

Dimensions in mm

Fig. P6.35

Dimensions in mm

Fig. P6.36

6.37 The vertical shear is 1200 lb in a beam having the cross section shown. Knowing that *d* = 4 in., determine the shearing stress (*a*) at point *a*, (*b*) at point *b*.

Fig. P6.37 and P6.38

6.38 The vertical shear is 1200 lb in a beam having the cross section shown. Determine (*a*) the distance *d* for which $\tau_a = \tau_b$, (*b*) the corresponding shearing stress at points *a* and *b*.

6.39 Knowing that a given vertical shear **V** causes a maximum shearing stress of 75 MPa in the hat-shaped extrusion shown, determine the corresponding shearing stress (*a*) at point *a*, (*b*) at point *b*.

Fig. P6.39

6.40 Knowing that a given vertical shear **V** causes a maximum shear-ing stress of 50 MPa in a thin-walled member having the cross section shown, determine the corresponding shearing stress (*a*) at point *a*, (*b*) at point *b*, (*c*) at point *c*.

Fig. P6.40

Fig. P6.41

Fig. P6.42

6.41 and 6.42 The extruded beam shown has a uniform wall thickness of $\frac{1}{8}$ in. Knowing that the vertical shear in the beam is 2 kips, determine the shearing stress at each of the five points indicated.

6.43 Three 20 × 450-mm steel plates are bolted to four L152 × 152 × 19.0 angles to form a beam with the cross section shown. The bolts have a 22-mm diameter and are spaced longitudinally every 125 mm. Knowing that the allowable average shearing stress in the bolts is 90 MPa, determine the largest permissible vertical shear in the beam. (*Given: $I_x = 1896 \times 10^6$ mm^4.*)

Fig. P6.43

6.44 A beam consists of three planks connected by steel bolts with a longitudinal spacing of 225 mm. Knowing that the shear in the beam is verti-cal and equal to 6 kN and that the allowable average shearing stress in each bolt is 60 MPa, determine the smallest permissible bolt diameter that can be used.

Fig. P6.44

6.45 and 6.46 Three planks are connected as shown by bolts of $\frac{3}{8}$-in. diameter spaced every 6 in. along the longitudinal axis of the beam. For a vertical shear of 2.5 kips, determine the average shearing stress in the bolts.

Fig. P6.45

Fig. P6.46

6.47 Three plates, each 12 mm thick, are welded together to form the section shown. For a vertical shear of 100 kN, determine the shear flow through the welded surfaces and sketch the shear flow in the cross section.

Fig. P6.47

6.48 A plate of 2-mm thickness is bent as shown and then used as a beam. For a vertical shear of 5 kN, determine the shearing stress at the five points indicated and sketch the shear flow in the cross section.

Fig. P6.48

6.49 A plate of $\frac{1}{4}$-in. thickness is corrugated as shown and then used as a beam. For a vertical shear of 1.2 kips, determine (*a*) the maximum shearing stress in the section, (*b*) the shearing stress at point *B*. Also sketch the shear flow in the cross section.

Fig. P6.49

6.50 A plate of thickness *t* is bent as shown and then used as a beam. For a vertical shear of 600 lb, determine (*a*) the thickness *t* for which the maximum shearing stress is 300 psi, (*b*) the corresponding shearing stress at point *E*. Also sketch the shear flow in the cross section.

Fig. P6.50

6.51 and 6.52

An extruded beam has a uniform wall thickness t. Denoting by $\mathbf{V}$ the vertical shear and by A the cross-sectional area of the beam, express the maximum shearing stress as $\tau_{max} = k(V/A)$ and determine the constant k for each of the two orientations shown.

(a)   (b)

Fig. P6.51

Fig. P6.52

6.53

The design of a beam calls for connecting two vertical rectangular $\frac{3}{8} \times 4$-in. plates by welding them to two horizontal $\frac{1}{2} \times 2$-in. plates as shown. For a vertical shear $\mathbf{V}$, determine the dimension a for which the shear flow through the welded surfaces is maximum.

Fig. P6.53

6.54

(a) Determine the shearing stress at point P of a thin-walled pipe of the cross section shown caused by a vertical shear $\mathbf{V}$. (b) Show that the maximum shearing stress occurs for $\theta = 90°$ and is equal to $2V/A$, where A is the cross-sectional area of the pipe.

Fig. P6.54

6.55

Consider the cantilever beam AB discussed in Sec. 6.8 and the portion $ACKJ$ of the beam that is located to the left of the transverse section CC' and above the horizontal plane JK, where K is a point at a distance $y < y_Y$ above the neutral axis (Fig. P6.55). (a) Recalling that $\sigma_x = \sigma_Y$ between C and E and $\sigma_x = (\sigma_Y/y_Y)y$ between E and K, show that the magnitude of the horizontal shearing force $\mathbf{H}$ exerted on the lower face of the portion of beam $ACKJ$ is

$$H = \frac{1}{2}b\sigma_Y\left(2c - y_Y - \frac{y^2}{y_Y}\right)$$

(b) Observing that the shearing stress at K is

$$\tau_{xy} = \lim_{\Delta A \to 0}\frac{\Delta H}{\Delta A} = \lim_{\Delta x \to 0}\frac{1}{b}\frac{\Delta H}{\Delta x} = \frac{1}{b}\frac{\partial H}{\partial x}$$

and recalling that y_Y is a function of x defined by Eq. (6.14), derive Eq. (6.15).

Fig. P6.55

6.56 For a beam made of two or more materials with different moduli of elasticity, show that Eq. (6.6)

$$\tau_{ave} = \frac{VQ}{It}$$

remains valid provided that both Q and I are computed by using the transformed section of the beam (see Sec. 4.6) and provided further that t is the actual width of the beam where τ_{ave} is computed.

6.57 and 6.58 A composite beam is made by attaching the timber and steel portions shown with bolts of 12-mm diameter spaced longitudinally every 200 mm. The modulus of elasticity is 10 GPa for the wood and 200 GPa for the steel. Determine the average shearing stress in the bolts caused by a vertical shearing force of 4 kN. (*Hint:* Use the method indicated in Prob. 6.56.)

Fig. P6.58

Fig. P6.57

6.59 and 6.60 A steel bar and an aluminum bar are bonded together as shown to form a composite beam. Knowing that the vertical shear in the beam is 4 kips and that the modulus of elasticity is 29×10^6 psi for the steel and 10.6×10^6 psi for the aluminum, determine (*a*) the average stress at the bonded surface, (*b*) the maximum stress in the beam. (*Hint:* Use the method indicated in Prob. 6.56.)

Fig. P6.60

Fig. P6.59

Fig. 6.42

*6.9. UNSYMMETRIC LOADING OF THIN-WALLED MEMBERS; SHEAR CENTER

Our analysis of the effects of transverse loadings in Chap. 5 and in the preceding sections of this chapter was limited to members possessing a vertical plane of symmetry and to loads applied in that plane. The members were observed to bend in the plane of loading (Fig. 6.42) and, in any given cross section, the bending couple **M** and the shear **V** (Fig. 6.43) were found to result in normal and shearing stresses defined, respectively, by the formulas

$$\sigma_x = -\frac{My}{I} \qquad (4.16)$$

and

$$\tau_{ave} = \frac{VQ}{It} \qquad (6.6)$$

$(V = P, M = Px)$

Fig. 6.43

$(V = P, M = Px)$

Fig. 6.44

In this section, the effects of transverse loadings on *thin-walled members that do not possess a vertical plane of symmetry* will be examined. Let us assume, for example, that the channel member of Fig. 6.42 has been rotated through 90° and that the line of action of **P** still passes through the centroid of the end section. The couple vector **M** representing the bending moment in a given cross section is still directed along a principal axis of the section (Fig. 6.44), and the neutral axis will coincide with that axis (cf. Sec. 4.13). Equation (4.16), therefore, is applicable and can be used to compute the normal stresses in the section. However, Eq. (6.6) cannot be used to determine the shearing stresses in the section, since this equation was derived for a member possessing a vertical plane of symmetry (cf. Sec. 6.7). Actually, the member will be observed to *bend and twist* under the applied load (Fig. 6.45), and the resulting distribution of shearing stresses will be quite different from that defined by Eq. (6.6).

The following question now arises: Is it possible to apply the vertical load **P** in such a way that the channel member of Fig. 6.45 will *bend without twisting* and, if so, where should the load **P** be applied? If the member bends without twisting, then the shearing stress at any point of a given cross section can be obtained from Eq. (6.6), where Q is the first moment of the shaded area with respect to the neutral axis (Fig. 6.46a), and the distribution of stresses will look as shown in Fig. 6.46b, with $\tau = 0$ at both A and E. We note that the shearing force ex-

Fig. 6.45

(a)

(b)

Fig. 6.46

erted on a small element of cross-sectional area $dA = t\,ds$ is $dF = \tau\,dA = \tau t\,ds$, or $dF = q\,ds$ (Fig. 6.47a), where q is the shear flow $q = \tau t = VQ/I$ at the point considered. The resultant of the shearing forces exerted on the elements of the upper flange AB of the channel is found to be a horizontal force $\mathbf{F}$ (Fig. 6.47b) of magnitude

$$F = \int_A^B q\,ds \qquad (6.17)$$

(a)　　(b)

Fig. 6.47

Because of the symmetry of the channel section about its neutral axis, the resultant of the shearing forces exerted on the lower flange DE is a force $\mathbf{F}'$ of the same magnitude as $\mathbf{F}$ but of opposite sense. We conclude that the resultant of the shearing forces exerted on the web BD must be equal to the vertical shear $\mathbf{V}$ in the section:

$$V = \int_B^D q\,ds \qquad (6.18)$$

We now observe that the forces $\mathbf{F}$ and $\mathbf{F}'$ form a couple of moment Fh, where h is the distance between the center lines of the flanges AB and DE (Fig. 6.48a). This couple can be eliminated if the vertical shear $\mathbf{V}$ is moved to the left through a distance e such that the moment of $\mathbf{V}$ about B is equal to Fh (Fig. 6.48b). We write $Ve = Fh$ or

$$e = \frac{Fh}{V} \qquad (6.19)$$

(a)　　(b)

Fig. 6.48

and conclude that, when the force $\mathbf{P}$ is applied at a distance e to the left of the center line of the web BD, the member bends in a vertical plane without twisting (Fig. 6.49).

The point O where the line of action of $\mathbf{P}$ intersects the axis of symmetry of the end section is called the *shear center* of that section. We note that, in the case of an oblique load $\mathbf{P}$ (Fig. 6.50a), the member will also be free of any twist if the load $\mathbf{P}$ is applied at the shear center of the section. Indeed, the load $\mathbf{P}$ can then be resolved into two components $\mathbf{P}_z$ and $\mathbf{P}_y$ (Fig. 6.50b) corresponding respectively to the loading conditions of Figs. 6.42 and 6.49, neither of which causes the member to twist.

Fig. 6.49

Fig. 6.50

EXAMPLE 6.05

Determine the shear center O of a channel section of uniform thickness (Fig. 6.51), knowing that $b = 4$ in., $h = 6$ in., and $t = 0.15$ in.

Assuming that the member does not twist, we first determine the shear flow q in flange AB at a distance s from A (Fig. 6.52). Recalling Eq. (6.5) and observing that the first moment Q of the shaded area with respect to the neutral axis is $Q = (st)(h/2)$, we write

$$q = \frac{VQ}{I} = \frac{Vsth}{2I} \qquad (6.20)$$

Fig. 6.51

where V is the vertical shear and I the moment of inertia of the section with respect to the neutral axis.

Recalling Eq. (6.17), we determine the magnitude of the shearing force $\mathbf{F}$ exerted on flange AB by integrating the shear flow q from A to B:

$$F = \int_0^b q\, ds = \int_0^b \frac{Vsth}{2I}\, ds = \frac{Vth}{2I} \int_0^b s\, ds$$

$$F = \frac{Vthb^2}{4I} \qquad (6.21)$$

The distance e from the center line of the web BD to the shear center O can now be obtained from Eq. (6.19):

$$e = \frac{Fh}{V} = \frac{Vthb^2}{4I} \frac{h}{V} = \frac{th^2b^2}{4I} \qquad (6.22)$$

The moment of inertia I of the channel section can be expressed as follows:

$$I = I_{web} + 2I_{flange}$$

$$= \frac{1}{12} th^3 + 2 \left[\frac{1}{12} bt^3 + bt \left(\frac{h}{2} \right)^2 \right]$$

Neglecting the term containing t^3, which is very small, we have

$$I = \frac{1}{12} th^3 + \frac{1}{2} tbh^2 = \frac{1}{12} th^2(6b + h) \qquad (6.23)$$

Fig. 6.52

Substituting this expression into (6.22), we write

$$e = \frac{3b^2}{6b + h} = \frac{b}{2 + \dfrac{h}{3b}} \qquad (6.24)$$

We note that the distance e does not depend upon t and can vary from 0 to $b/2$, depending upon the value of the ratio $h/3b$. For the given channel section, we have

$$\frac{h}{3b} = \frac{6 \text{ in.}}{3(4 \text{ in.})} = 0.5$$

and

$$e = \frac{4 \text{ in.}}{2 + 0.5} = 1.6 \text{ in.}$$

EXAMPLE 6.06

For the channel section of Example 6.05 determine the distribution of the shearing stresses caused by a 2.5-kip vertical shear **V** applied at the shear center O (Fig. 6.53).

Shearing stresses in flanges. Since **V** is applied at the shear center, there is no torsion, and the stresses in flange AB are obtained from Eq. (6.20) of Example 6.05. We have

$$\tau = \frac{q}{t} = \frac{VQ}{It} = \frac{Vh}{2I} s \qquad (6.25)$$

which shows that the stress distribution in flange AB is linear. Letting $s = b$ and substituting for I from Eq. (6.23), we obtain the value of the shearing stress at B:

$$\tau_B = \frac{Vhb}{2(\frac{1}{12}th^2)(6b + h)} = \frac{6Vb}{th(6b + h)} \qquad (6.26)$$

Letting $V = 2.5$ kips, and using the given dimensions, we have

$$\tau_B = \frac{6(2.5 \text{ kips})(4 \text{ in.})}{(0.15 \text{ in.})(6 \text{ in.})(6 \times 4 \text{ in.} + 6 \text{ in.})}$$

$$= 2.22 \text{ ksi}$$

Shearing stresses in web. The distribution of the shearing stresses in the web BD is parabolic, as in the case of a W-beam, and the maximum stress occurs at the neutral axis. Computing the first moment of the upper half of the cross section with respect to the neutral axis (Fig. 6.54), we write

$$Q = bt(\tfrac{1}{2}h) + \tfrac{1}{2}ht(\tfrac{1}{4}h) = \tfrac{1}{8}ht(4b + h) \qquad (6.27)$$

Substituting for I and Q from (6.23) and (6.27), respectively, into the expression for the shearing stress, we have

$$\tau_{max} = \frac{VQ}{It} = \frac{V(\frac{1}{8}ht)(4b + h)}{\frac{1}{12}th^2(6b + h)t} = \frac{3V(4b + h)}{2th(6b + h)}$$

or, with the given data,

$$\tau_{max} = \frac{3(2.5 \text{ kips})(4 \times 4 \text{ in.} + 6 \text{ in.})}{2(0.15 \text{ in.})(6 \text{ in.})(6 \times 4 \text{ in.} + 6 \text{ in.})}$$

$$= 3.06 \text{ ksi}$$

Distribution of stresses over the section. The distribution of the shearing stresses over the entire channel section has been plotted in Fig. 6.55.

Fig. 6.53

Fig. 6.54

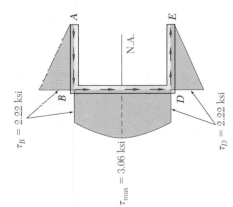

Fig. 6.55

EXAMPLE 6.07

For the channel section of Example 6.05, and neglecting stress concentrations, determine the maximum shearing stress caused by a 2.5-kip vertical shear **V** applied at the centroid C of the section, which is located 1.143 in. to the right of the center line of the web BD (Fig. 6.56).

Equivalent force-couple system at shear center. The shear center O of the cross section was determined in Example 6.05 and found to be at a distance $e = 1.6$ in. to the left of the center line of the web BD. We replace the shear **V** (Fig. 6.57a) by an equivalent force-couple system at the shear center O (Fig. 6.57b). This system consists of a 2.5-kip force **V** and of a torque **T** of magnitude

$$T = V(OC) = (2.5 \text{ kips})(1.6 \text{ in.} + 1.143 \text{ in.})$$
$$= 6.86 \text{ kip} \cdot \text{in.}$$

Stresses due to bending. The 2.5-kip force **V** causes the member to bend, and the corresponding distribution of shearing stresses in the section (Fig. 6.57c) was determined in Example 6.06. We recall that the maximum value of the stress due to this force was found to be

$$(\tau_{\text{max}})_{\text{bending}} = 3.06 \text{ ksi}$$

Stresses due to twisting. The torque **T** causes the member to twist, and the corresponding distribution of stresses is shown in Fig. 6.57d. We recall from Sec. 3.12 that the membrane analogy shows that, in a thin-walled member of uniform thickness, the stress caused by a torque **T** is maximum along

the edge of the section. Using Eqs. (3.45) and (3.43) with

$$a = 4 \text{ in.} + 6 \text{ in.} + 4 \text{ in.} = 14 \text{ in.}$$
$$b = t = 0.15 \text{ in.} \qquad b/a = 0.0107$$

we have

$$c_1 = \tfrac{1}{3}(1 - 0.630b/a) = \tfrac{1}{3}(1 - 0.630 \times 0.0107) = 0.331$$

$$(\tau_{\text{max}})_{\text{twisting}} = \frac{T}{c_1 a b^2} = \frac{6.86 \text{ kip} \cdot \text{in.}}{(0.331)(14 \text{ in.})(0.15 \text{ in.})^2} = 65.8 \text{ ksi}$$

Combined stresses. The maximum stress due to the combined bending and twisting occurs at the neutral axis, on the inside surface of the web, and is

$$\tau_{\text{max}} = 3.06 \text{ ksi} + 65.8 \text{ ksi} = 68.9 \text{ ksi}$$

Fig. 6.56

Fig. 6.57

Turning our attention to thin-walled members possessing no plane of symmetry, we now consider the case of an angle shape subjected to a vertical load **P.** If the member is oriented in such a way that the load **P** is perpendicular to one of the principal centroidal axes C_z of the cross section, the couple vector **M** representing the bending moment in a given section will be directed along C_z (Fig. 6.58), and the neutral axis will coincide with that axis (cf. Sec. 4.13). Equation (4.16), therefore, is applicable and can be used to compute the normal stresses in the section. We now propose to determine where the load **P** should be applied if Eq. (6.6) is to define the shearing stresses in the section, i.e., if the member is to *bend without twisting.*

Let us *assume* that the shearing stresses in the section are defined by Eq. (6.6). As in the case of the channel member considered earlier, the elementary shearing forces exerted on the section can be expressed as $dF = q \, ds$, with $q = VQ/I$, where Q represents a first moment with respect to the neutral axis (Fig. 6.59a). We note that the resultant of the

Fig. 6.58

Fig. 6.59

shearing forces exerted on portion OA of the cross section is a force $\mathbf{F_1}$ directed along OA, and that the resultant of the shearing forces exerted on portion OB is a force $\mathbf{F_2}$ along OB (Fig. 6.59b). Since both $\mathbf{F_1}$ and $\mathbf{F_2}$ pass through point O at the corner of the angle, it follows that their own resultant, which is the shear **V** in the section, must also pass through O (Fig. 6.59c). We conclude that the member will not be twisted if the line of action of the load **P** passes through the corner O of the section in which it is applied.

The same reasoning can be applied when the load **P** is perpendicular to the other principal centroidal axis C_y of the angle section. And, since any load **P** applied at the corner O of a cross section can be resolved into components perpendicular to the principal axes, it follows that the member will not be twisted if each load is applied at the corner O of a cross section. We thus conclude that O is the shear center of the section.

Angle shapes with one vertical and one horizontal leg are encountered in many structures. It follows from the preceding discussion that such members will not be twisted if vertical loads are applied along the

Fig. 6.60

center line of their vertical leg. We note from Fig. 6.60 that the resultant of the elementary shearing forces exerted on the vertical portion OA
of a given section will be equal to the shear $\mathbf{V}$, while the resultant of
the shearing forces on the horizontal portion OB will be zero:

$$\int_O^A q\,ds = V \qquad \int_O^B q\,ds = 0$$

Fig. 6.61

This does *not* mean, however, that there will be no shearing stress in
the horizontal leg of the member. By resolving the shear $\mathbf{V}$ into components perpendicular to the principal centroidal axes of the section and
computing the shearing stress at every point, we would verify that τ is
zero at only one point between O and B (see Sample Prob. 6.6).

Another type of thin-walled member frequently encountered in
practice is the Z shape. While the cross section of a Z shape does not
possess any axis of symmetry, it does possess a *center of symmetry* O
(Fig. 6.61). This means that, to any point H of the cross section corresponds another point H' such that the segment of straight line HH' is
bisected by O. Clearly, the center of symmetry O coincides with the
centroid of the cross section. As you will see presently, point O is also
the shear center of the cross section.

Fig. 6.62

As we did earlier in the case of an angle shape, we assume that the
loads are applied in a plane perpendicular to one of the principal axes
of the section, so that this axis is also the neutral axis of the section
(Fig. 6.62). We further assume that the shearing stresses in the section
are defined by Eq. (6.6), i.e., that the member is bent without being
twisted. Denoting by Q the first moment about the neutral axis of portion AH of the cross section, and by Q' the first moment of portion EH',
we note that $Q' = -Q$. Thus the shearing stresses at H and H' have
the same magnitude and the same direction, and the shearing forces exerted on small elements of area dA located respectively at H and H' are
equal forces that have equal and opposite moments about O (Fig. 6.63).
Since this is true for any pair of symmetric elements, it follows that the
resultant of the shearing forces exerted on the section has a zero moment about O. This means that the shear $\mathbf{V}$ in the section is directed
along a line that passes through O. Since this analysis can be repeated
when the loads are applied in a plane perpendicular to the other principal axis, we conclude that point O is the shear center of the section.

Fig. 6.63

SAMPLE PROBLEM 6.6

Determine the distribution of shearing stresses in the thin-walled angle shape DE of uniform thickness t for the loading shown.

SOLUTION

Shear Center. We recall from Sec. 6.9 that the shear center of the cross section of a thin-walled angle shape is located at its corner. Since the load **P** is applied at D, it causes bending but no twisting of the shape.

Principal Axes. We locate the centroid C of a given cross section AOB. Since the y' axis is an axis of symmetry, the y' and z' axes are the principal centroidal axes of the section. We recall that for the parallelogram shown $I_{nn} = \frac{1}{12}bh^3$ and $I_{mm} = \frac{1}{3}bh^3$. Considering each leg of the section as a parallelogram, we now determine the centroidal moments of inertia $I_{y'}$ and $I_{z'}$:

$$I_{y'} = 2\left[\frac{1}{3}\left(\frac{t}{\cos 45°}\right)(a\cos 45°)^3\right] = \frac{1}{3}ta^3$$

$$I_{z'} = 2\left[\frac{1}{12}\left(\frac{t}{\cos 45°}\right)(a\cos 45°)^3\right] = \frac{1}{12}ta^3$$

Superposition. The shear **V** in the section is equal to the load **P**. We resolve it into components parallel to the principal axes.

Shearing Stresses Due to $V_{y'}$. We determine the shearing stress at point e of coordinate y:

$$\bar{y}' = \frac{1}{2}(a+y)\cos 45° - \frac{1}{2}a\cos 45° = \frac{1}{2}y\cos 45°$$

$$Q = t(a-y)\bar{y}' = \frac{1}{2}t(a-y)y\cos 45°$$

$$\tau_1 = \frac{V_{y'}Q}{I_{z'}t} = \frac{(P\cos 45°)[\frac{1}{2}t(a-y)y\cos 45°]}{(\frac{1}{12}ta^3)t} = \frac{3P(a-y)y}{ta^3}$$

The shearing stress at point f is represented by a similar function of z.

Shearing Stresses Due to $V_{z'}$. We again consider point e:

$$\bar{z}' = \frac{1}{2}(a+y)\cos 45°$$

$$Q = (a-y)t\bar{z}' = \frac{1}{2}(a^2 - y^2)t\cos 45°$$

$$\tau_2 = \frac{V_{z'}Q}{I_{y'}t} = \frac{(P\cos 45°)[\frac{1}{2}(a^2-y^2)t\cos 45°]}{(\frac{1}{3}ta^3)t} = \frac{3P(a^2-y^2)}{4ta^3}$$

The shearing stress at point f is represented by a similar function of z.

Combined Stresses. *Along the Vertical Leg.* The shearing stress at point e is

$$\tau_e = \tau_2 + \tau_1 = \frac{3P(a^2-y^2)}{4ta^3} + \frac{3P(a-y)y}{ta^3} = \frac{3P(a-y)}{4ta^3}[(a+y)+4y]$$

$$\tau_e = \frac{3P(a-y)(a+5y)}{4ta^3} \blacktriangledown$$

The shearing stress at point f is

$$\tau_f = \tau_2 - \tau_1 = \frac{3P(a^2-z^2)}{4ta^3} - \frac{3P(a-z)z}{ta^3} = \frac{3P(a-z)}{4ta^3}[(a+z)-4z]$$

$$\tau_f = \frac{3P(a-z)(a-3z)}{4ta^3} \blacktriangledown$$

Along the Horizontal Leg. The shearing stress at point f is

PROBLEMS

6.61 through 6.64   Determine the location of the shear center O of a thin-walled beam of uniform thickness having the cross section shown.

Fig. P6.61

Fig. P6.62

Fig. P6.63

Fig. P6.64

6.65 and 6.66   An extruded beam has the cross section shown. Determine (a) the location of the shear center O, (b) the distribution of the shearing stresses caused by a 50-kN vertical shearing force applied at O.

Fig. P6.65

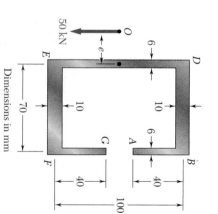

Fig. P6.66

6.67 and 6.68 An extruded beam has the cross section shown. Determine (a) the location of the shear center O, (b) the distribution of the shearing stresses caused by a 25-kip vertical shearing force applied at O.

Fig. *P6.67*

Fig. *P6.68*

6.69 through 6.74 Determine the location of the shear center O of a thin-walled beam of uniform thickness having the cross section shown.

Fig. **P6.69**

Fig. **P6.70**

Fig. *P6.71*

Fig. **P6.72**

Fig. *P6.73*

Fig. *P6.74*

6.75 and 6.76 A thin-walled beam of uniform thickness has the cross
section shown. Determine the dimension a for which the shear center O of the
cross section is located at the point indicated.

Fig. P6.75

Fig. P6.76

Fig. P6.77 and P6.78

6.77 A thin-walled beam of uniform thickness has the cross section
shown. Determine the location of the shear center O of the cross section, knowing that $h = 80$ mm.

6.78 A thin-walled beam of uniform thickness has the cross section
shown. Determine the dimension h for which the shear center O of the cross
section is located at a distance $e = 25$ mm from the center of the flange AB.

6.79 For the angle shape and loading of Sample Prob. 6.6, check that
$\int q \, dz = 0$ along the horizontal leg of the angle and $\int q \, dy = P$ along its vertical leg.

6.80 For the angle shape and loading of Sample Prob. 6.6, (a) determine
the points where the shearing stress is maximum and the corresponding values
of the stress, (b) verify that the points obtained are located on the neutral axis
corresponding to the given loading.

*__6.81__ A cantilever beam AB, consisting of half of a thin-walled pipe of
30-mm mean radius and 6-mm wall thickness, is subjected to a 1200-N vertical load. Knowing that the line of action of the load passes through the centroid C of the cross section of the beam, determine (a) the equivalent force-
couple system at the shear center of the cross section, (b) the maximum stress
in the beam. (*Hint:* The shear center O of this cross section was shown in Prob.
6.74 to be located twice as far from its vertical diameter as its centroid C.)

*__6.82__ Solve Prob. 6.81, assuming that the thickness of the beam is re-
duced to 5 mm.

*__6.83__ The cantilever beam shown at the top of the next page consists of
an angle shape of $\frac{3}{8}$-in. thickness. For the given loading, determine the location
and magnitude of the largest shearing stress along line $A'B'$ in the horizontal
leg of the angle shape. The x' and y' axes are the principal centroidal axes of
the cross section and the corresponding moments of inertia are $I_{x'} = 115.7$ in^4
and $I_{y'} = 12.61$ in^4.

1200 N

Fig. P6.81

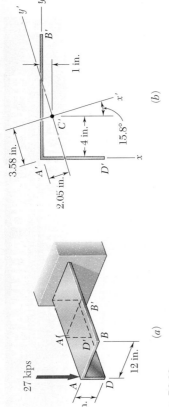

(a)

(b)

3.58 in.

2.05 m.

A'

C'

D'

B'

x'

y'

x

y

15.8°

4 in.

1 in.

Fig. P6.83

*6.84   For the cantilever beam and loading of Prob. 6.83, determine the location and magnitude of the largest shearing stress along line $A'D'$ in the vertical leg of the angle shape.

*6.85   Show that for the given angle section and shear **V**, (a) the distribution of the shearing stresses in each leg is parabolic, with $\tau = 0$ at A and B, (b) the shearing stress in the horizontal leg is also zero at point E of abscissa $x = a/3$, and is maximum at point F of abscissa $x = 2a/3$. (*Hint:* The resultant of the shearing forces in the horizontal leg must be zero.)

Fig. P6.85

*6.86   For the angle section and shear **V** of Prob. 6.85, show that the shearing stress in the vertical leg is maximum at point H of ordinate $y = 2b/(4 + \beta)$, where β is the ratio A_h/A_v, of the cross-sectional areas of the horizontal and vertical legs. (*Hint:* Points F and H, where the shearing stress is maximum, are located on the neutral axis, and this axis passes through the centroid C of the cross section.)

*6.87   The cantilever beam shown consists of a Z shape of $\frac{1}{4}$-in. thickness. For the given loading, determine the distribution of the shearing stresses along line $A'B'$ in the upper horizontal leg of the Z shape. The x' and y' axes are the principal centroidal axes of the cross section and the corresponding moments of inertia are $I_x = 166.3$ in⁴ and $I_y = 13.61$ in⁴.

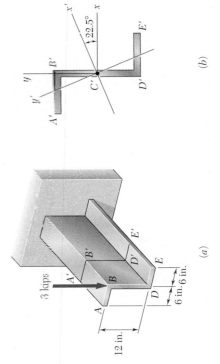

(a)

(b)

3 kips

12 in.

A

B

D

E

6 in.

6 in.

A'

B'

D'

E'

C'

x'

y'

x

y

22.5°

Fig. P6.87

*6.88   For the cantilever beam and loading of Prob. 6.87, determine the distribution of the shearing stresses along line $B'D'$ in the vertical web of the Z shape.

This chapter was devoted to the analysis of beams and thin-walled members under transverse loadings.

In Sec. 6.1 we considered a small element located in the vertical plane of symmetry of a beam under a transverse loading (Fig. 6.2) and found that normal stresses σ_x and shearing stresses τ_{xy} were exerted on the transverse faces of that element, while shearing stresses τ_{yx} equal in magnitude to τ_{xy} were exerted on its horizontal faces.

In Sec. 6.2 we considered a prismatic beam AB with a vertical plane of symmetry supporting various concentrated and distributed loads (Fig. 6.5). At a distance x from end A we detached from the

Stresses on a beam element

Fig. 6.2

Fig. 6.5

beam an element $CDD'C'$ of length Δx extending across the width of the beam from the upper surface of the beam to a horizontal plane located at a distance y_1 from the neutral axis (Fig. 6.6). We found

Horizontal shear in a beam

Fig. 6.6

that the magnitude of the shearing force $\Delta \mathbf{H}$ exerted on the lower face of the beam element was

$$\Delta H = \frac{VQ}{I} \Delta x \qquad (6.4)$$

where V = vertical shear in the given transverse section
Q = first moment with respect to the neutral axis of the shaded portion $\mathfrak{a}$ of the section
I = centroidal moment of inertia of the entire cross-sectional area

The *horizontal shear per unit length*, or *shear flow*, which was denoted by the letter q, was obtained by dividing both members of Eq. (6.4) by Δx:

$$q = \frac{\Delta H}{\Delta x} = \frac{VQ}{I} \qquad (6.5)$$

Shear flow

Dividing both members of Eq. (6.4) by the area ΔA of the horizontal face of the element and observing that $\Delta A = t \, \Delta x$, where t is the width of the element at the cut, we obtained in Sec. 6.3 the following expression for the *average shearing stress* on the horizontal face of the element

$$\tau_{ave} = \frac{VQ}{It} \qquad (6.6)$$

Shearing stresses in a beam

We further noted that, since the shearing stresses τ_{xy} and τ_{yx} exerted respectively on a transverse and a horizontal plane through D' are equal, the expression in (6.6) also represents the average value of τ_{xy} along the line $D_1' D_2'$ (Fig. 6.12).

Fig. 6.12

In Secs. 6.4 and 6.5 we analyzed the shearing stresses in a beam of rectangular cross section. We found that the distribution of stresses is parabolic and that the maximum stress, which occurs at the center of the section, is

$$\tau_{max} = \frac{3}{2} \frac{V}{A} \qquad (6.10)$$

Shearing stresses in a beam of rectangular cross section

where A is the area of the rectangular section. For wide-flange beams, we found that a good approximation of the maximum shearing stress can be obtained by dividing the shear V by the cross-sectional area of the web.

In Sec. 6.6 we showed that Eqs. (6.4) and (6.5) could still be used to determine, respectively, the longitudinal shearing force ΔH and the shear flow q exerted on a beam element if the element was bounded by an arbitrary curved surface instead of a horizontal plane (Fig. 6.24). This made it possible for us in Sec. 6.7 to extend the use

Longitudinal shear on curved surface

Fig. 6.24

of Eq. (6.6) to the determination of the average shearing stress in
thin-walled members such as wide-flange beams and box beams, in
the flanges of such members, and in their webs (Fig. 6.32).

**Shearing stresses in
thin-walled members**

(a)

(b)

Fig. 6.32

Plastic deformations

In Sec. 6.8 we considered the effect of plastic deformations on
the magnitude and distribution of shearing stresses. From Chap. 4
we recalled that once plastic deformation has been initiated, additional loading causes plastic zones to penetrate into the elastic core
of a beam. After demonstrating that shearing stresses can occur only
in the elastic core of a beam, we noted that both an increase in loading and the resulting decrease in the size of the elastic core contribute
to an increase in shearing stresses.

**Unsymmetric loading
Shear center**

In Sec. 6.9 we considered prismatic members that are *not* loaded
in their plane of symmetry and observed that, in general, both bending and twisting will occur. You learned to locate the point O of the
cross section, known as the *shear center*, where the loads should be
applied if the member is to bend without twisting (Fig. 6.49) and
found that if the loads are applied at that point, the following equations remain valid:

$$\sigma_x = -\frac{My}{I} \qquad \tau_{ave} = \frac{VQ}{It} \qquad (4.16, 6.6)$$

Using the principle of superposition, you also learned to determine
the stresses in unsymmetric thin-walled members such as channels,
angles, and extruded beams [Example 6.07 and Sample Prob. 6.6]

Fig. 6.49

REVIEW PROBLEMS

6.89 Three boards, each 50 mm thick, are nailed together to form a beam that is subjected to a 1200-N vertical shear. Knowing that the allowable shearing force in each nail is 600 N, determine the largest permissible spacing s between the nails.

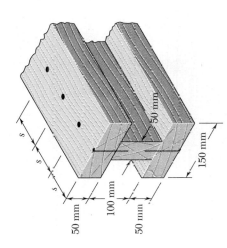

Fig. P6.89

6.90 The American Standard rolled-steel beam shown has been reinforced by attaching to it two 16 × 200-mm plates, using bolts of 18-mm diameter spaced longitudinally every 120 mm. Knowing that the allowable average shearing stress in the bolts is 90 MPa, determine the largest permissible shearing force.

Fig. P6.90

6.91 For the beam and loading shown, consider section n-n and determine the shearing stress at (a) point a, (b) point b.

Fig. P6.91

6.92 For the beam and loading shown, consider section *n-n* and determine (*a*) the largest shearing stress in that section, (*b*) the shearing stress at point *a*.

Fig. P6.92

6.93 The built-up timber beam shown is subjected to a 6-kN vertical shear. Knowing that the longitudinal spacing of the nails is *s* = 60 mm and that each nail is 90 mm long, determine the shearing force in each nail.

Fig. P6.93

6.94 The built-up beam shown was made by gluing together several wooden planks. Knowing that the beam is subjected to a 1200-lb vertical shear, determine the average shearing stress in the glued joint (*a*) at *A*, (*b*) at *B*.

Dimensions in inches

Fig. P6.94

6.95 A beam consists of three planks connected as shown by $\frac{3}{8}$-in.-diameter bolts spaced every 12 in. along the longitudinal axis of the beam. Knowing that the beam is subjected to a 2500-lb vertical shear, determine the maximum shearing stress in the bolts.

Fig. P6.95

COMPUTER PROBLEMS

The following problems are designed to be solved with a computer.

6.C1 A timber beam is to be designed to support a distributed load and up to two concentrated loads as shown. One of the dimensions of its uniform rectangular cross section has been specified and the other is to be determined so that the maximum normal stress and the maximum shearing stress in the beam will not exceed given allowable values σ_{all} and τ_{all}. Measuring x from end A and using SI units, write a computer program to calculate for successive cross sections, from $x = 0$ to $x = L$ and using given increments Δx, the shear, the bending moment, and the smallest value of the unknown dimension that satisfies in that section (1) the allowable normal stress requirement, (2) the allowable shearing stress requirement. Use this program to design the beams of uniform cross section of the following problems, assuming $\sigma_{all} = 12$ MPa and $\tau_{all} = 825$ kPa, and using the increments indicated: (a) Prob. 5.75 ($\Delta x = 0.1$ m), (b) Prob. 5.76 ($\Delta x = 0.2$ m).

Fig. P6.C1

6.C2 A cantilever timber beam AB of length L and of the uniform rectangular section shown supports a concentrated load $\mathbf{P}$ at its free end and a uniformly distributed load w along its entire length. Write a computer program to determine the length L and the width b of the beam for which both the maximum normal stress and the maximum shearing stress in the beam reach their largest allowable values. Assuming $\sigma_{all} = 1.8$ ksi and $\tau_{all} = 120$ psi, use this program to determine the dimensions L and b when (a) $P = 1000$ lb and $w = 0$, (b) $P = 0$ and $w = 12.5$ lb/in., (c) $P = 500$ lb and $w = 12.5$ lb/in.

Fig. P6.C2

6.C3 A beam having the cross section shown is subjected to a vertical shear $\mathbf{V}$. Write a computer program that, for loads and dimensions expressed in either SI or U.S. customary units, can be used to calculate the shearing stress along the line between any two adjacent rectangular areas forming the cross section. Use this program to solve (a) Prob. 6.10, (b) Prob. 6.11, (c) Prob. 6.21, (d) Prob. 6.23.

Fig. P6.C3

6.C4 A plate of uniform thickness t is bent as shown into a shape with a vertical plane of symmetry and is then used as a beam. Write a computer program that, for loads and dimensions expressed in either SI or U.S. customary units, can be used to determine the distribution of shearing stresses caused by a vertical shear **V**. Use this program (a) to solve Prob. 6.49, (b) to find the shearing stress at point E for the shape and load of Prob. 6.50, assuming a thickness $t = \frac{1}{4}$ in.

Fig. P6.C4

6.C5 The cross section of an extruded beam is symmetric with respect to the x axis and consists of several straight segments as shown. Write a computer program that, for loads and dimensions expressed in either SI or U.S. customary units, can be used to determine (a) the location of the shear center O, (b) the distribution of shearing stresses caused by a vertical force applied at O. Use this program to solve Probs. 6.65, 6.68, 6.69, and 6.70.

Fig. P6.C5

6.C6 A thin-walled beam has the cross section shown. Write a computer program that, for dimensions expressed in either SI or U.S. customary units, can be used to determine the location of the shear center O of the cross section. Use this program to solve Prob. 6.100.

Fig. P6.C6

7

Transformations of Stress and Strain

In the test setup shown the linear strain in the top surface of the bar is being measured by an electrical strain gage cemented to the top surface. This chapter deals with stresses and strains in structures and machine components.

7.1. INTRODUCTION

We saw in Sec. 1.12 that the most general state of stress at a given point Q may be represented by six components. Three of these components, σ_x, σ_y, and σ_z, define the normal stresses exerted on the faces of a small cubic element centered at Q and of the same orientation as the coordinate axes (Fig. 7.1a), and the other three, τ_{xy}, τ_{yz}, and τ_{zx},[†] the components of the shearing stresses on the same element. As we remarked at the time, the same state of stress will be represented by a different set of components if the coordinate axes are rotated (Fig. 7.1b). We propose in the first part of this chapter to determine how the components of stress are transformed under a rotation of the coordinate axes. The second part of the chapter will be devoted to a similar analysis of the transformation of the components of strain.

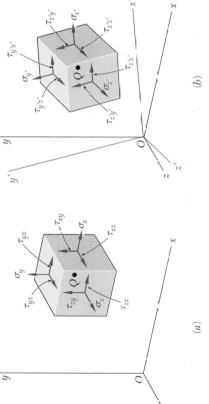

(a)

(b)

Fig. 7.1

Our discussion of the transformation of stress will deal mainly with *plane stress*, i.e., with a situation in which two of the faces of the cubic element are free of any stress. If the z axis is chosen perpendicular to these faces, we have $\sigma_z = \tau_{zx} = \tau_{zy} = 0$, and the only remaining stress components are σ_x, σ_y, and τ_{xy} (Fig. 7.2). Such a situation occurs in a thin plate subjected to forces acting in the midplane of the plate (Fig. 7.3). It also occurs on the free surface of a structural element or machine component, i.e., at any point of the surface of that element or component that is not subjected to an external force (Fig. 7.4).

Fig. 7.2

Fig. 7.4

Fig. 7.3

[†] We recall that $\tau_{yx} = \tau_{xy}$, $\tau_{zy} = \tau_{yz}$, and $\tau_{xz} = \tau_{zx}$.

Considering in Sec. 7.2 a state of plane stress at a given point Q characterized by the stress components $\sigma_x, \sigma_y,$ and τ_{xy} associated with the element shown in Fig. 7.5a, you will learn to determine the components $\sigma_{x'}, \sigma_{y'},$ and $\tau_{x'y'}$ associated with that element after it has been rotated through an angle θ about the z axis (Fig. 7.5b). In Sec. 7.3, you will determine the value θ_p of θ for which the stresses $\sigma_{x'}$ and $\sigma_{y'}$ are, respectively, maximum and minimum; these values of the normal stress are the *principal stresses* at point Q, and the faces of the corresponding element define the *principal planes of stress* at that point. You will also determine the value θ_s of the angle of rotation for which the shearing stress is maximum, as well as the value of that stress.

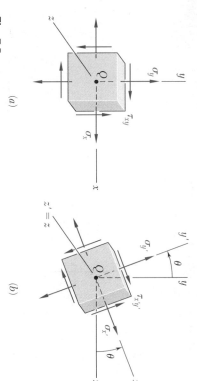

(a)

(b)

Fig. 7.5

In Sec. 7.4, an alternative method for the solution of problems involving the transformation of plane stress, based on the use of *Mohr's circle*, will be presented.

In Sec. 7.5, the *three-dimensional state of stress* at a given point will be considered and a formula for the determination of the normal stress on a plane of arbitrary orientation at that point will be developed. In Sec. 7.6, you will consider the rotations of a cubic element about each of the principal axes of stress and note that the corresponding transformations of stress can be described by three different Mohr's circles. You will also observe that, in the case of a state of *plane stress* at a given point, the maximum value of the shearing stress obtained earlier by considering rotations in the plane of stress does not necessarily represent the maximum shearing stress at that point. This will bring you to distinguish between *in-plane* and *out-of-plane* maximum shearing stresses.

Yield criteria for ductile materials under plane stress will be developed in Sec. 7.7. To predict whether a material will yield at some critical point under given loading conditions, you will determine the principal stresses σ_a and σ_b at that point and check whether $\sigma_a, \sigma_b,$ and the yield strength σ_Y of the material satisfy some criterion. Two criteria in common use are: the *maximum-shearing-strength criterion* and the *maximum-distortion-energy criterion*. In Sec. 7.8, *fracture criteria* for brittle materials under plane stress will be developed in a similar fashion; they will involve the principal stresses σ_a and σ_b at some critical point and the ultimate strength σ_U of the material. Two criteria will be discussed: the *maximum-normal-stress criterion* and *Mohr's criterion*.

Thin-walled pressure vessels provide an important application of the analysis of plane stress. In Sec. 7.9, we will discuss stresses in both cylindrical and spherical pressure vessels (Figs. 7.6 and 7.7).

Fig. 7.6

Fig. 7.7

Sections 7.10 and 7.11 will be devoted to a discussion of the *transformation of plane strain* and to *Mohr's circle for plane strain*. In Sec. 7.12, we will consider the three-dimensional analysis of strain and see how Mohr's circles can be used to determine the maximum shearing strain at a given point. Two particular cases are of special interest and should not be confused: the case of *plane strain* and the case of *plane stress*.

Finally, in Sec. 7.13, we discuss the use of *strain gages* to measure the normal strain on the surface of a structural element or machine component. You will see how the components ϵ_x, ϵ_y, and γ_{xy} characterizing the state of strain at a given point can be computed from the measurements made with three strain gages forming a *strain rosette*.

7.2. TRANSFORMATION OF PLANE STRESS

Let us assume that a state of plane stress exists at point Q (with $\sigma_z = \tau_{zx} = \tau_{zy} = 0$), and that it is defined by the stress components σ_x, σ_y, and τ_{xy} associated with the element shown in Fig. 7.5a. We propose to determine the stress components $\sigma_{x'}$, $\sigma_{y'}$, and $\tau_{x'y'}$ associated with the element after it has been rotated through an angle θ about the z axis (Fig. 7.5b), and to express these components in terms of σ_x, σ_y, τ_{xy}, and θ.

(a)

(b)

Fig. 7.5 (repeated)

In order to determine the normal stress $\sigma_{x'}$ and the shearing stress $\tau_{x'y'}$ exerted on the face perpendicular to the x' axis, we consider a prismatic element with faces respectively perpendicular to the x, y, and x' axes (Fig. 7.8a). We observe that, if the area of the oblique face is denoted by ΔA, the areas of the vertical and horizontal faces are respectively equal to $\Delta A \cos\theta$ and $\Delta A \sin\theta$. It follows that the *forces* exerted on the three faces are as shown in Fig. 7.8b. (No forces are ex-

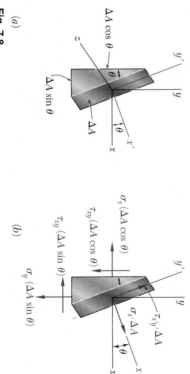

(a)

Fig. 7.8

(b)

erted on the triangular faces of the element, since the corresponding normal and shearing stresses have all been assumed equal to zero.) Using components along the x' and y' axes, we write the following equilibrium equations:

$$\sum F_{x'} = 0: \quad \sigma_{x'} \Delta A - \sigma_x (\Delta A \cos\theta) \cos\theta - \tau_{xy} (\Delta A \cos\theta) \sin\theta$$
$$-\sigma_y (\Delta A \sin\theta) \sin\theta - \tau_{xy} (\Delta A \sin\theta) \cos\theta = 0$$

$$\sum F_{y'} = 0: \quad \tau_{x'y'} \Delta A + \sigma_x (\Delta A \cos\theta) \sin\theta - \tau_{xy} (\Delta A \cos\theta) \cos\theta$$
$$-\sigma_y (\Delta A \sin\theta) \cos\theta + \tau_{xy} (\Delta A \sin\theta) \sin\theta = 0$$

Solving the first equation for $\sigma_{x'}$ and the second for $\tau_{x'y'}$, we have

$$\sigma_{x'} = \sigma_x \cos^2 \theta + \sigma_y \sin^2 \theta + 2\tau_{xy} \sin \theta \cos \theta \qquad (7.1)$$

$$\tau_{x'y'} = -(\sigma_x - \sigma_y) \sin \theta \cos \theta + \tau_{xy}(\cos^2 \theta - \sin^2 \theta) \qquad (7.2)$$

Recalling the trigonometric relations

$$\sin 2\theta = 2 \sin \theta \cos \theta \qquad \cos 2\theta = \cos^2 \theta - \sin^2 \theta \qquad (7.3)$$

and

$$\cos^2 \theta = \frac{1 + \cos 2\theta}{2} \qquad \sin^2 \theta = \frac{1 - \cos 2\theta}{2} \qquad (7.4)$$

we write Eq. (7.1) as follows:

$$\sigma_{x'} = \sigma_x \frac{1 + \cos 2\theta}{2} + \sigma_y \frac{1 - \cos 2\theta}{2} + \tau_{xy} \sin 2\theta$$

or

$$\sigma_{x'} = \frac{\sigma_x + \sigma_y}{2} + \frac{\sigma_x - \sigma_y}{2} \cos 2\theta + \tau_{xy} \sin 2\theta \qquad (7.5)$$

Using the relations (7.3), we write Eq. (7.2) as

$$\tau_{x'y'} = -\frac{\sigma_x - \sigma_y}{2} \sin 2\theta + \tau_{xy} \cos 2\theta \qquad (7.6)$$

The expression for the normal stress $\sigma_{y'}$ is obtained by replacing θ in Eq. (7.5) by the angle $\theta + 90°$ that the y' axis forms with the x axis. Since $\cos(2\theta + 180°) = -\cos 2\theta$ and $\sin(2\theta + 180°) = -\sin 2\theta$, we have

$$\sigma_{y'} = \frac{\sigma_x + \sigma_y}{2} - \frac{\sigma_x - \sigma_y}{2} \cos 2\theta - \tau_{xy} \sin 2\theta \qquad (7.7)$$

Adding Eqs. (7.5) and (7.7) member to member, we obtain

$$\sigma_{x'} + \sigma_{y'} = \sigma_x + \sigma_y \qquad (7.8)$$

Since $\sigma_z = \sigma_{z'} = 0$, we thus verify in the case of plane stress that the sum of the normal stresses exerted on a cubic element of material is independent of the orientation of that element.[†]

†Cf. footnote on page 88.

7.3. PRINCIPAL STRESSES; MAXIMUM SHEARING STRESS

The equations (7.5) and (7.6) obtained in the preceding section are the parametric equations of a circle. This means that, if we choose a set of rectangular axes and plot a point M of abscissa $\sigma_{x'}$ and ordinate $\tau_{x'y'}$ for any given value of the parameter θ, all the points thus obtained will lie on a circle. To establish this property we eliminate θ from Eqs. (7.5) and (7.6); this is done by first transposing $(\sigma_x + \sigma_y)/2$ in Eq. (7.5) and squaring both members of the equation, then squaring both members of Eq. (7.6), and finally adding member to member the two equations obtained in this fashion. We have

$$\left(\sigma_{x'} - \frac{\sigma_x + \sigma_y}{2}\right)^2 + \tau_{x'y'}^2 = \left(\frac{\sigma_x - \sigma_y}{2}\right)^2 + \tau_{xy}^2 \qquad (7.9)$$

Setting

$$\sigma_{\text{ave}} = \frac{\sigma_x + \sigma_y}{2} \qquad \text{and} \qquad R = \sqrt{\left(\frac{\sigma_x - \sigma_y}{2}\right)^2 + \tau_{xy}^2} \qquad (7.10)$$

we write the identity (7.9) in the form

$$(\sigma_{x'} - \sigma_{\text{ave}})^2 + \tau_{x'y'}^2 = R^2 \qquad (7.11)$$

which is the equation of a circle of radius R centered at the point C of abscissa σ_{ave} and ordinate 0 (Fig. 7.9). It can be observed that, due to the symmetry of the circle about the horizontal axis, the same result would have been obtained if, instead of plotting M, we had plotted a point N of abscissa $\sigma_{x'}$ and ordinate $-\tau_{x'y'}$ (Fig. 7.10). This property will be used in Sec. 7.4.

The two points A and B where the circle of Fig. 7.9 intersects the horizontal axis are of special interest: Point A corresponds to the maximum value of the normal stress $\sigma_{x'}$, while point B corresponds to its

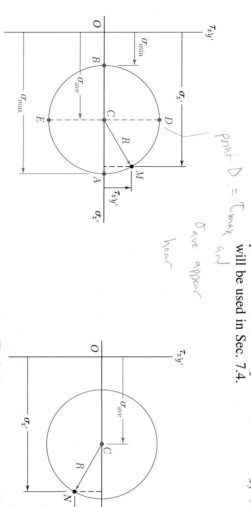

Fig. 7.9

Fig. 7.10

minimum value. Besides, both points correspond to a zero value of the shearing stress $\tau_{x'y'}$. Thus, the values θ_p of the parameter θ which correspond to points A and B can be obtained by setting $\tau_{x'y'} = 0$ in Eq. (7.6). We write†

from 7.6

$$\tan 2\theta_p = \frac{2\tau_{xy}}{\sigma_x - \sigma_y} \qquad (7.12)$$

This equation defines two values $2\theta_p$ that are 180° apart, and thus two values θ_p that are 90° apart. Either of these values can be used to determine the orientation of the corresponding element (Fig. 7.11). The planes containing the faces of the element obtained in this way are called

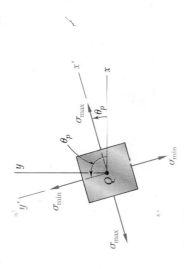

Fig. 7.11

the *principal planes of stress* at point Q, and the corresponding values σ_{max} and σ_{min} of the normal stress exerted on these planes are called the *principal stresses* at Q. Since the two values θ_p defined by Eq. (7.12) were obtained by setting $\tau_{x'y'} = 0$ in Eq. (7.6), it is clear that no shearing stress is exerted on the principal planes.

We observe from Fig. 7.9 that

$$\sigma_{max} = \sigma_{ave} + R \qquad \text{and} \qquad \sigma_{min} = \sigma_{ave} - R \qquad (7.13)$$

Substituting for σ_{ave} and R from Eq. (7.10), we write

$$\sigma_{max, min} = \frac{\sigma_x + \sigma_y}{2} \pm \sqrt{\left(\frac{\sigma_x - \sigma_y}{2}\right)^2 + \tau_{xy}^2} \qquad (7.14)$$

Unless it is possible to tell by inspection which of the two principal planes is subjected to σ_{max} and which is subjected to σ_{min}, it is necessary to substitute one of the values θ_p into Eq. (7.5) in order to determine which of the two corresponds to the maximum value of the normal stress.

Referring again to the circle of Fig. 7.9, we note that the points D and E located on the vertical diameter of the circle correspond to the largest numerical value of the shearing stress $\tau_{x'y'}$. Since the abscissa of points D and E is $\sigma_{ave} = (\sigma_x + \sigma_y)/2$, the values θ_s of the parameter θ corresponding to these points are obtained by setting $\sigma_{x'} = (\sigma_x + \sigma_y)/2$

†This relation can also be obtained by differentiating $\sigma_{x'}$ in Eq. (7.5) and setting the derivative equal to zero: $d\sigma_{x'}/d\theta = 0$.

Positive Shear Stress

Negative Shear Stress

in Eq. (7.5). It follows that the sum of the last two terms in that equation must be zero. Thus, for $\theta = \theta_s$, we write†

$$\frac{\sigma_x - \sigma_y}{2} \cos 2\theta_s + \tau_{xy} \sin 2\theta_s = 0$$

or

$$\tan 2\theta_s = -\frac{\sigma_x - \sigma_y}{2\tau_{xy}} \tag{7.15}$$

This equation defines two values $2\theta_s$ which are $180°$ apart, and thus two values θ_s which are $90°$ apart. Either of these values can be used to determine the orientation of the element corresponding to the maximum shearing stress (Fig. 7.12). Observing from Fig. 7.9 that the maximum value of the shearing stress is equal to the radius R of the circle, and recalling the second of Eqs. (7.10), we write

$$\tau_{max} = \sqrt{\left(\frac{\sigma_x - \sigma_y}{2}\right)^2 + \tau_{xy}^2} \tag{7.16}$$

Fig. 7.12

As observed earlier, the normal stress corresponding to the condition of maximum shearing stress is

$$\sigma' = \sigma_{ave} = \frac{\sigma_x + \sigma_y}{2} \tag{7.17}$$

Comparing Eqs. (7.12) and (7.15), we note that $\tan 2\theta_s$ is the negative reciprocal of $\tan 2\theta_p$. This means that the angles $2\theta_s$ and $2\theta_p$ are $90°$ apart and, therefore, that the angles θ_s and θ_p are $45°$ apart. We thus conclude that *the planes of maximum shearing stress are at 45° to the principal planes.* This confirms the results obtained earlier in Sec. 1.12 in the case of a centric axial loading (Fig. 1.40) and in Sec. 3.4 in the case of a torsional loading (Fig. 3.20.)

We should be aware that our analysis of the transformation of plane stress has been limited to rotations *in the plane of stress.* If the cubic element of Fig. 7.7 is rotated about an axis other than the z axis, its faces may be subjected to shearing stresses larger than the stress defined by Eq. (7.16). As you will see in Sec. 7.5, this occurs when the principal stresses defined by Eq. (7.14) have the same sign, i.e., when they are either both tensile or both compressive. In such cases, the value given by Eq. (7.16) is referred to as the maximum *in-plane* shearing stress.

†This relation may also be obtained by differentiating $\tau_{x'y'}$ in Eq. (7.6) and setting the derivative equal to zero: $d\tau_{x'y'}/d\theta = 0$.

EXAMPLE 7.01

For the state of plane stress shown in Fig. 7.13, determine (a) the principal planes, (b) the principal stresses, (c) the maximum shearing stress and the corresponding normal stress.

on this plane there are no shearing stresses

Fig. 7.13

(a) Principal Planes. Following the usual sign convention, we write the stress components as

$$\sigma_x = +50 \text{ MPa} \qquad \sigma_y = -10 \text{ MPa} \qquad \tau_{xy} = +40 \text{ MPa}$$

Substituting into Eq. (7.12), we have

$$\tan 2\theta_p = \frac{2\tau_{xy}}{\sigma_x - \sigma_y} = \frac{2(+40)}{50 - (-10)} = \frac{80}{60}$$

$$2\theta_p = 53.1° \quad \text{and} \quad 180° + 53.1° = 233.1°$$
$$\theta_p = 26.6° \quad \text{and} \quad 116.6°$$

(b) Principal Stresses. Formula (7.14) yields

$$\sigma_{\text{max, min}} = \frac{\sigma_x + \sigma_y}{2} \pm \sqrt{\left(\frac{\sigma_x - \sigma_y}{2}\right)^2 + \tau_{xy}^2}$$
$$= 20 \pm \sqrt{(30)^2 + (40)^2}$$
$$\sigma_{\text{max}} = 20 + 50 = 70 \text{ MPa}$$
$$\sigma_{\text{min}} = 20 - 50 = -30 \text{ MPa}$$

The principal planes and principal stresses are sketched in Fig. 7.14. Making $\theta = 26.6°$ in Eq. (7.5), we check that the normal stress exerted on face *BC* of the element is the maximum stress:

$$\sigma_{x'} = \frac{50}{2} + \frac{10}{2} + \frac{50 + 10}{2}\cos 53.1° + 40 \sin 53.1°$$
$$= 20 + 30 \cos 53.1° + 40 \sin 53.1° = 70 \text{ MPa} = \sigma_{\text{max}}$$

Fig. 7.14

(c) Maximum Shearing Stress. Formula (7.16) yields

$$\tau_{\text{max}} = \sqrt{\left(\frac{\sigma_x - \sigma_y}{2}\right)^2 + \tau_{xy}^2} = \sqrt{(30)^2 + (40)^2} = 50 \text{ MPa}$$

Since σ_{max} and σ_{min} have opposite signs, the value obtained for τ_{max} actually represents the maximum value of the shearing stress at the point considered. The orientation of the planes of maximum shearing stress and the sense of the shearing stresses are best determined by passing a section along the diagonal plane *AC* of the element of Fig. 7.14. Since the faces *AB* and *BC* of the element are contained in the principal planes, the diagonal plane *AC* must be one of the planes of maximum shearing stress (Fig. 7.15). Furthermore, the equilibrium conditions for the prismatic element *ABC* require that the shear-

shearing stresses

$\theta_s \Leftarrow$ *angle of plane of τ_{max}*

$$\theta_s = \theta_p - 45° = -18.4°$$

Fig. 7.15

ing stress exerted on *AC* be directed as shown. The cubic element corresponding to the maximum shearing stress is shown in Fig. 7.16. The normal stress on each of the four faces of the element is given by Eq. (7.17):

$$\sigma' = \sigma_{\text{ave}} = \frac{\sigma_x + \sigma_y}{2} = \frac{50 - 10}{2} = 20 \text{ MPa}$$

this isn't Mohr circle method (those will be less normal)

$$\sigma_{\text{ave}} = \frac{\sigma_x + \sigma_y}{2}$$

Fig. 7.16

SAMPLE PROBLEM 7.1

A single horizontal force **P** of magnitude 150 lb is applied to end *D* of lever *ABD*. Knowing that portion *AB* of the lever has a diameter of 1.2 in., determine (*a*) the normal and shearing stresses on an element located at point *H* and having sides parallel to the *x* and *y* axes, (*b*) the principal planes and the principal stresses at point *H*.

SOLUTION

Force-Couple System. We replace the force **P** by an equivalent force-couple system at the center *C* of the transverse section containing point *H*:

$$P = 150 \text{ lb} \qquad T = (150 \text{ lb})(18 \text{ in.}) = 2.7 \text{ kip} \cdot \text{in.}$$
$$M_x = (150 \text{ lb})(10 \text{ in.}) = 1.5 \text{ kip} \cdot \text{in.}$$

***a.* Stresses σ_x, σ_y, τ_{xy} at Point *H*.** Using the sign convention shown in Fig. 7.2, we determine the sense and the sign of each stress component by carefully examining the sketch of the force-couple system at point *C*:

$$\sigma_x = 0 \qquad \sigma_y = +\frac{Mc}{I} = +\frac{(1.5 \text{ kip} \cdot \text{in.})(0.6 \text{ in.})}{\frac{1}{4}\pi (0.6 \text{ in.})^4} \qquad \sigma_y = +8.84 \text{ ksi} \blacktriangleleft$$

$$\tau_{xy} = +\frac{Tc}{J} = +\frac{(2.7 \text{ kip} \cdot \text{in.})(0.6 \text{ in.})}{\frac{1}{2}\pi (0.6 \text{ in.})^4} \qquad \tau_{xy} = +7.96 \text{ ksi} \blacktriangleleft$$

We note that the shearing force **P** does not cause any shearing stress at point *H*.

***b.* Principal Planes and Principal Stresses.** Substituting the values of the stress components into Eq. (7.12), we determine the orientation of the principal planes:

$$\tan 2\theta_p = \frac{2\tau_{xy}}{\sigma_x - \sigma_y} = \frac{2(7.96)}{0 - 8.84} = -1.80$$

$$2\theta_p = -61.0° \qquad \text{and} \qquad 180° - 61.0° = +119°$$

$$\theta_p = -30.5° \qquad \text{and} \qquad +59.5° \blacktriangleleft$$

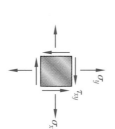

Substituting into Eq. (7.14), we determine the magnitudes of the principal stresses:

$$\sigma_{\text{max, min}} = \frac{\sigma_x + \sigma_y}{2} \pm \sqrt{\left(\frac{\sigma_x - \sigma_y}{2}\right)^2 + \tau_{xy}^2}$$
$$= \frac{0 + 8.84}{2} \pm \sqrt{\left(\frac{0 - 8.84}{2}\right)^2 + (7.96)^2} = +4.42 \pm 9.10$$

$$\sigma_{\text{max}} = +13.52 \text{ ksi} \blacktriangleleft$$
$$\sigma_{\text{min}} = -4.68 \text{ ksi} \blacktriangleleft$$

Considering face *ab* of the element shown, we make $\theta_p = -30.5°$ in Eq. (7.5) and find $\sigma_{x'} = -4.68$ ksi. We conclude that the principal stresses are as shown.

PROBLEMS

7.1 through 7.4 For the given state of stress, determine the normal and shearing stresses exerted on the oblique face of the shaded triangular element shown. Use a method of analysis based on the equilibrium of that element, as was done in the derivations of Sec. 7.2.

Fig. P7.1

Fig. P7.2

Fig. P7.3

Fig. P7.4

7.5 through 7.8 For the given state of stress, determine (*a*) the principal planes, (*b*) the principal stresses.

Fig. *P7.5* and P7.9

Fig. *P7.6* and P7.10

Fig. P7.7 and P7.11

Fig. P7.8 and *P7.12*

7.9 through 7.12 For the given state of stress, determine (*a*) the orientation of the planes of maximum in-plane shearing stress, (*b*) the maximum in-plane shearing stress, (*c*) the corresponding normal stress.

7.13 through 7.16 For the given state of stress, determine the normal and shearing stresses after the element shown has been rotated through (*a*) 25° clockwise, (*b*) 10° counterclockwise.

Fig. P7.13

Fig. *P7.14*

Fig. P7.15

Fig. P7.16

7.17 and 7.18 The grain of a wooden member forms an angle of 15° with the vertical. For the state of stress shown, determine (a) the in-plane shearing stress parallel to the grain, (b) the normal stress perpendicular to the grain.

Fig. P7.17

Fig. P7.18

7.19 The centric force **P** is applied to a short post as shown. Knowing that the stresses on plane a-a are σ = −15 ksi and τ = 5 ksi, determine the angle β that plane a-a forms with the horizontal, (b) the maximum compressive stress in the post.

Fig. P7.19

7.20 Two members of uniform cross section 50 × 80 mm are glued together along plane a-a, that forms an angle of 25° with the horizontal. Knowing that the allowable stresses for the glued joint are σ = 800 kPa and τ = 600 kPa, determine the largest axial load **P** that can be applied.

Fig. P7.20

7.21 Two steel plates of uniform cross section 10 × 80 mm are welded together as shown. Knowing that centric 100-kN forces are applied to the welded plates and that β = 25°, determine (a) the in-plane shearing stress parallel to the weld, (b) the normal stress perpendicular to the weld.

7.22 Two steel plates of uniform cross section 10 × 80 mm are welded together as shown. Knowing that centric 100-kN forces are applied to the welded plates and that the in-plane shearing stress parallel to the weld is 30 MPa, determine (a) the angle β, (b) the corresponding normal stress perpendicular to the weld.

Fig. P7.21 and P7.22

7.23   The steel pipe AB has a 102-mm outer diameter and a 6-mm wall thickness. Knowing that arm CD is rigidly attached to the pipe, determine the principal stresses and the maximum shearing stress at point H.

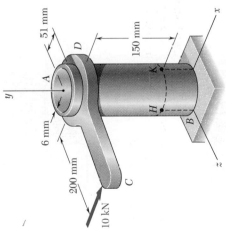

Fig. P7.23 and P7.24

7.24   The steel pipe AB has a 102-mm outer diameter and a 6-mm wall thickness. Knowing that arm CD is rigidly attached to the pipe, determine the principal stresses and the maximum shearing stress at point K.

7.25   A 400-lb vertical force is applied at D to a gear attached to the solid one-inch diameter shaft AB. Determine the principal stresses and the maximum shearing stress at point H located as shown on top of the shaft.

Fig. P7.25

7.26   A mechanic uses a crowfoot wrench to loosen a bolt at E. Knowing that the mechanic applies a vertical 24-lb force at A, determine the principal stresses and the maximum shearing stress at point H located as shown on top of the $\frac{3}{4}$-in. diameter shaft.

Fig. P7.26

7.27   For the state of plane stress shown, determine the largest value of σ_y for which the maximum in-plane shearing stress is equal to or less than 15 ksi.

Fig. P7.27

7.28 For the state of plane stress shown, determine (*a*) the largest value of τ_{xy} for which the maximum in-plane shearing stress is equal or less than 12 ksi, (*b*) the corresponding principal stresses.

Fig. P7.28

7.29 Determine the range of values of σ_x for which the maximum in-plane shearing stress is equal to or less than 50 MPa.

Fig. P7.29

7.30 For the state of plane stress shown, determine (*a*) the value of τ_{xy} for which the in-plane shearing stress parallel to the weld is zero, (*b*) the corresponding principal stresses.

Fig. P7.30

7.4. MOHR'S CIRCLE FOR PLANE STRESS

The circle used in the preceding section to derive some of the basic formulas relating to the transformation of plane stress was first introduced by the German engineer Otto Mohr (1835–1918) and is known as *Mohr's circle* for plane stress. As you will see presently, this circle can be used to obtain an alternative method for the solution of the various problems considered in Secs. 7.2 and 7.3. This method is based on simple geometric considerations and does not require the use of specialized formulas. While originally designed for graphical solutions, it lends itself well to the use of a calculator.

Consider a square element of a material subjected to plane stress (Fig. 7.17a), and let σ_x, σ_y, and τ_{xy} be the components of the stress exerted on the element. We plot a point X of coordinates σ_x and $-\tau_{xy}$, and a point Y of coordinates σ_y and $+\tau_{xy}$ (Fig. 7.17b). If τ_{xy} is positive, as assumed in Fig. 7.17a, point X is located below the σ axis and point Y above, as shown in Fig. 7.17b. If τ_{xy} is negative, X is located above the σ axis and Y below. Joining X and Y by a straight line, we define the point C of intersection of line XY with the σ axis and draw the circle of center C and diameter XY. Noting that the abscissa of C and the radius of the circle are respectively equal to the quantities σ_{ave} and R defined by Eqs. (7.10), we conclude that the circle obtained is Mohr's circle for plane stress. Thus the abscissas of points A and B where the circle intersects the σ axis represent respectively the principal stresses $\sigma_{\max}$ and $\sigma_{\min}$ at the point considered.

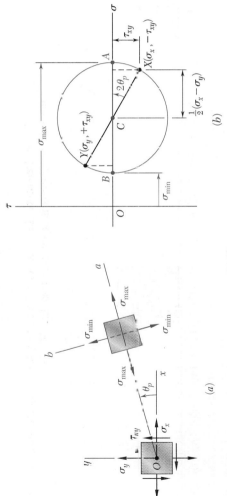

Clockwise rotation is positive

Fig. P7.17

(a)

(b)

We also note that, since $\tan (XCA) = 2\tau_{xy}/(\sigma_x - \sigma_y)$, the angle XCA is equal in magnitude to one of the angles $2\theta_p$ that satisfy Eq. (7.12). Thus, the angle θ_p that defines in Fig. 7.17a the orientation of the principal plane corresponding to point A in Fig. 7.17b can be obtained by dividing in half the angle XCA measured on Mohr's circle. We further observe that if $\sigma_x > \sigma_y$ and $\tau_{xy} > 0$, as in the case considered here, the rotation which brings CX into CA is counterclockwise. But, in that case, the angle θ_p obtained from Eq. (7.12) and defining the direction of the normal Oa to the principal plane is positive; thus the rotation bringing Ox into Oa is also counterclockwise. We conclude that the senses of rotation in both parts of Fig. 7.17 are the same; if a counterclockwise rotation through $2\theta_p$ is required to bring CX into CA on Mohr's circle, a counterclockwise rotation through θ_p will bring Ox into Oa in Fig. 7.17a.[†]

[†]This is due to the fact that we are using the circle of Fig. 7.10 rather than the circle of Fig. 7.9 as Mohr's circle.

Since Mohr's circle is uniquely defined, the same circle can be obtained by considering the stress components $\sigma_{x'}$, $\sigma_{y'}$, and $\tau_{x'y'}$, corresponding to the x' and y' axes as shown in Fig. 7.18a. The point X' of coordinates $\sigma_{x'}$ and $-\tau_{x'y'}$, and the point Y' of coordinates $\sigma_{y'}$ and $+\tau_{x'y'}$, are therefore located on Mohr's circle, and the angle $X'CA$ in Fig. 7.18b must be equal to twice the angle $x'Oa$ in Fig. 7.18a. Since, as noted

Fig. 7.18

before, the angle xOa is twice the angle XCA, it follows that the angle XCX' in Fig. 7.18b is twice the angle xOx' in Fig. 7.18a. Thus the diameter $X'Y'$ defining the normal and shearing stresses $\sigma_{x'}$, $\sigma_{y'}$, and $\tau_{x'y'}$ can be obtained by rotating the diameter XY through an angle equal to twice the angle θ formed by the x' and x axes in Fig. 7.18a. We note that the rotation that brings the diameter XY into the diameter $X'Y'$ in Fig. 7.18b has the same sense as the rotation that brings the xy axes into the $x'y'$ axes in Fig. 7.18a.

The property we have just indicated can be used to verify the fact that the planes of maximum shearing stress are at 45° to the principal planes. Indeed, we recall that points D and E on Mohr's circle correspond to the planes of maximum shearing stress, while A and B correspond to the principal planes (Fig. 7.19b). Since the diameters AB and DE of Mohr's circle are at 90° to each other, it follows that the faces of the corresponding elements are at 45° to each other (Fig. 7.19a).

Fig. 7.19

The construction of Mohr's circle for plane stress is greatly simplified if we consider separately each face of the element used to define the stress components. From Figs. 7.17 and 7.18 we observe that, when the shearing stress exerted *on a given face* tends to rotate the element *clockwise*, the point on Mohr's circle corresponding to that face is located *above* the σ axis. When the shearing stress on a given face tends to rotate the element *counterclockwise*, the point corresponding to that face is located *below* the σ axis (Fig. 7.20).† As far as the normal stresses are concerned, the usual convention holds, i.e., a tensile stress is considered as positive and is plotted to the right, while a compressive stress is considered as negative and is plotted to the left.

(a) Clockwise ——— Above

(b) Counterclockwise ——— Below

Fig. 7.20

---

† The following jingle is helpful in remembering this convention. "In the kitchen, the *clock* is above, and the *counter* is below."

EXAMPLE 7.02

For the state of plane stress already considered in Example 7.01, (a) construct Mohr's circle, (b) determine the principal stresses, (c) determine the maximum shearing stress and the corresponding normal stress.

(a) Construction of Mohr's Circle. We note from Fig. 7.21a that the normal stress exerted on the face oriented toward the x axis is tensile (positive) and that the shearing stress exerted on that face tends to rotate the element counterclockwise. Point X of Mohr's circle, therefore, will be plotted to the right of the vertical axis and below the horizontal axis (Fig. 7.21b). A similar inspection of the normal stress and shearing stress exerted on the upper face of the element shows that point Y should be plotted to the left of the vertical axis and above the horizontal axis. Drawing the line XY, we obtain the center C of Mohr's circle; its abscissa is

$$\sigma_{ave} = \frac{\sigma_x + \sigma_y}{2} = \frac{50 + (-10)}{2} = 20 \text{ MPa}$$

Since the sides of the shaded triangle are

$$CF = 50 - 20 = 30 \text{ MPa} \qquad \text{and} \qquad FX = 40 \text{ MPa}$$

the radius of the circle is

$$R = CX = \sqrt{(30)^2 + (40)^2} = 50 \text{ MPa}$$

(a)

Fig. 7.21

(b) Principal Planes and Principal Stresses. The principal stresses are

$$\sigma_{max} = OA = OC + CA = 20 + 50 = 70\ MPa$$

$$\sigma_{min} = OB = OC - BC = 20 - 50 = -30\ MPa$$

Recalling that the angle ACX represents $2\theta_p$ (Fig. 7.21b), we write

$$\tan 2\theta_p = \frac{FX}{CF} = \frac{40}{30}$$

$$2\theta_p = 53.1° \qquad \theta_p = 26.6°$$

Since the rotation which brings CX into CA in Fig. 7.22b is counterclockwise, the rotation which brings Ox into the axis Oa corresponding to σ_{max} in Fig. 7.22a is also counterclockwise.

(c) Maximum Shearing Stress. Since a further rotation of 90° counterclockwise brings CA into CD in Fig. 7.22b, a further rotation of 45° counterclockwise will bring the axis Oa into the axis Od corresponding to the maximum shearing stress in Fig. 7.22a. We note from Fig. 7.22b that $\tau_{max} = R = 50\ MPa$ and that the corresponding normal stress is $\sigma' = \sigma_{ave} = 20\ MPa$. Since point D is located above the σ axis in Fig. 7.22b, the shearing stresses exerted on the faces perpendicular to Od in Fig. 7.22a must be directed so that they will tend to rotate the element clockwise.

Fig. 7.21b (repeated)

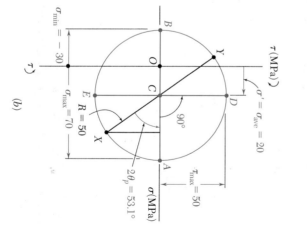

Fig. 7.22

Mohr's circle provides a convenient way of checking the results obtained earlier for stresses under a centric axial loading (Sec. 1.12) and under a torsional loading (Sec. 3.4). In the first case (Fig. 7.23a), we have $\sigma_x = P/A$, $\sigma_y = 0$, and $\tau_{xy} = 0$. The corresponding points X and Y define a circle of radius $R = P/2A$ that passes through the origin of

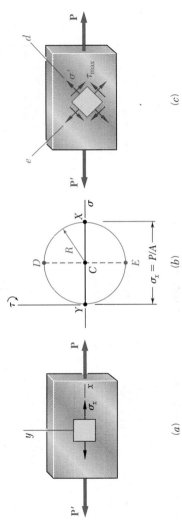

(a)   (b)   (c)

Fig. 7.23. Mohr's circle for centric axial loading.

coordinates (Fig. 7.23b). Points D and E yield the orientation of the planes of maximum shearing stress (Fig. 7.23c), as well as the values of τ_{max} and of the corresponding normal stresses σ':

$$\tau_{max} = \sigma' = R = \frac{P}{2A} \qquad (7.18)$$

In the case of torsion (Fig. 7.24a), we have $\sigma_x = \sigma_y = 0$ and $\tau_{xy} = \tau_{max} = Tc/J$. Points X and Y, therefore, are located on the τ axis,

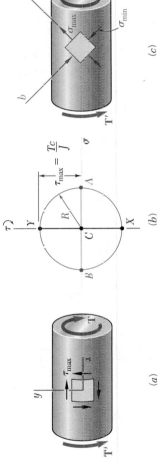

(a)   (b)   (c)

Fig. 7.24. Mohr's circle for torsional loading.

and Mohr's circle is a circle of radius $R = Tc/J$ centered at the origin (Fig. 7.24b). Points A and B define the principal planes (Fig. 7.24c) and the principal stresses:

$$\sigma_{max,min} = \pm R = \pm \frac{Tc}{J} \qquad (7.19)$$

SAMPLE PROBLEM 7.2

For the state of plane stress shown, determine (a) the principal planes and the principal stresses, (b) the stress components exerted on the element obtained by rotating the given element counterclockwise through 30°.

SOLUTION

Construction of Mohr's Circle. We note that on a face perpendicular to the x axis, the normal stress is tensile and the shearing stress tends to rotate the element clockwise; thus we plot X at a point 100 units to the right of the vertical axis and 48 units above the horizontal axis. In a similar fashion, we examine the stress components on the upper face and plot point $Y(60, -48)$. Joining points X and Y by a straight line, we define the center C of Mohr's circle. The abscissa of C, which represents σ_{ave}, and the radius R of the circle can be measured directly or calculated as follows:

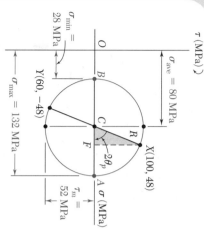

$$\sigma_{ave} = OC = \tfrac{1}{2}(\sigma_x + \sigma_y) = \tfrac{1}{2}(100 + 60) = 80 \text{ MPa}$$
$$R = \sqrt{(CF)^2 + (FX)^2} = \sqrt{(20)^2 + (48)^2} = 52 \text{ MPa}$$

a. Principal Planes and Principal Stresses. We rotate the diameter XY clockwise through $2\theta_p$ until it coincides with the diameter AB. We have

$$\tan 2\theta_p = \frac{XF}{CF} = \frac{48}{20} = 2.4 \qquad 2\theta_p = 67.4° \downarrow \quad \theta_p = 33.7° \downarrow \quad \blacktriangle$$

The principal stresses are represented by the abscissas of points A and B:

$$\sigma_{max} = OA = OC + CA = 80 + 52 \qquad \sigma_{max} = +132 \text{ MPa} \quad \blacktriangle$$
$$\sigma_{min} = OB = OC - BC = 80 - 52 \qquad \sigma_{min} = +28 \text{ MPa} \quad \blacktriangle$$

Since the rotation that brings XY into AB is clockwise, the rotation that brings Ox into the axis Oa corresponding to σ_{max} is also clockwise; we obtain the orientation shown for the principal planes.

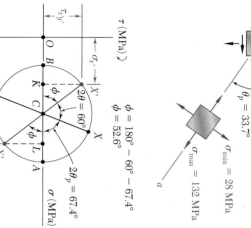

b. Stress Components on Element Rotated 30°↖. Points X' and Y' on Mohr's circle that correspond to the stress components on the rotated element are obtained by rotating XY counterclockwise through $2\theta = 60°$. We find

$$\phi = 180° - 60° - 67.4° \qquad\qquad \phi = 52.6° \quad \blacktriangle$$
$$\sigma_{x'} = OK = OC - KC = 80 - 52 \cos 52.6° \qquad \sigma_{x'} = +48.4 \text{ MPa} \quad \blacktriangle$$
$$\sigma_{y'} = OL = OC + CL = 80 + 52 \cos 52.6° \qquad \sigma_{y'} = +111.6 \text{ MPa} \quad \blacktriangle$$
$$\tau_{x'y'} = KX' = 52 \sin 52.6° \qquad\qquad \tau_{x'y'} = 41.3 \text{ MPa} \quad \blacktriangle$$

Since X' is located above the horizontal axis, the shearing stress on the face perpendicular to Ox' tends to rotate the element clockwise.

SAMPLE PROBLEM 7.3

A state of plane stress consists of a tensile stress $\sigma_0 = 8$ ksi exerted on vertical surfaces and of unknown shearing stresses. Determine (a) the magnitude of the shearing stress τ_0 for which the largest normal stress is 10 ksi, (b) the corresponding maximum shearing stress.

SOLUTION

Construction of Mohr's Circle. We assume that the shearing stresses act in the senses shown. Thus, the shearing stress τ_0 on a face perpendicular to the x axis tends to rotate the element clockwise and we plot the point X of coordinates 8 ksi and τ_0 above the horizontal axis. Considering a horizontal face of the element, we observe that $\sigma_y = 0$ and that τ_0 tends to rotate the element counterclockwise; thus, we plot point Y at a distance τ_0 below O. We note that the abscissa of the center C of Mohr's circle is

$$\sigma_{ave} = \tfrac{1}{2}(\sigma_x + \sigma_y) = \tfrac{1}{2}(8 + 0) = 4 \text{ ksi}$$

The radius R of the circle is determined by observing that the maximum normal stress, $\sigma_{max} = 10$ ksi, is represented by the abscissa of point A and writing

$$\sigma_{max} = \sigma_{ave} + R$$
$$10 \text{ ksi} = 4 \text{ ksi} + R \qquad R = 6 \text{ ksi}$$

a. Shearing Stress τ_0. Considering the right triangle CFX, we find

$$\cos 2\theta_p = \frac{CF}{CX} = \frac{CF}{R} = \frac{4 \text{ ksi}}{6 \text{ ksi}} \qquad 2\theta_p = 48.2° \downarrow \qquad \theta_p = 24.1° \downarrow$$

$$\tau_0 = FX = R \sin 2\theta_p = (6 \text{ ksi}) \sin 48.2° \qquad \tau_0 = 4.47 \text{ ksi} \blacktriangleleft$$

b. Maximum Shearing Stress. The coordinates of point D of Mohr's circle represent the maximum shearing stress and the corresponding normal stress.

$$\tau_{max} = R = 6 \text{ ksi} \qquad \tau_{max} = 6 \text{ ksi} \blacktriangleleft$$

$$2\theta_s = 90° - 2\theta_p = 90° - 48.2° = 41.8° \nwarrow \qquad \theta_s = 20.9° \nwarrow$$

The maximum shearing stress is exerted on an element that is oriented as shown in Fig. a. (The element upon which the principal stresses are exerted is also shown.)

Note. If our original assumption regarding the sense of τ_0 was reversed, we would obtain the same circle and the same answers, but the orientation of the elements would be as shown in Fig. b.

(a)

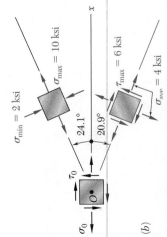

(b)

PROBLEMS

7.31 Solve Probs. 7.7 and 7.11, using Mohr's circle.

7.32 Solve Probs. 7.8 and 7.12, using Mohr's circle.

7.33 Solve Prob. 7.9, using Mohr's circle.

7.34 Solve Prob. 7.10, using Mohr's circle.

7.35 Solve Prob. 7.13, using Mohr's circle.

7.36 Solve Prob. 7.14, using Mohr's circle.

7.37 Solve Prob. 7.15, using Mohr's circle.

7.38 Solve Prob. 7.16, using Mohr's circle.

7.39 Solve Prob. 7.17, using Mohr's circle.

7.40 Solve Prob. 7.18, using Mohr's circle.

7.41 Solve Prob. 7.19, using Mohr's circle.

7.42 Solve Prob. 7.20, using Mohr's circle.

7.43 Solve Prob. 7.21, using Mohr's circle.

7.44 Solve Prob. 7.22, using Mohr's circle.

7.45 Solve Prob. 7.23, using Mohr's circle.

7.46 Solve Prob. 7.24, using Mohr's circle.

7.47 Solve Prob. 7.25, using Mohr's circle.

7.48 Solve Prob. 7.26, using Mohr's circle.

7.49 Solve Prob. 7.27, using Mohr's circle.

7.50 Solve Prob. 7.28, using Mohr's circle.

7.51 Solve Prob. 7.29, using Mohr's circle.

7.52 Solve Prob. 7.30, using Mohr's circle.

7.53 Solve Prob. 7.30, using Mohr's circle and assuming that the weld forms an angle of 60° with the horizontal.

7.54 and 7.55 Determine the principal planes and the principal stresses for the state of plane stress resulting from the superposition of the two states of stress shown.

Fig. P7.54

Fig. P7.55

7.56 and 7.57 Determine the principal planes and the principal stresses for the state of plane stress resulting from the superposition of the two states of stress shown.

Fig. *P7.56*

Fig. *P7.57*

7.58 For the state of stress shown, determine the range of values of θ for which the normal stress $\sigma_{x'}$ is equal to or less than 100 MPa.

Fig. P7.58 and P7.59

7.59 For the state of stress shown, determine the range of values of θ for which the normal stress $\sigma_{x'}$ is equal to or less than 50 MPa.

7.60 For the state of stress shown, determine the range of values of θ for which the magnitude of the shearing stress $\tau_{x'y'}$ is equal to or less than 8 ksi.

Fig. P7.60

7.61 For the element shown, determine the range of values of τ_{xy} for which the maximum tensile stress is equal to or less than 60 MPa.

Fig. P7.61 and P7.62

7.62 For the element shown, determine the range of values of τ_{xy} for which the maximum in plane shearing stress is equal to or less than 150 MPa.

7.63 For the state of stress shown it is known that the normal and shearing stresses are directed as shown and that $\sigma_x = 14$ ksi, $\sigma_y = 9$ ksi, and $\sigma_{min} = 5$ ksi. Determine (a) the orientation of the principal planes, (b) the principal stress σ_{max}, (c) the maximum inplane shearing stress.

Fig. P7.63

Fig. P7.64

7.64 The Mohr circle shown corresponds to the state of stress given in Fig. 7.5a and b. Noting that $\sigma_{x'} = OC + (CX') \cos (2\theta_p - 2\theta)$ and that $\tau_{x'y'} = (CX') \sin (2\theta_p - 2\theta)$, derive the expressions for $\sigma_{x'}$ and $\tau_{x'y'}$ given in Eqs. (7.5) and (7.6), respectively. [Hint: Use $\sin(A + B) = \sin A \cos B + \cos A \sin B$ and $\cos (A + B) = \cos A \cos B - \sin A \sin B$.]

7.65 (a) Prove that the expression $\sigma_{x'}\sigma_{y'} - \tau_{x'y'}^2$, where $\sigma_{x'}$, $\sigma_{y'}$, and $\tau_{x'y'}$ are components of stress along the rectangular axes x' and y', is independent of the orientation of these axes. Also, show that the given expression represents the square of the tangent drawn from the origin of the coordinates to Mohr's circle. (b) Using the invariance property established in part a, express the shearing stress τ_{xy} in terms of σ_x, σ_y, and the principal stresses σ_{max} and σ_{min}.

7.5. GENERAL STATE OF STRESS

In the preceding sections, we have assumed a state of plane stress with $\sigma_z = \tau_{zx} = \tau_{zy} = 0$, and have considered only transformations of stress associated with a rotation about the z axis. We will now consider the general state of stress represented in Fig. 7.1a and the transformation of stress associated with the rotation of axes shown in Fig. 7.1b. However, our analysis will be limited to the determination of the *normal stress* σ_n on a plane of arbitrary orientation.

Consider the tetrahedron shown in Fig. 7.25. Three of its faces are parallel to the coordinate planes, while its fourth face, *ABC*, is perpendicular to the line *QN*. Denoting by ΔA the area of face *ABC*, and by λ_x, λ_y, λ_z the direction cosines of line *QN*, we find that the areas of the faces perpendicular to the x, y, and z axes are, respectively, $(\Delta A)\lambda_x$, $(\Delta A)\lambda_y$, and $(\Delta A)\lambda_z$. If the state of stress at point Q is defined by the stress components σ_x, σ_y, σ_z, τ_{xy}, τ_{yz} and τ_{zx}, then the *forces* exerted on the faces parallel to the coordinate planes can be obtained by multiplying the appropriate stress components by the area of each face (Fig. 7.26). On the other hand, the forces exerted on face *ABC* consist of a normal force of magnitude $\sigma_n \Delta A$ directed along *QN*, and of a shearing force of magnitude $\tau \Delta A$ perpendicular to *QN* but of otherwise unknown direction. Note that, since *QBC*, *QCA*, and *QAB* respectively face the negative x, y, and z axes, the forces exerted on them must be shown with negative senses.

Fig. 7.25

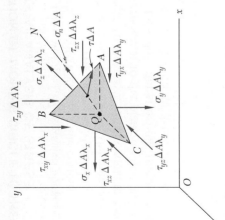

Fig. 7.26

We now express that the sum of the components along QN of all the forces acting on the tetrahedron is zero. Observing that the component along QN of a force parallel to the x axis is obtained by multiplying the magnitude of that force by the direction cosine λ_x, and that the components of forces parallel to the y and z axes are obtained in a similar way, we write

$$\sum F_n = 0: \qquad \sigma_n\, \Delta A - (\sigma_x\, \Delta A\, \lambda_x)\lambda_x - (\tau_{xy}\, \Delta A\, \lambda_x)\lambda_y - (\tau_{xz}\, \Delta A\, \lambda_x)\lambda_z$$
$$-(\tau_{yx}\, \Delta A\, \lambda_y)\lambda_x - (\sigma_y\, \Delta A\, \lambda_y)\lambda_y - (\tau_{yz}\, \Delta A\, \lambda_y)\lambda_z$$
$$-(\tau_{zx}\, \Delta A\, \lambda_z)\lambda_x - (\tau_{zy}\, \Delta A\, \lambda_z)\lambda_y - (\sigma_z\, \Delta A\, \lambda_z)\lambda_z = 0$$

Dividing through by ΔA and solving for σ_n, we have

$$\sigma_n = \sigma_x\lambda_x^2 + \sigma_y\lambda_y^2 + \sigma_z\lambda_z^2 + 2\tau_{xy}\lambda_x\lambda_y + 2\tau_{yz}\lambda_y\lambda_z + 2\tau_{zx}\lambda_z\lambda_x \qquad (7.20)$$

We note that the expression obtained for the normal stress σ_n is a *quadratic form* in λ_x, λ_y, and λ_z. It follows that we can select the coordinate axes in such a way that the right-hand member of Eq. (7.20) reduces to the three terms containing the squares of the direction cosines[†]. Denoting these axes by a, b, and c, the corresponding normal stresses by σ_a, σ_b, and σ_c, and the direction cosines of QN with respect to these axes by λ_a, λ_b, and λ_c, we write

$$\sigma_n = \sigma_a\lambda_a^2 + \sigma_b\lambda_b^2 + \sigma_c\lambda_c^2 \qquad (7.21)$$

The coordinate axes a, b, c are referred to as the *principal axes of stress*. Since their orientation depends upon the state of stress at Q, and thus upon the position of Q, they have been represented in Fig. 7.27 as attached to Q. The corresponding coordinate planes are known as the *principal planes of stress*, and the corresponding normal stresses σ_a, σ_b, and σ_c as the *principal stresses at* Q.[‡]

Fig. 7.27

[†] In Sec. 9.16 of F. P. Beer and E. R. Johnston, *Vector Mechanics for Engineers*, 6th ed., McGraw-Hill Book Company, 1988, a similar quadratic form is found to represent the moment of inertia of a rigid body with respect to an arbitrary axis. It is shown in Sec. 9.17 that this form is associated with a *quadric surface*, and that reducing the quadratic form to terms containing only the squares of the direction cosines is equivalent to determining the principal axes of that surface.

[‡] For a discussion of the determination of the principal planes of stress and of the principal stresses, see S. P. Timoshenko and J. N. Goodier, Theory of Elasticity, 3d ed., McGraw-Hill Book Company, 1970, sec. 77.

7.6. APPLICATION OF MOHR'S CIRCLE TO THE THREE-DIMENSIONAL ANALYSIS OF STRESS

If the element shown in Fig. 7.27 is rotated about one of the principal axes at Q, say the c axis (Fig. 7.28), the corresponding transformation of stress can be analyzed by means of Mohr's circle as if it were a transformation of plane stress. Indeed, the shearing stresses exerted on the faces perpendicular to the c axis remain equal to zero, and the normal stress σ_c is perpendicular to the plane ab in which the transformation takes place and, thus, does not affect this transformation. We therefore use the circle of diameter AB to determine the normal and shearing stresses exerted on the faces of the element as it is rotated about the c axis (Fig. 7.29). Similarly, circles of diameter BC and CA can be used to determine the stresses on the element as it is rotated about the a and b axes, respectively. While our analysis will be limited to rotations about

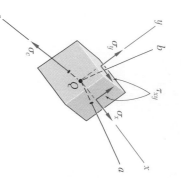

Fig. 7.28

the principal axes, it could be shown that any other transformation of axes would lead to stresses represented in Fig. 7.29 by a point located within the shaded area. Thus, the radius of the largest of the three circles yields the maximum value of the shearing stress at point Q. Noting that the diameter of that circle is equal to the difference between σ_{max} and σ_{min}, we write

$$\tau_{max} = \tfrac{1}{2}|\sigma_{max} - \sigma_{min}| \qquad (7.22)$$

where σ_{max} and σ_{min} represent the *algebraic* values of the maximum and minimum stresses at point Q.

Let us now return to the particular case of *plane stress*, which was discussed in Secs. 7.2 through 7.4. We recall that, if the x and y axes are selected in the plane of stress, we have $\sigma_z = \tau_{zx} = \tau_{zy} = 0$. This means that the z axis, i.e., the axis perpendicular to the plane of stress, is one of the three principal axes of stress. In a Mohr-circle diagram, this axis corresponds to the origin O, where $\sigma = \tau = 0$. We also recall that the other two principal axes correspond to points A and B where Mohr's circle for the xy plane intersects the σ axis. If A and B are located on opposite sides of the origin O (Fig. 7.30), the corresponding principal stresses represent

Fig. 7.29

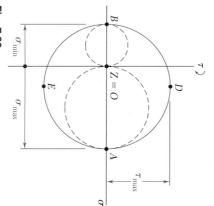

Fig. 7.30

the maximum and minimum normal stresses at point Q, and the maximum shearing stress is equal to the maximum "in-plane" shearing stress. As noted in Sec. 7.3, the planes of maximum shearing stress correspond to points D and E of Mohr's circle and arc at $45°$ to the principal planes corresponding to points A and B. They are, therefore, the shaded diagonal planes shown in Figs. 7.31a and b.

Fig. 7.31

(a) (b)

Fig. 7.32

If, on the other hand, A and B are on the same side of O, that is, if σ_a and σ_b have the same sign, then the circle defining σ_{max}, σ_{min}, and τ_{max} is *not* the circle corresponding to a transformation of stress within the xy plane. If $\sigma_a > \sigma_b > 0$, as assumed in Fig. 7.32, we have $\sigma_{max} = \sigma_a$, $\sigma_{min} = 0$, and τ_{max} is equal to the radius of the circle defined by points O and A, that is, $\tau_{max} = \frac{1}{2}\sigma_{max}$. We also note that the normals Qd' and Qe' to the planes of maximum shearing stress are obtained by rotating the axis Qa through $45°$ within the za plane. Thus, the planes of maximum shearing stress are the shaded diagonal planes shown in Figs. 7.33a and b.

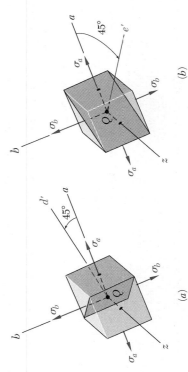

Fig. 7.33

(a) (b)

EXAMPLE 7.03

For the state of plane stress shown in Fig. 7.34, determine (a) the three principal planes and principal stresses, (b) the maximum shearing stress.

Fig. 7.34

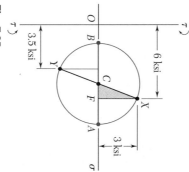

Fig. 7.35

(a) Principal Planes and Principal Stresses. We construct Mohr's circle for the transformation of stress in the xy plane (Fig. 7.35). Point X is plotted 6 units to the right of the τ axis and 3 units above the σ axis (since the corresponding shearing stress tends to rotate the element clockwise).

Point Y is plotted 3.5 units to the right of the τ axis and 3 units below the σ axis. Drawing the line XY, we obtain the center C of Mohr's circle for the xy plane; its abscissa is

$$\sigma_{ave} = \frac{\sigma_x + \sigma_y}{2} = \frac{6 + 3.5}{2} = 4.75 \text{ ksi}$$

Since the sides of the right triangle CFX are CF = 6 − 4.75 = 1.25 ksi and FX = 3 ksi, the radius of the circle is

$$R = CX = \sqrt{(1.25)^2 + (3)^2} = 3.25 \text{ ksi}$$

The principal stresses in the plane of stress are

$$\sigma_a = OA = OC + CA = 4.75 + 3.25 = 8.00 \text{ ksi}$$
$$\sigma_b = OB = OC - BC = 4.75 - 3.25 = 1.50 \text{ ksi}$$

Since the faces of the element that are perpendicular to the z axis are free of stress, these faces define one of the principal planes, and the corresponding principal stress is $\sigma_z = 0$. The other two principal planes are defined by points A and B on Mohr's circle. The angle θ_p through which the element should be rotated about the z axis to bring its faces to coincide with these planes (Fig. 7.36) is half the angle ACX. We have

$$\tan 2\theta_p = \frac{FX}{CF} = \frac{3}{1.25}$$
$$2\theta_p = 67.4° \curvearrowleft \qquad \theta_p = 33.7° \curvearrowleft$$

Fig. 7.36

(b) Maximum Shearing Stress. We now draw the circles of diameter OB and OA, which correspond respectively to rotations of the element about the a and b axes (Fig. 7.37). We note that the maximum shearing stress is equal to the radius of the circle of diameter OA. We thus have

$$\tau_{max} = \tfrac{1}{2}\sigma_a = \tfrac{1}{2}(8.00 \text{ ksi}) = 4.00 \text{ ksi}$$

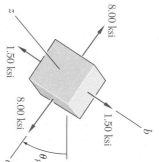

Fig. 7.37

Since points D′ and E′, which define the planes of maximum shearing stress, are located at the ends of the vertical diameter of the circle corresponding to a rotation about the b axis, the faces of the element of Fig. 7.36 can be brought to coincide with the planes of maximum shearing stress through a rotation of 45° about the b axis.

*7.7. YIELD CRITERIA FOR DUCTILE MATERIALS UNDER PLANE STRESS

Structural elements and machine components made of a ductile material are usually designed so that the material will not yield under the expected loading conditions. When the element or component is under uniaxial stress (Fig. 7.38), the value of the normal stress σ_x which will cause the material to yield can be obtained readily from a tensile test conducted

Fig. 7.38

on a specimen of the same material, since the test specimen and the structural element or machine component are in the same state of stress. Thus, regardless of the actual mechanism that causes the material to yield, we can state that the element or component will be safe as long as $\sigma_x < \sigma_Y$, where σ_Y is the yield strength of the test specimen.

On the other hand, when a structural element or machine component is in a state of plane stress (Fig. 7.39a), it is found convenient to use one of the methods developed earlier to determine the principal stresses σ_a and σ_b at any given point (Fig. 7.39b). The material can then be regarded as being in a state of biaxial stress at that point. Since this state is different from the state of uniaxial stress found in a specimen subjected to a tensile test, it is clearly not possible to predict directly from such a test whether or not the structural element or machine component under investigation will fail. Some criterion regarding the actual mechanism of failure of the material must first be established, which will make it possible to compare the effects of both states of stress on the material. The purpose of this section is to present the two yield criteria most frequently used for ductile materials.

Maximum-Shearing-Stress Criterion. This criterion is based on the observation that yield in ductile materials is caused by slippage of the material along oblique surfaces and is due primarily to shearing stresses (cf. Sec. 2.3). According to this criterion, a given structural component is safe as long as the maximum value τ_{max} of the shearing stress in that component remains smaller than the corresponding value of the shearing stress in a tensile-test specimen of the same material as the specimen starts to yield.

Recalling from Sec. 1.11 that the maximum value of the shearing stress under a centric axial load is equal to half the value of the corresponding normal, axial stress, we conclude that the maximum shearing stress in a tensile-test specimen is $\frac{1}{2}\sigma_Y$ as the specimen starts to yield. On the other hand, we saw in Sec. 7.6 that, for plane stress, the maximum value τ_{max} of the shearing stress is equal to $\frac{1}{2}|\sigma_{max}|$ if the principal stresses are either both positive or both negative, and to $\frac{1}{2}|\sigma_{max} - \sigma_{min}|$ if the maximum stress is positive and the minimum stress

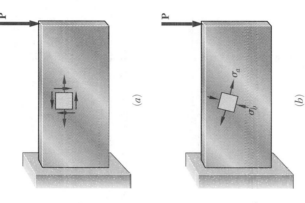

(a)

(b)

Fig. 7.39

negative. Thus, if the principal stresses σ_a and σ_b have the same sign, the maximum-shearing-stress criterion gives

$$|\sigma_a| < \sigma_Y \qquad |\sigma_b| < \sigma_Y \tag{7.23}$$

If the principal stresses σ_a and σ_b have opposite signs, the maximum-shearing-stress criterion yields

$$|\sigma_a - \sigma_b| < \sigma_Y \tag{7.24}$$

The relations obtained have been represented graphically in Fig. 7.40. Any given state of stress will be represented in that figure by a point of coordinates σ_a and σ_b, where σ_a and σ_b are the two principal stresses. If this point falls within the area shown in the figure, the structural component is safe. If it falls outside this area, the component will fail as a result of yield in the material. The hexagon associated with the initiation of yield in the material is known as *Tresca's hexagon* after the French engineer Henri Edouard Tresca (1814–1885).

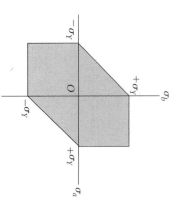

Fig. 7.40

Maximum-Distortion-Energy Criterion. This criterion is based on the determination of the distortion energy in a given material, i.e., of the energy associated with changes in shape in that material (as opposed to the energy associated with changes in volume in the same material). According to this criterion, also known as the *von Mises criterion*, after the German-American applied mathematician Richard von Mises (1883–1953), a given structural component is safe as long as the maximum value of the distortion energy per unit volume in that material remains smaller than the distortion energy per unit volume required to cause yield in a tensile-test specimen of the same material. As you will see in Sec. 11.6, the distortion energy per unit volume in an isotropic material under plane stress is

$$u_d = \frac{1}{6G}(\sigma_a^2 - \sigma_a\sigma_b + \sigma_b^2) \tag{7.25}$$

where σ_a and σ_b are the principal stresses and G the modulus of rigidity. In the particular case of a tensile-test specimen which is starting to yield, we have $\sigma_a = \sigma_Y$, $\sigma_b = 0$, and $(u_d)_Y = \sigma_Y^2/6G$. Thus, the maximum-distortion-energy criterion indicates that the structural component is safe as long as $u_d < (u_d)_Y$, or

$$\sigma_a^2 - \sigma_a\sigma_b + \sigma_b^2 < \sigma_Y^2 \tag{7.26}$$

i.e., as long as the point of coordinates σ_a and σ_b falls within the area shown in Fig. 7.41. This area is bounded by the ellipse of equation

$$\sigma_a^2 - \sigma_a\sigma_b + \sigma_b^2 = \sigma_Y^2 \tag{7.27}$$

which intersects the coordinate axes at $\sigma_a = \pm\sigma_Y$ and $\sigma_b = \pm\sigma_Y$. We can verify that the major axis of the ellipse bisects the first and third quadrants and extends from $A(\sigma_a = \sigma_b = \sigma_Y)$ to $B(\sigma_a = \sigma_b = -\sigma_Y)$, while its minor axis extends from $C(\sigma_a = -\sigma_b = -0.577\sigma_Y)$ to $D(\sigma_a = -\sigma_b = 0.577\sigma_Y)$.

The maximum-shearing-stress criterion and the maximum-distortion-energy criterion are compared in Fig. 7.42. We note that the ellipse passes through the vertices of the hexagon. Thus, for the states of stress represented by these six points, the two criteria give the same results.

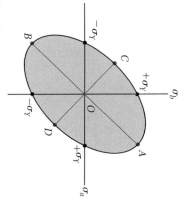

Fig. 7.41

For any other state of stress, the maximum-shearing-stress criterion is more conservative than the maximum-distortion-energy criterion, since the hexagon is located within the ellipse.

A state of stress of particular interest is that associated with yield in a torsion test. We recall from Fig. 7.24 of Sec. 7.4 that, for torsion, $\sigma_{min} = -\sigma_{max}$; thus, the corresponding points in Fig. 7.42 are located on the bisector of the second and fourth quadrants. It follows that yield occurs in a torsion test when $\sigma_a = -\sigma_b = \pm 0.5\sigma_Y$ according to the maximum-shearing-stress criterion, and when $\sigma_a = -\sigma_b = \pm 0.577\sigma_Y$ according to the maximum-distortion-energy criterion. But, recalling again Fig. 7.24, we note that σ_a and σ_b must be equal in magnitude to τ_{max}, that is, to the value obtained from a torsion test for the yield strength τ_Y of the material. Since the values of the yield strength σ_Y in tension and of the yield strength τ_Y in shear are given for various ductile materials in Appendix B, we can compute the ratio τ_Y/σ_Y for these materials and verify that the values obtained range from 0.55 to 0.60. Thus, the maximum-distortion-energy criterion appears somewhat more accurate than the maximum-shearing-stress criterion as far as predicting yield in torsion is concerned.

*7.8. FRACTURE CRITERIA FOR BRITTLE MATERIALS UNDER PLANE STRESS

As we saw in Chap. 2, brittle materials are characterized by the fact that, when subjected to a tensile test, they fail suddenly through rupture—or fracture—without any prior yielding. When a structural element or machine component made of a brittle material is under uniaxial tensile stress, the value of the normal stress that causes it to fail is equal to the ultimate strength σ_U of the material as determined from a tensile test, since both the tensile-test specimen and the element or component under investigation are in the same state of stress. However, when a structural element or machine component is in a state of plane stress, it is found convenient to first determine the principal stresses σ_a and σ_b at any given point, and to use one of the criteria indicated in this section to predict whether or not the structural element or machine component will fail.

Maximum-Normal-Stress Criterion. According to this criterion, a given structural component fails when the maximum normal stress in that component reaches the ultimate strength σ_U obtained from the tensile test of a specimen of the same material. Thus, the structural component will be safe as long as the absolute values of the principal stresses σ_a and σ_b are both less than σ_U:

$$|\sigma_a| < \sigma_U \qquad |\sigma_b| < \sigma_U \qquad (7.28)$$

The maximum-normal-stress criterion can be expressed graphically as shown in Fig. 7.43. If the point obtained by plotting the values σ_a and σ_b of the principal stresses falls within the square area shown in the figure, the structural component is safe. If it falls outside that area, the component will fail.

The maximum-normal-stress criterion, also known as *Coulomb's criterion*, after the French physicist Charles Augustin de Coulomb (1736–1806), suffers from an important shortcoming, since it is based

Fig. 7.42

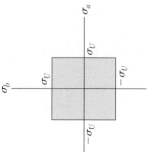

Fig. 7.43

on the assumption that the ultimate strength of the material is the same in tension and in compression. As we noted in Sec. 2.3, this is seldom the case, because of the presence of flaws in the material, such as microscopic cracks or cavities, which tend to weaken the material in tension, while not appreciably affecting its resistance to compressive failure. Besides, this criterion makes no allowance for effects other than those of the normal stresses on the failure mechanism of the material.[†]

Mohr's Criterion. This criterion, suggested by the German engineer Otto Mohr, can be used to predict the effect of a given state of plane stress on a brittle material, when results of various types of tests are available for that material.

Let us first assume that a tensile test and a compressive test have been conducted on a given material, and that the values σ_{UT} and σ_{UC} of the ultimate strength in tension and in compression have been determined for that material. The state of stress corresponding to the rupture of the tensile-test specimen can be represented on a Mohr-circle diagram by the circle intersecting the horizontal axis at O and σ_{UT} (Fig. 7.45a). Similarly, the state of stress corresponding to the failure of the compressive-test specimen can be represented by the circle intersecting the horizontal axis at O and σ_{UC}. Clearly, a state of stress represented by a circle entirely contained in either of these circles will be safe. Thus, if both principal stresses are positive, the state of stress is safe as long as $\sigma_a < \sigma_{UT}$ and $\sigma_b < \sigma_{UT}$; if both principal stresses are negative, the state of stress is safe as long as $|\sigma_a| < |\sigma_{UC}|$ and $|\sigma_b| < |\sigma_{UC}|$. Plotting the point of coordinates σ_a and σ_b (Fig. 7.45b), we verify that the state of stress is safe as long as that point falls within one of the square areas shown in that figure.

In order to analyze the cases when σ_a and σ_b have opposite signs, we now assume that a torsion test has been conducted on the material and that its ultimate strength in shear, τ_U, has been determined. Drawing the circle centered at O representing the state of stress corresponding to the failure of the torsion-test specimen (Fig. 7.46a), we observe that any state of stress represented by a circle entirely contained in that circle is also safe. Mohr's criterion is a logical extension of this obser-

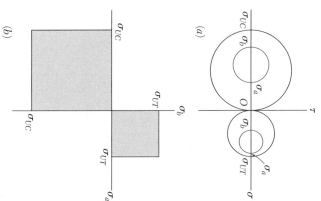

(a)

(b)

Fig. 7.45

Fig. 7.44

[†]Another failure criterion known as the *maximum-normal-strain criterion*, or *Saint-Venant's criterion*, was widely used during the nineteenth century. According to this criterion, a given structural component is safe as long as the maximum value of the normal strain in that component remains smaller than the value ϵ_U of the strain at which a tensile-test specimen of the same material will fail. But, as will be shown in Sec. 7.12, the strain is maximum along one of the principal axes of stress, if the deformation is elastic and the material homogeneous and isotropic. Thus, denoting by ϵ_a and ϵ_b the values of the normal strain along the principal axes in the plane of stress, we write

$$|\epsilon_a| < \epsilon_U \qquad |\epsilon_b| < \epsilon_U \qquad (7.29)$$

Making use of the generalized Hooke's law (Sec. 2.12), we could express these relations in terms of the principal stresses σ_a and σ_b and the ultimate strength σ_U of the material. We would find that, according to the maximum-normal-strain criterion, the structural component is safe as long as the point obtained by plotting σ_a and σ_b falls within the area shown in Fig. 7.44 where ν is Poisson's ratio for the given material.

Fig. 7.47

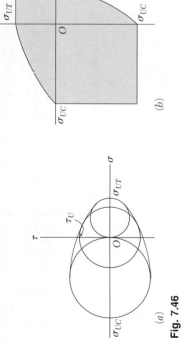

Fig. 7.46

vation: According to Mohr's criterion, a state of stress is safe if it is represented by a circle located entirely within the area bounded by the envelope of the circles corresponding to the available data. The remaining portions of the principal-stress diagram can now be obtained by drawing various circles tangent to this envelope, determining the corresponding values of σ_a and σ_b, and plotting the points of coordinates σ_a and σ_b (Fig. 7.46b).

More accurate diagrams can be drawn when additional test results, corresponding to various states of stress, are available. If, on the other hand, the only available data consists of the ultimate strengths σ_{UT} and σ_{UC}, the envelope in Fig. 7.46a is replaced by the tangents AB and $A'B'$ to the circles corresponding respectively to failure in tension and failure in compression (Fig. 7.47a). From the similar triangles drawn in that figure, we note that the abscissa of the center C of a circle tangent to AB and $A'B'$ is linearly related to its radius R. Since $\sigma_a = OC + R$ and $\sigma_b = OC - R$, it follows that σ_a and σ_b are also linearly related. Thus, the shaded area corresponding to this simplified Mohr's criterion is bounded by straight lines in the second and fourth quadrants (Fig. 7.47b).

Note that in order to determine whether a structural component will be safe under a given loading, the state of stress should be calculated at all critical points of the component, i.e., at all points where stress concentrations are likely to occur. This can be done in a number of cases by using the stress-concentration factors given in Figs. 2.64, 3.32, 4.31, and 4.32. There are many instances, however, when the theory of elasticity must be used to determine the state of stress at a critical point.

Special care should be taken when *macroscopic cracks* have been detected in a structural component. While it can be assumed that the test specimen used to determine the ultimate tensile strength of the material contained the same type of flaws (i.e., *microscopic* cracks or cavities) as the structural component under investigation, the specimen was certainly free of any detectable macroscopic cracks. When a crack is detected in a structural component, it is necessary to determine whether that crack will tend to propagate under the expected loading condition and cause the component to fail, or whether it will remain stable. This requires an analysis involving the energy associated with the growth of the crack. Such an analysis is beyond the scope of this text and should be carried out by the methods of fracture mechanics.

SAMPLE PROBLEM 7.4

The state of plane stress shown occurs at a critical point of a steel machine component. As a result of several tensile tests, it has been found that the tensile yield strength is $\sigma_Y = 250$ MPa for the grade of steel used. Determine the factor of safety with respect to yield, using (a) the maximum-shearing-stress criterion, and (b) the maximum-distortion-energy criterion.

SOLUTION

Mohr's Circle. We construct Mohr's circle for the given state of stress and find

$$\sigma_{ave} = OC = \tfrac{1}{2}(\sigma_x + \sigma_y) = \tfrac{1}{2}(80 - 40) = 20 \text{ MPa}$$

$$\tau_m = R = \sqrt{(CF)^2 + (FX)^2} = \sqrt{(60)^2 + (25)^2} = 65 \text{ MPa}$$

Principal Stresses

$$\sigma_a = OC + CA = 20 + 65 = +85 \text{ MPa}$$
$$\sigma_b = OC - BC = 20 - 65 = -45 \text{ MPa}$$

a. Maximum-Shearing-Stress Criterion. Since for the grade of steel used the tensile strength is $\sigma_Y = 250$ MPa, the corresponding shearing stress at yield is

$$\tau_Y = \tfrac{1}{2}\sigma_Y = \tfrac{1}{2}(250 \text{ MPa}) = 125 \text{ MPa}$$

For $\tau_m = 65$ MPa: $F.S. = \dfrac{\tau_Y}{\tau_m} = \dfrac{125 \text{ MPa}}{65 \text{ MPa}}$ $F.S. = 1.92$ ◀

b. Maximum-Distortion-Energy Criterion. Introducing a factor of safety into Eq. (7.26), we write

$$\sigma_a^2 - \sigma_a\sigma_b + \sigma_b^2 = \left(\frac{\sigma_Y}{F.S.}\right)^2$$

For $\sigma_a = +85$ MPa, $\sigma_b = -45$ MPa, and $\sigma_Y = 250$ MPa, we have

$$(85)^2 - (85)(-45) + (45)^2 = \left(\frac{250}{F.S.}\right)^2$$

$$114.3 = \frac{250}{F.S.}$$ $F.S. = 2.19$ ◀

Comment. For a ductile material with $\sigma_Y = 250$ MPa, we have drawn the hexagon associated with the maximum-shearing-stress criterion and the ellipse associated with the maximum-distortion-energy criterion. The given state of plane stress is represented by point H of coordinates $\sigma_a = 85$ MPa and $\sigma_b = -45$ MPa. We note that the straight line drawn through points O and H intersects the hexagon at point T and the ellipse at point M. For each criterion, the value obtained for F.S. can be verified by measuring the line segments indicated and computing their ratios:

$$(a)\ F.S. = \frac{OT}{OH} = 1.92 \qquad (b)\ F.S. = \frac{OM}{OH} = 2.19$$

PROBLEMS

7.66 For the state of plane stress shown, determine the maximum shearing stress when (*a*) $\sigma_y = 20$ MPa, (*b*) $\sigma_y = 140$ MPa. (*Hint:* Consider both in-plane and out-of-plane shearing stresses.)

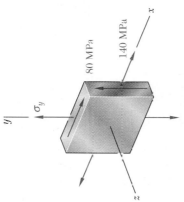

Fig. P7.66 and P7.67

7.67 For the state of plane stress shown, determine the maximum shearing stress when (*a*) $\sigma_y = 40$ MPa, (*b*) $\sigma_y = 120$ MPa. (*Hint:* Consider both in-plane and out-of-plane shearing stresses.)

7.68 For the state of plane stress shown, determine the maximum shearing stress when (*a*) $\sigma_x = 6$ ksi and $\sigma_y = 18$ ksi, (*b*) $\sigma_x = 14$ ksi and $\sigma_y = 2$ ksi. (*Hint:* Consider both in-plane and out-of-plane shearing stresses.)

7.69 For the state of plane stress shown, determine the maximum shearing stress when (*a*) $\sigma_x = 0$ and $\sigma_y = 12$ ksi, (*b*) $\sigma_x = 21$ ksi and $\sigma_y = 9$ ksi. (*Hint:* Consider both in-plane and out-of-plane shearing stresses.)

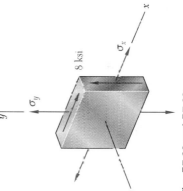

Fig. P7.68 and P7.69

7.70 and 7.71 For the state of stress shown, determine the maximum shearing stress when (*a*) $\sigma_z = 0$, (*b*) $\sigma_z = +45$ MPa, (*c*) $\sigma_z = -45$ MPa.

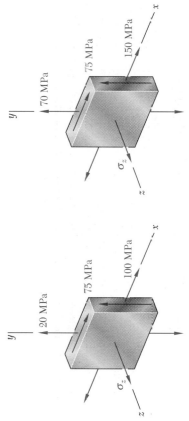

Fig. P7.70

Fig. P7.71

457

7.72 and 7.73 For the state of stress shown, determine the maximum shearing stress when (a) $\sigma_z = +4$ ksi, (b) $\sigma_z = -4$ ksi, (c) $\sigma_z = 0$.

Fig. P7.72

Fig. P7.73

7.74 For the state of stress shown, determine two values of σ_y for which the maximum shearing stress is 75 MPa.

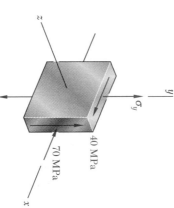

Fig. P7.74

7.75 For the state of stress shown, determine two values of σ_y for which the maximum shearing stress is 7.5 ksi.

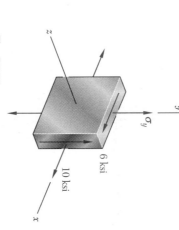

Fig. P7.75

7.76 For the state of stress shown, determine the value of τ_{xy} for which the maximum shearing stress is 80 MPa.

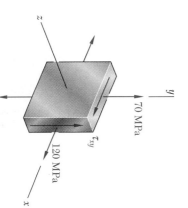

Fig. P7.76

7.77 For the state of stress shown, determine the value of τ_{xy} for which the maximum shearing stress is (a) 9 ksi, (b) 12 ksi.

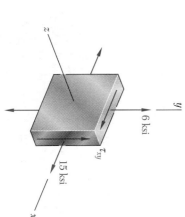

Fig. P7.77

7.78 For the state of stress shown, determine two values of σ_y for which the maximum shearing stress is 80 MPa.

Fig. P7.78

7.79 For the state of stress shown, determine the range of values of τ_{xy} for which the maximum shearing stress is equal to or less than 60 MPa.

Fig. P7.79

*7.80 For the state of stress of Prob. 7.66, determine (a) the value of σ_y for which the maximum shearing stress is as small as possible, (b) the corresponding value of the shearing stress.

7.81 The state of plane stress shown occurs in a machine component made of a steel with $\sigma_Y = 45$ ksi. Using the maximum-distortion-energy criterion, determine whether yield occurs when (a) $\tau_{xy} = 9$ ksi, (b) $\tau_{xy} = 18$ ksi, (c) $\tau_{xy} = 20$ ksi. If yield does not occur, determine the corresponding factor of safety.

Fig. P7.81

7.82 Solve Prob. 7.81, using the maximum-shearing-stress criterion.

7.83 The state of plane stress shown occurs in a machine component made of a steel with $\sigma_Y = 325$ MPa. Using the maximum-shearing-stress criterion, determine whether yield occurs when (a) $\sigma_0 = 200$ MPa, (b) $\sigma_0 = 240$ MPa, (c) $\sigma_0 = 280$ MPa. If yield does not occur, determine the corresponding factor of safety.

Fig. P7.83

7.84 Solve Prob. 7.83, using the maximum-distortion-energy criterion.

7.85 The 38-mm-diameter shaft AB is made of a grade of steel for which the yield strength is $\sigma_Y = 250$ MPa. Using the maximum-shearing-stress criterion, determine the magnitude of the torque **T** for which yield occurs when $P = 240$ kN.

Fig. P7.85

7.86 Solve Prob. 7.85, using the maximum-distortion-energy criterion.

7.87 The 1.5-in.-diameter shaft AB is made of a grade of steel for which the yield strength is $\sigma_Y = 42$ ksi. Using the maximum-shearing-stress criterion, determine the magnitude of the torque $\mathbf{T}$ for which yield occurs when $P = 60$ kips.

Fig. P7.87

7.88 Solve Prob. 7.87, using the maximum-distortion-energy criterion.

7.89 and 7.90 The state of plane stress shown is expected to occur in a cast-iron machine base. Knowing that for the grade of cast iron used $\sigma_{UT} = 160$ MPa and $\sigma_{UC} = 320$ MPa and using Mohr's criterion, determine whether rupture of the component will occur.

Fig. P7.89

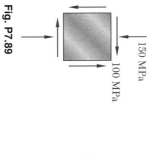

Fig. P7.90

7.91 and 7.92 The state of plane stress shown is expected to occur in an aluminum casting. Knowing that for the aluminum alloy used $\sigma_{UT} = 10$ ksi and $\sigma_{UC} = 30$ ksi and using Mohr's criterion, determine whether rupture of the component will occur.

Fig. P7.91

Fig. P7.92

7.93 The state of plane stress shown will occur at a critical point in a cast pipe made of an aluminum alloy for which $\sigma_{UT} = 75$ MPa and $\sigma_{UC} = 150$ MPa. Using Mohr's criterion, determine the shearing stress τ_0 for which failure should be expected.

Fig. P7.93

7.94 The state of plane stress shown will occur in an aluminum casting that is made of an alloy for which $\sigma_{UT} = 10$ ksi and $\sigma_{UC} = 25$ ksi. Using Mohr's criterion, determine the shearing stress τ_0 for which failure should be expected.

8 ksi

τ_0

Fig. P7.94

7.95 The cast-aluminum rod shown is made of an alloy for which $\sigma_{UT} = 8$ ksi and $\sigma_{UC} = 16$ ksi. Using Mohr's criterion, determine the magnitude of the torque **T** for which rupture should be expected.

1.25 in.

T

6 kips

A

B

Fig. P7.95

7.96 The cast-aluminum rod shown is made of an alloy for which $\sigma_{UT} = 70$ MPa and $\sigma_{UC} = 175$ MPa. Knowing that the magnitude T of the applied torques is slowly increased and using Mohr's criterion, determine the shearing stress τ_0 that should be expected at rupture.

T′

τ_0

T

Fig. P7.96

7.97 A machine component is made of a grade of cast iron for which $\sigma_{UT} = 8$ ksi and $\sigma_{UC} = 20$ ksi. For each of the states of plane stress shown, and using Mohr's criterion, determine the normal stress σ_0 at which rupture of the component should be expected.

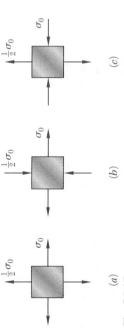

$\frac{1}{2}\sigma_0$ σ_0

$\frac{1}{2}\sigma_0$ σ_0

$\frac{1}{2}\sigma_0$ σ_0

(a)   (b)   (c)

Fig. P7.97

7.9. STRESSES IN THIN-WALLED PRESSURE VESSELS

Thin-walled pressure vessels provide an important application of the analysis of plane stress. Since their walls offer little resistance to bending, it can be assumed that the internal forces exerted on a given portion of wall are tangent to the surface of the vessel (Fig. 7.48). The resulting stresses on an element of wall will thus be contained in a plane tangent to the surface of the vessel.

Our analysis of stresses in thin-walled pressure vessels will be limited to the two types of vessels most frequently encountered: cylindrical pressure vessels and spherical pressure vessels (Figs. 7.49 and 7.50).

Fig. 7.48

Fig. 7.49

Fig. 7.50

Consider a cylindrical vessel of inner radius r and wall thickness t containing a fluid under pressure (Fig. 7.51). We propose to determine the stresses exerted on a small element of wall with sides respectively parallel and perpendicular to the axis of the cylinder. Because of the axisymmetry of the vessel and its contents, it is clear that no shearing stress is exerted on the element. The normal stresses σ_1 and σ_2 shown in Fig. 7.51 are therefore principal stresses. The stress σ_1 is known as the *hoop stress*, because it is the type of stress found in hoops used to hold together the various slats of a wooden barrel, and the stress σ_2 is called the *longitudinal stress*.

In order to determine the hoop stress σ_1, we detach a portion of the vessel and its contents bounded by the xy plane and by two planes parallel to the yz plane at a distance Δx from each other (Fig. 7.52). The forces parallel to the z axis acting on the free body defined in this fashion consist of the elementary internal forces $\sigma_1\, dA$ on the wall sections, and of the elementary pressure forces $p\, dA$ exerted on the portion of fluid included in the free body. Note that p denotes the *gage pressure* of the fluid, i.e., the excess of the inside pressure over the outside atmospheric pressure. The resultant of the internal forces $\sigma_1\, dA$ is equal to the product of σ_1 and of the cross-sectional area $2t\, \Delta x$ of the wall, while the resultant of the pressure forces $p\, dA$ is equal to the product of p and of the area $2r\, \Delta x$. Writing the equilibrium equation $\Sigma F_z = 0$, we have

Fig. 7.51

Fig. 7.52

$$\Sigma F_z = 0: \qquad \sigma_1(2t\,\Delta x) - p(2r\,\Delta x) = 0$$

and, solving for the hoop stress σ_1,

$$\sigma_1 = \frac{pr}{t} \tag{7.30}$$

To determine the longitudinal stress σ_2, we now pass a section perpendicular to the x axis and consider the free body consisting of the portion of the vessel and its contents located to the left of the section

ϕ be thin-walled :

$$\frac{r}{t} \geqslant 10$$

Fig. 7.53

(Fig. 7.53). The forces acting on this free body are the elementary internal forces $\sigma_2\,dA$ on the wall section and the elementary pressure forces $p\,dA$ exerted on the portion of fluid included in the free body. Noting that the area of the fluid section is πr^2 and that the area of the wall section can be obtained by multiplying the circumference $2\pi r$ of the cylinder by its wall thickness t, we write the equilibrium equation†.

$$\Sigma F_x = 0: \qquad \sigma_2(2\pi rt) - p(\pi r^2) = 0$$

and, solving for the longitudinal stress σ_2,

$$\sigma_2 = \frac{pr}{2t} \tag{7.31}$$

We note from Eqs. (7.30) and (7.31) that the hoop stress σ_1 is twice as large as the longitudinal stress σ_2:

$$\sigma_1 = 2\sigma_2 \tag{7.32}$$

---

†Using the mean radius of the wall section, $r_m = r + \frac{1}{2}t$, in computing the resultant of the forces on that section, we would obtain a more accurate value of the longitudinal stress, namely,

$$\sigma_2 = \frac{pr}{2t}\,\frac{1}{1 + \dfrac{t}{2r}} \tag{7.31'}$$

However, for a thin-walled pressure vessel, the term $t/2r$ is sufficiently small to allow the use of Eq. (7.31) for engineering design and analysis. If a pressure vessel is not thin-walled (i.e., if $t/2r$ is not small), the stresses σ_1 and σ_2 vary across the wall and must be determined by the methods of the theory of elasticity.

Drawing Mohr's circle through the points A and B which correspond respectively to the principal stresses σ_1 and σ_2 (Fig. 7.54), and recalling that the maximum in-plane shearing stress is equal to the radius of this circle, we have

Fig. 7.54

$$\tau_{max(in\ plane)} = \tfrac{1}{2}\sigma_2 = \frac{pr}{4t} \tag{7.33}$$

This stress corresponds to points D and E and is exerted on an element obtained by rotating the original element of Fig. 7.51 through 45° *within the plane* tangent to the surface of the vessel. The maximum shearing stress in the wall of the vessel, however, is larger. It is equal to the radius of the circle of diameter OA and corresponds to a rotation of 45° about a longitudinal axis and *out of the plane of stress.*† We have

$$\tau_{max} = \sigma_2 = \frac{pr}{2t} \tag{7.34}$$

We now consider a spherical vessel of inner radius r and wall thickness t, containing a fluid under a gage pressure p. For reasons of symmetry, the stresses exerted on the four faces of a small element of wall must be equal (Fig. 7.55). We have

Fig. 7.55

$$\sigma_1 = \sigma_2 \tag{7.35}$$

To determine the value of the stress, we pass a section through the center C of the vessel and consider the free body consisting of the portion of the vessel and its contents located to the left of the section (Fig. 7.56). The equation of equilibrium for this free body is the same as for the free body of Fig. 7.53. We thus conclude that, for a spherical vessel,

Fig. 7.56

$$\sigma_1 = \sigma_2 = \frac{pr}{2t} \tag{7.36}$$

Since the principal stresses σ_1 and σ_2 are equal, Mohr's circle for transformations of stress within the plane tangent to the surface of the vessel reduces to a point (Fig. 7.57); we conclude that the in-plane normal stress is constant and that the in-plane maximum shearing stress is zero. The maximum shearing stress in the wall of the vessel, however, is not zero; it is equal to the radius of the circle of diameter OA and corresponds to a rotation of 45° out of the plane of stress. We have

Fig. 7.57

$$\tau_{max} = \tfrac{1}{2}\sigma_1 = \frac{pr}{4t} \tag{7.37}$$

†It should be observed that, while the third principal stress is zero on the outer surface of the vessel, it is equal to $-p$ on the inner surface, and is represented by a point $C(-p, 0)$ on a Mohr-circle diagram. Thus, close to the inside surface of the vessel, the maximum shearing stress is equal to the radius of a circle of diameter CA, and we have

$$\tau_{max} = \tfrac{1}{2}(\sigma_1 + p) = \frac{pr}{2t}\left(1 + \frac{t}{r}\right)$$

For a thin-walled vessel, however, the term t/r is small, and we can neglect the variation of τ_{max} across the wall section. This remark also applies to spherical pressure vessels.

SAMPLE PROBLEM 7.5

A compressed-air tank is supported by two cradles as shown; one of the cradles is designed so that it does not exert any longitudinal force on the tank. The cylindrical body of the tank has a 30-in. outer diameter and is fabricated from a $\frac{3}{8}$-in. steel plate by butt welding along a helix that forms an angle of $25°$ with a transverse plane. The end caps are spherical and have a uniform wall thickness of $\frac{5}{16}$ in. For an internal gage pressure of 180 psi, determine (a) the normal stress and the maximum shearing stress in the spherical caps, (b) the stresses in directions perpendicular and parallel to the helical weld.

30 in.

8 ft

$25°$

SOLUTION

a. Spherical Cap.
Using Eq. (7.36), we write

$p = 180$ psi, $t = \frac{5}{16}$ in. $= 0.3125$ in., $r = 15 - 0.3125 = 14.688$ in.

$$\sigma_1 = \sigma_2 = \frac{pr}{2t} = \frac{(180 \text{ psi})(14.688 \text{ in.})}{2(0.3125 \text{ in.})} \qquad \sigma = 4230 \text{ psi} \blacktriangledown$$

We note that for stresses in a plane tangent to the cap, Mohr's circle reduces to a point (A, B) on the horizontal axis and that all in-plane shearing stresses are zero. On the surface of the cap the third principal stress is zero and corresponds to point O. On a Mohr's circle of diameter AO, point D' represents the maximum shearing stress; it occurs on planes at $45°$ to the plane tangent to the cap.

$$\tau_{max} = \frac{1}{2}(4230 \text{ psi}) \qquad \tau_{max} = 2115 \text{ psi} \blacktriangledown$$

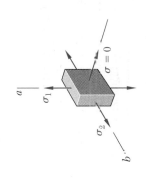

b. Cylindrical Body of the Tank.
We first determine the hoop stress σ_1 and the longitudinal stress σ_2. Using Eqs. (7.30) and (7.32), we write

$p = 180$ psi, $t = \frac{3}{8}$ in. $- 0.375$ in., $r = 15 \qquad 0.375 = 14.625$ in.

$$\sigma_1 = \frac{pr}{t} = \frac{(180 \text{ psi})(14.625 \text{ in.})}{0.375 \text{ in.}} = 7020 \text{ psi} \qquad \sigma_2 = \frac{1}{2}\sigma_1 = 3510 \text{ psi}$$

$$\sigma_{ave} = \frac{1}{2}(\sigma_1 + \sigma_2) = 5265 \text{ psi} \qquad R = \frac{1}{2}(\sigma_1 - \sigma_2) = 1755 \text{ Psi}$$

$\sigma_1 = 7020$ psi

$\sigma_2 = 3510$ psi

Stresses at the Weld.
Noting that both the hoop stress and the longitudinal stress are principal stresses, we draw Mohr's circle as shown.

An element having a face parallel to the weld is obtained by rotating the face perpendicular to the axis Ob counterclockwise through $25°$. Therefore, on Mohr's circle we locate the point X' corresponding to the stress components on the weld by rotating radius CB counterclockwise through $2\theta = 50°$.

$\sigma_w = \sigma_{ave} - R \cos 50° = 5265 - 1755 \cos 50° \qquad \sigma_w = +4140 \text{ psi} \blacktriangledown$

$\tau_w = R \sin 50° = 1755 \sin 50° \qquad \tau_w = 1344 \text{ psi} \blacktriangledown$

Since X' is below the horizontal axis, τ_w tends to rotate the element counter-clockwise.

$\sigma_w = 4140$ psi

$\tau_w = 1344$ psi

Weld

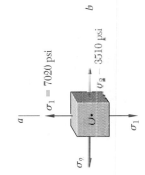

τ

$\sigma_1 = 7020$ psi

$\sigma_{ave} = 5265$ psi

$\sigma_2 = 3510$ psi

$2\theta = 50°$

$R = 1755$ psi

σ

σ_w

τ_w

PROBLEMS

7.98 Determine the normal stress in a basketball of 9.5-in. outer diameter and 0.125-in. wall thickness that is inflated to a gage pressure of 9 psi.

7.99 A spherical gas container made of steel has an 18-ft outer diameter and a wall thickness of $\frac{7}{8}$ in. Knowing that the internal pressure is 60 psi, determine the maximum normal stress and the maximum shearing stress in the container.

7.100 The maximum gage pressure is known to be 8 MPa in a spherical steel pressure vessel having a 250-mm outer diameter and a 6-mm wall thickness. Knowing that the ultimate stress in the steel used is $\sigma_U = 400$ MPa, determine the factor of safety with respect to tensile failure.

7.101 A spherical pressure vessel of 900-mm outer diameter is to be fabricated from a steel having an ultimate stress $\sigma_U = 400$ MPa. Knowing that a factor of safety of 4 is desired and that the gage pressure can reach 3.5 MPa, determine the smallest wall thickness that should be used.

7.102 A spherical gas container having an inner diameter of 5 m and a wall thickness of 24 mm is made of a steel for which $E = 200$ GPa and $\nu = 0.29$. Knowing that the gage pressure in the container is increased from zero to 1.8 MPa, determine (a) the maximum normal stress in the container, (b) the increase in the diameter of the container.

7.103 A spherical pressure vessel has an inner diameter of 3 m and a wall thickness of 12 mm. Knowing that for the steel used $E = 200$ GPa, and $\nu = 0.29$, determine (a) the allowable gage pressure, (b) the corresponding increase in the diameter of the vessel.

7.104 When filled to capacity, the unpressurized storage tank shown contains water to a height of 15.5 m above its base. Knowing that the lower portion of the tank has a wall thickness of 16 mm, determine the maximum normal stress and the maximum shearing stress in the tank. (Density of water = 1000 kg/m³.)

7.105 Determine the largest internal pressure that can be applied to a cylindrical tank of 5.5-ft outer diameter and $\frac{5}{8}$-in. wall thickness if the ultimate normal stress of the steel used is 65 ksi and a factor of safety of 5.0 is desired.

8 m

16 m

Fig. P7.104

7.106 The storage tank shown contains liquefied propane under a pressure of 210 psi at a temperature of 100°F. Knowing that the tank has an outer diameter of 12.6 in. and a wall thickness of 0.11 in, determine the maximum normal stress and the maximum shearing stress in the tank.

Fig. 7.106

7.107 The bulk storage tank shown in Fig. 7.49 has an outer diameter of 3.3 m and a wall thickness of 18 mm. At a time when the internal pressure of the tank is 1.5 MPa, determine the maximum normal stress and the maximum shearing stress in the tank.

7.108 A steel penstock has a 36-in. outer diameter, a 0.5-in. wall thickness, and connects a reservoir at A with a generating station at B. Knowing that the specific weight of water is 62.4 lb/ft³, determine the maximum normal stress and the maximum shearing stress in the penstock under static conditions.

500 ft

36 in.

Fig. P7.108 and *P7.109*

7.109 A steel penstock of 36-in. outer diameter connects a reservoir at A with a generating station at B. Knowing that the specific weight of water is 62.4 lb/ft³ and that the allowable normal stress in the steel is 12.5 ksi, determine the smallest wall thickness that can be used for the penstock.

7.110 The cylindrical portion of the compressed air tank shown is fabricated of 6-mm-thick plate welded along a helix forming an angle $\beta = 30°$ with the horizontal. Knowing that the allowable stress normal to the weld is 75 MPa, determine the largest gage pressure that can be used in the tank.

500 mm

β

1.5 m

Fig. P7.110 and P7.111

7.111 The cylindrical portion of the compressed air tank shown is fabricated of 6-mm-thick plate welded along a helix forming an angle $\beta = 30°$ with the horizontal. Determine the gage pressure that will cause a shearing stress parallel to the weld of 30 MPa.

7.112 The pressure tank shown has a $\frac{3}{8}$-in. wall thickness and butt-welded seams forming an angle $\beta = 20°$ with a transverse plane. For a gage pressure of 85 psi, determine (*a*) the normal stress perpendicular to the weld, (*b*) the shearing stress parallel to the weld.

Fig. P7.112, P7.113, and P7.114

7.113 The pressure tank shown has a $\frac{3}{8}$-in. wall thickness and butt-welded seams forming an angle β with a transverse plane. Determine the range of values of β that can be used if the shearing stress parallel to the weld is not to exceed 1350 psi when the gage pressure is 85 psi.

7.114 The pressure tank shown has a $\frac{3}{8}$-in. wall thickness and butt-welded seams forming an angle β with a transverse plane. Determine the largest allowable gage pressure, knowing that the allowable normal stress perpendicular to the weld is 18 ksi and the allowable shearing stress parallel to the weld is 10 ksi.

7.115 The pipe shown was fabricated by welding strips of plate along a helix forming an angle β with a transverse plane. Determine the largest value of β that can be used if the normal stress perpendicular to the weld is not to be larger than 85 percent of the maximum stress in the pipe.

Fig. P7.115 and P7.116

7.116 The pipe shown has an outer diameter of 600 mm and was fabricated by welding strips of 10-mm-thick plate along a helix forming an angle $\beta = 25°$ with a transverse plane. Knowing that the ultimate normal stress perpendicular to the weld is 450 MPa and that a factor of safety of 6.0 is desired, determine the largest allowable gage pressure that can be used.

7.117 Square plates, each of 0.5-in. thickness, can be bent and welded together in either of the two ways shown to form the cylindrical portion of a compressed-air tank. Knowing that the allowable normal stress perpendicular to the weld is 12 ksi, determine the largest allowable gage pressure in each case.

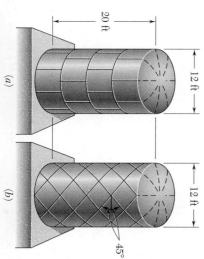

Fig. P7.117

7.118 A torque of magnitude $T = 12$ kN · m is applied to the end of a tank containing compressed air under a pressure of 8 MPa. Knowing that the tank has a 180-mm inner diameter and a 12-mm wall thickness, determine the maximum normal stress and the maximum shearing stress in the tank.

Fig. P7.118 and P7.119

7.119 The tank shown has a 180-mm inner diameter and a 12-mm wall thickness. Knowing that the tank contains compressed air under a pressure of 8 MPa, determine the magnitude T of the applied torque for which the maximum normal stress in the tank is 75 MPa.

7.120 A pressure vessel of 250-mm inner diameter and 6-mm wall thickness is fabricated from a 1.2-m section of spirally welded pipe AB and is equipped with two rigid end plates. The gage pressure inside the vessel is 2 MPa and 45-kN centric axial forces **P** and **P'** are applied to the end plates. Determine (a) the normal stress perpendicular to the weld, (b) the shearing stress parallel to the weld.

Fig. P7.120

7.121 Solve Prob. 7.120, assuming that the magnitude P of the two forces is increased to 120 kN.

7.122 The cylindrical tank AB has an 8-in. inner diameter and a 0.32-in. wall thickness. Knowing that the pressure inside the tank is 600 psi, determine the maximum normal stress and the maximum shearing stress at point K.

7.123 Solve Prob. 7.122, assuming that the 9-kip force applied at point D is directed vertically downward.

7.124 A brass ring of 160-mm outer diameter fits exactly inside a steel ring of 160-mm inner diameter when the temperature of both rings is 5°C. Knowing that the temperature of the rings is then raised to 55°C, determine (a) the tensile stress in the steel ring, (b) the corresponding pressure exerted by the brass ring on the steel ring.

7.125 Solve Prob. 7.124, assuming that the thickness of the brass ring is $t_b = 6$ mm.

Fig. P7.122

STEEL
$t_s = 4$ mm
$E_s = 210$ GPa
$\alpha_s = 12 \times 10^{-6}$/°C

BRASS
$t_b = 4$ mm
$E_b = 105$ GPa
$\alpha_b = 20 \times 10^{-6}$/°C

Fig. P7.124

*7.10. TRANSFORMATION OF PLANE STRAIN

Transformations of *strain* under a rotation of the coordinate axes will now be considered. Our analysis will first be limited to states of *plane strain*, i.e., to situations where the deformations of the material take place within parallel planes, and are the same in each of these planes. If the z axis is chosen perpendicular to the planes in which the deformations take place, we have $\epsilon_z = \gamma_{zx} = \gamma_{zy} = 0$, and the only remaining strain components are ϵ_x, ϵ_y, and γ_{xy}. Such a situation occurs in a plate subjected along its edges to uniformly distributed loads and restrained from expanding or contracting laterally by smooth, rigid, and fixed supports (Fig. 7.58). It would also be found in a bar of infinite length subjected on its sides to uniformly distributed loads and, by reason of symmetry, the elements located in a given transverse plane cannot move out of that plane. This idealized model shows that, in the actual case of a long bar subjected to uniformly distributed transverse loads (Fig. 7.59), a state of plane strain exists in any given transverse section that is not located too close to either end of the bar.†

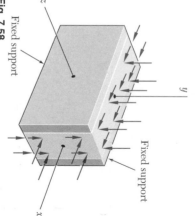

Fixed support
Fixed support

Fig. 7.58

Fig. 7.59

Let us assume that a state of plane strain exists at point Q (with $\epsilon_z = \gamma_{zx} = \gamma_{zy} = 0$), and that it is defined by the strain components ϵ_x, ϵ_y, and γ_{xy} associated with the x and y axes. As we know from Secs. 2.12 and 2.14, this means that a square element of center Q, with sides of length Δs respectively parallel to the x and y axes, is deformed into a parallelogram with sides of length respectively equal to $\Delta s\,(1+\epsilon_x)$ and $\Delta s\,(1+\epsilon_y)$, forming angles of $\frac{\pi}{2}-\gamma_{xy}$ and $\frac{\pi}{2}+\gamma_{xy}$ with each other (Fig. 7.60). We recall that, as a result of the deformations of the other elements located in the xy plane, the element considered may also undergo a rigid-body motion, but such a motion is irrelevant to the determination of the strains at point Q and will be ignored in this analysis. Our purpose is to determine in terms of ϵ_x, ϵ_y, γ_{xy}, and θ the strain components $\epsilon_{x'}$, $\epsilon_{y'}$, and $\gamma_{x'y'}$ associated with the frame of reference $x'y'$ obtained by rotating the x and y axes through the angle θ. As shown in Fig. 7.61, these new strain components define the parallelogram into which a square with sides respectively parallel to the x' and y' axes is deformed.

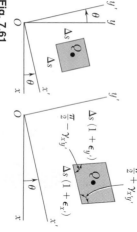

Fig. 7.60

Fig. 7.61

†It should be observed that a state of *plane strain* and a state of *plane stress* (cf. Sec. 7.1) do not occur simultaneously, except for ideal materials with a Poisson ratio equal to zero. The constraints placed on the elements of the plate of Fig. 7.58 and of the bar of Fig. 7.59 result in a stress σ_z different from zero. On the other hand, in the case of the plate of Fig. 7.3, the absence of any lateral restraint results in $\sigma_z = 0$ and $\epsilon_z \neq 0$.

We first derive an expression for the normal strain $\epsilon(\theta)$ along a line AB forming an arbitrary angle θ with the x axis. To do so, we consider the right triangle ABC, which has AB for hypotenuse (Fig. 7.62a), and the oblique triangle $A'B'C'$ into which triangle ABC is deformed (Fig. 7.62b). Denoting by Δs the length of AB, we express the length of $A'B'$ as $\Delta s[1 + \epsilon(\theta)]$. Similarly, denoting by Δx and Δy the lengths of sides AC and CB, we express the lengths of $A'C'$ and $C'B'$ as $\Delta x(1 + \epsilon_x)$ and $\Delta y(1 + \epsilon_y)$, respectively. Recalling from Fig. 7.60 that the right angle at C in Fig. 7.62a deforms into an angle equal to $\frac{\pi}{2} + \gamma_{xy}$ in Fig. 7.62b, and applying the law of cosines to triangle $A'B'C'$, we write

$$(A'B')^2 = (A'C')^2 + (C'B')^2 - 2(A'C')(C'B')\cos\left(\frac{\pi}{2} + \gamma_{xy}\right)$$

$$(\Delta s)^2[1 + \epsilon(\theta)]^2 = (\Delta x)^2(1 + \epsilon_x)^2 + (\Delta y)^2(1 + \epsilon_y)^2$$
$$- 2(\Delta x)(1 + \epsilon_x)(\Delta y)(1 + \epsilon_y)\cos\left(\frac{\pi}{2} + \gamma_{xy}\right) \quad (7.38)$$

But from Fig. 7.62a we have

$$\Delta x = (\Delta s)\cos\theta \qquad \Delta y = (\Delta s)\sin\theta \qquad (7.39)$$

and we note that, since γ_{xy} is very small,

$$\cos\left(\frac{\pi}{2} + \gamma_{xy}\right) = -\sin\gamma_{xy} \approx -\gamma_{xy} \qquad (7.40)$$

Substituting from Eqs. (7.39) and (7.40) into Eq. (7.38), recalling that $\cos^2\theta + \sin^2\theta = 1$, and neglecting second-order terms in $\epsilon(\theta)$, ϵ_x, ϵ_y, and γ_{xy}, we write

$$\epsilon(\theta) = \epsilon_x\cos^2\theta + \epsilon_y\sin^2\theta + \gamma_{xy}\sin\theta\cos\theta \qquad (7.41)$$

Equation (7.41) enables us to determine the normal strain $\epsilon(\theta)$ in any direction AB in terms of the strain components ϵ_x, ϵ_y, γ_{xy}, and the angle θ that AB forms with the x axis. We check that, for $\theta = 0$, Eq. (7.41) yields $\epsilon(0) = \epsilon_x$ and that, for $\theta = 90°$, it yields $\epsilon(90°) = \epsilon_y$. On the other hand, making $\theta = 45°$ in Eq. (7.41), we obtain the normal strain in the direction of the bisector OB of the angle formed by the x and y axes (Fig. 7.63). Denoting this strain by ϵ_{OB}, we write

$$\epsilon_{OB} = \epsilon(45°) = \tfrac{1}{2}(\epsilon_x + \epsilon_y + \gamma_{xy}) \qquad (7.42)$$

Solving Eq. (7.42) for γ_{xy}, we have

$$\gamma_{xy} = 2\epsilon_{OB} - (\epsilon_x + \epsilon_y) \qquad (7.43)$$

This relation makes it possible to express the *shearing strain* associated with a given pair of rectangular axes in terms of the *normal strains* measured along these axes and their bisector. It will play a fundamental role in our present derivation and will also be used in Sec. 7.13 in connection with the experimental determination of shearing strains.

Fig. 7.62

Fig. 7.63

Recalling that the main purpose of this section is to express the strain components associated with the frame of reference $x'y'$ of Fig. 7.61 in terms of the angle θ and the strain components ϵ_x, ϵ_y, and γ_{xy} associated with the x and y axes, we note that the normal strain $\epsilon_{x'}$ along the x' axis is given by Eq. (7.41). Using the trigonometric relations (7.3) and (7.4), we write this equation in the alternative form

$$\epsilon_{x'} = \frac{\epsilon_x + \epsilon_y}{2} + \frac{\epsilon_x - \epsilon_y}{2}\cos 2\theta + \frac{\gamma_{xy}}{2}\sin 2\theta \qquad (7.44)$$

Replacing θ by $\theta + 90°$, we obtain the normal strain along the y' axis. Since $\cos(2\theta + 180°) = -\cos 2\theta$ and $\sin(2\theta + 180°) = -\sin 2\theta$, we have

$$\epsilon_{y'} = \frac{\epsilon_x + \epsilon_y}{2} - \frac{\epsilon_x - \epsilon_y}{2}\cos 2\theta - \frac{\gamma_{xy}}{2}\sin 2\theta \qquad (7.45)$$

Adding Eqs. (7.44) and (7.45) member to member, we obtain

$$\epsilon_{x'} + \epsilon_{y'} = \epsilon_x + \epsilon_y \qquad (7.46)$$

Since $\epsilon_z = \epsilon_{z'} = 0$, we thus verify in the case of plane strain that the sum of the normal strains associated with a cubic element of material is independent of the orientation of that element.†

Replacing now θ by $\theta + 45°$ in Eq. (7.44), we obtain an expression for the normal strain along the bisector OB' of the angle formed by the x' and y' axes. Since $\cos(2\theta + 90°) = -\sin 2\theta$ and $\sin(2\theta + 90°) = \cos 2\theta$, we have

$$\epsilon_{OB'} = \frac{\epsilon_x + \epsilon_y}{2} - \frac{\epsilon_x - \epsilon_y}{2}\sin 2\theta + \frac{\gamma_{xy}}{2}\cos 2\theta \qquad (7.47)$$

Writing Eq. (7.43) with respect to the x' and y' axes, we express the shearing strain $\gamma_{x'y'}$ in terms of the normal strains measured along the x' and y' axes and the bisector OB':

$$\gamma_{x'y'} = 2\epsilon_{OB'} - (\epsilon_{x'} + \epsilon_{y'}) \qquad (7.48)$$

Substituting from Eqs. (7.46) and (7.47) into (7.48), we obtain

$$\gamma_{x'y'} = -(\epsilon_x - \epsilon_y)\sin 2\theta + \gamma_{xy}\cos 2\theta \qquad (7.49)$$

Equations (7.44), (7.45), and (7.49) are the desired equations defining the transformation of plane strain under a rotation of axes in the plane of strain. Dividing all terms in Eq. (7.49) by 2, we write this equation in the alternative form

$$\frac{\gamma_{x'y'}}{2} = -\frac{\epsilon_x - \epsilon_y}{2}\sin 2\theta + \frac{\gamma_{xy}}{2}\cos 2\theta \qquad (7.49')$$

and observe that Eqs. (7.44), (7.45), and (7.49') for the transformation of plane strain closely resemble the equations derived in Sec. 7.2 for the transformation of plane stress. While the former may be obtained from the latter by replacing the normal stresses by the corresponding normal strains, it should be noted, however, that the shearing stresses τ_{xy} and $\tau_{x'y'}$ should be replaced by *half* of the corresponding shearing strains, i.e., by $\frac{1}{2}\gamma_{xy}$ and $\frac{1}{2}\gamma_{x'y'}$, respectively.

†Cf. footnote on page 88.

*7.11. MOHR'S CIRCLE FOR PLANE STRAIN

Since the equations for the transformation of plane strain are of the same form as the equations for the transformation of plane stress, the use of Mohr's circle can be extended to the analysis of plane strain. Given the strain components ϵ_x, ϵ_y, and γ_{xy} defining the deformation represented in Fig. 7.60, we plot a point $X(\epsilon_x, -\tfrac{1}{2}\gamma_{xy})$ of abscissa equal to the normal strain ϵ_x and of ordinate equal to minus half the shearing strain γ_{xy}, and a point $Y(\epsilon_y, +\tfrac{1}{2}\gamma_{xy})$ (Fig. 7.64). Drawing the diameter XY, we define the center C of Mohr's circle for plane strain. The abscissa of C and the radius R of the circle are respectively equal to

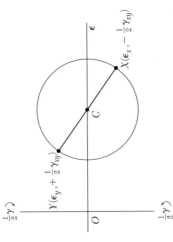

Fig. 7.64

$$\epsilon_{ave} = \frac{\epsilon_x + \epsilon_y}{2} \qquad \text{and} \qquad R = \sqrt{\left(\frac{\epsilon_x - \epsilon_y}{2}\right)^2 + \left(\frac{\gamma_{xy}}{2}\right)^2} \qquad (7.50)$$

We note that if γ_{xy} is positive, as assumed in Fig. 7.60, points X and Y are plotted, respectively, below and above the horizontal axis in Fig. 7.60. But, in the absence of any overall rigid-body rotation, the side of the element in Fig. 7.60 that is associated with ϵ_x is observed to rotate counterclockwise, while the side associated with ϵ_y is observed to rotate clockwise. Thus, if the shear deformation causes a given side to rotate *clockwise*, the corresponding point on Mohr's circle for plane strain is plotted *above* the horizontal axis, and if the deformation causes the side to rotate *counterclockwise*, the corresponding point is plotted *below* the horizontal axis. We note that this convention matches the convention used to draw Mohr's circle for plane stress.

Points A and B where Mohr's circle intersects the horizontal axis correspond to the *principal strains* ϵ_{max} and ϵ_{min} (Fig. 7.65a). We find

$$\epsilon_{max} = \epsilon_{ave} + R \qquad \text{and} \qquad \epsilon_{min} = \epsilon_{ave} - R \qquad (7.51)$$

Fig. 7.65

where ϵ_{ave} and R are defined by Eqs. (7.50). The corresponding value θ_p of the angle θ is obtained by observing that the shearing strain is zero for A and B. Setting $\gamma_{x'y'} = 0$ in Eq. (7.49), we have

$$\tan 2\theta_p = \frac{\gamma_{xy}}{\epsilon_x - \epsilon_y} \qquad (7.52)$$

The corresponding axes a and b in Fig. 7.65b are the *principal axes of strain*. The angle θ_p, which defines the direction of the principal axis Oa in Fig. 7.65b corresponding to point A in Fig. 7.65a, is equal to half of the angle XCA measured on Mohr's circle, and the rotation that brings Ox into Oa has the same sense as the rotation that brings the diameter XY of Mohr's circle into the diameter AB.

We recall from Sec. 2.14 that, in the case of the elastic deformation of a homogeneous, isotropic material, Hooke's law for shearing stress and strain applies and yields $\tau_{xy} = G\gamma_{xy}$ for any pair of rectangular x and y axes. Thus, $\gamma_{xy} = 0$ when $\tau_{xy} = 0$, which indicates that the principal axes of strain coincide with the principal axes of stress.

Fig. 7.61 (repeated)

The maximum in-plane shearing strain is defined by points D and E in Fig. 7.65a. It is equal to the diameter of Mohr's circle. Recalling the second of Eqs. (7.50), we write

$$\gamma_{\text{max(in plane)}} = 2R = \sqrt{(\epsilon_x - \epsilon_y)^2 + \gamma_{xy}^2} \qquad (7.53)$$

Finally, we note that the points X' and Y' that define the components of strain corresponding to a rotation of the coordinate axes through an angle θ (Fig. 7.61) are obtained by rotating the diameter XY of Mohr's circle in the same sense through an angle 2θ (Fig. 7.66).

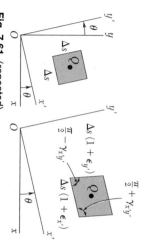

Fig. 7.66

EXAMPLE 7.04

In a material in a state of plane strain, it is known that the horizontal side of a 10×10-mm square elongates by 4 μm, while its vertical side remains unchanged, and that the angle at the lower left corner increases by 0.4×10^{-3} rad (Fig. 7.67). Determine (a) the principal axes and principal strains, (b) the maximum shearing strain and the corresponding normal strain.

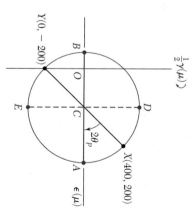

Fig. 7.67

(a) Principal Axes and Principal Strains. We first determine the coordinates of points X and Y on Mohr's circle for strain. We have

$$\epsilon_x = \frac{+4 \times 10^{-6} \text{m}}{10 \times 10^3 \text{m}} = +400 \ \mu \qquad \epsilon_y = 0 \qquad \left| \frac{\gamma_{xy}}{2} \right| = 200 \ \mu$$

Since the side of the square associated with ϵ_x rotates *clockwise*, point X of coordinates ϵ_x and $|\gamma_{xy}/2|$ is plotted *above* the horizontal axis. Since $\epsilon_y = 0$ and the corresponding side ro-

tates *counterclockwise*, point Y is plotted directly *below* the origin (Fig. 7.68). Drawing the diameter XY, we determine the center C of Mohr's circle and its radius R. We have

$$OC = \frac{\epsilon_x + \epsilon_y}{2} = 200 \ \mu \qquad OY = 200 \ \mu$$

$$R = \sqrt{(OC)^2 + (OY)^2} = \sqrt{(200 \ \mu)^2 + (200 \ \mu)^2} = 283 \ \mu$$

The principal strains are defined by the abscissas of points A and B. We write

$$\epsilon_a = OA = OC + R = 200 \ \mu + 283 \ \mu = 483 \ \mu$$
$$\epsilon_b = OB = OC - R = 200 \ \mu - 283 \ \mu = -83 \ \mu$$

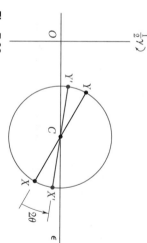

Fig. 7.68

The principal axes Oa and Ob are shown in Fig. 7.69. Since $OC = OY$, the angle at C in triangle OCY is 45°. Thus, the angle $2\theta_p$ which brings XY into AB is 45°⤸ and the angle θ_p bringing Ox into Oa is 22.5°⤸.

(b) Maximum Shearing Strain. Points D and E define the maximum in-plane shearing strain which, since the principal strains have opposite signs, is also the actual maximum shearing strain (see Sec. 7.12). We have

$$\frac{\gamma_{max}}{2} = R = 283\,\mu \qquad \gamma_{max} = 566\,\mu$$

The corresponding normal strains are both equal to

$$\epsilon' = OC = 200\,\mu$$

The axes of maximum shearing strain are shown in Fig. 7.70.

Fig. 7.69

Fig. 7.70

*7.12. THREE-DIMENSIONAL ANALYSIS OF STRAIN

We saw in Sec. 7.5 that, in the most general case of stress, we can determine three coordinate axes a, b, and c, called the principal axes of stress. A small cubic element with faces respectively perpendicular to these axes is free of shearing stresses (Fig. 7.27); i.e., we have $\tau_{ab} = \tau_{bc} = \tau_{ca} = 0$. As recalled in the preceding section, Hooke's law for shearing stress and strain applies when the deformation is elastic and the material homogeneous and isotropic. It follows that, in such a case, $\gamma_{ab} = \gamma_{bc} = \gamma_{ca} = 0$, i.e., the axes a, b, and c are also *principal axes of strain*. A small cube of side equal to unity, centered at Q and with faces respectively perpendicular to the principal axes, is deformed into a rectangular parallelepiped of sides $1 + \epsilon_a$, $1 + \epsilon_b$, and $1 + \epsilon_c$ (Fig. 7.71).

Fig. 7.27 (repeated)

Fig. 7.71

Fig. 7.71 (repeated)

Fig. 7.72

If the element of Fig. 7.71 is rotated about one of the principal axes at Q, say the c axis (Fig. 7.72), the method of analysis developed earlier for the transformation of plane strain can be used to determine the strain components ϵ_x, ϵ_y, and γ_{xy} associated with the faces perpendicular to the c axis, since the derivation of this method did not involve any of the other strain components.† We can, therefore, draw Mohr's circle through the points A and B corresponding to the principal axes a and b (Fig. 7.73). Similarly, circles of diameters BC and CA can be used to analyze the transformation of strain as the element is rotated about the a and b axes, respectively.

The three-dimensional analysis of strain by means of Mohr's circle is limited here to rotations about principal axes (as was the case for the analysis of stress) and is used to determine the maximum shearing strain γ_{max} at point Q. Since γ_{max} is equal to the diameter of the largest of the three circles shown in Fig. 7.73, we have

$$\boxed{\gamma_{max} = |\epsilon_{max} - \epsilon_{min}|} \qquad (7.54)$$

where ϵ_{max} and ϵ_{min} represent the *algebraic* values of the maximum and minimum strains at point Q.

Returning to the particular case of *plane strain*, and selecting the x and y axes in the plane of strain, we have $\epsilon_z = \gamma_{zx} = \gamma_{zy} = 0$. Thus, the z axis is one of the three principal axes at Q, and the corresponding

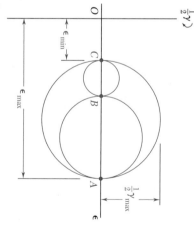

Fig. 7.73

---

†We note that the other four faces of the element remain rectangular and that the edges parallel to the c axis remain unchanged.

point in the Mohr-circle diagram is the origin O, where $\epsilon = \gamma = 0$. If the points A and B that define the principal axes within the plane of strain fall on opposite sides of O (Fig. 7.74a), the corresponding principal strains represent the maximum and minimum normal strains at point Q, and the maximum shearing strain is equal to the maximum inplane shearing strain corresponding to points D and E. If, on the other hand, A and B are on the same side of O (Fig. 7.74b), that is, if ϵ_a and ϵ_b have the same sign, then the maximum shearing strain is defined by points D' and F' on the circle of diameter OA, and we have $\gamma_{max} = \epsilon_{max}$.

We now consider the particular case of *plane stress* encountered in a thin plate or on the free surface of a structural element or machine component (Sec. 7.1). Selecting the x and y axes in the plane of stress, we have $\sigma_z = \tau_{zx} = \tau_{zy} = 0$ and verify that the z axis is a principal axis of stress. As we saw earlier, if the deformation is elastic and if the material is homogeneous and isotropic, it follows from Hooke's law that $\gamma_{zx} = \gamma_{zy} = 0$; thus, the z axis is also a principal axis of strain, and Mohr's circle can be used to analyze the transformation of strain in the xy plane. However, as we shall see presently, it does *not* follow from Hooke's law that $\epsilon_z = 0$; indeed, a state of plane stress does not, in general, result in a state of plane strain.†

Denoting by a and b the principal axes within the plane of stress, and by c the principal axis perpendicular to that plane, we let $\sigma_x = \sigma_a$, $\sigma_y = \sigma_b$, and $\sigma_z = 0$ in Eqs. (2.28) for the generalized Hooke's law (Sec. 2.12) and write

$$\epsilon_a = \frac{\sigma_a}{E} - \frac{\nu\sigma_b}{E} \tag{7.55}$$

$$\epsilon_b = -\frac{\nu\sigma_a}{E} + \frac{\sigma_b}{E} \tag{7.56}$$

$$\epsilon_c = -\frac{\nu}{E}(\sigma_a + \sigma_b) \tag{7.57}$$

Adding Eqs. (7.55) and (7.56) member to member, we have

$$\epsilon_a + \epsilon_b = \frac{1 - \nu}{E}(\sigma_a + \sigma_b) \tag{7.58}$$

Solving Eq. (7.58) for $\sigma_a + \sigma_b$ and substituting into Eq. (7.57), we write

$$\epsilon_c = -\frac{\nu}{1 - \nu}(\epsilon_a + \epsilon_b) \tag{7.59}$$

The relation obtained defines the third principal strain in terms of the "in-plane" principal strains. We note that, if B is located between A and C on the Mohr-circle diagram (Fig. 7.75), the maximum shearing strain is equal to the diameter CA of the circle corresponding to a rotation about the b axis, out of the plane of stress.

$\left(\frac{1}{2}\gamma\right)$

ϵ

(a)

$\left(\frac{1}{2}\gamma\right)$

ϵ

$\epsilon_{min} = 0 \quad F'$

$\epsilon_{max} = \epsilon_u$

(b)

Fig. 7.74

$\left(\frac{1}{2}\gamma\right)$

ϵ

Fig. 7.75

†See footnote on page 470.

EXAMPLE 7.05

As a result of measurements made on the surface of a machine component with strain gages oriented in various ways, it has been established that the principal strains on the free surface are $\epsilon_a = +400 \times 10^{-6}$ in./in. and $\epsilon_b = -50 \times 10^{-6}$ in./in. Knowing that Poisson's ratio for the given material is $\nu = 0.30$, determine (a) the maximum in-plane shearing strain, (b) the true value of the maximum shearing strain near the surface of the component.

(a) *Maximum In-Plane Shearing Strain.* We draw Mohr's circle through the points A and B corresponding to the given principal strains (Fig. 7.76). The maximum in-plane shearing strain is defined by points D and E and is equal to the diameter of Mohr's circle:

$$\gamma_{\text{max(in plane)}} = 400 \times 10^{-6} + 50 \times 10^{-6} = 450 \times 10^{-6} \text{ rad}$$

Fig. 7.76

(b) *Maximum Shearing Strain.* We first determine the third principal strain ϵ_c. Since we have a state of plane *stress* on the surface of the machine component, we use Eq. (7.59) and write

$$\epsilon_c = -\frac{\nu}{1 - \nu}(\epsilon_a + \epsilon_b)$$
$$= -\frac{0.30}{0.70}(400 \times 10^{-6} - 50 \times 10^{-6}) = -150 \times 10^{-6} \text{ in./in.}$$

Drawing Mohr's circles through A and C and through B and C (Fig. 7.77), we find that the maximum shearing strain is equal to the diameter of the circle of diameter CA:

$$\gamma_{\text{max}} = 400 \times 10^{-6} + 150 \times 10^{-6} = 550 \times 10^{-6} \text{ rad}$$

We note that, even though ϵ_a and ϵ_b have opposite signs, the maximum in-plane shearing strain does not represent the true maximum shearing strain.

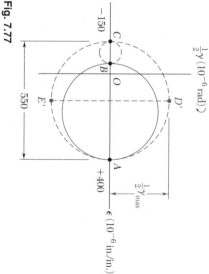

Fig. 7.77

*7.13. MEASUREMENTS OF STRAIN; STRAIN ROSETTE

The normal strain can be determined in any given direction on the surface of a structural element or machine component by scribing two gage marks A and B across a line drawn in the desired direction and measuring the length of the segment AB before and after the load has been applied. If L is the undeformed length of AB and δ its deformation, the normal strain along AB is $\epsilon_{AB} = \delta/L$.

A more convenient and more accurate method for the measurement of normal strains is provided by electrical strain gages. A typical electrical strain gage consists of a length of thin wire arranged as shown in Fig. 7.78 and cemented to two pieces of paper. In order to measure the strain ϵ_{AB} of a given material in the direction AB, the gage is cemented to the surface of the material, with the wire folds running parallel to AB. As the material elongates, the wire increases in length and decreases in diameter, causing the electrical resistance of the gage to increase. By measuring the current passing through a properly calibrated gage, the

Fig. 7.78

strain ϵ_{AB} can be determined accurately and continuously as the load is increased.

The strain components ϵ_x and ϵ_y can be determined at a given point of the free surface of a material by simply measuring the normal strain along x and y axes drawn through that point. Recalling Eq. (7.43) of Sec. 7.10, we note that a third measurement of normal strain, made along the bisector OB of the angle formed by the x and y axes, enables us to determine the shearing strain γ_{xy} as well (Fig. 7.79):

$$\gamma_{xy} = 2\epsilon_{OB} - (\epsilon_x + \epsilon_y) \qquad (7.43)$$

Fig. 7.79

It should be noted that the strain components ϵ_x, ϵ_y, and γ_{xy} at a given point could be obtained from normal strain measurements made along *any three lines* drawn through that point (Fig. 7.80). Denoting respectively by θ_1, θ_2, and θ_3 the angle each of the three lines forms with the x axis, by ϵ_1, ϵ_2, and ϵ_3 the corresponding strain measurements, and substituting into Eq. (7.41), we write the three equations

$$\epsilon_1 = \epsilon_x \cos^2 \theta_1 + \epsilon_y \sin^2 \theta_1 + \gamma_{xy} \sin \theta_1 \cos \theta_1$$
$$\epsilon_2 = \epsilon_x \cos^2 \theta_2 + \epsilon_y \sin^2 \theta_2 + \gamma_{xy} \sin \theta_2 \cos \theta_2 \qquad (7.60)$$
$$\epsilon_3 = \epsilon_x \cos^2 \theta_3 + \epsilon_y \sin^2 \theta_3 + \gamma_{xy} \sin \theta_3 \cos \theta_3$$

which can be solved simultaneously for ϵ_x, ϵ_y, and γ_{xy}.†

The arrangement of strain gages used to measure the three normal strains ϵ_1, ϵ_2, and ϵ_3 is known as a *strain rosette*. The rosette used to measure normal strains along the x and y axes and their bisector is referred to as a 45° rosette. Another rosette frequently used is the 60° rosette (see Sample Prob. 7.7).

Fig. 7.80

† It should be noted that the free surface on which the strain measurements are made is in a state of *plane stress*, while Eqs. (7.41) and (7.43) were derived for a state of *plane strain*. However, as observed earlier, the normal to the free surface is a principal axis of strain and the derivations given in sec. 7.10 remain valid.

SAMPLE PROBLEM 7.6

A cylindrical storage tank used to transport gas under pressure has an inner diameter of 24 in. and a wall thickness of $\frac{3}{4}$ in. Strain gages attached to the surface of the tank in transverse and longitudinal directions indicate strains of 255×10^{-6} and 60×10^{-6} in./in. respectively. Knowing that a torsion test has shown that the modulus of rigidity of the material used in the tank is $G = 11.2 \times 10^6$ psi, determine (a) the gage pressure inside the tank, (b) the principal stresses and the maximum shearing stress in the wall of the tank.

SOLUTION

a. **Gage Pressure Inside Tank.** We note that the given strains are the principal strains at the surface of the tank. Plotting the corresponding points A and B, we draw Mohr's circle for strain. The maximum in-plane shearing strain is equal to the diameter of the circle.

$$\gamma_{\text{max(in plane)}} = \epsilon_1 - \epsilon_2 = 255 \times 10^{-6} - 60 \times 10^{-6} = 195 \times 10^{-6} \text{ rad}$$

From Hooke's law for shearing stress and strain, we have

$$\tau_{\text{max(in plane)}} = G\gamma_{\text{max(in plane)}}$$
$$= (11.2 \times 10^6 \text{ psi})(195 \times 10^{-6} \text{ rad})$$
$$= 2184 \text{ psi} = 2.184 \text{ ksi}$$

Substituting this value and the given data in Eq. (7.33), we write

$$\tau_{\text{max(in plane)}} = \frac{pr}{4t} \qquad 2184 \text{ psi} = \frac{p(12 \text{ in.})}{4(0.75 \text{ in.})}$$

Solving for the gage pressure p, we have

$$p = 546 \text{ psi} \quad \blacktriangleleft$$

b. **Principal Stresses and Maximum Shearing Stress.** Recalling that, for a thin-walled cylindrical pressure vessel, $\sigma_1 = 2\sigma_2$, we draw Mohr's circle for stress and obtain

$$\sigma_2 = 2\tau_{\text{max(in plane)}} = 2(2.184 \text{ ksi}) = 4.368 \text{ ksi} \qquad \sigma_2 = 4.37 \text{ ksi} \quad \blacktriangleleft$$
$$\sigma_1 = 2\sigma_2 = 2(4.368 \text{ ksi}) \qquad\qquad\qquad \sigma_1 = 8.74 \text{ ksi} \quad \blacktriangleleft$$

The maximum shearing stress is equal to the radius of the circle of diameter OA and corresponds to a rotation of $45°$ about a longitudinal axis.

$$\tau_{\text{max}} = \frac{1}{2}\sigma_1 = \sigma_2 = 4.368 \text{ ksi} \qquad \tau_{\text{max}} = 4.37 \text{ ksi} \quad \blacktriangleleft$$

SAMPLE PROBLEM 7.7

Using a 60° rosette, the following strains have been determined at point Q on the surface of a steel machine base:

$$\epsilon_1 = 40\ \mu \qquad \epsilon_2 = 980\ \mu \qquad \epsilon_3 = 330\ \mu$$

Using the coordinate axes shown, determine at point Q, (a) the strain components ϵ_x, ϵ_y, and γ_{xy}, (b) the principal strains, (c) the maximum shearing strain. (Use $\nu = 0.29$.)

SOLUTION

a. Strain Components ϵ_x, ϵ_y, γ_{xy}. For the coordinate axes shown

$$\theta_1 = 0 \qquad \theta_2 = 60° \qquad \theta_3 = 120°$$

Substituting these values into Eqs. (7.60), we have

$$\epsilon_1 = \epsilon_x(1) \qquad\qquad + \epsilon_y(0) \qquad\qquad + \gamma_{xy}(0)(1)$$
$$\epsilon_2 = \epsilon_x(0.500)^2 \quad + \epsilon_y(0.866)^2 + \gamma_{xy}(0.866)(0.500)$$
$$\epsilon_3 = \epsilon_x(-0.500)^2 + \epsilon_y(0.866)^2 + \gamma_{xy}(0.866)(\ 0.500)$$

Solving these equations for ϵ_x, ϵ_y, and γ_{xy}, we obtain

$$\epsilon_x = \epsilon_1 \qquad \epsilon_y = \tfrac{1}{3}(2\epsilon_2 + 2\epsilon_3 - \epsilon_1) \qquad \gamma_{xy} = \frac{\epsilon_2 - \epsilon_3}{0.866}$$

Substituting the given values for ϵ_1, ϵ_2, and ϵ_3, we have

$$\epsilon_x = 40\ \mu \qquad \epsilon_y = \tfrac{1}{3}[2(980) + 2(330) - 40]$$
$$\gamma_{xy} = (980 - 330)/0.866$$

$$\epsilon_y = +860\ \mu \qquad\blacktriangledown$$
$$\gamma_{xy} = 750\ \mu \qquad\blacktriangledown$$

These strains are indicated on the element shown.

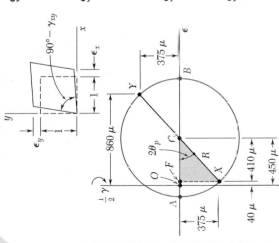

b. Principal Strains. We note that the side of the element associated with ϵ_x rotates counterclockwise, thus we plot point X below the horizontal axis, i.e., $X(40, -375)$. We then plot $Y(860, +375)$ and draw Mohr's circle.

$$\epsilon_{ave} = \tfrac{1}{2}(860\ \mu + 40\ \mu) = 450\ \mu$$
$$R = \sqrt{(375\ \mu)^2 + (410\ \mu)^2} - 556\ \mu$$
$$\tan 2\theta_p = \frac{375\ \mu}{410\ \mu} \qquad 2\theta_p = 42.4° \qquad \theta_p = 21.2° \downarrow$$

Points A and B correspond to the principal strains. We have

$$\epsilon_a = \epsilon_{ave} - R = 450\ \mu - 556\ \mu \qquad \epsilon_a = -106\ \mu \qquad\blacktriangledown$$
$$\epsilon_b = \epsilon_{ave} + R = 450\ \mu + 556\ \mu \qquad \epsilon_b = +1006\ \mu \qquad\blacktriangledown$$

Since $\sigma_z = 0$ on the surface, we use Eq. (7.59) to find the principal strain ϵ_c:

$$\epsilon_c = -\frac{\nu}{1-\nu}(\epsilon_a + \epsilon_b) = -\frac{0.29}{1 - 0.29}(-106\ \mu + 1006\ \mu) \quad \epsilon_c = -368\ \mu \qquad\blacktriangledown$$

c. Maximum Shearing Strain. Plotting point C and drawing Mohr's circle through points B and C, we obtain point D' and write

$$\tfrac{1}{2}\gamma_{max} = \tfrac{1}{2}(1006\ \mu + 368\ \mu) \qquad \gamma_{max} = 1374\ \mu \qquad\blacktriangledown$$

PROBLEMS

7.126 through 7.129 For the given state of plane strain, use the methods of Sec. 7.10 to determine the state of strain associated with axes x' and y' rotated through the given angle θ.

Fig. P7.126 through P7.133

	ϵ_x	ϵ_y	γ_{xy}	θ
7.126 and 7.130	$-720\ \mu$	0	$+300\ \mu$	$30°\swarrow$
7.127 and 7.131	0	$+320\ \mu$	$-100\ \mu$	$30°\nearrow$
7.128 and 7.132	$-800\ \mu$	$+450\ \mu$	$+200\ \mu$	$25°\nearrow$
7.129 and 7.133	$-500\ \mu$	$+250\ \mu$	0	$15°\nwarrow$

7.130 through 7.133 For the given state of plane strain, use Mohr's circle to determine the state of strain associated with axes x' and y' rotated through the angle θ.

7.134 through 7.137 The following state of strain has been measured on the surface of a thin plate. Knowing that the surface of the plate is unstressed, determine (a) the direction and magnitude of the principal strains, (b) the maximum in-plane shearing strain, (c) the maximum shearing strain. (Use $\nu = \frac{1}{3}$)

	ϵ_x	ϵ_y	γ_{xy}
7.134	$+160\ \mu$	$-480\ \mu$	$-600\ \mu$
7.135	$-260\ \mu$	$-60\ \mu$	$+480\ \mu$
7.136	$-40\ \mu$	$+760\ \mu$	$+960\ \mu$
7.137	$-300\ \mu$	$-200\ \mu$	$+175\ \mu$

7.138 through 7.141 For the given state of plane strain, use Mohr's circle to determine (a) the orientation and magnitude of the principal strains, (b) the maximum in-plane strain, (c) the maximum shearing strain.

	ϵ_x	ϵ_y	γ_{xy}
7.138	$-90\ \mu$	$-130\ \mu$	$+150\ \mu$
7.139	$+375\ \mu$	$+75\ \mu$	$+125\ \mu$
7.140	$+400\ \mu$	$+200\ \mu$	$+375\ \mu$
7.141	$+60\ \mu$	$+240\ \mu$	$-50\ \mu$

7.142 The strains determined by use of the rosette shown during the test of a rocker arm are:

$$\epsilon_1 = +600 \ \mu \qquad \epsilon_2 = +450 \ \mu \qquad \epsilon_3 = -75 \ \mu$$

Determine (*a*) the in-plane principal strains, (*b*) the in-plane maximum shearing strain.

Fig. P7.142

7.143 Determine the strain ϵ_x, knowing that the following strains have been determined by use of the rosette shown:

$$\epsilon_1 = +720 \times 10^{-6} \ \text{in./in.} \qquad \epsilon_2 = -180 \times 10^{-6} \ \text{in./in.}$$
$$\epsilon_3 = +120 \times 10^{-6} \ \text{in./in.}$$

Fig. P7.143

7.144 The rosette shown has been used to determine the following strains at a point on the surface of a crane hook:

$$\epsilon_1 = +420 \ \mu \qquad \epsilon_2 = -45 \ \mu \qquad \epsilon_4 = +165 \ \mu$$

(*a*) What should be the reading of gage 3? (*b*) Determine the principal strains and the maximum in-plane shearing strain.

Fig. P7.144

7.145 Determine the largest in-plane normal strain, knowing that the following strains have been obtained by use of the rosette shown:

$$\epsilon_1 = -50 \times 10^{-6} \ \text{in./in.} \qquad \epsilon_2 = +360 \times 10^{-6} \ \text{in./in.}$$
$$\epsilon_3 = +315 \times 10^{-6} \ \text{in./in.}$$

Fig. P7.145

7.146 Show that the sum of the three strain measurements made with a 60° rosette is independent of the orientation of the rosette and equal to

$$\epsilon_1 + \epsilon_2 + \epsilon_3 = 3\epsilon_{\text{ave}}$$

where ϵ_{ave} is the abscissa of the center of the corresponding Mohr's circle for strain.

Fig. P7.146

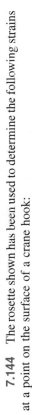

7.147 Using a 45° rosette, the strains ϵ_1, ϵ_2, and ϵ_3 have been determined at a given point. Using Mohr's circle, show that the principal strains are

$$\epsilon_{max,\ min} = \frac{1}{2}(\epsilon_1 + \epsilon_3) \pm \frac{1}{\sqrt{2}}[(\epsilon_1 - \epsilon_2)^2 + (\epsilon_2 - \epsilon_3)^2]^{1/2}$$

(*Hint:* The shaded triangles are congruent.)

Fig. P7.147

7.148 The given state of plane stress is known to exist on the surface of a machine component. Knowing that $E = 200$ GPa and $G = 77$ GPa, determine the direction and magnitude of the three principal strains (*a*) by determining the corresponding state of strain [use Eq. (2.43), page 94, and Eq. (2.38), page 91] and then using Mohr's circle for strain, (*b*) by using Mohr's circle for stress to determine the principal planes and principal stresses and then determining the corresponding strains.

Fig. P7.148

7.149 The following state of strain has been determined on the surface of a cast-iron machine element:

$$\epsilon_1 = -720 \times 10^{-6}\ \text{in./in.} \qquad \epsilon_2 = -400 \times 10^{-6}\ \text{in./in.}$$

$$\gamma = +660 \times 10^{-6}\ \text{rad}$$

Knowing that $E = 10 \times 10^6$ psi and $G = 4 \times 10^6$ psi, determine the principal planes and the principal stresses (*a*) by determining the corresponding state of plane stress [use Eq. (2.36), page 91; Eq. (2.43), page 94; and the first two equations of Prob. 2.75, page 101] and then using Mohr's circle for stress, (*b*) by using Mohr's circle for strain to determine the orientation and magnitude of the principal strains and then determining the corresponding stresses.

7.150 A single strain gage forming an angle $\beta = 30°$ with the vertical is used to determine the gage pressure in the cylindrical steel tank shown. The cylindrical wall of the tank is $\frac{3}{8}$ in. thick, has a 36-in. inner diameter, and is made of a steel with $E = 29 \times 10^6$ psi and $\nu = 0.30$. Determine the pressure in the tank corresponding to a gage reading of 220×10^{-6} in./in.

Fig. P7.150

7.151 Solve Prob. 7.150, assuming that the gage forms an angle $\beta = 60°$ with the vertical.

7.152 A single strain gage is cemented to a solid 96-mm-diameter aluminum shaft at an angle $\beta = 20°$ with a line parallel to the axis of the shaft. Knowing that $G = 27$ GPa, determine the torque **T** corresponding to a gage reading of 400 μ.

Fig. P7.152

7.153 Solve Prob. 7.152, assuming that the gage forms an angle $\beta = 60°$ with a line parallel to the axis of the shaft.

7.154 A centric axial force **P** and a horizontal force $\mathbf{Q}_x$ are both applied at point C of the rectangular bar shown. A 45° strain rosette on the surface of the bar at point A indicates the following strains:

$$\epsilon_1 = -75 \times 10^{-6} \text{ in./in.} \quad \epsilon_2 = +300 \times 10^{-6} \text{ in./in.} \quad \epsilon_3 = +250 \times 10^{-6} \text{ in./in.}$$

Knowing that $E = 29 \times 10^6$ psi and $\nu = 0.30$, determine the magnitudes of **P** and $\mathbf{Q}_x$.

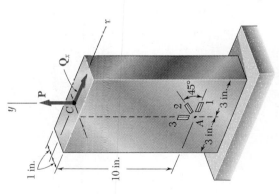

Fig. P7.154

7.155 Solve Prob. 7.154, assuming that the rosette at point A indicates the following strains:

$$\epsilon_1 = -60 \times 10^{-6} \text{ in./in.} \quad \epsilon_2 = +410 \times 10^{-6} \text{ in./in.} \quad \epsilon_3 = +200 \times 10^{-6} \text{ in./in.}$$

The first part of this chapter was devoted to a study of the *transformation of stress* under a rotation of axes and to its application to the solution of engineering problems, and the second part to a similar study of the *transformation of strain*.

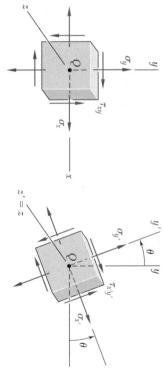

Fig. 7.5

Considering first a state of *plane stress* at a given point Q [Sec. 7.2] and denoting by σ_x, σ_y, and τ_{xy} the stress components associated with the element shown in Fig. 7.5a, we derived the following formulas defining the components $\sigma_{x'}$, $\sigma_{y'}$, and $\tau_{x'y'}$ associated with that element after it had been rotated through an angle θ about the z axis (Fig. 7.5b):

$$\sigma_{x'} = \frac{\sigma_x + \sigma_y}{2} + \frac{\sigma_x - \sigma_y}{2}\cos 2\theta + \tau_{xy}\sin 2\theta \qquad (7.5)$$

$$\sigma_{y'} = \frac{\sigma_x + \sigma_y}{2} - \frac{\sigma_x - \sigma_y}{2}\cos 2\theta - \tau_{xy}\sin 2\theta \qquad (7.7)$$

$$\tau_{x'y'} = -\frac{\sigma_x - \sigma_y}{2}\sin 2\theta + \tau_{xy}\cos 2\theta \qquad (7.6)$$

In Sec. 7.3, we determined the values θ_p of the angle of rotation which correspond to the maximum and minimum values of the normal stress at point Q. We wrote

$$\tan 2\theta_p = \frac{2\tau_{xy}}{\sigma_x - \sigma_y} \qquad (7.12)$$

The two values obtained for θ_p are 90° apart (Fig. 7.11) and define the *principal planes of stress* at point Q. The corresponding values

Fig. 7.11

Transformation of plane stress

of the normal stress are called the *principal stresses* at Q; we obtained

$$\sigma_{max, min} = \frac{\sigma_x + \sigma_y}{2} \pm \sqrt{\left(\frac{\sigma_x - \sigma_y}{2}\right)^2 + \tau_{xy}^2} \quad (7.14)$$

We also noted that the corresponding value of the shearing stress is zero. Next, we determined the values θ_s of the angle θ for which the largest value of the shearing stress occurs. We wrote

$$\tan 2\theta_s = -\frac{\sigma_x - \sigma_y}{2\tau_{xy}} \quad (7.15)$$

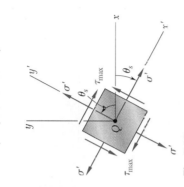

Fig. 7.12

The two values obtained for θ_s are 90° apart (Fig. 7.12). We also noted that the planes of maximum shearing stress are at 45° to the principal planes. The maximum value of the shearing stress for a rotation *in the plane of stress is*

$$\tau_{max} = \sqrt{\left(\frac{\sigma_x - \sigma_y}{2}\right)^2 + \tau_{xy}^2} \quad (7.16)$$

and the corresponding value of the normal stresses is

$$\sigma' = \sigma_{ave} = \frac{\sigma_x + \sigma_y}{2} \quad (7.17)$$

We saw in Sec. 7.4 that *Mohr's circle* provides an alternative method, based on simple geometric considerations, for the analysis

(a)

Fig. 7.17

(b)

of the transformation of plane stress. Given the state of stress shown in black in Fig. 7.17a, we plot point X of coordinates σ_x, $-\tau_{xy}$ and point Y of coordinates σ_y, $+\tau_{xy}$ (Fig. 7.17b). Drawing the circle of diameter XY, we obtain Mohr's circle. The abscissas of the points of intersection A and B of the circle with the horizontal axis represent the principal stresses, and the angle of rotation bringing the diameter XY into AB is twice the angle θ_p defining the principal planes in Fig. 7.17a, with both angles having the same sense. We also noted that diameter DE defines the maximum shearing stress and the orientation of the corresponding plane (Fig. 7.19b) [Example 7.02, Sample Probs. 7.2 and 7.3].

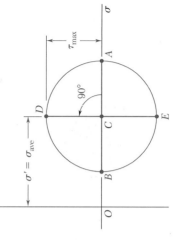

Fig. 7.19b

General state of stress

Considering a *general state of stress* characterized by six stress components [Sec. 7.5], we showed that the normal stress on a plane of arbitrary orientation can be expressed as a quadratic form of the direction cosines of the normal to that plane. This proves the existence of three *principal axes of stress* and three *principal stresses* at any given point. Rotating a small cubic element about each of the three principal axes [Sec. 7.6], we drew the corresponding Mohr's circles which yield the values of σ_{max}, σ_{min}, and τ_{max} (Fig. 7.29). In the particular case of *plane stress*, and if the x and y axes are selected in the plane of stress, point C coincides with the origin O. If A and B are located on opposite sides of O, the maximum shearing stress is equal to the maximum "in-plane" shearing stress as determined in Secs. 7.3 or 7.4. If A and B are located on the same side of O, this will not be the case. If $\sigma_a > \sigma_b > 0$, for instance the maximum shearing stress is equal to $\frac{1}{2}\sigma_a$ and corresponds to a rotation out of the plane of stress (Fig. 7.32).

Fig. 7.29

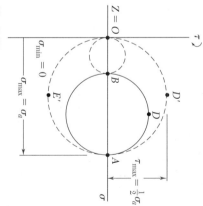

Fig. 7.32

Yield criteria for ductile materials

Yield criteria for ductile materials under plane stress were developed in Sec. 7.7. To predict whether a structural or machine component will fail at some critical point due to yield in the material, we first determine the principal stresses σ_a and σ_b at that point for the given loading condition. We then plot the point of coordinates σ_a and σ_b. If this point falls within a certain area, the component is safe; if it falls outside, the component will fail. The area used with the maximum-shearing-stress criterion is shown in Fig. 7.40 and 7.41. We note that both areas depend upon the value of the yield strength σ_Y of the material.

Fig. 7.40

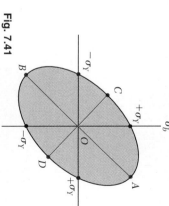

Fig. 7.41

Fracture criteria for brittle materials under plane stress were developed in Sec. 7.8 in a similar fashion. The most commonly used is *Mohr's criterion*, which utilizes the results of various types of test available for a given material. The shaded area shown in Fig. 7.47b is used when the ultimate strengths σ_{UT} and σ_{UC} have been deter-

Fracture criteria for brittle materials

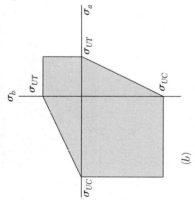

(b)

Fig. 7.47b

mined, respectively, from a tension and a compression test. Again, the principal stresses σ_a and σ_b are determined at a given point of the structural or machine component being investigated. If the corresponding point falls within the shaded area, the component is safe; if it falls outside, the component will rupture.

In Sec. 7.9, we discussed the stresses in *thin-walled pressure vessels* and derived formulas relating the stresses in the walls of the vessels and the *gage pressure p* in the fluid they contain. In the case of a *cylindrical vessel* of inside radius r and thickness t (Fig. 7.51), we obtained the following expressions for the *hoop stress* σ_1 and the *longitudinal stress* σ_2:

Cylindrical pressure vessels

Fig 7.51

$$\sigma_1 = \frac{pr}{t} \qquad \sigma_2 = \frac{pr}{2t} \qquad \text{(7.30, 7.31)}$$

We also found that the *maximum shearing* stress occurs out of the plane of stress and is

$$\tau_{max} = \sigma_2 = \frac{pr}{2t} \qquad \text{(7.34)}$$

In the case of a *spherical vessel* of inside radius r and thickness t (Fig. 7.55), we found that the two principal stresses are equal:

Spherical pressure vessels

Fig. 7.55

$$\sigma_1 = \sigma_2 = \frac{pr}{2t} \qquad \text{(7.36)}$$

Again, the *maximum shearing* stress occurs out of the plane of stress; it is

$$\tau_{max} = \tfrac{1}{2}\sigma_1 = \frac{pr}{4t} \qquad \text{(7.37)}$$

Transformation of plane strain

The last part of the chapter was devoted to the *transformation of strain*. In Secs. 7.10 and 7.11, we discussed the transformation of *plane strain* and introduced *Mohr's circle for plane strain*. The discussion was similar to the corresponding discussion of the transformation of stress, except that, where the shearing stress τ was used, we now used $\frac{1}{2}\gamma$, that is, *half the shearing strain*. The formulas obtained for the transformation of strain under a rotation of axes through an angle θ were

$$\epsilon_{x'} = \frac{\epsilon_x + \epsilon_y}{2} + \frac{\epsilon_x - \epsilon_y}{2}\cos 2\theta + \frac{\gamma_{xy}}{2}\sin 2\theta \quad (7.44)$$

$$\epsilon_{y'} = \frac{\epsilon_x + \epsilon_y}{2} - \frac{\epsilon_x - \epsilon_y}{2}\cos 2\theta - \frac{\gamma_{xy}}{2}\sin 2\theta \quad (7.45)$$

$$\gamma_{x'y'} = -(\epsilon_x - \epsilon_y)\sin 2\theta + \gamma_{xy}\cos 2\theta \quad (7.49)$$

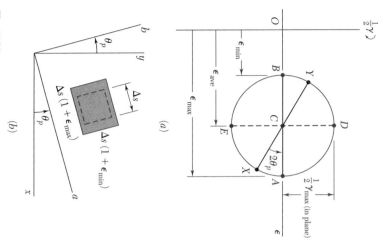

Mohr's circle for strain

Using Mohr's circle for strain (Fig. 7.65), we also obtained the following relations defining the angle of rotation θ_p corresponding to the *principal axes of strain* and the values of the *principal strains* ϵ_{max} and ϵ_{min}:

$$\tan 2\theta_p = \frac{\gamma_{xy}}{\epsilon_x - \epsilon_y} \quad (7.52)$$

$$\epsilon_{max} = \epsilon_{ave} + R \quad \text{and} \quad \epsilon_{min} = \epsilon_{ave} - R \quad (7.51)$$

where

$$\epsilon_{ave} = \frac{\epsilon_x + \epsilon_y}{2} \quad \text{and} \quad R = \sqrt{\left(\frac{\epsilon_x - \epsilon_y}{2}\right)^2 + \left(\frac{\gamma_{xy}}{2}\right)^2} \quad (7.50)$$

The *maximum shearing strain* for a rotation in the plane of strain was found to be

$$\gamma_{max(in\ plane)} = 2R = \sqrt{(\epsilon_x - \epsilon_y)^2 + \gamma_{xy}^2} \quad (7.53)$$

Section 7.12 was devoted to the three-dimensional analysis of strain, with application to the determination of the maximum shearing strain in the particular cases of plane strain and plane stress. In the case of *plane stress*, we also found that the principal strain ϵ_c in a direction perpendicular to the plane of stress could be expressed as follows in terms of the "in-plane" principal strains ϵ_a and ϵ_b:

$$\epsilon_c = -\frac{\nu}{1-\nu}(\epsilon_a + \epsilon_b) \quad (7.59)$$

Strain gages. Strain rosette

Finally, we discussed in Sec.7.13 the use of *strain gages* to measure the normal strain on the surface of a structural element or machine component. Considering a *strain rosette* consisting of three gages aligned along lines forming respectively, angles θ_1, θ_2, and θ_3 with the x axis (Fig. 7.80), we wrote the following relations among the measurements ϵ_1, ϵ_2, ϵ_3 of the gages and the components ϵ_x, ϵ_y, γ_{xy} characterizing the state of strain at that point:

$$\epsilon_1 = \epsilon_x \cos^2\theta_1 + \epsilon_y \sin^2\theta_1 + \gamma_{xy} \sin\theta_1 \cos\theta_1$$
$$\epsilon_2 = \epsilon_x \cos^2\theta_2 + \epsilon_y \sin^2\theta_2 + \gamma_{xy} \sin\theta_2 \cos\theta_2 \quad (7.60)$$
$$\epsilon_3 = \epsilon_x \cos^2\theta_3 + \epsilon_y \sin^2\theta_3 + \gamma_{xy} \sin\theta_3 \cos\theta_3$$

These equations can be solved for ϵ_x, ϵ_y, and γ_{xy}, once ϵ_1, ϵ_2, and ϵ_3 have been determined.

Fig. 7.65

Fig. 7.80

REVIEW PROBLEMS

7.156 The state of stress shown occurs in a steel member made of a grade of steel with a tensile yield strength of 270 MPa. Determine the factor of safety with respect to yield strength, using (a) the maximum-shearing-stress criterion, (b) the maximum distortion-energy criterion.

Fig. P7.156

7.157 A spherical pressure tank has a 1.2-m outer diameter and a uniform wall thickness of 10 mm. Knowing that the gage pressure is 1.25 MPa in the tank, determine (a) the maximum normal stress, (b) the maximum shearing stress, (c) the normal strain on the surface of the tank. (Use $E = 200$ GPa and $\nu = 0.30$.)

7.158 The strains determined by the use of a rosette attached as shown to the surface of a structural member are:

$$\epsilon_1 = 200 \times 10^{-6} \text{ in./in.} \qquad \epsilon_2 = 425 \times 10^{-6} \text{ in./in.} \qquad \epsilon_3 = 480 \times 10^{-6} \text{ in./in.}$$

Determine (a) the orientation and magnitude of the principal strains in the plane of the rosette, (b) the maximum in-plane shearing strain.

Fig. P7.158

7.159 For a state of plane stress it is known that the normal and shearing stresses are directed as shown and that $\sigma_x = 5$ ksi, $\sigma_y = 12$ ksi, and $\sigma_{\max} = 18$ ksi. Determine (a) the orientation of the principal planes, (b) the maximum in-plane shearing stress.

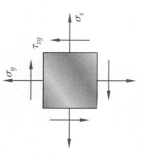

Fig. P7.159

7.160 The compressed-air tank *AB* has an inner diameter of 450 mm and a uniform wall thickness of 6 mm. Knowing that the gage pressure in the tank is 1.2 MPa, determine the maximum normal stress and the maximum in-plane shearing stress at points *a* and *b* on the top of the tank.

Fig. P7.160

7.161 Solve Prob. 7.160, assuming that the 5-kN force applied at *D* is directed vertically upward.

7.162 The steel pressure tank shown has a 30-in. inner diameter and a $\frac{3}{8}$-in. wall thickness. Knowing that the butt-welded seams form an angle of 50° with the longitudinal axis of the tank and that the gage pressure in the tank is 200 psi, determine (*a*) the normal stress perpendicular to the weld, (*b*) the shearing stress parallel to the weld.

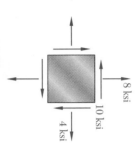

Fig. P7.162

7.163 A square *ABCD* of 2.4-in. side is scribed on the surface of a thin plate while the plate is unloaded. After the plate is loaded, the lengths of sides *AB* and *AD* are observed to have increased, respectively, by 540×10^{-6} in. and 900×10^{-6} in., while the angle *DAB* is observed to have decreased by 360×10^{-6} rad. Knowing that $\nu = \frac{1}{3}$, determine (*a*) the orientation and magnitude of the principal strains, (*b*) the maximum in-plane shearing strain, (*c*) the maximum shearing strain.

Fig. P7.163

7.164 For the state of plane stress shown, determine (*a*) the principal planes, (*b*) the principal stresses, (*c*) the maximum shearing stress.

Fig. P7.164

7.165 The grain of a wooden member forms an angle of 15° with the vertical. For the state of plane stress shown, determine (a) the in-plane shearing stress parallel to the grain, (b) the normal stress perpendicular to the grain.

Fig. P7.165

7.166 A cylindrical steel pressure tank has a 26-in. inner diameter and a uniform $\frac{1}{4}$-in. wall thickness. Knowing that the ultimate stress of the steel used is 65 ksi, determine the maximum allowable gage pressure if a factor of safety of 5.0 must be maintained.

7.167 For the state of stress shown, determine the maximum shearing stress when (a) $\sigma_y = +72$ MPa, (b) $\sigma_y = -72$ MPa.

Fig. P7.167

COMPUTER PROBLEMS

The following problems are to be solved with a computer.

7.C1 A state of plane stress is defined by the stress components σ_x, σ_y, and τ_{xy} associated with the element shown in Fig. P7.C1a. (a) Write a computer program that can be used to calculate the stress components $\sigma_{x'}$, $\sigma_{y'}$, and $\tau_{x'y'}$ associated with the element after it has rotated through an angle θ about the z axis (Fig. P7.C1b). (b) Use this program to solve Probs. 7.13 through 7.16.

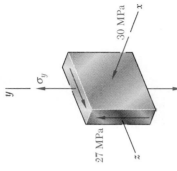

Fig. P7.C1

7.C2 A state of plane stress is defined by the stress components σ_x, σ_y, and τ_{xy} associated with the element shown in Fig. P7.C1a. (*a*) Write a computer program that can be used to determine the principal axes, the principal stresses, the maximum in-plane shearing stress, and the maximum shearing stress. (*b*) Use this program to solve Probs. 7.7, 7.11, 7.66, and 7.67.

7.C3 (*a*) Write a computer program that, for a given state of plane stress and a given yield strength of a ductile material, can be used to determine whether the material will yield. The program should use both the maximum-shearing-strength criterion and the maximum-distortion-energy criterion. It should also print the values of the principal stresses and, if the material does not yield, calculate the factor of safety. (*b*) Use this program to solve Probs. 7.81 through 7.84.

7.C4 (*a*) Write a computer program based on Mohr's fracture criterion for brittle materials that, for a given state of plane stress and given values of the ultimate strength of the material in tension and in compression, can be used to determine whether rupture will occur. The program should also print the values of the principal stresses. (*b*) Use this program to solve Probs. 7.91 and 7.92 and to check the answers given for Probs. 7.93 and 7.94.

7.C5 A state of plane strain is defined by the strain components ϵ_x, ϵ_y, and γ_{xy} associated with the *x* and *y* axes. (*a*) Write a computer program that can be used to calculate the strain components $\epsilon_{x'}$, $\epsilon_{y'}$, and $\gamma_{x'y'}$ associated with the frame of reference $x'y'$ obtained by rotating the *x* and *y* axes through an angle θ. (*b*) Use this program to solve Probs. 7.126 through 7.129.

Fig. P7.C5

7.C6 A state of strain is defined by the strain components ϵ_x, ϵ_y, and γ_{xy} associated with the *x* and *y* axes. (*a*) Write a computer program that can be used to determine the orientation and magnitude of the principal strains, the maximum in-plane shearing strain, and the maximum shearing strain. (*b*) Use this program to solve Probs. 7.134 through 7.137.

7.C7 A state of plane strain is defined by the strain components ϵ_x, ϵ_y, and γ_{xy} measured at a point. (*a*) Write a computer program that can be used to determine the orientation and magnitude of the principal strains, the maximum in-plane shearing strain, and the maximum shearing strain. (*b*) Use this program to solve Probs. 7.138 through 7.141.

7.C8 A rosette consisting of three gages forming, respectively, angles θ_1, θ_2, and θ_3 with the *x* axis is attached to the free surface of a machine component made of a material with a given Poisson's ratio ν. (*a*) Write a computer program that, for given readings ϵ_1, ϵ_2, and ϵ_3 of the gages, can be used to calculate the strain components associated with the *x* and *y* axes and to determine the orientation and magnitude of the three principal strains, the maximum in-plane shearing strain, and the maximum shearing strain. (*b*) Use this program to solve Probs. 7.142 through 7.145.

Principal Stresses under a Given Loading

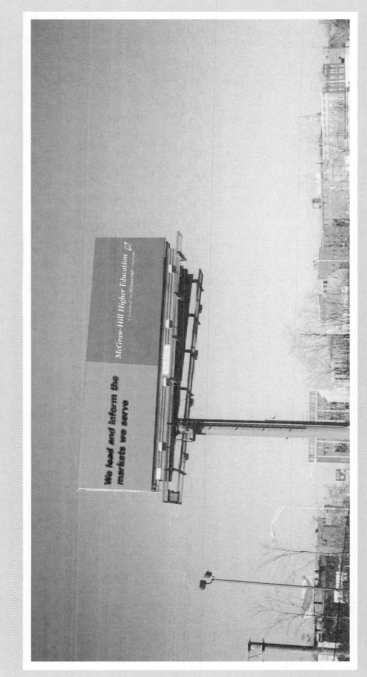

Due to gravity and wind load, the post supporting the sign shown is subjected simultaneously to compression, bending, and torsion. In this chapter you will learn to determine the stresses created by such combined loadings in structures and machine components.

Fig. 8.1

(a)

(b)

Fig. 8.2

(a)

(b)

*8.1. INTRODUCTION

In the first part of this chapter, you will apply to the design of beams and shafts the knowledge that you acquired in Chap. 7 on the transformation of stresses. In the second part of the chapter, you will learn how to determine the principal stresses in structural members and machine elements under given loading conditions.

In Chap. 5 you learned to calculate the maximum normal stress σ_m occurring in a beam under a transverse loading (Fig. 8.1a) and check whether this value exceeded the allowable stress σ_{all} for the given material. If it did, the design of the beam was not acceptable. While the danger for a brittle material is actually to fail in tension, the danger for a ductile material is to fail in shear (Fig. 8.1b). The fact that $\sigma_m > \sigma_{all}$ indicates that $|M|_{max}$ is too large for the cross section selected, but does not provide any information on the actual mechanism of failure. Similarly, the fact that $\tau_m > \tau_{all}$ simply indicates that $|V|_{max}$ is too large for the cross section selected. While the danger for a ductile material is actually to fail in shear (Fig. 8.2a), the danger for a brittle material is to fail in tension under the principal stresses (Fig. 8.2b). The distribution of the principal stresses in a beam will be discussed in Sec. 8.2.

Depending upon the shape of the cross section of the beam and the value of the shear V in the critical section where $|M| = |M|_{max}$, it may happen that the largest value of the normal stress will not occur at the top or bottom of the section, but at some other point within the section. As you will see in Sec. 8.2, a combination of large values of σ_x and τ_{xy} near the junction of the web and the flanges of a W-beam or an S-beam can result in a value of the principal stress σ_{max} (Fig. 8.3) that is larger than the value of σ_m on the surface of the beam.

Fig. 8.3

Section 8.3 will be devoted to the design of transmission shafts subjected to transverse loads as well as to torques. The effect of both the normal stresses due to bending and the shearing stresses due to torsion will be taken into account.

In Sec. 8.4 you will learn to determine the stresses at a given point K of a body of arbitrary shape subjected to a combined loading. First, you will reduce the given loading to forces and couples in the section containing K. Next, you will calculate the normal and shearing stresses at K. Finally, using one of the methods for the transformation of stresses that you learned in Chap. 7, you will determine the principal planes, principal stresses, and maximum shearing stress at K.

*8.2. PRINCIPAL STRESSES IN A BEAM

Consider a prismatic beam AB subjected to some arbitrary transverse loading (Fig. 8.4). We denote by V and M, respectively, the shear and bending moment in a section through a given point C. We recall from Chaps. 5 and 6 that, within the elastic limit, the stresses exerted on a small element with faces perpendicular, respectively, to the x and y axes reduce to the normal stresses $\sigma_m = Mc/I$ if the element is at the free surface of the beam, and to the shearing stresses $\tau_m = VQ/It$ if the element is at the neutral surface (Fig. 8.5).

Fig. 8.4

Fig. 8.6

At any other point of the cross section, an element of material is subjected simultaneously to the normal stresses

$$\sigma_x = -\frac{My}{I} \qquad (8.1)$$

where y is the distance from the neutral surface and I the centroidal moment of inertia of the section, and to the shearing stresses

$$\tau_{xy} = -\frac{VQ}{It} \qquad (8.2)$$

where Q is the first moment about the neutral axis of the portion of the cross-sectional area located above the point where the stresses are computed, and t the width of the cross section at that point. Using either of the methods of analysis presented in Chap. 7, we can obtain the principal stresses at any point of the cross section (Fig. 8.6).

The following question now arises: Can the maximum normal stress $\sigma_{\max}$ at some point within the cross section be larger than the value of $\sigma_m = Mc/I$ computed at the surface of the beam? If it can, then the determination of the largest normal stress in the beam will involve a great deal more than the computation of $|M|_{\max}$ and the use of Eq. (8.1). We can obtain an answer to this question by investigating the distribution

Fig. 8.5

of the principal stresses in a narrow rectangular cantilever beam subjected to a concentrated load **P** at its free end (Fig. 8.7). We recall from Sec. 6.5 that the normal and shearing stresses at a distance x from the load **P** and a distance y above the neutral surface are given, respectively, by Eq. (6.13) and Eq. (6.12). Since the moment of inertia of the cross section is

$$I = \frac{bh^3}{12} = \frac{(b h)(2c)^2}{12} = \frac{Ac^2}{3}$$ (8.3)

where A is the cross-sectional area and c the half-depth of the beam, we write

$$\sigma_x = \frac{Pxy}{I} = \frac{Pxy}{\frac{1}{3}Ac^2} = 3\frac{Pxy}{Ac^2}$$

and

$$\tau_{xy} = \frac{3}{2}\frac{P}{A}\left(1 - \frac{y^2}{c^2}\right)$$ (8.4)

Fig. 8.7

Using the method of Sec. 7.3 or Sec. 7.4, the value of σ_{max} can be determined at any point of the beam. Figure 8.8 shows the results of the computation of the ratios σ_{max}/σ_m and σ_{min}/σ_m in two sections of the beam, corresponding respectively to $x = 2c$ and $x = 8c$. In each

y/c	$x = 2c$			$x = 8c$		
	σ_{min}/σ_m	σ_{max}/σ_m		σ_{min}/σ_m	σ_{max}/σ_m	
1.0	0	1.000		0	1.000	
0.8	−0.010	0.810		−0.001	0.801	
0.6	−0.040	0.640		−0.003	0.603	
0.4	−0.090	0.490		−0.007	0.407	
0.2	−0.160	0.360		−0.017	0.217	
0	−0.250	0.250		−0.063	0.063	
−0.2	−0.360	0.160		−0.217	0.017	
−0.4	−0.490	0.090		−0.407	0.007	
−0.6	−0.640	0.040		−0.603	0.003	
−0.8	−0.810	0.010		−0.801	0.001	
−1.0	−1.000	0		−1.000	0	

Fig. 8.8. Distribution of principal stresses in two transverse sections of a rectangular cantilever beam supporting a single concentrated load.

section, these ratios have been determined at 11 different points, and the orientation of the principal axes has been indicated at each point.[†]

It is clear that σ_{max} does not exceed σ_m in either of the two sections considered in Fig. 8.8 and that, if it does exceed σ_m elsewhere, it will be in sections close to the load $\mathbf{P}$, where σ_m is small compared to τ_m.[‡] But, for sections close to the load $\mathbf{P}$, Saint-Venant's principle does not apply, Eqs. (8.3) and (8.4) cease to be valid, except in the very unlikely case of a load distributed parabolically over the end section (cf. Sec. 6.5), and more advanced methods of analysis taking into account the effect of stress concentrations should be used. We thus conclude that, for beams of rectangular cross section, and within the scope of the theory presented in this text, the maximum normal stress can be obtained from Eq. (8.1).

In Fig. 8.8 the directions of the principal axes were determined at 11 points in each of the two sections considered. If this analysis were extended to a larger number of sections and a larger number of points in each section, it would be possible to draw two orthogonal systems of curves on the side of the beam (Fig. 8.9). One system would consist of curves tangent to the principal axes corresponding to σ_{max} and the other of curves tangent to the principal axes corresponding to σ_{min}. The curves obtained in this manner are known as the *stress trajectories*. A trajectory of the first group (solid lines) defines at each of its points the direction of the largest tensile stress, while a trajectory of the second group (dashed lines) defines the direction of the largest compressive stress.[§]

The conclusion we have reached for beams of rectangular cross section, that the maximum normal stress in the beam can be obtained from Eq. (8.1), remains valid for many beams of nonrectangular cross section. However, when the width of the cross section varies in such a way that large shearing stresses τ_{xy} will occur at points close to the surface of the beam, where σ_x is also large, a value of the principal stress σ_{max} larger than σ_m may result at such points. One should be particularly aware of this possibility when selecting W-beams or S-beams, and calculate the principal stress σ_{max} at the junctions b and d of the web with the flanges of the beam (Fig. 8.10). This is done by determining σ_x and τ_{xy} at that point from Eqs. (8.1) and (8.2), respectively, and using either of the methods of analysis of Chap. 7 to obtain σ_{max} (see Sample Prob. 8.1). An alternative procedure consists of using for τ_{xy} the maximum value of the shearing stress in the section, $\tau_{max} = V/A_{web}$, given by Eq. (6.11) of Sec. 6.4. This leads to a slightly larger, and thus conservative, value of the principal stress σ_{max} at the junction of the web with the flanges of the beam (see Sample Prob. 8.2).

P Tensile

Compressive

Fig. 8.9. Stress trajectories.

Fig. 8.10

[†]See Prob. 8.C2, which refers to the program used to obtain the results shown in Fig. 8.8.

[‡]As will be verified in Prob. 8.C2, σ_{max} exceeds σ_m if $x \leq 0.544c$.

[§]A brittle material, such as concrete, will fail in tension along planes that are perpendicular to the tensile-stress trajectories. Thus, to be effective, steel reinforcing bars should be placed so that they intersect these planes. On the other hand, stiffeners attached to the web of a plate girder will be effective in preventing buckling only if they intersect planes perpendicular to the compressive-stress trajectories.

*8.3. DESIGN OF TRANSMISSION SHAFTS

When we discussed the design of transmission shafts in Sec. 3.7, we considered only the stresses due to the torques exerted on the shafts. However, if the power is transferred to and from the shaft by means of gears or sprocket wheels (Fig. 8.11a), the forces exerted on the gear teeth or sprockets are equivalent to force-couple systems applied at the centers of the corresponding cross sections (Fig. 8.11b). This means that the shaft is subjected to a transverse loading, as well as to a torsional loading.

Fig. 8.11

The shearing stresses produced in the shaft by the transverse loads are usually much smaller than those produced by the torques and will be neglected in this analysis.† The normal stresses due to the transverse loads, however, may be quite large and, as you will see presently, their contribution to the maximum shearing stress τ_{max} should be taken into account.

---

†For an application where the shearing stresses produced by the transverse loads must be considered, see Probs. 8.21 and 8.22.

Consider the cross section of the shaft at some point C. We represent the torque $\mathbf{T}$ and the bending couples $\mathbf{M}_y$ and $\mathbf{M}_z$ acting respectively in a horizontal and a vertical plane by the couple vectors shown (Fig. 8.12a). Since any diameter of the section is a principal axis of inertia for the section, we can replace $\mathbf{M}_y$ and $\mathbf{M}_z$ by their resultant $\mathbf{M}$ (Fig. 8.12b) in order to compute the normal stresses σ_x exerted on the section. We thus find that σ_x is maximum at the end of the diameter perpendicular to the vector representing $\mathbf{M}$ (Fig. 8.13). Recalling that the values of the normal stresses at that point are, respectively, $\sigma_m = Mc/I$ and zero, while the shearing stress is $\tau_m = Tc/J$, we plot the corresponding points X and Y on a Mohr-circle diagram (Fig. 8.14) and determine the value of the maximum shearing stress:

$$\tau_{max} = R = \sqrt{\left(\frac{\sigma_m}{2}\right)^2 + (\tau_m)^2} = \sqrt{\left(\frac{Mc}{2I}\right)^2 + \left(\frac{Tc}{J}\right)^2}$$

Recalling that, for a circular or annular cross section, $2I = J$, we write

$$\tau_{max} = \frac{c}{J}\sqrt{M^2 + T^2} \qquad (8.5)$$

It follows that the minimum allowable value of the ratio J/c for the cross section of the shaft is

$$\frac{J}{c} = \frac{\left(\sqrt{M^2 + T^2}\right)_{max}}{\tau_{all}} \qquad (8.6)$$

where the numerator in the right-hand member of the expression obtained represents the maximum value of $\sqrt{M^2 + T^2}$ in the shaft, and τ_{all} the allowable shearing stress. Expressing the bending moment M in terms of its components in the two coordinate planes, we can also write

$$\frac{J}{c} = \frac{\left(\sqrt{M_y^2 + M_z^2 + T^2}\right)_{max}}{\tau_{all}} \qquad (8.7)$$

Equations (8.6) and (8.7) can be used to design both solid and hollow circular shafts and should be compared with Eq. (3.22) of Sec. 3.7, which was obtained under the assumption of a torsional loading only.

The determination of the maximum value of $\sqrt{M_y^2 + M_z^2 + T^2}$ will be facilitated if the bending-moment diagrams corresponding to M_y and M_z are drawn, as well as a third diagram representing the values of T along the shaft (see Sample Prob. 8.3).

Fig. 8.12

Fig. 8.13

Fig. 8.14

502

SAMPLE PROBLEM 8.1

160 kN
L = 375 mm
A'
A

A 160-kN force is applied as shown at the end of a W200 × 52 rolled-steel beam. Neglecting the effect of fillets and of stress concentrations, determine whether the normal stresses in the beam satisfy a design specification that they be equal to or less than 150 MPa at section A-A'.

SOLUTION

160 kN
0.375 m
V_A
M_A

Shear and Bending Moment.
At section A-A', we have
$$M_A = (160 \text{ kN})(0.375 \text{ m}) = 60 \text{ kN} \cdot \text{m}$$
$$V_A = 160 \text{ kN}$$

12.6 mm
$c = 103$ mm
204 mm
$y_b = 90.4$ mm
206 mm
7.9 mm

$I = 52.7 \times 10^{-6} \text{ m}^4$
$S = 512 \times 10^{-6} \text{ m}^3$

Normal Stresses on Transverse Plane.
Referring to the table of *Properties of Rolled-Steel Shapes* in Appendix C, we obtain the data shown and then determine the stresses σ_a and σ_b.

At point a:
$$\sigma_a = \frac{M_A}{S} = \frac{60 \text{ kN} \cdot \text{m}}{512 \times 10^{-6} \text{ m}^3} = 117.2 \text{ MPa}$$

At point b:
$$\sigma_b = \sigma_a \frac{y_b}{c} = (117.2 \text{ MPa})\frac{90.4 \text{ mm}}{103 \text{ mm}} = 102.9 \text{ MPa}$$

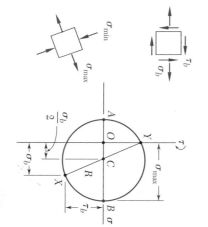

12.6 mm
103 mm
204 mm
96.7 mm
a
b
c
σ_b
σ_a

We note that all normal stresses on the transverse plane are less than 150 MPa.

Shearing Stresses on Transverse Plane
At point a:
$$Q = 0 \quad \tau_a = 0$$

At point b:
$$Q = (204 \times 12.6)(96.7) = 248.6 \times 10^3 \text{ mm}^3 = 248.6 \times 10^{-6} \text{ m}^3$$
$$\tau_b = \frac{V_A Q}{It} = \frac{(160 \text{ kN})(248.6 \times 10^{-6} \text{ m}^3)}{(52.7 \times 10^{-6} \text{ m}^4)(0.0079 \text{ m})} = 95.5 \text{ MPa}$$

τ_b
σ_b
σ_{min}
σ_{max}

Principal Stress at Point b.
The state of stress at point b consists of the normal stress $\sigma_b = 102.4$ MPa and the shearing stress $\tau_b = 95.5$ MPa. We draw Mohr's circle and find

$$\sigma_{max} = \frac{1}{2}\sigma_b + R = \frac{1}{2}\sigma_b + \sqrt{\left(\frac{1}{2}\sigma_b\right)^2 + \tau_b^2}$$
$$= \frac{102.9}{2} + \sqrt{\left(\frac{102.9}{2}\right)^2 + (95.5)^2}$$
$$\sigma_{max} = 159.9 \text{ MPa}$$

A
O
C
R
X
Y
B
τ
$\frac{\sigma_b}{2}$
σ_b
σ_{max}
τ_b
a
b

The specification, $\sigma_{max} \leq 150$ MPa, is *not* satisfied ▲

P
L = 874 mm
W200 × 52
a
b
c

Comment. For this beam and loading, the principal stress at point b is 36% larger than the normal stress at point a. For $L \geq 874$ mm, the maximum normal stress would occur at point a.

SAMPLE PROBLEM 8.2

The overhanging beam AB supports a uniformly distributed load of 3.2 kips/ft and a concentrated load of 20 kips at C. Knowing that for the grade of steel to be used $\sigma_{all} = 24$ ksi and $\tau_{all} = 14.5$ ksi, select the wide-flange shape which should be used.

Req'd S (section mod) to not exceed this

SOLUTION

Reactions at A and D. We draw the free-body diagram of the beam. From the equilibrium equations $\Sigma M_D = 0$ and $\Sigma M_A = 0$ we find the values of R_A and R_D shown in the diagram.

Shear and Bending-Moment Diagrams. Using the methods of Secs. 5.2 and 5.3, we draw the diagrams and observe that

$$|M|_{max} = 239.4 \text{ kip} \cdot \text{ft} = 2873 \text{ kip} \cdot \text{in.} \qquad |V|_{max} = 43 \text{ kips}$$

Section Modulus. For $|M|_{max} = 2873$ kip $\cdot$ in. and $\sigma_{all} = 24$ ksi, the minimum acceptable section modulus of the rolled-steel shape is

$$S_{min} = \frac{|M|_{max}}{\sigma_{all}} = \frac{2873 \text{ kip} \cdot \text{in.}}{24 \text{ ksi}} = 119.7 \text{ in}^3$$

Selection of Wide-Flange Shape. From the table of *Properties of Rolled-Steel Shapes* in Appendix C, we compile a list of shapes which have a section modulus larger than S_{min} and are also the lightest shape in a given depth group.

depth ← *weight*

Shape	S (in³)
W24 × 68	154
W21 × 62	127
W18 × 76	146
W16 × 77	134
W14 × 82	123
W12 × 96	131

We now select the lightest shape available, namely   **W21 × 62** ▼

Shearing Stress. Assuming that the maximum shear is uniformly distributed over the web area of a W21 × 62, we write

used for

designing purposes only

$$\tau_m = \frac{V_{max}}{A_{web}} = \frac{43 \text{ kips}}{8.40 \text{ in}^2} = 5.12 \text{ ksi} < 14.5 \text{ ksi} \qquad \text{(OK)}$$

Principal Stress at Point b. We check that the maximum principal stress at point b in the critical section where M is maximum does not exceed $\sigma_{all} = 24$ ksi. We write

$$\sigma_a = \frac{M_{max}}{S} = \frac{2873 \text{ kip} \cdot \text{in.}}{127 \text{ in}^3} = 22.6 \text{ ksi}$$

$$\sigma_b = \sigma_a \frac{y_b}{c} = (22.6 \text{ ksi}) \frac{9.88 \text{ in.}}{10.50 \text{ in.}} = 21.3 \text{ ksi}$$

$$\tau_b = \frac{V}{A_{web}} = \frac{12.2 \text{ kips}}{8.40 \text{ in}^2} = 1.45 \text{ ksi}$$

We draw Mohr's circle and find

$$\sigma_{max} = \tfrac{1}{2}\sigma_b + R = \frac{21.3 \text{ ksi}}{2} + \sqrt{\left(\frac{21.3 \text{ ksi}}{2}\right)^2 + (1.45 \text{ ksi})^2}$$

$$\sigma_{max} = 21.4 \text{ ksi} \le 24 \text{ ksi} \qquad \text{(OK)} \qquad \blacktriangledown$$

20 kips

3.2 kips/ft

$t_w = 0.400$ in.
W21 × 62
$S = 127$ in³
$A_{web} = t_w d = 8.40$ in²
$d = 21$ in.

$\sigma_a = 22.6$ ksi
$\sigma_b = 21.3$ ksi
$t_f = 0.615$ in.
10.5 in.
9.88 in.

$\tau_b = 1.45$ ksi
$\sigma_b = 21.3$ ksi
$\sigma_b = 21.3$ ksi
$\sigma_{max} = 21.4$ ksi
$\tau_b = 1.45$ ksi

41 kips 59 kips
9 ft   11 ft   5 ft

16 kips
(12.2 kips) (40)
(239.4) (−279.4)
43 kips
−7.8 kips
41 kips
V

moment here is small
x to use
M max, So we also analyze the V here

239.4 kip · ft
−40 kip · ft
M

SAMPLE PROBLEM 8.3

The solid shaft AB rotates at 480 rpm and transmits 30 kW from the motor M to machine tools connected to gears G and H. Knowing that $\tau_{all} = 50$ MPa, determine the smallest permissible diameter for shaft AB.

A $r_C = 60$ $r_D = 80$

Dimensions in mm

$r_C = 0.060$ m
$r_D = 0.080$ m
$r_E = 0.160$ m

$\mathbf{T}_C = 398$ N · m
$\mathbf{T}_D = 199$ N · m
$\mathbf{T}_E = 597$ N · m

$\mathbf{F}_C = 6.63$ kN
$\mathbf{F}_D = 2.49$ kN
$\mathbf{F}_E = 3.73$ kN

SOLUTION

Torques Exerted on Gears. Observing that $f = 480$ rpm = 8 Hz, we determine the torque exerted on gear E:

$$T_E = \frac{P}{2\pi f} = \frac{30 \text{ kW}}{2\pi(8 \text{ Hz})} = 597 \text{ N} \cdot \text{m}$$

The corresponding tangential force acting on the gear is

$$F_E = \frac{T_E}{r_E} = \frac{597 \text{ N} \cdot \text{m}}{0.16 \text{ m}} = 3.73 \text{ kN}$$

A similar analysis of gears C and D yields

$$T_C = \frac{20 \text{ kW}}{2\pi(8 \text{ Hz})} = 398 \text{ N} \cdot \text{m} \qquad F_C = 6.63 \text{ kN}$$

$$T_D = \frac{10 \text{ kW}}{2\pi(8 \text{ Hz})} = 199 \text{ N} \cdot \text{m} \qquad F_D = 2.49 \text{ kN}$$

We now replace the forces on the gears by equivalent force-couple systems.

Bending-Moment and Torque Diagrams

$\mathbf{F}_E = 3.73$ kN

M_z
186 N · m 373 N · m
A C D E B 560 N · m
0.6 m 0.2 m
0.932 kN 2.80 kN

$=$

$\mathbf{F}_C = 6.63$ kN
$\mathbf{F}_D = 2.49$ kN

M_y
A C D E B
580 N · m
1160 N · m
1244 N · m
6.22 kN 0.2 m 0.4 m 2.90 kN

$+$

$\mathbf{T}_C = 398$ N · m
$\mathbf{T}_D = 199$ N · m
$\mathbf{T}_E = 597$ N · m

T
398 N · m
A C D E B
597 N · m

Critical Transverse Section. By computing $\sqrt{M_y^2 + M_z^2 + T^2}$ at all potentially critical sections, we find that its maximum value occurs just to the right of D:

$$(\sqrt{M_y^2 + M_z^2 + T^2})_{max} = \sqrt{(1160)^2 + (373)^2 + (597)^2} = 1357 \text{ N} \cdot \text{m}$$

Diameter of Shaft. For $\tau_{all} = 50$ MPa, Eq. (7.32) yields

$$\frac{J}{c} = \frac{(\sqrt{M_y^2 + M_z^2 + T^2})_{max}}{\tau_{all}} = \frac{1357 \text{ N} \cdot \text{m}}{50 \text{ MPa}} = 27.14 \times 10^{-6} \text{ m}^3$$

For a solid circular shaft of radius c, we have

$$\frac{J}{c} = \frac{\pi}{2} c^3 = 27.14 \times 10^{-6} \qquad c = 0.02585 \text{ m} = 25.85 \text{ mm}$$

$$\text{Diameter} = 2c = 51.7 \text{ mm} \blacktriangleleft$$

PROBLEMS

8.1 An overhanging W250 × 58 rolled-steel beam supports two loads as shown. Knowing that $P = 400$ kN, $a = 0.25$ m, and $\sigma_{all} = 250$ MPa, determine (a) the maximum value of the normal stress σ_m in the beam, (b) the maximum value of the principal stress σ_{max} at the junction of a flange and the web, (c) whether the specified shape is acceptable as far as these two stresses are concerned.

Fig. P8.1

8.2 Solve Prob. 8.1, assuming that $P = 200$ kN and $a = 0.5$ m.

8.3 An overhanging W36 × 300 rolled-steel beam supports a load **P** as shown. Knowing that $P = 320$ kips, $a = 100$ in., and $\sigma_{all} = 29$ ksi, determine (a) the maximum value of the normal stress σ_m in the beam, (b) the maximum value of the principal stress σ_{max} at the junction of a flange and the web, (c) whether the specified shape is acceptable as far as these two stresses are concerned.

Fig. P8.3

8.4 Solve Prob. 8.3, assuming that $P = 400$ kips and $a = 80$ in.

8.5 and 8.6 (a) Knowing that $\sigma_{all} = 160$ MPa and $\tau_{all} = 100$ MPa, select the most economical metric wide-flange shape that should be used to support the loading shown. (b) Determine the values to be expected for σ_m, τ_m, and the principal stress σ_{max} at the junction of a flange and the web of the selected beam.

Fig. P8.5

Fig. P8.6

8.7 and 8.8 (a) Knowing that $\sigma_{all} = 24$ ksi and $\tau_{all} = 14.5$ ksi, select the most economical wide-flange shape that should be used to support the loading shown. (b) Determine the values to be expected for σ_m, τ_m, and the principal stress σ_{max} at the junction of a flange and the web of the selected beam.

Fig. P8.7

Fig. P8.8

8.9 through 8.14 Each of the following problems refers to a rolled-steel shape selected in a problem of Chap. 5 to support a given loading at a minimal cost while satisfying the requirement $\sigma_m \leq \sigma_{all}$. For the selected design, determine (a) the actual value of σ_m in the beam, (b) the maximum value of the principal stress σ_{max} at the junction of a flange and the web.

   8.9 Loading of Prob. 5.81 and selected W410 × 60 shape.
   8.10 Loading of Prob. 5.86 and selected S510 × 98.3 shape.
   8.11 Loading of Prob. 5.83 and selected W27 × 84 shape.
   8.12 Loading of Prob. 5.84 and selected W18 × 50 shape.
   8.13 Loading of Prob. 5.87 and selected S12 × 31.8 shape.
   8.14 Loading of Prob. 5.88 and selected S15 × 42.9 shape.

   8.15 Determine the smallest allowable diameter of the solid shaft ABCD, knowing that $\tau_{all} = 60$ MPa and that the radius of disk B is $r = 80$ mm.

   8.16 Determine the smallest allowable diameter of the solid shaft ABCD, knowing that $\tau_{all} = 60$ MPa and that the radius of disk B is $r = 120$ mm.

Fig. P8.15 and P8.16

   8.17 Using the notation of Sec. 8.3 and neglecting the effect of shearing stresses caused by transverse loads, show that the maximum normal stress in a cylindrical shaft can be expressed as

$$\sigma_{max} = \frac{c}{J}[(M_y^2 + M_z^2)^{\frac{1}{2}} + (M_y^2 + M_z^2 + T^2)^{\frac{1}{2}}]_{max}$$

   8.18 Use the expression given in Prob. 8.17 to determine the maximum normal stress in the solid shaft AB, knowing that its diameter is 36 mm.

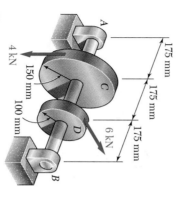

Fig. P8.18

   8.19 The vertical force $\mathbf{P}_1$ and the horizontal force $\mathbf{P}_2$ are applied as shown to disks welded to the solid shaft AD. Knowing that the diameter of the shaft is 1.75 in. and that $\tau_{all} = 8$ ksi, determine the largest permissible magnitude of the force $\mathbf{P}_2$.

   8.20 Solve Prob. 8.19, assuming that the solid shaft AD has been replaced by a hollow shaft of the same material and of inner diameter 1.50 in. and outer diameter 1.75 in.

Fig. P8.19

8.21 It was stated in Sec. 8.3 that the shearing stresses produced in a shaft by the transverse loads are usually much smaller than those produced by the torques. In the preceding problems their effect was ignored and it was assumed that the maximum shearing stress in a given section occurred at point H (Fig. P8.21a) and was equal to the expression obtained in Eq. (8.5), namely,

$$\tau_H = \frac{c}{J} \sqrt{M^2 + T^2}$$

Show that the maximum shearing stress at point K (Fig. P8.21b), where the effect of the shear V is greatest, can be expressed as

$$\tau_K = \frac{c}{J} \sqrt{(M \cos \beta)^2 + (\tfrac{2}{3} cV + T)^2}$$

where β is the angle between the vectors $\mathbf{V}$ and $\mathbf{M}$. It is clear that the effect of the shear V cannot be ignored when $\tau_K \geq \tau_H$. (*Hint.* Only the component of $\mathbf{M}$ along $\mathbf{V}$ contributes to the shearing stress at K.)

8.22 Assuming that the magnitudes of the forces applied to disks A and C of Prob. 8.19 are, respectively, $P_1 = 1080$ lb and $P_2 = 810$ lb, and using the expressions given in Prob. 8.21, determine the values of τ_H and τ_K in a section (*a*) just to the left of B, (*b*) just to the left of C.

8.23 The solid shaft ABC and the gears shown are used to transmit 10 kW from the motor M to a machine tool connected to gear D. Knowing that the motor rotates at 240 rpm and that $\tau_{\text{all}} = 60$ MPa, determine the smallest permissible diameter of shaft ABC.

8.24 Assuming that shaft ABC of Prob. 8.23 is hollow and has an outer diameter of 50 mm, determine the largest permissible inner diameter of the shaft.

8.25 The solid shaft AB rotates at 600 rpm and transmits 80 kW from the motor M to a machine tool connected to gear F. Knowing that $\tau_{\text{all}} = 60$ MPa, determine the smallest permissible diameter of shaft AB.

8.26 Solve Prob. 8.25, assuming that shaft AB rotates at 720 rpm.

Fig. P8.21

Fig. P8.23

Fig. P8.25

8.27 The solid shafts *ABC* and *DEF* and the gears shown are used to transmit 20 hp from the motor *M* to a machine tool connected to shaft *DEF*. Knowing that the motor rotates at 240 rpm and that τ_{all} = 7.5 ksi, determine the smallest permissible diameter of (*a*) shaft *ABC*, (*b*) shaft *DEF*.

Fig. P8.27

8.28 Solve Prob. 8.27, assuming that the motor rotates at 360 rpm.

8.29 The solid shaft *AB* rotates at 450 rpm and transmits 20 kW from the motor *M* to machine tools connected to gears *F* and *G*. Knowing that τ_{all} = 55 MPa and assuming that 8 kW is taken off at gear *F* and 12 kW is taken off at gear *G*, determine the smallest permissible diameter of shaft *AB*.

Fig. P8.29

8.30 Solve Prob. 8.29, assuming that the entire 20 kW is taken off at gear *G*.

*8.4. STRESSES UNDER COMBINED LOADINGS

In Chaps. 1 and 2 you learned to determine the stresses caused by a centric axial load. In Chap. 3, you analyzed the distribution of stresses in a cylindrical member subjected to a twisting couple. In Chap. 4, you determined the stresses caused by bending couples and, in Chaps. 5 and 6, the stresses produced by transverse loads. As you will see presently, you can combine the knowledge you have acquired to determine the stresses in slender structural members or machine components under fairly general loading conditions.

Consider, for example, the bent member *ABDE* of circular cross section, that is subjected to several forces (Fig. 8.15). In order to determine the stresses produced at points *H* or *K* by the given loads, we first pass a section through these points and determine the force-couple system at the centroid *C* of the section that is required to maintain the equilibrium of portion *ABC*.† This system represents the internal forces

Fig. 8.15

---

†The force-couple system at *C* can also be defined as equivalent to the forces acting on the portion of the member located to the right of the section (see Example 8.01).

Fig. 8.16

(a)   (b)

Fig. 8.17

(a)   (b)

Fig. 8.18

Fig. 8.19

Fig. 8.20

in the section and, in general, consists of three force components and three couple vectors that will be assumed directed as shown (Fig. 8.16).

The force **P** is a centric axial force that produces normal stresses in the section. The couple vectors M_y and M_z cause the member to bend and also produce normal stresses in the section. They have therefore been grouped with the force **P** in part *a* of Fig. 8.17 and the sums σ_x of the normal stresses they produce at points *H* and *K* have been shown in part *a* of Fig. 8.18. These stresses can be determined as shown in Sec. 4.14.

On the other hand, the twisting couple **T** and the shearing forces V_y and V_z produce shearing stresses in the section. The sums τ_{xy} and τ_{xz} of the components of the shearing stresses they produce at points *H* and *K* have been shown in part *b* of Fig. 8.18 and can be determined as indicated in Secs. 3.4 and 6.3.† The normal and shearing stresses shown in parts *a* and *b* of Fig. 8.18 can now be combined and displayed at points *H* and *K* on the surface of the member (Fig. 8.19).

The principal stresses and the orientation of the principal planes at points *H* and *K* can be determined from the values of σ_x, τ_{xy}, and τ_{xz} at each of these points by one of the methods presented in Chap. 7 (Fig. 8.20). The values of the maximum shearing stress at each of these points and the corresponding planes can be found in a similar way.

The results obtained in this section are valid only to the extent that the conditions of applicability of the superposition principle (Sec. 2.12) and of Saint-Venant's principle (Sec. 2.17) are met. This means that the stresses involved must not exceed the proportional limit of the material, that the deformations due to one of the loadings must not affect the determination of the stresses due to the others, and that the section used in your analysis must not be too close to the points of application of the given forces. It is clear from the first of these requirements that the method presented here cannot be applied to plastic deformations.

---

†Note that your present knowledge allows you to determine the effect of the twisting couple **T** only in the cases of circular shafts, of members with a rectangular cross section (Sec. 3.12), or of thin-walled hollow members (Sec. 3.13).

EXAMPLE 8.01

Two forces $\mathbf{P}_1$ and $\mathbf{P}_2$, of magnitude $P_1 = 15$ kN and $P_2 = 18$ kN, are applied as shown to the end A of bar AB, which is welded to a cylindrical member BD of radius $c = 20$ mm (Fig. 8.21). Knowing that the distance from A to the axis of member BD is $a = 50$ mm and assuming that all stresses remain below the proportional limit of the material, determine (a) the normal and shearing stresses at point K of the transverse section of member BD located at a distance $b = 60$ mm from end B, (b) the principal axes and principal stresses at K, (c) the maximum shearing stress at K.

Internal Forces in Given Section. We first replace the forces $\mathbf{P}_1$ and $\mathbf{P}_2$ by an equivalent system of forces and couples applied at the center C of the section containing point K (Fig. 8.22). This system, which represents the internal forces in the section, consists of the following forces and couples:

1. A centric axial force $\mathbf{F}$ equal to the force $\mathbf{P}_1$, of magnitude

$$F = P_1 = 15 \text{ kN}$$

2. A shearing force $\mathbf{V}$ equal to the force $\mathbf{P}_2$, of magnitude

$$V = P_2 = 18 \text{ kN}$$

3. A twisting couple $\mathbf{T}$ of torque T equal to the moment of $\mathbf{P}_2$ about the axis of member BD:

$$T = P_2 a = (18 \text{ kN})(50 \text{ mm}) = 900 \text{ N} \cdot \text{m}$$

4. A bending couple $\mathbf{M}_y$, of moment M_y equal to the moment of $\mathbf{P}_1$ about a vertical axis through C:

$$M_y = P_1 a = (15 \text{ kN})(50 \text{ mm}) = 750 \text{ N} \cdot \text{m}$$

5. A bending couple $\mathbf{M}_z$, of moment M_z equal to the moment of $\mathbf{P}_2$ about a transverse, horizontal axis through C:

$$M_z = P_2 b = (18 \text{ kN})(60 \text{ mm}) = 1080 \text{ N} \cdot \text{m}$$

The results obtained are shown in Fig. 8.23.

a. Normal and Shearing Stresses at Point K.
Each of the forces and couples shown in Fig. 8.23 can produce a normal or shearing stress at point K. Our purpose is to compute separately each of these stresses, and then to add the normal stresses and add the shearing stresses. But we must first determine the geometric properties of the section.

Geometric Properties of the Section. We have

$$A = \pi c^2 = \pi (0.020 \text{ m})^2 = 1.257 \times 10^{-3} \text{ m}^2$$
$$I_y = I_z = \tfrac{1}{4}\pi c^4 = \tfrac{1}{4}\pi (0.020 \text{ m})^4 = 125.7 \times 10^{-9} \text{ m}^4$$
$$J_C = \tfrac{1}{2}\pi c^4 = \tfrac{1}{2}\pi (0.020 \text{ m})^4 = 251.3 \times 10^{-9} \text{ m}^4$$

We also determine the first moment Q and the width t of the area of the cross section located above the z axis. Recalling that $\bar{y} = 4c/3\pi$ for a semicircle of radius c, we have

$$Q = A\bar{y} = \left(\tfrac{1}{2}\pi c^2\right)\left(\frac{4c}{3\pi}\right) = \tfrac{2}{3}c^3 = \tfrac{2}{3}(0.020 \text{ m})^3$$
$$= 5.33 \times 10^{-6} \text{ m}^3$$

and

$$t = 2c = 2(0.020 \text{ m}) = 0.040 \text{ m}$$

Normal Stresses. We observe that normal stresses are produced at K by the centric force $\mathbf{F}$ and the bending couple $\mathbf{M}_y$, but that the couple $\mathbf{M}_z$ does not produce any stress at K, since K is located on the neutral axis corresponding to that couple. Determining each sign from Fig. 8.23, we write

$$\sigma_x = -\frac{F}{A} + \frac{M_y c}{I_y} = -11.9 \text{ MPa} + \frac{(750 \text{ N} \cdot \text{m})(0.020 \text{ m})}{125.7 \times 10^{-9} \text{ m}^4}$$
$$= -11.9 \text{ MPa} + 119.3 \text{ MPa}$$
$$\sigma_x = +107.4 \text{ MPa}$$

$b = 60$ mm

$a = 50$ mm

$\mathbf{P}_2 = 18$ kN

$\mathbf{P}_1 = 15$ kN

Fig. 8.21

Fig. 8.22

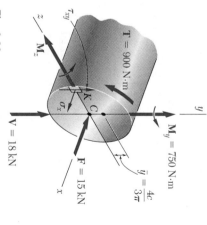

$\mathbf{T} = 900$ N·m

$\mathbf{M}_y = 750$ N·m

$\mathbf{V} = 18$ kN

$\mathbf{F} = 15$ kN

$\bar{y} = \dfrac{4c}{3\pi}$

Fig. 8.23

$\sigma_x = +107.4$ MPa

$\tau_{xy} = -52.5$ MPa

Fig. 8.24

Shearing Stresses. These consist of the shearing stress $(\tau_{xy})_V$ due to the vertical shear **V** and of the shearing stress $(\tau_{xy})_{twist}$ caused by the torque **T**. Recalling the values obtained for Q, t, I_z and J_C, we write

$$(\tau_{xy})_V = +\frac{VQ}{I_z t} = +\frac{(18 \times 10^3 \text{ N})(5.33 \times 10^{-6} \text{ m}^3)}{(125.7 \times 10^{-9} \text{ m}^4)(0.040 \text{ m})}$$
$$= +19.1 \text{ MPa}$$

$$(\tau_{xy})_{twist} = -\frac{Tc}{J_C} = -\frac{(900 \text{ N} \cdot \text{m})(0.020 \text{ m})}{251.3 \times 10^{-9} \text{ m}^4} = -71.6 \text{ MPa}$$

Adding these two expressions, we obtain τ_{xy} at point K.

$$\tau_{xy} = (\tau_{xy})_V + (\tau_{xy})_{twist} = +19.1 \text{ MPa} - 71.6 \text{ MPa}$$
$$\tau_{xy} = -52.5 \text{ MPa}$$

In Fig. 8.24, the normal stress σ_x and the shearing stresses and τ_{xy} have been shown acting on a square element located at K on the surface of the cylindrical member. Note that shearing stresses acting on the longitudinal sides of the element have been included.

b. Principal Planes and Principal Stresses at Point K. We can use either of the two methods of Chap. 7 to determine the principal planes and principal stresses at K. Selecting Mohr's circle, we plot point X of coordinates $\sigma_x = +107.4$ MPa and $-\tau_{xy} = +52.5$ MPa and point Y of coordinates $\sigma_y = 0$ and $+\tau_{xy} = -52.5$ MPa and draw the circle of diameter XY (Fig. 8.25). Observing that

$$OC = CD = \tfrac{1}{2}(107.4) = 53.7 \text{ MPa} \qquad DX = 52.5 \text{ MPa}$$

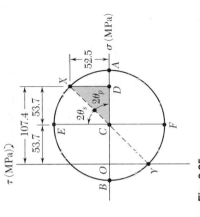

Fig. 8.25

we determine the orientation of the principal planes:

$$\tan 2\theta_p = \frac{DX}{CD} = \frac{52.5}{53.7} = 0.97765 \qquad 2\theta_p = 44.4° \,\rangle$$

$$\theta_p = 22.2° \,\rangle$$

We now determine the radius of the circle,

$$R = \sqrt{(53.7)^2 + (52.5)^2} = 75.1 \text{ MPa}$$

and the principal stresses,

$$\sigma_{max} = OC + R = 53.7 + 75.1 = 128.8 \text{ MPa}$$
$$\sigma_{min} = OC - R = 53.7 - 75.1 = -21.4 \text{ MPa}$$

The results obtained are shown in Fig. 8.26.

$\sigma_{max} = 128.8$ MPa

$\sigma_{min} = -21.4$ MPa

Fig. 8.26

c. Maximum Shearing Stress at Point K. This stress corresponds to points E and F in Fig. 8.25. We have

$$\tau_{max} = CE = R = 75.1 \text{ MPa}$$

Observing that $2\theta_s = 90° - 2\theta_p = 90° - 44.4° = 45.6°$, we conclude that the planes of maximum shearing stress form an angle $\theta_p = 22.8° \,\rangle$ with the horizontal. The corresponding element is shown in Fig. 8.27. Note that the normal stresses acting on this element are represented by OC in Fig. 8.25 and are thus equal to $+53.7$ MPa.

$\tau_{max} = 75.1$ MPa

$\sigma = 53.7$ MPa

Fig. 8.27

SAMPLE PROBLEM 8.4

A horizontal 500-lb force acts at point D of crankshaft AB which is held in static equilibrium by a twisting couple $\mathbf{T}$ and by reactions at A and B. Knowing that the bearings are self-aligning and exert no couples on the shaft, determine the normal and shearing stresses at points H, J, K, and L located at the ends of the vertical and horizontal diameters of a transverse section located 2.5 in. to the left of bearing B.

SOLUTION

Free Body, Entire Crankshaft. $A = B = 250$ lb

$$+\,\circlearrowleft\,\Sigma M_x = 0: \qquad -(500\text{ lb})(1.8\text{ in.}) + T = 0 \qquad T = 900\text{ lb} \cdot \text{in.}$$

Internal Forces in Transverse Section. We replace the reaction $\mathbf{B}$ and the twisting couple $\mathbf{T}$ by an equivalent force-couple system at the center C of the transverse section containing H, J, K, and L.

$$V = B = 250\text{ lb} \qquad T = 900\text{ lb} \cdot \text{in.}$$
$$M_y = (250\text{ lb})(2.5\text{ in.}) = 625\text{ lb} \cdot \text{in.}$$

The geometric properties of the 0.9-in.-diameter section are

$$A = \pi(0.45\text{ in.})^2 = 0.636\text{ in}^2 \qquad I = \tfrac{1}{4}\pi(0.45\text{ in.})^4 = 32.2 \times 10^{-3}\text{ in}^4$$
$$J = \tfrac{1}{2}\pi(0.45\text{ in.})^4 = 64.4 \times 10^{-3}\text{ in}^4$$

Stresses Produced by Twisting Couple T. Using Eq. (3.8), we determine the shearing stresses at points H, J, K, and L and show them in Fig. (a).

$$\tau = \frac{Tc}{J} = \frac{(900\text{ lb} \cdot \text{in.})(0.45\text{ in.})}{64.4 \times 10^{-3}\text{ in}^4} = 6290\text{ psi}$$

Stresses Produced by Shearing Force V. The shearing force $\mathbf{V}$ produces no shearing stresses at points J and L. At points H and K we first compute Q for a semicircle about a vertical diameter and then determine the shearing stress produced by the shear force $V = 250$ lb. These stresses are shown in Fig. (b).

$$Q = \left(\tfrac{1}{2}\pi c^2\right)\left(\frac{4c}{3\pi}\right) = \tfrac{2}{3}c^3 = \tfrac{2}{3}(0.45\text{ in.})^3 = 60.7 \times 10^{-3}\text{ in}^3$$
$$\tau = \frac{VQ}{It} = \frac{(250\text{ lb})(60.7 \times 10^{-3}\text{ in}^3)}{(32.2 \times 10^{-3}\text{ in}^4)(0.9\text{ in.})} = 524\text{ psi}$$

Stresses Produced by the Bending Couple M_y. Since the bending couple $\mathbf{M}_y$ acts in a horizontal plane, it produces no stresses at H and K. Using Eq. (4.15), we determine the normal stresses at points J and L and show them in Fig. (c).

$$\sigma = \frac{|M_y|c}{I} = \frac{(625\text{ lb} \cdot \text{in.})(0.45\text{ in.})}{32.2 \times 10^{-3}\text{ in}^4} = 8730\text{ psi}$$

Summary. We add the stresses shown and obtain the total normal and shearing stresses at points H, J, K, and L.

(a)
$\tau = 6290$ psi

(b)
$\tau = 6290$ psi $\tau = 524$ psi $\tau = 0$

(c)
$\sigma = 0$ $\sigma = 8730$ psi $K\,\sigma = 0$

$\tau = 6290$ psi $\tau = 5770$ psi $\tau = 6810$ psi $\tau = 6290$ psi $\sigma = 8730$ psi $\sigma = 8730$ psi

$A = 250$ lb
$B = 250$ lb
$\mathbf{M}_y = 625$ lb $\cdot$ in.
$\mathbf{V} = 250$ lb
$\mathbf{T} = 900$ lb $\cdot$ in.
0.9-in. diameter

SAMPLE PROBLEM 8.5

Three forces are applied as shown at points A, B, and D of a short steel post. Knowing that the horizontal cross section of the post is a 40×140-mm rectangle, determine the principal stresses, principal planes and maximum shearing stress at point H.

SOLUTION

Internal Forces in Section EFG. We replace the three applied forces by an equivalent force-couple system at the center C of the rectangular section EFG. We have

$$V_x = -30\ \text{kN} \quad P = 50\ \text{kN} \quad V_z = -75\ \text{kN}$$
$$M_x = (50\ \text{kN})(0.130\ \text{m}) - (75\ \text{kN})(0.200\ \text{m}) = -8.5\ \text{kN} \cdot \text{m}$$
$$M_y = 0 \quad M_z = (30\ \text{kN})(0.100\ \text{m}) = 3\ \text{kN} \cdot \text{m}$$

We note that there is no twisting couple about the y axis. The geometric properties of the rectangular section are

$$A = (0.040\ \text{m})(0.140\ \text{m}) = 5.6 \times 10^{-3}\ \text{m}^2$$
$$I_x = \tfrac{1}{12}(0.040\ \text{m})(0.140\ \text{m})^3 = 9.15 \times 10^{-6}\ \text{m}^4$$
$$I_z = \tfrac{1}{12}(0.140\ \text{m})(0.040\ \text{m})^3 = 0.747 \times 10^{-6}\ \text{m}^4$$

Normal Stress at H. We note that normal stresses σ_y are produced by the centric force $\mathbf{P}$ and by the bending couples $\mathbf{M}_x$ and $\mathbf{M}_z$. We determine the sign of each stress by carefully examining the sketch of the force-couple system at C.

$$\sigma_y = +\frac{P}{A} + \frac{|M_z|a}{I_z} - \frac{|M_x|b}{I_x}$$
$$= \frac{50\ \text{kN}}{5.6 \times 10^{-3}\ \text{m}^2} + \frac{(3\ \text{kN} \cdot \text{m})(0.020\ \text{m})}{0.747 \times 10^{-6}\ \text{m}^4} - \frac{(8.5\ \text{kN} \cdot \text{m})(0.025\ \text{m})}{9.15 \times 10^{-6}\ \text{m}^4}$$
$$= 8.93\ \text{MPa} + 80.3\ \text{MPa} - 23.2\ \text{MPa}$$
$$\sigma_y = 66.0\ \text{MPa} \quad \triangledown$$

Shearing Stress at H. Considering first the shearing force $\mathbf{V}_x$, we note that $Q = 0$ with respect to the z axis, since H is on the edge of the cross section. Thus $\mathbf{V}_x$ produces no shearing stress at H. The shearing force $\mathbf{V}_z$ does produce a shearing stress at H and we write

$$Q = A_1\bar{y}_1 = [(0.040\ \text{m})(0.045\ \text{m})](0.0475\ \text{m}) = 85.5 \times 10^{-6}\ \text{m}^3$$
$$\tau_{yz} = \frac{V_z Q}{I_x t} = \frac{(75\ \text{kN})(85.5 \times 10^{-6}\ \text{m}^3)}{(9.15 \times 10^{-6}\ \text{m}^4)(0.040\ \text{m})} \qquad \tau_{yz} = 17.52\ \text{MPa} \quad \triangledown$$

Principal Stresses, Principal Planes, and Maximum Shearing Stress at H. We draw Mohr's circle for the stresses at point H

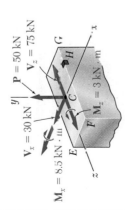

$$\tan 2\theta_p = \frac{17.52}{33.0} \qquad 2\theta_p = 27.96° \qquad \theta_p = 13.98° \quad \blacktriangledown$$
$$R = \sqrt{(33.0)^2 + (17.52)^2} = 37.4\ \text{MPa} \qquad \tau_{max} = 37.4\ \text{MPa} \quad \blacktriangledown$$
$$\sigma_{max} = OA = OC + R = 33.0 + 37.4 \qquad \sigma_{max} = 70.4\ \text{MPa} \quad \blacktriangledown$$
$$\sigma_{min} = OB = OC - R = 33.0 - 37.4 \qquad \sigma_{min} = -7.4\ \text{MPa} \quad \blacktriangledown$$

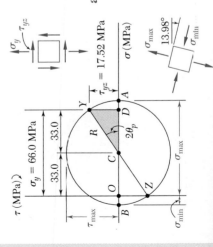

PROBLEMS

8.31 The cantilever beam AB has a rectangular cross section of 150 × 200 mm. Knowing that the tension in cable BD is 10.4 kN and neglecting the weight of the beam, determine the normal and shearing stresses at the three points indicated.

Fig. P8.31

8.32 A 6-kip force is applied to the machine element AB as shown. Knowing that the uniform thickness of the element is 0.8 in., determine the normal and shearing stresses at (a) point a, (b) point b, (c) point c.

Fig. P8.32 and P8.33

8.33 A 6-kip force is applied to the machine element AB as shown. Knowing that the uniform thickness of the element is 0.8 in., determine the normal and shearing stresses at (a) point d, (b) point e, (c) point f.

8.34 through 8.36 Member AB has a uniform rectangular cross section of 10 × 24 mm. For the loading shown, determine the normal and shearing stresses at (a) point H, (b) point K.

Fig. P8.34

Fig. P8.35

Fig. P8.36

8.37   Two forces are applied to the bar shown. At point *a*, determine (*a*) the principal stresses and principal planes, (*b*) the maximum shearing stress.

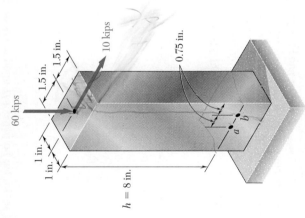

Fig. P8.37 and P8.38

8.38   Two forces are applied to the bar shown. At point *b*, determine (*a*) the principal stresses and principal planes, (*b*) the maximum shearing stress.

8.39   The billboard shown weighs 8000 lb and is supported by a structural tube that has a 15-in. outer diameter and a 0.5-in. wall thickness. At a time when the resultant of the wind pressure is 3 kips located at the center *C* of the billboard, determine the normal and shearing stresses at point *H*.

8.40   The steel pipe *AB* has a 100-mm outer diameter and an 8-mm wall thickness. Knowing that the tension in the cable is 40 kN, determine the normal and shearing stresses at point *H*.

Fig. P8.39

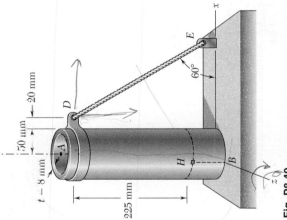

Fig. P8.40

8.41   The axle of a small truck is acted upon by the forces and couple shown. Knowing that the diameter or the axle is 1.42 in., determine the normal and shearing stresses at point H located on the top of the axle.

Fig. P8.41

8.42   A 1.5-kip force and a 9-kip in. couple are applied at the top of the 2.5-in.-diameter cast-iron post shown. Determine the normal and shearing stresses at (*a*) point H, (*b*) point K.

Fig. P8.42

8.43   For the truck axle and loading of Prob. 8.41, determine the principal stresses and the maximum shearing stress at point H.

8.44   For the post and loading of Prob. 8.42, determine the principal stresses and the maximum shearing stress at (*a*) point H, (*b*) point K.

8.45   The steel pipe AB has a 72-mm outer diameter and a 5-mm wall thickness. Knowing that arm CDE is rigidly attached to the pipe, determine the principal stresses, principal planes, and maximum shearing stress at point H.

Fig. P8.45 and P8.46

8.46   The steel pipe AB has a 72-mm outer diameter and a 5-mm wall thickness. Knowing that arm CDE is rigidly attached to the pipe, determine the principal stresses and the maximum shearing stress at point K.

8.47 Three forces are applied to a 4-in.-diameter plate that is attached to the solid 1.8-in. diameter shaft *AB*. At point *H*, determine (*a*) the principal stresses and principal planes, (*b*) the maximum shearing stress.

Fig. P8.47

Fig. P8.48

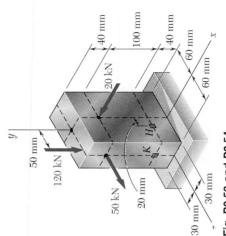

Fig. *P8.49* and *P8.50*

Fig. P8.53 and P8.54

8.48 A 13-kN force is applied as shown to the 60-mm-diameter cast-iron post *ABD*. At point *H*, determine (*a*) the principal stresses and principal planes, (*b*) the maximum shearing stress.

8.49 Three forces are applied to the cantilever beam shown. Determine the normal and shearing stresses at point *H*.

8.50 Three forces are applied to the cantilever beam shown. Determine the normal and shearing stresses at point *K*.

8.51 For the beam and loading of Prob. 8.49, determine the principal stresses and the maximum shearing stress at point *H*.

8.52 For the beam and loading of Prob. 8.50, determine the principal stresses and the maximum shearing stress at point *K*.

8.53 Three forces are applied to a steel post as shown. Determine the normal and shearing stresses at point *H*.

8.54 Three forces are applied to a steel post as shown. Determine the normal and shearing stresses at point *K*.

8.55 Two forces are applied to the small post *BD* as shown. Knowing that the vertical portion of the post has a cross section of 1.5×2.4 in., determine the principal stresses, principal planes, and maximum shearing stress at point *H*.

8.56 Solve Prob 8.55, assuming that the magnitude of the 6000-lb force is reduced to 1500 lb.

Fig. P8.55

8.57 Three forces are applied to the machine component *ABD* as shown. Knowing that the cross section containing point *H* is a 20×40-mm rectangle, determine the principal stresses and the maximum shearing stress at point *H*.

8.58 Solve Prob. 8.57, assuming that the magnitude of the 2.5-kN force is increased to 10 kN.

Fig. P8.57

8.59 Three steel plates, each 13 mm thick, are welded together to form a cantilever beam. For the loading shown, determine the normal and shearing stresses at points *a* and *b*.

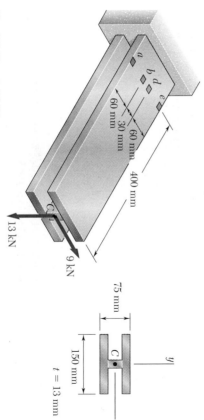

Fig. P8.59 and P8.60

8.60 Three steel plates, each 13 mm thick, are welded together to form a cantilever beam. For the loading shown, determine the normal and shearing stresses at points *d* and *e*.

8.61 For the beam and loading of Prob. 8.59, determine the principal stresses and the maximum shearing stress at points *a* and *b*.

8.62 For the beam and loading of Prob. 8.60, determine the principal stresses and the maximum shearing stress at points *d* and *e*.

8.63 Two forces are applied to a W8 × 28 rolled-steel beam as shown. Determine the principal stresses, principal planes, and maximum shearing stress at point *a*.

8.64 Two forces are applied to a W8 × 28 rolled-steel beam as shown. Determine the principal stresses, principal planes, and maximum shearing stress at point *b*.

Fig. P8.63 and P8.64

8.65 Two forces, $\mathbf{P}_1$ and $\mathbf{P}_2$, are applied as shown in directions perpendicular to the longitudinal axis of a W310 × 60 beam. Knowing that $\mathbf{P}_1 = 25$ kN and $\mathbf{P}_2 = 24$ kN, determine the principal stresses and the maximum shearing stress at point *a*.

8.66 Two forces, $\mathbf{P}_1$ and $\mathbf{P}_2$, are applied as shown in directions perpendicular to the longitudinal axis of a W310 × 60 beam. Knowing that $\mathbf{P}_1 = 25$ kN and $\mathbf{P}_2 = 24$ kN, determine the principal stresses and the maximum shearing stress at point *b*.

Fig. P8.65 and P8.66

8.67 A force **P** is applied to a cantilever beam by means of a cable attached to a bolt located at the center of the free end of the beam. Knowing that **P** acts in a direction perpendicular to the longitudinal axis of the beam, determine (*a*) the normal stress at point *a* in terms of P, b, h, l, and β, (*b*) the values of β for which the normal stress at *a* is zero.

Fig. P8.67

8.68 A vertical force **P** is applied at the center of the free end of cantilever beam AB. (*a*) If the beam is installed with the web vertical ($\beta = 0$) and with its longitudinal axis AB horizontal, determine the magnitude of the force **P** for which the normal stress at point a is $+120$ MPa. (*b*) Solve part a, assuming that the beam is installed with $\beta = 3°$.

$l = 1.25$ m
β
P
W250 × 44.8
A
B
a

Fig. P8.68

Fig. P8.69

A
$\dfrac{d}{2}$
d
a
B
P

*8.69 A 500-lb force **P** is applied to a wire that is wrapped around the bar AB as shown. Knowing that the cross section of the bar is a square of side $d = 0.75$ in., determine the principal stresses and the maximum shearing stress at point a.

*8.70 A vertical 20-kN force is applied to end A of the bar AB, which is welded to an extruded aluminum tube. Knowing that the tube has a uniform wall thickness of 6 mm, determine the shearing stress at points a, b, and c.

*8.71 For the tube and loading of Prob. 8.70, determine the principal stresses and the maximum shearing stress at point b.

*8.72 Knowing that the structural tube shown has a uniform wall thickness of 0.3 in., determine the principal stresses, principal planes, and maximum shearing stress at (*a*) point H, (*b*) point K.

20 kN
A
100 mm
80 mm
B
25 mm
50 mm
150 mm
50 mm
6 mm
a
b
c

Fig. P8.70

2 in.
K
4 in.
10 in.
H
6 in.
3 in.
0.15 in.
9 kips

Fig. P8.72

REVIEW AND SUMMARY
FOR CHAPTER 8

This chapter was devoted to the determination of the principal stresses in beams, transmission shafts, and bodies of arbitrary shape subjected to combined loadings.

We first recalled in Sec. 8.2 the two fundamental relations derived in Chaps. 5 and 6 for the normal stress σ_x and the shearing stress τ_{xy} at any given point of a cross section of a prismatic beam,

$$\sigma_x = -\frac{My}{I} \qquad \tau_{xy} = -\frac{VQ}{It} \qquad (8.1,\ 8.2)$$

where V = shear in the section
M = bending moment in the section
y = distance of the point from the neutral surface
I = centroidal moment of inertia of the cross section
Q = first moment about the neutral axis of the portion of the cross section located above the given point
t = width of the cross section at the given point

Using one of the methods presented in Chap. 7 for the transformation of stresses, we were able to obtain the principal planes and principal stresses at the given point (Fig. 8.6).

We investigated the distribution of the principal stresses in a narrow rectangular cantilever beam subjected to a concentrated load **P** at its free end and found that in any given transverse section—except close to the point of application of the load—the maximum principal stress σ_{max} did not exceed the maximum normal stress σ_m occurring at the surface of the beam.

While this conclusion remains valid for many beams of nonrectangular cross section, it may not hold for W-beams or S-beams, where σ_{max} at the junctions b and d of the web with the flanges of the beam (Fig. 8.10) may exceed the value of σ_m occurring at points a and e. Therefore, the design of a rolled-steel beam should include the computation of the maximum principal stress at these points. (See Sample Probs. 8.1 and 8.2.)

Principal planes and principal stresses in a beam

Fig. 8.6

Fig. 8.10

Design of transmission shafts
under transverse loads

In Sec. 8.3, we considered the design of *transmission shafts* subjected to *transverse loads* as well as to torques. Taking into account
the effect of both the normal stresses due to the bending moment M
and the shearing stresses due to the torque T in any given transverse
section of a cylindrical shaft (either solid or hollow), we found that
the minimum allowable value of the ratio J/c for the cross section
was

$$\frac{J}{c} = \frac{(\sqrt{M^2 + T^2})_{\max}}{\tau_{\text{all}}} \qquad (8.6)$$

Stresses under general
loading conditions

In preceding chapters, you learned to determine the stresses in
prismatic members caused by axial loadings (Chaps. 1 and 2), torsion (Chap. 3), bending (Chap. 4), and transverse loadings (Chaps.
5 and 6). In the second part of this chapter (Sec. 8.4), we combined
this knowledge to determine stresses under more general loading conditions.

For instance, to determine the stresses at point H or K of the bent
member shown in Fig. 8.15, we passed a section through these points
and replaced the applied loads by an equivalent force-couple system
at the centroid C of the section (Fig. 8.16). The normal and shearing stresses produced at H or K by each of the forces and couples
applied at C were determined and then combined to obtain the resulting normal stress σ_x and the resulting shearing stresses τ_{xy} and
τ_{xz} at H or K. Finally, the principal stresses, the orientation of the
principal planes, and the maximum shearing stress at point H or K
were determined by one of the methods presented in Chap. 7 from
the values obtained for σ_x, τ_{xy}, and τ_{xz}.

Fig. 8.15

Fig. 8.16

REVIEW PROBLEMS

8.73 (*a*) Knowing that $\sigma_{all} = 165$ MPa and $\tau_{all} = 100$ MPa, select the most economical wide-flange shape that should be used to support the loading shown. (*b*) Determine the values to be expected for σ_m, τ_m, and the principal stress σ_{max} at the junction of a flange and the web of the selected beam.

Fig. P8.73

8.74 Knowing that the shear and bending moment in a given section of a W21 × 101 rolled-steel beam are, respectively, 120 kips and 300 kip · ft, determine the values in that section of (*a*) the maximum normal stress σ_m, (*b*) the principal stress σ_{max} at the junction of a flange and the web.

8.75 The 6 kN force is vertical and the force **P** is parallel to the z axis. Knowing that $\tau_{all} = 60$ MPa, determine the smallest permissible diameter of the solid shaft AD.

Fig. P8.75

8.76 The two 500-lb forces are vertical and the force **P** is parallel to the z axis. Knowing that $\tau_{all} = 8$ ksi, determine the smallest permissible diameter of the solid shaft AE.

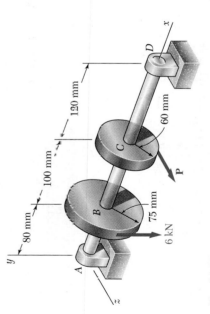

Fig. P8.76

8.77   The solid shaft *AE* rotates at 600 rpm and transmits 60 hp from the motor *M* to machine tools connected to gears *G* and *H*. Knowing that $\tau_{\text{all}} = 8$ ksi and that 40 hp is taken off at gear *G* and 20 hp is taken off at gear *H*, determine the smallest permissible diameter of shaft *AE*.

Fig. P8.77

8.78   The motor *M* rotates at 300 rpm and transmits 30 kW to the solid shaft *AB* through a flexible connection. Half of this power is transferred to a machine tool connected to gear *E* and the other half to a machine tool connected to gear *F*. Knowing that $\tau_{\text{all}} = 60$ MPa, determine the smallest permissible diameter of shaft *AB*.

Fig. P8.78

8.79   Several forces are applied to the pipe assembly shown. Knowing that each section of pipe has inner and outer diameters respectively equal to 36 mm and 42 mm, determine the normal and shearing stresses at point *H* located at the top of the outer surface of the pipe.

Fig. P8.79

Dimensions in mm

8.80 A vertical force **P** of magnitude 250 N is applied to the crank at point A. Knowing that the shaft BDE has a diameter of 18 mm, determine the principal stresses and the maximum shearing stress at point H located at the top of the shaft, 50 mm to the right of support D.

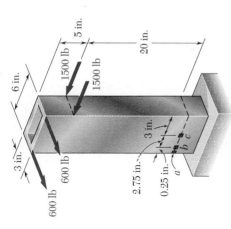

Fig. P8.80

8.81 Knowing that the structural tube shown has a uniform wall thickness of 0.25 in., determine the normal and shearing stresses at the three points indicated.

Fig. P8.81

8.82 For the post and loading shown, determine the principal stresses, principal planes, and maximum shearing stress at point H.

8.83 For the post and loading shown, determine the principal stresses, principal planes, and maximum shearing stress at point K.

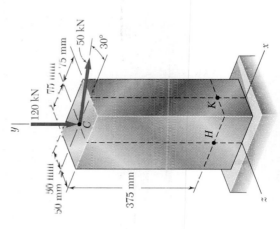

Fig. P8.82 and P8.83

8.84 Forces are applied at points A and B of the solid cast-iron bracket shown. Knowing that the bracket has a diameter of 0.8 in., determine the principal stresses and the maximum shearing stress (a) at point H, (b) at point K.

Fig. P8.84

COMPUTER PROBLEMS

The following problems are designed to be solved with a computer.

8.C1 Let us assume that the shear V and the bending moment M have been determined in a given section of a rolled-steel beam. Write a computer program to calculate in that section, from the data available in Appendix C, (*a*) the maximum normal stress σ_m, (*b*) the principal stress σ_{max} at the junction of a flange and the web. Use this program to solve parts *a* and *b* of the following problems:

(1) Prob. 8.1 (Use $V = 400$ kN and $M = 100$ kN · m)
(2) Prob. 8.2 (Use $V = 200$ kN and $M = 100$ kN · m)
(3) Prob. 8.3 (Use $V = 320$ kips and $M = 32 \times 10^3$ kip · in.)
(4) Prob. 8.74.

8.C2 A cantilever beam AB with a rectangular cross section of width b and depth $2c$ supports a single concentrated load P at its end A. Write a computer program to calculate, for any values of x/c and y/c, (*a*) the ratios σ_{max}/σ_m and σ_{min}/σ_m, where σ_{max} and σ_{min} are the principal stresses at point $K(x, y)$ and σ_m the maximum normal stress in the same transverse section, (*b*) the angle θ_p that the principal planes at K form with a transverse and a horizontal plane through K. Use this program to check the values shown in Fig. 8.8 and to verify that σ_m exceeds σ_m if $x \leq 0.544c$, as indicated in the second footnote on page 499.

Fig. P8.C2

8.C3 Disks $D_1, D_2, \ldots, D_n$ are attached as shown in Fig. P8.C3 to the solid shaft AB of length L, uniform diameter d, and allowable shearing stress τ_{all}. Forces $P_1, P_2, \ldots, P_n$ of known magnitude (except for one of them) are applied to the disks, either at the top or bottom of a vertical diameter, or at the left or right end of a horizontal diameter. Denoting by r_i the radius of disk D_i and by c_i its distance from the support at A, write a computer program to calculate (*a*) the magnitude of the unknown force P_i, (*b*) the smallest permissible value of the diameter d of shaft AB. Use this program to solve Probs. 8.75 and 8.76.

Fig. P8.C3

8.C4 The solid shaft AB of length L, uniform diameter d, and allowable shearing stress τ_{all} rotates at a given speed expressed in rpm (Fig. P8.C4). Gears $G_1, G_2, \ldots, G_n$ are attached to the shaft and each of these gears meshes with another gear (not shown), either at the top or bottom of its vertical diameter, or at the left or right end of its horizontal diameter. One of these other gears is connected to a motor and the rest of them to various machine tools. Denoting by r_i the radius of gear G_i, by c_i its distance from the support at A, and by P_i the power transmitted to that gear ($+$ sign) or taken off that gear ($-$ sign), write a computer program to calculate the smallest permissible value of the diameter d of shaft AB. Use this program to solve Probs. 8.25, 8.29, and 8.77.

Fig. P8.C4

8.C5 Write a computer program that can be used to calculate the normal and shearing stresses at points with given coordinates y and z located on the surface of a machine part having a rectangular cross section. The internal forces are known to be equivalent to the force-couple system shown. Write the program so that the loads and dimensions can be expressed in either SI or U.S. customary units. Use this program to solve (*a*) Prob. 8.50, (*b*) Prob. 8.53.

Fig. P8.C5

8.C6 Member AB has a rectangular cross section of 10×24 mm. For the loading shown, write a computer program that can be used to determine the normal and shearing stresses at points H and K for values of d from 0 to 120 mm, using 15-mm increments. Use this program to solve Prob. 8.35.

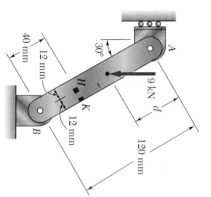

Fig. P8.C6

*8.C7 The structural tube shown has a uniform wall thickness of 0.3 in. A 9-kip force is applied to a bar (not shown) that is welded to the end of the tube. Write a computer program that can be used to determine, for any given value of c, the principal stresses, principal planes, and maximum shearing stress at point H for values of d from -3 in. to 3 in., using one-inch increments. Use this program to solve Prob. 8.72*a*.

Fig. P8.C7

Deflection of Beams

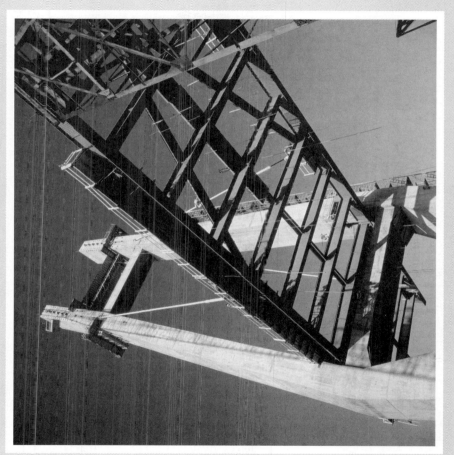

The photo shows a cable-stayed bridge during construction. The design of beams in the bridge deck is based on both strength considerations and deflection evaluations.

$[y_A = 0]$
$[\theta_A = 0]$

(a) Cantilever beam

$[y_A = 0]$ $[y_B = 0]$

(b) Simply supported beam

Fig. 9.1

9.1. INTRODUCTION

In the preceding chapter we learned to design beams for strength. In this chapter we will be concerned with another aspect in the design of beams, namely, the determination of the *deflection*. Of particular interest is the determination of the *maximum deflection* of a beam under a given loading, since the design specifications of a beam will generally include a maximum allowable value for its deflection. Also of interest is that a knowledge of the deflections is required to analyze *indeterminate beams*. These are beams in which the number of reactions at the supports exceeds the number of equilibrium equations available to determine these unknowns.

We saw in Sec. 4.4 that a prismatic beam subjected to pure bending is bent into an arc of circle and that, within the elastic range, the curvature of the neutral surface can be expressed as

$$\frac{1}{\rho} = \frac{M}{EI} \tag{4.21}$$

where M is the bending moment, E the modulus of elasticity, and I the moment of inertia of the cross section about its neutral axis.

When a beam is subjected to a transverse loading, Eq. (4.21) remains valid for any given transverse section, provided that Saint-Venant's principle applies. However, both the bending moment and the curvature of the neutral surface will vary from section to section. Denoting by x the distance of the section from the left end of the beam, we write

$$\frac{1}{\rho} = \frac{M(x)}{EI} \tag{9.1}$$

The knowledge of the curvature at various points of the beam will enable us to draw some general conclusions regarding the deformation of the beam under loading (Sec. 9.2).

To determine the slope and deflection of the beam at any given point, we first derive the following second-order linear differential equation, which governs the *elastic curve* characterizing the shape of the deformed beam (Sec. 9.3):

$$\frac{d^2y}{dx^2} = \frac{M(x)}{EI}$$

If the bending moment can be represented for all values of x by a single function $M(x)$, as in the case of the beams and loadings shown in Fig. 9.1, the slope $\theta = dy/dx$ and the deflection y at any point of the beam may be obtained through two successive integrations. The two constants of integration introduced in the process will be determined from the boundary conditions indicated in the figure.

However, if different analytical functions are required to represent the bending moment in various portions of the beam, different differ-

$[x = L, y_2 = 0]$

$[x = \frac{1}{4}L, \; \theta_1 = \theta_2]$

$[x = \frac{1}{4}L, \; y_1 = y_2]$

$[x = 0, y_1 = 0]$

Fig. 9.2

ential equations will also be required, leading to different functions defining the elastic curve in the various portions of the beam. In the case of the beam and loading of Fig. 9.2, for example, two differential equations are required, one for the portion of beam AD and the other for the portion DB. The first equation yields the functions θ_1 and y_1, and the second the functions θ_2 and y_2. Altogether, four constants of integration must be determined; two will be obtained by writing that the deflection is zero at A and B, and the other two by expressing that the portions of beam AD and DB have the same slope and the same deflection at D.

You will observe in Sec. 9.4 that in the case of a beam supporting a distributed load $w(x)$, the elastic curve can be obtained directly from $w(x)$ through four successive integrations. The constants introduced in this process will be determined from the boundary conditions imposed by the supports.

In Sec. 9.5, we will discuss *statically indeterminate beams* where the reactions at the supports involve four or more unknowns. The three equilibrium equations must be supplemented with equations obtained from the boundary conditions imposed by the supports.

The method described earlier for the determination of the elastic curve when several functions are required to represent the bending moment M can be quite laborious, since it requires matching slopes and deflections at every transition point. You will see in Sec. 9.6 that the use of *singularity functions* (previously discussed in Sec. 5.5) considerably simplifies the determination of θ and y at any point of the beam.

The next part of the chapter (Secs. 9.7 and 9.8) is devoted to the *method of superposition*, which consists of determining separately, and then adding, the deflection and slope caused by the various loads applied to a beam. This procedure can be facilitated by the use of the table in Appendix D, which gives the slopes and deflections of beams for various loadings and types of support.

In Sec. 9.9, certain geometric properties of the elastic curve will be used to determine the deflection and slope of a beam at a given point. Instead of expressing the bending moment as a function $M(x)$ and integrating this function analytically, the diagram representing the variation of M/EI over the length of the beam will be drawn and two *moment-area theorems* will be derived. The *first moment-area theorem* will enable us to calculate the angle between the tangents to the beam at two points; the *second moment-area theorem* will be used to calculate the vertical distance from a point on the beam to a tangent at a second point.

The moment-area theorems will be used in Sec. 9.10 to determine the slope and deflection at selected points of cantilever beams and beams with symmetric loadings. In Sec. 9.11 you will find that in many cases the areas and moments of areas defined by the M/EI diagram may be more easily determined if you draw the *bending-moment diagram by parts*. As you study the moment-area method, you will observe that this method is particularly effective in the case of *beams of variable cross section*.

Beams with unsymmetric loadings and overhanging beams will be considered in Sec. 9.12. Since for an unsymmetric loading the maximum deflection does not occur at the center of a beam, you will learn in Sec. 9.13 how to locate the point where the tangent is horizontal in order to determine the *maximum deflection*. Section 9.14 will be devoted to the solution of problems involving *statically indeterminate beams*.

9.2. DEFORMATION OF A BEAM UNDER TRANSVERSE LOADING

At the beginning of this chapter, we recalled Eq. (4.21) of Sec. 4.4, which relates the curvature of the neutral surface and the bending moment in a beam in pure bending. We pointed out that this equation remains valid for any given transverse section of a beam subjected to a transverse loading, provided that Saint-Venant's principle applies. However, both the bending moment and the curvature of the neutral surface will vary from section to section. Denoting by x the distance of the section from the left end of the beam, we write

$$\frac{1}{\rho} = \frac{M(x)}{EI} \tag{9.1}$$

Consider, for example, a cantilever beam AB of length L subjected to a concentrated load $\mathbf{P}$ at its free end A (Fig. 9.3a). We have $M(x) = -Px$ and, substituting into (9.1),

$$\frac{1}{\rho} = -\frac{Px}{EI}$$

which shows that the curvature of the neutral surface varies linearly with x, from zero at A, where ρ_A itself is infinite, to $-PL/EI$ at B, where $|\rho_B| = EI/PL$ (Fig. 9.3b).

Consider now the overhanging beam AD of Fig. 9.4a that supports two concentrated loads as shown. From the free-body diagram of the beam (Fig. 9.4b), we find that the reactions at the supports are $R_A = 1$ kN and $R_C = 5$ kN, respectively, and draw the corresponding bending-moment diagram (Fig. 9.5a). We note from the diagram that M, and thus the curvature of the beam, are both zero at each end of the beam, and also at a point E located at $x = 4$ m. Between A and E the bending moment is positive and the beam is concave upward; between

(a)

(b)

Fig. 9.3

(a)

(b)

Fig. 9.4

Fig. 9.5

Fig. 9.6

E and D the bending moment is negative and the beam is concave downward (Fig. 9.5b). We also note that the largest value of the curvature (i.e., the smallest value of the radius of curvature) occurs at the support C, where $|M|$ is maximum.

From the information obtained on its curvature, we get a fairly good idea of the shape of the deformed beam. However, the analysis and design of a beam usually require more precise information on the *deflection* and the *slope* of the beam at various points. Of particular importance is the knowledge of the *maximum deflection* of the beam. In the next section Eq. (9.1) will be used to obtain a relation between the deflection y measured at a given point Q on the axis of the beam and the distance x of that point from some fixed origin (Fig. 9.6). The relation obtained is the equation of the *elastic curve*, i.e., the equation of the curve into which the axis of the beam is transformed under the given loading (Fig. 9.6b).†

9.3. EQUATION OF THE ELASTIC CURVE

We first recall from elementary calculus that the curvature of a plane curve at a point $Q(x,y)$ of the curve can be expressed as

$$\frac{1}{\rho} = \frac{\dfrac{d^2y}{dx^2}}{\left[1 + \left(\dfrac{dy}{dx}\right)^2\right]^{3/2}} \qquad (9.2)$$

where dy/dx and d^2y/dx^2 are the first and second derivatives of the function $y(x)$ represented by that curve. But, in the case of the elastic curve of a beam, the slope dy/dx is very small, and its square is negligible compared to unity. We write, therefore,

$$\frac{1}{\rho} = \frac{d^2y}{dx^2} \qquad (9.3)$$

Substituting for $1/\rho$ from (9.3) into (9.1), we have

$$\boxed{\frac{d^2y}{dx^2} = \frac{M(x)}{EI}} \qquad (9.4)$$

The equation obtained is a second-order linear differential equation; it is the governing differential equation for the elastic curve.

†It should be noted that, in this chapter, y represents a vertical displacement, while it was used in previous chapters to represent the distance of a given point in a transverse section from the neutral axis of that section.

The product EI is known as the *flexural rigidity* and, if it varies along the beam, as in the case of a beam of varying depth, we must express it as a function of x before proceeding to integrate Eq. (9.4). However, in the case of a prismatic beam, which is the case considered here, the flexural rigidity is constant. We may thus multiply both members of Eq. (8.4) by EI and integrate in x. We write

$$EI\frac{dy}{dx} = \int_0^x M(x)\,dx + C_1 \qquad (9.5)$$

where C_1 is a constant of integration. Denoting by $\theta(x)$ the angle, measured in radians, that the tangent to the elastic curve at Q forms with the horizontal (Fig. 9.7), and recalling that this angle is very small, we have

$$\frac{dy}{dx} = \tan\theta \approx \theta(x)$$

Fig. 9.7

Thus, we write Eq. (9.5) in the alternative form

$$EI\,\theta(x) = \int_0^x M(x)\,dx + C_1 \qquad (9.5')$$

Integrating both members of Eq. (9.5) in x, we have

$$EI\,y = \int_0^x \left[\int_0^x M(x)\,dx + C_1 \right] dx + C_2$$

$$EI\,y = \int_0^x dx \int_0^x M(x)\,dx + C_1 x + C_2 \qquad (9.6)$$

where C_2 is a second constant, and where the first term in the right-hand member represents the function of x obtained by integrating twice in x the bending moment $M(x)$. If it were not for the fact that the constants C_1 and C_2 are as yet undetermined, Eq. (9.6) would define the deflection of the beam at any given point Q, and Eq. (9.5) or (9.5') would similarly define the slope of the beam at Q.

The constants C_1 and C_2 are determined from the *boundary conditions* or, more precisely, from the conditions imposed on the beam by its supports. Limiting our analysis in this section to *statically determinate beams*, i.e., to beams supported in such a way that the reactions at the supports can be obtained by the methods of statics, we note that only three types of beams need to be considered here (Fig. 9.8): (*a*) the *simply supported beam*, (*b*) the *overhanging beam*, and (*c*) the *cantilever beam*.

In the first two cases, the supports consist of a pin and bracket at A and of a roller at B, and require that the deflection be zero at each of these points. Letting first $x = x_A$, $y = y_A = 0$ in Eq. (9.6), and then $x = x_B$, $y = y_B = 0$ in the same equation, we obtain two equations that can be solved for C_1 and C_2. In the case of the cantilever beam (Fig. 9.8c), we note that both the deflection and the slope at A must be zero. Letting $x = x_A$, $y = y_A = 0$ in Eq. (9.6), and $x = x_A$, $\theta = \theta_A = 0$ in Eq. (9.5'), we obtain again two equations which can be solved for C_1 and C_2.

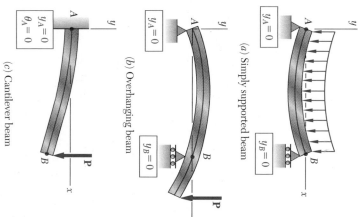

(a) Simply supported beam

(b) Overhanging beam

(c) Cantilever beam

Fig. 9.8 Boundary conditions for statically determinate beams.

EXAMPLE 9.01

The cantilever beam AB is of uniform cross section and carries a load $\mathbf{P}$ at its free end A (Fig. 9.9). Determine the equation of the elastic curve and the deflection and slope at A.

Fig. 9.9

Fig. 9.10

Using the free-body diagram of the portion AC of the beam (Fig. 9.10), where C is located at a distance x from end A, we find

$$M = -Px \qquad (9.7)$$

Substituting for M into Eq. (9.4) and multiplying both members by the constant EI, we write

$$EI \frac{d^2y}{dx^2} = -Px$$

Integrating in x, we obtain

$$EI \frac{dy}{dx} = -\frac{1}{2}Px^2 + C_1 \qquad (9.8)$$

We now observe that at the fixed end B we have $x = L$ and $\theta = dy/dx = 0$ (Fig. 9.11). Substituting these values into (9.8) and solving for C_1, we have

$$C_1 = \frac{1}{2}PL^2$$

which we carry back into (9.8):

$$EI \frac{dy}{dx} = -\frac{1}{2}Px^2 + \frac{1}{2}PL^2 \qquad (9.9)$$

Integrating both members of Eq. (9.9), we write

$$EI\, y = -\frac{1}{6}Px^3 + \frac{1}{2}PL^2x + C_2 \qquad (9.10)$$

But, at B we have $x = L$, $y = 0$. Substituting into (9.10), we have

$$0 = -\frac{1}{6}PL^3 + \frac{1}{2}PL^3 + C_2$$

$$C_2 = -\frac{1}{3}PL^3$$

Carrying the value of C_2 back into Eq. (9.10), we obtain the equation of the elastic curve:

$$EI\, y = -\frac{1}{6}Px^3 + \frac{1}{2}PL^2x - \frac{1}{3}PL^3 \qquad (9.11)$$

or

$$y = -\frac{P}{6EI}\left(x^3 + 3L^2x - 2L^3\right) \qquad (9.11')$$

The deflection and slope at A are obtained by letting $x = 0$ in Eqs. (9.11) and (9.9). We find

$$y_A = -\frac{PL^3}{3EI} \qquad \text{and} \qquad \theta_A = \left(\frac{dy}{dx}\right)_A = \frac{PL^2}{2EI}$$

$$[x = L,\ \theta = 0]$$
$$[x = L,\ y = 0]$$

Fig. 9.11

EXAMPLE 9.02

The simply supported prismatic beam AB carries a uniformly distributed load w per unit length (Fig. 9.12). Determine the equation of the elastic curve and the maximum deflection of the beam.

Fig. 9.12

Drawing the free-body diagram of the portion AD of the beam (Fig. 9.13) and taking moments about D, we find that

$$M = \frac{1}{2}wLx - \frac{1}{2}wx^2 \qquad (9.12)$$

Substituting for M into Eq. (9.4) and multiplying both members of this equation by the constant EI, we write

$$EI \frac{d^2y}{dx^2} = -\frac{1}{2}wx^2 + \frac{1}{2}wLx \qquad (9.13)$$

Integrating twice in x, we have

$$EI \frac{dy}{dx} = -\frac{1}{6}wx^3 + \frac{1}{4}wLx^2 + C_1 \qquad (9.14)$$

$$EI\, y = -\frac{1}{24}wx^4 + \frac{1}{12}wLx^3 + C_1x + C_2 \qquad (9.15)$$

$$R_A = \frac{1}{2}wL$$

Fig. 9.13

@ $x = 0$, $y = 0$
@ $x = L$, $y = 0$

by these $\rightarrow$ $C_2 = 0$, $C_1 = -\frac{1}{24}wL^3$

Observing that $y = 0$ at both ends of the beam (Fig. 9.14), we first let $x = 0$ and $y = 0$ in Eq. (9.15) and obtain $C_2 = 0$. We then make $x = L$ and $y = 0$ in the same equation and write

$$0 = -\tfrac{1}{24} wL^4 + \tfrac{1}{12} wL^4 + C_1 L$$
$$C_1 = -\tfrac{1}{24} wL^3$$

Carrying the values of C_1 and C_2 back into Eq. (9.15), we obtain the equation of the elastic curve:

$$EI\, y = -\tfrac{1}{24} wx^4 + \tfrac{1}{12} wL x^3 - \tfrac{1}{24} wL^3 x$$

or

$$y = \frac{w}{24EI}(-x^4 + 2Lx^3 - L^3 x) \qquad (9.16)$$

Substituting into Eq. (9.14) the value obtained for C_1, we check that the slope of the beam is zero for $x = L/2$ and that the elastic curve has a minimum at the midpoint C of the beam (Fig. 9.15). Letting $x = L/2$ in Eq. (9.16), we have

$$y_C = \frac{w}{24EI}\left(-\frac{L^4}{16} + 2L\frac{L^3}{8} - L^3\frac{L}{2}\right) = -\frac{5wL^4}{384EI}$$

So, what is y @ center?

The maximum deflection or, more precisely, the maximum absolute value of the deflection, is thus

$$|y|_{max} = \frac{5wL^4}{384EI}$$

Fig. 9.14

$[x = 0, y = 0]$ $[x = L, y = 0]$

Fig. 9.15

In each of the two examples considered so far, only one free-body diagram was required to determine the bending moment in the beam. As a result, a single function of x was used to represent M throughout the beam. This, however, is not generally the case. Concentrated loads, reactions at supports, or discontinuities in a distributed load will make it necessary to divide the beam into several portions, and to represent the bending moment by a different function $M(x)$ in each of these portions of beam (Fig. 9.16). Each of the functions $M(x)$ will then lead to a different expression for the slope $\theta(x)$ and for the deflection $y(x)$. Since each of the expressions obtained for the deflection must contain two constants of integration, a large number of constants will have to be determined. As you will see in the next example, the required additional boundary conditions can be obtained by observing that, while the shear and bending moment can be discontinuous at several points in a beam, the *deflection* and the *slope* of the beam *cannot be discontinuous* at any point.

Fig. 9.16 A different function $M(x)$ is required in each portion of the cantilever arms.

EXAMPLE 9.03

For the prismatic beam and the loading shown (Fig. 9.17), determine the slope and deflection at point D.

We must divide the beam into two portions, AD and DB, and determine the function $y(x)$ which defines the elastic curve for each of these portions.

Fig. 9.17

1. From A to D (x < L/4). We draw the free-body diagram of a portion of beam AE of length $x < L/4$ (Fig. 9.18). Taking moments about E, we have

$$M_1 = \frac{3P}{4}x \qquad (9.17)$$

or, recalling Eq. (9.4),

$$EI\frac{d^2y_1}{dx^2} = \frac{3}{4}Px \qquad (9.18)$$

where $y_1(x)$ is the function which defines the elastic curve for portion AD of the beam. Integrating in x, we write

$$EI\,\theta_1 = EI\frac{dy_1}{dx} = \frac{3}{8}Px^2 + C_1 \qquad (9.19)$$

$$EI\,y_1 = \frac{1}{8}Px^3 + C_1x + C_2 \qquad (9.20)$$

This is 4 eqn, 4 b.c. → moment area method, we'll try so much shit, too much shit

Fig. 9.18

2. From D to B (x > L/4). We now draw the free-body diagram of a portion of beam AE of length $x > L/4$ (Fig. 9.19) and write

$$M_2 - \frac{3P}{4}x - P\left(x - \frac{L}{4}\right) \qquad (9.21)$$

or, recalling Eq. (9.4) and rearranging terms,

$$EI\frac{d^2y_2}{dx^2} = -\frac{1}{4}Px + \frac{1}{4}PL \qquad (9.22)$$

where $y_2(x)$ is the function which defines the elastic curve for portion DB of the beam. Integrating in x, we write

$$EI\,\theta_2 = EI\frac{dy_2}{dx} = -\frac{1}{8}Px^2 + \frac{1}{4}PLx + C_3 \qquad (9.23)$$

$$EI\,y_2 = -\frac{1}{24}Px^3 + \frac{1}{8}PLx^2 + C_3x + C_4 \qquad (9.24)$$

Fig. 9.19

Determination of the Constants of Integration. The conditions that must be satisfied by the constants of integration have been summarized in Fig. 9.20. At the support A,

$[x = 0, y_1 = 0]$

$\left[x - \frac{1}{4}L, \theta_1 = \theta_2\right]$
$\left[x = \frac{1}{4}L, y_1 = y_2\right]$

$[x = L, y_2 = 0]$

Fig. 9.20

where the deflection is defined by Eq. (9.20), we must have $x = 0$ and $y_1 = 0$. At the support B, where the deflection is defined by Eq. (9.24), we must have $x = L$ and $y_2 = 0$. Also, the fact that there can be no sudden change in deflection or in slope at point D requires that $y_1 = y_2$ and $\theta_1 = \theta_2$ when $x = L/4$. We have therefore:

$[x = 0, y_1 = 0]$, Eq. (9.20): $\qquad 0 = C_2 \qquad (9.25)$

$[x = L, y_2 = 0]$, Eq. (9.24): $\qquad 0 = \frac{1}{12}PL^3 + C_3L + C_4 \qquad (9.26)$

$[x = L/4, \theta_1 = \theta_2]$, Eqs. (9.19) and (9.23):

$$\frac{3}{128}PL^2 + C_1 = \frac{7}{128}PL^2 + C_3 \qquad (9.27)$$

$[x = L/4, y_1 = y_2]$, Eqs. (9.20) and (9.24):

$$\frac{PL^3}{512} + C_1\frac{L}{4} = \frac{11PL^3}{1536} + C_3\frac{L}{4} + C_4 \qquad (9.28)$$

Solving these equations simultaneously, we find

$$C_1 = \frac{7PL^2}{128}, \quad C_2 = 0, \quad C_3 = -\frac{11PL^2}{128}, \quad C_4 = \frac{PL^3}{384}$$

Substituting for C_1 and C_2 into Eqs. (9.19) and (9.20), we write that for $x \le L/4$,

$$EI\,\theta_1 = \frac{3}{8}Px^2 - \frac{7PL^2}{128} \qquad (9.29)$$

$$EI\,y_1 = \frac{1}{8}Px^3 - \frac{7PL^2}{128}x \qquad (9.30)$$

Letting $x = L/4$ in each of these equations, we find that the slope and deflection at point D are, respectively,

$$\theta_D = -\frac{PL^2}{32EI} \quad \text{and} \quad y_D = -\frac{3PL^3}{256EI}$$

We note that, since $\theta_D \neq 0$, the deflection at D is not the maximum deflection of the beam.

(a) Cantilever beam

$[V_A = 0]$
$[M_B = 0]$

$[y_A = 0]$　$[y_B = 0]$
$[\theta_A = 0]$　$[M_A = 0]$

$[y_A = 0]$　$[y_B = 0]$
$[M_A = 0]$　$[M_B = 0]$

(b) Simply supported beam

Fig. 9.21 Boundary conditions for beams carrying a distributed load.

*9.4. DIRECT DETERMINATION OF THE ELASTIC CURVE FROM THE LOAD DISTRIBUTION

We saw in Sec. 9.3 that the equation of the elastic curve can be obtained by integrating twice the differential equation

$$\frac{d^2y}{dx^2} = \frac{M(x)}{EI} \qquad (9.4)$$

where $M(x)$ is the bending moment in the beam. We now recall from Sec. 5.3 that, when a beam supports a distributed load $w(x)$, we have $dM/dx = V$ and $dV/dx = -w$ at any point of the beam. Differentiating both members of Eq. (9.4) with respect to x and assuming EI to be constant, we have therefore

$$\frac{d^3y}{dx^3} = \frac{1}{EI}\frac{dM}{dx} = \frac{V(x)}{EI} \qquad (9.31)$$

and, differentiating again,

$$\frac{d^4y}{dx^4} = \frac{1}{EI}\frac{dV}{dx} = -\frac{w(x)}{EI}$$

We conclude that, when a prismatic beam supports a distributed load $w(x)$, its elastic curve is governed by the fourth-order linear differential equation

$$\frac{d^4y}{dx^4} = -\frac{w(x)}{EI} \qquad (9.32)$$

Multiplying both members of Eq. (9.32) by the constant EI and integrating four times, we write

$$EI\frac{d^4y}{dx^4} = -w(x)$$

$$EI\frac{d^3y}{dx^3} = V(x) = -\int w(x)\,dx + C_1$$

$$EI\frac{d^2y}{dx^2} = M(x) = -\int dx \int w(x)\,dx + C_1x + C_2$$

$$EI\frac{dy}{dx} = EI\,\theta(x) = -\int dx \int dx \int w(x)\,dx + \frac{1}{2}C_1x^2 + C_2x + C_3$$

$$EI\,y(x) = -\int dx \int dx \int dx \int w(x)\,dx + \frac{1}{6}C_1x^3 + \frac{1}{2}C_2x^2 + C_3x + C_4$$

$$\qquad\qquad (9.33)$$

The four constants of integration can be determined from the boundary conditions. These conditions include (a) the conditions imposed on the deflection or slope of the beam by its supports (cf. Sec. 9.3), and (b) the condition that V and M be zero at the free end of a cantilever beam, or that M be zero at both ends of a simply supported beam (cf. Sec. 5.3). This has been illustrated in Fig. 9.21.

The method presented here can be used effectively with cantilever or simply supported beams carrying a distributed load. In the case of over-hanging beams, however, the reactions at the supports will cause discon-tinuities in the shear, i.e., in the third derivative of y, and different func-tions would be required to define the elastic curve over the entire beam.

Fig. 9.22

Fig. 9.23

EXAMPLE 9.04

The simply supported prismatic beam AB carries a uniformly distributed load w per unit length (Fig. 9.22). Determine the equation of the elastic curve and the maximum deflection of the beam. (This is the same beam and loading as in Example 9.02.)

Since w = constant, the first three of Eqs. (9.33) yield

$$EI \frac{d^4y}{dx^4} = -w$$

$$EI \frac{d^3y}{dx^3} = V(x) = -wx + C_1$$

$$EI \frac{d^2y}{dx^2} = M(x) = -\frac{1}{2}wx^2 + C_1 x + C_2 \quad (9.34)$$

Noting that the boundary conditions require that $M = 0$ at both ends of the beam (Fig. 9.23), we first let $x = 0$ and $M = 0$ in Eq. (9.34) and obtain $C_2 = 0$. We then make $x = L$ and $M = 0$ in the same equation and obtain $C_1 = \frac{1}{2}wL$.

Carrying the values of C_1 and C_2 back into Eq. (9.34), and integrating twice, we write

$$EI \frac{d^2y}{dx^2} = -\frac{1}{2}wx^2 + \frac{1}{2}wLx$$

$$EI \frac{dy}{dx} = -\frac{1}{6}wx^3 + \frac{1}{4}wLx^2 + C_3$$

$$EI\, y = -\frac{1}{24}wx^4 + \frac{1}{12}wLx^3 + C_3 x + C_4 \quad (9.35)$$

But the boundary conditions also require that $y = 0$ at both ends of the beam. Letting $x = 0$ and $y = 0$ in Eq. (9.35), we obtain $C_4 = 0$; letting $x = L$ and $y = 0$ in the same equation,

$[x = 0, M = 0]$
$[x = 0, y = 0]$
$[x = L, M = 0]$
$[x = L, y = 0]$

we write

$$0 = -\frac{1}{24}wL^4 + \frac{1}{12}wL^4 + C_3 L$$
$$C_3 = -\frac{1}{24}wL^3.$$

Carrying the values of C_3 and C_4 back into Eq. (9.35) and dividing both members by EI, we obtain the equation of the elastic curve:

$$y = \frac{w}{24EI}(-x^4 + 2Lx^3 - L^3 x) \quad (9.36)$$

The value of the maximum deflection is obtained by mak-ing $x = L/2$ in Eq. (9.36). We have

$$|y|_{max} = \frac{5wL^4}{384EI}$$

9.5. STATICALLY INDETERMINATE BEAMS

In the preceding sections, our analysis was limited to statically determinate beams. Consider now the prismatic beam AB (Fig. 9.24a), which has a fixed end at A and is supported by a roller at B. Drawing the free-body diagram of the beam (Fig. 9.24b), we note that the reactions involve four unknowns, while only three equilibrium equations are available, namely

$$\sum F_x = 0 \quad \sum F_y = 0 \quad \sum M_A = 0 \qquad (9.37)$$

(a)

Fig. 9.24

(b)

Since only A_x can be determined from these equations, we conclude that the beam is *statically indeterminate*.

However, we recall from Chaps. 2 and 3 that, in a statically indeterminate problem, the reactions can be obtained by considering the *deformations* of the structure involved. We should, therefore, proceed with the computation of the slope and deformation along the beam. Following the method used in Sec. 9.3, we first express the bending moment $M(x)$ at any given point of AB in terms of the distance x from A, the given load, and the unknown reactions. Integrating in x, we obtain expressions for θ and y which contain two additional unknowns, namely the constants of integration C_1 and C_2. But altogether six unknowns are available to determine the reactions and the constants C_1 and C_2; they are the three equilibrium equations (9.37) and the three equations expressing that the boundary conditions are satisfied, i.e., that the slope and deflection at A are zero, and that the deflection at B is zero (Fig. 9.25). Thus, the reactions at the supports can be determined, and the equation of the elastic curve can be obtained.

$$[x = 0, \theta = 0]$$
$$[x = 0, y = 0]$$
$$[x = L, y = 0]$$

Fig. 9.25

Determine the reactions at the supports for the prismatic beam of Fig. 9.24a.

Equilibrium Equations. From the free-body diagram of Fig. 9.24b we write

$$\xrightarrow{+} \sum F_x = 0: \quad A_x = 0$$
$$+\uparrow \sum F_y = 0: \quad A_y + B - wL = 0$$
$$+\circlearrowleft \sum M_A = 0: \quad M_A + BL - \tfrac{1}{2}wL^2 = 0 \qquad (9.38)$$

Equation of Elastic Curve. Drawing the free-body diagram of a portion of beam AC (Fig. 9.26), we write

Fig. 9.26

$$+\circlearrowleft \sum M_C = 0: \quad M + \tfrac{1}{2}wx^2 + M_A - A_y x = 0 \qquad (9.39)$$

Solving Eq. (9.39) for M and carrying into Eq. (9.4), we write

$$EI \frac{d^2 y}{dx^2} = -\frac{1}{2}wx^2 + A_y x - M_A$$

Integrating in x, we have

$$EI\,\theta = EI\frac{dy}{dx} = -\frac{1}{6}wx^3 + \frac{1}{2}A_y x^2 - M_A x + C_1 \qquad (9.40)$$

$$EI\,y = -\frac{1}{24}wx^4 + \frac{1}{6}A_y x^3 - \frac{1}{2}M_A x^2 + C_1 x + C_2 \qquad (9.41)$$

Referring to the boundary conditions indicated in Fig. 9.25, we make $x = 0$, $\theta = 0$ in Eq. (9.40), $x = 0$, $y = 0$ in Eq. (9.41), and conclude that $C_1 = C_2 = 0$. Thus, we rewrite Eq. (9.41) as follows:

$$EI\,y = -\frac{1}{24}wx^4 + \frac{1}{6}A_y x^3 - \frac{1}{2}M_A x^2 \qquad (9.42)$$

But the third boundary condition requires that $y = 0$ for $x = L$. Carrying these values into (9.42), we write

$$0 = -\frac{1}{24}wL^4 + \frac{1}{6}A_y L^3 - \frac{1}{2}M_A L^2$$

or

$$3M_A - A_y L + \frac{1}{4}wL^2 = 0 \qquad (9.43)$$

Solving this equation simultaneously with the three equilibrium equations (9.38), we obtain the reactions at the supports:

$$A_x = 0 \qquad A_y = \tfrac{5}{8}wL \qquad M_A = \tfrac{1}{8}wL^2 \qquad B = \tfrac{3}{8}wL$$

In the example we have just considered, there was one redundant reaction, i.e., there was one more reaction than could be determined from the equilibrium equations alone. The corresponding beam is said to be *statically indeterminate to the first degree*. Another example of a beam indeterminate to the first degree is provided in Sample Prob. 9.3. If the beam supports are such that two reactions are redundant (Fig. 9.27a), the beam is said to be *indeterminate to the second degree*. While there are now five unknown reactions (Fig. 9.27b), we find that four equations may be obtained from the boundary conditions (Fig. 9.27c). Thus, altogether seven equations are available to determine the five reactions and the two constants of integration.

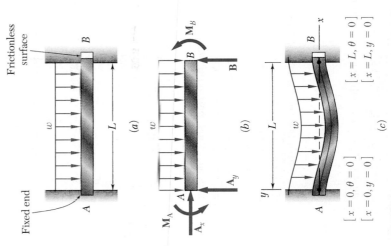

Fig. 9.27

SAMPLE PROBLEM 9.1

The overhanging steel beam ABC carries a concentrated load $\mathbf{P}$ at end C. For portion AB of the beam, (a) derive the equation of the elastic curve, (b) determine the maximum deflection, (c) evaluate y_{max} for the following data:

W 14 × 68 $I = 723$ in^4 $E = 29 \times 10^6$ psi

$P = 50$ kips $L = 15$ ft $= 180$ in. $a = 4$ ft $= 48$ in.

SOLUTION

Free-Body Diagrams.   Reactions: $\mathbf{R}_A = Pa/L\downarrow$ $\mathbf{R}_B = P(1 + a/L)\uparrow$
Using the free-body diagram of the portion of beam AD of length x, we find

$$M = -P\frac{a}{L}x \qquad (0 < x < L)$$

Differential Equation of the Elastic Curve.   We use Eq. (9.4) and write

$$EI\frac{d^2y}{dx^2} = -P\frac{a}{L}x$$

Noting that the flexural rigidity EI is constant, we integrate twice and find

$$EI\frac{dy}{dx} = -\frac{1}{2}P\frac{a}{L}x^2 + C_1 \tag{1}$$

$$EI\,y = -\frac{1}{6}P\frac{a}{L}x^3 + C_1 x + C_2 \tag{2}$$

Determination of Constants.   For the boundary conditions shown, we have

$[x = 0, y = 0]$:   From Eq. (2), we find $C_2 = 0$

$[x = L, y = 0]$:   Again using Eq. (2), we write

$$EI(0) = -\frac{1}{6}P\frac{a}{L}L^3 + C_1 L \qquad C_1 = +\frac{1}{6}PaL$$

a. Equation of the Elastic Curve.   Substituting for C_1 and C_2 into Eqs. (1) and (2), we have

$$EI\frac{dy}{dx} = -\frac{1}{2}P\frac{a}{L}x^2 + \frac{1}{6}PaL \qquad \frac{dy}{dx} = \frac{PaL}{6EI}\left[1 - 3\left(\frac{x}{L}\right)^2\right]$$

$$EI\,y = -\frac{1}{6}P\frac{a}{L}x^3 + \frac{1}{6}PaL.x \qquad y = \frac{PaL^2}{6EI}\left[\frac{x}{L} - \left(\frac{x}{L}\right)^3\right] \tag{4} \blacktriangleleft$$

b. Maximum Deflection in Portion AB.   The maximum deflection y_{max} occurs at point E where the slope of the elastic curve is zero. Setting $dy/dx = 0$ in Eq. (3), we determine the abscissa x_m of point E:

$$0 = \frac{PaL}{6EI}\left[1 - 3\left(\frac{x_m}{L}\right)^2\right] \qquad x_m = \frac{L}{\sqrt{3}} = 0.577L$$

We substitute $x_m/L = 0.577$ into Eq. (4) and have

$$y_{max} = \frac{PaL^2}{6EI}\left[(0.577) - (0.577)^3\right] \quad y_{max} = 0.0642\frac{PaL^2}{EI} \blacktriangleleft$$

c. Evaluation of y_{max}.   For the data given, the value of y_{max} is

$$y_{max} = 0.0642\frac{(50 \text{ kips})(48 \text{ in.})(180 \text{ in.})^2}{(29 \times 10^6 \text{ psi})(723 \text{ in}^4)} \qquad y_{max} = 0.238 \text{ in.} \blacktriangleleft$$

SAMPLE PROBLEM 9.2

For the beam and loading shown, determine (*a*) the equation of the elastic curve, (*b*) the slope at end A, (*c*) the maximum deflection.

SOLUTION

Differential Equation of the Elastic Curve. From Eq. (9.32),

$$EI\frac{d^4y}{dx^4} = -w(x) = -w_0 \sin\frac{\pi x}{L} \qquad (1)$$

Integrate Eq. (1) twice:

$$EI\frac{d^3y}{dx^3} = V = +w_0\frac{L}{\pi}\cos\frac{\pi x}{L} + C_1 \qquad (2)$$

$$EI\frac{d^2y}{dx^2} = M = +w_0\frac{L^2}{\pi^2}\sin\frac{\pi x}{L} + C_1 x + C_2 \qquad (3)$$

Boundary Conditions:

$[x = 0, M = 0]$: From Eq. (3), we find $C_2 = 0$

$[x = L, M = 0]$: Again using Eq. (3), we write

$$0 = w_0\frac{L^2}{\pi^2}\sin\pi + C_1 L \qquad C_1 = 0$$

Thus:

$$EI\frac{d^2y}{dx^2} = +w_0\frac{L^2}{\pi^2}\sin\frac{\pi x}{L} \qquad (4)$$

Integrate Eq. (4) twice:

$$EI\frac{dy}{dx} = EI\theta = -w_0\frac{L^3}{\pi^3}\cos\frac{\pi x}{L} + C_3 \qquad (5)$$

$$EI\,y = -w_0\frac{L^4}{\pi^4}\sin\frac{\pi x}{L} + C_3 x + C_4 \qquad (6)$$

Boundary Conditions:

$[x = 0, y = 0]$: Using Eq. (6), we find $C_4 = 0$

$[x = L, y = 0]$: Again using Eq. (6), we find $C_3 = 0$

***a*. Equation of Elastic Curve**

$$EIy = -w_0\frac{L^4}{\pi^4}\sin\frac{\pi x}{L} \qquad \blacktriangledown$$

***b*. Slope at End A.** For $x = 0$, we have

$$EI\theta_A = -w_0\frac{L^3}{\pi^3}\cos 0$$

$$\theta_A = \frac{w_0 L^3}{\pi^3 EI} \qquad \blacktriangledown$$

***c*. Maximum Deflection.** For $x = \frac{1}{2}L$,

$$EIy_{max} = -w_0\frac{L^4}{\pi^4}\sin\frac{\pi}{2}$$

$$y_{max} = \frac{w_0 L^4}{\pi^4 EI} \qquad \blacktriangledown$$

543

SAMPLE PROBLEM 9.3

For the uniform beam AB, (a) determine the reaction at A, (b) derive the equation of the elastic curve, (c) determine the slope at A. (Note that the beam is statically indeterminate to the first degree.)

SOLUTION

Bending Moment. Using the free body shown, we write

$$+\gamma\Sigma M_D = 0: \quad R_A x - \frac{1}{2}\left(\frac{w_0 x^2}{L}\right)\frac{x}{3} - M = 0 \quad M = R_A x - \frac{w_0 x^3}{6L}$$

Noting that the flexural rigidity EI is constant, we integrate twice and find

$$EI\frac{dy}{dx} = EI\theta = \frac{1}{2}R_A x^2 - \frac{w_0 x^4}{24L} + C_1 \tag{1}$$

$$EI\,y = \frac{1}{6}R_A x^3 - \frac{w_0 x^5}{120L} + C_1 x + C_2 \tag{2}$$

Differential Equation of the Elastic Curve. We use Eq. (9.4) and write

$$EI\frac{d^2 y}{dx^2} = R_A x - \frac{w_0 x^3}{6L}$$

Boundary Conditions. The three boundary conditions that must be satisfied are shown on the sketch

$$[x = 0, y = 0]: \quad C_2 = 0 \tag{3}$$

$$[x = L, \theta = 0]: \quad \frac{1}{2}R_A L^2 - \frac{w_0 L^3}{24} + C_1 = 0 \tag{4}$$

$$[x = L, y = 0]: \quad \frac{1}{6}R_A L^3 - \frac{w_0 L^4}{120} + C_1 L + C_2 = 0 \tag{5}$$

a. Reaction at A. Multiplying Eq. (4) by L, subtracting Eq. (5) member by member from the equation obtained, and noting that $C_2 = 0$, we have

$$\frac{1}{3}R_A L^3 - \frac{1}{30}w_0 L^4 = 0 \qquad R_A = \frac{1}{10}w_0 L\uparrow \blacktriangleleft$$

We note that the reaction is independent of E and I. Substituting $R_A = \frac{1}{10}w_0 L$ into Eq. (4), we have

$$\frac{1}{2}\left(\frac{1}{10}w_0 L\right)L^2 - \frac{1}{24}w_0 L^3 + C_1 = 0 \qquad C_1 = -\frac{1}{120}w_0 L^3$$

b. Equation of the Elastic Curve. Substituting for R_A, C_1, and C_2 into Eq. (2), we have

$$EI\,y = \frac{1}{6}\left(\frac{1}{10}w_0 L\right)x^3 - \frac{w_0 x^5}{120L} - \left(\frac{1}{120}w_0 L^3\right)x$$

$$y = \frac{w_0}{120EIL}(-x^5 + 2L^2 x^3 - L^4 x) \blacktriangleleft$$

c. Slope at A. We differentiate the above equation with respect to x:

$$\theta = \frac{dy}{dx} = \frac{w_0}{120EIL}(-5x^4 + 6L^2 x^2 - L^4)$$

Making $x = 0$, we have

$$\theta_A = -\frac{w_0 L^3}{120EI} \qquad \theta_A = \frac{w_0 L^3}{120EI} \text{↘} \blacktriangleleft$$

$[x = 0, y = 0]$

$[x = L, \theta = 0]$

$[x = L, y = 0]$

PROBLEMS

In the following problems assume that the flexural rigidity EI of each beam is constant.

9.1 through 9.4 For the loading shown, determine (*a*) the equation of the elastic curve for the cantilever beam AB, (*b*) the deflection at the free end, (*c*) the slope at the free end.

Fig. P9.1

Fig. P9.2

Fig. P9.3

Fig. P9.4

9.5 For the beam and loading shown, determine (*a*) the equation of the elastic curve for portion AB of the beam, (*b*) the slope at A, (*c*) the slope at B.

Fig. P9.5

9.6 For the beam and loading shown, determine (*a*) the equation of the elastic curve for portion BC of the beam, (*b*) the deflection at midspan, (*c*) the slope at B.

Fig. P9.6

9.7 For the cantilever beam and loading shown, determine (*a*) the equation of the elastic curve for portion *AB* of the beam, (*b*) the deflection at *B*, (*c*) the slope at *B*.

9.8 For the beam shown with load **P**, determine (*a*) the equation of the elastic curve for portion *AC* of the beam, (*b*) the slope at *A*, (*c*) the deflection at *C*.

Fig. P9.8

9.9 and 9.10 For the beam and loading shown, (*a*) express the magnitude and location of the maximum deflection in terms of w_0, *L*, *E*, and *I*. (*b*) Calculate the value of the maximum deflection, assuming that beam *AB* is a W18 × 50 rolled shape and that $w_0 = 4.5$ kips/ft, $L = 18$ ft, and $E = 29 \times 10^6$ psi.

Fig. P9.9

Fig. P9.10

9.11 (*a*) Determine the location and magnitude of the maximum deflection of beam *AB*. (*b*) Assuming that beam *AB* is a W360 × 64, $L = 3.5$ m, and $E = 200$ GPa, calculate the maximum allowable value of the applied moment M_0 if the maximum deflection is not to exceed 1 mm.

Fig. P9.11

9.12 (*a*) Determine the location and magnitude of the maximum absolute deflection in *AB* between *A* and the center of the beam. (*b*) Assuming that beam *AB* is a W460 × 113, $M_0 = 224$ kN · m and $E = 200$ GPa, determine the maximum allowable length *L* of the beam if the maximum deflection is not to exceed 1.2 mm.

Fig. P9.12

9.13 and 9.14 For the beam and loading shown, determine the deflection at point *C*. Use $E = 200$ GPa.

$M_0 = 38$ kN · m

$a = 0.8$ m

$L = 3.2$ m

W100 × 19.3

Fig. P9.13

P = 20 kN

$a = 1$ m

$L = 3$ m

W150 × 18.0

Fig. P9.14

9.15 Knowing that beam AE is an S200 × 27.4 rolled shape and that $P = 17.5$ kN, $L = 2.5$ m, $a = 0.8$ m, and $E = 200$ GPa, determine (*a*) the equation of the elastic curve for portion BD, (*b*) the deflection at the center C of the beam.

Fig. P9.15

Fig. P9.16

9.16 Uniformly distributed loads are applied to beam AE as shown. (*a*) Selecting the x axis through the centers A and E of the end sections of the beam, determine the equation of the elastic curve for portion AB of the beam. (*b*) Knowing that the beam is a W200 × 35.9 rolled shape and that $L = 3$ m, $w = 5$ kN/m, and $E = 200$ GPa, determine the distance of the center of the beam from the x axis.

9.17 through 9.20 For the beam and loading shown, determine the reaction at the roller support.

Fig. P9.17

Fig. P9.19

Fig. P9.18

Fig. P9.20

9.21 For the beam shown, determine the reaction at the roller support when $w_0 = 6$ kips/ft.

Fig. P9.21

9.22 For the beam shown, determine the reaction at the roller support when $w_0 = 15$ kN/m.

Fig. P9.22

9.23 through 9.26 Determine the reaction at the roller support and draw the bending moment diagram for the beam and loading shown.

Fig. P9.23

Fig. P9.24

Fig. P9.25

Fig. P9.26

9.27 and 9.28 Determine the reaction at the roller support and the deflection at point C.

Fig. P9.27

Fig. P9.28

9.29 and 9.30 Determine the reaction at point D, knowing that a is equal to L/3.

Fig. P9.29

Fig. P9.30

9.31 and 9.32 Determine the reaction at A and draw the bending moment diagram for the beam and loading shown.

Fig. P9.31

Fig. P9.32

9.33 For the beam and loading shown, determine (*a*) the equation of the elastic curve, (*b*) the slope at end *A*, (*c*) the deflection at the midpoint of the span.

$$w = 4w_0 \left[\frac{x}{L} - \frac{x^2}{L^2} \right]$$

Fig. P9.33

9.34 For the beam and loading shown, determine (*a*) the equation of the elastic curve, (*b*) the slope at the free end, (*c*) the deflection at the free end.

$$w = w_0 \cos \frac{\pi x}{2L}$$

Fig. P9.34

*9.6. USING SINGULARITY FUNCTIONS TO DETERMINE THE SLOPE AND DEFLECTION OF A BEAM

Reviewing the work done so far in this chapter, we note that the integration method provides a convenient and effective way of determining the slope and deflection at any point of a prismatic beam, *as long as the bending moment can be represented by a single analytical function M(x)*. However, when the loading of the beam is such that two different functions are needed to represent the bending moment over the entire length of the beam, as in Example 9.03 (Fig. 9.17), four constants of integration are required, and an equal number of equations, expressing continuity conditions at point *D*, as well as boundary conditions at the supports *A* and *B*, must be used to determine these constants. If three or more functions were needed to represent the bending moment, additional constants and a corresponding number of additional equations would be required, resulting in rather lengthy computations. Such would be the case for the beam shown in Fig. 9.28. In this section these computations will be simplified through the use of the singularity functions discussed in Sec. 5.5.

Fig. 9.28 In this roof structure, each of the joists applies a concentrated load to the beam that supports it.

Let us consider again the beam and loading of Example 9.03 (Fig. 9.17) and draw the free-body diagram of that beam (Fig. 9.29). Using

Fig. 9.29

the appropriate singularity function, as explained in Sec. 5.5, to represent the contribution to the shear of the concentrated load **P**, we write

$$V(x) = \frac{3P}{4} - P\langle x - \tfrac{1}{4}L \rangle^0$$

Integrating in x and recalling from Sec. 5.5 that in the absence of any concentrated couple, the expression obtained for the bending moment will not contain any constant term, we have

$$M(x) = \frac{3P}{4}x - P\langle x - \tfrac{1}{4}L \rangle \qquad (9.44)$$

Substituting for $M(x)$ from (9.44) into Eq. (9.4), we write

$$EI\frac{d^2y}{dx^2} = \frac{3P}{4}x - P\langle x - \tfrac{1}{4}L \rangle \qquad (9.45)$$

and, integrating in x,

$$EI\theta = EI\frac{dy}{dx} = \frac{3}{8}Px^2 - \frac{1}{2}P\langle x - \tfrac{1}{4}L \rangle^2 + C_1 \qquad (9.46)$$

$$EI\,y = \frac{1}{8}Px^3 - \frac{1}{6}P\langle x - \tfrac{1}{4}L \rangle^3 + C_1x + C_2 \qquad (9.47)\dagger$$

The constants C_1 and C_2 can be determined from the boundary conditions shown in Fig. 9.30. Letting $x = 0$, $y = 0$ in Eq. (9.47), we have

$$0 = 0 - \frac{1}{6}P\langle 0 - \tfrac{1}{4}L \rangle^3 + 0 + C_2$$

which reduces to $C_2 = 0$, since any bracket containing a negative quantity is equal to zero. Letting now $x = L$, $y = 0$, and $C_2 = 0$ in Eq. (9.47), we write

$$0 = \frac{1}{8}PL^3 - \frac{1}{6}P\langle \tfrac{3}{4}L \rangle^3 + C_1L$$

$$[x = 0, y = 0] \qquad [x = L, y = 0]$$

Fig. 9.30

---

†The continuity conditions for the slope and deflection at D are "built-in" in Eqs. (9.46) and (9.47). Indeed, the difference between the expressions for the slope θ_1 in AD and the slope θ_2 in DB is represented by the term $-\frac{1}{2}P(x - \frac{1}{4}L)^2$ in Eq. (9.46), and this term is equal to zero at D. Similarly, the difference between the expressions for the deflection y_1 in AD and the deflection y_2 in DB is represented by the term $-\frac{1}{6}P(x - \frac{1}{4}L)^3$ in Eq. (9.47), and this term is also equal to zero at D.

Since the quantity between brackets is positive, the brackets can be replaced by ordinary parentheses. Solving for C_1, we have

$$C_1 = \frac{7PL^2}{128}$$

We check that the expressions obtained for the constants C_1 and C_2 are the same that were found earlier in Sec. 9.3. But the need for additional constants C_3 and C_4 has now been eliminated, and we do not have to write equations expressing that the slope and the deflection are continuous at point D.

EXAMPLE 9.06

For the beam and loading shown (Fig. 9.31a) and using singularity functions, (a) express the slope and deflection as functions of the distance x from the support at A, (b) determine the deflection at the midpoint D. Use $E = 200$ GPa and $I = 6.87 \times 10^{-6}$ m^4.

Fig. 9.31

(a) We note that the beam is loaded and supported in the same manner as the beam of Example 5.05. Referring to that example, we recall that the given distributed loading was replaced by the two equivalent open-ended loadings shown in Fig. 9.31b and that the following expressions were obtained for the shear and bending moment:

$$V(x) = -1.5\langle x - 0.6 \rangle^1 + 1.5\langle x - 1.8 \rangle^1 \\ + 2.6 - 1.2\langle x - 0.6 \rangle^0$$

$$M(x) = -0.75\langle x - 0.6 \rangle^2 + 0.75\langle x - 1.8 \rangle^2 \\ + 2.6x - 1.2\langle x - 0.6 \rangle^1 - 1.44\langle x - 2.6 \rangle^0$$

Integrating the last expression twice, we obtain

$$EI\theta = -0.25\langle x - 0.6 \rangle^3 + 0.25\langle x - 1.8 \rangle^3 \\ + 1.3x^2 - 0.6\langle x - 0.6 \rangle^2 - 1.44\langle x - 2.6 \rangle^1 + C_1 \quad (9.48)$$

$$EIy = -0.0625\langle x - 0.6 \rangle^4 + 0.0625\langle x - 1.8 \rangle^4 \\ + 0.4333x^3 - 0.2\langle x - 0.6 \rangle^3 - 0.72\langle x - 2.6 \rangle^2 \\ + C_1x + C_2 \quad (9.49)$$

The constants C_1 and C_2 can be determined from the boundary conditions shown in Fig. 9.32. Letting $x = 0$, $y = 0$ in Eq. (9.49) and noting that all the brackets contain negative

$$[x = 0, \ y = 0] \quad [x = 3.6, \ y = 0]$$

Fig. 9.32

quantities and, therefore, are equal to zero, we conclude that $C_2 = 0$. Letting now $x = 3.6$, $y = 0$, and $C_2 = 0$ in Eq. (9.49), we write

$$0 = -0.0625\langle 3.0 \rangle^4 + 0.0625\langle 1.8 \rangle^4 \\ + 0.4333(3.6)^3 - 0.2\langle 3.0 \rangle^3 - 0.72\langle 1.0 \rangle^2 + C_1(3.6) + 0$$

Since all the quantities between brackets are positive, the brackets can be replaced by ordinary parentheses. Solving for C_1, we find $C_1 = -2.692$.

(b) Substituting for C_1 and C_2 into Eq. (9.49) and making $x = x_D = 1.8$ m, we find that the deflection at point D is defined by the relation

$$EIy_D = -0.0625\langle 1.2 \rangle^4 + 0.0625\langle 0 \rangle^4 \\ + 0.4333(1.8)^3 - 0.2(1.2)^3 - 0.72\langle -0.8 \rangle^2 - 2.692(1.8)$$

The last bracket contains a negative quantity and, therefore, is equal to zero. All the other brackets contain positive quantities and can be replaced by ordinary parentheses. We have

$$EIy_D = -0.0625(1.2)^4 + 0.0625(0)^4 \\ + 0.4333(1.8)^3 - 0.2(1.2)^3 - 0 - 2.692(1.8) = -2.794$$

Recalling the given numerical values of E and I, we write

$$(200 \text{ GPa})(6.87 \times 10^{-6} \text{ m}^4)y_D = -2.794 \text{ kN} \cdot \text{m}^3$$

$$y_D = -13.64 \times 10^{-3} \text{ m} = -2.03 \text{ mm}$$

SAMPLE PROBLEM 9.4

For the prismatic beam and loading shown, determine (a) the equation of the elastic curve, (b) the slope at A, (c) the maximum deflection.

SOLUTION

Bending Moment. The equation defining the bending moment of the beam was obtained in Sample Prob. 5.9. Using the modified loading diagram shown, we had [Eq. (3)]:

$$M(x) = -\frac{w_0}{3L}x^3 + \frac{2w_0}{3L}\langle x - \tfrac{1}{2}L\rangle^3 + \tfrac{1}{4}w_0 Lx$$

***a.* Equation of the Elastic Curve.** Using Eq. (9.4), we write

$$EI\frac{d^2y}{dx^2} = -\frac{w_0}{3L}x^3 + \frac{2w_0}{3L}\langle x - \tfrac{1}{2}L\rangle^3 + \tfrac{1}{4}w_0 Lx \qquad (1)$$

and, integrating twice in x,

$$EI\,\theta = -\frac{w_0}{12L}x^4 + \frac{w_0}{6L}\langle x - \tfrac{1}{2}L\rangle^4 + \frac{w_0 L}{8}x^2 + C_1 \qquad (2)$$

$$EI\,y = -\frac{w_0}{60L}x^5 + \frac{w_0}{30L}\langle x - \tfrac{1}{2}L\rangle^5 + \frac{w_0 L}{24}x^3 + C_1 x + C_2 \qquad (3)$$

Boundary Conditions.

$[x = 0, y = 0]$: Using Eq. (3) and noting that each bracket $\langle\ \rangle$ contains a negative quantity and, thus, is equal to zero, we find $C_2 = 0$.

$[x = L, y = 0]$: Again using Eq. (3), we write

$$0 = -\frac{w_0 L^4}{60} + \frac{w_0}{30L}\left(\frac{L}{2}\right)^5 + \frac{w_0 L^4}{24} + C_1 L \qquad C_1 = -\frac{5}{192}w_0 L^3$$

Substituting C_1 and C_2 into Eqs. (2) and (3), we have

$$EI\,\theta = -\frac{w_0}{12L}x^4 + \frac{w_0}{6L}\langle x - \tfrac{1}{2}L\rangle^4 + \frac{w_0 L}{8}x^2 - \frac{5}{192}w_0 L^3 \qquad (4)$$

$$EI\,y = -\frac{w_0}{60L}x^5 + \frac{w_0}{30L}\langle x - \tfrac{1}{2}L\rangle^5 + \frac{w_0 L}{24}x^3 - \frac{5}{192}w_0 L^3 x \qquad (5)\ \blacktriangle$$

***b.* Slope at A.** Substituting $x = 0$ into Eq. (4), we find

$$EI\,\theta_A = -\frac{5}{192}w_0 L^3 \qquad \theta_A = \frac{5w_0 L^3}{192EI} \ \measuredangle \ \blacktriangle$$

***c.* Maximum Deflection.** Because of the symmetry of the supports and loading, the maximum deflection occurs at point C, where $x = \tfrac{1}{2}L$. Substituting into Eq. (5), we obtain

$$EI\,y_{\max} = w_0 L^4\left[-\frac{1}{60(32)} + 0 + \frac{1}{24(8)} - \frac{5}{192(2)}\right] = -\frac{w_0 L^4}{120}$$

$$y_{\max} = \frac{w_0 L^4}{120EI}\downarrow \ \blacktriangle$$

50 lb/ft

A

C

D
F
E

8 ft — 3 ft — 5 ft

1 in.

3 in.

160 lb

SAMPLE PROBLEM 9.5

The rigid bar DEF is welded at point D to the uniform steel beam AB. For the loading shown, determine (a) the equation of the elastic curve of the beam, (b) the deflection at the midpoint C of the beam. Use $E = 29 \times 10^6$ psi.

SOLUTION

Bending Moment. The equation defining the bending moment of the beam was obtained in Sample Prob. 5.10. Using the modified loading diagram shown and expressing x in feet, we had [Eq. (3)]:

$$M(x) = -25x^2 + 480x - 160\langle x - 11\rangle^1 - 480\langle x - 11\rangle^0 \quad \text{lb} \cdot \text{ft}$$

w

$w_0 = 50$ lb/ft

A   B x

$M_D = 480$ lb · ft $P = 160$ lb
$R_A = 480$ lb

R_B

— 11 ft — 5 ft —

a. Equation of the Elastic Curve. Using Eq. (8.4), we write

$$EI(d^2y/dx^2) = -25x^2 + 480x - 160\langle x - 11\rangle^1 - 480\langle x - 11\rangle^0 \quad \text{lb} \cdot \text{ft} \quad (1)$$

and, integrating twice in x,

$$EI\,\theta = -8.333x^3 + 240x^2 - 80\langle x - 11\rangle^2 - 480\langle x - 11\rangle^1 + C_1 \quad \text{lb} \cdot \text{ft}^2 \quad (2)$$

$$EI\,y = -2.083x^4 + 80x^3 - 26.67\langle x - 11\rangle^3 - 240\langle x - 11\rangle^2 \\ + C_1x + C_2 \quad \text{lb} \cdot \text{ft}^3 \quad (3)$$

Boundary Conditions.

$[x = 0, y = 0]$: Using Eq. (3) and noting that each bracket $\langle \ \rangle$ contains a negative quantity and, thus, is equal to zero, we find $C_2 = 0$.

$[x = 16$ ft, $y = 0]$: Again using Eq. (3) and noting that each bracket contains a positive quantity and, thus, can be replaced by a parenthesis, we write

$$0 = -2.083(16)^4 + 80(16)^3 - 26.67(5)^3 - 240(5)^2 + C_1(16)$$

$$C_1 = -11\,36 \times 10^3$$

y
$[x = 0, y = 0]$ $[x = 16$ ft, $y = 0]$

A   B x

— 16 ft —

Substituting the values found for C_1 and C_2 into Eq. (3), we have

$$EI\,y = -2.083x^4 + 80x^3 - 26.67\langle x - 11\rangle^3 - 240\langle x - 11\rangle^2 \\ -11.36 \times 10^3 x \quad \text{lb} \cdot \text{ft}^3 \quad (3')$$

To determine EI, we recall that $E = 29 \times 10^6$ psi and compute

$$I = \tfrac{1}{12}bh^3 = \tfrac{1}{12}(1\,\text{in.})(3\,\text{in.})^3 = 2.25\,\text{in}^4$$

$$EI = (29 \times 10^6\,\text{psi})(2.25\,\text{in}^4) = 65.25 \times 10^6\,\text{lb} \cdot \text{in}^2$$

However, since all previous computations have been carried out with feet as the unit of length, we write

$$EI = (65.25 \times 10^6\,\text{lb} \cdot \text{in}^2)(1\,\text{ft}/12\,\text{in.})^2 = 453.1 \times 10^3\,\text{lb} \cdot \text{ft}^2$$

b. Deflection at Midpoint C. Making $x = 8$ ft in Eq. (3'), we write

$$EI\,y_C = -2.083(8)^4 + 80(8)^3 - 26.67\langle -3\rangle^3 - 240\langle -3\rangle^2 - 11.36 \times 10^3(8)$$

Noting that each bracket is equal to zero and substituting for EI its numerical value, we have

$$(453.1 \times 10^3\,\text{lb} \cdot \text{ft}^2)y_C = -58.45 \times 10^3\,\text{lb} \cdot \text{ft}^3$$

and, solving for y_C: $y_C = -0.1290$ ft $y_C = -1.548$ in. ◀

Note that the deflection obtained *is not* the maximum deflection.

y

A y_C

C   B x

— 8 ft — 8 ft —

SAMPLE PROBLEM 9.6

For the uniform beam ABC, (a) express the reaction at A in terms of P, L, a, E, and I, (b) determine the reaction at A and the deflection under the load when $a = L/2$.

SOLUTION

Reactions. For the given vertical load **P** the reactions are as shown. We note that they are statically indeterminate.

Shear and Bending Moment. Using a step function to represent the contribution of **P** to the shear, we write

$$V(x) = R_A - P\langle x - a\rangle^0$$

Integrating in x, we obtain the bending moment:

$$M(x) = R_A x - P\langle x - a\rangle^1$$

Equation of the Elastic Curve. Using Eq. (9.4), we write

$$EI\frac{d^2y}{dx^2} = R_A x - P\langle x - a\rangle^1$$

Integrating twice in x,

$$EI\frac{dy}{dx} = EI\,\theta = \frac{1}{2}R_A x^2 - \frac{1}{2}P\langle x - a\rangle^2 + C_1$$

$$EI\,y = \frac{1}{6}R_A x^3 - \frac{1}{6}P\langle x - a\rangle^3 + C_1 x + C_2$$

Boundary Conditions. Noting that the bracket $\langle x - a\rangle$ is equal to zero for $x = 0$, and to $(L - a)$ for $x = L$, we write

$$[x = 0, y = 0]: \qquad C_2 = 0$$
$$[x = L, \theta = 0]: \qquad \frac{1}{2}R_A L^2 - \frac{1}{2}P(L - a)^2 + C_1 = 0 \qquad (1)$$
$$[x = L, y = 0]: \qquad \frac{1}{6}R_A L^3 - \frac{1}{6}P(L - a)^3 + C_1 L + C_2 = 0 \qquad (2)$$

$$[x = 0, y = 0]$$
$$[x = L, \theta = 0]$$

a. Reaction at A. Multiplying Eq. (2) by L, subtracting Eq. (3) member by member from the equation obtained, and noting that $C_2 = 0$, we have

$$\frac{1}{3}R_A L^3 - \frac{1}{6}P(L - a)^2[3L - (L - a)] = 0$$

$$\mathbf{R}_A = P\left(1 - \frac{a}{L}\right)^2\left(1 + \frac{a}{2L}\right)\uparrow \qquad \blacktriangle$$

We note that the reaction is independent of E and I.

b. Reaction at A and Deflection at B when $a = \frac{1}{2}L$. Making $a = \frac{1}{2}L$ in the expression obtained for R_A, we have

$$R_A = P(1 - \frac{1}{2})^2(1 + \frac{1}{4}) = 5P/16 \qquad \mathbf{R}_A = \frac{5}{16}P\uparrow \qquad \blacktriangle$$

Substituting $a = L/2$ and $R_A = 5P/16$ into Eq. (2) and solving for C_1, we find $C_1 = -PL^2/32$. Making $x = L/2$, $C_1 = -PL^2/32$, and $C_2 = 0$ in the expression obtained for y, we have

$$y_B = -\frac{7PL^3}{768EI} \qquad \qquad y_B = \frac{7PL^3}{768EI}\downarrow \qquad \blacktriangle$$

Note that the deflection obtained is *not* the maximum deflection.

PROBLEMS

Use singularity functions to solve the following problems and assume that the flexural rigidity EI of each beam is constant.

9.35 through 9.38 For the beam and loading shown, determine (*a*) the equation of the elastic curve, (*b*) the slope at end A, (*c*) the deflection of point C.

Fig. P9.36

Fig. P9.38

Fig. P9.35

Fig. P9.37

9.39 and 9.40 For the beam and loading shown, determine (*a*) the deflection at end A, (*b*) the deflection point C, (*c*) the slope at end D.

Fig. P9.39

Fig. P9.40

9.41 For the beam and loading shown, determine (*a*) the equation of the elastic curve, (*b*) the deflection at point B, (*c*) the deflection at point C.

Fig. P9.41

9.42 and 9.43 For the beam and loading shown, determine (*a*) the equation of the elastic curve, (*b*) the deflection at the midpoint C.

Fig. P9.42

Fig. P9.43

9.44 For the beam and loading shown, determine (*a*) the equation of the elastic curve, (*b*) the deflection at point A, (*c*) the deflection at point C.

Fig. P9.44

9.45 For the beam and loading shown, determine (*a*) the slope at end A (*b*) the deflection at the midpoint C. Use $E = 29 \times 10^6$ psi.

Fig. P9.45

9.46 For the beam and loading shown, determine (*a*) the slope at end A, (*b*) the deflection at point C. Use $E = 200$ GPa.

9.47 For the timber beam and loading shown, determine (*a*) the slope at end A, (*b*) the deflection at the midpoint C. Use $E = 1.6 \times 10^6$ psi.

Fig. P9.47

9.48 For the beam and loading shown, determine (*a*) the slope at end A, (*b*) the deflection at the midpoint C. Use $E = 200$ GPa.

Fig. P9.46

Fig. P9.48

9.49 and 9.50   For the beam and loading indicated, write a computer program and use it to calculate the slope and deflection of the beam at intervals ΔL, starting at point A and ending at the right-hand support.

    9.49   Beam and loading of Prob. 9.47 with $\Delta L = 3.0$ in.
    9.50   Beam and loading of Prob. 9.48 with $\Delta L = 0.3$ m.

9.51 through 9.54   For the beam and loading shown, determine (*a*) the reaction at the roller support, (*b*) the deflection at point C.

Fig. P9.51

Fig. P9.52

Fig. P9.53

Fig. P9.54

9.55 and 9.56   For the beam and loading shown, determine (*a*) the reaction at A, (*b*) the deflection at C. Use $E = 29 \times 10^6$ psi.

W12 × 22

Fig. P9.55

W14 × 22

Fig. P9.56

9.57   For the beam shown and knowing that $P = 40$ kN, determine (*a*) the reaction at E, (*b*) the deflection at C. Use $E = 200$ GPa.

W200 × 46.1

Fig. P9.57

9.58   For the beam and loading shown, determine (*a*) the reaction at C, (*b*) the deflection at B. Use $E = 200$ GPa.

W410 × 60

Fig. P9.58

9.59   For the beam and loading shown, determine (*a*) the reaction at A, (*b*) the slope at C.

Fig. P9.59

9.60   For the beam and loading shown, determine (a) the reaction at A, (b) the deflection at D.

Fig. P9.60

9.61 through 9.64   For the beam and loading indicated, determine the magnitude and location of the largest downward deflection.

9.61   Beam and loading of Prob. 9.45.
9.62   Beam and loading of Prob. 9.46.
9.63   Beam and loading of Prob. 9.47.
9.64   Beam and loading of Prob. 9.48.

9.65   The rigid bar BDE is welded at point B to the rolled steel beam AC. For the loading shown, determine (a) the slope at point A, (b) the deflection at point B. Use $E = 200$ GPa.

Fig. P9.65

9.66   Rigid bars are welded to the steel rod AD as shown. For the loading shown, determine (a) the deflection at point B, (b) the slope at end A. Use $E = 200$ GPa.

Fig. P9.66

9.7. METHOD OF SUPERPOSITION

When a beam is subjected to several concentrated or distributed loads, it is often found convenient to compute separately the slope and deflection caused by each of the given loads. The slope and deflection due to the combined loads are then obtained by applying the principle of superposition (Sec. 2.12) and adding the values of the slope or deflection corresponding to the various loads.

EXAMPLE 9.07

Determine the slope and deflection at D for the beam and loading shown (Fig. 9.33), knowing that the flexural rigidity of the beam is $EI = 100$ MN $\cdot$ m².

The slope and deflection at any point of the beam can be obtained by superposing the slopes and deflections caused respectively by the concentrated load and by the distributed load (Fig. 9.34).

Fig. 9.33

Fig. 9.34

Since the concentrated load in Fig. 9.34b is applied at quarter span, we can use the results obtained for the beam and loading of Example 9.03 and write

$$(\theta_D)_P = \frac{PL^2}{32EI} = -\frac{(150 \times 10^3)(8)^2}{32(100 \times 10^6)} = -3 \times 10^{-3} \text{ rad}$$

$$(y_D)_P = -\frac{3PL^3}{256EI} = -\frac{3(150 \times 10^3)(8)^3}{256(100 \times 10^6)} = -9 \times 10^{-3} \text{ m}$$
$$= -9 \text{ mm}$$

On the other hand, recalling the equation of the elastic curve obtained for a uniformly distributed load in Example 9.02, we express the deflection in Fig. 9.34c as

$$y = \frac{w}{24EI}(-x^4 + 2Lx^3 - L^3x) \qquad (9.50)$$

and, differentiating with respect to x,

$$\theta = \frac{dy}{dx} = \frac{w}{24EI}(-4x^3 + 6Lx^2 - L^3) \qquad (9.51)$$

Making $w = 20$ kN/m, $x = 2$ m, and $L = 8$ m in Eqs. (9.51) and (9.50), we obtain

$$(\theta_D)_w = \frac{20 \times 10^3}{24(100 \times 10^6)}(-352) = -2.93 \times 10^{-3} \text{ rad}$$

$$(y_D)_w = \frac{20 \times 10^3}{24(100 \times 10^6)}(-912) = -7.60 \times 10^{-3} \text{ m}$$
$$= -7.60 \text{ mm}$$

Combining the slopes and deflections produced by the concentrated and the distributed loads, we have

$$\theta_D = (\theta_D)_P + (\theta_D)_w = -3 \times 10^{-3} - 2.93 \times 10^{-3}$$
$$= -5.93 \times 10^{-3} \text{ rad}$$

$$y_D = (y_D)_P + (y_D)_w = -9 \text{ mm} - 7.60 \text{ mm} = -16.60 \text{ mm}$$

To facilitate the task of practicing engineers, most structural and mechanical engineering handbooks include tables giving the deflections and slopes of beams for various loadings and types of support. Such a table will be found in Appendix D. We note that the slope and deflection of the beam of Fig. 9.33 could have been determined from that table. Indeed, using the information given under cases 5 and 6, we could have expressed the deflection of the beam for any value $x \leq L/4$. Taking the derivative of the expression obtained in this way would have yielded the slope of the beam over the same interval. We also note that the slope at both ends of the beam can be obtained by simply adding the corre-

559

sponding values given in the table. However, the maximum deflection of the beam of Fig. 9.33 *cannot* be obtained by adding the maximum deflections of cases 5 and 6, since these deflections occur at different points of the beam.†

Fig. 9.35 The continuous beams supporting this highway overpass have three supports and are thus indeterminate.

9.8. APPLICATION OF SUPERPOSITION TO STATICALLY INDETERMINATE BEAMS

We often find it convenient to use the method of superposition to determine the reactions at the supports of a statically indeterminate beam. Considering first the case of a beam indeterminate to the first degree (cf. Sec. 9.5), such as the beam shown in Fig. 9.35, we follow the approach described in Sec. 9.5. We designate one of the reactions as redundant and eliminate or modify accordingly the corresponding support. The redundant reaction is then treated as an unknown load that, together with the other loads, must produce deformations which are compatible with the original supports. The slope or deflection at the point where the support has been modified or eliminated is obtained by computing separately the deformations caused by the given loads and by the redundant reaction, and by superposing the results obtained. Once the reactions at the supports have been found, the slope and deflection can be determined in the usual way at any other point of the beam.

†An approximate value of the maximum deflection of the beam can be obtained by plotting the values of y corresponding to various values of x. The determination of the exact location and magnitude of the maximum deflection would require setting equal to zero the expression obtained for the slope of the beam and solving this equation for x.

EXAMPLE 9.08

Determine the reactions at the supports for the prismatic beam and loading shown in Fig. 9.36. (This is the same beam and loading as in Example 9.05 of Sec. 9.5.)

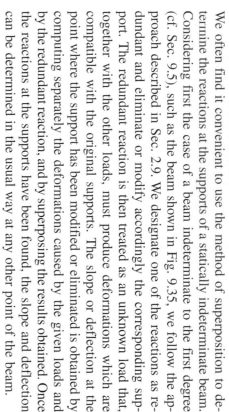

Fig. 9.36

We consider the reaction at B as redundant and release the beam from the support. The reaction $\mathbf{R}_B$ is now considered as an unknown load (Fig. 9.37a) and will be determined from the condition that the deflection of the beam at B must be zero.

The solution is carried out by considering separately the deflection $(y_B)_w$ caused at B by the uniformly distributed load w (Fig. 9.37b) and the deflection $(y_B)_R$ produced at the same point by the redundant reaction $\mathbf{R}_B$ (Fig. 9.37c).

Fig. 9.37

From the table of Appendix D (cases 2 and 1), we find that

$$(y_B)_w = -\frac{wL^4}{8EI} \qquad (y_B)_R = +\frac{R_B L^3}{3EI}$$

Writing that the deflection at B is the sum of these two quantities and that it must be zero, we have

$$y_B = (y_B)_w + (y_B)_R = 0$$

$$y_B = -\frac{wL^4}{8EI} + \frac{R_B L^3}{3EI} = 0$$

and, solving for R_B,

$$R_B = \frac{3}{8}wL \qquad \mathbf{R}_B = \frac{3}{8}wL \uparrow$$

Fig. 9.38

Drawing the free-body diagram of the beam (Fig. 9.38) and writing the corresponding equilibrium equations, we have

$$+\uparrow\sum F_y = 0: \qquad R_A + R_B - wL = 0 \qquad (9.52)$$
$$R_A = wL - R_B = wL - \frac{3}{8}wL = \frac{5}{8}wL$$
$$\mathbf{R}_A = \frac{5}{8}wL \uparrow$$

$$+\text{\rotatebox{90}{$\curvearrowleft$}}\sum M_A = 0: \quad M_A + R_B L - (wL)(\tfrac{1}{2}L) = 0 \qquad (9.53)$$
$$M_A = \tfrac{1}{2}wL^2 - R_B L = \tfrac{1}{2}wL^2 - \tfrac{3}{8}wL^2 = \tfrac{1}{8}wL^2$$
$$\mathbf{M}_A = \tfrac{1}{8}wL^2 \text{\rotatebox{90}{$\curvearrowright$}}$$

Alternative Solution. We may consider the couple exerted at the fixed end A as redundant and replace the fixed end by a pin-and-bracket support. The couple $\mathbf{M}_A$ is now considered as an unknown load (Fig. 9.39a) and will be deter-

Fig. 9.39

mined from the condition that the slope of the beam at A must be zero. The solution is carried out by considering separately the slope $(\theta_A)_w$ caused at A by the uniformly distributed load w (Fig. 9.39b) and the slope $(\theta_A)_M$ produced at the same point by the unknown couple $\mathbf{M}_A$ (Fig. 9.39c).

Using the table of Appendix D (cases 6 and 7), and noting that in case 7, A and B must be interchanged, we find that

$$(\theta_A)_w = -\frac{wL^3}{24EI} \qquad (\theta_A)_M = \frac{M_A L}{3EI}$$

Writing that the slope at A is the sum of these two quantities and that it must be zero, we have

$$\theta_A = (\theta_A)_w + (\theta_A)_M = 0$$

$$\theta_A = -\frac{wL^3}{25EI} + \frac{M_A L}{3EI} = 0$$

and, solving for M_A,

$$M_A = \tfrac{1}{8}wL^2 \qquad \mathbf{M}_A = \tfrac{1}{8}wL^2 \text{\rotatebox{90}{$\curvearrowright$}}$$

The values of R_A and R_B may then be found from the equilibrium equations (9.52) and (9.53).

The beam considered in the preceding example was indeterminate to the first degree. In the case of a beam indeterminate to the second degree (cf. Sec. 9.5), two reactions must be designated as redundant, and the corresponding supports must be eliminated or modified accordingly. The redundant reactions are then treated as unknown loads which, simultaneously and together with the other loads, must produce deformations which are compatible with the original supports. (See Sample Prob. 9.9.)

SAMPLE PROBLEM 9.7

For the beam and loading shown, determine the slope and deflection at point B.

SOLUTION

Principle of Superposition. The given loading can be obtained by superposing the loadings shown in the following "picture equation." The beam AB is, of course, the same in each part of the figure.

For each of the loadings I and II, we now determine the slope and deflection at B by using the table of *Beam Deflections and Slopes* in Appendix D.

Loading I

$$(\theta_B)_I = -\frac{wL^3}{6EI} \qquad (y_B)_I = -\frac{wL^4}{8EI}$$

Loading II

$$(\theta_C)_{II} = +\frac{w(L/2)^3}{6EI} = +\frac{wL^3}{48EI} \qquad (y_C)_{II} = +\frac{w(L/2)^4}{8EI} = +\frac{wL^4}{128EI}$$

In portion CB, the bending moment for loading II is zero and thus the elastic curve is a straight line.

$$(\theta_B)_{II} = (\theta_C)_{II} = +\frac{wL^3}{48EI}$$

$$(y_B)_{II} = (y_C)_{II} + (\theta_C)_{II}\left(\frac{L}{2}\right)$$

$$= \frac{wL^4}{128EI} + \frac{wL^3}{48EI}\left(\frac{L}{2}\right) = +\frac{7wL^4}{384EI}$$

Slope at Point B

$$\theta_B = (\theta_B)_I + (\theta_B)_{II} = -\frac{wL^3}{6EI} + \frac{wL^3}{48EI} = -\frac{7wL^3}{48EI} \qquad \theta_B = \frac{7wL^3}{48EI} \; \blacktriangle$$

Deflection at B

$$y_B = (y_B)_I + (y_B)_{II} = -\frac{wL^4}{8EI} + \frac{7wL^4}{384EI} = -\frac{41wL^4}{384EI} \qquad y_B = \frac{41wL^4}{384EI} \downarrow \; \blacktriangle$$

SAMPLE PROBLEM 9.8

For the uniform beam and loading shown, determine (a) the reaction at each support, (b) the slope at end A.

SOLUTION

Principle of Superposition. The reaction $\mathbf{R}_B$ is designated as redundant and considered as an unknown load. The deflections due to the distributed load and to the reaction $\mathbf{R}_B$ are considered separately as shown below.

For each loading the deflection at point B is found by using the table of *Beam Deflections and Slopes* in Appendix D.

Distributed Loading. We use case 6, Appendix D

$$y = -\frac{w}{24EI}(x^4 - 2Lx^3 + L^3x)$$

At point B, $x = \frac{2}{3}L$:

$$(y_B)_w = -\frac{w}{24EI}\left[\left(\frac{2}{3}L\right)^4 - 2L\left(\frac{2}{3}L\right)^3 + L^3\left(\frac{2}{3}L\right)\right] = -0.01132\frac{wL^4}{EI}$$

Redundant Reaction Loading. From case 5, Appendix D, with $a = \frac{2}{3}L$ and $b = \frac{1}{3}L$, we have

$$(y_B)_R = -\frac{Pa^2b^2}{3EIL} = +\frac{R_D}{3EIL}\left(\frac{2}{3}L\right)^2\left(\frac{L}{3}\right)^2 = 0.01646\frac{R_BL^3}{EI}$$

a. Reactions at Supports. Recalling that $y_B = 0$, we write

$$y_B = (y_B)_w + (y_B)_R$$

$$0 = -0.01132\frac{wL^4}{EI} + 0.01646\frac{R_BL^3}{EI} \qquad \mathbf{R}_B = 0.688wL \uparrow \quad \blacktriangleleft$$

Since the reaction R_B is now known, we may use the methods of statics to determine the other reactions: $\quad \mathbf{R}_A = 0.271wL \uparrow \qquad \mathbf{R}_C = 0.0413wL \uparrow \quad \blacktriangleleft$

b. Slope at End A. Referring again to Appendix D, we have

Distributed Loading. $\quad (\theta_A)_w = -\dfrac{wL^3}{24EI} = -0.04167\dfrac{wL^3}{EI}$

Redundant Reaction Loading. For $P = -R_B = -0.688wL$ and $b = \frac{1}{3}L$

$$(\theta_A)_R = -\frac{Pb(L^2 - b^2)}{6EIL} = +\frac{0.688wL}{6EIL}\left(\frac{L}{3}\right)\left[L^2 - \left(\frac{L}{3}\right)^2\right] = 0.03398\frac{wL^3}{EI}$$

Finally, $\quad \theta_A = (\theta_A)_w + (\theta_A)_R$

$$\theta_A = -0.04167\frac{wL^3}{EI} + 0.03398\frac{wL^3}{EI} = -0.00769\frac{wL^3}{EI} \qquad \theta_A = 0.00769\frac{wL^3}{EI} \searrow \quad \blacktriangleleft$$

$\mathbf{R}_A = 0.271\,wL \qquad \mathbf{R}_B = 0.688\,wL \qquad \mathbf{R}_C = 0.0413\,wL$

SAMPLE PROBLEM 9.9

For the beam and loading shown, determine the reaction at the fixed support C.

SOLUTION

Principle of Superposition. Assuming the axial force in the beam to be zero, the beam ABC is indeterminate to the second degree and we choose two reaction components as redundant, namely, the vertical force $\mathbf{R}_C$ and the couple $\mathbf{M}_C$. The deformations caused by the given load $\mathbf{P}$, the force $\mathbf{R}_C$, and the couple $\mathbf{M}_C$ will be considered separately as shown.

For each load, the slope and deflection at point C will be found by using the table of *Beam Deflections and Slopes* in Appendix D.

Load P. We note that, for this loading, portion BC of the beam is straight.

$$(\theta_C)_P = (\theta_B)_P = -\frac{Pa^2}{2EI} \qquad (y_C)_P = (y_B)_P + (\theta_B)_P b$$

$$= -\frac{Pa^3}{3EI} - \frac{Pa^2}{2EI}b = -\frac{Pa^2}{6EI}(2a + 3b)$$

Force R_C

$$(\theta_C)_R = +\frac{R_C L^2}{2EI} \qquad (y_C)_R = +\frac{R_C L^3}{3EI}$$

Couple M_C

$$(\theta_C)_M = +\frac{M_C L}{EI} \qquad (y_C)_M = +\frac{M_C L^2}{2EI}$$

Boundary Conditions. At end C the slope and deflection must be zero:

$[x = L, \theta_C = 0]$: $\theta_C = (\theta_C)_P + (\theta_C)_R + (\theta_C)_M$

$$0 = -\frac{Pa^2}{2EI} + \frac{R_C L^2}{2EI} + \frac{M_C L}{EI} \qquad (1)$$

$[x = L, y_C = 0]$: $y_C = (y_C)_P + (y_C)_R + (y_C)_M$

$$0 = -\frac{Pa^2}{6EI}(2a + 3b) + \frac{R_C L^3}{3EI} + \frac{M_C L^2}{2EI} \qquad (2)$$

Reaction Components at C. Solving simultaneously Eqs. (1) and (2), we find after reductions

$$R_C = +\frac{Pa^2}{L^3}(a + 3b) \qquad \mathbf{R}_C = \frac{Pa^2}{L^3}(a + 3b) \uparrow \quad \blacktriangle$$

$$M_C = -\frac{Pa^2 b}{L^2} \qquad \mathbf{M}_C = \frac{Pa^2 b}{L^2} \downarrow \quad \blacktriangle$$

Using the methods of statics, we can now determine the reaction at A.

$$\mathbf{M}_A = \frac{Pab^2}{L^2}$$

$$\mathbf{R}_A = \frac{Pb^2}{L^3}(3a + b)$$

PROBLEMS

Use the method of superposition to solve the following problems and assume that the flexural rigidity EI of each beam is constant.

9.67 and 9.68 For the beam and loading shown, determine (*a*) the deflection at point C, (*b*) the slope at end A.

Fig. P9.67

Fig. P9.68

Fig. P9.70

9.69 and 9.70 For the beam and loading shown, determine (*a*) the deflection at the midpoint C, (*b*) the slope at end A.

Fig. P9.69

9.71 and 9.72 For the cantilever beam and loading shown, determine the slope and deflection at the free end.

Fig. P9.71

Fig. P9.72

9.73 and 9.74 For the cantilever beam and loading shown, determine the slope and deflection at point C.

Fig. P9.73

Fig. P9.74

9.75 For the W360 × 39 beam and loading shown, determine (*a*) the slope at end *A*, (*b*) the deflection at point *C*. Use *E* = 200 GPa.

Fig. P9.75

8 kN/m

35 kN

1.3 m ─ 2.6 m

A ─ *C* ─ *B*

9.76 For the W410 × 46.1 beam and loading shown, determine (*a*) the slope at end *A*, (*b*) the deflection at point *C*. Use *E* = 200 GPa.

Fig. P9.76

80 kN · m

140 kN

80 kN · m

2.5 m ─ 2.5 m

A ─ *C* ─ *B*

9.77 For the cantilever beam shown, determine the slope and deflection at end *A*. Use *E* = 29 × 10⁶ psi.

9.78 For the cantilever beam shown, determine the slope and deflection at point *B*. Use *E* = 29 × 10⁶ psi.

Fig. P9.77 and P9.78

1 kip

1 kip/ft

A ─ *B* ─ *C*

2 ft ─ 3 ft

2.0 in.

4.0 in.

9.79 For the cantilever beam shown, determine the slope and deflection at end *C*. Use *E* = 200 GPa.

9.80 For the cantilever beam shown, determine the slope and deflection at point *B*. Use *E* = 200 GPa.

Fig. P9.79 and P9.80

3 kN

3 kN

0.75 m ─ 0.5 m

A ─ *B* ─ *C*

S100 × 11.5

9.81 and 9.82 For the uniform beam shown, determine (*a*) the reaction at *A*, (*b*) the reaction at *B*.

Fig. P9.81

P

P

A ─ *C* ─ *D* ─ *B*

L/3 ─ L/3 ─ L/3

Fig. P9.82

w

A ─ *C* ─ *B*

L/2 ─ L/2

9.83 and 9.84 For the uniform beam shown, determine the reaction at each of the three supports.

Fig. P9.83

Fig. P9.84

9.85 and 9.86 For the beam shown, determine the reaction at B.

Fig. P9.85

Fig. P9.86

9.87 The two beams shown have the same cross section and are joined by a hinge at C. For the loading shown, determine (a) the slope at point A, (b) the deflection at point B. Use E = 29 × 10⁶ psi.

Fig. P9.87

9.88 A central beam BD is joined by hinges to two cantilever beams AB and DE. All beams have the same cross section. For the loading shown, determine the largest allowable value of w if the deflection at C is not to exceed 3 mm. Use E = 200 GPa.

Fig. P9.88

9.89 Beam AC rests on the cantilever beam DE, as shown. Knowing that a W410 × 38.8 rolled-steel shape is used for each beam, determine for the loading shown (a) the deflection at point B, (b) the deflection at point D. Use E = 200 GPa.

Fig. P9.89

9.90 Beam AD rests on beam EF as shown. Knowing that a W12 × 26 rolled-steel shape is used for each beam, determine for the loading shown the deflection at points B and C. Use E = 29 × 10⁶ psi.

Fig. P9.90

9.91 For the loading shown, and knowing that beams AB and DE have the same flexural rigidity, determine the reaction (a) at B, (b) at E.

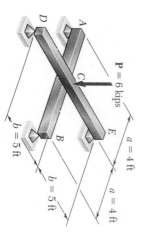

Fig. P9.91

9.92 Knowing that the rod ABC and the wire BD are both made of steel, determine (a) the deflection at B, (b) the reaction at A. Use $E = 200$ GPa.

Fig. P9.92

9.93 Before the load $\mathbf{P}$ was applied, a gap $\delta_0 = 0.5$ mm existed between the cantilever beam AC and the support at B. Knowing that $E = 200$ GPa, determine the magnitude of $\mathbf{P}$ for which the deflection at C is 1 mm.

Fig. P9.93

9.94 Before the 60 kip · ft couple was applied, a gap, $\delta_0 = 0.05$ in., existed between the W16 × 26 beam and the support at C. Knowing that $E = 29 \times 10^6$ psi, determine the reaction at each support after the couple is applied.

Fig. P9.94

9.95 A $\frac{5}{8}$-in.-diameter rod ABC has been bent into the shape shown. Determine the deflection of end C after the 30-lb force is applied. Use $E = 29 \times 10^6$ psi and $G = 11.2 \times 10^6$ psi.

Fig. P9.95

9.96 Two 24-mm-diameter aluminum rods are welded together to form the T-shaped hanger shown. Knowing that $E = 70$ GPa and $G = 26$ GPa, determine the deflection at (a) end A, (b) end B.

Fig. P9.96

*9.9. MOMENT-AREA THEOREMS

In Sec. 9.2 through Sec. 9.6 we used a mathematical method based on the integration of a differential equation to determine the deflection and slope of a beam at any given point. The bending moment was expressed as a function $M(x)$ of the distance x measured along the beam, and two successive integrations led to the functions $\theta(x)$ and $y(x)$ representing, respectively, the slope and deflection at any point of the beam. In this section you will see how geometric properties of the elastic curve can be used to determine the deflection and slope of a beam at a specific point (Fig. 9.40).

Consider a beam AB subjected to some arbitrary loading (Fig. 9.41a). We draw the diagram representing the variation along the beam of the quantity M/EI obtained by dividing the bending moment M by the flexural rigidity EI (Fig. 9.41b). We note that, except for a difference in the scales of ordinates, this diagram will be the same as the bending-moment diagram if the flexural rigidity of the beam is constant.

Recalling Eq. (9.4) of Sec. 9.3, and the fact that $dy/dx = \theta$, we write

$$\frac{d\theta}{dx} = \frac{d^2y}{dx^2} = \frac{M}{EI}$$

or

$$d\theta = \frac{M}{EI}\,dx \qquad (9.54)\dagger$$

Considering two arbitrary points C and D on the beam and integrating both members of Eq. (9.54) from C to D, we write

$$\int_{\theta_C}^{\theta_D} d\theta = \int_{x_C}^{x_D} \frac{M}{EI}\,dx$$

or

$$\theta_D - \theta_C = \int_{x_C}^{x_D} \frac{M}{EI}\,dx \qquad (9.55)$$

where θ_C and θ_D denote the slope at C and D, respectively (Fig. 9.41c). But the right-hand member of Eq. (9.55) represents the area under the (M/EI) diagram between C and D, and the left-hand member the angle between the tangents to the elastic curve at C and D (Fig. 9.41d). Denoting this angle by $\theta_{D/C}$, we have

$$\theta_{D/C} = \text{area under } (M/EI) \text{ diagram} \\ \text{between } C \text{ and } D \qquad (9.56)$$

This is the *first moment-area theorem.*

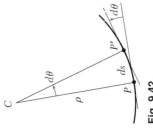

Fig. 9.40 The deflections of the beams supporting the floors of a building should be taken into account in the design process.

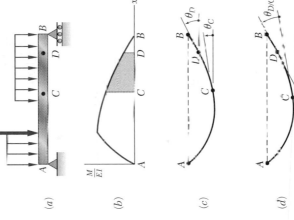

(a)

(b)

(c)

(d)

Fig. 9.41

Fig. 9.42

---

†This relation can also be derived without referring to the results obtained in Sec. 9.3, by noting that the angle $d\theta$ formed by the tangents to the elastic curve at P and P' is also the angle formed by the corresponding normals to that curve (Fig. 9.42). We thus have $d\theta = ds/\rho$ where ds is the length of the arc PP' and ρ the radius of curvature at P. Substituting for $1/\rho$ from Eq. (4.21), and noting that, since the slope at P is very small, ds is equal in first approximation to the horizontal distance dx between P and P', we write

$$d\theta = \frac{M}{EI}\,dx \qquad (9.54)$$

We note that the angle $\theta_{D/C}$ and the area under the (M/EI) diagram have the same sign. In other words, a positive area (i.e., an area located above the x axis) corresponds to a counterclockwise rotation of the tangent as we move from C to D, and a negative area corresponds to a clockwise rotation.

Fig. 9.43

Let us now consider two points P and P' located between C and D, and at a distance dx from each other (Fig. 9.43). The tangents to the elastic curve drawn at P and P' intercept a segment of length dt on the vertical through point C. Since the slope θ at P and the angle $d\theta$ formed by the tangents at P and P' are both small quantities, we can assume that dt is equal to the arc of circle of radius x_1 subtending the angle $d\theta$. We have, therefore,

$$dt = x_1 \, d\theta$$

or, substituting for $d\theta$ from Eq. (9.54),

$$dt = x_1 \frac{M}{EI} dx \qquad (9.57)$$

Fig. 9.44

We now integrate Eq. (9.57) from C to D. We note that, as point P describes the elastic curve from C to D, the tangent at P sweeps the vertical through C from C to E. The integral of the left-hand member is thus equal to the vertical distance from C to the tangent at D. This distance is denoted by $t_{C/D}$ and is called the *tangential deviation of C with respect to D*. We have, therefore,

$$t_{C/D} = \int_{x_C}^{x_D} x_1 \frac{M}{EI} dx \qquad (9.58)$$

We now observe that $(M/EI) \, dx$ represents an element of area under the (M/EI) diagram, and $x_1 \,(M/EI) \, dx$ the first moment of that element with respect to a vertical axis through C (Fig. 9.44). The right-hand member in Eq. (9.58), thus, represents the first moment with respect to that axis of the area located under the (M/EI) diagram between C and D.

We can, therefore, state the second moment-area theorem as follows: *The tangential deviation $t_{C/D}$ of C with respect to D is equal to the first moment with respect to a vertical axis through C of the area under the (M/EI) diagram between C and D.*

Recalling that the first moment of an area with respect to an axis is equal to the product of the area and of the distance from its centroid to that axis, we may also express the second moment-area theorem as follows:

$$t_{C/D} = (\text{area between } C \text{ and } D) \, \bar{x}_1 \qquad (9.59)$$

where the area refers to the area under the (M/EI) diagram, and where $\bar{x}_1$ is the distance from the centroid of the area to the vertical axis through C (Fig. 9.45a).

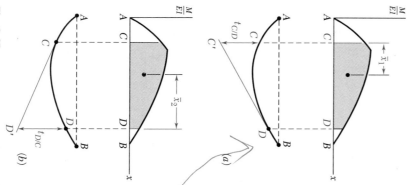

Fig. 9.45

Care should be taken to distinguish between the tangential deviation of C with respect to D, denoted by $t_{C/D}$, and the tangential deviation of D with respect to C, which is denoted by $t_{D/C}$. The tangential deviation $t_{D/C}$ represents the vertical distance from D to the tangent to the elastic curve at C, and is obtained by multiplying the area under the (M/EI) diagram by the distance $\bar{x}_2$ from its centroid to the vertical axis through D (Fig. 9.45b):

$$t_{D/C} = (\text{area between } C \text{ and } D)\,\bar{x}_2 \qquad (9.60)$$

We note that, if an area under the (M/EI) diagram is located above the x axis, its first moment with respect to a vertical axis will be positive; if it is located below the x axis, its first moment will be negative. We check from Fig. 9.45, that a point with a *positive* tangential deviation is located *above* the corresponding tangent, while a point with a *negative* tangential deviation would be located *below* that tangent.

*9.10. APPLICATION TO CANTILEVER BEAMS AND BEAMS WITH SYMMETRIC LOADINGS

We recall that the first moment-area theorem derived in the preceding section defines the angle $\theta_{D/C}$ *between* the tangents at two points C and D of the elastic curve. Thus, the angle θ_D that the tangent at D forms with the horizontal, i.e., the slope at D, can be obtained only if the slope at C is known. Similarly, the second moment-area theorem defines the vertical distance of one point of the elastic curve from the tangent at another point. The tangential deviation $t_{D/C}$, therefore, will help us to locate point D only if the tangent at C is known. We conclude that the two moment area theorems can be applied effectively to the determination of slopes and deflections only if a certain *reference tangent* to the elastic curve has first been determined.

In the case of a *cantilever beam* (Fig. 9.46), the tangent to the elastic curve at the fixed end A is known and can be used as the reference tangent. Since $\theta_A = 0$, the slope of the beam at any point D is $\theta_D = \theta_{D/A}$ and can be obtained by the first moment-area theorem. On the other hand, the deflection y_D of point D is equal to the tangential deviation $t_{D/A}$ measured from the horizontal reference tangent at A and can be obtained by the second moment-area theorem.

In the case of a simply supported beam AB with a *symmetric loading* (Fig. 9.47a) or in the case of an overhanging symmetric beam with a symmetric loading (see Sample Prob. 9.11), the tangent at the center C of the beam must be horizontal by reason of symmetry and can be used as the reference tangent (Fig. 9.47b). Since $\theta_C = 0$, the slope at the support B is $\theta_B = \theta_{B/C}$ and can be obtained by the first moment-area theorem. We also note that $|y|_{\max}$ is equal to the tangential deviation $t_{B/C}$ and can, therefore, be obtained by the second moment-area theorem. The slope at any other point D of the beam (Fig. 9.47c) is found in a similar fashion, and the deflection at D can be expressed as

$$y_D = t_{D/C} - t_{B/C}.$$

Fig. 9.46

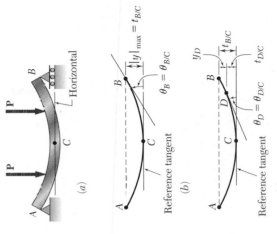

Fig. 9.47

EXAMPLE 9.09

Determine the slope and deflection at end B of the prismatic cantilever beam AB when it is loaded as shown (Fig. 9.48), knowing that the flexural rigidity of the beam is $EI = 10$ MN · m².

Fig. 9.48

We first draw the free-body diagram of the beam (Fig. 9.49a). Summing vertical components and moments about A, we find that the reaction at the fixed end A consists of a 50 kN upward vertical force $\mathbf{R}_A$ and a 60 kN · m counterclockwise couple $\mathbf{M}_A$. Next, we draw the bending-moment diagram (Fig. 9.49b) and determine from similar triangles the distance x_D from the end A to the point D of the beam where $M = 0$:

$$\frac{x_D}{60} = \frac{3 - x_D}{90} = \frac{3}{150} \qquad x_D = 1.2 \text{ m}$$

Dividing by the flexural rigidity EI the values obtained for M, we draw the (M/EI) diagram (Fig. 9.50) and compute the areas corresponding respectively to the segments AD and DB, assigning a positive sign to the area located above the x axis, and a negative sign to the area located below that axis. Using the first moment-area theorem, we write

$$\theta_{B/A} = \theta_B - \theta_A = \text{area from } A \text{ to } B = A_1 + A_2$$
$$= -\tfrac{1}{2}(1.2 \text{ m})(6 \times 10^{-3} \text{ m}^{-1})$$
$$\quad + \tfrac{1}{2}(1.8 \text{ m})(9 \times 10^{-3} \text{ m}^{-1})$$
$$= -3.6 \times 10^{-3} + 8.1 \times 10^{-3}$$
$$= +4.5 \times 10^{-3} \text{ rad}$$

and, since $\theta_A = 0$,

$$\theta_B = +4.5 \times 10^{-3} \text{ rad}$$

Using now the second moment-area theorem, we write that the tangential deviation $t_{B/A}$ is equal to the first moment about a vertical axis through B of the total area between A and B. Expressing the moment of each partial area as the product of that area and of the distance from its centroid to the axis through B, we have

$$t_{B/A} = A_1(2.6 \text{ m}) + A_2(0.6 \text{ m})$$
$$= (-3.6 \times 10^{-3})(2.6 \text{ m}) + (8.1 \times 10^{-3})(0.6 \text{ m})$$
$$= -9.36 \text{ mm} + 4.86 \text{ mm} = -4.50 \text{ mm}$$

Since the reference tangent at A is horizontal, the deflection at B is equal to $t_{B/A}$ and we have

$$y_B = t_{B/A} = -4.50 \text{ mm}$$

The deflected beam has been sketched in Fig. 9.51.

Fig. 9.49

Fig. 9.50

Fig. 9.51

*9.11. BENDING-MOMENT DIAGRAMS BY PARTS

In many applications the determination of the angle $\theta_{D/C}$ and of the tangential deviation $t_{D/C}$ is simplified if the effect of each load is evaluated independently. A separate (M/EI) diagram is drawn for each load, and the angle $\theta_{D/C}$ is obtained by adding algebraically the areas under the various diagrams. Similarly, the tangential deviation $t_{D/C}$ is obtained by adding the first moments of these areas about a vertical axis through D. A bending-moment or (M/EI) diagram plotted in this fashion is said to be *drawn by parts*.

When a bending-moment or (M/EI) diagram is drawn by parts, the various areas defined by the diagram consist of simple geometric shapes, such as rectangles, triangles, and parabolic spandrels. For convenience, the areas and centroids of these various shapes have been indicated in Fig. 9.52.

Shape		Area	c
Rectangle		bh	$\dfrac{b}{2}$
Triangle		$\dfrac{bh}{2}$	$\dfrac{b}{3}$
Parabolic spandrel	$y = kx^2$	$\dfrac{bh}{3}$	$\dfrac{b}{4}$
Cubic spandrel	$y = kx^3$	$\dfrac{bh}{4}$	$\dfrac{b}{5}$
General spandrel	$y = kx^n$	$\dfrac{bh}{n+1}$	$\dfrac{b}{n+2}$

Fig. 9.52 Areas and centroids of common shapes.

EXAMPLE 9.10

Determine the slope and deflection at end B of the prismatic beam of Example 9.09, drawing the bending-moment diagram by parts.

Fig. 9.53

We replace the given loading by the two equivalent loadings shown in Fig. 9.53, and draw the corresponding bending-moment and (M/EI) diagrams from right to left, starting at the free end B.

Applying the first moment-area theorem, and recalling that $\theta_A = 0$, we write

$$\theta_B = \theta_{B/A} = A_1 + A_2$$
$$= (9 \times 10^{-3} \text{ m}^{-1})(3 \text{ m}) - \tfrac{1}{2}(15 \times 10^{-3} \text{ m}^{-1})(3 \text{ m})$$
$$= 27 \times 10^{-3} - 22.5 \times 10^{-3} = 4.5 \times 10^{-3} \text{ rad}$$

Applying the second moment-area theorem, we compute the first moment of each area about a vertical axis through B and write

$$y_B = t_{B/A} = A_1(1.5 \text{ m}) + A_2(2 \text{ m})$$
$$= (27 \times 10^{-3})(1.5 \text{ m}) - (22.5 \times 10^{-3})(2 \text{ m})$$
$$= 40.5 \text{ mm} - 45 \text{ mm} = -4.5 \text{ mm}$$

It is convenient, in practice, to group into a single drawing the two portions of the (M/EI) diagram (Fig. 9.54).

Fig. 9.54

EXAMPLE 9.11

For the prismatic beam AB and the loading shown (Fig. 9.55), determine the slope at a support and the maximum deflection.

Fig. 9.55

$$|y|_{max} = t_{A/C}$$

Reference tangent

$$\theta_A = -\theta_{C/A}$$

Fig. 9.56

Fig. 9.57

Fig. 9.58

We first sketch the deflected beam (Fig. 9.56). Since the tangent at the center C of the beam is horizontal, it will be used as the reference tangent, and we have $|y|_{max} = t_{A/C}$. On the other hand, since $\theta_C = 0$, we write

$$\theta_{C/A} = \theta_C - \theta_A = -\theta_A \qquad \text{or} \qquad \theta_A = -\theta_{C/A}$$

From the free-body diagram of the beam (Fig. 9.57), we find that

$$R_A = R_B = wa$$

Next, we draw the shear and bending-moment diagrams for the portion AC of the beam. We draw these diagrams by parts, considering separately the effects of the reaction $\mathbf{R}_A$ and of the distributed load. However, for convenience, the two parts of each diagram have been plotted together (Fig. 9.58). We recall from Sec. 5.3 that, the distributed load being uniform, the corresponding parts of the shear and bending-moment diagrams will he, respectively, linear and parabolic. The area and centroid of the triangle and of the parabolic spandrel can he obtained by referring to Fig. 9.52. The areas of the triangle and spandrel are found to be, respectively,

$$A_1 = \frac{1}{2}(2a)\left(\frac{2wa^2}{EI}\right) = \frac{2wa^3}{EI}$$

and

$$A_2 = -\frac{1}{3}(a)\left(\frac{wa^2}{2EI}\right) = -\frac{wa^3}{6EI}$$

Applying the first moment-area theorem, we write

$$\theta_{C/A} = A_1 + A_2 = \frac{2wa^3}{EI} - \frac{wa^3}{6EI} = \frac{11wa^3}{6EI}$$

Recalling from Figs. 9.55 and 9.56 that $a = \frac{1}{4}L$ and $\theta_A = -\theta_{C/A}$, we have

$$\theta_A = -\frac{11wa^3}{6EI} = -\frac{11wL^3}{384EI}$$

Applying now the second moment-area theorem, we write

$$t_{A/C} = A_1\frac{4a}{3} + A_2\frac{7a}{4} = \left(\frac{2wa^3}{EI}\right)\frac{4a}{3} + \left(-\frac{wa^3}{6EI}\right)\frac{7a}{4} = \frac{19wa^4}{8EI}$$

and

$$|y|_{max} = t_{A/C} = \frac{19wa^4}{8EI} = \frac{19wL^4}{2048EI}$$

The prismatic rods AD and DB are welded together to form the cantilever beam ADB. Knowing that the flexural rigidity is EI in portion AD of the beam and $2EI$ in portion DB, determine, for the loading shown, the slope and deflection at end A.

SOLUTION

(M/EI) Diagram. We first draw the bending-moment diagram for the beam and then obtain the (M/EI) diagram by dividing the value of M at each point of the beam by the corresponding value of the flexural rigidity.

Reference Tangent. We choose the horizontal tangent at the fixed end B as the reference tangent. Since $\theta_B = 0$ and $y_B = 0$, we note that

$$\theta_A = -\theta_{B/A} \qquad y_A = t_{A/B}$$

$y_A = t_{A/B}$

$-\theta_{B/A}$

Reference tangent

θ_A

Slope at A. Dividing the (M/EI) diagram into the three triangular portions shown, we write

$$A_1 = -\frac{1}{2}\frac{Pa}{EI}a = -\frac{Pa^2}{2EI}$$

$$A_2 = -\frac{1}{2}\frac{Pa}{2EI}a = -\frac{Pa^2}{4EI}$$

$$A_3 = -\frac{1}{2}\frac{3Pa}{2EI}a = -\frac{3Pa^2}{4EI}$$

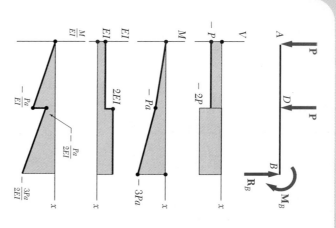

Using the first moment-area theorem, we have

$$\theta_{B/A} = A_1 + A_2 + A_3 = -\frac{Pa^2}{2EI} - \frac{Pa^2}{4EI} - \frac{3Pa^2}{4EI} = -\frac{3Pa^2}{2EI}$$

$$\theta_A = -\theta_{B/A} = +\frac{3Pa^2}{2EI} \qquad \theta_A = \frac{3Pa^2}{2EI} \; \measuredangle \quad \blacktriangleright$$

Deflection at A. Using the second moment-area theorem, we have

$$y_A = t_{A/B} = A_1\left(\frac{2}{3}a\right) + A_2\left(\frac{4}{3}a\right) + A_3\left(\frac{5}{3}a\right)$$

$$= \left(-\frac{Pa^2}{2EI}\right)\frac{2a}{3} + \left(-\frac{Pa^2}{4EI}\right)\frac{4a}{3} + \left(-\frac{3Pa^2}{4EI}\right)\frac{5a}{3}$$

$$y_A = -\frac{23Pa^3}{12EI} \qquad y_A = \frac{23Pa^3}{12EI} \downarrow \quad \blacktriangleright$$

SAMPLE PROBLEM 9.11

For the prismatic beam and loading shown, determine the slope and deflection at end E.

SOLUTION

(M/EI) Diagram. From a free-body diagram of the beam, we determine the reactions and then draw the shear and bending-moment diagrams. Since the flexural rigidity of the beam is constant, we divide each value of M by EI and obtain the (M/EI) diagram shown.

Reference Tangent. Since the beam and its loading are symmetric with respect to the midpoint C, the tangent at C is horizontal and is used as the reference tangent. Referring to the sketch, we observe that, since $\theta_C = 0$,

$$\theta_E = \theta_C + \theta_{E/C} = \theta_{E/C} \qquad (1)$$
$$y_E = t_{E/C} - t_{D/C} \qquad (2)$$

Slope at E. Referring to the (M/EI) diagram and using the first moment-area theorem, we write

$$A_1 = -\frac{wa^2}{2EI}\left(\frac{L}{2}\right) = -\frac{wa^2L}{4EI}$$

$$A_2 = -\frac{1}{3}\left(\frac{wa^2}{2EI}\right)(a) = -\frac{wa^3}{6EI}$$

Using Eq. (1), we have

$$\theta_E = \theta_{E/C} = A_1 + A_2 = -\frac{wa^2L}{4EI} - \frac{wa^3}{6EI}$$

$$\theta_E = -\frac{wa^2}{12EI}(3L + 2a)$$

$$\boxed{\theta_E = -\frac{wa^2}{12EI}(3L + 2a)\;\searrow} \qquad \blacktriangledown$$

Deflection at E. Using the second moment-area theorem, we write

$$t_{D/C} = A_1\frac{L}{4} = \left(-\frac{wa^2L}{4EI}\right)\frac{L}{4} = -\frac{wa^2L^2}{16EI}$$

$$t_{E/C} = A_1\left(a + \frac{L}{4}\right) + A_2\left(\frac{3a}{4}\right)$$
$$= \left(-\frac{wa^2L}{4EI}\right)\left(a + \frac{L}{4}\right) + \left(-\frac{wa^3}{6EI}\right)\left(\frac{3a}{4}\right)$$
$$= -\frac{wa^3L}{4EI} - \frac{wa^2L^2}{16EI} - \frac{wa^4}{8EI}$$

Using Eq. (2), we have

$$y_E = t_{E/C} - t_{D/C} = -\frac{wa^3L}{4EI} - \frac{wa^4}{8EI}$$

$$y_E = -\frac{wa^3}{8EI}(2L + a)$$

$$\boxed{y_E = \frac{wa^3}{8EI}(2L + a)\;\downarrow} \qquad \blacktriangledown$$

577

Use the moment-area method to solve the following problems.

9.97 through 9.100 For the uniform cantilever beam and loading shown, determine (a) the slope at the free end, (b) the deflection at the free end.

Fig. P9.97

Fig. P9.98

Fig. P9.99

Fig. P9.100

9.101 and 9.102 For the uniform cantilever beam and loading shown, determine (a) the slope at point B, (b) the deflection at point A.

Fig. P9.101

Fig. P9.102

9.103 For the cantilever beam and loading shown, determine (a) the slope at point A, (b) the deflection at point A. Use $E = 200$ GPa.

Fig. P9.103

9.104 For the cantilever beam and loading shown, determine (a) the slope at point A, (b) the deflection at point A. Use $E = 29 \times 10^6$ psi.

Fig. P9.104

9.105 For the cantilever beam and loading shown, determine (*a*) the slope at point *C*, (*b*) the deflection at point *C*.

Fig. P9.105

9.106 Two C6 × 8.2 channels are welded back to back and loaded as shown. Knowing that $E = 29 \times 10^6$ psi, determine (*a*) the slope at *D*, (*b*) the deflection at *D*.

Fig. P9.106

9.107 For the cantilever beam and loading shown, determine the deflection and slope at end *D* caused by the couple $\mathbf{M}_0$.

Fig. P9.107

9.108 For the cantilever beam and loading shown, determine the deflection at (*a*) point *B*, (*b*) point *C*.

Fig. P9.108

9.109 Two cover plates are welded to the rolled-steel beam as shown. Using $E = 29 \times 10^6$ psi, determine (*a*) the slope at end *C*, (*b*) the deflection at end *C*.

Fig. P9.109

9.110 Two cover plates are welded to the rolled-steel beam as shown. Using $E = 200$ GPa, determine (*a*) the slope at end *A*, (*b*) the deflection at end *A*.

Fig. P9.110

9.111 through 9.114 For the prismatic beam and loading shown, determine (*a*) the slope at end *A*, (*b*) the deflection at the center *C* of the beam.

Fig. P9.111

Fig. P9.112

Fig. P9.113

Fig. P9.114

9.115 For the beam and loading shown, determine (*a*) the slope at end *A*, (*b*) the deflection at the midpoint *C* of the beam. Use $E = 200$ GPa.

$w_0 = 54$ kN/m

W460 × 52

Fig. P9.115

9.116 For the beam and loading shown, determine (*a*) the slope at end *A*, (*b*) the deflection at the midpoint *C* of the beam. Use $E = 29 \times 10^6$ psi.

1.6 kips   1.6 kips

Fig. P9.116

9.117 and 9.118 For the beam and loading shown, determine (*a*) the slope at end *A*, (*b*) the deflection at the midpoint *C* of the beam.

Fig. P9.117

Fig. P9.118

Fig. P9.119

9.119 For the beam and loading shown, determine (*a*) the slope at end *A*, (*b*) the deflection at the midpoint *C* of the beam. Use $E = 29 \times 10^6$ psi.

9.120 For the beam and loading shown and knowing that $w = 8$ kN/m, determine (*a*) the slope at end *A*, (*b*) the deflection at the midpoint *C* of the beam. Use $E = 200$ GPa.

Fig P9.120

9.121 For the beam and loading shown, determine (*a*) the slope at end *A*, (*b*) the deflection at point *A*. Use $E = 29 \times 10^6$ psi.

9.122 Knowing that $P = 8$ kN, determine (*a*) the slope at end *A*, (*b*) the deflection at the midpoint *C* of the beam. Use $E = 200$ GPa.

Fig. P9.122

9.123 For the beam and loading of Prob. 9.120, determine the value of w for which the deflection is zero at the midpoint *C* of the beam. Use $E = 200$ GPa.

9.124 For the beam and loading of Prob. 9.122, determine the magnitude of the forces **P** for which the deflection is zero at end *A*. Use $E = 200$ GPa.

*9.125 A uniform rod *AE* is supported at two points *B* and *D*. Determine the distance *a* from the ends of the rod to the points of support if the downward deflections of points *A*, *C*, and *E* are to be equal.

*9.126 A uniform rod *AE* is supported at two points *B* and *D*. Determine the distance *a* for which the slope at ends *A* and *E* is to be zero.

Fig. P9.121

Fig. P9.125 and P9.126

*9.12. APPLICATION OF MOMENT-AREA THEOREMS TO BEAMS WITH UNSYMMETRIC LOADINGS

We saw in Sec. 9.10 that, when a simply supported or overhanging beam carries a symmetric load, the tangent at the center C of the beam is horizontal and can be used as the reference tangent. When a simply supported or overhanging beam carries an unsymmetric load, it is generally not possible to determine by inspection the point of the beam where the tangent is horizontal. Other means must then be found for locating a reference tangent, i.e., a tangent of known slope to be used in applying either of the two moment-area theorems.

It is usually most convenient to select the reference tangent at one of the beam supports. Considering, for example, the tangent at the support A of the simply supported beam AB (Fig. 9.59a), we determine its slope by computing the tangential deviation $t_{B/A}$ of the support B with respect to A, and dividing $t_{B/A}$ by the distance L between the supports. Recalling that the tangential deviation of a point located above the tangent is positive, we write

$$\theta_A = -\frac{t_{B/A}}{L} \qquad (9.61)$$

Once the slope of the reference tangent has been found, the slope θ_D of the beam at any point D (Fig. 9.60) can be determined by using the first moment-area theorem to obtain $\theta_{D/A}$, and then writing

$$\theta_D = \theta_A + \theta_{D/A} \qquad (9.62)$$

The tangential deviation $t_{D/A}$ of D with respect to the support A can be obtained from the second moment-area theorem. We note that $t_{D/A}$ is equal to the segment ED (Fig. 9.61) and represents the vertical distance of D *from the reference tangent*. On the other hand, the deflection y_D of point D represents the vertical distance of D *from the horizontal line AB* (Fig. 9.62). Since y_D is equal in magnitude to the segment FD, it can be expressed as the difference between EF and ED (Fig. 9.63). Observing from the similar triangles AFE and ABH that

$$\frac{EF}{x} = \frac{HB}{L} \qquad \text{or} \qquad EF = \frac{x}{L} t_{B/A}$$

and recalling the sign conventions used for deflections and tangential deviations, we write

$$y_D = ED - EF = t_{D/A} - \frac{x}{L} t_{B/A} \qquad (9.63)$$

Fig. 9.59

(a)

(b)

Fig. 9.60

Fig. 9.61

Fig. 9.62

Fig. 9.63

EXAMPLE 9.12

For the prismatic beam and loading shown (Fig. 9.64), determine the slope and deflection at point D.

Fig. 9.64

Reference Tangent at Support A. We compute the reactions at the supports and draw the (M/EI) diagram (Fig. 9.65). We determine the tangential deviation $t_{B/A}$ of the support B with respect to the support A by applying the second moment-area theorem and computing the moments about a vertical axis through B of the areas A_1 and A_2. We have

$$A_1 = \frac{1}{2}\frac{L}{4}\frac{3PL}{16EI} = \frac{3PL^2}{128EI} \qquad A_2 = \frac{1}{2}\frac{3L}{4}\frac{3PL}{16EI} = \frac{9PL^2}{128EI}$$

Fig. 9.65

$$t_{B/A} = A_1\left(\frac{L}{12} + \frac{3L}{4}\right) + A_2\left(\frac{L}{2}\right)$$

$$= \frac{3PL^2}{128EI}\frac{10L}{12} + \frac{9PL^2}{128EI}\frac{L}{2} = \frac{7PL^3}{128EI}$$

The slope of the reference tangent at A (Fig. 9.66) is

$$\theta_A = -\frac{t_{B/A}}{L} = -\frac{7PL^2}{128EI}$$

Fig. 9.66

Slope at D. Applying the first moment-area theorem from A to D, we write

$$\theta_{D/A} = A_1 = \frac{3PL^2}{128EI}$$

Thus, the slope at D is

$$\theta_D = \theta_A + \theta_{D/A} = -\frac{7PL^2}{128EI} + \frac{3PL^2}{128EI} = -\frac{PL^2}{32EI}$$

Deflection at D. We first determine the tangential deviation $DE = t_{D/A}$ by computing the moment of the area A_1 about a vertical axis through D:

$$DE = t_{D/A} = A_1\left(\frac{L}{12}\right) = \frac{3PL^2}{128EI}\frac{L}{12} = \frac{PL^3}{512EI}$$

The deflection at D is equal to the difference between the segments DE and EF (Fig. 9.66). We have

$$y_D = DE - EF = t_{D/A} - \tfrac{1}{4}t_{B/A}$$
$$= \frac{PL^3}{512EI} - \frac{1}{4}\frac{7PL^3}{128EI}$$
$$y_D = -\frac{3PL^3}{256EI} = -0.01172PL^3/EI$$

*9.13. MAXIMUM DEFLECTION

When a simply supported or overhanging beam carries an unsymmetric load, the maximum deflection generally does not occur at the center of the beam. This will be the case for the beams forming the frame of the flatbed trailer shown in Fig. 9.67, which is being used to transport a wheeled loader.

Fig. 9.67 The deflections of the beams used for the trailer must be reviewed for different possible positions of the load.

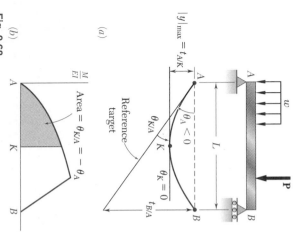

(a)

(b)

Fig. 9.68

To determine the maximum deflection of such a beam, we should locate the point K of the beam where the tangent is horizontal, and compute the deflection at that point.

Our analysis must begin with the determination of a reference tangent at one of the supports. If support A is selected, the slope θ_A of the tangent at A is obtained by the method indicated in the preceding section, i.e., by computing the tangential deviation $t_{B/A}$ of support B with respect to A and dividing that quantity by the distance L between the two supports.

Since the slope θ_K at point K is zero (Fig. 9.68a), we must have

$$\theta_{K/A} = \theta_K - \theta_A = 0 - \theta_A = -\theta_A$$

Recalling the first moment-area theorem, we conclude that point K may be determined by measuring under the (M/EI) diagram an area equal to $\theta_{K/A} = -\theta_A$ (Fig. 9.68b).

Observing that the maximum deflection $|y|_{max}$ is equal to the tangential deviation $t_{A/K}$ of support A with respect to K (Fig. 9.68a), we can obtain $|y|_{max}$ by computing the first moment with respect to the vertical axis through A of the area between A and K (Fig. 9.68b).

EXAMPLE 9.13

Determine the maximum deflection of the beam of Example 9.12.

Determination of Point K Where Slope Is Zero. We recall from Example 9.12 that the slope at point D, where the load is applied, is negative. It follows that point K, where the slope is zero, is located between D and the support B (Fig. 9.69). Our computations, therefore, will be simplified if we relate the slope at K to the slope at B, rather than to the slope at A.

Since the slope at A has already been determined in Example 9.12, the slope at B is obtained by writing

$$\theta_B = \theta_A + \theta_{B/A} = \theta_A + A_1 + A_2$$

$$\theta_B = -\frac{7PL^2}{128EI} + \frac{3PL^2}{128EI} + \frac{9PL^2}{128EI} = \frac{5PL^2}{128EI}$$

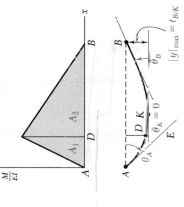

Observing that the bending moment at a distance u from end B is $M = \frac{1}{4}Pu$ (Fig. 9.70a), we express the area A' located between K and B under the (M/EI) diagram (Fig. 9.70b) as

$$A' = \frac{1}{2}\frac{Pu}{4EI}u = \frac{Pu^2}{8EI}$$

By the first moment-area theorem, we have

$$\theta_{B/K} = \theta_B - \theta_K = A'$$

and, since $\theta_K = 0$, $\theta_B = A'$

Substituting the values obtained for θ_B and A', we write

$$\frac{5PL^2}{128EI} = \frac{Pu^2}{8EI}$$

and, solving for u,

$$u = \frac{\sqrt{5}}{4}L = 0.559L$$

Thus, the distance from the support A to point K is

$$AK = L - 0.559L = 0.441L$$

Maximum Deflection. The maximum deflection $|y|_{max}$ is equal to the tangential deviation $t_{B/K}$ and, thus, to the first moment of the area A' about a vertical axis through B (Fig. 9.70b). We write

$$|y|_{max} = t_{B/K} = A'\left(\frac{2u}{3}\right) = \frac{Pu^2}{8EI}\left(\frac{2u}{3}\right) = \frac{Pu^3}{12EI}$$

Substituting the value obtained for u, we have

$$|y|_{max} = \frac{P}{12EI}\left(\frac{\sqrt{5}}{4}L\right)^3 = 0.01456PL^3/EI$$

Fig. 9.69

Fig. 9.70

*9.14. USE OF MOMENT-AREA THEOREMS WITH STATICALLY INDETERMINATE BEAMS

The reactions at the supports of a statically indeterminate beam can be determined by the moment-area method in much the same way that was described in Sec. 9.8. In the case of a beam indeterminate to the first degree, for example, we designate one of the reactions as redundant and eliminate or modify accordingly the corresponding support. The redundant reaction is then treated as an unknown load, which, together with the other loads, must produce deformations that are compatible with the original supports. The compatibility condition is usually expressed by writing that the tangential deviation of one support with respect to another either is zero or has a predetermined value.

Two separate free-body diagrams of the beam are drawn. One shows *the given loads and the corresponding reactions at the supports that have not been eliminated;* the other shows *the redundant reaction and the corresponding reactions* at the supports (see Example 9.14). An M/EI diagram is then drawn for each of the two loadings, and the desired tangential deviations are obtained by the second moment-area theorem. Superposing the results obtained, we express the required compatibility condition and determine the redundant reaction. The other reactions are obtained from the free-body diagram of beam.

Once the reactions at the supports have been determined, the slope and deflection may be obtained by the moment-area method at any other point of the beam.

Fig. 9.71

EXAMPLE 9.14

Determine the reaction at the supports for the prismatic beam and loading shown (Fig. 9.71).

We consider the couple exerted at the fixed end A as redundant and replace the fixed end by a pin-and-bracket support. The couple M_A is now considered as an unknown load (Fig. 9.72a) and will be determined from the condition that the tangent to the beam at A must be horizontal. It follows that this tangent must pass through the support B and, thus, that the tangential deviation $t_{B/A}$ of B with respect to A must be zero. The solution is carried out by computing separately the tangential deviation $(t_{B/A})_w$ caused by the uniformly distributed load w (Fig. 9.72b) and the tangential deviation $(t_{B/A})_M$ produced by the unknown couple M_A (Fig. 9.72c).

Fig. 9.72

Considering first the free-body diagram of the beam under the known distributed load w (Fig. 9.73a), we determine the corresponding reactions at the supports A and B. We have

$$(\mathbf{R}_A)_1 = (\mathbf{R}_B)_1 = \tfrac{1}{2}wL\uparrow \qquad (9.64)$$

We can now draw the corresponding shear and (M/EI) diagrams (Figs. 9.73b and c). Observing that M/EI is represented by an arc of parabola, and recalling the formula, $A = \tfrac{2}{3}bh$, for the area under a parabola, we compute the first moment of this area about a vertical axis through B and write

$$(t_{B/A})_w = A_1\left(\frac{L}{2}\right) = \left(\frac{2}{3}L\,\frac{wL^2}{8EI}\right)\left(\frac{L}{2}\right) = \frac{wL^4}{24EI} \qquad (9.65)$$

Considering next the free-body diagram of the beam when it is subjected to the unknown couple $\mathbf{M}_A$ (Fig. 9.74a), we determine the corresponding reactions at A and B:

$$(\mathbf{R}_A)_2 = \frac{M_A}{L}\uparrow \qquad (\mathbf{R}_B)_2 = \frac{M_A}{L}\downarrow \qquad (9.66)$$

Drawing the corresponding (M/EI) diagram (Fig. 9.74b), we apply again the second moment-area theorem and write

$$(t_{B/A})_M = A_2\left(\frac{2L}{3}\right) = \left(-\frac{1}{2}L\,\frac{M_A}{EI}\right)\left(\frac{2L}{3}\right) = -\frac{M_AL^2}{3EI} \qquad (9.67)$$

Combining the results obtained in (9.65) and (9.67), and expressing that the resulting tangential deviation $t_{B/A}$ must be zero (Fig. 9.72), we have

$$t_{B/A} = (t_{B/A})_w + (t_{B/A})_M = 0$$
$$\frac{wL^4}{24EI} - \frac{M_AL^2}{3EI} = 0$$

and, solving for M_A,

$$M_A = +\tfrac{1}{8}wL^2 \qquad \mathbf{M}_A = \tfrac{1}{8}wL^2 \curvearrowright$$

Substituting for M_A into (9.66), and recalling (9.64), we obtain the values of R_A and R_B:

$$R_A = (R_A)_1 + (R_A)_2 = \tfrac{1}{2}wL + \tfrac{1}{8}wL = \tfrac{5}{8}wL$$
$$R_B = (R_B)_1 + (R_B)_2 = \tfrac{1}{2}wL - \tfrac{1}{8}wL = \tfrac{3}{8}wL$$

Fig. 9.73

Fig. 9.74

In the example we have just considered, there was a single redundant reaction, i.e., the beam was *statically indeterminate to the first degree*. The *moment-area theorems* can also be used when there are additional redundant reactions. As discussed in Sec. 9.5, it is then necessary to write additional equations. Thus for a beam that is *statically indeterminate to the second degree*, it would be necessary to select two redundants and write two equations considering the *deformations* of the structure involved.

SAMPLE PROBLEM 9.12

For the beam and loading shown, (a) determine the deflection at end A. (b) evaluate y_A for the following data:

$$W10 \times 33: I = 170 \text{ in}^4 \qquad E = 29 \times 10^6 \text{ psi}$$
$$a = 3 \text{ ft} = 36 \text{ in.} \qquad L = 5.5 \text{ ft} = 66 \text{ in.}$$
$$w = 13.5 \text{ kips/ft} = 1125 \text{ lb/in.}$$

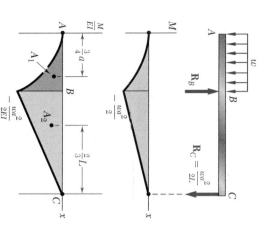

SOLUTION

(M/EI) **Diagram.** We first draw the bending-moment diagram. Since the flexural rigidity EI is constant, we obtain the (M/EI) diagram shown, which consists of a parabolic spandrel of area A_1 and a triangle of area A_2.

$$A_1 = \frac{1}{3}\left(-\frac{wa^2}{2EI}\right)a = -\frac{wa^3}{6EI}$$

$$A_2 = \frac{1}{2}\left(-\frac{wa^2}{2EI}\right)L = -\frac{wa^2 L}{4EI}$$

Reference Tangent at B. The reference tangent is drawn at point B as shown. Using the second moment-area theorem, we determine the tangential deviation of C with respect to B:

$$t_{C/B} = A_2 \frac{2L}{3} = \left(-\frac{wa^2 L}{4EI}\right)\frac{2L}{3} = -\frac{wa^2 L^2}{6EI}$$

From the similar triangles $A''A'B$ and $CC'B$, we find

$$A''A' = t_{C/B}\left(\frac{a}{L}\right) = -\frac{wa^2 L^2}{6EI}\left(\frac{a}{L}\right) = -\frac{wa^3 L}{6EI}$$

Again using the second moment-area theorem, we write

$$t_{A/B} = A_1 \frac{3a}{4} = \left(-\frac{wa^3}{6EI}\right)\frac{3a}{4} = -\frac{wa^4}{8EI}$$

a. Deflection at End A

$$y_A = A''A' + t_{A/B} = -\frac{wa^3 L}{6EI} - \frac{wa^4}{8EI} = -\frac{wa^4}{8EI}\left(\frac{4}{3}\frac{L}{a}+1\right)$$

$$y_A = \frac{wa^4}{8EI}\left(1 + \frac{4}{3}\frac{L}{a}\right) \blacktriangleleft$$

b. Evaluation of y_A. Substituting the data given, we write

$$y_A = \frac{(1125 \text{ lb/in.})(36 \text{ in.})^4}{8(29 \times 10^6 \text{ lb/in}^2)(170 \text{ in}^4)}\left(1 + \frac{4}{3}\frac{66 \text{ in.}}{36 \text{ in.}}\right)$$

$$y_A = 0.1650 \text{ in.} \downarrow \blacktriangleleft$$

SAMPLE PROBLEM 9.13

For the beam and loading shown, determine the magnitude and location of the largest deflection. Use $E = 200$ GPa.

W230 × 22.3

$w = 25$ kN/m

$a = 1.4$ m

$b = 2.2$ m

$L = 3.6$ m

SOLUTION

Reactions. Using the free-body diagram of the entire beam, we find

$$\mathbf{R}_A = 16.81 \text{ kN}\uparrow \qquad \mathbf{R}_B = 38.2 \text{ kN}\uparrow$$

(M/EI) Diagram. We draw the (M/EI) diagram by parts, considering separately the effects of the reaction $\mathbf{R}_A$ and of the distributed load. The areas of the triangle and of the spandrel are

$$A_1 = \frac{1}{2}\frac{R_A L}{EI}L = \frac{R_A L^2}{2EI} \qquad A_2 = \frac{1}{3}\left(-\frac{wb^2}{2EI}\right)b = -\frac{wb^3}{6EI}$$

Reference Tangent. The tangent to the beam at support A is chosen as the reference tangent. Using the second moment-area theorem, we determine the tangential deviation $t_{B/A}$ of support B with respect to support A:

$$t_{B/A} = A_1\frac{L}{3} \mid A_2\frac{b}{4} = \left(\frac{R_A L^2}{2EI}\right)\frac{L}{3} + \left(-\frac{wb^3}{6EI}\right)\frac{b}{4} = \frac{R_A L^3}{6EI} - \frac{wb^4}{24EI}$$

Slope at A

$$\theta_A = -\frac{t_{B/A}}{L} = -\left(\frac{R_A L^2}{6EI} - \frac{wb^4}{24EIL}\right) \tag{1}$$

Largest Deflection. The largest deflection occurs at point K, where the slope of the beam is zero. We write therefore

$$\theta_K = \theta_A + \theta_{K/A} = 0 \tag{2}$$

But

$$\theta_{K/A} = A_3 + A_4 = \frac{R_A x_m^2}{2EI} - \frac{w}{6EI}(x_m - a)^3 \tag{3}$$

We substitute for θ_A and $\theta_{K/A}$ from Eqs. (1) and (3) into Eq. (2):

$$-\left(\frac{R_A L^2}{6EI} - \frac{wb^4}{24EIL}\right) + \left[\frac{R_A x_m^2}{2EI} - \frac{w}{6EI}(x_m - a)^3\right] = 0$$

Substituting the numerical data, we have

$$-29.53\frac{10^3}{EI} + 8.405x_m^2\frac{10^3}{EI} - 4.167(x_m - 1.4)^3\frac{10^3}{EI} = 0$$

Solving by trial and error for x_m, we find

$$x_m = 1.890 \text{ m} \qquad \blacktriangledown$$

Computing the moments of A_3 and A_4 about a vertical axis through A, we have

$$|y|_m = t_{A/K} = A_3\frac{2x_m}{3} + A_4\left[a + \frac{3}{4}(x_m - a)\right]$$
$$= \frac{R_A x_m^3}{3EI} - \frac{wa}{6EI}(x_m - a)^3 - \frac{w}{8EI}(x_m - a)^4$$

Using the given data, $R_A = 16.81$ kN, and $I = 28.9 \times 10^{-6}$ m⁴, we find

$$y_m = 6.39 \text{ mm}\downarrow \qquad \blacktriangledown$$

Reference tangent

SAMPLE PROBLEM 9.14

For the uniform beam and loading shown, determine the reaction at B.

SOLUTION

The beam is indeterminate to the first degree. We choose the reaction $\mathbf{R}_B$ as redundant and consider separately the distributed loading and the redundant re-action loading. We next select the tangent at A as the reference tangent. From the similar triangles ABB' and ACC', we find that

$$\frac{t_{C/A}}{L} = \frac{t_{B/A}}{\frac{2}{3}L} \qquad t_{C/A} = \frac{3}{2}t_{B/A} \qquad (1)$$

Distributed Loading. Considering the (M/EI) diagram from end A to an arbitrary point X, we write

$$(t_{X/A})_w = A_1 \frac{x}{3} + A_2 \frac{x}{4} = \left(\frac{1}{2}\frac{wLx}{2EI} x\right)\frac{x}{3} + \left(-\frac{1}{3}\frac{wx^2}{2EI} x\right)\frac{x}{4} = \frac{wx^3}{24EI}(2L - x)$$

Letting successively $x = L$ and $x = \frac{2}{3}L$, we have

$$(t_{C/A})_w = \frac{wL^4}{24EI} \qquad (t_{B/A})_w = \frac{4}{243}\frac{wL^4}{EI}$$

Redundant Reaction Loading

$$(t_{C/A})_R = A_3 \frac{L}{9} + A_4 \frac{L}{3} = \left(\frac{1}{2}\frac{R_BL}{3EI}L\right)\frac{L}{9} + \left(-\frac{1}{2}\frac{R_BL}{3EI}L\right)\frac{L}{3} = -\frac{4}{81}\frac{R_BL^3}{EI}$$

$$(t_{B/A})_R = A_5 \frac{2L}{9} = \left[-\frac{1}{2}\frac{2R_BL}{9EI}\left(\frac{2L}{3}\right)\right]\frac{2L}{9} = -\frac{4}{243}\frac{R_BL^3}{EI}$$

Combined Loading. Adding the results obtained, we write

$$t_{C/A} = \frac{wL^4}{24EI} - \frac{4}{81}\frac{R_BL^3}{EI} \qquad t_{B/A} = \frac{4}{243}(wL^4 - R_BL^3)$$

Reaction at B. Substituting for $t_{C/A}$ and $t_{B/A}$ into Eq. (1), we have

$$\left(\frac{wL^4}{24EI} - \frac{4}{81}\frac{R_BL^3}{EI}\right) = \frac{3}{2}\left[\frac{4}{243}\frac{(wL^4 - R_BL^3)}{EI}\right]$$

$$R_B = 0.6875wL \qquad \mathbf{R}_B = 0.688wL \uparrow \blacktriangleleft$$

PROBLEMS

Use the moment-area method to solve the following problems.

9.127 through 9.130 For the prismatic beam and loading shown, determine (*a*) the deflection at point *D*, (*b*) the slope at end *A*.

Fig. P9.127

Fig. P9.128

Fig. P9.130

9.131 For the beam and loading shown, determine (*a*) the slope at point *A*, (*b*) the deflection at point *C*. Use $E = 29 \times 10^6$ psi.

Fig. P9.129

Fig. P9.131

9.132 and 9.133 For the beam and loading shown, determine (*a*) the slope at point *A*, (*b*) the deflection at point *D*. Use $E = 200$ GPa.

Fig. P9.132

Fig. P9.133

591

9.134   For the timber beam and loading shown, determine (*a*) the slope at point *A*, (*b*) the deflection at point *D*. Use $E = 1.5 \times 10^6$ psi.

Fig. P9.134

9.135 and 9.136   For the beam and loading shown, determine (*a*) the slope at point *A*, (*b*) the deflection at point *D*.

Fig. P9.135

Fig. P9.136

9.137   For the beam and loading shown, determine (*a*) the slope at point *C*, (*b*) the deflection at point *D*. Use $E = 200$ GPa.

Fig. P9.137

9.138   For the beam and loading shown, determine (*a*) the slope at point *B*, (*b*) the deflection at point *A*. Use $E = 29 \times 10^6$ psi.

Fig. P9.138

9.139   For the beam and loading shown, determine (*a*) the slope at point *D*, (*b*) the deflection at point *E*. Use $E = 29 \times 10^6$ psi.

Fig. P9.139

9.140   Knowing the beam *AD* is made of a solid steel bar, determine the (*a*) slope at point *B*, (*c*) the deflection at point *A*. Use $E = 200$ GPa.

Fig. P9.140

9.141 and 9.142 For the beam and loading shown, determine (*a*) the slope at end *A*, (*b*) the slope at end *B*, (*c*) the deflection at the midpoint *C*.

Fig. P9.141

Fig. P9.142

9.143 For the beam and loading shown, determine the magnitude and location of the maximum deflection.

Fig. P9.143

9.144 through 9.147 For the beam and loading shown, determine the magnitude and location of the largest downward deflection.

    9.144 Beam and loading of Prob. 9.129
    9.145 Beam and loading of Prob. 9.130
    9.146 Beam and loading of Prob. 9.132
    9.147 Beam and loading of Prob. 9.133

9.148 For the beam and loading of Prob. 9.135, determine the magnitude and location of the largest upward deflection in span *AC*.

9.149 For the beam and loading of Prob. 9.136, determine the magnitude and location of the largest downward deflection in span *AB*.

*9.150 The cantilever *AB* is a beam of constant strength. It has a rectangular cross section of uniform width *b* and variable depth *h*. Express the deflection at end *A* in terms of *P*, *L*, and the flexural rigidity EI_0 at *B*. (*Hint:* Since the beam is of constant strength, Mc/I has a constant value along *AB*.)

Fig. P9.150

9.151 through 9.154 For the beam and loading shown, determine the reaction at the roller support.

Fig. P9.151

Fig. P9.152

Fig. P9.153

Fig. P9.154

9.155 and 9.156   For the beam and loading shown, determine the reaction at each support.

Fig. P9.155

Fig. P9.156

9.157 and 9.158   Determine the reaction at the roller support and draw the bending moment diagram for the beam and loading shown.

Fig. P9.157

Fig. P9.158

9.159   For the beam and loading shown, determine the spring constant k for which the bending moment at B is $M_B = -wL^2/10$.

9.160   For the beam and loading shown, determine the spring constant k for which the force in the spring is equal to one-third of the total load on the beam.

Fig. P9.159 and P9.160

REVIEW AND SUMMARY
FOR CHAPTER 9

This chapter was devoted to the determination of slopes and deflections of beams under transverse loadings. Two approaches were used. First we used a mathematical method based on the method of integration of a differential equation to get the slopes and deflections at any point along the beam. We then used the *moment-area method* to find the slopes and deflections at a given point along the beam. Particular emphasis was placed on the computation of the maximum deflection of a beam under a given loading. We also applied these methods for determining deflections to the analysis of *indeterminate beams*, those in which the number of reactions at the supports exceeds the number of equilibrium equations available to determine these unknowns.

We noted in Sec. 9.2 that Eq. (4.21) of Sec. 4.4, which relates the curvature $1/\rho$ of the neutral surface and the bending moment M in a prismatic beam in pure bending, can be applied to a beam under a transverse loading, but that both M and $1/\rho$ will vary from section to section. Denoting by x the distance from the left end of the beam, we wrote

$$\frac{1}{\rho} = \frac{M(x)}{EI} \qquad (9.1)$$

This equation enabled us to determine the radius of curvature of the neutral surface for any value of x and to draw some general conclusions regarding the shape of the deformed beam.

In Sec. 9.3, we discussed how to obtain a relation between the deflection y of a beam, measured at a given point Q, and the distance x of that point from some fixed origin (Fig. 9.6b). Such a relation defines the *elastic curve* of a beam. Expressing the curvature $1/\rho$ in terms of the derivatives of the function $y(x)$ and substituting into (9.1), we obtained the following second-order linear differential equation:

$$\frac{d^2y}{dx^2} = \frac{M(x)}{EI} \qquad (9.4)$$

Integrating this equation twice, we obtained the following expressions defining the slope $\theta(x) = dy/dx$ and the deflection $y(x)$, respectively:

$$EI\frac{dy}{dx} = \int_0^x M(x)\,dx + C_1 \qquad (9.5)$$

$$EI\,y = \int_0^x dx \int_0^x M(x)\,dx + C_1 x + C_2 \qquad (9.6)$$

The product EI is known as the *flexural rigidity* of the beam; C_1 and C_2 are two constants of integration that can be determined from the *boundary conditions* imposed on the beam by its supports (Fig. 9.8) [Example 9.01]. The maximum deflection can then be obtained by determining the value of x for which the slope is zero and the corresponding value of y [Example 9.02, Sample Prob. 9.1].

Deformation of a beam under transverse loading

Fig. 9.6b

Boundary conditions

(a) Simply supported beam   (b) Overhanging beam

$y_A = 0$ $y_B = 0$ $y_A = 0$ $y_B = 0$

(c) Cantilever beam

$y_A = 0$
$\theta_A = 0$

Fig. 9.8 Boundary conditions for statically determinate beams.

Elastic curve defined by
different functions

$[x = 0, y_1 = 0]$

$[x = \frac{1}{4}L, \theta_1 = \theta_2]$
$[x = \frac{1}{4}L, y_1 = y_2]$

$[x = L, y_2 = 0]$

Fig. 9.20

When the loading is such that different analytical functions are
required to represent the bending moment in various portions of the
beam, then different differential equations are also required, leading
to different functions representing the slope $\theta(x)$ and the deflection
$y(x)$ in the various portions of the beam. In the case of the beam and
loading considered in Example 9.03 (Fig. 9.20), two differential
equations were required, one for the portion AD and the
other for the portion DB. The first equation yielded the functions θ_1
and y_1, and the second the functions θ_2 and y_2. Altogether, four constants of integration had to be determined; two were obtained by
writing that the deflections at A and B were zero, and the other two
by expressing that the portions of beam AD and DB had the same
slope and the same deflection at D.

We observed in Sec. 9.4 that in the case of a beam supporting a
distributed load $w(x)$, the elastic curve can be determined directly
from $w(x)$ through four successive integrations yielding V, M, θ, and
y in that order. For the cantilever beam of Fig. 9.21a and the simply
supported beam of Fig. 9.21b, the resulting four constants of integration can be determined from the four boundary conditions indicated in each part of the figure [Example 9.04, Sample Prob. 9.2].

$[y_A = 0]$
$[\theta_A = 0]$

$[V_A = 0]$
$[M_B = 0]$

(a) Cantilever beam

$[y_A = 0]$
$[M_A = 0]$

$[y_B = 0]$
$[M_B = 0]$

(b) Simply supported beam

Fig. 9.21 Boundary conditions for beams carrying a distributed load.

Statically indeterminate beams

In Sec. 9.5, we discussed *statically indeterminate beams*, i.e.,
beams supported in such a way that the reactions at the supports involved four or more unknowns. Since only three equilibrium equations are available to determine these unknowns, the equilibrium
equations had to be supplemented by equations obtained from the
boundary conditions imposed by the supports. In the case of the beam
of Fig. 9.24, we noted that the reactions at the supports involved four
unknowns, namely, M_A, A_x, A_y, and B. Such a beam is said to be *indeterminate to the first degree*. (If five unknowns were involved, the

(a)
Fig. 9.24

(b)

beam would be indeterminate to the *second degree*.) Expressing the bending moment $M(x)$ in terms of the four unknowns and integrating twice [Example 9.05], we determined the slope $\theta(x)$ and the deflection $y(x)$ in terms of the same unknowns and the constants of integration C_1 and C_2. The six unknowns involved in this computation were obtained by solving simultaneously the three equilibrium equations for the free body of Fig. 9.24b and the three equations expressing that $\theta = 0$, $y = 0$ for $x = 0$, and that $y = 0$ for $x = L$ (Fig. 9.25) [see also Sample Prob. 9.3].

$[x = 0, \theta = 0]$
$[x = 0, y = 0]$
$[x = L, y = 0]$

Fig. 9.25

The integration method provides an effective way for determining the slope and deflection at any point of a prismatic beam, as long as the bending moment M can be represented by a single analytical function. However, when several functions are required to represent M over the entire length of the beam, this method can become quite laborious, since it requires matching slopes and deflections at every transition point. We saw in Sec. 9.6 that the use of *singularity functions* (previously introduced in Sec. 5.5) considerably simplifies the determination of θ and y at any point of the beam. Considering again

Use of singularity functions

Fig. 9.17

the beam of Example 9.03 (Fig. 9.17) and drawing its free-body diagram (Fig. 9.29), we expressed the shear at any point of the beam as

$$V(x) = \frac{3P}{4} - P\langle x - \tfrac{1}{4}L\rangle^0$$

where the step function $\langle x - \tfrac{1}{4}L\rangle^0$ is equal to zero when the quantity inside the brackets $\langle\ \rangle$ is negative, and equal to one otherwise. Integrating three times, we obtained successively

$$M(x) = \frac{3P}{4}x - P\langle x - \tfrac{1}{4}L\rangle \tag{9.44}$$

$$EI\,\theta = EI\frac{dy}{dx} = \frac{3}{8}Px^2 - \frac{1}{2}P\langle x - \tfrac{1}{4}L\rangle^2 + C_1 \tag{9.46}$$

$$EI\,y = \frac{1}{8}Px^3 - \frac{1}{6}P\langle x - \tfrac{1}{4}L\rangle^3 + C_1x + C_2 \tag{9.47}$$

where the brackets $\langle\ \rangle$ should be replaced by zero when the quantity inside is negative, and by ordinary parentheses otherwise. The constants C_1 and C_2 were determined from the boundary conditions shown in Fig. 9.30 [Example 9.06; Sample Probs. 9.4, 9.5, and 9.6].

Fig. 9.29

$[x = 0, y = 0]$ $[x = L, y = 0]$

Fig. 9.30

The next section was devoted to the *method of superposition*, which consists of determining separately, and then adding, the slope and deflection caused by the various loads applied to a beam [Sec. 9.7]. This procedure was facilitated by the use of the table of Appendix D, which gives the slopes and deflections of beams for various loadings and types of support [Example 9.07, Sample Prob. 9.7].

Statically indeterminate beams by superposition

The method of superposition can be used effectively with *statically indeterminate beams* [Sec. 9.8]. In the case of the beam of Example 9.08 (Fig. 9.36), which involves four unknown reactions and is thus indeterminate to the *first degree*, the reaction at *B* was con-

Fig. 9.36

sidered as *redundant* and the beam was released from that support. Treating the reaction R_B as an unknown load and considering separately the deflections caused at *B* by the given distributed load and by R_B, we wrote that the sum of these deflections was zero (Fig. 9.37). The equation obtained was solved for R_B [see also Sample Prob. 9.8]. In the case of a beam indeterminate to the *second degree*, i.e., with reactions at the supports involving five unknowns, two reactions must be designated as redundant, and the corresponding supports must be eliminated or modified accordingly [Sample Prob. 9.9].

$y_B = 0$

Fig. 9.37

(a)   =   (b)   +   (c)

Method of superposition

First moment-area theorem

We next studied the determination of deflections and slopes of beams using the *moment-area method*. In order to derive the *moment-area theorems* [Sec. 9.9], we first drew a diagram representing the variation along the beam of the quantity M/EI obtained by dividing the bending moment M by the flexural rigidity EI (Fig. 9.41). We then derived the *first moment-area theorem*, which may be stated as follows: *The area under the (M/EI) diagram between two points is equal to the angle between the tangents to the elastic curve drawn at these points.* Considering tangents at *C* and *D*, we wrote

$$\theta_{D/C} = \text{area under } (M/EI) \text{ diagram between } C \text{ and } D \qquad (9.56)$$

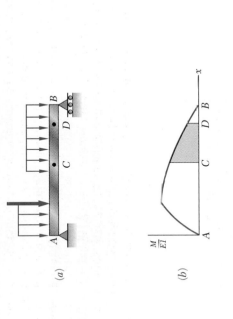

Fig. 9.41   First moment-area theorem

Fig. 9.46   Second moment-area theorem

Second moment-area theorem

Again using the (M/EI) diagram and a sketch of the deflected beam (Fig. 9.45), we drew a tangent at point D and considered the vertical distance $t_{C/D}$, which is called the tangential deviation of C with respect to D. We then derived the second moment-area theorem, which may be stated as follows: The tangential deviation $t_{C/D}$ of C with respect to D is equal to the first moment with respect to a vertical axis through C of the area under the (M/EI) diagram between C and D. We were careful to distinguish between the tangential deviation of C with respect to D (Fig. 9.45a).

$$t_{C/D} = \text{(area between } C \text{ and } D)\, \bar{x}_1 \qquad (9.59)$$

and the tangential deviation of D with respect to C (Fig. 9.45b):

$$t_{D/C} = \text{(area between } C \text{ and } D)\, \bar{x}_2 \qquad (9.60)$$

Cantilever Beams

Beams with symmetric loadings

In Sec. 9.10 we learned to determine the slope and deflection at points of *cantilever beams* and beams with *symmetric loadings*. For cantilever beams, the tangent at the fixed support is horizontal (Fig. 9.46); and for symmetrically loaded beams, the tangent is horizontal at the midpoint C of the beam (Fig. 9.47). Using the horizontal tangent at the midpoint C of the beam (Fig. 9.47). Using the horizontal tangent as a *reference tangent*, we were able to determine slopes and deflections by using, respectively, the first and second moment-area theorems [Example 9.09, Sample Probs. 9.10 and 9.11]. We noted that to find a deflection that is not a tangential deviation (Fig. 9.47c), it is necessary to first determine which tangential deviations can be combined to obtain the desired deflection.

Fig. 9.46

(a)

Fig. 9.47

(b)

(c)

Bending-moment diagram by parts

In many cases the application of the moment-area theorems is simplified if we consider the effect of each load separately [Sec. 9.11]. To do this we drew the (M/EI) *diagram by parts* by drawing a separate (M/EI) diagram for each load. The areas and the moments of areas under the several diagrams could then be added to determine slopes and tangential deviations for the original beam and loading [Examples 9.10 and 9.11].

Unsymmetric loadings

In Sec. 9.12 we expanded the use of the moment-area method to cover beams with *unsymmetric loadings*. Observing that locating a horizontal tangent is usually not possible, we selected a reference tangent at one of the beam supports, since the slope of that tangent can be readily determined. For example, for the beam and loading shown in Fig. 9.59, the slope of the tangent at A can be obtained by computing the tangential deviation $t_{B/A}$ and dividing it by the dis-

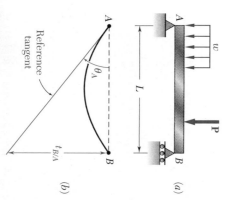

Fig. 9.59

tance L between the supports A and B. Then, using both moment-area theorems and simple geometry, we could determine the slope and deflection at any point of the beam [Example 9.12, Sample Prob. 9.12].

The *maximum deflection* of an unsymmetrically loaded beam generally does not occur at midspan. The approach indicated in the preceding paragraph was used to determine point K where the maximum deflection occurs and the magnitude of that deflection [Sec. 9.13]. Observing that the slope at K is zero (Fig. 9.68), we concluded

Fig. 9.68

Maximum deflection

that $\theta_{K/A} = -\theta_A$. Recalling the first moment-area theorem, we determined the location of K by measuring under the (M/EI) diagram an area equal to $\theta_{K/A}$. The maximum deflection was then obtained by computing the tangential deviation $t_{A/K}$ [Sample Probs. 9.12 and 9.13].

In the last section of the chapter [Sec. 9.14] we applied the moment-area method to the analysis of *statically indeterminate beams*. Since the reactions for the beam and loading shown in Fig. 9.71 cannot be determined by statics alone, we designated one of the reactions of the beam as redundant (M_A in Fig 9.72a) and considered the redundant reaction as an unknown load. The tangential deviation of B with respect to A was considered separately for the distributed load (Fig. 9.72b) and for the redundant reaction (Fig. 9.72c). Expressing that under the combined action of the distributed load and of the couple M_A the tangential deviation of B with respect to A must be zero, we wrote

$$t_{B/A} = (t_{B/A})_w + (t_{B/A})_M = 0$$

From this expression we determined the magnitude of the redundant reaction M_A [Example 9.14, Sample Prob. 9.14].

Statically indeterminate beams

Fig. 9.71

REVIEW PROBLEMS

9.161 For the cantilever beam and loading shown, determine (*a*) the deflection at point *B*, (*b*) the slope at point *B*.

Fig. P9.161

9.162 For the beam and loading shown, determine (*a*) the slope at point *A*, (*b*) the deflection at point *D*.

Fig. P9.162

9.163 The rigid bars *BF* and *DH* are welded to the rolled-steel beam *AE* as shown. Knowing that *c* = 0.4 m, determine for the loading shown (*a*) the deflection at point *B*, (*b*) the deflection at the midpoint *C* of the beam. Use *E* = 200 GPa.

W100 × 19.3

Fig. P9.163

9.164 For the beam and loading shown, determine the deflection at point *A*.

Fig. P9.164

9.165 For the beam and loading shown, determine (*a*) the reaction at *C*, (*b*) the deflection at point *B*. Use *E* = 29 × 10⁶ psi.

W12 × 40

Fig. P9.165

Fig. P9.166

9.166 For the loading shown, knowing that beams *AC* and *BD* have the same flexural rigidity, determine the reaction at *B*.

9.167 Beam *DE* rests on the cantilever beam *AC* as shown. Knowing that a square rod of side 10 mm is used for each beam, determine the deflection at end *C* if the 25-N · m couple is applied (*a*) to end *E* of beam *DE*, (*b*) to end *C* of beam *AC*. Use *E* = 200 GPa.

Fig. P9.167

9.168 For the beam and loading shown, determine the value of *P* for which the deflection is zero at end *A* of the beam. Use *E* = 29 × 10⁶ psi.

9.169 For the beam and loading shown, determine the deflection (*a*) at point *D*, (*b*) at point *E*.

9.170 For the beam and loading shown, determine the magnitude and location of the largest downward deflection.

9.171 For the beam and loading shown, determine (*a*) the value of *a* for which the slope at end *A* is zero, (*b*) the corresponding deflection at point *C*.

Fig. P9.171

Fig. P9.168

Fig. P9.169 and P9.170

9.172 A hydraulic jack can be used to raise point *B* of the cantilever beam *ABC*. The beam was originally straight, horizontal, and unloaded. A 20-kN load was then applied at point *C*, causing this point to move down. Determine (*a*) how much point *B* should be raised to return point *C* to its original position, (*b*) the final value of the reaction at *B*. Use *E* = 200 GPa.

Fig. P9.172

COMPUTER PROBLEMS

The following problems are designed to be solved with a computer.

9.C1 Several concentrated loads can be applied to the cantilever beam AB. Write a computer program to calculate the slope and deflection of beam AB from $x = 0$ to $x = L$, using given increments Δx. Apply this program with increments $\Delta x = 50$ mm to the beam and loading of Probs. 9.79 and 9.80.

Fig. P9.C1

9.C2 The 22-ft beam AB consists of a W21 × 62 rolled-steel shape and supports a 3.5 kips/ft distributed load as shown. Write a computer program and use it to calculate for values of a from 0 to 22 ft, using 1-ft increments, (*a*) the slope and deflection at D, (*b*) the location and magnitude of the maximum deflection. Use $E = 29 \times 10^6$ psi.

Fig. P9.C2

9.C3 The cantilever beam AB carries the distributed loads shown. Write a computer program to calculate the slope and deflection of beam AB from $x = 0$ to $x = L$ using given increments Δx. Apply this program with increments $\Delta x = 100$ mm, assuming that $L = 2.4$ m, $w = 36$ kN/m, and (*a*) $a = 0.6$ m, (*b*) $a = 1.2$ m, (*c*) $a = 1.8$ m. Use $E = 200$ GPa.

W250 × 32.7

Fig. P9.C3

9.C4 The simply supported beam AB is of constant flexural rigidity EI and carries several concentrated loads as shown. Using the *Method of Integration*, write a computer program to calculate the slope and deflection at points along the beam from $x = 0$ to $x = L$ using given increments Δx. Apply this program to the beam and loading of (*a*) Prob. 9.14 with $\Delta x = 0.25$ m, (*b*) Prob. 9.15 with $\Delta x = 0.05$ m, (*c*) Prob. 9.132 with $\Delta x = 0.25$ m.

Fig. P9.C4

Fig. P9.C5

9.C5 The supports of beam AB consist of a fixed support at end A and a roller located at point D. Write a computer program to calculate the slope and deflection at the free end of the beam for values of a from 0 to L using given increments Δa. Apply this program to calculate the slope and deflection at point B for each of the following cases:

	L	Δa	w	E	Shape
(a)	12 ft	0.5 ft	1.6 klps/ft	29×10^6 psi	W16 × 57
(b)	3 m	0.2 m	18 kN/m	200 GPa	W460 × 113

9.C6 For the beam and loading shown, use the *Moment-Area Method* to write a computer program to calculate the slope and deflection at points along the beam from $x = 0$ to $x = L$ using given increments Δx. Apply this program to calculate the slope and deflection at each concentrated load for the beam of (a) Prob. 9.76 with $\Delta x = 0.5$ m, (b) Prob. 9.116 with $\Delta x = 3$ in., (c) Prob. 9.119 with $\Delta x = 0.5$ ft.

Fig. P9.C6

9.C7 Two 52-kN loads are maintained 2.5 m apart as they are moved slowly across beam AB. Write a computer program to calculate the deflection at the midpoint C of the beam for values of x from 0 to 9 m, using 0.5-m increments. Use $E = 200$ GPa.

Fig. P9.C7

9.C8 A uniformly distributed load w and several concentrated loads P_i may be applied to the cantilever beam AB. Write a computer program to determine the reaction at the roller support and apply this program to the beam and loading of (a) Prob. 9.57a, (b) Prob. 9.58a.

Fig. P9.C8

10

Columns

A steel wide-flange column is being tested in the five-million-pound universal testing machine at Lehigh University, Bethlehem, Pennsylvania. The analysis and design of members supporting axial compressive loads will be discussed in this chapter.

10.1. INTRODUCTION

In the preceding chapters, we had two primary concerns: (1) the strength of the structure, i.e., its ability to support a specified load without experiencing excessive stress; (2) the ability of the structure to support a specified load without undergoing unacceptable deformations. In this chapter, our concern will be with the stability of the structure, i.e., with its ability to support a given load without experiencing a sudden change in its configuration. Our discussion will relate chiefly to columns, i.e., to the analysis and design of vertical prismatic members supporting axial loads.

In Sec. 10.2, the stability of a simplified model of a column, consisting of two rigid rods connected by a pin and a spring and supporting a load **P**, will first be considered. You will observe that if its equilibrium is disturbed, this system will return to its original equilibrium position as long as P does not exceed a certain value P_{cr}, called the *critical load*. However, if $P > P_{cr}$, the system will move away from its original position and settle in a new position of equilibrium. In the first case, the system is said to be *stable*, and in the second case, it is said to be *unstable*.

In Sec. 10.3, you will begin the study of the *stability of elastic columns* by considering a pin-ended column subjected to a centric axial load. *Euler's formula* for the critical load of the column will be derived and from that formula the corresponding critical normal stress in the column will be determined. By applying a factor of safety to the critical load, you will be able to determine the allowable load that can be applied to a pin-ended column.

In Sec. 10.4, the analysis of the stability of columns with different end conditions will be considered. You will simplify these analyses by learning how to determine the *effective length* of a column, i.e., the length of a pin-ended column having the same critical load.

In Sec. 10.5, you will consider columns supporting eccentric axial loads; these columns have transverse deflections for all magnitudes of the load. An expression for the maximum deflection under a given load will be derived and used to determine the maximum normal stress in the column. Finally, the *secant formula* which relates the average and maximum stresses in a column will be developed.

In the first sections of the chapter, each column is initially assumed to be a straight homogeneous prism. In the last part of the chapter, you will consider real columns which are designed and analyzed using empirical formulas set forth by professional organizations. In Sec. 10.6, formulas will be presented for the allowable stress in columns made of steel, aluminum, or wood and subjected to a centric axial load. In the last section of the chapter (Sec. 10.7), the design of columns under an eccentric axial load will be considered.

10.2. STABILITY OF STRUCTURES

Suppose we are to design a column AB of length L to support a given load **P** (Fig. 10.1). The column will be pin-connected at both ends and we assume that **P** is a centric axial load. If the cross-sectional area A of the column is selected so that the value $\sigma = P/A$ of the stress on a transverse section is less than the allowable stress σ_{all} for the material used, and if the deformation $\delta = PL/AE$ falls within the given specifications, we might conclude that the column has been properly designed. However, it may happen that, as the load is applied, the column will *buckle*; instead of remaining straight, it will suddenly become sharply curved (Fig. 10.2). Figure 10.3 shows a column similar to that in the opening photo of this chapter after it has been loaded so that it is no longer straight; the column has buckled. Clearly, a column that buckles under the load it is to support is not properly designed.

Fig. 10.1

Fig. 10.2

Fig. 10.3 Buckled column

Before getting into the actual discussion of the stability of elastic columns, some insight will be gained on the problem by considering a simplified model consisting of two rigid rods AC and BC connected at C by a pin and a torsional spring of constant K (Fig. 10.4).

Fig. 10.4

Fig. 10.5

If the two rods and the two forces **P** and **P′** are perfectly aligned, the system will remain in the position of equilibrium shown in Fig. 10.5*a* as long as it is not disturbed. But suppose that we move *C* slightly to the right, so that each rod now forms a small angle $\Delta\theta$ with the vertical (Fig. 10.5*b*). Will the system return to its original equilibrium position, or will it move further away from that position? In the first case, the system is said to be *stable*, and in the second case, it is said to be *unstable*.

To determine whether the two-rod system is stable or unstable, we consider the forces acting on rod *AC* (Fig. 10.6). These forces consist of two couples, namely the couple formed by **P** and **P′**, of moment $P(L/2) \sin \Delta\theta$, which tends to move the rod away from the vertical, and the couple **M** exerted by the spring, which tends to bring the rod back into its original vertical position. Since the angle of deflection of the spring is $2 \Delta\theta$, the moment of the couple **M** is $M = K(2 \Delta\theta)$. If the moment of the second couple is larger than the moment of the first couple, the system tends to return to its original equilibrium position; the system is stable. If the moment of the first couple is larger than the moment of the second couple, the system tends to move away from its original equilibrium position; the system is unstable. The value of the load for which the two couples balance each other is called the *critical load* and is denoted by P_{cr}. We have

$$P_{cr}(L/2) \sin \Delta\theta = K(2 \Delta\theta) \qquad (10.1)$$

or, since $\sin \Delta\theta \approx \Delta\theta$,

$$P_{cr} = 4K/L \qquad (10.2)$$

Clearly, the system is stable for $P < P_{cr}$, that is, for values of the load smaller than the critical value, and unstable for $P > P_{cr}$.

Let us assume that a load $P > P_{cr}$ has been applied to the two rods of Fig. 10.4 and that the system has been disturbed. Since $P > P_{cr}$, the system will move further away from the vertical and, after some oscil-

Fig. 10.6

lations, will settle into a new equilibrium position (Fig. 10.7a). Considering the equilibrium of the free body AC (Fig. 10.7b), we obtain an equation similar to Eq. (10.1), but involving the finite angle θ, namely

$$P(L/2)\sin\theta = K(2\theta)$$

or

$$\frac{PL}{4K} = \frac{\theta}{\sin\theta} \qquad (10.3)$$

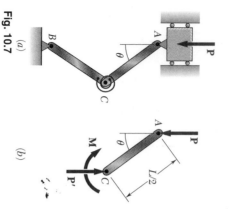

(a)

(b)

Fig. 10.7

The value of θ corresponding to the equilibrium position represented in Fig. 10.7 is obtained by solving Eq. (10.3) by trial and error. But we observe that, for any positive value of θ, we have $\sin\theta < \theta$. Thus, Eq. (10.3) yields a value of θ different from zero only when the left-hand member of the equation is larger than one. Recalling Eq. (10.2), we note that this is indeed the case here, since we have assumed $P > P_{cr}$. But, if we had assumed $P < P_{cr}$, the second equilibrium position shown in Fig. 10.7 would not exist and the only possible equilibrium position would be the position corresponding to $\theta = 0$. We thus check that, for $P < P_{cr}$, the position $\theta = 0$ must be stable.

This observation applies to structures and mechanical systems in general, and will be used in the next section, where the stability of elastic columns will be discussed.

10.3. EULER'S FORMULA FOR PIN-ENDED COLUMNS

Returning to the column AB considered in the preceding section (Fig. 10.1), we propose to determine the critical value of the load $\mathbf{P}$, i.e., the value P_{cr} of the load for which the position shown in Fig. 10.1 ceases to be stable. If $P > P_{cr}$, the slightest misalignment or disturbance will cause the column to buckle, i.e., to assume a curved shape as shown in Fig. 10.2.

Fig. 10.1
(repeated)

Fig. 10.2
(repeated)

Our approach will be to determine the conditions under which the configuration of Fig. 10.2 is possible. Since a column can be considered as a beam placed in a vertical position and subjected to an axial load, we proceed as in Chap. 9 and denote by x the distance from end

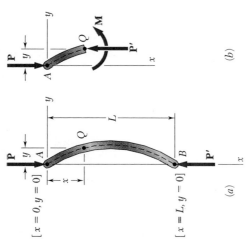

Fig. 10.8

A of the column to a given point Q of its elastic curve, and by y the deflection of that point (Fig. 10.8a). It follows that the x axis will be vertical and directed downward, and the y axis horizontal and directed to the right. Considering the equilibrium of the free body AQ (Fig. 10.8b), we find that the bending moment at Q is $M = -Py$. Substituting this value for M in Eq. (9.4) of Sec. 9.3, we write

$$\frac{d^2y}{dx^2} = \frac{M}{EI} = -\frac{P}{EI}y \qquad (10.4)$$

or, transposing the last term,

$$\frac{d^2y}{dx^2} + \frac{P}{EI}y = 0 \qquad (10.5)$$

This equation is a linear, homogeneous differential equation of the second order with constant coefficients. Setting

$$p^2 = \frac{P}{EI} \qquad (10.6)$$

we write Eq. (10.5) in the form

$$\frac{d^2y}{dx^2} + p^2y = 0 \qquad (10.7)$$

which is the same as that of the differential equation for simple harmonic motion except, of course, that the independent variable is now the distance x instead of the time t. The general solution of Eq. (10.7) is

$$y = A \sin px + B \cos px \qquad (10.8)$$

as we easily check by computing d^2y/dx^2 and substituting for y and d^2y/dx^2 into Eq. (10.7),

Recalling the boundary conditions that must be satisfied at ends A and B of the column (Fig. 10.8a), we first make $x = 0, y = 0$ in Eq. (10.8) and find that $B = 0$. Substituting next $x = L, y = 0$, we obtain

$$A \sin pL = 0 \qquad (10.9)$$

This equation is satisfied either if $A = 0$, or if $\sin pL = 0$. If the first of these conditions is satisfied, Eq. (10.8) reduces to $y = 0$ and the column is straight (Fig. 10.1). For the second condition to be satisfied, we must have $pL = n\pi$ or, substituting for p from (10.6) and solving for P,

$$P = \frac{n^2\pi^2 EI}{L^2} \qquad (10.10)$$

The smallest of the values of P defined by Eq. (10.10) is that corresponding to $n = 1$. We thus have

$$P_{cr} = \frac{\pi^2 EI}{L^2} \qquad (10.11)$$

The expression obtained is known as *Euler's formula*, after the Swiss mathematician Leonhard Euler (1707–1783). Substituting this expression for P into Eq. (10.6) and the value obtained for p into Eq. (10.8), and recalling that $B = 0$, we write

$$y = A \sin \frac{\pi x}{L} \qquad (10.12)$$

which is the equation of the elastic curve after the column has buckled (Fig. 10.2). We note that the value of the maximum deflection, $y_m = A$, is indeterminate. This is due to the fact that the differential equation (10.5) is a linearized approximation of the actual governing differential equation for the elastic curve.†

If $P < P_{cr}$ the condition $\sin pL = 0$ cannot be satisfied, and the solution given by Eq. (10.12) does not exist. We must then have $A = 0$, and the only possible configuration for the column is a straight one. Thus, for $P < P_{cr}$ the straight configuration of Fig. 10.1 is stable.

In the case of a column with a circular or square cross section, the moment of inertia I of the cross section is the same about any centroidal axis, and the column is as likely to buckle in one plane as another, except for the restraints that can be imposed by the end connections. For other shapes of cross section, the critical load should be computed by making $I = I_{min}$ in Eq. (10.11); if buckling occurs, it will take place in a plane perpendicular to the corresponding principal axis of inertia.

The value of the stress corresponding to the critical load is called the *critical stress* and is denoted by σ_{cr}. Recalling Eq. (10.11) and setting $I = Ar^2$, where A is the cross-sectional area and r its radius of gyration, we have

$$\sigma_{cr} = \frac{P_{cr}}{A} = \frac{\pi^2 E A r^2}{AL^2}$$

or

$$\sigma_{cr} = \frac{\pi^2 E}{(L/r)^2} \qquad (10.13)$$

The quantity L/r is called the *slenderness ratio* of the column. It is clear, in view of the remark of the preceding paragraph, that the minimum value of the radius of gyration r should be used in computing the slenderness ratio and the critical stress in a column.

Equation (10.13) shows that the critical stress is proportional to the modulus of elasticity of the material, and inversely proportional to the square of the slenderness ratio of the column. The plot of σ_{cr} versus L/r is shown in Fig. 10.9 for structural steel, assuming $E = 200$ GPa and $\sigma_Y = 250$ MPa. We should keep in mind that no factor of safety has been used in plotting σ_{cr}. We also note that, if the value obtained for σ_{cr} from Eq. (10.13) or from the curve of Fig. 10.9 is larger than the yield strength σ_Y, this value is of no interest to us, since the column will yield in compression and cease to be elastic before it has a chance to buckle.

†We recall that the equation $d^2y/dx^2 = M/EI$ was obtained in Sec. 9.3 by assuming that the slope dy/dx of the beam could be neglected and that the exact expression given in Eq. (9.3) for the curvature of the beam could be replaced by $1/\rho = d^2y/dx^2$.

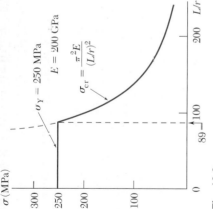

σ (MPa)

$\sigma_Y = 250$ MPa

$E = 200$ GPa

$\sigma_{cr} = \dfrac{\pi^2 E}{(L/r)^2}$

Fig. 10.9

Our analysis of the behavior of a column has been based so far on the assumption of a perfectly aligned centric load. In practice, this is seldom the case, and in Sec. 10.5 the effect of the eccentricity of the loading is taken into account. This approach will lead to a smoother transition from the buckling failure of long, slender columns to the compression failure of short, stubby columns. It will also provide us with a more realistic view of the relation between the slenderness ratio of a column and the load that causes it to fail.

EXAMPLE 10.01

A 2-m-long pin-ended column of square cross section is to be made of wood. Assuming $E = 13$ GPa, $\sigma_{all} = 12$ MPa, and using a factor of safety of 2.5 in computing Euler's critical load for buckling, determine the size of the cross section if the column is to safely support (a) a 100-kN load, (b) a 200-kN load.

(a) For the 100-kN Load. Using the given factor of safety, we make

$$P_{cr} = 2.5(100 \text{ kN}) = 250 \text{ kN} \qquad L = 2 \text{ m} \qquad E = 13 \text{ GPa}$$

in Euler's formula (10.11) and solve for I. We have

$$I = \frac{P_{cr}L^2}{\pi^2 E} = \frac{(250 \times 10^3 \text{ N})(2 \text{ m})^2}{\pi^2(13 \times 10^9 \text{ Pa})} = 7.794 \times 10^{-6} \text{ m}^4$$

Recalling that, for a square of side a, we have $I = a^4/12$, we write

$$\frac{a^4}{12} = 7.794 \times 10^{-6} \text{ m}^4 \qquad a = 98.3 \text{ mm} \approx 100 \text{ mm}$$

We check the value of the normal stress in the column:

$$\sigma = \frac{P}{A} = \frac{100 \text{ kN}}{(0.100 \text{ m})^2} = 10 \text{ MPa}$$

Since σ is smaller than the allowable stress, a 100×100-mm cross section is acceptable.

(b) For the 200-kN Load. Solving again Eq. (10.11) for I, but making now $P_{cr} = 2.5(200) = 500$ kN, we have

$$I = 15.588 \times 10^{-6} \text{ m}^4$$

$$\frac{a^4}{12} = 15.588 \times 10^{-6} \qquad a = 116.95 \text{ mm}$$

The value of the normal stress is

$$\sigma = \frac{P}{A} = \frac{200 \text{ kN}}{(0.11695 \text{ m})^2} = 14.62 \text{ MPa}$$

Since this value is larger than the allowable stress, the dimension obtained is not acceptable, and we must select the cross section on the basis of its resistance to compression. We write

$$A = \frac{P}{\sigma_{all}} = \frac{200 \text{ kN}}{12 \text{ MPa}} = 16.67 \times 10^{-3} \text{ m}^2$$

$$a^2 = 16.67 \times 10^{-3} \text{ m}^2 \qquad a = 129.1 \text{ mm}$$

A 130×130-mm cross section is acceptable.

10.4. EXTENSION OF EULER'S FORMULA TO COLUMNS WITH OTHER END CONDITIONS

Euler's formula (10.11) was derived in the preceding section for a column that was pin-connected at both ends. Now the critical load P_{cr} will be determined for columns with different end conditions.

In the case of a column with one free end A supporting a load **P** and one fixed end B (Fig. 10.10a), we observe that the column will behave as the upper half of a pin-connected column (Fig. 10.10b). The critical load for the column of Fig. 10.10a is thus the same as for the pin-ended column of Fig. 10.10b and can be obtained from Euler's

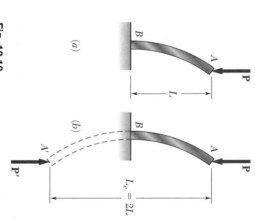

(a)   (b)

Fig. 10.10

formula (10.11) by using a column length equal to twice the actual length L of the given column. We say that the *effective length* L_e of the column of Fig. 10.10 is equal to $2L$ and substitute $L_e = 2L$ in Euler's formula:

$$P_{cr} = \frac{\pi^2 EI}{L_e^2}$$

(10.11')

The critical stress is found in a similar way from the formula

$$\sigma_{cr} = \frac{\pi^2 E}{(L_e/r)^2}$$

(10.13')

The quantity L_e/r is referred to as the *effective slenderness ratio* of the column and, in the case considered here, is equal to $2L/r$.

Consider next a column with two fixed ends A and B supporting a load **P** (Fig. 10.11). The symmetry of the supports and of the loading about a horizontal axis through the midpoint C requires that the shear at C and the horizontal components of the reactions at A and B be zero (Fig. 10.12). It follows that the restraints imposed upon the upper half AC of the column by the support at A and by the lower half CB are

Fig. 10.11

identical (Fig. 10.13). Portion AC must thus be symmetric about its midpoint D, and this point must be a point of inflection, where the bending moment is zero. A similar reasoning shows that the bending moment at the midpoint E of the lower half of the column must also be zero (Fig. 10.14a). Since the bending moment at the ends of a pin-ended column is zero, it follows that the portion DE of the column of Fig. 10.14a must behave as a pin-ended column (Fig. 10.14b). We thus conclude that the effective length of a column with two fixed ends is $L_e = L/2$.

Fig. 10.12

Fig. 10.13

Fig. 10.14

In the case of a column with one fixed end B and one pin-connected end A supporting a load $\mathbf{P}$ (Fig. 10.15), we must write and solve the differential equation of the elastic curve to determine the effective length of the column. From the free-body diagram of the entire column (Fig. 10.16), we first note that a transverse force $\mathbf{V}$ is exerted at end A, in addition to the axial load $\mathbf{P}$, and that $\mathbf{V}$ is statically indeterminate. Considering now the free-body diagram of a portion AQ of the column (Fig. 10.17), we find that the bending moment at Q is

$$M = -Py - Vx$$

Fig. 10.15

Fig. 10.16

Fig. 10.17

Substituting this value into Eq. (9.4) of Sec. 9.3, we write

$$\frac{d^2y}{dx^2} = \frac{M}{EI} = -\frac{P}{EI}y - \frac{V}{EI}x$$

Transposing the term containing y and setting

$$p^2 = \frac{P}{EI} \qquad (10.6)$$

as we did in Sec. 10.3, we write

$$\frac{d^2y}{dx^2} + p^2y = -\frac{V}{EI}x \qquad (10.14)$$

This equation is a linear, nonhomogeneous differential equation of the second order with constant coefficients. Observing that the left-hand members of Eqs. (10.7) and (10.14) are identical, we conclude that the general solution of Eq. (10.14) can be obtained by adding a particular solution of Eq. (10.14) to the solution (10.8) obtained for Eq. (10.7). Such a particular solution is easily seen to be

$$y = -\frac{V}{p^2EI}x$$

or, recalling (10.6),

$$y = -\frac{V}{P}x \qquad (10.15)$$

Adding the solutions (10.8) and (10.15), we write the general solution of Eq. (10.14) as

$$y = A \sin px + B \cos px - \frac{V}{P}x \qquad (10.16)$$

The constants A and B, and the magnitude V of the unknown transverse force $\mathbf{V}$ are obtained from the boundary conditions indicated in Fig. (10.16). Making first $x = 0, y = 0$ in Eq. (10.16), we find that $B = 0$. Making next $x = L, y = 0$, we obtain

$$A \sin pL = \frac{V}{P}L \qquad (10.17)$$

Finally, computing

$$\frac{dy}{dx} = Ap \cos px - \frac{V}{P}$$

and making $x = L, dy/dx = 0$, we have

$$Ap \cos pL = \frac{V}{P} \qquad (10.18)$$

Fig. 10.16 (repeated)

Dividing (10.17) by (10.18) member by member, we conclude that a solution of the form (10.16) can exist only if

$$\tan pL = pL \qquad (10.19)$$

Solving this equation by trial and error, we find that the smallest value of pL which satisfies (10.19) is

$$pL = 4.4934 \qquad (10.20)$$

Carrying the value of p defined by Eq. (10.20) into Eq. (10.6) and solving for P, we obtain the critical load for the column of Fig. 10.15

$$P_{cr} = \frac{20.19EI}{L^2} \qquad (10.21)$$

Fig. 10.15 (repeated)

The effective length of the column is obtained by equating the right-hand members of Eqs. (10.11') and (10.21):

$$\frac{\pi^2 EI}{L_e^2} = \frac{20.19EI}{L^2}$$

Solving for L_e, we find that the effective length of a column with one fixed end and one pin-connected end is $L_e = 0.699L \approx 0.7L$.

The effective lengths corresponding to the various end conditions considered in this section are shown in Fig. 10.18.

(a) One fixed end, one free end $\quad L_e = 2L$

(b) Both ends pinned $\quad L_e = L$

(c) One fixed end, one pinned end $\quad L_e = 0.7L$

(d) Both ends fixed $\quad L_e = 0.5L$

Fig. 10.18 Effective length of column for various end conditions.

SAMPLE PROBLEM 10.1

An aluminum column of length L and rectangular cross section has a fixed end B and supports a centric load at A. Two smooth and rounded fixed plates restrain end A from moving in one of the vertical planes of symmetry of the column, but allow it to move in the other plane. (a) Determine the ratio a/b of the two sides of the cross section corresponding to the most efficient design against buckling. (b) Design the most efficient cross section for the column, knowing that $L = 20$ in., $E = 10.1 \times 10^6$ psi, $P = 5$ kips, and that a factor of safety of 2.5 is required.

SOLUTION

Buckling in xy Plane. Referring to Fig. 10.18, we note that the effective length of the column with respect to buckling in this plane is $L_e = 0.7L$. The radius of gyration r_z of the cross section is obtained by writing

$$I_x = \tfrac{1}{12}ba^3 \qquad A = ab$$

and, since $I_z = Ar_z^2$, $\qquad r_z^2 = \dfrac{I_z}{A} = \dfrac{\tfrac{1}{12}ba^3}{ab} = \dfrac{a^2}{12} \qquad r_z = a/\sqrt{12}$

The effective slenderness ratio of the column with respect to buckling in the xy plane is

$$\frac{L_e}{r_z} = \frac{0.7L}{a/\sqrt{12}} \tag{1}$$

Buckling in xz Plane. The effective length of the column with respect to buckling in this plane is $L_e = 2L$, and the corresponding radius of gyration is $r_y = b/\sqrt{12}$. Thus,

$$\frac{L_e}{r_y} = \frac{2L}{b/\sqrt{12}} \tag{2}$$

a. Most Efficient Design. The most efficient design is that for which the critical stresses corresponding to the two possible modes of buckling are equal. Referring to Eq. (10.13′), we note that this will be the case if the two values obtained above for the effective slenderness ratio are equal. We write

$$\frac{0.7L}{a/\sqrt{12}} = \frac{2L}{b/\sqrt{12}}$$

and, solving for the ratio a/b, $\qquad \dfrac{a}{b} = \dfrac{0.7}{2} \qquad \dfrac{a}{b} = 0.35 \quad \blacktriangleleft$

b. Design for Given Data. Since $F.S. = 2.5$ is required,

$$P_{cr} = (F.S.)P = (2.5)(5 \text{ kips}) = 12.5 \text{ kips}$$

Using $a = 0.35b$, we have $A = ab = 0.35b^2$ and

$$\sigma_{cr} = \frac{P_{cr}}{A} = \frac{12{,}500 \text{ lb}}{0.35b^2}$$

Making $L = 20$ in. in Eq. (2), we have $L_e/r_y = 138.6/b$. Substituting for E, L_e/r, and σ_{cr} into Eq. (10.13′), we write

$$\sigma_{cr} = \frac{\pi^2 E}{(L_e/r)^2}$$

$$\frac{12{,}500 \text{ lb}}{0.35b^2} = \frac{\pi^2 (10.1 \times 10^6 \text{ psi})}{(138.6/b)^2}$$

$$b = 1.620 \text{ in.} \qquad a = 0.35b = 0.567 \text{ in.} \quad \blacktriangleleft$$

PROBLEMS

10.1 Knowing that the torsional spring at B is of constant K and that the bar AB is rigid, determine the critical load P_{cr}.

10.2 Knowing that the spring at A is of constant k and that the bar AB is rigid, determine the critical load P_{cr}.

Fig. P10.1

Fig. P10.2

Fig. P10.3

Fig. P10.4

10.3 Two rigid bars AC and BC are connected as shown to a spring of constant k. Knowing that the spring can act in either tension or compression, determine the critical load P_{cr} for the system.

10.4 Two rigid bars AC and BC are connected by a pin at C as shown. Knowing that the torsional spring at B is of constant K, determine the critical load P_{cr} for the system.

10.5 The rigid bar AD is attached to two springs of constant k and is in equilibrium in the position shown. Knowing that the equal and opposite loads $\mathbf{P}$ and $\mathbf{P}'$ remain horizontal, determine the magnitude P_{cr} of the critical load for the system.

Fig. P10.5

10.6 A frame consists of four L-shaped members connected by four torsional springs, each of constant K. Knowing that equal loads $\mathbf{P}$ are applied at points A and D as shown, determine the critical value P_{cr} of the loads applied to the frame.

Fig. P10.6

10.7 The rigid rod AB is attached to a hinge at A and to two springs, each of constant $k = 2$ kip/in., that can act in either tension or compression. Knowing that $h = 2$ ft, determine the critical load.

Fig. P10.7

10.8 If $m = 125$ kg, $h = 700$ mm, and the constant of each spring is $k = 2.8$ kN/m, determine the range of values of the distance d for which the equilibrium of the rigid rod AB is stable in the position shown. Each spring can act in either tension or compression.

Fig. P10.8

10.9 Determine the critical load of a round wooden dowel that is 48 in. long and has a diameter of (a) 0.375 in., (b) 0.5 in. Use $E = 1.6 \times 10^6$ psi.

10.10 Determine the critical load of a steel tube that is 5 m long and has a 100-mm outer diameter and a 16-mm wall thickness. Use $E = 200$ GPa.

Fig. P10.10

P

P

1.1 m

1.1 m

20 mm

A

B

C

D

d

d

Brass
$E = 120$ GPa
$\rho = 8740$ kg/m³

Aluminum
$E = 70$ GPa
$\rho = 2710$ kg/m³

Fig. P10.11

10.11 Determine (*a*) the critical load for the brass strut, (*b*) the dimension *d* for which the aluminum strut will have the same critical load, (*c*) the weight of the aluminum strut as a percent of the weight of the brass strut.

10.12 A compression member of 20-in. effective length consists of a solid 1-in.-diameter aluminum rod. In order to reduce the weight of the member by 25%, the solid rod is replaced by a hollow rod of the cross section shown. Determine (*a*) the percent reduction in the critical load, (*b*) the value of the critical load for the hollow rod. Use $E = 10.6 \times 10^6$ psi.

0.5 m.

1.0 in.

1.0 in.

Fig. P10.12

60 mm

40 mm

60 mm

Fig. P10.13

10.13 Two brass rods used as compression members, each of 3-m effective length, have the cross sections shown. (*a*) Determine the wall thickness of the hollow square rod for which the rods have the same cross-sectional area. (*b*) Using $E = 105$ GPa, determine the critical load of each rod.

10.14 A column of effective length *L* can be made by gluing together identical planks in either of the arrangements shown. Determine the ratio of the critical load using the arrangement *a* to the critical load using the arrangement *b*.

(*b*)

d

d/3

(*a*)

Fig. *P10.14*

10.15 A compression member of 7-m effective length is made by welding together two L152 × 102 × 12.7 angles as shown. Using $E = 200$ GPa, determine the allowable centric load for the member if a factor of safety of 2.2 is required.

152 mm

102 mm

102 mm

Fig. P10.15

10.16 A column of 26-ft effective length is made from half a W16 × 40 rolled-steel shape. Knowing that the centroid of the cross section is located as shown, determine the factor of safety if the allowable centric load is 20 kips. Use $E = 29 \times 10^6$ psi.

8.005 in.

1.81 in.

Fig. P10.16

10.17 A column of 22-ft effective length is to be made by welding two 9 × 0.5-in. plates to a W8 × 35 as shown. Determine the allowable centric load if a factor of safety of 2.3 is required. Use $E = 29 \times 10^6$ psi.

4.5 in.

4.5 in.

4.5 in.

Fig. P10.17

10.18 A column of 3-m effective length is to be made by welding together two C130 × 13 rolled-steel channels. Using $E = 200$ GPa, determine for each arrangement shown the allowable centric load if a factor of safety of 2.4 is required.

(a)   (b)

Fig. P10.18

10.19 Knowing that $P = 5.2$ kN, determine the factor of safety for the structure shown. Use $E = 200$ GPa and consider only buckling in the plane of the structure.

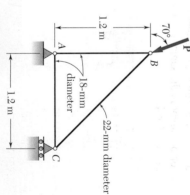

1.2 m

70°

18-mm diameter

22-mm diameter

1.2 m

P

A

B

C

Fig. P10.19

10.20 Members *AB* and *CD* are 30-mm-diameter steel rods, and members *BC* and *AD* are 22-mm-diameter steel rods. When the turnbuckle is tightened, the diagonal member *AC* is put in tension. Knowing that a factor of safety with respect to buckling of 2.75 is required, determine the largest allowable tension in *AC*. Use $E = 200$ GPa and consider only buckling in the plane of the structure.

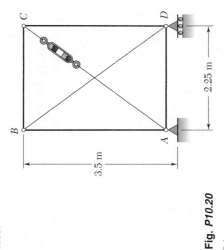

Fig. P10.20

10.21 Each of the five struts consists of an aluminum tube that has a 32-mm outer diameter and a 4-mm wall thickness. Using $E = 70$ GPa and a factor of safety of 2.3, determine the allowable load $\mathbf{P}_0$ for each support condition shown.

Fig. P10.21

10.22 Two columns are used to support a block weighing 3.25 kips in each of the four ways shown. (*a*) Knowing that the column of Fig. (1) is made of steel with a 1.25-in.-diameter, determine the factor of safety with respect to buckling for the loading shown. (*b*) Determine the diameter of each of the other columns for which the factor of safety is the same as the factor of safety obtained in part *a*. Use $E = 29 \times 10^6$ psi.

Fig. P10.22

10.23 A 25-mm-square aluminum strut is maintained in the position shown by a pin support at A and by sets of rollers at B and C that prevent rotation of the strut in the plane of the figure. Knowing that $L_{AB} = 1.0$ m, $L_{BC} = 1.25$ m, and $L_{CD} = 0.5$ m, determine the allowable load **P** using a factor of safety with respect to buckling of 2.8. Consider only buckling in the plane of the figure and use $E = 75$ GPa.

10.24 A 32-mm-square aluminum strut is maintained in the position shown by a pin support at A and by sets of rollers at B and C that prevent rotation of the strut in the plane of the figure. Knowing that $L_{AB} = 1.4$ m, determine (a) the largest values of L_{BC} and L_{CD} that can be used if the allowable load **P** is to be as large as possible, (b) the magnitude of the corresponding allowable load if the factor of safety is 2.8. Consider only buckling in the plane of the figure and use $E = 72$ GPa.

Fig. P10.23 and P10.24

10.25 The aluminum column ABC has a uniform rectangular cross section and is braced in the xz plane at its midpoint C. (a) Determine the ratio b/d for which the factor of safety is the same with respect to buckling in the xz and yz planes. (b) Using the ratio found in part a, design the cross section of the column so that the factor of safety will be 2.7 when $P = 1.2$ kips, $L = 24$ in., and $E = 10.6 \times 10^6$ psi.

10.26 The aluminum column ABC has a uniform rectangular cross section with $b = \frac{1}{2}$ in. and $d = \frac{7}{8}$ in. The column is braced in the xz plane at its midpoint C and carries a centric load **P** of magnitude 1.1 kips. Knowing that a factor of safety of 2.5 is required, determine the largest allowable length L. Use $E = 10.6 \times 10^6$ psi.

Fig. P10.25 and P10.26

10.27 The uniform brass bar AB has a rectangular cross section and is supported by pins and brackets as shown. Each end of the bar can rotate freely about a horizontal axis through the pin, but rotation about a vertical axis is prevented by the brackets. (a) Determine the ratio b/d for which the factor of safety is the same about the horizontal and vertical axes. (b) Determine the factor of safety if $P = 1.8$ kips, $L = 7$ ft, $d = 1.5$ in., and $E = 15 \times 10^6$ psi.

Fig. P10.27

10.28 Column AB carries a centric load $\mathbf{P}$ of magnitude 72 kN. Cables BC and BD are taut and prevent motion of point B in the xz plane. Using Euler's formula and a factor of safety of 2.3, and neglecting the tension in the cables, determine the maximum allowable length L. Use $E = 200$ GPa.

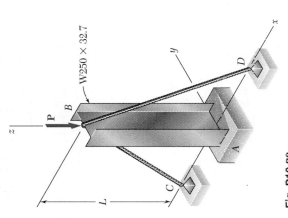

Fig. P10.28

*10.5. ECCENTRIC LOADING; THE SECANT FORMULA

In this section the problem of column buckling will be approached in a different way, by observing that the load $\mathbf{P}$ applied to a column is never perfectly centric. Denoting by e the eccentricity of the load, i.e., the distance between the line of action $\mathbf{P}$ and the axis of the column (Fig. 10.19a), we replace the given eccentric load by a centric force $\mathbf{P}$ and a couple $\mathbf{M}_A$ of moment $M_A = Pe$ (Fig. 10.19b). It is clear that, no matter how small the load $\mathbf{P}$ and the eccentricity e, the couple $\mathbf{M}_A$ will cause some bending of the column (Fig. 10.20). As the eccentric load

Fig. 10.19

Fig. 10.20

is increased, both the couple M_A and the axial force $\mathbf{P}$ increase, and both cause the column to bend further. Viewed in this way, the problem of buckling is not a question of determining how long the column can remain straight and stable under an increasing load, but rather how much the column can be permitted to bend under the increasing load, if the allowable stress is not to be exceeded and if the deflection y_{max} is not to become excessive.

We first write and solve the differential equation of the elastic curve, proceeding in the same manner as we did earlier in Secs. 10.3 and 10.4. Drawing the free-body diagram of a portion AQ of the column and choosing the coordinate axes as shown (Fig. 10.21), we find that the bending moment at Q is

Fig. 10.21

$$M = -Py - M_A = -Py - Pe \qquad (10.22)$$

Substituting the value of M into Eq. (9.4) of Sec. 9.3, we write

$$\frac{d^2y}{dx^2} = \frac{M}{EI} = -\frac{P}{EI}y - \frac{Pe}{EI}$$

Transposing the term containing y and setting

$$p^2 = \frac{P}{EI} \qquad (10.6)$$

as done earlier, we write

$$\frac{d^2y}{dx^2} + p^2y = -p^2e \qquad (10.23)$$

Since the left-hand member of this equation is the same as that of Eq. (10.7), which was solved in Sec. 10.3, we write the general solution of Eq. (10.23) as

$$y = A \sin px + B \cos px - e \qquad (10.24)$$

where the last term is a particular solution of Eq. (10.23).

The constants A and B are obtained from the boundary conditions shown in Fig. 10.22. Making first $x = 0, y = 0$ in Eq. (10.24), we have

$[x = 0, y = 0]$

$[x = L, y = 0]$

Fig. 10.22

$$B = e$$

Making next $x = L, y = 0$, we write

$$A \sin pL = e(1 - \cos pL)$$

Recalling that

$$\sin pL = 2 \sin \frac{pL}{2} \cos \frac{pL}{2}$$

and

$$1 - \cos pL = 2 \sin^2 \frac{pL}{2}$$

and substituting into Eq. (10.25), we obtain, after reductions,

$$A = e \tan \frac{pL}{2}$$

(10.25)

Substituting for A and B into Eq. (10.24), we write the equation of the elastic curve:

$$y = e\left(\tan\frac{pL}{2}\sin px + \cos px - 1\right) \qquad (10.26)$$

The value of the maximum deflection is obtained by setting $x = L/2$ in Eq. (10.26). We have

$$y_{max} = e\left(\tan\frac{pL}{2}\sin\frac{pL}{2} + \cos\frac{pL}{2} - 1\right)$$

$$= e\left(\frac{\sin^2\dfrac{pL}{2} + \cos^2\dfrac{pL}{2}}{\cos\dfrac{pL}{2}} - 1\right)$$

$$y_{max} = e\left(\sec\frac{pL}{2} - 1\right) \qquad (10.27)$$

Recalling Eq. (10.6), we write

$$y_{max} = e\left[\sec\left(\sqrt{\frac{P}{EI}}\frac{L}{2}\right) - 1\right] \qquad (10.28)$$

We note from the expression obtained that y_{max} becomes infinite when

$$\sqrt{\frac{P}{EI}}\frac{L}{2} = \frac{\pi}{2} \qquad (10.29)$$

While the deflection does not actually become infinite, it nevertheless becomes unacceptably large, and P should not be allowed to reach the critical value which satisfies Eq. (10.29). Solving (10.29) for P, we have

$$P_{cr} = \frac{\pi^2 EI}{L^2} \qquad (10.30)$$

which is the value that we obtained in Sec. 10.3 for a column under a centric load. Solving (10.30) for EI and substituting into (10.28), we can express the maximum deflection in the alternative form

$$y_{max} = e\left(\sec\frac{\pi}{2}\sqrt{\frac{P}{P_{cr}}} - 1\right) \qquad (10.31)$$

The maximum stress σ_{max} occurs in the section of the column where the bending moment is maximum, i.e., in the transverse section through the midpoint C, and can be obtained by adding the normal stresses due, respectively, to the axial force and the bending couple exerted on that section (cf. Sec. 4.12). We have

$$\sigma_{max} = \frac{P}{A} + \frac{M_{max}c}{I} \qquad (10.32)$$

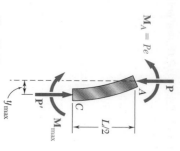

Fig. 10.23

From the free-body diagram of the portion AC of the column (Fig. 10.23), we find that

$$M_{max} = Py_{max} + M_A = P(y_{max} + e)$$

Substituting this value into (10.32) and recalling that $I = Ar^2$, we write

$$\sigma_{max} = \frac{P}{A}\left[1 + \frac{(y_{max} + e)c}{r^2}\right] \qquad (10.33)$$

Substituting for y_{max} the value obtained in (10.28), we write

$$\sigma_{max} = \frac{P}{A}\left[1 + \frac{ec}{r^2}\sec\left(\sqrt{\frac{P}{EI}}\frac{L}{2}\right)\right] \qquad (10.34)$$

An alternative form for σ_{max} is obtained by substituting for y_{max} from (10.31) into (10.33). We have

$$\sigma_{max} = \frac{P}{A}\left(1 + \frac{ec}{r^2}\sec\frac{\pi}{2}\sqrt{\frac{P}{P_{cr}}}\right) \qquad (10.35)$$

The equation obtained can be used with any end conditions, as long as the appropriate value is used for the critical load (cf. Sec. 10.4).

We note that, since σ_{max} does not vary linearly with the load P, the principle of superposition does not apply to the determination of the stress due to the simultaneous application of several loads; the resultant load must first be computed, and then Eq. (10.34) or Eq. (10.35) can be used to determine the corresponding stress. For the same reason, any given factor of safety should be applied to the load, and not to the stress.

Making $I = Ar^2$ in Eq. (10.34) and solving for the ratio P/A in front of the bracket, we write

$$\frac{P}{A} = \frac{\sigma_{max}}{1 + \dfrac{ec}{r^2}\sec\left(\dfrac{1}{2}\sqrt{\dfrac{P}{EA}}\dfrac{L_e}{r}\right)} \qquad (10.36)$$

where the effective length is used to make the formula applicable to various end conditions. This formula is referred to *as the secant formula*; it defines the force per unit area, P/A, that causes a specified

maximum stress σ_{max} in a column of given effective slenderness ratio, L_e/r, for a given value of the ratio ec/r^2, where e is the eccentricity of the applied load. We note that, since P/A appears in both members, it is necessary to solve a transcendental equation by trial and error to obtain the value of P/A corresponding to a given column and loading condition.

Equation (10.36) was used to draw the curves shown in Fig. 10.24a and b for a steel column, assuming the values of E and σ_Y shown in the figure. These curves make it possible to determine the load per unit area P/A, which causes the column to yield for given values of the ratios L_e/r and ec/r^2.

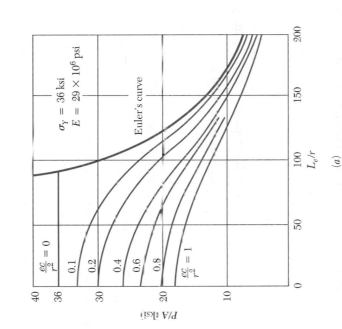

Fig. 10.24 Load per unit area, P/A, causing yield in column.

We note that, for small values of L_e/r, the secant is almost equal to 1 in Eq. (10.36), and P/A can be assumed equal to

$$\frac{P}{A} = \frac{\sigma_{max}}{1 + \dfrac{ec}{r^2}} \qquad (10.37)$$

a value that could be obtained by neglecting the effect of the lateral deflection of the column and using the method of Sec. 4.12. On the other hand, we note from Fig. 10.24 that, for large values of L_e/r, the curves corresponding to the various values of the ratio ec/r^2 get very close to Euler's curve defined by Eq. (10.13'), and thus that the effect of the eccentricity of the loading on the value of P/A becomes negligible. The secant formula is chiefly useful for intermediate values of L_e/r. However, to use it effectively, we should know the value of the eccentricity e of the loading, and this quantity, unfortunately, is seldom known with any degree of accuracy.

SAMPLE PROBLEM 10.2

The uniform column AB consists of an 8-ft section of structural tubing having the cross section shown. (a) Using Euler's formula and a factor of safety of two, determine the allowable centric load for the column and the corresponding normal stress. (b) Assuming that the allowable load, found in part a, is applied as shown at a point 0.75 in. from the geometric axis of the column, determine the horizontal deflection of the top of the column and the maximum normal stress in the column. Use $E = 29 \times 10^6$ psi.

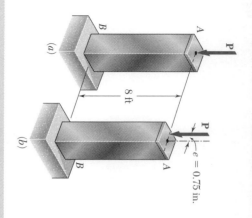

(a)

(b)

8 ft

$e = 0.75$ in.

$A = 3.54$ in^2
$I = 8.00$ in^4
$r = 1.50$ in.
$c = 2.00$ in.

4 in.

4 in.

SOLUTION

Effective Length. Since the column has one end fixed and one end free, its effective length is

$$L_e = 2(8 \text{ ft}) = 16 \text{ ft} = 192 \text{ in.}$$

Critical Load. Using Euler's formula, we write

$$P_{cr} = \frac{\pi^2 EI}{L_e^2} = \frac{\pi^2(29 \times 10^6 \text{ psi})(8.00 \text{ in}^4)}{(192 \text{ in.})^2} \qquad P_{cr} = 62.1 \text{ kips}$$

a. Allowable Load and Stress. For a factor of safety of 2, we find

$$P_{all} = \frac{P_{cr}}{F.S.} = \frac{62.1 \text{ kips}}{2} \qquad P_{all} = 31.1 \text{ kips} \blacktriangleleft$$

and

$$\sigma = \frac{P_{all}}{A} = \frac{31.1 \text{ kips}}{3.54 \text{ in}^2} \qquad \sigma = 8.79 \text{ ksi} \blacktriangleleft$$

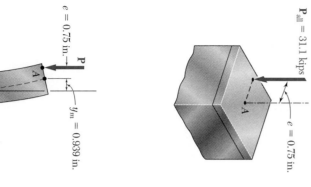

$e = 0.75$ in.

$\mathbf{P}_{all} = 31.1$ kips

b. Eccentric Load. We observe that column AB and its loading are identical to the upper half of the column of Fig. 10.19 which was used in the derivation of the secant formulas; we conclude that the formulas of Sec. 10.5 apply directly to the case considered here. Recalling that $P_{all}/P_{cr} = \frac{1}{2}$ and using Eq. (10.31), we compute the horizontal deflection of point A:

$$y_m = e\left[\sec\left(\frac{\pi}{2}\sqrt{\frac{P}{P_{cr}}}\right) - 1\right] = (0.75 \text{ in.})\left[\sec\left(\frac{\pi}{2\sqrt{2}}\right) - 1\right]$$

$$= (0.75 \text{ in.})(2.252 - 1) \qquad y_m = 0.939 \text{ in.} \blacktriangleleft$$

The maximum normal stress is obtained from Eq. (10.35):

$$\sigma_m = \frac{P}{A}\left[1 + \frac{ec}{r^2}\sec\left(\frac{\pi}{2}\sqrt{\frac{P}{P_{cr}}}\right)\right]$$

$$= \frac{31.1 \text{ kips}}{3.54 \text{ in}^2}\left[1 + \frac{(0.75 \text{ in.})(2 \text{ in.})}{(1.50 \text{ in.})^2}\sec\left(\frac{\pi}{2\sqrt{2}}\right)\right]$$

$$= (8.79 \text{ ksi})[1 + 0.667(2.252)] \qquad \sigma_m = 22.0 \text{ ksi} \blacktriangleleft$$

$\mathbf{P}$

$\mathbf{P}$

A

B

$y_m = 0.939$ in.

PROBLEMS

10.29 An axial load **P** is applied to the 1.25-in.-square aluminum bar *ABC* as shown. When *P* = 3.8 kips, the horizontal deflection at end *C* is 0.16 in. Using $E = 10.1 \times 10^6$ psi, determine (*a*) the eccentricity *e* of the load, (*b*) the maximum stress in the rod.

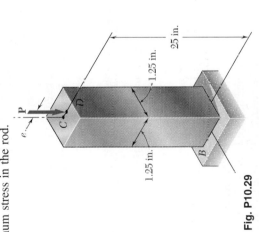

Fig. P10.29

10.30 The line of action of the 310 kN axial load is parallel to the geometric axis of the column *AB* and intersects the *x* axis at *x* = *e*. Using $E = 200$ GPa, determine (*a*) the eccentricity *e* when the deflection of the midpoint *C* of the column is 9 mm, (*b*) the corresponding maximum stress in the column.

10.31 The axial load **P** is applied at a point located on the *x* axis at a distance *e* from the geometric axis of the rolled-steel column *BC*. When *P* = 82 kips, the horizontal deflection of the top of the column is 0.20 in. Using $E = 29 \times 10^6$ psi, determine (*a*) the eccentricity *e* of the load, (*b*) the maximum stress in the column.

Fig. P10.30

Fig. P10.31

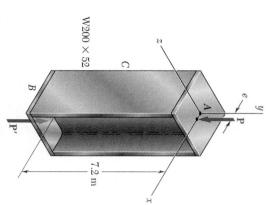

Fig. P10.33

10.32 An axial load **P** is applied to the 32-mm-diameter steel rod *AB* as shown. For *P* = 37 kN and *e* = 1.2 mm, determine (*a*) the deflection at the midpoint *C* of the rod, (*b*) the maximum stress in the rod. Use *E* = 200 GPa.

Fig. P10.32

10.33 The line of action of the axial load **P** of magnitude 270 kN is parallel to the geometric axis of the column *AB* and intersects the *x* axis at *e* = 14 mm. Using *E* = 200 GPa, determine (*a*) the deflection of the midpoint *C* of the column, (*b*) the maximum stress in the column.

10.34 Solve Prob. 10.33 if the load **P** is applied parallel to the geometric axis of the column *AB* so that it intersects the *x* axis at *e* = 21 mm.

10.35 An axial load **P** is applied at a point *D* that is 0.25 in. from the geometric axis of the square aluminum bar *BC*. Using *E* = 10.1 × 10⁶ psi, determine (*a*) the load **P** for which the horizontal deflection of end *C* is 0.50 in., (*b*) the corresponding maximum stress in the column.

Fig. P10.35

10.36 A brass pipe having the cross section shown has an axial load **P** applied 5 mm from its geometric axis. Using *E* = 120 GPa, determine (*a*) the load **P** for which the horizontal deflection at the midpoint *C* is 5 mm, (*b*) the corresponding maximum stress in the column.

Fig. P10.36

10.37 Solve Prob. 10.36, assuming that the axial load **P** is applied 10 mm from the geometric axis of the column.

Fig. P10.38

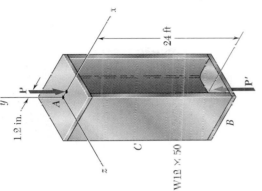

Fig. P10.40

10.38 An axial load **P** is applied at a point located on the x axis at a distance $e = 12$ mm from the geometric axis of the W310 × 60 rolled-steel column BC. Assuming that $L = 3.5$ m and using $E = 200$ GPa, determine (*a*) the load **P** for which the horizontal deflection at end C is 15 mm, (*b*) the corresponding maximum stress in the column.

10.39 Solve Prob. 10.38, assuming that $L = 4.5$ m.

10.40 The line of action of an axial load **P** is parallel to the geometric axis of the column AB and intersects the x axis at $x = 1.2$ in. Using $E = 29 \times 10^6$ psi, determine (*a*) the load **P** for which the horizontal deflection of the midpoint C of the column is 0.8 in., (*b*) the corresponding maximum stress in the column.

10.41 The steel bar AB has a $\frac{3}{8} \times \frac{3}{8}$-in.-square cross section and is held by pins that are a fixed distance apart and are located at a distance $e = 0.03$ in. from the geometric axis of the bar. Knowing that at temperature T_0 the pins are in contact with the bar and that the force in the bar is zero, determine the increase in temperature for which the bar will just make contact with point C if $d = 0.01$ in. Use $E = 29 \times 10^6$ psi and the coefficient of thermal expansion $\alpha = 6.5 \times 10^{-6}/°F$.

Fig. P10.41

10.42 For the bar of Prob. 10.41, determine the required distance d for which the bar will just make contact with point C when the temperature increases by 120°F.

$A = 3400 \text{ mm}^2$
$I = 7.93 \times 10^{-6} \text{ m}^4$
$r = 48.3 \text{ mm}$

Fig. P10.43 and P10.44

10.43 A 3.5-m-long steel tube having the cross section and properties shown is used as a column. For the grade of steel used $\sigma_Y = 250$ MPa and $E = 200$ GPa. Knowing that a factor of safety of 2.6 with respect to permanent deformation is required, determine the allowable load **P** when the eccentricity e is (a) 15 mm, (b) 7.5 mm. (*Hint:* Since the factor of safety must be applied to the load **P**, not to the stress, use Fig. 10.24 to determine P_Y).

10.44 Solve Prob. 10.43, assuming that the length of the steel tube is increased to 5 m.

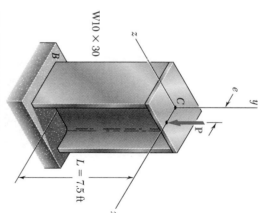

Fig. P10.45 and P10.46

10.45 An axial load **P** is applied to the W10 × 30 rolled-steel column *BC* that is free at its top *C* and fixed at its base *B*. Knowing that the eccentricity of the load is $e = 0.5$ in. and that for the grade of steel used $\sigma_Y = 36$ ksi and $E = 29 \times 10^6$ psi, determine (a) the magnitude of **P** of the allowable load when a factor of safety of 2.4 with respect to permanent deformation is required, (b) the ratio of the load found in part a to the magnitude of the allowable centric load for the column. (See hint of Prob. 10.43.)

10.46 Solve Prob. 10.45, assuming that the length of the column is reduced to 5.0 ft.

10.47 A 55-kip axial load **P** is applied to a W8 × 24 rolled-steel column *BC* that is free at its top *C* and fixed at its base *B*. Knowing that the eccentricity of the load is $e = 0.25$ in., determine the largest permissible length *L* if the allowable stress in the column is 14 ksi. Use $E = 29 \times 10^6$ psi.

10.48 A 26-kip axial load **P** is applied to a W6 × 12 rolled-steel column *BC* that is free at its top *C* and fixed at its base *B*. Knowing that the eccentricity of the load is $e = 0.25$ in., determine the largest permissible length *L* if the allowable stress in the column is 14 ksi. Use $E = 29 \times 10^6$ psi.

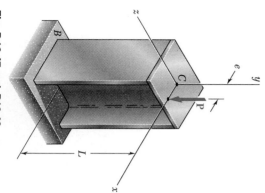

Fig. P10.47 and P10.48

10.49 Axial loads of magnitude $P = 84$ kN are applied parallel to the geometric axis of a W200 × 22.5 rolled-steel column AB and intersect the x axis at a distance e from its geometric axis. Knowing that the allowable stress is $\sigma_{all} = 75$ MPa and that $E = 200$ GPa, determine the largest permissible length L when (a) $e = 5$ mm, (b) $e = 12$ mm.

10.50 Axial loads of magnitude $P = 580$ kN are applied parallel to the geometric axis of a W250 × 80 rolled-steel column AB and intersect the x axis at a distance e from its geometric axis. Knowing that the allowable stress is $\sigma_{all} = 75$ MPa and that $E = 200$ GPa, determine the largest permissible length L when (a) $e = 5$ mm, (b) $e = 10$ mm.

Fig. P10.49 and P10.50

10.51 A 12-kip axial load is applied with an eccentricity $e = 0.375$ in. to the circular steel rod BC that is free at its top C and fixed at its base B. Knowing that the stock of rods available for use have diameters in increments of $\frac{1}{8}$ in. from 1.5 in. to 3.0 in., determine the lightest rod that can be used if $\sigma_{all} = 15$ ksi. Use $E = 29 \times 10^6$ psi.

10.52 Solve Prob. 10.51, assuming that the 12-kip axial load will be applied to the rod with an eccentricity $e = \frac{1}{2}d$.

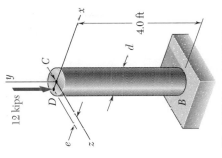

Fig. P10.51 and P10.52

10.53 An axial load of magnitude $P = 220$ kN is applied at a point located on the x axis at a distance $e = 6$ mm from the geometric axis of the wide-flange column BC. Knowing that $E = 200$ GPa, chose the lightest W200 shape that can be used if $\sigma_{all} = 120$ MPa.

10.54 Solve Prob. 10.53, assuming that the magnitude of the axial load is $P = 345$ kN.

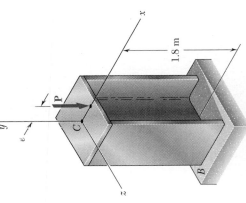

Fig. P10.53 and P10.54

10.55 Axial loads of magnitude $P = 175$ kN are applied parallel to the geometric axis of a W250 × 44.8 rolled-steel column AB and intersect the x axis at a distance $e = 12$ mm from its geometric axis. Knowing that $\sigma_Y = 250$ MPa and $E = 200$ GPa, determine the factor of safety with respect to yield. (*Hint:* Since the factor of safety must be applied to the load **P**, not to the stresses, use Fig. 10.24 to determine P_Y.)

10.56 Solve Prob. 10.55, assuming that $e = 16$ mm and $P = 155$ kN.

Fig. P10.55 and 10.56

10.6. DESIGN OF COLUMNS UNDER A CENTRIC LOAD

In the preceding sections, we have determined the critical load of a column by using Euler's formula, and we have investigated the deformations and stresses in eccentrically loaded columns by using the secant formula. In each case we assumed that all stresses remained below the proportional limit and that the column was initially a straight homogeneous prism. Real columns fall short of such an idealization, and in practice the design of columns is based on empirical formulas that reflect the results of numerous laboratory tests.

Over the last century, many steel columns have been tested by applying to them a centric axial load and increasing the load until failure occurred. The results of such tests are represented in Fig. 10.25 where, for each of many tests, a point has been plotted with its ordinate equal to the normal stress σ_{cr} at failure, and its abscissa equal to the corresponding value of the effective slenderness ratio, L_e/r. Although there is considerable scatter in the test results, regions corresponding to three types of failure can be observed. For long columns, where L_e/r is large, failure is closely predicted by Euler's formula, and the value of σ_{cr} is

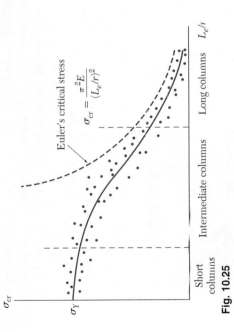

Fig. 10.25

observed to depend on the modulus of elasticity E of the steel used, but
not on its yield strength σ_Y. For very short columns and compression
blocks, failure occurs essentially as a result of yield, and we have
$\sigma_{cr} \approx \sigma_Y$. Columns of intermediate length comprise those cases where
failure is dependent on both σ_Y and E. In this range, column failure is
an extremely complex phenomenon, and test data have been used
extensively to guide the development of specifications and design
formulas.

Empirical formulas that express an allowable stress or critical stress
in terms of the effective slenderness ratio were first introduced over a
century ago, and since then have undergone a continuous process of re-
finement and improvement. Typical empirical formulas used to approximate test data are shown in Fig. 10.26. It is not always feasible to
use a single formula for all values of L_e/r. Most design specifications

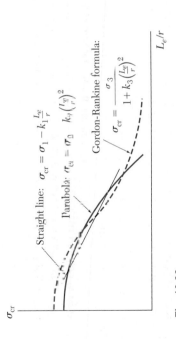

Fig. 10.26

use different formulas, each with a definite range of applicability. In
each case we must check that the formula we propose to use is applicable for the value of L_e/r for the column involved. Furthermore, we
must determine whether the formula provides the value of the critical
stress for the column, in which case we must apply the appropriate factor of safety, or whether it provides directly an allowable stress.

Specific formulas for the design of steel, aluminum, and wood columns under centric loading will now be considered. Figure 10.27 shows examples of columns that would be designed using these formulas. The design for the three different materials using *Allowable Stress Design* is first presented. This is followed with the formulas needed for the design of steel columns based on *Load and Resistance Factor Design*.†

(a)

(b)

Fig. 10.27 The water tank in (a) is supported by steel columns and the building in construction in (b) is framed with wood columns.

Structural Steel—Allowable Stress Design. The formulas most widely used for the allowable stress design of steel columns under a centric load are found in the *Allowable Stress Design Specifications* of the American Institute of Steel Construction.‡ A parabolic expression will be used to predict σ_{all} for columns of short and intermediate lengths, and an Euler-type relation will be used for long columns. These relations are developed in two steps:

1. First a curve representing the variation of σ_{cr} with L/r is obtained (Fig. 10.28). It is important to note that this curve does not incorporate any factor of safety. The portion *AB* of this curve is an arc of parabola defined by an equation of the form

$$\sigma_{cr} = \sigma_0 - k\left(\frac{L}{r}\right)^2 \qquad (10.38)$$

while portion *BE* is part of Euler's curve *DBE* defined by the equation

$$\sigma_{cr} = \frac{\pi^2 E}{(L/r)^2} \qquad (10.39)$$

We note that, since $\sigma_{cr} = \sigma_Y$ for $L/r = 0$, the constant σ_0 in Eq. (10.38) must be equal to σ_Y. On the other hand, it is assumed that at point *B* where the parabola joins Euler's curve, the critifications that at point *B* where the parabola joins Euler's curve, the crit-

† In specific design formulas, the letter *L* will always refer to the effective length of the column.

‡ *Manual of Steel Construction, Allowable Stress Design,* 9th ed., American Institute of Steel Construction, New York, 1989.

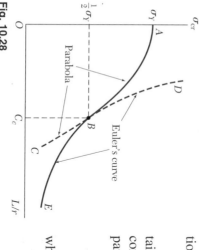

Fig. 10.28

ical stress is equal to half of the yield stress. Denoting by C_c the value of L/r at that point, Eq. (10.38) yields therefore

$$\tfrac{1}{2}\sigma_Y = \sigma_Y - kC_c^2$$

and $k = \sigma_Y/2C_c^2$. Substituting for σ_0 and k into Eq. (10.38), we find that

$$\text{for } L/r \le C_c: \quad \sigma_{\text{cr}} = \sigma_Y\left[1 - \frac{(L/r)^2}{2C_c^2}\right] \qquad (10.40)$$

and, recalling (10.39), that

$$\text{for } L/r \ge C_c: \quad \sigma_{\text{cr}} = \frac{\pi^2 E}{(L/r)^2} \qquad (10.41)$$

Making $\sigma_{\text{cr}} = \tfrac{1}{2}\sigma_Y$ and $L/r = C_c$ in Eq. (10.41), we find that

$$C_c^2 = \frac{2\pi^2 E}{\sigma_Y} \qquad (10.42)$$

2. A factor of safety must be introduced to obtain the final AISC design formulas defining σ_{all} as a function of L/r. For $L/r \ge C_c$, that is, for long columns, a constant factor of safety of 1.92 is used. Dividing the value obtained in Eq. (10.41) for σ_{cr} by this factor of safety, and noting that the AISC specifications state that L/r preferably should not exceed 200, we write[†]

$$\frac{L}{r} \ge C_c: \quad \sigma_{\text{all}} = \frac{\sigma_{\text{cr}}}{F.S.} = \frac{\pi^2 E}{1.92(L/r)^2} \qquad (10.43)$$

For short and intermediate-length columns, the following formula is used to determine the factor of safety:

$$F.S. = \frac{5}{3} + \frac{3}{8}\frac{L/r}{C_c} - \frac{1}{8}\left(\frac{L/r}{C_c}\right)^3 \qquad (10.44)$$

Dividing the expression obtained in (10.40) for σ_{cr} by this factor of safety, we write

$$\frac{L}{r} < C_c: \quad \sigma_{\text{all}} = \frac{\sigma_{\text{cr}}}{F.S.} = \frac{\sigma_Y}{F.S.}\left[1 - \frac{1}{2}\left(\frac{L/r}{C_c}\right)^2\right] \qquad (10.45)$$

The formulas obtained can be used with SI or U.S. customary units.

We observe that, by using Eqs. (10.42), (10.43), (10.44), and (10.45), we can determine the allowable axial stress for a given grade of steel and any given allowable value of L/r. The procedure is to first compute C_c by substituting the given value of σ_Y into Eq (10.42). For values of L/r larger than C_c, we then use Eq. (10.43) to determine σ_{all}, and for values of L/r smaller than C_c, we determine σ_{all} from Eqs. (10.44) and (10.45). For the convenience of the designer, values of the allowable stress have been tabulated in the AISC *Manual of Steel Construction* for different grades of steel and for all values of L/r from 1 to 200. The variation of σ_{all} with L/r is shown in Fig. 10.29 for different grades of structural steel.

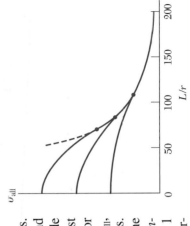

Fig. 10.29

[†]The AISC actually specifies $\sigma_{\text{all}} = 12\pi^2 E/23(L/r)^2$.

EXAMPLE 10.02

Determine the longest unsupported length L for which the S100 × 11.5 rolled-steel compression member AB can safely carry the centric load shown (Fig. 10.30). Assume $\sigma_Y = 250$ MPa and $E = 200$ GPa.

Fig. 10.30

From Appendix C we find that, for an S100 × 11.5 shape,

$$A = 1460 \text{ mm}^2 \qquad r_x = 41.6 \text{ mm} \qquad r_y = 14.8 \text{ mm}$$

Fig. 10.31

If the 60-kN load is to be safely supported, we must have

$$\sigma_{all} = \frac{P}{A} = \frac{60 \times 10^3 \text{N}}{1460 \times 10^{-6} \text{ m}^2} = 41.1 \times 10^6 \text{ Pa} \quad (10.42)$$

On the other hand, for the given yield strength, Eq. (10.42) gives

$$C_c^2 = \frac{2\pi^2 E}{\sigma_Y} = \frac{2\pi^2(200 \times 10^9)}{250 \times 10^6} = 15.79 \times 10^3$$

$$C_c = 125.7$$

Assuming $L/r \geq C_c$, we use Eq. (10.43) and write

$$\sigma_{all} = \frac{\pi^2 E}{1.92(L/r)^2} = \frac{\pi^2(200 \times 10^9 \text{ Pa})}{1.92(L/r)^2} = \frac{1.028 \times 10^{12} \text{ Pa}}{(L/r)^2}$$

Equating this expression to the required value of σ_{all}, we write

$$\frac{1.028 \times 10^{12} \text{ Pa}}{(L/r)^2} = 41.1 \times 10^6 \text{ Pa} \qquad L/r = 158.2$$

Our assumption that $L/r \geq C_c$ was correct. Choosing the smaller of the two radii of gyration, we have

$$\frac{L}{r_y} = \frac{L}{14.8 \times 10^{-3} \text{ m}} = 158.2 \qquad L = 2.34 \text{ m}$$

Aluminum. Many aluminum alloys are available for use in structural and machine construction. For most columns the specifications of the Aluminum Association† provide two formulas for the allowable stress in columns under centric loading. The variation of σ_{all} with L/r defined by these formulas is shown in Fig. 10.31. We note that for short columns a linear relation between σ_{all} with L/r is used and for long columns an Euler-type formula is used. Specific formulas for use in the design of buildings and similar structures are given below in both SI and U.S. customary units for two commonly used alloys.

Alloy 6061-T6:

$$L/r < 66: \quad \sigma_{all} = [20.2 - 0.126(L/r)] \text{ ksi} \quad (10.46)$$
$$= [139 - 0.868(L/r)] \text{ MPa} \quad (10.46')$$

$$L/r \geq 66: \quad \sigma_{all} = \frac{51{,}000 \text{ ksi}}{(L/r)^2} = \frac{351 \times 10^3 \text{ MPa}}{(L/r)^2} \quad (10.47)$$

Alloy 2014-T6:

$$L/r < 55: \quad \sigma_{all} = [30.7 - 0.23(L/r)] \text{ ksi} \quad (10.48)$$
$$= [212 - 1.585(L/r)] \text{ MPa} \quad (10.48')$$

$$L/r \geq 55: \quad \sigma_{all} = \frac{54{,}000 \text{ ksi}}{(L/r)^2} = \frac{372 \times 10^3 \text{ MPa}}{(L/r)^2} \quad (10.49)$$

†*Specifications and Guidelines for Aluminum Structures*, Aluminum Association, Inc., Washington D.C., 2000.

Wood. For the design of wood columns the specifications of the American Forest and Paper Association[†] provides a single equation that can be used to obtain the allowable stress for short, intermediate, and long columns under centric loading. For a column with a *rectangular* cross section of sides b and d, where $d < b$, the variation of σ_{all} with L/d is shown in Fig. 10.32.

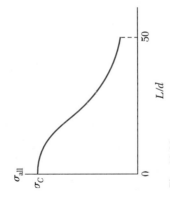

Fig. 10.32

For solid columns made from a single piece of wood or made by gluing laminations together, the allowable stress σ_{all} is

$$\sigma_{\text{all}} = \sigma_C C_P \qquad (10.50)$$

where σ_C is the adjusted allowable stress for compression parallel to the grain.[‡] Adjustments used to obtain σ_C are included in the specifications to account for different variations, such as in the load duration. The column stability factor C_P accounts for the column length and is defined by the following equation:

$$C_P = \frac{1 + (\sigma_{CE}/\sigma_C)}{2c} - \sqrt{\left[\frac{1 + (\sigma_{CE}/\sigma_C)}{2c}\right]^2 - \frac{\sigma_{CE}/\sigma_C}{c}} \qquad (10.51)$$

The parameter c accounts for the type of column, and it is equal to 0.8 for sawn lumber columns and 0.90 for glued laminated wood columns. The value of σ_{CE} is defined as

$$\sigma_{CE} = \frac{K_{CE}E}{(L/d)^2} \qquad (10.52)$$

where K_{CE} is the Euler buckling coefficient. Values of K_{CE} are 0.300 for columns made from a single piece of structurally graded wood and 0.418 for glued laminated columns. Columns in which L/d exceeds 50 are not permitted by the *National Design Specification for Wood Construction*.

[†]*National Design Specification for Wood Construction,* American Forest and Paper Association, American Wood Council, Washington, D.C., 1997.

[‡]In the *National Design Specification for Wood Construction,* the symbol F is used for stresses.

EXAMPLE 10.03

Knowing that column AB (Fig. 10.33) has an effective length of 14 ft, and that it must safely carry a 32-kip load, design the column using a square glued laminated cross section. The modulus of elasticity for the wood is $E = 1.6 \times 10^6$ psi, and the adjusted allowable stress for compression parallel to the grain is $\sigma_C = 1060$ psi.

We note that $c = 0.90$ and that $K_{CE} = 0.418$ for glued laminated wood columns. We must compute the value of σ_{CE}. Using Eq. (10.52) we write

$$\sigma_{CE} = \frac{K_{CE}E}{(L/d)^2} = \frac{0.418(1.6 \times 10^6 \text{ psi})}{(168 \text{ in.}/d)^2} = 23.696 d^2 \text{ psi}$$

We then use Eq. (10.51) to express the column stability factor in terms of d, with $(\sigma_{CE}/\sigma_C) = (23.696 d^2/1.060 \times 10^3) = 22.35 \times 10^{-3} d^2$,

$$C_P = \frac{1 + (\sigma_{CE}/\sigma_C)}{2c} - \sqrt{\left[\frac{1 + (\sigma_{CE}/\sigma_C)}{2c} \right]^2 - \frac{\sigma_{CE}/\sigma_C}{c}}$$

$$= \frac{1 + 22.35 \times 10^{-3} d^2}{2(0.90)} - \sqrt{\left[\frac{1 + 22.35 \times 10^{-3} d^2}{2(0.90)} \right]^2 - \frac{22.35 \times 10^{-3} d^2}{0.90}}$$

Since the column must carry 32 kips, which is equal to $\sigma_C d^2$, we use Eq. (10.50) to write

$$\sigma_{all} = \frac{32 \text{ kips}}{d^2} = \sigma_C C_P = 1.060 C_P$$

Solving this equation for C_P and substituting the value obtained into the previous equation, we write

$$\frac{30.19}{d^2} = \frac{1 + 22.35 \times 10^{-3} d^2}{2(0.90)} - \sqrt{\left[\frac{1 + 22.35 \times 10^{-3} d^2}{2(0.90)} \right]^2 - \frac{22.35 \times 10^{-3} d^2}{0.90}}$$

Solving for d by trial and error yields $d = 6.44$ in.

Fig. 10.33

*Structural Steel—Load and Resistance Factor Design. As we saw in Sec. 1.13, an alternative method of design is based on the determination of the load at which the structure ceases to be useful. Design is based on the inequality given by Eq. (1.26):

$$\gamma_D P_D + \gamma_L P_L \le \phi P_U \qquad (1.26)$$

The approach used for the design of steel columns under a centric load as found in the Load and Resistance Factor Design Specifications of the American Institute of Steel Construction† is similar to that in the Allowable Stress Design Specifications. Two equations are used to obtain the *ultimate load* P_U, one for short and intermediate-length columns and another for long columns. To determine P_U, the ratio L/r is first converted to a slenderness parameter λ_C, defined as

$$\lambda_C = \frac{L}{r\pi}\sqrt{\frac{\sigma_Y}{E}} \qquad (10.53)$$

Short and intermediate-length columns are those with $\lambda_C \le 1.5$ and long columns are those with $\lambda_C \ge 1.5$.

The ultimate load for short and intermediate-length columns ($\lambda_C \le 1.5$) is

$$P_U = A(0.658^{\lambda_C^2})\sigma_Y \qquad (10.54)$$

The Euler relation forms the basis for the determination of the ultimate load for long columns. Thus, for long columns ($\lambda_C \ge 1.5$),

$$P_U = A\left(\frac{0.877}{\lambda_C^2}\right)\sigma_Y \qquad (10.55)$$

These equations can be used with SI or U.S. customary units.

We observe that, by using Eqs. (10.53), (10.54), and (10.55) with Eq. (1.26), we can determine if the design is acceptable. The procedure is to first determine the slenderness parameter λ_C from Eq. (10.53). For values of L/r smaller than λ_C, the *ultimate load* P_U for use with (1.26) is obtained from Eq. (10.54), and for values of L/r larger than λ_C, the *ultimate load* P_U is obtained from Eq. (10.55). We can then use P_U in Eq. (1.26) to determine if the design is acceptable. The Load and Resistance Factor Design Specifications of the American Institute of Steel Construction specify that the *resistance factor* ϕ is 0.85.

Note: The design formulas presented in this section are intended to provide examples of different design approaches. These formulas do not provide all the requirements that are needed for many designs, and the student should refer to the appropriate design specifications before attempting actual designs.

†*Manual of Steel Construction, Load & Resistance Factor Design*, 2nd ed., American Institute of Steel Construction, New York, 1994.

SAMPLE PROBLEM 10.3

Column AB consists of a W10 × 39 rolled-steel shape made of a grade of steel for which $\sigma_Y = 36$ ksi and $E = 29 \times 10^6$ psi. Determine the allowable centric load $\mathbf{P}$ (a) if the effective length of the column is 24 ft in all directions, (b) if bracing is provided to prevent the movement of the midpoint C in the xz plane. (Assume that the movement of point C in the yz plane is not affected by the bracing.)

(a)

(b)

W10 × 39
$A = 11.5$ in²
$r_x = 4.27$ in.
$r_y = 1.98$ in.

SOLUTION

We first compute the value of C_c corresponding to the given yield strength $\sigma_Y = 36$ ksi.

$$C_c^2 = \frac{2\pi^2 E}{\sigma_Y} = \frac{2\pi^2(29 \times 10^6 \text{ psi})}{36 \text{ ksi}} = 15.90 \times 10^3 \qquad C_c = 126.1$$

a. Effective Length = 24 ft. Since $r_y < r_x$, buckling will take place in the xz plane. For $L = 24$ ft and $r = r_y = 1.98$ in. the slenderness ratio is

$$\frac{L}{r_y} = \frac{(24 \times 12) \text{ in.}}{1.98 \text{ in.}} = \frac{288 \text{ in.}}{1.98 \text{ in.}} = 145.5$$

Since $L/r > C_c$, we use Eq. (10.43), with $E = 29 \times 10^6$ psi

$$\sigma_{all} = \frac{\pi^2 E}{1.92(L/r)^2} = \frac{\pi^2(29 \times 10^6 \text{ psi})}{1.92(145.5)^2} = 7.04 \text{ ksi}$$

$$P_{all} = \sigma_{all}A = (7.04 \text{ ksi})(11.5 \text{ in}^2) \qquad P_{all} = 81.0 \text{ kips} \blacktriangleleft$$

b. Bracing at Midpoint C. Since bracing prevents movement of point C in the xz plane but not in the yz plane, we must compute the slenderness ratio corresponding to buckling in each plane and determine which is larger.

xz Plane: Effective length = 12 ft = 144 in., $r = r_y = 1.98$ in.

$$L/r = (144 \text{ in.})/(1.98 \text{ in.}) = 72.7$$

yz Plane: Effective length = 24 ft = 288 in., $r = r_x = 4.27$ in.

$$L/r = (288 \text{ in.})/(4.27 \text{ in.}) = 67.4$$

Since the larger slenderness ratio corresponds to a smaller allowable load, we choose $L/r = 72.7$. Since $L/r < C_c$, the column is of intermediate length and we use Eqs. (10.44) and (10.45). We first compute the factor of safety, then the allowable stress, and finally the allowable load $\mathbf{P}$.

$$F.S. = \frac{5}{3} + \frac{3}{8}\left(\frac{L/r}{C_c}\right) - \frac{1}{8}\left(\frac{L/r}{C_c}\right)^3 = \frac{5}{3} + \frac{3}{8}\left(\frac{72.7}{126.1}\right) - \frac{1}{8}\left(\frac{72.7}{126.1}\right)^3 = 1.86$$

$$\sigma_{all} = \frac{\sigma_Y}{F.S.}\left[1 - \frac{1}{2}\left(\frac{L/r}{C_c}\right)^2\right] = \frac{36 \text{ ksi}}{1.86}\left[1 - \frac{1}{2}\left(\frac{72.7}{126.1}\right)^2\right] = 16.14 \text{ ksi}$$

$$P_{all} = \sigma_{all}A = (16.14 \text{ ksi})(11.5 \text{ in}^2) \qquad P_{all} = 185.6 \text{ kips} \blacktriangleleft$$

Buckling in xz plane

Buckling in yz plane

SAMPLE PROBLEM 10.4

Using the aluminum alloy 2014-T6, determine the smallest diameter rod which can be used to support the centric load $P = 60$ kN if (a) $L = 750$ mm, (b) $L = 300$ mm.

SOLUTION

For the cross section of a solid circular rod, we have

$$I = \frac{\pi}{4}c^4 \qquad A = \pi c^2 \qquad r = \sqrt{\frac{I}{A}} = \sqrt{\frac{\pi c^4/4}{\pi c^2}} = \frac{c}{2}$$

a. **Length of 750 mm.** Since the diameter of the rod is not known, a value of L/r must be assumed; we *assume* that $L/r > 55$ and use Eq. (10.49). For the centric load **P**, we have $\sigma = P/A$ and write

$$\frac{P}{A} = \sigma_{all} = \frac{372 \times 10^3 \text{ MPa}}{(L/r)^2}$$

$$\frac{60 \times 10^3 \text{ N}}{\pi c^2} = \frac{372 \times 10^9 \text{ Pa}}{\left(\dfrac{0.750 \text{ m}}{c/2}\right)^2}$$

$$c^4 = 115.5 \times 10^{-9} \text{ m}^4 \qquad c = 18.44 \text{ mm}$$

For $c = 18.44$ mm, the slenderness ratio is

$$\frac{L}{r} = \frac{L}{c/2} = \frac{750 \text{ mm}}{(18.44 \text{ mm})/2} = 81.3 > 55$$

Our assumption is correct, and for $L = 750$ mm, the required diameter is

$$d = 2c = 2(18.44 \text{ mm}) \qquad d = 36.9 \text{ mm} \blacktriangledown$$

b. **Length of 300 mm.** We again *assume* that $L/r > 55$. Using Eq. (11.49), and following the procedure used in part *a*, we find that $c = 11.66$ mm and $L/r = 51.5$. Since L/r is less than 55, our assumption is wrong; we now assume that $L/r < 55$ and use Eq. (10.48) for the design of this rod.

$$\frac{P}{A} = \sigma_{all} = \left[212 - 1.585\left(\frac{L}{r}\right)\right] \text{MPa}$$

$$\frac{60 \times 10^3 \text{ N}}{\pi c^2} = \left[212 - 1.585\left(\frac{0.3 \text{ m}}{c/2}\right)\right]10^6 \text{ Pa}$$

$$c = 12.00 \text{ mm}$$

For $c = 12.00$ mm, the slenderness ratio is

$$\frac{L}{r} = \frac{L}{c/2} = \frac{300 \text{ mm}}{(12.00 \text{ mm})/2} = 50$$

Our second assumption that $L/r < 55$ is correct. For $L = 300$ mm, the required diameter is

$$d = 2c = 2(12.00 \text{ mm}) \qquad d = 24.0 \text{ mm} \blacktriangledown$$

10.57 Using allowable stress design, determine the allowable centric load for a column of 6.5-m effective length that is made from the following rolled-steel shape: (a) W250 × 49.1, (b) W250 × 80. Use $\sigma_Y = 250$ MPa and $E = 200$ GPa.

10.58 A W8 × 31 rolled-steel shape is used to form a column of 21-ft effective length. Using allowable stress design, determine the allowable centric load if the yield strength of the grade of steel used is (a) $\sigma_Y = 36$ ksi, (b) $\sigma_Y = 50$ ksi. Use $E = 29 × 10^6$ psi.

10.59 A steel pipe having the cross section shown is used as a column. Using allowable stress design, determine the allowable centric load if the effective length of the column is (a) 18 ft, (b) 26 ft. Use $\sigma_Y = 36$ ksi and $E = 29 × 10^6$ psi.

$t = 0.28$ in.

6.0 in.

Fig. P10.59

10.60 A column is made from half of a W360 × 216 rolled-steel shape, with the geometric properties as shown. Using allowable stress design, determine the allowable centric load if the effective length of the column is (a) 4.0 m, (b) 6.5 m. Use $\sigma_Y = 345$ MPa and $E = 200$ GPa.

$A = 13.8 × 10^3$ mm^2
$I_x = 26.0 × 10^6$ mm^4
$I_y = 142.0 × 10^6$ mm^4

Fig. P10.60

10.61 A column having an effective length of 3.5 m is made of sawn lumber with a 114 × 140-mm cross section. Knowing that for the grade of wood used the adjusted allowable stress for compression parallel to the grain is $\sigma_C = 7.6$ MPa and that $E = 10$ GPa, determine the maximum allowable centric load for the column.

10.62 A sawn lumber column with a 7.5 × 5.5-in. cross section has an 18-ft effective length. Knowing that for the grade of wood used the adjusted allowable stress for compression parallel to the grain is $\sigma_C = 1220$ psi and that $E = 1.3 × 10^6$ psi, determine the maximum allowable centric load for the column.

10.63 A compression member has the cross section shown and an effective length of 5 ft. Knowing that the aluminum alloy used is 2014-T6, determine the allowable centric load.

$t = 0.375$ in.

4.0 in.

4.0 in.

Fig. P10.63

Fig. P10.64

10.64 A compression member has the cross section shown and an effective length of 1.55 m. Knowing that the aluminum alloy used is 6061-T6, determine the allowable centric load.

10.65 A column of 6.4-m effective length is obtained by connecting four 89 × 89 × 9.5-mm steel angles with lacing bars as shown. Using allowable stress design, determine the allowable centric load for the column. Use $\sigma_Y = 345$ MPa and $E = 200$ GPa.

Fig. *P10.65*

10.66 A column of 21-ft effective length is obtained by connecting two C10 × 20 steel channels with lacing bars as shown. Using allowable stress design, determine the allowable centric load for the column. Use $\sigma_Y = 36$ ksi and $E = 29 \times 10^6$ psi.

Fig. P10.66

10.67 A compression member of 2.3-m effective length is obtained by bolting together two 127 × 76 × 12.7-mm steel angles as shown. Using allowable stress design, determine the allowable centric load for the column. Use $\sigma_Y = 250$ MPa and $E = 200$ GPa.

Fig. *P10.67*

10.68 A column of 23-ft effective length is obtained by welding two $\frac{3}{8}$-in. steel plates to a W10 × 33 rolled-steel shape as shown. Using allowable stress design, determine the allowable centric load for the column. Use $\sigma_Y = 50$ ksi and $E = 29 \times 10^6$ psi.

Fig. P10.68

10.69 A rectangular column with a 4.4-m effective length is made of glued laminated wood. Knowing that for the grade of wood used the adjusted allowable stress for compression parallel to the grain is $\sigma_C = 8.3$ MPa and that $E = 10$ GPa, determine the maximum allowable centric load for the column.

Fig. P10.69

10.70 An aluminum structural tube is reinforced by riveting two plates to its as shown for use as a column of 1.7-m effective length. Knowing that all material is aluminum alloy 2014-T6, determine the maximum allowable centric load.

Fig. P10.70

10.71 A 280-kN centric load is applied to the column shown, which is free at its top A and fixed at its base B. Using aluminum alloy 2014-T6, select the smallest square cross section that can be used.

Fig. P10.71

10.72 A 16-kip centric load must be supported by an aluminum column as shown. Using the aluminum alloy 6061-T6, determine the minimum dimension b that can be used.

Fig. P10.72

10.73 An aluminum tube of 90-mm outer diameter is to carry a centric load of 120 kN. Knowing that the stock of tubes available for use are made of alloy 2014-T6 and with wall thickness in increments of 3 mm from 6 mm to 15 mm, determine the lightest tube that can be used.

Fig. P10.73

Fig. P10.74

Fig. P10.75

10.74 An 18-kip centric load is applied to a rectangular sawn lumber column of 22-ft effective length. Using sawn lumber for which the adjusted allowable stress for compression parallel to the grain is $\sigma_C = 1050$ psi and knowing that $E = 10 \times 10^6$ psi, determine the smallest cross section that can be used.

10.75 The glued laminated column shown is free at its top A and fixed at its base B. Using wood that has an adjusted allowable stress for compression parallel to the grain $\sigma_C = 9.2$ MPa and a modulus of elasticity $E = 12$ GPa, determine the smallest cross section that can support a centric load of 62 kN.

10.76 A column of 2.1-m effective length is to be made by gluing together laminated wood pieces of 25×150-mm cross section. Knowing that the adjusted allowable stress for compression parallel to the grain is $\sigma_C = 7.7$ MPa and that $E = 10.5$ GPa, determine the number of wood pieces that must be used to support the centric load shown when (a) $P = 52$ kN, (b) $P = 108$ kN.

Fig. P10.76

10.77 A column of 5.6-m effective length must carry a centric load of 2750 kN. Knowing that $\sigma_Y = 250$ MPa and $E = 200$ GPa, use allowable stress design to select the wide-flange shape of 360-mm nominal depth that should be used.

10.78 A column of 4.6-m effective length must carry a centric load of 525 kN. Knowing that $\sigma_Y = 345$ MPa and $E = 200$ GPa, use allowable stress design to select the wide-flange shape of 200-mm nominal depth that should be used.

10.79 A column of 22.5-ft effective length must carry a centric load of 288 kips. Using allowable stress design, select the wide-flange shape of 14-in. nominal depth that should be used. Use $\sigma_Y = 50$ ksi and $E = 29 \times 10^6$ psi.

10.80 A column of 17-ft effective length must carry a centric load of 235 kips. Using allowable stress design, select the wide-flange shape of 10-in. nominal depth that should be used. Use $\sigma_Y = 36$ ksi and $E = 29 \times 10^6$ psi.

10.81 A centric load **P** must be supported by the steel bar AB. Using allowable stress design, determine the smallest dimension d of the cross section that can be used when (*a*) $P = 108$ kN, (*b*) $P = 166$ kN. Use $\sigma_Y = 250$ MPa and $E = 200$ GPa.

Fig. P10.81

10.82 Two $3\frac{1}{2} \times 2\frac{1}{2}$-in. angles are bolted together as shown for use as a column of 8-ft effective length to carry a centric load of 41 kips. Knowing that the angles available have thicknesses of $\frac{1}{4}$, $\frac{3}{8}$, and $\frac{1}{2}$ in., use allowable stress design to determine the lightest angles that can be used. Use $\sigma_Y = 36$ ksi and $E = 29 \times 10^6$ psi.

Fig. P10.82

10.83 Two $3\frac{1}{2} \times 2\frac{1}{2}$-in. angles are bolted together as shown for use as a column of 6-ft effective length to carry a centric load of 54 kips. Knowing that the angles available have thicknesses of $\frac{1}{4}$, $\frac{3}{8}$, and $\frac{1}{2}$ in., use allowable stress design to determine the lightest angles that can be used. Use $\sigma_Y = 36$ ksi and $E = 29 \times 10^6$ psi.

Fig. P10.83

10.84 A square structural tube having the cross section shown is used as a column of 3.1-m effective length to carry a centric load of 129 kN. Knowing that the tubes available for use are made with wall thicknesses of 3.2, 4.8, 6.4, and 7.9 mm, use allowable stress design to determine the lightest tube that can be used. Use $\sigma_Y = 250$ MPa and $E = 200$ GPa.

Fig. P10.84

Fig. P10.85

*10.85** A rectangular tube having the cross section shown is used as a column of 4.5-m effective length. Knowing that $\sigma_Y = 250$ MPa and $E = 200$ MPa, use load and resistance factor design to determine the largest centric live load that can be applied if the centric dead load is 140 kN. The dead load factor is $\gamma_D = 1.2$, the live load factor $\gamma_L = 1.6$, and the resistance factor $\phi = 0.85$.

*10.86** A column with a 19.5-ft effective length supports a centric load, with ratio of dead to live load equal to 1.35. The dead load factor is $\gamma_D = 1.2$, the live load factor $\gamma_L = 1.6$, and the resistance factor $\phi = 0.85$. Use load and resistance factor design to determine the allowable centric dead and live loads if the column is made of the following rolled-steel shape: (a) W10 × 39, (b) W14 × 68. Use $E = 29 \times 10^6$ psi and $\sigma_Y = 50$ ksi.

*10.87** The structural tube having the cross section shown is used as a column of 15-ft effective length to carry a centric dead load of 51 kips and a centric live load of 58 kips. Knowing that the tubes available for use are made with wall thicknesses in increments of $\frac{1}{16}$ in. from $\frac{3}{16}$ in. to $\frac{3}{8}$ in., use load and resistance factor design to determine the lightest tube that can be used. Use $\sigma_Y = 36$ ksi and $E = 29 \times 10^6$ psi. The dead load factor is $\gamma_D = 1.2$, the live load factor $\gamma_L = 1.6$, and the resistance factor $\phi = 0.85$.

Fig. P10.87

*10.88** A column of 5.5-m effective length must carry a centric dead load of 310 kN and a centric live load of 375 kN. Knowing that $\sigma_Y = 250$ MPa and $E = 200$ GPa, use load and resistance factor design to select the wide-flange shape of 310-mm nominal depth that should be used. The dead load factor is $\gamma_D = 1.2$, the live load factor $\gamma_L = 1.6$ and the resistance factor $\phi = 0.85$.

10.7. DESIGN OF COLUMNS UNDER AN ECCENTRIC LOAD

In this section, the design of columns subjected to an eccentric load will be considered. You will see how the empirical formulas developed in the preceding section for columns under a centric load can be modified and used when the load P applied to the column has an eccentricity e which is known.

We first recall from Sec. 4.12 that an eccentric axial load $\mathbf{P}$ applied in a plane of symmetry of the column can be replaced by an equivalent system consisting of a centric load $\mathbf{P}$ and a couple $\mathbf{M}$ of moment $M = Pe$, where e is the distance from the line of action of the load to the longitudinal axis of the column (Fig. 10.34). The normal stresses exerted on a transverse section of the column can then be obtained by superposing the stresses due, respectively, to the centric load $\mathbf{P}$ and to the couple $\mathbf{M}$ (Fig. 10.35), provided that the section considered is not too close to either end of the column, and as long as the stresses involved do not exceed the proportional limit of the material. The normal stresses due to the eccentric load $\mathbf{P}$ can thus be expressed as

$$\sigma = \sigma_{\text{centric}} + \sigma_{\text{bending}} \qquad (10.56)$$

Fig. 10.34

Fig. 10.35

Recalling the results obtained in Sec. 4.12, we find that the maximum compressive stress in the column is

$$\sigma_{\max} = \frac{P}{A} + \frac{Mc}{I} \qquad (10.57)$$

In a properly designed column, the maximum stress defined by Eq. (10.57) should not exceed the allowable stress for the column. Two alternative approaches can be used to satisfy this requirement, namely, the *allowable-stress method* and the *interaction method*.

a. Allowable-Stress Method. This method is based on the assumption that the allowable stress for an eccentrically loaded column is the same as if the column were centrically loaded. We must have, therefore, $\sigma_{\max} \leq \sigma_{\text{all}}$, where σ_{all} is the allowable stress under a centric load, or substituting for $\sigma_{\max}$ from Eq. (10.57)

$$\frac{P}{A} + \frac{Mc}{I} \leq \sigma_{\text{all}} \qquad (10.58)$$

The allowable stress is obtained from the formulas of Sec. 10.6 which, for a given material, express σ_{all} as a function of the slenderness ratio of the column. The major engineering codes require that the largest value of the slenderness ratio of the column be used to determine the allowable stress, whether or not this value corresponds to the actual plane of bending. This requirement sometimes results in an overly conservative design.

EXAMPLE 10.04

A column with a 2-in.-square cross section and 28-in. effective length is made of the aluminum alloy 2014-T6. Using the allowable-stress method, determine the maximum load P that can be safely supported with an eccentricity of 0.8 in.

We first compute the radius of gyration r using the given data

$$A = (2\text{ in.})^2 = 4\text{ in}^2 \qquad I = \tfrac{1}{12}(2\text{ in.})^4 = 1.333\text{ in}^4$$

$$r = \sqrt{\frac{I}{A}} = \sqrt{\frac{1.333\text{ in}^4}{4\text{ in}^2}} = 0.5773\text{ in.}$$

We next compute $L/r = (28\text{ in.})/(0.5773\text{ in.}) = 48.50$. Since $L/r < 55$, we use Eq. (10.48) to determine the allowable stress for the aluminum column subjected to a centric load. We have

$$\sigma_{\text{all}} = [30.7 - 0.23(48.50)] = 19.55\text{ ksi}$$

We now use Eq. (10.58) with $M = Pe$ and $c = \tfrac{1}{2}(2\text{ in.}) = 1$ in. to determine the allowable load:

$$\frac{P}{4\text{ in}^2} + \frac{P(0.8\text{ in.})(1\text{ in.})}{1.333\text{ in}^4} \le 19.55\text{ ksi}$$

$$P \le 22.3\text{ kips}$$

The maximum load that can be safely applied is $P = 22.3$ kips.

Fig. 10.36

b. Interaction Method. We recall that the allowable stress for a column subjected to a centric load (Fig. 10.36a) is generally smaller than the allowable stress for a column in pure bending (Fig. 10.36b), since the former takes into account the possibility of buckling. Therefore, when we use the allowable-stress method to design an eccentrically loaded column and write that the sum of the stresses due to the centric load **P** and the bending couple **M** (Fig. 10.36c) must not exceed the allowable stress for a centrically loaded column, the resulting design is generally overly conservative. An improved method of design can be developed by rewriting Eq. 10.58 in the form

$$\frac{P/A}{\sigma_{\text{all}}} + \frac{Mc/I}{\sigma_{\text{all}}} \le 1 \qquad (10.59)$$

and substituting for σ_{all} in the first and second terms the values of the allowable stress which correspond, respectively, to the centric loading of Fig. 10.36a and to the pure bending of Fig. 10.36b. We have

$$\frac{P/A}{(\sigma_{\text{all}})_{\text{centric}}} + \frac{Mc/I}{(\sigma_{\text{all}})_{\text{bending}}} \le 1 \qquad (10.60)$$

The type of formula obtained is known as an *interaction formula*.

We note that, when $M = 0$, the use of this formula results in the design of a centrically loaded column by the method of Sec. 10.6. On the other hand, when $P = 0$, the use of the formula results in the design of a beam in pure bending by the method of Chap. 4. When P and M are both different from zero, the interaction formula results in a design that takes into account the capacity of the member to resist bending as well as axial loading. In all cases, $(\sigma_{\text{all}})_{\text{centric}}$ will be determined by using the largest slenderness ratio of the column, regardless of the plane in which bending takes place.[†]

[†]This procedure is required by all major codes for the design of steel, aluminum, and timber compression members. In addition, many specifications call for the use of an additional factor in the second term of Eq. (10.60); this factor takes into account the additional stresses resulting from the deflection of the column due to bending.

When the eccentric load **P** is not applied in a plane of symmetry of the column, it causes bending about both of the principal axes of the cross section. We recall from Sec. 4.14 that the load **P** can then be replaced by a centric load **P** and two couples represented by the couple

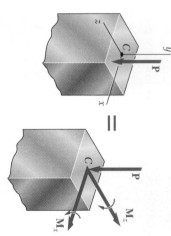

vectors $\mathbf{M}_x$ and $\mathbf{M}_z$ shown in Fig. 10.37. The interaction formula to be used in this case is

Fig. 10.37

$$\frac{P/A}{(\sigma_{all})_{centric}} + \frac{|M_x|z_{max}/I_x}{(\sigma_{all})_{bending}} + \frac{|M_z|x_{max}/I_z}{(\sigma_{all})_{bending}} \leq 1 \qquad (10.61)$$

EXAMPLE 10.05

Use the interaction method to determine the maximum load P that can be safely supported by the column of Example 10.04 with an eccentricity of 0.8 in. The allowable stress in bending is 24 ksi.

The value of $(\sigma_{all})_{centric}$ has already been determined in Example 10.04. We have

$$(\sigma_{all})_{centric} = 19.55 \text{ ksi} \qquad (\sigma_{all})_{bending} = 24 \text{ ksi}$$

Substituting these values into Eq. (10.60), we write

$$\frac{P/A}{19.55 \text{ ksi}} + \frac{Mc/I}{24 \text{ ksi}} \leq 1.0$$

Using the numerical data from Example 11.04, we write

$$\frac{P/4}{19.55 \text{ ksi}} + \frac{P(0.8)(1.0)/1.333}{24 \text{ ksi}} \leq 1.0$$

$$P \leq 26.5 \text{ kips}$$

The maximum load that can be safely applied is thus $P = 26.5$ kips.

SAMPLE PROBLEM 10.5

Using the allowable-stress method, determine the largest load $\mathbf{P}$ that can be safely carried by a W310 × 74 steel column of 4.5-m effective length. Use $E = 200$ GPa and $\sigma_Y = 250$ MPa.

W310 × 74

$A = 9480$ mm^2
$r_x = 132$ mm
$r_y = 49.7$ mm
$S_x = 1060 \times 10^3$ mm^3

SOLUTION

The largest slenderness ratio of the column is $L/r_y = (4.5\text{ m})/(0.0497\text{ m})$ $= 90.5$. Using Eq. (10.42) with $E = 200$ GPa and $\sigma_Y = 250$ MPa, we find that $C_c = 125.7$. Since $L/r < C_c$, we use Eqs. (10.44) and (10.45) and find

$$\sigma_{all} = (\nu_{all})_{centric} = 97.9 \text{ MPa}$$

For the given column and loading, we have

$$\frac{P}{A} = \frac{P}{9.48 \times 10^{-3}\text{ m}^2} \qquad \frac{Mc}{I} = \frac{M}{S} = \frac{P(0.200\text{ m})}{1.060 \times 10^{-3}\text{ m}^3}$$

Substituting into Eq. (10.58), we write

$$\frac{P}{A} + \frac{Mc}{I} < \sigma_{all}$$

$$\frac{P}{9.48 \times 10^{-3}\text{ m}^2} + \frac{P(0.200\text{ m})}{1.060 \times 10^{-3}\text{ m}^3} \leq 97.9 \text{ MPa} \qquad P \leq 333 \text{ kN}$$

The largest allowable load $\mathbf{P}$ is thus

$$\mathbf{P} = 333 \text{ kN} \downarrow \quad \blacktriangledown$$

SAMPLE PROBLEM 10.6

Using the interaction method, solve Sample Prob. 10.5. Assume $(\sigma_{all})_{bending}$ $= 150$ MPa.

SOLUTION

Using Eq. (10.60), we write

$$\frac{P/A}{(\sigma_{all})_{centric}} + \frac{Mc/I}{(\sigma_{all})_{bending}} \leq 1$$

Substituting the given allowable bending stress and the allowable centric stress found in Sample Prob. 10.5, as well as the other given data, we have

$$\frac{P/(9.48 \times 10^{-3}\text{ m}^2)}{97.9 \times 10^6 \text{ Pa}} + \frac{P(0.200\text{ m})/(1.060 \times 10^{-3}\text{ m}^3)}{150 \times 10^6 \text{ Pa}} \leq 1$$

$$P \leq 428 \text{ kN}$$

The largest allowable load $\mathbf{P}$ is thus

$$\mathbf{P} = 428 \text{ kN} \downarrow \quad \blacktriangledown$$

SAMPLE PROBLEM 10.7

A steel column having an effective length of 16 ft is loaded eccentrically as shown. Using the interaction method, select the wide-flange shape of 8-in. nominal depth that should be used. Assume $E = 29 \times 10^6$ psi and $\sigma_Y = 36$ ksi, and use an allowable stress in bending of 22 ksi.

P = 85 kips

5 in.

C

SOLUTION

So that we can select a trial section, we use the allowable-stress method with $\sigma_{all} = 22$ ksi and write

$$\sigma_{all} = \frac{P}{A} + \frac{Mc}{I_x} = \frac{P}{A} + \frac{Mc}{Ar_x^2}$$

From Appendix C we observe for shapes of 8-in. nominal depth that $c \approx 4$ in. and $r_x \approx 3.5$ in. Substituting into Eq. (1), we have

$$22 \text{ ksi} = \frac{85 \text{ kips}}{A} + \frac{(425 \text{ kip} \cdot \text{in.})(4 \text{ in.})}{A(3.5 \text{ in.})^2} \qquad A \approx 10.2 \text{ in}^2$$

We select for a first trial shape: W8 × 35.

Trial 1: W8 × 35. The allowable stresses are

Allowable Bending Stress: (see data) $(\sigma_{all})_{bending} = 22$ ksi

Allowable Centric Stress: The largest slenderness ratio is $L/r_y = (192 \text{ in.})/(2.03 \text{ in.}) = 94.6$. For $E = 29 \times 10^6$ psi and $\sigma_Y = 36$ ksi, Eq. (10.42) yields $C_c = 126.1$ and we note that $L/r < C_c$. Using Eqs. (10.44) and (10.45) we find

$$(\sigma_{all})_{centric} = 13.65 \text{ ksi}$$

For the W8 × 35 trial shape, we have

$$\frac{P}{A} = \frac{85 \text{ kips}}{10.3 \text{ in}^2} = 8.25 \text{ ksi} \qquad \frac{Mc}{I} = \frac{M}{S_x} = \frac{425 \text{ kip} \cdot \text{in.}}{31.2 \text{ in}^3} = 13.62 \text{ ksi}$$

With this data we find that the left-hand member of Eq. (10.60) is

$$\frac{P/A}{(\sigma_{all})_{centric}} + \frac{Mc/I}{(\sigma_{all})_{bending}} = \frac{8.25 \text{ ksi}}{13.65 \text{ ksi}} + \frac{13.62 \text{ ksi}}{22 \text{ ksi}} = 1.223$$

Since $1.223 > 1.000$, the requirement expressed by the interaction formula is not satisfied; we must select a larger trial shape.

Trial 2: W8 × 48. Following the procedure used in trial 1, we write

$$\frac{L}{r_y} = \frac{192 \text{ in.}}{2.08 \text{ in.}} = 92.3 \qquad (\sigma_{all})_{centric} = 13.93 \text{ ksi}$$

$$\frac{P}{A} = \frac{85 \text{ kips}}{14.1 \text{ in}^2} = 6.03 \text{ ksi} \qquad \frac{Mc}{I} = \frac{M}{S_x} = \frac{425 \text{ kip} \cdot \text{in.}}{43.3 \text{ in}^3} = 9.82 \text{ ksi}$$

Substituting into Eq. (10.60) gives

$$\frac{P/A}{(\sigma_{all})_{centric}} + \frac{Mc/I}{(\sigma_{all})_{bending}} = \frac{6.03 \text{ ksi}}{13.93 \text{ ksi}} + \frac{9.82 \text{ ksi}}{22 \text{ ksi}} = 0.879 < 1.000$$

The W8 × 48 shape is satisfactory but may be unnecessarily large.

Trial 3: W8 × 40. Following again the same procedure, we find that the interaction formula is not satisfied.

Selection of Shape. The shape to be used is **W8 × 48** ◀

P = 85 kips

5 in.

M = (85 kips)(5 in.)
  = 425 kip · in.

P = 85 kips

W8 × 35

A = 10.3 in²
r_x = 3.51 in.
r_y = 2.03 in.
S_x = 31.2 in³
L = 16 ft = 192 in.

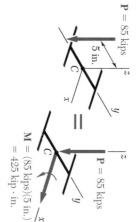

W8 × 48

A = 14.1 in²
r_x = 3.61 in.
r_y = 2.08 in.
S_x = 43.3 in³
L = 16 ft = 192 in.

W8 × 40

A = 11.7 in²
r_x = 3.53 in.
r_y = 2.04 in.
S_x = 35.5 in³
L = 16 ft = 192 in.

PROBLEMS

10.89 A sawn lumber column with a 125-mm-square cross section and a 3.6-m effective length is made of a grade of wood that has an adjusted allowable stress $\sigma_C = 9.2$ MPa for compression parallel to the grain and a modulus of elasticity $E = 12$ GPa. Using the allowable-stress method, determine the maximum load **P** that can be safely supported with an eccentricity of 50 mm.

10.90 Solve Prob. 10.89 using the interaction method and an allowable stress in bending of 12.8 MPa.

10.91 An eccentric load is applied at a point 1 in. from the geometric axis of a 2.2-in.-diameter rod made of a steel for which $\sigma_Y = 36$ ksi and $E = 29 \times 10^6$ psi. Using the allowable-stress method, determine the allowable load **P**.

10.92 Solve Prob. 10.91, assuming that the load is applied at a point 1.6 in. from the geometric axis and that the effective length is 33 in.

Fig. P10.91

10.93 A column of 5.5 m effective length is made of the aluminum alloy 2014-T6, for which the allowable stress in bending is 220 MPa. Using the interaction method, determine the allowable load **P**, knowing that the eccentricity is (*a*) $e = 0$, (*b*) $e = 40$ mm.

10.94 Solve Prob. 10.93, assuming that the effective length of the column is 3.0 m.

Fig. P10.93

10.95 An eccentric load $P = 10.8$ kips is applied at a point 0.8 in. from the geometric axis of a 2-in.-diameter rod made of the aluminum alloy 6061-T6. Using the interaction method and an allowable stress in bending of 21 ksi, determine the largest allowable effective length L that can be used.

Fig. P10.95

10.96 Solve Prob. 10.95, assuming that the aluminum alloy used is 2014-T6 and that the allowable stress in bending is 26 ksi.

10.97 A rectangular column is made of sawn lumber that has an adjusted allowable stress for compression parallel to the grain $\sigma_C = 8.3$ MPa and a modulus of elasticity $E = 11.1$ GPa. Using the allowable-stress method, determine the largest allowable effective length L that can be used.

Fig. P10.97

10.98 Solve Prob. 10.97, assuming that $P = 85$ kN.

10.99 A column of 14-ft effective length consists of a section of steel tubing having the cross section shown. Using the allowable-stress method, determine the maximum allowable eccentricity e if (*a*) $P = 55$ kips, (*b*) $P = 35$ kips. Use $\sigma_Y = 36$ ksi and $E = 29 \times 10^6$ psi.

Fig. P10.99

10.100 Solve Prob. 10.99, assuming that the effective length of the column is increased to 18 ft and that (*a*) $P = 28$ kips, (*b*) $P = 18$ kips.

10.101 The compression member AB is made of a steel for which $\sigma_Y = 250$ MPa and $E = 200$ GPa. It is free at its top A and fixed at its base B. Using the allowable-stress method, determine the largest allowable eccentricity e_x knowing that (*a*) $e_y = 0$, (*b*) $e_y = 8$ mm.

10.102 The compression member AB is made of a steel for which $\sigma_Y = 250$ MPa and $E = 200$ GPa. It is free at its top A and fixed at its base B. Using the interaction method with an allowable bending stress equal to 120 MPa and knowing that the eccentricities e_x and e_y are equal, determine their largest allowable common value.

Fig. P10.101 and P10.102

10.103 A sawn lumber column of rectangular cross section has a 7.2-ft effective length and supports a 9.2-kip load as shown. The sizes available for use have *b* equal to 3.5, 5.5, 7.5, and 9.5 in. The grade of wood has an adjusted allowable stress for compression parallel to the grain $\sigma_C = 1180$ psi and $E = 1.2 \times 10^6$ psi. Using the allowable-stress method, determine the lightest section that can be used.

9.2 kips

$e = 1.6$ in.

C

D

b

7.5 in.

Fig. P10.103

10.104 Solve Prob. 10.103, assuming that $e = 3.2$ in.

10.105 A 32-kN vertical load **P** is applied at the midpoint of one edge of the square cross section of the aluminum compression member *AB* that is free at its top *A* and fixed at its base *B*. Knowing that the alloy used is 6061-T6 and using the allowable-stress method, determine the smallest allowable dimension *d*.

P = 32 kN

A

D

d

d

1.2 m

B

Fig. P10.105

10.106 Solve Prob. 10.105, assuming that the vertical load **P** is applied at a corner of the square cross section of the compression member *AB*.

10.107 A compression member made of steel has a 720-mm effective length and must support the 198-kN load **P** as shown. For the material used $\sigma_Y = 250$ MPa and $E = 200$ GPa. Using the interaction method with an allowable bending stress equal to 150 MPa, determine the smallest dimension *d* of the cross section that can be used.

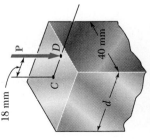

P

18 mm

D

C

40 mm

d

Fig. P10.107

10.108 Solve Prob. 10.107, assuming that the effective length is 1.62 m and that the magnitude *P* of the eccentric load is 128 kN.

10.109   The eccentric load **P** has a magnitude of 85 kN and is applied at a point located at a distance $e = 30$ mm from the geometric axis of a rod made of the aluminum alloy 6016-T6. Using the interaction method with a 140-MPa allowable stress in bending, determine the smallest diameter d that can be used.

Fig. P10.111

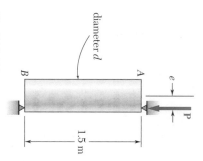

Fig. P10.109

10.110   Solve Prob. 10.109, using the allowable-stress method and assuming that the aluminum alloy used is 2014-T6.

10.111   A steel compression member of 5.8-m effective length is to support a 296-kN eccentric load **P**. Using the interaction method, select the wide-flange shape of 200-mm nominal depth that should be used. Use $E = 200$ GPa, $\sigma_Y = 250$ MPa, and $\sigma_{all} = 150$ MPa in bending.

10.112   A steel column of 7.2-m effective length is to support an 83-kN eccentric load **P** at a point D located on the x axis as shown. Using the allowable-stress method, select the wide-flange shape of 250-mm nominal depth that should be used. Use $E = 200$ GPa and $\sigma_Y = 250$ MPa.

Fig. P10.112

10.113 A steel column of 21-ft effective length must carry a load of 82 kips with an eccentricity of 2.1 in. as shown. Using the interaction method, select the wide-flange shape of 12-in. nominal depth that should be used. Use $E = 29 \times 10^6$ psi, $\sigma_Y = 36$ ksi, and $\sigma_{all} = 22$ ksi in bending.

Fig. P10.113

10.114 A 43-kip axial load P is applied to the rolled-steel column BC at a point on the x axis at a distance $e = 2.5$ in. from the geometric axis of the column. Using the allowable-stress method, select the wide-flange shape of 8-in. nominal depth that should be used. Use $F = 29 \times 10^6$ psi and $\sigma_Y = 36$ ksi.

Fig. P10.114

Fig. P10.115

10.115 A steel tube of 80-mm outer diameter is to carry a 93-kN load P with an eccentricity of 20 mm. The tubes available for use are made with wall thicknesses in increments of 3 mm from 6 mm to 15 mm. Using the allowable-stress method, determine the lightest tube that can be used. Assume $E = 200$ GPa and $\sigma_Y = 250$ MPa.

10.116 Solve Prob. 10.115, using the interaction method with $P = 165$ kN, $e = 15$ mm, and an allowable stress in bending of 150 MPa.

Critical load

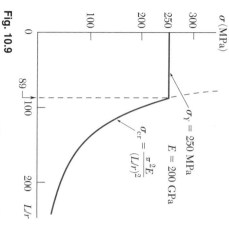

(a)

$[x = 0, y = 0]$

$[x = L, y = 0]$

(b)

Fig. 10.8

Euler's formula

σ (MPa)

300

250

200

100

0

$\sigma_Y = 250$ MPa

$E = 200$ GPa

$\sigma_{cr} = \dfrac{\pi^2 E}{(L/r)^2}$

89

100

200

L/r

Fig. 10.9

Slenderness ratio

This chapter was devoted to the design and analysis of columns, i.e., prismatic members supporting axial loads. In order to gain insight into the behavior of columns, we first considered in Sec. 10.2 the equilibrium of a simple model and found that for values of the load P exceeding a certain value P_{cr}, called the *critical load*, two equilibrium positions of the model were possible: the original position with zero transverse deflections and a second position involving deflections that could be quite large. This led us to conclude that the first equilibrium position was unstable for $P > P_{cr}$ and stable for $P < P_{cr}$, since in the latter case it was the only possible equilibrium position.

In Sec. 10.3, we considered a pin-ended column of length L and of constant flexural rigidity EI subjected to an axial centric load P. Assuming that the column had buckled (Fig. 10.8), we noted that the bending moment at point Q was equal to $-Py$ and wrote

$$\frac{d^2y}{dx^2} = \frac{M}{EI} = -\frac{P}{EI}y \qquad (10.4)$$

Solving this differential equation, subject to the boundary conditions corresponding to a pin-ended column, we determined the smallest load P for which buckling can take place. This load, known as the *critical load* and denoted by P_{cr}, is given by *Euler's formula*:

$$P_{cr} = \frac{\pi^2 EI}{L^2} \qquad (10.11)$$

where L is the length of the column. For this load or any larger load, the equilibrium of the column is unstable and transverse deflections will occur.

Denoting the cross-sectional area of the column by A and its radius of gyration by r, we determined the critical stress σ_{cr} corresponding to the critical load P_{cr}:

$$\sigma_{cr} = \frac{\pi^2 E}{(L/r)^2} \qquad (10.13)$$

The quantity L/r is called the *slenderness ratio* and we plotted σ_{cr} as a function of L/r (Fig. 10.9). Since our analysis was based on stresses remaining below the yield strength of the material, we noted that the column would fail by yielding when $\sigma_{cr} > \sigma_Y$.

In Sec. 10.4, we discussed the critical load of columns with various end conditions and wrote

$$P_{cr} = \frac{\pi^2 EI}{L_e^2} \qquad (10.11')$$

where L_e is the *effective length* of the column, i.e., the length of an equivalent pin-ended column. The effective lengths of several columns with various end conditions were calculated and shown in Fig. 10.18 on page 617.

Effective length

In Sec. 10.5, we considered columns supporting an *eccentric axial load*. For a pin-ended column subjected to a load P applied with an eccentricity e, we replaced the load by a centric axial load and a couple of moment $M_A = Pe$ (Figs. 10.19a and 10.20), and derived the following expression for the maximum transverse deflection:

$$y_{max} = e\left[\sec\left(\sqrt{\frac{P}{EI}}\frac{L}{2}\right) - 1\right] \qquad (10.28)$$

Eccentric axial load. Secant formula.

Fig. 10.19a

Fig. 10.20

We then determined the maximum stress in the column, and from the expression obtained for that stress, we derived the *secant formula*:

$$\frac{P}{A} = \frac{\sigma_{max}}{1 + \dfrac{ec}{r^2}\sec\left(\dfrac{1}{2}\sqrt{\dfrac{P}{EA}}\dfrac{L_e}{r}\right)} \qquad (10.36)$$

This equation can be solved for the force per unit area, P/A, that causes a specified maximum stress σ_{max} in a pin-ended column or any other column of effective slenderness ratio L_e/r.

In the first part of the chapter we considered each column as a straight homogeneous prism. Since imperfections exist in all real columns, the *design of real columns* is done by using empirical formulas based on laboratory tests and set forth in specifications codes issued by professional organizations. In Sec. 10.6, we discussed the design of *centrically loaded columns* made of steel, aluminum, or wood. For each material, the design of the column was based on formulas expressing the allowable stress as a function of the slenderness ratio L/r of the column. For structural steel, we also discussed the alternative method of *Load and Resistance Factor Design*.

Design of real columns

Centrically loaded columns

In the last section of the chapter [Sec. 10.7], we studied two methods used for the design of columns under an *eccentric* load. The first method was the *allowable-stress method*, a conservative method in which it is assumed that the allowable stress is the same as if the column were centrically loaded. The allowable-stress method requires that the following inequality be satisfied:

$$\frac{P}{A} + \frac{Mc}{I} \le \sigma_{all} \qquad (10.58)$$

Eccentrically loaded columns

Allowable-stress method

The second method was the *interaction method*, a method used in most modern specifications. In this method the allowable stress for a centrically loaded column is used for the portion of the total stress due to the axial load and the allowable stress in bending for the stress due to bending. Thus, the inequality to be satisfied is

$$\frac{P/A}{(\sigma_{all})_{centric}} + \frac{Mc/I}{(\sigma_{all})_{bending}} \le 1 \qquad (10.60)$$

Interaction method

10.117 A column of 3.5-m effective length is made by welding together two 89 × 64 × 6.4-mm angles as shown. Using $E = 200$ GPa, determine the allowable centric load if a factor of safety of 2.8 is required.

6.4 mm

89 mm

64 mm

Fig. P10.117

10.118 Member AB consists of a single C130 × 10.4 steel channel of length 2.5 m. Knowing that the pins at A and B pass through the centroid of the cross section of the channel, determine the factor of safety for the load shown with respect to buckling in the plane of the figure when $\theta = 30°$. Use Euler's formula with $E = 200$ GPa.

2.5 m

6.8 kN

θ

Fig. P10.118

10.119 Supports A and B of the pin-ended column shown are at a fixed distance L from each other. Knowing that at a temperature T_0 the force in the column is zero and that buckling occurs when the temperature is $T_1 = T_0 + \Delta T$, express ΔT in terms of b, L, and the coefficient of thermal expansion α.

L

b

B

A

Fig. P10.119

10.120 Knowing that a factor of safety of 2.6 is required, determine the largest load **P** that can be applied to the structure shown. Use $E = 200$ GPa and consider only buckling in the plane of the structure.

15-mm diameter

20-mm diameter

A

B

C

P

0.5 m

1 m

0.5 m

Fig. P10.120

10.121 The steel rod *BC* is attached to the rigid bar *AB* and to the fixed support at *C*. Knowing that $G = 11.2 \times 10^6$ psi, determine the diameter of rod *BC* for which the critical load P_{cr} of the system is 80 lb.

Fig. P10.121

10.122 An axial load **P** of magnitude 560 kN is applied at a point on the *x* axis at a distance $e = 8$ mm from the geometric axis of the W200 × 46.1 rolled-steel column *BC*. Using $E = 200$ GPa, determine (*a*) the horizontal deflection of end *C*, (*b*) the maximum stress in the column.

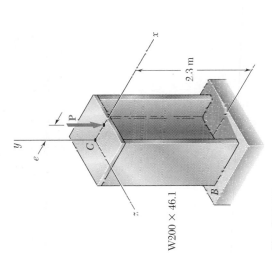

Fig. P10.122

10.123 A column with the cross section shown has a 13.5-ft effective length. Knowing that $\sigma_Y = 36$ ksi and $E = 29 \times 10^6$ psi, use the AISC allowable stress design formulas to determine the largest centric load that can be applied to the column.

Fig. P10.123

10.124 A column of 4.5-m effective length must carry a centric load of 900 kN. Knowing that $\sigma_Y = 345$ MPa and $E = 200$ GPa and using the AISC allowable stress design formulas, select the wide-flange shape of 250-mm nominal depth that should be used.

10.125 Bar AB is free at end A and fixed at base B. Determine the allowable centric load **P** if the aluminum alloy used is (*a*) 6061-T6, (*b*) 2014-T6.

Fig. P10.125

10.126 A sawn lumber column of 5.0×7.5-in. cross section has an effective length of 8.5 ft. The grade of wood used has an adjusted allowable stress for compression parallel to the grain $\sigma_C = 1180$ psi and a modulus of elasticity $E = 1.2 \times 10^6$ psi. Using the allowable-stress method, determine the largest eccentric load **P** that can be applied when (*a*) $e = 0.5$ in., (*b*) $e = 1.0$ in.

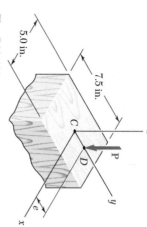

Fig. P10.126

10.127 Two $4 \times 3 \times \frac{3}{8}$-in. steel angles are welded together to form the column AB. An axial load **P** of magnitude 14 kips is applied at point D. Using the allowable-stress method, determine the largest allowable length L. Assume $E = 29 \times 10^6$ psi and $\sigma_Y = 36$ ksi.

Fig. 10.127

10.128 A compression member of rectangular cross section has an effective length of 36 in. and is made of the aluminum alloy 2014-T6 for which the allowable stress in bending is 24 ksi. Using the interaction method, determine the smallest dimension d of the cross section that can be used when $e = 0.4$ in.

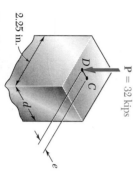

Fig. P10.128

COMPUTER PROBLEMS

The following problems are designed to be solved with a computer.

10.C1 A solid steel rod having an effective length of 500 mm is to be used as a compression strut to carry a centric load **P**. For the grade of steel used $E = 200$ GPa and $\sigma_Y = 245$ MPa. Knowing that a factor of safety of 2.8 is required and using Euler's formula, write a computer program and use it to calculate the allowable centric load P_{all} for values of the radius of the rod from 6 mm to 24 mm, using 2-mm increments.

10.C2 An aluminum bar is fixed at end A and supported at end B so that it is free to rotate about a horizontal axis through the pin. Rotation about a vertical axis at end B is prevented by the brackets. Knowing that $E = 10.1 \times 10^6$ psi, use Euler's formula with a factor of safety of 2.5 to determine the allowable centric load **P** for values of b from 0.75 in. to 1.5 in., using 0.125-in. increments.

Fig. P10.C2

10.C3 The pin-ended members AB and BC consist of sections of aluminum pipe of 120-mm outer diameter and 10-mm wall thickness. Knowing that a factor of safety of 3.5 is required, determine the mass m of the largest block that can be supported by the cable arrangement shown for values of h from 4 m to 8 m, using 0.25-m increments. Use $E = 70$ GPa and consider only buckling in the plane of the structure.

Fig. P10.C3

10.C4 An axial load **P** is applied at a point located on the x axis at a distance $e = 0.5$ in. from the geometric axis of the W8 × 40 rolled-steel column AB. Using $E = 29 \times 10^6$ psi, write a computer program and use it to calculate for values of P from 25 to 75 kips, using 5-kip increments, (a) the horizontal deflection at the midpoint C, (b) the maximum stress in the column.

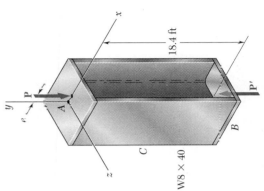

Fig. P10.C4

10.C5 A column of effective length L is made from a rolled-steel shape and carries a centric axial load **P**. The yield strength for the grade of steel used is denoted by σ_Y, the modulus of elasticity by E, the cross-sectional area of the selected shape by A, and its smallest radius of gyration by r. Using the AISC design formulas for allowable stress design, write a computer program that can be used with either SI or U.S. customary units to determine the allowable load **P**. Use this program to solve (a) Prob. 10.57, (b) Prob. 10.58, (c) Prob. 10.60.

10.C6 A column of effective length L is made from a rolled-steel shape and is loaded eccentrically as shown. The yield strength of the grade of steel used is denoted by σ_Y, the allowable stress in bending by σ_{all}, the modulus of elasticity by E, the cross-sectional area of the selected shape by A, and its smallest radius of gyration by r. Write a computer program that can be used with either SI or U.S. customary units to determine the allowable load **P**, using either the allowable-stress method or the interaction method. Use this program to check the given answer for (a) Prob. 10.111, (b) Prob. 10.112, (c) Prob. 10.113.

Fig. P10.C6

11

Energy Methods

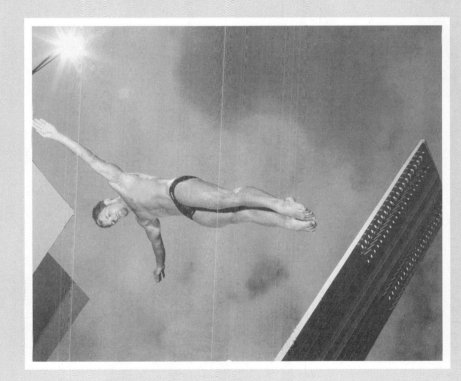

As the diver comes down on the diving board the potential energy due to his elevation above the board will be converted into strain energy due to the bending of the board. The normal and shearing stresses resulting from energy loadings will be determined in this chapter.

11.1. INTRODUCTION

In the previous chapter we were concerned with the relations existing between forces and deformations under various loading conditions. Our analysis was based on two fundamental concepts, the concept of *stress* (Chap. 1) and the concept of *strain* (Chap. 2). A third important concept, the concept of *strain energy*, will now be introduced.

In Sec. 11.2, the *strain energy* of a member will be defined as the increase in energy associated with the deformation of the member. You will see that the strain energy is equal to the work done by a slowly increasing load applied to the member. The *strain-energy density* of a material will be defined as the strain energy per unit volume; it will be seen that it is equal to the area under the stress-strain diagram of the material (Sec. 11.3). From the stress-strain diagram of a material two additional properties will be defined, namely, the *modulus of toughness* and the *modulus of resilience* of the material.

In Sec. 11.4 the elastic strain energy associated with *normal stresses* will be discussed, first in members under axial loading and then in members in bending. Later you will consider the elastic strain energy associated with *shearing stresses* such as occur in torsional loadings of shafts and in transverse loadings of beams (Sec. 11.5). Strain energy for a *general state of stress* will be considered in Sec. 11.6, where the *maximum-distortion-energy criterion* for yielding will be derived.

The effect of *impact loading* on members will be considered in Sec. 11.7. You will learn to calculate both the *maximum stress* and the *maximum deflection* caused by a moving mass impacting on a member. Properties that increase the ability of a structure to withstand impact loads effectively will be discussed in Sec. 11.8.

In Sec. 11.9 the elastic strain of a member subjected to a *single concentrated load* will be calculated, and in Sec. 11.10 the deflection at the point of application of a single load will be determined.

The last portion of the chapter will be devoted to the determination of the strain energy of structures subjected to *several loads* (Sec. 11.11). *Castigliano's theorem* will be derived in Sec. 11.12 and used in Sec. 11.13 to determine the deflection at a given point of a structure subjected to several loads. In the last section Castigliano's theorem will be applied to the analysis of indeterminate structures (Sec. 11.14).

11.2. STRAIN ENERGY

Consider a rod BC of length L and uniform cross-sectional area A, which is attached at B to a fixed support, and subjected at C to a *slowly increasing* axial load $\mathbf{P}$ (Fig. 11.1). As we noted in Sec. 2.2, by plotting the magnitude P of the load against the deformation x of the rod, we obtain a certain load-deformation diagram (Fig. 11.2) that is characteristic of the rod BC.

Fig. 11.1

Fig. 11.2

Let us now consider the work dU done by the load **P** as the rod elongates by a small amount dx. This *elementary work* is equal to the product of the magnitude P of the load and of the small elongation dx. We write

$$dU = P\,dx \tag{11.1}$$

and note that the expression obtained is equal to the element of area of width dx located under the load-deformation diagram (Fig. 11.3). The *total work* U done by the load as the rod undergoes a deformation x_1 is thus

$$U = \int_0^{x_1} P\,dx$$

and is equal to the area under the load-deformation diagram between $x = 0$ and $x = x_1$.

The work done by the load **P** as it is slowly applied to the rod must result in the increase of some energy associated with the deformation of the rod. This energy is referred to as the *strain energy* of the rod. We have, by definition,

$$\text{Strain energy} = U = \int_0^{x_1} P\,dx \tag{11.2}$$

We recall that work and energy should be expressed in units obtained by multiplying units of length by units of force. Thus, if SI metric units are used, work and energy are expressed in N · m; this unit is called a *joule* (J). If U.S. customary units are used, work and energy are expressed in ft · lb or in in · lb.

In the case of a linear and elastic deformation, the portion of the load-deformation diagram involved can be represented by a straight line of equation $P = kx$ (Fig. 11.4). Substituting for P in Eq. (11.2), we have

$$U = \int_0^{x_1} kx\,dx = \tfrac{1}{2}kx_1^2$$

or

$$U = \tfrac{1}{2}P_1 x_1 \tag{11.3}$$

where P_1 is the value of the load corresponding to the deformation x_1.

Fig. 11.3

Fig. 11.4

(a) $U = 0$ $\sigma = 0$ $T = \frac{1}{2} m v_0^2$

(b) $U = U_m$ $\sigma = \sigma_m$ $v = 0$ $T = 0$

Fig. 11.5

The concept of strain energy is particularly useful in the determination of the effects of impact loadings on structures or machine components. Consider, for example, a body of mass m moving with a velocity v_0 which strikes the end B of a rod AB (Fig. 11.5a). Neglecting the inertia of the elements of the rod, and assuming no dissipation of energy during the impact, we find that the maximum strain energy U_m acquired by the rod (Fig. 11.5b) is equal to the original kinetic energy $T = \frac{1}{2} m v_0^2$ of the moving body. We then determine the value P_m of the static load which would have produced the same strain energy in the rod, and obtain the value σ_m of the largest stress occurring in the rod by dividing P_m by the cross-sectional area of the rod.

11.3. STRAIN-ENERGY DENSITY

As we noted in Sec. 2.2, the load-deformation diagram for a rod BC depends upon the length L and the cross-sectional area A of the rod. The strain energy U defined by Eq. (11.2), therefore, will also depend upon the dimensions of the rod. In order to eliminate the effect of size from our discussion and direct our attention to the properties of the material, the strain energy per unit volume will be considered. Dividing the strain energy U by the volume $V = AL$ of the rod (Fig. 11.1), and using Eq. (11.2), we have

$$\frac{U}{V} = \int_0^{x_1} \frac{P}{A} \frac{dx}{L}$$

Recalling that P/A represents the normal stress σ_x in the rod, and x/L the normal strain ϵ_x, we write

$$\frac{U}{V} = \int_0^{\epsilon_1} \sigma_x \, d\epsilon_x$$

where ϵ_1 denotes the value of the strain corresponding to the elongation x_1. The strain energy per unit volume, U/V, is referred to as the *strain-energy density* and will be denoted by the letter u. We have, therefore,

$$\text{Strain-energy density} = u = \int_0^{\epsilon_1} \sigma_x \, d\epsilon_x \qquad (11.4)$$

The strain-energy density u is expressed in units obtained by dividing units of energy by units of volume. Thus, if SI metric units are used, the strain-energy density is expressed in J/m^3 or its multiples kJ/m^3 and MJ/m^3; if U.S. customary units are used, it is expressed in in $\cdot$ lb/in^3.†

† We note that $1 \, J/m^3$ and $1 \, Pa$ are both equal to $1 \, N/m^2$, while $1 \, \text{in} \cdot \text{lb/in}^3$ and $1 \, \text{psi}$ are both equal to $1 \, \text{lb/in}^2$. Thus, strain-energy density and stress are dimensionally equal and could be expressed in the same units.

Referring to Fig. 11.6, we note that the strain-energy density u is equal to the area under the stress-strain curve, measured from $\epsilon_x = 0$ to $\epsilon_x = \epsilon_1$. If the material is unloaded, the stress returns to zero, but there is a permanent deformation represented by the strain ϵ_p, and only the portion of the strain energy per unit volume corresponding to the triangular area is recovered. The remainder of the energy spent in deforming the material is dissipated in the form of heat.

Fig. 11.6

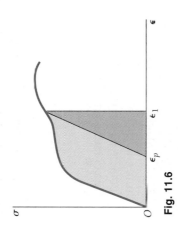

Fig. 11.7

The value of the strain-energy density obtained by setting $\epsilon_1 = \epsilon_R$ in Eq. (11.4), where ϵ_R is the strain at rupture, is known as the *modulus of toughness* of the material. It is equal to the area under the entire stress-strain diagram (Fig. 11.7) and represents the energy per unit volume required to cause the material to rupture. It is clear that the toughness of a material is related to its ductility as well as to its ultimate strength (Sec. 2.3), and that the capacity of a structure to withstand an impact load depends upon the toughness of the material used (Fig. 11.8).

If the stress σ_x remains within the proportional limit of the material, Hooke's law applies and we write

$$\sigma_x = E\epsilon_x \qquad (11.5)$$

Substituting for σ_x from (11.5) into (11.4), we have

$$u = \int_0^{\epsilon_1} E\epsilon_x \, d\epsilon_x = \frac{E\epsilon_1^2}{2} \qquad (11.6)$$

or, using Eq. (11.5) to express ϵ_1 in terms of the corresponding stress σ_1,

$$u = \frac{\sigma_1^2}{2E} \qquad (11.7)$$

The value u_Y of the strain-energy density obtained by setting $\sigma_1 = \sigma_Y$ in Eq. (11.7), where σ_Y is the yield strength, is called the *modulus of resilience* of the material. We have

$$u_Y = \frac{\sigma_Y^2}{2E} \qquad (11.8)$$

Fig. 11.8 The railroad coupler is made of a ductile steel which has a large modulus of toughness.

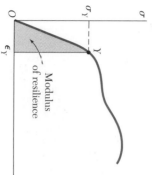

Fig. 11.9

The modulus of resilience is equal to the area under the straight-line portion OY of the stress-strain diagram (Fig. 11.9) and represents the energy per unit volume that the material can absorb without yielding. The capacity of a structure to withstand an impact load without being permanently deformed clearly depends upon the resilience of the material used.

Since the modulus of toughness and the modulus of resilience represent characteristic values of the strain-energy density of the material considered, they are both expressed in J/m³ or its multiples if SI units are used, and in in · lb/in³ if U.S. customary units are used.†

11.4. ELASTIC STRAIN ENERGY FOR NORMAL STRESSES

Since the rod considered in the preceding section was subjected to uniformly distributed stresses σ_x, the strain-energy density was constant throughout the rod and could be defined as the ratio U/V of the strain energy U and the volume V of the rod. In a structural element or machine part with a nonuniform stress distribution, the strain-energy density u can be defined by considering the strain energy of a small element of material of volume ΔV and writing

$$u = \lim_{\Delta V \to 0} \frac{\Delta U}{\Delta V}$$

or

$$u = \frac{dU}{dV} \tag{11.9}$$

The expression obtained for u in Sec. 11.3 in terms of σ_x and ϵ_x remains valid, i.e., we still have

$$u = \int_0^{\epsilon_x} \sigma_x \, d\epsilon_x \tag{11.10}$$

but the stress σ_x, the strain ϵ_x, and the strain-energy density u will generally vary from point to point.

For values of σ_x within the proportional limit, we may set $\sigma_x = E\epsilon_x$ in Eq. (11.10) and write

$$u = \frac{1}{2} E \epsilon_x^2 = \frac{1}{2} \sigma_x \epsilon_x = \frac{1}{2} \frac{\sigma_x^2}{E} \tag{11.11}$$

The value of the strain energy U of a body subjected to uniaxial normal stresses can be obtained by substituting for u from Eq. (11.11) into Eq. (11.9) and integrating both members. We have

$$U = \int \frac{\sigma_x^2}{2E} dV \tag{11.12}$$

The expression obtained is valid only for elastic deformations and is referred to as the *elastic strain energy* of the body.

†However, referring to the footnote on page 672, we note that the modulus of toughness and the modulus of resilience could be expressed in the same units as stress.

Strain Energy under Axial Loading. We recall from Sec. 2.17 that, when a rod is subjected to a centric axial loading, the normal stresses σ_x can be assumed uniformly distributed in any given transverse section. Denoting by A the area of the section located at a distance x from the end B of the rod (Fig. 11.10), and by P the internal force in that section, we write $\sigma_x = P/A$. Substituting for σ_x into Eq. (11.12), we have

$$U = \int \frac{P^2}{2EA^2}dV$$

or, setting $dV = A\,dx$,

$$U = \int_0^L \frac{P^2}{2AE}dx$$

Fig. 11.10

(11.13)

In the case of a rod of uniform cross section subjected at its ends to equal and opposite forces of magnitude P (Fig. 11.11), Eq. (11.13) yields

$$U = \frac{P^2L}{2AE}$$

(11.14)

Fig. 11.11

EXAMPLE 11.01

A rod consists of two portions BC and CD of the same material and same length, but of different cross sections (Fig. 11.12). Determine the strain energy of the rod when it is subjected to a centric axial load $\mathbf{P}$, expressing the result in terms of P, L, E, the cross-sectional area A of portion CD, and the ratio n of the two diameters.

Fig. 11.12

We use Eq. (11.14) to compute the strain energy of each of the two portions, and add the expressions obtained:

$$U_n = \frac{P^2(\frac{1}{2}L)}{2AE} + \frac{P^2(\frac{1}{2}L)}{2(n^2A)E} = \frac{P^2L}{4AE}\left(1 + \frac{1}{n^2}\right)$$

or

$$U_n = \frac{1+n^2}{2n^2}\frac{P^2L}{2AE}$$

(11.15)

We check that, for $n = 1$, we have

$$U_1 = \frac{P^2L}{2AE}$$

which is the expression given in Eq. (11.14) for a rod of length L and uniform cross section of area A. We also note that, for $n > 1$, we have $U_n < U_1$; for example, when $n = 2$, we have $U_2 = (\frac{5}{8})U_1$. Since the maximum stress occurs in portion CD of the rod and is equal to $\sigma_{max} = P/A$, it follows that, for a given allowable stress, increasing the diameter of portion BC of the rod results in a *decrease* of the overall energy-absorbing capacity of the rod. Unnecessary changes in cross-sectional area should therefore be avoided in the design of members that may be subjected to loadings, such as impact loadings, where the energy-absorbing capacity of the member is critical.

EXAMPLE 11.02

A load **P** is supported at B by two rods of the same material and of the same uniform cross section of area A (Fig. 11.13). Determine the strain energy of the system.

Fig. 11.13

Denoting by F_{BC} and F_{BD}, respectively, the forces in members BC and BD, and recalling Eq. (11.14), we express the strain energy of the system as

$$U = \frac{F_{BC}^2(BC)}{2AE} + \frac{F_{BD}^2(BD)}{2AE}$$

But we note from Fig. 11.13 that

$$BC = 0.6l \qquad BD = 0.8l$$

and from the free-body diagram of pin B and the corresponding force triangle (Fig. 11.14) that

Fig. 11.14

$$F_{BC} = +0.6P \qquad F_{BD} = -0.8P$$

Substituting into Eq. (11.16), we have

$$U = \frac{P^2[(0.6)^3 + (0.8)^3]}{2AE} = 0.364 \frac{P^2 l}{AE}$$

(11.16)

Strain Energy in Bending. Consider a beam AB subjected to a given loading (Fig. 11.15), and let M be the bending moment at a distance x from end A. Neglecting for the time being the effect of shear, and taking into account only the normal stresses $\sigma_x = My/I$, we substitute this expression into Eq. (11.12) and write

$$U = \int \frac{\sigma_x^2}{2E} dV = \int \frac{M^2 y^2}{2EI^2} dV$$

Fig. 11.15

Setting $dV = dA\,dx$, where dA represents an element of the cross-sectional area, and recalling that $M^2/2EI^2$ is a function of x alone, we have

$$U = \int_0^L \frac{M^2}{2EI^2} \left(\int y^2\,dA \right) dx$$

Recalling that the integral within the parentheses represents the moment of inertia I of the cross section about its neutral axis, we write

$$U = \int_0^L \frac{M^2}{2EI} dx$$

(11.17)

EXAMPLE 11.03

Determine the strain energy of the prismatic cantilever beam AB (Fig. 11.16), taking into account only the effect of the normal stresses.

The bending moment at a distance x from end A is $M = -Px$. Substituting this expression into Eq. (11.17), we write

$$U = \int_0^L \frac{P^2 x^2}{2EI} dx = \frac{P^2 L^3}{6EI}$$

Fig. 11.16

11.5. ELASTIC STRAIN ENERGY FOR SHEARING STRESSES

When a material is subjected to plane shearing stresses τ_{xy}, the strain-energy density at a given point can be expressed as

$$u = \int_0^{\gamma_{xy}} \tau_{xy} \, d\gamma_{xy} \qquad (11.18)$$

where γ_{xy} is the shearing strain corresponding to τ_{xy} (Fig. 11.17a). We note that the strain-energy density u is equal to the area under the shearing-stress-strain diagram (Fig. 11.17b).

For values of τ_{xy} within the proportional limit, we have $\tau_{xy} = G\gamma_{xy}$, where G is the modulus of rigidity of the material. Substituting for τ_{xy} into Eq. (11.18) and performing the integration, we write

$$u = \frac{1}{2} G\gamma_{xy}^2 = \frac{1}{2} \tau_{xy}\gamma_{xy} = \frac{\tau_{xy}^2}{2G} \qquad (11.19)$$

The value of the strain energy U of a body subjected to plane shearing stresses can be obtained by recalling from Sec. 11.4 that

$$u = \frac{dU}{dV} \qquad (11.9)$$

Substituting for u from Eq. (11.19) into Eq. (11.9) and integrating both members, we have

$$U = \int \frac{\tau_{xy}^2}{2G} dV \qquad (11.20)$$

This expression defines the elastic strain associated with the shear deformations of the body. Like the similar expression obtained in Sec. 11.4 for uniaxial normal stresses, it is valid only for elastic deformations.

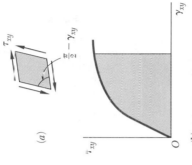

(a)

(b)

Fig. 11.17

Strain Energy in Torsion. Consider a shaft BC of length L subjected to one or several twisting couples. Denoting by J the polar moment of inertia of the cross section located at a distance x from B (Fig. 11.18), and by T the internal torque in that section, we recall that the shearing stresses in the section are $\tau_{xy} = Tp/J$. Substituting for τ_{xy} into Eq. (11.20), we have

$$U = \int \frac{\tau_{xy}^2}{2G}\,dV = \int \frac{T^2\rho^2}{2GJ^2}\,dV$$

Fig. 11.18

Setting $dV = dA\,dx$, where dA represents an element of the cross-sectional area, and observing that $T^2/2GJ^2$ is a function of x alone, we write

$$U = \int_0^L \frac{T^2}{2GJ^2}\left(\int \rho^2\,dA\right)dx \qquad (11.21)$$

Recalling that the integral within the parentheses represents the polar moment of inertia J of the cross section, we have

$$U = \int_0^L \frac{T^2}{2GJ}\,dx \qquad (11.21)$$

In the case of a shaft of uniform cross section subjected at its ends to equal and opposite couples of magnitude T (Fig. 11.19), Eq. (11.21) yields

$$\boxed{U = \frac{T^2L}{2GJ}} \qquad (11.22)$$

Fig. 11.19

EXAMPLE 11.04

A circular shaft consists of two portions BC and CD of the same material and same length, but of different cross sections (Fig. 11.20). Determine the strain energy of the shaft when it is subjected to a twisting couple $\mathbf{T}$ at end D, expressing the result in terms of T, L, G, the polar moment of inertia J of the smaller cross section, and the ratio n of the two diameters.

Fig. 11.20

We use Eq. (11.22) to compute the strain energy of each of the two portions of shaft, and add the expressions obtained. Noting that the polar moment of inertia of portion BC is equal to n^4J, we write

$$U_n = \frac{T^2(\tfrac{1}{2}L)}{2GJ} + \frac{T^2(\tfrac{1}{2}L)}{2G(n^4J)} = \frac{T^2L}{4GJ}\left(1 + \frac{1}{n^4}\right)$$

or

$$U_n = \frac{1 + n^4}{2n^4}\,\frac{T^2L}{2GJ} \qquad (11.23)$$

We check that, for $n = 1$, we have

$$U_1 = \frac{T^2L}{2GJ}$$

which is the expression given in Eq. (11.22) for a shaft of length L and uniform cross section. We also note that, for $n > 1$, we have $U_n < U_1$; for example, when $n = 2$, we have $U_2 = (\tfrac{17}{32})U_1$. Since the maximum shearing stress occurs in the portion CD of the shaft and is proportional to the torque T, we note as we did earlier in the case of the axial loading of a rod that, for a given allowable stress, increasing the diameter of portion BC of the shaft results in a *decrease* of the overall energy-absorbing capacity of the shaft.

Strain Energy under Transverse Loading. In Sec. 11.4 we obtained an expression for the strain energy of a beam subjected to a transverse loading. However, in deriving that expression we took into account only the effect of the normal stresses due to bending and neglected the effect of the shearing stresses. In Example 11.05 both types of stresses will be taken into account.

EXAMPLE 11.05

Determine the strain energy of the rectangular cantilever beam AB (Fig. 11.21), taking into account the effect of both normal and shearing stresses.

We first recall from Example 11.03 that the strain energy due to the normal stresses σ_x is

$$U_\sigma = \frac{P^2 L^3}{6EI}$$

To determine the strain energy U_τ due to the shearing stresses τ_{xy}, we recall Eq. (6.9) of Sec. 6.4 and find that, for a beam with a rectangular cross section of width b and depth h,

$$\tau_{xy} = \frac{3}{2}\frac{V}{A}\left(1 - \frac{y^2}{c^2}\right) = \frac{3}{2}\frac{P}{bh}\left(1 - \frac{y^2}{c^2}\right)$$

Substituting for τ_{xy} into Eq. (11.20), we write

$$U_\tau = \frac{1}{2G}\left(\frac{3}{2}\frac{P}{bh}\right)^2 \int\left(1 - \frac{y^2}{c^2}\right)^2 dV$$

or, setting $dV = b\,dy\,dx$, and after reductions,

$$U_\tau = \frac{9P^2}{8Gbh^2}\int_{-c}^{c}\left(1 - 2\frac{y^2}{c^2} + \frac{y^4}{c^4}\right)dy\int_{0}^{L}dx$$

Performing the integrations, and recalling that $c = h/2$, we have

$$U_\tau = \frac{9P^2 L}{8Gbh^2}\left[y - \frac{2}{3}\frac{y^3}{c^2} + \frac{1}{5}\frac{y^5}{c^4}\right]_{-c}^{+c} = \frac{3P^2 L}{5Gbh} = \frac{3P^2 L}{5GA}$$

The total strain energy of the beam is thus

$$U = U_\sigma + U_\tau = \frac{P^2 L^3}{6EI} + \frac{3P^2 L}{5GA}$$

or, noting that $I/A = h^2/12$ and factoring the expression for U_σ,

$$U = \frac{P^2 L^3}{6EI}\left(1 + \frac{3Eh^2}{10GL^2}\right) = U_\sigma\left(1 + \frac{3Eh^2}{10GL^2}\right) \qquad (11.24)$$

Fig. 11.21

Recalling from Sec. 2.14 that $G \geq E/3$, we conclude that the parenthesis in the expression obtained is less than $1 + 0.9(h/L)^2$ and, thus, that the relative error is less than $0.9(h/L)^2$ when the effect of shear is neglected. For a beam with a ratio h/L less than $\frac{1}{10}$, the percentage error is less than 0.9%. It is therefore customary in engineering practice to neglect the effect of shear in computing the strain energy of slender beams.

11.6. STRAIN ENERGY FOR A GENERAL STATE OF STRESS

In the preceding sections, we determined the strain energy of a body in a state of uniaxial stress (Sec. 11.4) and in a state of plane shearing stress (Sec. 11.5). In the case of a body in a general state of stress characterized by the six stress components $\sigma_x, \sigma_y, \sigma_z, \tau_{xy}, \tau_{yz},$ and τ_{zx} the strain-energy density can be obtained by adding the expressions given in Eqs. (11.10) and (11.18), as well as the four other expressions obtained through a permutation of the subscripts.

In the case of the elastic deformation of an isotropic body, each of the six stress-strain relations involved is linear, and the strain-energy density can be expressed as

$$u = \tfrac{1}{2}(\sigma_x \epsilon_x + \sigma_y \epsilon_y + \sigma_z \epsilon_z + \tau_{xy}\gamma_{xy} + \tau_{yz}\gamma_{yz} + \tau_{zx}\gamma_{zx}) \qquad (11.25)$$

Recalling the relations (2.38) obtained in Sec. 2.14, and substituting for the strain components into (11.25), we have, for the most general state of stress at a given point of an elastic isotropic body,

$$u = \frac{1}{2E}[\sigma_x^2 + \sigma_y^2 + \sigma_z^2 - 2\nu(\sigma_x\sigma_y + \sigma_y\sigma_z + \sigma_z\sigma_x)]$$

$$+ \frac{1}{2G}(\tau_{xy}^2 + \tau_{yz}^2 + \tau_{zx}^2) \qquad (11.26)$$

If the principal axes at the given point are used as coordinate axes, the shearing stresses become zero and Eq. (11.26) reduces to

$$u = \frac{1}{2E}[\sigma_a^2 + \sigma_b^2 + \sigma_c^2 - 2\nu(\sigma_a\sigma_b + \sigma_b\sigma_c + \sigma_c\sigma_a)] \qquad (11.27)$$

where $\sigma_a, \sigma_b,$ and σ_c are the principal stresses at the given point.

We now recall from Sec. 7.7 that one of the criteria used to predict whether a given state of stress will cause a ductile material to yield, namely, the maximum-distortion-energy criterion, is based on the determination of the energy per unit volume associated with the distortion, or change in shape, of that material. Let us, therefore, attempt to separate the strain-energy density u at a given point into two parts, a part u_v associated with a change in volume of the material at that point, and a part u_d associated with a distortion, or change in shape, of the material at the same point. We write

$$u = u_v + u_d \qquad (11.28)$$

In order to determine u_v and u_d, we introduce the *average value* $\bar{\sigma}$ of the principal stresses at the point considered,

$$\bar{\sigma} = \frac{\sigma_a + \sigma_b + \sigma_c}{3} \qquad (11.29)$$

and set

$$\sigma_a = \bar{\sigma} + \sigma_a' \qquad \sigma_b = \bar{\sigma} + \sigma_b' \qquad \sigma_c = \bar{\sigma} + \sigma_c' \qquad (11.30)$$

σ'_a

σ'_b

σ'_c

(c)

$\overline{\sigma}$

$\overline{\sigma}$

$\overline{\sigma}$

(b)

σ_a

σ_b

σ_c

(a)

Fig. 11.22

Thus, the given state of stress (Fig. 11.22a) can be obtained by superposing the states of stress shown in Fig. 11.22b and c. We note that the state of stress described in Fig. 11.22b tends to change the volume of the element of material, but not its shape, since all the faces of the element are subjected to the same stress $\overline{\sigma}$. On the other hand, it follows from Eqs. (11.29) and (11.30) that

$$\sigma'_a + \sigma'_b + \sigma'_c = 0 \qquad (11.31)$$

which indicates that some of the stresses shown in Fig. 11.22c are tensile and others compressive. Thus, this state of stress tends to change the shape of the element. However, it does not tend to change its volume. Indeed, recalling Eq. (2.31) of Sec. 2.13, we note that the dilatation e (i.e., the change in volume per unit volume) caused by this state of stress is

$$e = \frac{1 - 2\nu}{E}(\sigma'_a + \sigma'_b + \sigma'_c)$$

or $e = 0$, in view of Eq. (11.31). We conclude from these observations that the portion u_v of the strain-energy density must be associated with the state of stress shown in Fig. 11.22b, while the portion u_d must be associated with the state of stress shown in Fig. 11.22c.

It follows that the portion u_v of the strain-energy density corresponding to a change in volume of the element can be obtained by substituting $\overline{\sigma}$ for each of the principal stresses in Eq. (11.27). We have

$$u_v = \frac{1}{2E}[3\overline{\sigma}^2 - 2\nu(3\overline{\sigma}^2)] = \frac{3(1 - 2\nu)}{2E}\overline{\sigma}^2 \qquad (11.32)$$

or, recalling Eq. (11.29),

$$u_v = \frac{1 - 2\nu}{6E}(\sigma_a + \sigma_b + \sigma_c)^2$$

The portion of the strain-energy density corresponding to the distortion of the element is obtained by solving Eq. (11.28) for u_d and substituting for u and u_v from Eqs. (11.27) and (11.32), respectively. We write

$$u_d = u - u_v = \frac{1}{6E}[3(\sigma_a^2 + \sigma_b^2 + \sigma_c^2) - 6v(\sigma_a\sigma_b + \sigma_b\sigma_c + \sigma_c\sigma_a)$$

$$- (1 - 2v)(\sigma_a + \sigma_b + \sigma_c)^2]$$

Expanding the square and rearranging terms, we have

$$u_d = \frac{1 + v}{6E}[(\sigma_a^2 - 2\sigma_a\sigma_b + \sigma_b^2) + (\sigma_b^2 - 2\sigma_b\sigma_c + \sigma_c^2)$$

$$+ (\sigma_c^2 - 2\sigma_c\sigma_a + \sigma_a^2)]$$

Noting that each of the parentheses inside the bracket is a perfect square, and recalling from Eq. (2.43) of Sec. 2.15 that the coefficient in front of the bracket is equal to $1/12G$, we obtain the following expression for the portion u_d of the strain-energy density, i.e., for the distortion energy per unit volume,

$$u_d = \frac{1}{12G}[(\sigma_a - \sigma_b)^2 + (\sigma_b - \sigma_c)^2 + (\sigma_c - \sigma_a)^2] \quad (11.33)$$

In the case of *plane stress*, and assuming that the c axis is perpendicular to the plane of stress, we have $\sigma_c = 0$ and Eq. (11.33) reduces to

$$u_d = \frac{1}{6G}(\sigma_a^2 - \sigma_a\sigma_b + \sigma_b^2) \quad (11.34)$$

Considering the particular case of a tensile-test specimen, we note that, at yield, we have $\sigma_a = \sigma_Y, \sigma_b = 0$, and thus $(u_d)_Y = \sigma_Y^2/6G$. The maximum-distortion-energy criterion for plane stress indicates that a given state of stress is safe as long as $u_d < (u_d)_Y$ or, substituting for u_d from Eq. (11.34), as long as

$$\sigma_a^2 - \sigma_a\sigma_b + \sigma_b^2 < \sigma_Y^2 \quad (7.26)$$

which is the condition stated in Sec. 7.7 and represented graphically by the ellipse of Fig. 7.41. In the case of a general state of stress, the expression (11.33) obtained for u_d should be used. The maximum-distortion-energy criterion is then expressed by the condition.

$$(\sigma_a - \sigma_b)^2 + (\sigma_b - \sigma_c)^2 + (\sigma_c - \sigma_a)^2 < 2\sigma_Y^2 \quad (11.35)$$

which indicates that a given state of stress is safe if the point of coordinates $\sigma_a, \sigma_b, \sigma_c$ is located within the surface defined by the equation

$$(\sigma_a - \sigma_b)^2 + (\sigma_b - \sigma_c)^2 + (\sigma_c - \sigma_a)^2 = 2\sigma_Y^2 \quad (11.36)$$

This surface is a circular cylinder of radius $\sqrt{2/3}\,\sigma_Y$ with an axis of symmetry forming equal angles with the three principal axes of stress.

SAMPLE PROBLEM 11.1

$\frac{3}{4}$-in. diameter

5 ft

During a routine manufacturing operation, rod AB must acquire an elastic strain energy of 120 in · lb. Using $E = 29 \times 10^6$ psi, determine the required yield strength of the steel if the factor of safety with respect to permanent deformation is to be five.

SOLUTION

Factor of Safety.   Since a factor of safety of five is required, the rod should be designed for a strain energy of

$$U = 5(120 \text{ in} \cdot \text{lb}) = 600 \text{ in} \cdot \text{lb}$$

Strain-Energy Density.   The volume of the rod is

$$V = AL = \frac{\pi}{4}(0.75 \text{ in.})^2(60 \text{ in.}) = 26.5 \text{ in}^3$$

Since the rod is of uniform cross section, the required strain-energy density is

$$u = \frac{U}{V} = \frac{600 \text{ in} \cdot \text{lb}}{26.5 \text{ in}^3} = 22.6 \text{ in} \cdot \text{lb/in}^3$$

Yield Strength.   We recall that the modulus of resilience is equal to the strain-energy density when the maximum stress is equal to σ_Y. Using Eq. (11.8), we write

$$u = \frac{\sigma_Y^2}{2E}$$

$$22.6 \text{ in} \cdot \text{lb/in}^3 = \frac{\sigma_Y^2}{2(29 \times 10^6 \text{ psi})}$$

$$\sigma_Y = 36.2 \text{ ksi} \quad \blacktriangleleft$$

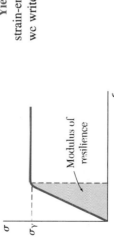

Modulus of resilience

Comment.   It is important to note that, since energy loads are not linearly related to the stresses they produce, factors of safety associated with energy loads should be applied to the energy loads and not to the stresses.

683

SAMPLE PROBLEM 11.2

(a) Taking into account only the effect of normal stresses due to bending, determine the strain energy of the prismatic beam AB for the loading shown. (b) Evaluate the strain energy, knowing that the beam is a W10 × 45, $P = 40$ kips, $L = 12$ ft, $a = 3$ ft, $b = 9$ ft, and $E = 29 \times 10^6$ psi.

SOLUTION

Bending Moment. Using the free-body diagram of the entire beam, we determine the reactions

$$R_A = \frac{Pb}{L}\uparrow \qquad R_B = \frac{Pa}{L}\uparrow$$

For portion AD of the beam, the bending moment is

$$M_1 = \frac{Pb}{L}x$$

For portion DB, the bending moment at a distance v from end B is

$$M_2 = \frac{Pa}{L}v$$

a. Strain Energy. Since strain energy is a scalar quantity, we add the strain energy of portion AD to that of portion DB to obtain the total strain energy of the beam. Using Eq. (11.17), we write

$$U = U_{AD} + U_{DB}$$
$$= \int_0^a \frac{M_1^2}{2EI}dx + \int_0^b \frac{M_2^2}{2EI}dv$$
$$= \frac{1}{2EI}\int_0^a \left(\left(\frac{Pb}{L}x\right)^2 dx + \frac{1}{2EI}\int_0^b \left(\frac{Pa}{L}v\right)^2 dv\right.$$
$$= \frac{1}{2EI}\frac{P^2}{L^2}\left(\frac{b^2a^3}{3} + \frac{a^2b^3}{3}\right) = \frac{P^2a^2b^2}{6EIL^2}(a+b)$$

or, since $(a+b) = L$,

$$U = \frac{P^2a^2b^2}{6EIL} \quad \blacktriangleleft$$

b. Evaluation of the Strain Energy. The moment of inertia of a W10 × 45 rolled-steel shape is obtained from Appendix C and the given data is restated using units of kips and inches.

$P = 40$ kips	$L = 12$ ft $= 144$ in.
$a = 3$ ft $= 36$ in.	$b = 9$ ft $= 108$ in.
$E = 29 \times 10^6$ psi $= 29 \times 10^3$ ksi	$I = 248$ in^4

Substituting into the expression for U, we have

$$U = \frac{(40 \text{ kips})^2(36 \text{ in.})^2(108 \text{ in.})^2}{6(29 \times 10^3 \text{ ksi})(248 \text{ in}^4)(144 \text{ in.})} \qquad U = 3.89 \text{ in} \cdot \text{kips} \quad \blacktriangleleft$$

PROBLEMS

11.1 Determine the modulus of resilience for each of the following grades of structural steel:

- (a) ASTM A709 Grade 50: $\sigma_Y = 50$ ksi
- (b) ASTM A913 Grade 65: $\sigma_Y = 65$ ksi
- (c) ASTM A709 Grade 100: $\sigma_Y = 100$ ksi

11.2 Determine the modulus of resilience for each of the following aluminum alloys:

- (a) 1100-H14: $E = 70$ GPa, $\sigma_Y = 55$ MPa
- (b) 2014-T6: $E = 72$ GPa, $\sigma_Y = 220$ MPa
- (c) 6061-Y6: $E = 69$ GPa, $\sigma_Y = 140$ MPa

11.3 Determine the modulus of resilience for each of the following metals:

- (a) Stainless steel
  AISI 302 (annealed): $E = 190$ GPa, $\sigma_Y = 260$ MPa
- (b) Stainless steel
  AISI 302 (cold-rolled): $E = 190$ GPa, $\sigma_Y = 520$ MPa
- (c) Malleable cast iron: $E = 165$ GPa, $\sigma_Y = 230$ MPa

11.4 Determine the modulus of resilience for each of the following alloys:

- (a) Titanium: $E = 16.5 \times 10^6$ psi, $\sigma_Y = 120$ ksi
- (b) Magnesium: $E = 6.5 \times 10^6$ psi, $\sigma_Y = 29$ ksi
- (c) Cupronickel (annealed): $E = 20 \times 10^6$ psi, $\sigma_Y = 16$ ksi

11.5 The stress-strain diagram shown has been drawn from data obtained during a tensile test of an aluminum alloy. Using $E = 72$ GPa, determine (a) the modulus of resilience of the alloy, (b) the modulus of toughness of the alloy.

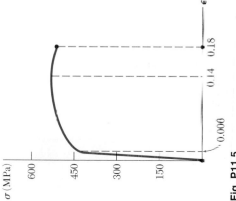

Fig. P11.5

11.6 The stress-strain diagram shown has been drawn from data obtained during the tensile test of a specimen of structural steel. Using $E = 29 \times 10^6$ psi, determine (a) the modulus of resilience of the steel, (b) the modulus of toughness of the steel.

Fig. P11.6

11.7 The load-deformation diagram shown has been drawn from data obtained during the tensile test of a specimen of structural steel. Knowing that the cross-sectional area of the specimen is 250 mm² and that the deformation was measured using a 500-mm gage length, determine (*a*) the modulus of resilience of the steel, (*b*) the modulus of toughness of the steel.

P (kN)

100
75
50
25

0.6   8.6   78   96 δ (mm)

Fig. P11.7

500 mm

δ

P'   P

Fig. P11.8

P (kips)

40
30
20
10

0.104   1.85 δ (in.)

16 in.

δ

P'   P

11.8 The load-deformation diagram shown has been drawn from data obtained during the tensile test of a 0.75-in.-diameter rod of an aluminum alloy. Knowing that the deformation was measured using a 16-in. gage length, determine (*a*) the modulus of resilience of the alloy, (*b*) the modulus of toughness of the alloy.

11.9 Using $E = 29 \times 10^6$ psi, determine (*a*) the strain energy of the steel rod ABC when $P = 8$ kips, (*b*) the corresponding strain energy density in portions AB and BC of the rod.

3 ft

2 ft

$\frac{3}{4}$ in.

$\frac{5}{8}$ in.

A   B   C   P

Fig. P11.9

11.10 A 30-in. length of aluminum pipe of cross-sectional area 1.85 in² is welded to a fixed support A and to a rigid cap B. The steel rod EF, of 0.75-in. diameter, is welded to cap B. Knowing that the modulus of elasticity is 29×10^6 psi for steel and 10.6×10^6 psi for aluminum, determine (*a*) the total strain energy of the system when $P = 10$ kips, (*b*) the corresponding strain-energy density in the pipe CD and in the rod EF.

B
D   E   C
30 in.

48 in.

A

F

P

Fig. P11.10

11.11 Solve Prob. 11.10 when $P = 8$ kips.

11.12 Using $E = 200$ GPa, determine (*a*) the strain energy of the steel rod ABC when $P = 25$ kN, (*b*) the corresponding strain-energy density of portions AB and BC of the rod.

11.13 Rod ABC is made of a steel for which the yield strength is $\sigma_Y = 250$ MPa and the modulus of elasticity is $E = 200$ GPa. Determine, for the loading shown, the maximum strain energy that can be acquired by the rod without causing any permanent deformation.

Fig. P11.12 and *P11.13*

11.14 Rods AB and BC are made of a steel for which the yield strength is $\sigma_Y = 300$ MPa and the modulus of elasticity is $E = 200$ GPa. Determine the maximum strain energy that can be acquired by the assembly without causing any permanent deformation when the length a of rod AB is (*a*) 2 m, (*b*) 4 m.

Fig. *P11.14*

11.15 Rod AB is made of a steel for which the yield strength is $\sigma_Y = 65$ ksi and the modulus of elasticity is $E = 29 \times 10^6$ psi; rod BC is made of an aluminum alloy for which $\sigma_Y = 40$ ksi and $E = 10.6 \times 10^6$ psi. Determine the maximum strain energy that can be acquired by the composite rod ABC without causing any permanent deformation.

Fig. P11.15

11.16 Rod AB is made of a steel for which the yield strength is $\sigma_Y = 300$ MPa and the modulus of elasticity is $E = 200$ GPa. Knowing that a strain energy of 10 J must be acquired by the rod when the axial load **P** is applied, determine the diameter of the rod for which the factor of safety with respect to permanent deformation is six.

Fig. P11.16

11.17 Rod ABC is made of a steel for which the yield strength is $\sigma_Y = 65$ ksi and the modulus of elasticity is $E = 29 \times 10^6$ psi. Knowing that a strain energy of 90 in · lb must be acquired by the rod as the axial load **P** is applied, determine the factor of safety of the rod with respect to permanent deformation when $a = 18$ in.

Fig. P11.17

11.18 Solve Prob. 11.17, assuming that $a = 30$ in.

11.19 Show by integration that the strain energy of the tapered rod AB
is

$$U = \frac{1}{4} \frac{P^2 L}{E A_{\min}}$$

where $A_{\min}$ is the cross-sectional area at end B.

Fig. P11.19

Fig. P11.20

11.20 Solve Prob. 11.19, using the stepped rod shown as an approximation of the tapered rod. What is the percentage error in the answer obtained?

11.21 Using $E = 10.6 \times 10^6$ psi, determine by approximate means the maximum strain energy that can be acquired by the aluminum rod shown if the allowable normal stress is $\sigma_{\text{all}} = 22$ ksi.

Fig. P11.21

11.22 through 11.25 In the truss shown, all members are made of the same material and have the uniform cross-sectional area indicated. Determine the strain energy of the truss when the load **P** is applied.

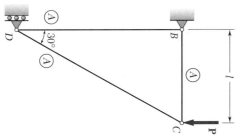

Fig. P11.22

Fig. P11.23

Fig. P11.24

Fig. P11.25

11.26 In the truss shown, all members are made of aluminum and have the uniform cross-sectional area indicated. Using $E = 72$ GPa, determine the strain energy of the truss for the loading shown.

Fig. P11.26

11.27 Solve Prob. 11.26, assuming that the 120-kN load is removed.

11.28 through 11.31 Taking into account only the effect of normal stresses, determine the strain energy of the prismatic beam AB for the loading shown.

Fig. P11.28

Fig. P11.30 and P11.32

Fig. P11.29

Fig. P11.31 and P11.33

11.32 Assuming that the prismatic beam AB has a rectangular cross section, show that for the given loading the maximum value of the strain-energy density in the beam is

$$u_{max} = 15\frac{U}{V}$$

where U is the strain energy of the beam and V is its volume.

11.33 Assuming that the prismatic beam AB has a rectangular cross section, show that for the given loading the maximum value of the strain-energy density in the beam is

$$u_{max} = \frac{45}{8}\frac{U}{V}$$

where U is the strain energy of the beam and V is its volume.

11.34 and 11.35   Using $E = 200$ GPa, determine the strain energy due to bending for the steel beam and loading shown.

Fig. P11.34

W310 × 74

160 kN

2.4 m
4.8 m
2.4 m

A   C   B

Fig. P11.35

W310 × 74

80 kN   80 kN

1.6 m   1.6 m   1.6 m
4.8 m

A   D   E   B

11.36   Using $E = 29 \times 10^6$ psi, determine the strain energy due to bending for the steel beam and loading shown.

Fig. P11.36

S8 × 18.4

8 kips

6 ft   3 ft

A   D   B

11.37   Using $E = 1.8 \times 10^6$ psi, determine the strain energy due to bending for the timber beam and loading shown.

11.38   Rod AC is made of aluminum ($G = 73$ GPa) and is subjected to a torque $\mathbf{T}$ applied at end C. Knowing that portion BC of the rod is hollow and has an inner diameter of 16 mm, determine the strain energy of the rod for a maximum shearing stress of 120 MPa.

Fig. P11.37

16 kip · ft

4 ft   6 ft
10 ft

A   D   B

$3\frac{1}{2}$ in.

$9\frac{1}{2}$ in.

Fig. P11.38

24-mm diameter

400 mm   500 mm

A   (B)   C

T

11.39   In the assembly shown torques $\mathbf{T}_A$ and $\mathbf{T}_B$ are exerted on disks A and B respectively. Knowing that both shafts are solid and made of aluminum ($G = 73$ GPa), determine the total energy acquired by the assembly.

Fig. P11.39

$T_A = 300$ N · m

$T_B = 400$ N · m

30 mm

46 mm

0.9 m   0.75 m

A   B   C

11.40 The ship at A has just started to drill for oil on the ocean floor at a depth of 5000 ft. The steel drill pipe has an outer diameter of 8 in. and a uniform wall thickness of 0.5 in. Knowing that the top of the drill pipe rotates through two complete revolutions before the drill bit at B starts to operate and using $G = 11.2 \times 10^6$ psi, determine the maximum strain energy acquired by the drill pipe.

Fig. P11.40

11.41 The design specifications for the steel shaft AB require that the shaft acquire a strain energy of 300 in · lb as the 25-kip · in. torque is applied. Using $G = 11.2 \times 10^6$ psi, determine (a) the largest inner diameter of the shaft that can be used, (b) the corresponding maximum shearing stress in the shaft.

Fig. P11.41

11.42 Show by integration that the strain energy in the tapered rod AB is

$$U = \frac{7}{48} \frac{T^2 L}{GJ_{min}}$$

where J_{min} is the polar moment of inertia of the rod at end B.

Fig. P11.42

11.43 The state of stress shown occurs in a machine component made of a grade of steel for which $\sigma_Y = 65$ ksi. Using the maximum-distortion-energy criterion, determine the range of values of σ_Y for which the factor of safety associated with the yield strength is equal to or larger than 2.2.

11.44 The state of stress shown occurs in a machine component made of a grade of steel for which $\sigma_Y = 65$ ksi. Using the maximum-distortion-energy criterion, determine the factor of safety associated with the yield strength when (a) $\sigma_Y = +16$ ksi, (b) $\sigma_Y = -16$ ksi.

Fig. P11.43 and P11.44

11.45 The state of stress shown occurs in a machine component made of a brass for which $\sigma_Y = 160$ MPa. Using the maximum-distortion-energy criterion, determine whether yield occurs when (a) $\sigma_z = +45$ MPa, (b) $\sigma_z = -45$ MPa.

Fig. P11.45 and P11.46

11.46 The state of stress shown occurs in a machine component made of a brass for which $\sigma_Y = 160$ MPa. Using the maximum-distortion-energy criterion, determine the range of values of σ_z for which yield does not occur.

11.47 Determine the strain energy of the prismatic beam AB, taking into account the effect of both normal and shearing stresses.

Fig. P11.47

11.48 For the state of stress shown in Fig. a, determine the stresses in an element oriented as shown in Fig. b. Compare the strain energy density in the given state first by using Fig. a and then by using Fig. b. Equating the two results obtained, show that

$$G = \frac{E}{2(1 + \nu)}$$

Fig. P11.48

*11.49 A vibration isolation support is made by bonding a rod A, of radius R_1, and a tube B, of inner radius R_2, to a hollow rubber cylinder. Denoting by G the modulus of rigidity of the rubber, determine the strain energy of the hollow rubber cylinder for the loading shown.

Fig. P11.49

11.7. IMPACT LOADING

Consider a rod BD of uniform cross section which is hit at its end B by a body of mass m moving with a velocity $\mathbf{v}_0$ (Fig. 11.23a). As the rod deforms under the impact (Fig. 11.23b), stresses develop within the rod and reach a maximum value σ_m. After vibrating for a while, the rod will come to rest, and all stresses will disappear. Such a sequence of events is referred to as an *impact loading* (Fig. 11.24).

In order to determine the maximum value σ_m of the stress occurring at a given point of a structure subjected to an impact loading, we are going to make several simplifying assumptions.

First, we assume that the kinetic energy $T = \frac{1}{2}mv_0^2$ of the striking body is transferred entirely to the structure and, thus, that the strain energy U_m corresponding to the maximum deformation x_m is

$$U_m = \tfrac{1}{2}mv_0^2 \qquad (11.37)$$

This assumption leads to the following two specific requirements:

1. No energy should be dissipated during the impact.
2. The striking body should not bounce off the structure and retain part of its energy. This, in turn, necessitates that the inertia of the structure be negligible, compared to the inertia of the striking body.

In practice, neither of these requirements is satisfied, and only part of the kinetic energy of the striking body is actually transferred to the structure. Thus, assuming that all of the kinetic energy of the striking body is transferred to the structure leads to a conservative design of that structure.

We further assume that the stress-strain diagram obtained from a static test of the material is also valid under impact loading. Thus, for an elastic deformation of the structure, we can express the maximum value of the strain energy as

$$U_m = \int \frac{\sigma_m^2}{2E}dV \qquad (11.38)$$

In the case of the uniform rod of Fig. 11.23, the maximum stress σ_m has the same value throughout the rod, and we write $U_m = \sigma_m^2 V/2E$. Solving for σ_m and substituting for U_m from Eq. (11.37), we write

$$\sigma_m = \sqrt{\frac{2U_mE}{V}} = \sqrt{\frac{mv_0^2E}{V}} \qquad (11.39)$$

We note from the expression obtained that selecting a rod with a large volume V and a low modulus of elasticity E will result in a smaller value of the maximum stress σ_m for a given impact loading.

In most problems, the distribution of stresses in the structure is not uniform, and formula (11.39) does not apply. It is then convenient to determine the static load $\mathbf{P}_m$ which would produce the same strain energy as the impact loading, and compute from P_m the corresponding value σ_m of the largest stress occurring in the structure.

Area = A

x_m

$\mathbf{v} = 0$

(a)

(b)

Fig. 11.23

Fig. 11.24 Steam alternately lifts a weight inside the pile driver and then propels it downward. This delivers a large impact load to the pile which is being driven into the ground.

EXAMPLE 11.06

A body of mass m moving with a velocity $\mathbf{v}_0$ hits the end B of the nonuniform rod BCD (Fig. 11.25). Knowing that the diameter of portion BC is twice the diameter of portion CD, determine the maximum value σ_m of the stress in the rod.

Fig. P11.25

Making $n = 2$ in the expression (11.15) obtained in Example 11.01, we find that when rod BCD is subjected to a static load P_m, its strain energy is

$$U_m = \frac{5P_m^2 L}{16AE} \qquad (11.40)$$

where A is the cross-sectional area of portion CD of the rod. Solving Eq. (11.40) for P_m, we find that the static load that produces in the rod the same strain energy as the given impact loading is

$$P_m = \sqrt{\frac{16}{5}\frac{U_m AE}{L}}$$

where U_m is given by Eq. (11.37). The largest stress occurs in portion CD of the rod. Dividing P_m by the area A of that portion, we have

$$\sigma_m = \frac{P_m}{A} = \sqrt{\frac{16}{5}\frac{U_m E}{AL}} \qquad (11.41)$$

or, substituting for U_m from Eq. (11.37),

$$\sigma_m = \sqrt{\frac{8}{5}\frac{mv_0^2 E}{AL}} = 1.265\sqrt{\frac{mv_0^2 E}{AL}}$$

Comparing this value with the value obtained for σ_m in the case of the uniform rod of Fig. 11.24 and making $V = AL$ in Eq. (11.39), we note that the maximum stress in the rod of variable cross section is 26.5% larger than in the lighter uniform rod. Thus, as we observed earlier in our discussion of Example 11.01, increasing the diameter of portion BC of the rod results in a *decrease* of the energy-absorbing capacity of the rod.

EXAMPLE 11.07

A block of weight W is dropped from a height h onto the free end of the cantilever beam AB (Fig. 11.26). Determine the maximum value of the stress in the beam.

Fig. 11.26

As it falls through the distance h, the potential energy Wh of the block is transformed into kinetic energy. As a result of the impact, the kinetic energy in turn is transformed into strain energy. We have, therefore,†

$$U_m = Wh \qquad (11.42)$$

†The total distance through which the block drops is actually $h + y_m$, where y_m is the maximum deflection of the end of the beam. Thus, a more accurate expression for U_m (see Sample Prob. 11.3) is

$$U_m = W(h + y_m) \qquad (11.42')$$

However, when $h \gg y_m$, we may neglect y_m and use Eq. (11.42).

Recalling the expression obtained for the strain energy of the cantilever beam AB in Example 11.03 and neglecting the effect of shear, we write

$$U_m = \frac{P_m^2 L^3}{6EI}$$

Solving this equation for P_m, we find that the static force that produces in the beam the same strain energy is

$$P_m = \sqrt{\frac{6U_m EI}{L^3}} \qquad (11.43)$$

The maximum stress σ_m occurs at the fixed end B and is

$$\sigma_m = \frac{|M|c}{I} = \frac{P_m Lc}{I}$$

Substituting for P_m from (11.43), we write

$$\sigma_m = \sqrt{\frac{6U_m E}{L(I/c^2)}} \qquad (11.44)$$

or, recalling (11.42),

$$\sigma_m = \sqrt{\frac{6WhE}{L(I/c^2)}}$$

11.8. DESIGN FOR IMPACT LOADS

Let us now compare the values obtained in the preceding section for the maximum stress σ_m (a) in the rod of uniform cross section of Fig. 11.23, (b) in the rod of variable cross section of Example 11.06, and (c) in the cantilever beam of Example 11.07, assuming that the last has a circular cross section of radius c.

(a) We first recall from Eq. (11.39) that, if U_m denotes the amount of energy transferred to the rod as a result of the impact loading, the maximum stress in the rod of uniform cross section is

$$\sigma_m = \sqrt{\frac{2U_mE}{V}} \tag{11.45a}$$

where V is the volume of the rod.

(b) Considering next the rod of Example 11.06 and observing that the volume of the rod is

$$V = 4A(L/2) + A(L/2) = 5AL/2$$

we substitute $AL = 2V/5$ into Eq. (11.41) and write

$$\sigma_m = \sqrt{\frac{8U_mE}{V}} \tag{11.45b}$$

(c) Finally, recalling that $I = \frac{1}{4}\pi c^4$ for a beam of circular cross section, we note that

$$L(I/c^2) = L(\tfrac{1}{4}\pi c^4/c^2) = \tfrac{1}{4}(\pi c^2 L) = \tfrac{1}{4}V$$

where V denotes the volume of the beam. Substituting into Eq. (11.44), we express the maximum stress in the cantilever beam of Example 11.07 as

$$\sigma_m = \sqrt{\frac{24U_mE}{V}} \tag{11.45c}$$

We note that, in each case, the maximum stress σ_m is proportional to the square root of the modulus of elasticity of the material and inversely proportional to the square root of the volume of the member. Assuming all three members to have the same volume and to be of the same material, we also note that, for a given value of the absorbed energy, the uniform rod will experience the lowest maximum stress, and the cantilever beam the highest one.

This observation can be explained by the fact that, the distribution of stresses being uniform in case a, the strain energy will be uniformly distributed throughout the rod. In case b, on the other hand, the stresses in portion BC of the rod are only 25% as large as the stresses in portion CD. This uneven distribution of the stresses and of the strain energy results in a maximum stress σ_m twice as large as the corresponding stress in the uniform rod. Finally, in case c, where the cantilever beam is subjected to a transverse impact loading, the stresses vary linearly along the beam as well as across a transverse section. The very uneven resulting distribution of strain energy causes the maximum stress σ_m to be 3.46 times larger than if the same member had been loaded axially as in case a.

The properties noted in the three specific cases discussed in this section are quite general and can be observed in all types of structures and impact loadings. We thus conclude that a structure designed to withstand effectively an impact load should

1. Have a large volume
2. Be made of a material with a low modulus of elasticity and a high yield strength
3. Be shaped so that the stresses are distributed as evenly as possible throughout the structure

11.9. WORK AND ENERGY UNDER A SINGLE LOAD

When we first introduced the concept of strain energy at the beginning of this chapter, we considered the work done by an axial load $\mathbf{P}$ applied to the end of a rod of uniform cross section (Fig. 11.1). We defined the strain energy of the rod for an elongation x_1 as the work of the load $\mathbf{P}$ as it is slowly increased from 0 to the value P_1 corresponding to x_1. We wrote

$$\text{Strain energy} = U = \int_0^{x_1} P\,dx \qquad (11.2)$$

In the case of an elastic deformation, the work of the load $\mathbf{P}$, and thus the strain energy of the rod, were expressed as

$$U = \tfrac{1}{2}P_1x_1 \qquad (11.3)$$

Later, in Secs. 11.4 and 11.5, we computed the strain energy of structural members under various loading conditions by determining the strain-energy density u at every point of the member and integrating u over the entire member.

However, when a structure or member is subjected to a *single concentrated load*, it is possible to use Eq. (11.3) to evaluate its elastic strain energy, provided, of course, that the relation between the load and the resulting deformation is known. For instance, in the case of the cantilever beam of Example 11.03 (Fig. 11.27), we write

Fig. 11.27

$$U = \tfrac{1}{2}P_1y_1$$

and, substituting for y_1 the value obtained from the table of *Beam Deflections and Slopes* of Appendix D,

$$U = \frac{1}{2}P_1\!\left(\frac{P_1L^3}{3EI}\right) = \frac{P_1^2L^3}{6EI} \qquad (11.46)$$

A similar approach can be used to determine the strain energy of a structure or member subjected to a *single couple*. Recalling that the elementary work of a couple of moment M is $M\,d\theta$, where $d\theta$ is a small angle, we find, since M and θ are linearly related, that the elastic strain energy of a cantilever beam AB subjected to a single couple $\mathbf{M}_1$ at its end A (Fig. 11.28) can be expressed as

Fig. 11.28

$$U = \int_0^{\theta_1} M\,d\theta = \tfrac{1}{2}M_1\theta_1 \qquad (11.47)$$

where θ_1 is the slope of the beam at A. Substituting for θ_1 the value obtained from Appendix D, we write

$$U = \frac{1}{2}M_1\left(\frac{M_1 L}{EI}\right) = \frac{M_1^2 L}{2EI} \qquad (11.48)$$

In a similar way, the elastic strain energy of a uniform circular shaft AB of length L subjected at its end B to a single torque $\mathbf{T}_1$ (Fig. 11.29) can be expressed as

$$U = \int_0^{\phi_1} T \, d\phi = \frac{1}{2}T_1\phi_1 \qquad (11.49)$$

Fig. 11.29

Substituting for the angle of twist ϕ_1 from Eq. (3.16), we verify that

$$U = \frac{1}{2}T_1\left(\frac{T_1 L}{JG}\right) = \frac{T_1^2 L}{2JG}$$

as previously obtained in Sec. 11.5.

The method presented in this section may simplify the solution of many impact-loading problems. In Example 11.08, the crash of an automobile into a barrier (Fig. 11.30) is considered by using a simplified model consisting of a block and a simple beam.

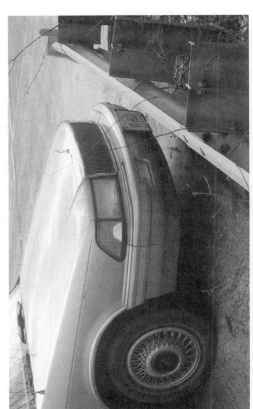

Fig. 11.30 As the automobile crashes into the barrier considerable energy will be dissipated as heat during the permanent deformation of the automobile and the barrier.

EXAMPLE 11.08

A block of mass m moving with a velocity $\mathbf{v}_0$ hits squarely the prismatic member AB at its midpoint C (Fig. 11.31). Determine (a) the equivalent static load P_m, (b) the maximum stress σ_m in the member, and (c) the maximum deflection x_m at point C.

(a) Equivalent Static Load. The maximum strain energy of the member is equal to the kinetic energy of the block before impact. We have

$$U_m = \tfrac{1}{2}mv_0^2 \qquad (11.50)$$

On the other hand, expressing U_m as the work of the equivalent horizontal static load as it is slowly applied at the midpoint C of the member, we write

$$U_m = \tfrac{1}{2}P_m x_m \qquad (11.51)$$

where x_m is the deflection of C corresponding to the static load P_m. From the table of *Beam Deflections and Slopes* of Appendix D, we find that

$$x_m = \frac{P_m L^3}{48EI} \qquad (11.52)$$

Substituting for x_m from (11.52) into (11.51), we write

$$U_m = \frac{1}{2}\frac{P_m^2 L^3}{48EI}$$

Solving for P_m and recalling Eq. (11.50), we find that the static load equivalent to the given impact loading is

$$P_m = \sqrt{\frac{96 U_m EI}{L^3}} = \sqrt{\frac{48 m v_0^2 EI}{L^3}} \qquad (11.53)$$

Fig. 11.31

(b) Maximum Stress. Drawing the free-body diagram of the member (Fig. 11.32), we find that the maximum value of the bending moment occurs at C and is $M_{max} = P_m L/4$. The maximum stress, therefore, occurs in a transverse section through C and is equal to

$$\sigma_m = \frac{M_{max}c}{I} = \frac{P_m L c}{4I}$$

Substituting for P_m from (11.53), we write

$$\sigma_m = \sqrt{\frac{3mv_0^2 EI}{L(I/c)^2}}$$

Fig. 11.32

(c) Maximum Deflection. Substituting into Eq. (11.52) the expression obtained for P_m in (11.53), we have

$$x_m = \frac{L^3}{48EI}\sqrt{\frac{48mv_0^2 EI}{L^3}} = \sqrt{\frac{mv_0^2 L^3}{48EI}}$$

11.10. DEFLECTION UNDER A SINGLE LOAD BY THE WORK-ENERGY METHOD

We saw in the preceding section that, if the deflection x_1 of a structure or member under a single concentrated load $\mathbf{P}_1$ is known, the corresponding strain energy U is obtained by writing

$$U = \tfrac{1}{2}P_1 x_1 \qquad (11.3)$$

A similar expression for the strain energy of a structural member under a single couple $\mathbf{M}_1$ is:

$$U = \tfrac{1}{2}M_1\theta_1 \qquad (11.47)$$

Conversely, if the strain energy U of a structure or member subjected to a single concentrated load $\mathbf{P}_1$ or couple $\mathbf{M}_1$ is known, Eq. (11.3) or (11.47) can be used to determine the corresponding deflection x_1 or angle θ_1. In order to determine the deflection under a single load applied to a structure consisting of several component parts, it is easier, rather than use one of the methods of Chap. 9, to first compute the strain energy of the structure by integrating the strain-energy density over its various parts, as was done in Secs. 11.4 and 11.5, and then use either Eq. (11.3) or Eq. (11.47) to obtain the desired deflection. Similarly, the angle of twist ϕ_1 of a composite shaft can be obtained by integrating the strain-energy density over the various parts of the shaft and solving Eq. (11.49) for ϕ_1.

It should be kept in mind that the method presented in this section can be used *only if the given structure is subjected to a single concentrated load or couple.* The strain energy of a structure subjected to several loads *cannot* be determined by computing the work of each load as if it were applied independently to the structure (see Sec. 11.11). We can also observe that, even if it were possible to compute the strain energy of the structure in this manner, only one equation would be available to determine the deflections corresponding to the various loads. In Secs. 11.12 and 11.13, another method based on the concept of strain energy is presented, one that can be used to determine the deflection or slope at a given point of a structure, even when that structure is subjected simultaneously to several concentrated loads, distributed loads, or couples.

EXAMPLE 11.09

A load $\mathbf{P}$ is supported at B by two uniform rods of the same cross-sectional area A (Fig. 11.33). Determine the vertical deflection of point B.

Fig. 11.33

The strain energy of the system under the given load was determined in Example 11.02. Equating the expression obtained for U to the work of the load, we write

$$U = 0.364\frac{P^2l}{AE} = \frac{1}{2}Py_B$$

and, solving for the vertical deflection of B,

$$y_B = 0.728\frac{Pl}{AE}$$

Remark. We should note that, once the forces in the two rods have been obtained (see Example 11.02), the deformations $\delta_{B/C}$ and $\delta_{B/D}$ of the rods could be obtained by the method of Chap. 2. Determining the vertical deflection of point B from these deformations, however, would require a careful geometric analysis of the various displacements involved. The strain-energy method used here makes such an analysis unnecessary.

EXAMPLE 11.10

Determine the deflection of end A of the cantilever beam AB (Fig. 11.34), taking into account the effect of (a) the normal stresses only, (b) both the normal and shearing stresses.

Fig. 11.34

the normal stresses was considered, we write

and, solving for y_A,

$$\frac{P^2L^3}{6EI} = \frac{1}{2}Py_A$$

$$y_A = \frac{PL^3}{3EI}$$

(b) Effect of Normal and Shearing Stresses. We now substitute for U the expression (11.24) obtained in Example 11.05, where the effects of both the normal and shearing stresses were taken into account. We have

$$\frac{P^2L^3}{6EI}\left(1 + \frac{3Eh^2}{10GL^2}\right) = \frac{1}{2}Py_A$$

and, solving for y_A,

$$y_A = \frac{PL^3}{3EI}\left(1 + \frac{3Eh^2}{10GL^2}\right)$$

We note that the relative error when the effect of shear is neglected is the same that was obtained in Example 11.05, i.e., less than $0.9(h/L)^2$. As we indicated then, this is less than 0.9% for a beam with a ratio h/L less than $\frac{1}{10}$.

(a) Effect of Normal Stresses. The work of the force $\mathbf{P}$ as it is slowly applied to A is

$$U = \frac{1}{2}Py_A$$

Substituting for U the expression obtained for the strain energy of the beam in Example 11.03, where only the effect of

EXAMPLE 11.11

A torque $\mathbf{T}$ is applied at the end D of shaft BCD (Fig. 11.35). Knowing that both portions of the shaft are of the same material and same length, but that the diameter of BC is twice the diameter of CD, determine the angle of twist for the entire shaft.

Fig. 11.35

The strain energy of a similar shaft was determined in Example 11.04 by breaking the shaft into its component parts BC and CD. Making $n = 2$ in Eq. (11.23), we have

$$U = \frac{17}{32}\frac{T^2L}{2GJ}$$

where G is the modulus of rigidity of the material and J the polar moment of inertia of portion CD of the shaft. Setting U equal to the work of the torque as it is slowly applied to end D, and recalling Eq. (11.49), we write

$$\frac{17}{32}\frac{T^2L}{2GJ} = \frac{1}{2}T\phi_{D/B}$$

and, solving for the angle of twist $\phi_{D/B}$,

$$\phi_{D/B} = \frac{17TL}{32GJ}$$

SAMPLE PROBLEM 11.3

The block D of mass m is released from rest and falls a distance h before it strikes the midpoint C of the aluminum beam AB. Using $E = 73$ GPa, determine (a) the maximum deflection of point C, (b) the maximum stress that occurs in the beam.

$m = 80$ kg

$h = 40$ mm

$L = 1$ m

40 mm

40 mm

Position 1

y_m

Position 2

From Appendix D

$P_m = \dfrac{48\,EI}{L^3}\,y_m$

$y_m = \dfrac{P_m L^3}{48\,EI}$

SOLUTION

Principle of Work and Energy. Since the block is released from rest, we note that in position *1* both the kinetic energy and the strain energy are zero. In position 2, where the maximum deflection y_m occurs, the kinetic energy is again zero. Referring to the table of *Beam Deflections and Slopes* of Appendix D, we find the expression for y_m shown. The strain energy of the beam in position 2 is

$$U_2 = \frac{1}{2} P_m y_m = \frac{1}{2}\frac{48EI}{L^3}y_m^2 \qquad U_2 = \frac{24EI}{L^3}y_m^2$$

We observe that the work done by the weight **W** of the block is $W(h + y_m)$. Equating the strain energy of the beam to the work done by **W**, we have

$$\frac{24EI}{L^3}y_m^2 = W(h + y_m) \tag{1}$$

a. Maximum Deflection of Point C. From the given data we have

$$EI = (73 \times 10^9 \text{ Pa})\tfrac{1}{12}(0.04 \text{ m})^4 = 15.573 \times 10^3 \text{ N} \cdot \text{m}^2$$

$$L = 1 \text{ m} \quad h = 0.040 \text{ m} \quad W = mg = (80 \text{ kg})(9.81 \text{ m/s}^2) = 784.8 \text{ N}$$

Substituting into Eq. (1), we obtain and solve the quadratic equation

$$(3'/3.8 \times 10^3)y_m^2 - 784.8y_m \qquad 31.39 - 0 \qquad y_m = 10.27 \text{ mm} \qquad \blacktriangledown$$

b. Maximum Stress. The value of P_m is

$$P_m = \frac{48EI}{L^3}\,y_m = \frac{48(15.573 \times 10^3 \text{ N} \cdot \text{m})}{(1 \text{ m})^3}(0.01027 \text{ m}) \qquad P_m = 7677 \text{ N}$$

Recalling that $\sigma_m = M_{\max} c/I$ and $M_{\max} = \tfrac{1}{4} P_m L$, we write

$$\sigma_m = \frac{(\tfrac{1}{4}P_m L)c}{I} = \frac{\tfrac{1}{4}(7677 \text{ N})(1 \text{ m})(0.020 \text{ m})}{\tfrac{1}{12}(0.040 \text{ m})^4} \qquad \sigma_m = 179.9 \text{ MPa} \qquad \blacktriangledown$$

An approximation for the work done by the weight of the block can be obtained by omitting y_m from the expression for the work and from the right-hand member of Eq. (1), as was done in Example 11.07. If this approximation is used here, we find $y_m = 9.16$ mm; the error is 10.8%. However, if an 8-kg block is dropped from a height of 400 mm, producing the same value of Wh, omitting y_m from the right-hand member of Eq. (1) results in an error of only 1.2%. A further discussion of this approximation is given in Prob. 11.70.

SAMPLE PROBLEM 11.4

Members of the truss shown consist of sections of aluminum pipe with the cross-sectional areas indicated. Using $E = 73$ GPa, determine the vertical deflection of point E caused by the load **P**.

SOLUTION

Axial Forces in Truss Members. The reactions are found by using the free-body diagram of the entire truss. We then consider in sequence the equilibrium of joints, E, C, D, and B. At each joint we determine the forces indicated by dashed lines. At joint B, the equation $\Sigma F_x = 0$ provides a check of our computations.

$A_x = 21P/8$

$A_y = P$

$B = 21P/8$

$\Sigma F_y = 0: F_{DE} = -\tfrac{17}{8}P \quad \Sigma F_x = 0: F_{AC} = +\tfrac{15}{8}P \quad \Sigma F_y = 0: F_{AD} = +\tfrac{5}{4}P$

$\Sigma F_x = 0: F_{CE} = +\tfrac{15}{8}P \quad \Sigma F_y = 0: F_{CD} = 0 \quad \Sigma F_x = 0: F_{BD} = -\tfrac{21}{8}P$

$\Sigma F_y = 0: F_{AB} = 0 \quad \Sigma F_x = 0:$ (Checks)

Strain Energy. Noting that E is the same for all members, we express the strain energy of the truss as follows

$$U = \sum \frac{F_i^2 L_i}{2A_i E} = \frac{1}{2E}\sum \frac{F_i^2 L_i}{A_i} \tag{1}$$

where F_i is the force in a given member as indicated in the following table and where the summation is extended over all members of the truss.

Member	F_i	L_i, m	A_i, m^2	$\dfrac{F_i^2 L_i}{A_i}$
AB	0	0.8	500×10^{-6}	0
AC	$+15P/8$	0.6	500×10^{-6}	$4\,219P^2$
AD	$+5P/4$	1.0	500×10^{-6}	$3\,125P^2$
BD	$-21P/8$	0.6	1000×10^{-6}	$4\,134P^2$
CD	0	0.8	1000×10^{-6}	0
CE	$+15P/8$	1.5	500×10^{-6}	$10\,547P^2$
DE	$-17P/8$	1.7	1000×10^{-6}	$7\,677P^2$

$$\sum \frac{F_i^2 L_i}{A_i} = 29\,700P^2$$

Returning to Eq. (1), we have

$$U = (1/2E)(29.7 \times 10^3 P^2).$$

Principle of Work-Energy. We recall that the work done by the load **P** as it is gradually applied is $\tfrac{1}{2}Py_E$. Equating the work done by **P** to the strain energy U and recalling that $E = 73$ GPa and $P = 40$ kN, we have

$$\frac{1}{2}Py_E = U \qquad \frac{1}{2}Py_E = \frac{1}{2E}(29.7 \times 10^3 P^2)$$

$$y_E = \frac{1}{E}(29.7 \times 10^3 P) = \frac{(29.7 \times 10^3)(40 \times 10^3)}{73 \times 10^9}$$

$$y_E = 16.27 \times 10^{-3}\ \text{m} \qquad y_E = 16.27\ \text{mm} \downarrow \quad \blacktriangleleft$$

PROBLEMS

11.50 The cylindrical block E has a speed $v_0 = 16$ ft/s when it strikes squarely the yoke BD that is attached to the $\frac{7}{8}$-in.-diameter rods AB and CD. Knowing that the rods are made of a steel for which $\sigma_Y = 50$ ksi and $E = 29 \times 10^6$ psi, determine the weight of the block E for which the factor of safety is five with respect to permanent deformation of the rods.

11.51 The 18-lb cylindrical block E has a horizontal velocity v_0 when it strikes squarely the yoke BD that is attached to the $\frac{7}{8}$-in.-diameter rods AB and CD. Knowing that the rods are made of a steel for which $\sigma_Y = 50$ ksi and $E = 29 \times 10^6$ psi, determine the maximum allowable speed v_0 if the rods are not to be permanently deformed.

Fig. P11.50 and P11.51

11.52 The uniform rod AB is made of a brass for which $\sigma_Y = 125$ MPa and $E = 105$ GPa. Collar D moves along the rod and has a speed $v_0 = 3$ m/s as it strikes a small plate attached to end B of the rod. Using a factor of safety of four, determine the largest allowable mass of the collar if the rod is not to be permanently deformed.

Fig. P11.52

11.53 Solve Prob. 11.52, assuming that the length of the brass rod is increased from 1.2 m to 2.4 m.

11.54 Collar D is released from rest in the position shown and is stopped by a small plate attached at end C of the vertical rod ABC. Determine the mass of the collar for which the maximum normal stress in portion BC is 125 MPa.

Fig. P11.54

11.55 Solve Prob. 11.54, assuming that both portions of rod ABC are made of aluminum.

11.56 The 100-lb collar G is released from rest in the position shown and is stopped by plate BDF that is attached to the $\frac{7}{8}$-in.-diameter steel rod CD and to the $\frac{5}{8}$-in.-diameter steel rods AB and EF. Knowing that for the grade of steel used $\sigma_{\text{all}} = 24$ ksi and $E = 29 \times 10^6$ psi, determine the largest allowable distance h.

Fig. P11.56

11.57 Solve Prob. 11.56, assuming that the $\frac{7}{8}$-in.-diameter steel rod CD is replaced by a $\frac{7}{8}$-in.-diameter rod made of a grade of aluminum for which $\sigma_{\text{all}} = 20$ ksi and $E = 10.6 \times 10^6$ psi.

Fig. P11.58

Fig. P11.58

11.58 The steel beam AB is struck squarely at its midpoint C by a 45-kg block moving horizontally with a speed $v_0 = 2$ m/s. Using $E = 200$ GPa, determine (a) the equivalent static load, (b) the maximum normal stress in the beam, (c) the maximum deflection of the midpoint C of the beam.

11.59 Solve Prob. 11.58, assuming that the W150 × 13.5 rolled-steel beam is rotated by 90° about its longitudinal axis so that its web is vertical.

11.60 The post AB consists of a steel pipe of 3.5-in. outer diameter and 0.3-in. wall thickness. A 15-lb block C moving horizontally with a velocity v_0 hits the post squarely at A. Using $E = 29 \times 10^6$ psi, determine the largest speed v_0 for which the maximum normal stress in the pipe does not exceed 24 ksi.

11.61 Solve Prob. 11.60, assuming that the post AB consists of a solid steel rod of 3.5-in. diameter.

11.62 and 11.63 The 2-kg block D is dropped from the position shown onto the end of a 16-mm-diameter rod. Knowing that $E = 200$ GPa, determine (a) the maximum deflection of end A, (b) the maximum bending moment in the rod, (c) the maximum normal stress in the rod.

11.64 The 50-lb block D is dropped from a height of 20 in. onto the steel beam AB. Knowing that $E = 29 \times 10^6$ psi, determine (a) the maximum deflection at point E, (b) the maximum normal stress in the beam.

11.65 A 160-lb diver jumps from a height of 20 in. onto end C of a diving board having the uniform cross section shown. Assuming that the diver's legs remain rigid and using $E = 1.8 \times 10^6$ psi, determine (a) the maximum deflection at point C, (b) the maximum normal stress in the board, (c) the equivalent static load.

Fig. P11.62

Fig. P11.63

Fig. P11.64

Fig. P11.65

Fig. P11.66

11.66 The 3-lb block D is released from rest in the position shown and strikes a steel bar AB having the uniform cross section shown. The bar is supported at each end by springs of constant 20 kips/in. Using $E = 29 \times 10^6$ psi, determine the maximum deflection at the midpoint of the bar.

11.67 Solve Prob 11.66, assuming that the constant of each spring is 40 kips/in.

11.68 A block of weight W is placed in contact with a beam at some given point D and released. Show that the resulting maximum deflection at point D is twice as large as the deflection due to a static load W applied at D.

11.69 A block of weight W is dropped from a height h onto the horizontal beam AB and hits it at point D. (a) Show that the maximum deflection y_m at point D can be expressed as

$$y_m = y_{st}\left(1 + \sqrt{1 + \frac{2h}{y_{st}}}\right)$$

where y_{st} represents the deflection at D caused by a static load W applied at that point and where the quantity in parentheses is referred to as the *impact factor*. (b) Compute the impact factor for the beam and impact loading of Prob. 11.62.

11.70 A block of weight W is dropped from a height h onto the horizontal beam AB and hits it at point D. (a) Denoting by y_m the exact value of the maximum deflection at D and by y_m' the value obtained by neglecting the effect of this deflection on the change in potential energy of the block, show that the absolute value of the relative error $(y_m' - y_m)/y_m$ never exceeds $y_m'/2h$. (b) Check the result obtained in part a by solving part a of Prob. 11.62 without taking y_m into account when determining the change in potential energy of the load, and comparing the answer obtained in this way with the exact answer to that problem.

11.71 and 11.72 Using the method of work and energy, determine the deflection at point D caused by the load **P**.

Fig. P11.69 and P11.70

Fig. P11.72

Fig. P11.71

11.73 and 11.74 Using the method of work and energy, determine the slope at point D caused by the couple M_0.

Fig. P11.73

Fig. P11.74

11.75 and 11.76 Using the method of work and energy, determine the deflection at point C caused by the load P.

Fig. P11.75

Fig. P11.76

Fig. P11.77

11.77 Using the method of work and energy, determine the slope at point B caused by the couple M_0.

Fig. P11.78

11.78 Using the method of work and energy, determine the slope at point A caused by the couple M_0.

11.79 The 12-mm-diameter steel rod ABC has been bent into the shape shown. Knowing that $E = 200$ GPa and $G = 77.2$ GPa, determine the deflection of end C caused by the 150-N force.

Fig. P11.79

Fig. P11.80

11.80 Two steel shafts, each of 0.75-in. diameter, are connected by the gears shown. Knowing that $G = 11.2 \times 10^6$ psi and that shaft DF is fixed at F, determine the angle through which end A rotates when a 750 lb · in. torque is applied at A. Ignore the strain energy due to the bending of the shafts.

11.81 The 20-mm-diameter steel rod AC is welded to the 20-mm-diameter steel shaft AB as shown. End C of rod AC is touching the rigid surface when a couple $\mathbf{T}_B$ is applied to a disk attached to shaft AB. Knowing that the bearings are self-aligning and exert no couples on the shaft, determine the angle of rotation of the disk when $T_B = 400$ N · m. Use $E = 200$ GPa and $G = 77.2$ GPa. Consider the strain energy due to both bending and twisting of shaft AB and to bending of arm CD.

Fig. P11.81

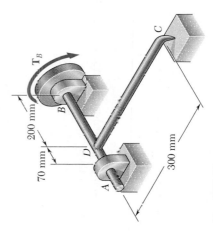

Fig. P11.82

11.82 A disk of radius a has been welded to end B of the solid steel shaft AB. A cable is then wrapped around the disk and a vertical force $\mathbf{P}$ is applied to end C of the cable. Knowing that the radius of the shaft is r and neglecting the deformations of the disk and of the cable, show that the deflection of point C caused by the application of $\mathbf{P}$ is

$$\delta_C = \frac{PL^3}{3EI}\left(1 + 1.5\frac{Er^2}{GL^2}\right)$$

11.83   The thin-walled hollow cylindrical member AB has a noncircular cross section of nonuniform thickness. Using the expression given in Eq. (3.53) of Sec. 3.13, and the expression for the strain-energy density given in Eq. (11.19) of Sec. 11.4, show that the angle of twist of member AB is

$$\phi = \frac{TL}{4\mathscr{A}^2 G} \oint \frac{ds}{t}$$

where ds is an element of the center line of the wall cross section and $\mathscr{A}$ is the area enclosed by that center line.

Fig. P11.83

11.84 and 11.85   Each member of the truss shown has a uniform cross-sectional area A. Using the method of work and energy, determine the horizontal deflection of the point of application of the load **P**.

Fig. P11.84

Fig. P11.85

11.86   Each member of the truss shown is made of steel; the cross-sectional area of member BC is 800 mm² and for all other members the cross-sectional area is 400 mm². Using $E = 200$ GPa, determine the deflection of point D caused by the 60-kN load shown.

Fig. P11.86

11.87   Each member of the truss shown is made of steel and has a uniform cross-sectional area of 3 in². Using $E = 29 \times 10^6$ psi, determine the vertical deflection of joint A caused by the application of the 16-kip load.

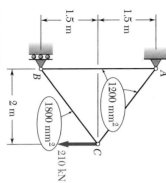

Fig. P11.87

11.88   Members of the truss shown are made of steel and have the cross-sectional areas shown. Using $E = 200$ GPa, determine the vertical deflection of joint C caused by the application of the 210-kN load.

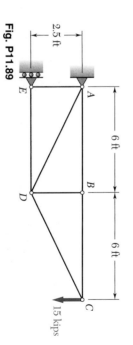

Fig. P11.88

11.89   Each member of the truss shown is made of steel and has a uniform cross-sectional area of 5 in². Using $E = 29 \times 10^6$ psi, determine the vertical deflection of joint C caused by the application of the 15-kip load.

Fig. P11.89

*11.11. WORK AND ENERGY UNDER SEVERAL LOADS

In this section, the strain energy of a structure subjected to several loads will be considered and will be expressed in terms of the loads and the resulting deflections.

Consider an elastic beam AB subjected to two concentrated loads $\mathbf{P}_1$ and $\mathbf{P}_2$. The strain energy of the beam is equal to the work of $\mathbf{P}_1$ and $\mathbf{P}_2$ as they are slowly applied to the beam at C_1 and C_2, respectively (Fig. 11.36). However, in order to evaluate this work, we must first express the deflections x_1 and x_2 in terms of the loads $\mathbf{P}_1$ and $\mathbf{P}_2$.

Let us assume that only $\mathbf{P}_1$ is applied to the beam (Fig. 11.37). We note that both C_1 and C_2 are deflected and that their deflections are proportional to the load P_1. Denoting these deflections by x_{11} and x_{21}, respectively, we write

Fig. 11.36

$$x_{11} = \alpha_{11}P_1 \qquad x_{21} = \alpha_{21}P_1 \tag{11.54}$$

where α_{11} and α_{21} are constants called *influence coefficients*. These constants represent the deflections of C_1 and C_2, respectively, when a unit load is applied at C_1 and are characteristics of the beam AB.

Let us now assume that only $\mathbf{P}_2$ is applied to the beam (Fig. 11.38). Denoting by x_{12} and x_{22}, respectively, the resulting deflections of C_1 and C_2, we write

Fig. 11.37

$$x_{12} = \alpha_{12}P_2 \qquad x_{22} = \alpha_{22}P_2 \tag{11.55}$$

where α_{12} and α_{22} are the influence coefficients representing the deflections of C_1 and C_2, respectively, when a unit load is applied at C_2. Applying the principle of superposition, we express the deflections x_1 and x_2 of C_1 and C_2 when both loads are applied (Fig. 11.36) as

Fig. 11.38

$$x_1 = x_{11} + x_{12} = \alpha_{11}P_1 + \alpha_{12}P_2 \tag{11.56}$$
$$x_2 = x_{21} + x_{22} = \alpha_{21}P_1 + \alpha_{22}P_2 \tag{11.57}$$

To compute the work done by $\mathbf{P}_1$ and $\mathbf{P}_2$, and thus the strain energy of the beam, it is convenient to assume that $\mathbf{P}_1$ is first applied slowly at C_1 (Fig. 11.39a). Recalling the first of Eqs. (11.54), we express the work of $\mathbf{P}_1$ as

(a)

(b)

Fig. 11.39

$$\tfrac{1}{2}P_1x_{11} = \tfrac{1}{2}P_1(\alpha_{11}P_1) = \tfrac{1}{2}\alpha_{11}P_1^2 \tag{11.58}$$

and note that $\mathbf{P}_2$ does no work while C_2 moves through x_{21}, since it has not yet been applied to the beam.

Now we slowly apply $\mathbf{P}_2$ at C_2 (Fig. 11.39b); recalling the second of Eqs. (11.55), we express the work of $\mathbf{P}_2$ as

$$\tfrac{1}{2}P_2x_{22} = \tfrac{1}{2}P_2(\alpha_{22}P_2) = \tfrac{1}{2}\alpha_{22}P_2^2 \tag{11.59}$$

But, as $\mathbf{P}_2$ is slowly applied at C_2, the point of application of $\mathbf{P}_1$ moves through x_{12} from C'_1 to C_1, and the load $\mathbf{P}_1$ does work. Since $\mathbf{P}_1$ is *fully*

(a) Load-displacement
diagram for C_1

(b) Load-displacement
diagram for C_2

Fig. 11.40

applied during this displacement (Fig. 11.40), its work is equal to $P_1 x_{12}$ or, recalling the first of Eqs. (11.55),

$$P_1 x_{12} = P_1(\alpha_{12} P_2) = \alpha_{12} P_1 P_2 \qquad (11.60)$$

Adding the expressions obtained in (11.58), (11.59), and (11.60), we express the strain energy of the beam under the loads $\mathbf{P}_1$ and $\mathbf{P}_2$ as

$$U = \tfrac{1}{2}(\alpha_{11} P_1^2 + 2\alpha_{12} P_1 P_2 + \alpha_{22} P_2^2) \qquad (11.61)$$

If the load $\mathbf{P}_2$ had first been applied to the beam (Fig. 11.41a), and then the load $\mathbf{P}_1$ (Fig. 11.41b), the work done by each load would have been as shown in Fig. 11.42. Calculations similar to those we have just carried out would lead to the following alternative expression for the strain energy of the beam:

$$U = \tfrac{1}{2}(\alpha_{22} P_2^2 + 2\alpha_{21} P_2 P_1 + \alpha_{11} P_1^2) \qquad (11.62)$$

Equating the right-hand members of Eqs. (11.61) and (11.62), we find that $\alpha_{12} = \alpha_{21}$, and thus conclude that the deflection produced at C_1 by a unit load applied at C_2 is equal to the deflection produced at C_2 by a unit load applied at C_1. This is known as *Maxwell's reciprocal theorem*, after the British physicist James Clerk Maxwell (1831–1879).

(a)

(b)

Fig. 11.41

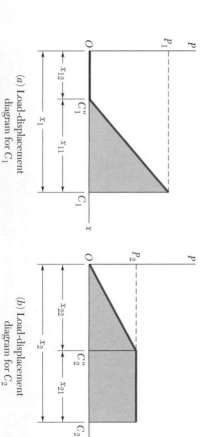

(a) Load-displacement
diagram for C_1

(b) Load-displacement
diagram for C_2

Fig. 11.42

While we are now able to express the strain energy U of a structure subjected to several loads as a function of these loads, we cannot use the method of Sec. 11.10 to determine the deflection of such a structure. Indeed, computing the strain energy U by integrating the strain-energy density u over the structure and substituting the expression obtained into (11.61) would yield only one equation, which clearly could not be solved for the various coefficients α.

*11.12. CASTIGLIANO'S THEOREM

We recall the expression obtained in the preceding section for the strain energy of an elastic structure subjected to two loads $\mathbf{P}_1$ and $\mathbf{P}_2$:

$$U = \tfrac{1}{2}(\alpha_{11}P_1^2 + 2\alpha_{12}P_1P_2 + \alpha_{22}P_2^2) \qquad (11.61)$$

where α_{11}, α_{12}, and α_{22} are the influence coefficients associated with the points of application C_1 and C_2 of the two loads. Differentiating both members of Eq. (11.61) with respect to P_1 and recalling Eq. (11.56), we write

$$\frac{\partial U}{\partial P_1} = \alpha_{11}P_1 + \alpha_{12}P_2 = x_1 \qquad (11.63)$$

Differentiating both members of Eq. (11.61) with respect to P_2, recalling Eq. (11.57), and keeping in mind that $\alpha_{12} = \alpha_{21}$, we have

$$\frac{\partial U}{\partial P_2} = \alpha_{12}P_1 + \alpha_{22}P_2 = x_2 \qquad (11.64)$$

More generally, if an elastic structure is subjected to n loads $\mathbf{P}_1, \mathbf{P}_2, \ldots, \mathbf{P}_n$, the deflection x_j of the point of application of $\mathbf{P}_j$, measured along the line of action of $\mathbf{P}_j$, can be expressed as the partial derivative of the strain energy of the structure with respect to the load $\mathbf{P}_j$. We write

$$\boxed{x_j = \frac{\partial U}{\partial P_j}} \qquad (11.65)$$

This is *Castigliano's theorem*, named after the Italian engineer Alberto Castigliano (1847–1884) who first stated it.[†]

[†]In the case of an elastic structure subjected to n loads $\mathbf{P}_1, \mathbf{P}_2, \ldots, \mathbf{P}_n$, the deflection of the point of application of $\mathbf{P}_j$, measured along the line of action of $\mathbf{P}_j$, can be expressed as

$$x_j = \sum_k \alpha_{jk}P_k \qquad (11.66)$$

and the strain energy of the structure is found to be

$$U = \tfrac{1}{2}\sum_i\sum_k \alpha_{ik}P_iP_k \qquad (11.67)$$

Differentiating U with respect to P_j, and observing that P_j is found in terms corresponding to either $i = j$ or $k = j$, we write

$$\frac{\partial U}{\partial P_j} = \frac{1}{2}\sum_k \alpha_{jk}P_k + \frac{1}{2}\sum_i \alpha_{ij}P_i$$

or, since $\alpha_{ij} = \alpha_{ji}$,

$$\frac{\partial U}{\partial P_j} = \frac{1}{2}\sum_k \alpha_{jk}P_k + \frac{1}{2}\sum_i \alpha_{ji}P_i = \sum_k \alpha_{jk}P_k$$

Recalling Eq. (11.66), we verify that

$$x_j = \frac{\partial U}{\partial P_j} \qquad (11.65)$$

Recalling that the work of a couple $\mathbf{M}$ is $\frac{1}{2}M\theta$, where θ is the angle of rotation at the point where the couple is slowly applied, we note that Castigliano's theorem may be used to determine the slope of a beam at the point of application of a couple $\mathbf{M}_j$. We have

$$\theta_j = \frac{\partial U}{\partial M_j} \qquad (11.68)$$

Similarly, the angle of twist ϕ_j in a section of a shaft where a torque $\mathbf{T}_j$ is slowly applied is obtained by differentiating the strain energy of the shaft with respect to T_j:

$$\phi_j = \frac{\partial U}{\partial T_j} \qquad (11.69)$$

*11.13. DEFLECTIONS BY CASTIGLIANO'S THEOREM

We saw in the preceding section that the deflection x_j of a structure at the point of application of a load $\mathbf{P}_j$ can be determined by computing the partial derivative $\partial U/\partial P_j$ of the strain energy U of the structure. As we recall from Secs. 11.4 and 11.5, the strain energy U is obtained by integrating or summing over the structure the strain energy of each element of the structure. The calculation by Castigliano's theorem of the deflection x_j is simplified if the differentiation with respect to the load P_j is carried out before the integration or summation.

In the case of a beam, for example, we recall from Sec. 11.4 that

$$U = \int_0^L \frac{M^2}{2EI} dx \qquad (11.17)$$

and determine the deflection x_j of the point of application of the load $\mathbf{P}_j$ by writing

$$x_j = \frac{\partial U}{\partial P_j} = \int_0^L \frac{M}{EI} \frac{\partial M}{\partial P_j} dx \qquad (11.70)$$

In the case of a truss consisting of n uniform members of length L_i, cross-sectional area A_i, and internal force F_i, we recall Eq. (11.14) and express the strain energy U of the truss as

$$U = \sum_{i=1}^n \frac{F_i^2 L_i}{2A_i E} \qquad (11.71)$$

The deflection x_j of the point of application of the load P_j is obtained by differentiating with respect to P_j each term of the sum. We write

$$x_j = \frac{\partial U}{\partial P_j} = \sum_{i=1}^n \frac{F_i L_i}{A_i E} \frac{\partial F_i}{\partial P_j} \qquad (11.72)$$

EXAMPLE 11.12

The cantilever beam AB supports a uniformly distributed load w and a concentrated load $\mathbf{P}$ as shown (Fig. 11.43). Knowing that $L = 2$ m, $w = 4$ kN/m, $P = 6$ kN, and $EI = 5$ MN $\cdot$ m^2, determine the deflection at A.

Fig. 11.43

The deflection y_A of the point A where the load $\mathbf{P}$ is applied is obtained from Eq. (11.70). Since $\mathbf{P}$ is vertical and directed downward, y_A represents a vertical deflection and is positive downward. We have

$$y_A = \frac{\partial U}{\partial P} = \int_0^L \frac{M}{EI} \frac{\partial M}{\partial P} dx \qquad (11.73)$$

The bending moment M at a distance x from A is

$$M = -(Px + \tfrac{1}{2} wx^2) \qquad (11.74)$$

and its derivative with respect to P is

$$\frac{\partial M}{\partial P} = -x$$

Substituting for M and $\partial M / \partial P$ into Eq. (11.73), we write

$$y_A = \frac{1}{EI} \int_0^L \left(Px^2 + \frac{1}{2} wx^3 \right) dx$$

$$y_A = \frac{1}{EI} \left(\frac{PL^3}{3} + \frac{wL^4}{8} \right) \qquad (11.75)$$

Substituting the given data, we have

$$y_A = \frac{1}{5 \times 10^6 \text{ N} \cdot \text{m}^2}$$
$$\left[\frac{(6 \times 10^3 \text{ N})(2 \text{ m})^3}{3} + \frac{(4 \times 10^3 \text{ N/m})(2 \text{ m})^4}{8} \right]$$

$$y_A = 4.8 \times 10^{-3} \text{ m} \qquad y_A - 4.8 \text{ mm} \downarrow$$

We note that the computation of the partial derivative $\partial M / \partial P$ *could not have been carried out if the numerical value of P had been substituted for P in the expression (11.74)* for the bending moment.

We can observe that the deflection x_j of a structure at a given point C_j can be obtained by the direct application of Castigliano's theorem only if a load $\mathbf{P}_j$ happens to be applied at C_j in the direction in which x_j is to be determined. When no load is applied at C_j, or when a load is applied in a direction other than the desired one, we can still obtain the deflection x_j by Castigliano's theorem if we use the following procedure: We apply a fictitious or "dummy" load $\mathbf{Q}_j$ at C_j in the direction in which the deflection x_j is to be determined and use Castigliano's theorem to obtain the deflection

$$x_j = \frac{\partial U}{\partial Q_j} \qquad (11.76)$$

due to $\mathbf{Q}_j$ and the actual loads. Making $Q_j = 0$ in Eq. (11.76) yields the deflection at C_j in the desired direction under the given loading.

The slope θ_j of a beam at a point C_j can be determined in a similar manner by applying a fictitious couple $\mathbf{M}_j$ at C_j, computing the partial derivative $\partial U / \partial M_j$, and making $M_j = 0$ in the expression obtained.

EXAMPLE 11.13

The cantilever beam AB supports a uniformly distributed load w (Fig. 11.44). Determine the deflection and slope at A.

Deflection at A. We apply a dummy downward load Q_A at A (Fig. 11.45) and write

$$y_A = \frac{\partial U}{\partial Q_A} = \int_0^L \frac{M}{EI} \frac{\partial M}{\partial Q_A} dx \qquad (11.77)$$

The bending moment M at a distance x from A is

$$M = -Q_A x - \frac{1}{2} w x^2 \qquad (11.78)$$

and its derivative with respect to Q_A is

$$\frac{\partial M}{\partial Q_A} = -x \qquad (11.79)$$

Substituting for M and $\partial M/\partial Q_A$ from (11.78) and (11.79) into (11.77), and making $Q_A = 0$, we obtain the deflection at A for the given loading:

$$y_A = \frac{1}{EI} \int_0^L \left(-\frac{1}{2} w x^2\right)(-x)\, dx = +\frac{w L^4}{8EI}$$

Since the dummy load was directed downward, the positive sign indicates that

$$y_A = \frac{w L^4}{8EI} \downarrow$$

Slope at A. We apply a dummy counterclockwise couple $\mathbf{M}_A$ at A (Fig. 11.46) and write

$$\theta_A = \frac{\partial U}{\partial M_A}$$

Recalling Eq. (11.17), we have

$$\theta_A = \frac{\partial}{\partial M_A} \int_0^L \frac{M^2}{2EI} dx = \int_0^L \frac{M}{EI} \frac{\partial M}{\partial M_A} dx \qquad (11.80)$$

The bending moment M at a distance x from A is

$$M = -M_A - \frac{1}{2} w x^2 \qquad (11.81)$$

and its derivative with respect to M_A is

$$\frac{\partial M}{\partial M_A} = -1 \qquad (11.82)$$

Substituting for M and $\partial M/\partial M_A$ from (11.81) and (11.82) into (11.80), and making $M_A = 0$, we obtain the slope at A for the given loading:

$$\theta_A = \frac{1}{EI} \int_0^L \left(-\frac{1}{2} w x^2\right)(-1)\, dx = +\frac{w L^3}{6EI}$$

Since the dummy couple was counterclockwise, the positive sign indicates that the angle θ_A is also counterclockwise:

$$\theta_A = \frac{w L^3}{6EI} \; \measuredangle$$

Fig. 11.44

Fig. 11.45

Fig. 11.46

EXAMPLE 11.14

A load **P** is supported at B by two rods of the same material and of the same cross-sectional area A (Fig. 11.47). Determine the horizontal and vertical deflection of point B.

Fig. 11.47

Fig. 11.48

We apply a dummy horizontal load **Q** at B (Fig. 11.48). From Castigliano's theorem we have

$$x_B = \frac{\partial U}{\partial Q} \qquad y_B = \frac{\partial U}{\partial P}$$

Recalling from Sec. 11.4 the expression (11.14) for the strain energy of a rod, we write

$$U = \frac{F_{BC}^2 (BC)}{2AE} + \frac{F_{BD}^2 (BD)}{2AE} \qquad (11.83)$$

where F_{BC} and F_{BD} represent the forces in BC and BD, respectively. We have, therefore,

$$x_B = \frac{\partial U}{\partial Q} = \frac{F_{BC}(BC)}{AE}\frac{\partial F_{BC}}{\partial Q} + \frac{F_{BD}(BD)}{AE}\frac{\partial F_{BD}}{\partial Q} \qquad (11.83)$$

and

$$y_B = \frac{\partial U}{\partial P} = \frac{F_{BC}(BC)}{AE}\frac{\partial F_{BC}}{\partial P} + \frac{F_{BD}(BD)}{AE}\frac{\partial F_{BD}}{\partial P} \qquad (11.84)$$

From the free-body diagram of pin B (Fig. 11.49), we obtain

$$F_{BC} = 0.6P + 0.8Q \qquad F_{BD} = -0.8P + 0.6Q \qquad (11.85)$$

Differentiating these expressions with respect to Q and P, we write

$$\frac{\partial F_{BC}}{\partial Q} = 0.8 \qquad \frac{\partial F_{BD}}{\partial Q} = 0.6$$

$$\frac{\partial F_{BC}}{\partial P} = 0.6 \qquad \frac{\partial F_{BD}}{\partial P} = -0.8 \qquad (11.86)$$

Fig. 11.49

Substituting from (11.85) and (11.86) into both (11.83) and (11.84), making $Q = 0$, and noting that $BC = 0.6l$ and $BD = 0.8l$, we obtain the horizontal and vertical deflections of point B under the given load **P**:

$$x_B = \frac{(0.6P)(0.6l)}{AE}(0.8) + \frac{(-0.8P)(0.8l)}{AE}(0.6)$$

$$= -0.096\frac{Pl}{AE}$$

$$y_B = \frac{(0.6P)(0.6l)}{AE}(0.6) + \frac{(-0.8P)(0.8l)}{AE}(-0.8)$$

$$= +0.728\frac{Pl}{AE}$$

Referring to the directions of the loads **Q** and **P**, we conclude that

$$x_B = 0.096\frac{Pl}{AE} \leftarrow \qquad y_B = 0.728\frac{Pl}{AE}\downarrow$$

We check that the expression obtained for the vertical deflection of B is the same that was found in Example 11.09.

*11.14. STATICALLY INDETERMINATE STRUCTURES

The reactions at the supports of a statically indeterminate elastic structure can be determined by Castigliano's theorem. In the case of a structure indeterminate to the first degree, for example, we designate one of the reactions as redundant and eliminate or modify accordingly the corresponding support. The redundant reaction is then treated as an unknown load that, together with the other loads, must produce deformations that are compatible with the original supports. We first calculate the strain energy U of the structure due to the combined action of the given loads and the redundant reaction. Observing that the partial derivative of U with respect to the redundant reaction represents the deflection (or slope) at the support that has been eliminated or modified, we then set this derivative equal to zero and solve the equation obtained for the redundant reaction.† The remaining reactions can be obtained from the equations of statics.

†This is in the case of a rigid support allowing no deflection. For other types of support, the partial derivative of U should be set equal to the allowed deflection.

EXAMPLE 11.15

Determine the reactions at the supports for the prismatic beam and loading shown (Fig. 11.50).

The beam is statically indeterminate to the first degree. We consider the reaction at A as redundant and release the beam from that support. The reaction $\mathbf{R}_A$ is now considered as an unknown load (Fig. 11.51) and will be determined from the condition that the deflection y_A at A must be zero. By Castigliano's theorem $y_A = \partial U / \partial R_A$, where U is the strain energy of the beam under the distributed load and the redundant reaction. Recalling Eq. (11.70), we write

$$y_A = \frac{\partial U}{\partial R_A} = \int_0^L \frac{M}{EI} \frac{\partial M}{\partial R_A} dx \qquad (11.87)$$

We now express the bending moment M for the loading of Fig. 11.51.

The bending moment at a distance x from A is

$$M = R_A x - \tfrac{1}{2} w x^2 \qquad (11.88)$$

and its derivative with respect to R_A is

$$\frac{\partial M}{\partial R_A} = x \qquad (11.89)$$

Substituting for M and $\partial M / \partial R_A$ from (11.88) and (11.89) into (11.87), we write

Fig. 11.50

Fig. 11.51

$$y_A = \frac{1}{EI} \int_0^L \left(R_A x^2 - \frac{1}{2} w x^3 \right) dx = \frac{1}{EI} \left(\frac{R_A L^3}{3} - \frac{w L^4}{8} \right)$$

Setting $y_A = 0$ and solving for R_A, we have

$$R_A = \tfrac{3}{8} w L \qquad \mathbf{R}_A = \tfrac{3}{8} w L \uparrow$$

From the conditions of equilibrium for the beam, we find that the reaction at B consists of the following force and couple:

$$\mathbf{R}_B = \tfrac{5}{8} w L \uparrow \qquad \mathbf{M}_B = \tfrac{1}{8} w L^2 \, \zeta$$

EXAMPLE 11.16

A load **P** is supported at B by three rods of the same material and the same cross-sectional area A (Fig. 11.52). Determine the force in each rod.

Fig. 11.52

The structure is statically indeterminate to the first degree. We consider the reaction at H as redundant and release rod BH from its support at H. The reaction $\mathbf{R}_H$ is now considered as an unknown load (Fig. 11.53) and will be determined from the condition that the deflection y_H of point H must be zero. By Castigliano's theorem $y_H = \partial U/\partial R_H$, where U is the strain energy of the three-rod system under the load **P** and the redundant reaction $\mathbf{R}_H$. Recalling Eq. (11.72), we write

$$y_H = \frac{F_{BC}(BC)}{AE}\frac{\partial F_{BC}}{\partial R_H} + \frac{F_{BD}(BD)}{AE}\frac{\partial F_{BD}}{\partial R_H}$$
$$+ \frac{F_{BH}(BH)}{AE}\frac{\partial F_{BH}}{\partial R_H} \qquad (11.90)$$

Fig. 11.53

We note that the force in rod BH is equal to R_H and write

$$F_{BH} = R_H \qquad (11.91)$$

Then, from the free-body diagram of pin B (Fig. 11.54), we obtain

$$F_{BC} = 0.6P - 0.6R_H \qquad F_{BD} = 0.8R_H - 0.8P \qquad (11.92)$$

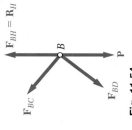

Fig. 11.54

Differentiating with respect to R_H the force in each rod, we write

$$\frac{\partial F_{BC}}{\partial R_H} = -0.6 \qquad \frac{\partial F_{BD}}{\partial R_H} = 0.8 \qquad \frac{\partial F_{BH}}{\partial R_H} = 1 \qquad (11.93)$$

Substituting from (11.91), (11.92), and (11.93) into (11.90), and noting that the lengths BC, BD, and BH are respectively equal to $0.6l$, $0.8l$, and $0.5l$, we write

$$y_H = \frac{1}{AE}[(0.6P - 0.6R_H)(0.6l)(-0.6)$$
$$+ (0.8R_H - 0.8P)(0.8l)(0.8) + R_H(0.5l)(1)]$$

Setting $y_H = 0$, we obtain

$$1.228R_H - 0.728P = 0$$

and, solving for R_H,

$$R_H = 0.593P$$

Carrying this value into Eqs. (11.91) and (11.92), we obtain the forces in the three rods:

$$F_{BC} = +0.244P \qquad F_{BD} = -0.326P \qquad F_{BH} = +0.593P$$

SAMPLE PROBLEM 11.5

For the truss and loading of Sample Prob. 11.4, determine the vertical deflection of joint C.

$P = 40$ kN

SOLUTION

Castigliano's Theorem. Since no vertical load is applied at joint C, we introduce the dummy load Q as shown. Using Castigliano's theorem, and denoting by F_i the force in a given member i caused by the combined loading of P and Q, we have, since $E = $ constant,

$$y_C = \sum \left(\frac{F_i L_i}{A_i E} \right) \frac{\partial F_i}{\partial Q} = \frac{1}{E} \sum \left(\frac{F_i L_i}{A_i} \right) \frac{\partial F_i}{\partial Q} \qquad (1)$$

Force in Members. Considering in sequence the equilibrium of joints E, C, B, and D, we determine the force in each member caused by load Q.

Joint E: $\quad F_{CE} = F_{DE} = 0$
Joint C: $\quad F_{AC} = 0;\ F_{CD} = -Q$
Joint B: $\quad F_{AB} = 0;\ F_{BD} = -\frac{3}{4}Q$

Joint D Force triangle

The force in each member caused by the load P was previously found in Sample Prob. 11.4. The total force in each member under the combined action of Q and P is shown in the following table. Forming $\partial F_i / \partial Q$ for each member, we then compute $(F_i L_i / A_i)(\partial F_i / \partial Q)$ as indicated in the table.

Member	F_i	$\partial F_i / \partial Q$	L_i, m	A_i, m²	$\left(\dfrac{F_i L_i}{A_i} \right) \dfrac{\partial F_i}{\partial Q}$
AB	0	0	0.8	500×10^{-6}	0
AC	$+15P/8$	0	0.6	500×10^{-6}	0
AD	$+5P/4 + 5Q/4$	$\frac{5}{4}$	1.0	500×10^{-6}	$+3125P + 3125Q$
BD	$-21P/8 - 3Q/4$	$-\frac{3}{4}$	0.6	1000×10^{-6}	$+1181P + 338Q$
CD	$-Q$	-1	0.8	1000×10^{-6}	$+800Q$
CE	$+15P/8$	0	1.5	500×10^{-6}	0
DE	$-17P/8$	0	1.7	1000×10^{-6}	0

$$\sum \left(\frac{F_i L_i}{A_i} \right) \frac{\partial F_i}{\partial Q} = 4306P + 4263Q$$

Deflection of C. Substituting into Eq. (1), we have

$$y_C = \frac{1}{E} \sum \left(\frac{F_i L_i}{A_i} \right) \frac{\partial F_i}{\partial Q} = \frac{1}{E}(4306P + 4263Q)$$

Since the load Q is not part of the original loading, we set $Q = 0$. Substituting the given data, $P = 40$ kN and $E = 73$ GPa, we find

$$y_C = \frac{4306(40 \times 10^3 \text{ N})}{73 \times 10^9 \text{ Pa}} = 2.36 \times 10^{-3} \text{ m} \qquad y_C = 2.36 \text{ mm} \downarrow \blacktriangleleft$$

SAMPLE PROBLEM 11.6

For the beam and loading shown, determine the deflection at point D. Use $E = 29 \times 10^6$ psi.

W10 × 15

$w = 1.8$ kips/ft

$a = 4.5$ ft
$b = 7.5$ ft
$L = 12$ ft

SOLUTION

Castigliano's Theorem. Since the given loading does not include a vertical load at point D, we introduce the dummy load $\mathbf{Q}$ as shown. Using Castigliano's theorem and noting that EI is constant, we write

$$y_D = \int \frac{M}{EI}\left(\frac{\partial M}{\partial Q}\right) dx = \frac{1}{EI}\int M\left(\frac{\partial M}{\partial Q}\right) dx \qquad (1)$$

The integration will be performed separately for portions AD and DB.

Reactions. Using the free-body diagram of the entire beam, we find

$$\mathbf{R}_A = \frac{wb^2}{2L} + Q\frac{b}{L}\uparrow \qquad \mathbf{R}_B = \frac{wb(a + \frac{1}{2}b)}{L} + Q\frac{a}{L}\uparrow$$

w

Portion AD of Beam. Using the free body shown, we find

$$M_1 = R_A x = \left(\frac{wb^2}{2L} + Q\frac{b}{L}\right) x \qquad \frac{\partial M_1}{\partial Q} = +\frac{bx}{L}$$

Substituting into Eq. (1) and integrating from A to D gives

$$\frac{1}{EI}\int M_1 \frac{\partial M_1}{\partial Q} dx = \frac{1}{EI}\int_0^a R_A x \left(\frac{bx}{L}\right) dx = \frac{R_A a^3 b}{3EIL}$$

We substitute for R_A and then set the dummy load Q equal to zero.

$$\frac{1}{EI}\int M_1 \frac{\partial M_1}{\partial Q} dx = \frac{wa^3 b^3}{6EIL^2} \qquad (2)$$

From A to D

wb $\frac{1}{2}b$

$a + \frac{1}{2}b$

$(x \le a)$

Portion DB of Beam. Using the free body shown, we find that the bending moment at a distance v from end B is

$$M_2 = R_B v - \frac{wv^2}{2} = \left[\frac{wb(a + \frac{1}{2}b)}{L} + Q\frac{a}{L}\right]v - \frac{wv^2}{2} \qquad \frac{\partial M_2}{\partial Q} = +\frac{av}{L}$$

Substituting into Eq. (1) and integrating from point B where $v = 0$, to point D where $v = b$, we write

$$\frac{1}{EI}\int M_2 \frac{\partial M_2}{\partial Q} dv = \frac{1}{EI}\int_0^b \left(R_B v - \frac{wv^2}{2}\right)\left(\frac{av}{L}\right) dv = \frac{R_B ab^3}{3EIL} - \frac{wab^4}{8EIL}$$

Substituting for R_B and setting $Q = 0$,

$$\frac{1}{EI}\int M_2 \frac{\partial M_2}{\partial Q} dv = \left[\frac{wb(a + \frac{1}{2}b)}{L}\right]\frac{ab^3}{3EIL} - \frac{wab^4}{8EIL} = \frac{R_B ab^3}{3EIL} - \frac{wab^4}{8EIL}$$

$$= \frac{5a^2 b^4 + ab^5}{24EIL^2} w \qquad (3)$$

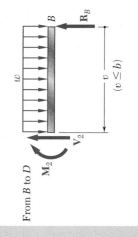

From B to D

w

$(v \le b)$

Deflection at Point D. Recalling Eqs. (1), (2), and (3), we have

$$y_D = \frac{wab^3}{24EIL^2}(4a^2 + 5ab + b^2) = \frac{wab^3}{24EIL^2}(4a + b)(a + b) = \frac{wab^3}{24EIL}(4a + b)$$

From Appendix C we find that $I = 68.9$ in⁴ for a W10 × 15. Substituting for I, w, a, b, and L their numerical values, we obtain

$$y_D = 0.262 \text{ in.} \downarrow \qquad \blacktriangleleft$$

SAMPLE PROBLEM 11.7

For the uniform beam and loading shown, determine the reactions at the supports.

SOLUTION

Castigliano's Theorem. The beam is indeterminate to the first degree and we choose the reaction $\mathbf{R}_A$ as redundant. Using Castigliano's theorem, we determine the deflection at A due to the combined action of $\mathbf{R}_A$ and the distributed load. Since EI is constant, we write

$$y_A = \int \frac{M}{EI}\left(\frac{\partial M}{\partial R_A}\right) dx = \frac{1}{EI}\int M\frac{\partial M}{\partial R_A}dx \qquad (1)$$

The integration will be performed separately for portions AB and BC of the beam. Finally, $\mathbf{R}_A$ is obtained by setting y_A equal to zero.

Free Body: Entire Beam. We express the reactions at B and C in terms of R_A and the distributed load

$$R_B = \frac{9}{4}wL - 3R_A \qquad R_C = 2R_A - \frac{3}{4}wL \qquad (2)$$

Portion AB of Beam. Using the free-body diagram shown, we find

$$M_1 = R_A x - \frac{wx^2}{2} \qquad \frac{\partial M_1}{\partial R_A} = x$$

Substituting into Eq. (1) and integrating from A to B, we have

$$\frac{1}{EI}\int M_1 \frac{\partial M_1}{\partial R_A}dx = \frac{1}{EI}\int_0^L \left(R_A x^2 - \frac{wx^3}{2} \right) dx = \frac{1}{EI}\left(\frac{R_A L^3}{3} - \frac{wL^4}{8} \right) \qquad (3)$$

Portion BC of Beam. We have

$$M_2 = \left(2R_A - \frac{3}{4}wL \right) v - \frac{wv^2}{2} \qquad \frac{\partial M_2}{\partial R_A} = 2v$$

Substituting into Eq. (1) and integrating from C, where $v = 0$, to B, where $v = \frac{1}{2}L$, we have

$$\frac{1}{EI}\int M_2 \frac{\partial M_2}{\partial R_A}dv = \frac{1}{EI}\int_0^{L/2}\left(4R_A v^2 - \frac{3}{2}wLv^2 - wv^3 \right) dv$$

$$= \frac{1}{EI}\left(\frac{R_A L^3}{6} - \frac{wL^4}{16} - \frac{wL^4}{64} \right) = \frac{1}{EI}\left(\frac{R_A L^3}{6} - \frac{5wL^4}{64} \right) \qquad (4)$$

Reaction at A. Adding the expressions obtained in (3) and (4), we determine y_A and set it equal to zero

$$y_A = \frac{1}{EI}\left(\frac{R_A L^3}{3} - \frac{wL^4}{8} \right) + \frac{1}{EI}\left(\frac{R_A L^3}{6} - \frac{5wL^4}{64} \right) = 0$$

$$R_A = \frac{13}{32}wL$$

Solving for R_A,

$$R_A = \frac{13}{32}wL \uparrow \quad \blacktriangleleft$$

Reactions at B and C. Substituting for R_A into Eqs. (2), we obtain

$$R_B = \frac{33}{32}wL\uparrow \qquad R_C = \frac{wL}{16}\uparrow \quad \blacktriangleleft$$

PROBLEMS

11.90 through 11.92 Using the information provided in Appendix D, compute the work of the loads as they are applied to the beam (*a*) if the load **P** is applied first, (*b*) if the couple M_0 is applied first.

Fig. 11.90

Fig. *P11.91*

Fig. P11.92

11.93 through 11.95 For the beam and loading shown, (*a*) compute the work of the loads as they are applied successively to the beam, using the information provided in Appendix D, (*b*) compute the strain energy of the beam by the method of Sec. 11.4 and show that it is equal to the work obtained in part *a*.

Fig. *P11.93*

Fig. P11.94

Fig. P11.95

11.96 and 11.97 For the prismatic beam shown, determine the deflection at point *D*.

Fig. P11.96 and P11.98

Fig. P11.97 and P11.99

11.98 and 11.99 For the prismatic beam shown, determine the slope at point *D*.

721.

11.100 and *11.101* For the prismatic beam shown, determine the slope at point A.

Fig. P11.100

Fig. *P11.101*

11.102 and 11.103 For the prismatic beam shown, determine the deflection at point D.

Fig. P11.102 and *P11.104*

Fig. P11.103 and P11.105

11.104 and 11.105 For the prismatic beam shown, determine the slope at point D.

11.106 For the beam and loading shown, determine the deflection at point B. Use E = 200 GPa.

11.107 For the beam and loading shown, determine the deflection at point B. Use E = 29 × 10³ ksi.

11.108 For the beam and loading shown, determine the deflection at point A. Use E = 29 × 10³ ksi.

Fig. P11.106

Fig. P11.107 and P11.108

11.109 For the beam and loading shown, determine the slope at end A. Use $E = 200$ GPa.

160 kN

W310 × 74

2.4 m — 2.4 m

4.8 m

Fig. P11.109

11.110 For the beam and loading shown, determine the deflection at point C. Use $E = 29 \times 10^3$ ksi.

11.111 For the beam and loading shown, determine the slope at end A. Use $E = 29 \times 10^3$ ksi.

8 kips

3 ft — 3 ft

6 ft — 3 ft

S8 × 18.4

Fig. P11.110 and P11.111

11.112 and 11.113 Each member of the truss shown is made of steel and has the cross-sectional area shown. Using $E = 200$ GPa, determine the deflection indicated.

11.112 Vertical deflection of joint C.
11.113 Horizontal deflection of joint C.

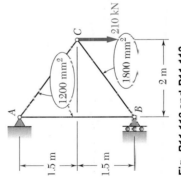

C

210 kN

1200 mm²

1800 mm²

2 m

1.5 m   1.5 m

Fig. P11.112 and P11.113

11.114 and 11.115 Each member of the truss shown is made of steel and has the cross-sectional area shown. Using $E = 29 \times 10^6$ psi, determine the deflection indicated.

11.114 Vertical deflection of joint C.
11.115 Horizontal deflection of joint C.

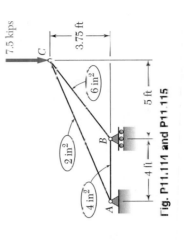

7.5 kips

3.75 ft

6 in²

2 in²

4 in²

5 ft

4 ft

Fig. P11.114 and P11.115

11.116 and 11.117 Each member of the truss shown is made of steel and has a cross-sectional area of 500 mm². Using $E = 200$ GPa, determine the deflection indicated.

11.116 Vertical deflection of joint B.
11.117 Horizontal deflection of joint B.

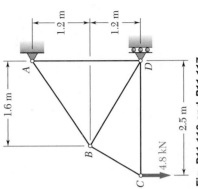

1.2 m   1.2 m

A   D

1.6 m

B

2.5 m

4.8 kN

C

Fig. P11.116 and P11.117

Fig. P11.118

*11.118 For the uniform rod and loading shown and using Castigliano's theorem, determine (*a*) the horizontal deflection of point *B*, (*b*) the vertical deflection of point *B*.

Fig. P11.119

11.119 Two rods *AB* and *BC* of the same flexural rigidity *EI* are welded together at *B*. For the loading shown, determine (*a*) the deflection of point *C*, (*b*) the slope of member *BC* at point *C*.

Fig. P11.120 and P11.121

11.120 A uniform rod of flexural rigidity *EI* is bent and loaded as shown. Determine (*a*) the horizontal deflection of point *D*, (*b*) the slope at point *D*.

11.121 A uniform rod of flexural rigidity *EI* is bent and loaded as shown. Determine (*a*) the vertical deflection of point *D*, (*b*) the slope of *BC* at point *C*.

Fig. P11.122 and P11.123

11.122 A uniform rod of flexural rigidity *EI* is bent and loaded as shown. Determine (*a*) the vertical deflection of point *A*, (*b*) the horizontal deflection of point *A*.

11.123 A uniform rod of flexural rigidity *EI* is bent and loaded as shown. Determine (*a*) the vertical deflection of point *B*, (*b*) the slope of *BC* at point *B*.

11.124 through 11.127 Determine the reaction at the roller support and draw the bending-moment diagram for the beam and loading shown.

Fig. P11.124

Fig. P11.125

Fig. P11.126

Fig. P11.127

11.128 Determine the reaction at the roller support and draw the bending-moment diagram for the beam and loading shown.

Fig. P11.128

11.129 For the uniform beam and loading shown, determine the reaction at each support.

Fig. P11.129

11.130 through 11.133 Three members of the same material and same cross-sectional area are used to support the load **P**. Determine the force in member *BC*.

Fig. P11.130

Fig. *P11.132*

Fig. P11.131

Fig. P11.133

11.134 **and** *11.135* Knowing that the eight members of the indeterminate truss shown have the same uniform cross-sectional area, determine the force in member *AB*.

Fig. P11.134

Fig. P11.135

This chapter was devoted to the study of strain energy and to the ways in which it can be used to determine the stresses and deformations in structures subjected to both static and impact loadings.

In Sec. 11.2 we considered a uniform rod subjected to a slowly increasing axial load **P** (Fig 11.1). We noted that the area under the

Fig. 11.1

Strain energy

load-deformation diagram (Fig 11.3) represents the work done by **P**. This work is equal to the *strain energy* of the rod associated with the deformation caused by the load **P:**

$$\text{Strain energy} = U = \int_{0}^{x_1} P \, dx \qquad (11.2)$$

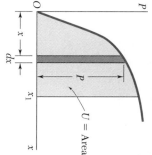

Fig. 11.3

Since the stress is uniform throughout the rod, we were able to divide the strain energy by the volume of the rod and obtain the strain energy per unit volume, which we defined as the *strain-energy density* of the material [Sec. 11.3]. We found that

$$\text{Strain-energy density} = u = \int_{0}^{\epsilon_1} \sigma_x \, d\epsilon_x \qquad (11.4)$$

Strain-energy density

and noted that the strain-energy density is equal to the area under the stress-strain diagram of the material (Fig. 11.6). As we saw in Sec. 11.4, Eq. (11.4) remains valid when the stresses are not uniformly distributed, but the strain-energy density will then vary from point to point. If the material is unloaded, there is a permanent strain ϵ_p and only the strain-energy density corresponding to the triangular area is recovered, the remainder of the energy having been dissipated in the form of heat during the deformation of the material.

The area under the entire stress-strain diagram was defined as the *modulus of toughness* and is a measure of the total energy that can be acquired by the material.

Fig. 11.6

Modulus of toughness

If the normal stress σ remains within the proportional limit of the material, the strain-energy density u is expressed as

$$u = \frac{\sigma^2}{2E}$$

The area under the stress-strain curve from zero strain to the strain ϵ_Y at yield (Fig. 11.9) is referred to as the *modulus of resilience* of the material and represents the energy per unit volume that the material can absorb without yielding. We wrote

$$u_Y = \frac{\sigma_Y^2}{2E} \qquad (11.8)$$

Modulus of resilience

Fig. 11.9

In Sec. 11.4 we considered the strain energy associated with *normal stresses*. We saw that if a rod of length L and *variable cross-sectional area A* is subjected at its end to a centric axial load **P,** the strain energy of the rod is

$$U = \int_0^L \frac{P^2}{2AE}\,dx \qquad (11.13)$$

Strain energy under axial load

If the rod is of *uniform cross section* of area A, the strain energy is

$$U = \frac{P^2 L}{2AE} \qquad (11.14)$$

We saw that for a beam subjected to transverse loads (Fig. 11.15) the strain energy associated with the normal stresses is

$$U = \int_0^L \frac{M^2}{2EI}\,dx \qquad (11.17)$$

Strain energy due to bending

Fig. 11.15

where M is the bending moment and EI the flexural rigidity of the beam.

The strain energy associated with *shearing stresses* was considered in Sec. 11.5. We found that the strain-energy density for a material in pure shear is

$$u = \frac{\tau_{xy}^2}{2G} \qquad (11.19)$$

Strain energy due to shearing stresses

where τ_{xy} is the shearing stress and G the modulus of rigidity of the material.

For a shaft of length L and uniform cross section subjected at its ends to couples of magnitude T (Fig. 11.19) the strain energy was found to be

$$U = \frac{T^2 L}{2GJ} \qquad (11.22)$$

Strain energy due to torsion

Fig. 11.19

where J is the polar moment of inertia of the cross-sectional area of the shaft.

General state of stress

In Sec. 11.6 we considered the strain energy of an elastic isotropic material under a general state of stress and expressed the strain-energy density at a given point in terms of the principal stresses σ_a, σ_b, and σ_c at that point:

$$u = \frac{1}{2E} [\sigma_a^2 + \sigma_b^2 + \sigma_c^2 - 2v(\sigma_a\sigma_b + \sigma_b\sigma_c + \sigma_c\sigma_a)] \quad (11.27)$$

The strain-energy density at a given point was divided into two parts: u_v, associated with a change in volume of the material at that point, and u_d, associated with a distortion of the material at the same point. We wrote $u = u_v + u_d$, where

$$u_v = \frac{1-2v}{6E}(\sigma_a + \sigma_b + \sigma_c)^2 \quad (11.32)$$

and

$$u_d = \frac{1}{12G}[(\sigma_a - \sigma_b)^2 + (\sigma_b - \sigma_c)^2 + (\sigma_c - \sigma_a)^2] \quad (11.33)$$

Using the expression obtained for u_d, we derived the maximum-distortion-energy criterion, which was used in Sec. 7.7 to predict whether a ductile material would yield under a given state of plane stress.

Impact loading

In Sec. 11.7 we considered the *impact loading* of an elastic structure being hit by a mass moving with a given velocity. We assumed that the kinetic energy of the mass is transferred entirely to the structure, and that the material used should have a low modulus of elasticity and a high yield strength [Sec. 11.8].

Equivalent static load

After discussing several examples, we noted that a structure designed to withstand effectively an impact load should be shaped in such a way that stresses are evenly distributed throughout the structure, and that the material used should have a low modulus of elasticity and a high yield strength [Sec. 11.8].

The strain energy of structural members subjected to a *single load* was considered in Sec. 11.9. In the case of the beam and loading of Fig. 11.27 we found that the strain energy of the beam is

$$U = \frac{P_1^2 L^3}{6EI} \quad (11.46)$$

Members subjected to a single load

Observing that the work done by the load **P** is equal to $\frac{1}{2}P_1 y_1$, we equated the work of the load and the strain energy of the beam and determined the deflection y_1 at the point of application of the load [Sec. 11.10 and Example 11.10].

Fig. 11.27

The method just described is of limited value, since it is restricted to structures subjected to a single concentrated load and to the determination of the deflection at the point of application of that load. In the remaining sections of the chapter, we presented a more general method, which can be used to determine deflections at various points of structures subjected to several loads.

In Sec. 11.11 we discussed the strain energy of a structure subjected to several loads, and in Sec. 11.12 introduced *Castigliano's theorem*, which states that the deflection x_j, of the point of application of a load $\mathbf{P}_j$ measured along the line of action of $\mathbf{P}_j$ is equal to the partial derivative of the strain energy of the structure with respect to the load $\mathbf{P}_j$. We wrote

$$x_j = \frac{\partial U}{\partial P_j} \qquad (11.65)$$

We also found that we could use Castigliano's theorem to determine the *slope* of a beam at the point of application of a couple $\mathbf{M}_j$ by writing

$$\theta_j = \frac{\partial U}{\partial M_j} \qquad (11.68)$$

and the *angle of twist* in a section of a shaft where a torque $\mathbf{T}_j$ is applied by writing

$$\phi_j = \frac{\partial U}{\partial T_j} \qquad (11.69)$$

In Sec. 11.13, Castigliano's theorem was applied to the determination of deflections and slopes at various points of a given structure. The use of "dummy" loads enabled us to include points where no actual load was applied. We also observed that the calculation of a deflection x_j was simplified if the differentiation with respect to the load P_j was carried out before the integration. In the case of a beam, recalling Eq. (11.17), we wrote

$$x_j = \frac{\partial U}{\partial P_j} = \int_0^L \frac{M}{EI}\frac{\partial M}{\partial P_j}\, dx \qquad (11.70)$$

Similarly, for a truss consisting of n members, the deflection x_j at the point of application of the load $\mathbf{P}_j$ was found by writing

$$x_j = \frac{\partial U}{\partial P_j} = \sum_{i=1}^n \frac{F_i L_i}{A_i E}\frac{\partial F_i}{\partial P_j} \qquad (11.72)$$

The chapter concluded [Sec. 11.14] with the application of Castigliano's theorem to the analysis of *statically indeterminate structures* [Sample Prob. 11.7, Examples 11.15 and 11.16].

REVIEW PROBLEMS

11.136 The steel bar ABC has a square cross section of side 0.75 in. and is subjected to a 50-lb load **P**. Using $E = 29 \times 10^6$ psi, determine the deflection of point C.

Fig. P11.136

11.137 The steel bars BE and AD have each a 5×15-mm cross section. Assuming that lever ABC is rigid and using $E = 200$ GPa, determine the deflection of point C.

11.138 The steel bars BE and AD have each a 5×15-mm cross section and the steel lever ABC has a square cross section of side 25 mm. Using $E = 200$ GPa, determine the deflection of point C.

Fig. P11.137 and P11.138

11.139 Two solid steel shafts are connected by the gears shown. Using $G = 11.2 \times 10^6$ psi, determine in each shaft the strain energy due to twisting when a 24-kip · in. torque is applied at D. (Ignore the strain energy due to bending of the shafts.)

11.140 For the shaft and gear system of Prob. 11.139, determine the angle through which end D rotates when $T = 24$ kip · in. (Ignore the strain energy due to bending of the shafts.)

11.141 (a) Determine the modulus of resilience of a grade of structural steel for which $\sigma_Y = 300$ MPa and $E = 200$ GPa. (b) Determine the required yield strength of an aluminum alloy for which $E = 72$ GPa if the modulus of resilience of the alloy is to be the same as that of the structural steel.

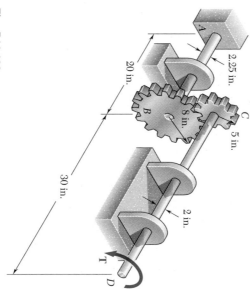

Fig. P11.139

11.142 A single 6-mm-diameter steel pin B is used to connect the steel strip DE to two aluminum strips, each of 20-mm width and 5-mm thickness. The modulus of elasticity is 200 GPa for the steel and 70 GPa for the aluminum. Knowing that for the pin at B the allowable shearing stress is $\tau_{all} = 85$ MPa, determine, for the loading shown, the maximum strain energy that can be acquired by the assembled strips.

Fig. P11.142

11.143 The 18-mm-diameter steel rod BC is attached to the lever AB and to the fixed support C. The uniform steel lever AB is 9 mm wide and 24 mm deep. Using $E = 200$ GPa, $G = 77$ GPa, and the method of work and energy, determine the deflection of point A.

Fig. P11.143

11.144 The 75-lb collar D is released from rest in the position shown and is stopped by a plate attached at end C of the vertical rod ABC. Knowing that $E = 29 \times 10^6$ psi for both portions of the rod, determine the distance h for which the maximum stress in the rod is 30 ksi.

11.145 The 75-lb collar D is released from rest when $h = 20$ in. and is stopped by a plate attached at end C of the vertical rod ABC. Knowing that $E = 29 \times 10^6$ psi for both portions of the rod, determine (a) the maximum deflection of end C, (b) the equivalent static load, (c) the maximum stress that occurs in the rod.

Fig. P11.144 and P11.145

11.146 The steel rod BC has a 24-mm diameter and the steel cable ABDCA has a 12-mm diameter. Using $E = 200$ GPa, determine the deflection of point D caused by the 12-kN load.

Fig. P11.146

11.147 The simply supported beam AB is struck squarely at D by a block of mass m moving horizontally with a velocity v_0. Show that the resulting maximum normal stress σ_m in the beam due to bending is independent of the location of point D.

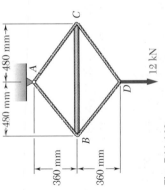

Fig. P11.147

COMPUTER PROBLEMS

The following problems are designed to be solved with a computer.

11.C1 A rod consisting of n elements, each of which is homogeneous and of uniform cross section, is subjected to a load $\mathbf{P}$ applied at its free end. The length of element i is denoted by L_i and its diameter by d_i. (a) Denoting by E the modulus of elasticity of the material used in the rod, write a computer program that can be used to determine the strain energy acquired by the rod and the deformation measured at the free end. (b) Use this program to determine the strain energy and deformation of the rods of Probs. 11.9 and 11.12.

Fig. P11.C1

11.C2 Two 0.75×6-in. cover plates are welded to a W8 $\times$ 18 rolled-steel beam as shown. The 1500-lb block is to be dropped from a height $h = 2$ in. onto the beam. (a) Write a computer program to calculate the maximum normal stress on transverse sections just to the left of D and at the center of the beam for values of a from 0 to 60 in., using 5-in. increments. (b) From the values considered in part a, select the distance a for which the maximum normal stress is as small as possible. Use $E = 29 \times 10^6$ psi.

Fig. P11.C2

11.C3 The 16-kg block D is dropped from a height h onto the free end of the steel bar AB. For the steel used $\sigma_{all} = 120$ MPa and $E = 200$ GPa. (a) Write a computer program to calculate the maximum allowable height h for values of the length L from 100 mm to 1.2 m, using 100-mm increments. (b) From the values considered in part a, select the length corresponding to the largest allowable height.

Fig. P11.C3

11.C4 The block D of mass $m = 8$ kg is dropped from a height $h = 750$ mm onto the rolled-steel beam AB. Knowing that $E = 200$ GPa, write a computer program to calculate the maximum deflection of point E and the maximum normal stress in the beam for values of a from 100 to 900 mm, using 100-mm increments.

Fig. P11.C4

W150 × 13.5

11.C5 The steel rods AB and BC are made of a steel for which $\sigma_Y = 300$ MPa and $E = 200$ GPa. (*a*) Write a computer program to calculate, for values of a from 0 to 6 m, using 1-m increments, the maximum strain energy that can be acquired by the assembly without causing any permanent deformation. (*b*) For each value of a considered, calculate the diameter of a uniform rod of length 6 m and of the same mass as the original assembly, and the maximum strain energy that could be acquired by this uniform rod without causing permanent deformation.

Fig. P11.C5

11.C6 A 160-lb diver jumps from a height of 20 in. onto end C of a diving board having the uniform cross section shown. Write a computer program to calculate for values of a from 10 to 50 in., using 10-in. increments, (*a*) the maximum deflection of point C, (*b*) the maximum bending moment in the board, (*c*) the equivalent static load. Assume that the diver's legs remain rigid and use $E = 1.8 \times 10^6$ psi.

Fig. P11.C6

Appendices

†Courtesy of the American Institute of Steel Construction, Chicago, Illinois.

A P P E N D I X

A

Moments of Areas

A.1. FIRST MOMENT OF AN AREA; CENTROID OF AN AREA

Consider an area A located in the xy plane (Fig. A.1). Denoting by x and y the coordinates of an element of area dA, we define the *first moment of the area A with respect to the x axis* as the integral

Fig. A.1

$$Q_x = \int_A y \, dA \qquad (A.1)$$

Similarly, the *first moment of the area A with respect to the y axis* is defined as the integral

$$Q_y = \int_A x \, dA \qquad (A.2)$$

We note that each of these integrals may be positive, negative, or zero, depending on the position of the coordinate axes. If SI units are used, the first moments Q_x and Q_y are expressed in m^3 or mm^3; if U.S. customary units are used, they are expressed in ft^3 or in^3.

The *centroid of the area A* is defined as the point C of coordinates $\bar{x}$ and $\bar{y}$ (Fig. A.2), which satisfy the relations

Fig. A.2

$$\int_A x \, dA = A\bar{x} \qquad \int_A y \, dA = A\bar{y} \qquad (A.3)$$

Comparing Eqs. (A.1) and (A.2) with Eqs. (A.3), we note that the first moments of the area A can be expressed as the products of the area and of the coordinates of its centroid:

$$Q_x = A\bar{y} \qquad Q_y = A\bar{x} \qquad (A.4)$$

When an area possesses an *axis of symmetry*, the first moment of the area with respect to that axis is zero. Indeed, considering the area A of Fig. A.3, which is symmetric with respect to the y axis, we observe that to every element of area dA of abscissa x corresponds an element of area dA' of abscissa $-x$. It follows that the integral in Eq. (A.2) is zero and, thus, that $Q_y = 0$. It also follows from the first of the relations (A.3) that $\bar{x} = 0$. Thus, if an area A possesses an axis of symmetry, its centroid C is located on that axis.

Fig. A.3

Since a rectangle possesses two axes of symmetry (Fig. A.4a), the centroid C of a rectangular area coincides with its geometric center. Similarly, the centroid of a circular area coincides with the center of the circle (Fig. A.4b).

When an area possesses a *center of symmetry* O, the first moment of the area about any axis through O is zero. Indeed, considering the area A of Fig. A.5, we observe that to every element of area dA of coordinates x and y corresponds an element of area dA' of coordinates $-x$ and $-y$. It follows that the integrals in Eqs. (A.1) and (A.2) are both zero, and that $Q_x = Q_y = 0$. It also follows from Eqs. (A.3) that $\bar{x} = \bar{y} = 0$, that is, the centroid of the area coincides with its center of symmetry.

When the centroid C of an area can be located by symmetry, the first moment of that area with respect to any given axis can be readily obtained from Eqs. (A.4). For example, in the case of the rectangular area of Fig. A.6, we have

$$Q_x = A\bar{y} = (bh)(\tfrac{1}{2}h) = \tfrac{1}{2}bh^2$$

and

$$Q_y = A\bar{x} = (bh)(\tfrac{1}{2}b) = \tfrac{1}{2}b^2h$$

In most cases, however, it is necessary to perform the integrations indicated in Eqs. (A.1) through (A.3) to determine the first moments and the centroid of a given area. While each of the integrals involved is actually a double integral, it is possible in many applications to select elements of area dA in the shape of thin horizontal or vertical strips, and thus to reduce the computations to integrations in a single variable. This is illustrated in Example A.01. Centroids of common geometric shapes are indicated in a table inside the back cover of this book.

(a)

(b)

Fig. A.4

Fig. A.5

Fig. A.6

For the triangular area of Fig. A.7, determine (*a*) the first moment Q_x of the area with respect to the *x* axis, (*b*) the ordinate $\bar{y}$ of the centroid of the area.

(a) First Moment Q_x. We select as an element of area a horizontal strip of length *u* and thickness *dy*, and note that all the points within the element are at the same distance *y* from the *x* axis (Fig. A.8). From similar triangles, we have

$$\frac{u}{b} = \frac{h - y}{h} \qquad u = b\,\frac{h - y}{h}$$

and

$$dA = u\,dy = b\,\frac{h - y}{h}\,dy$$

The first moment of the area with respect to the *x* axis is

$$Q_x = \int_A y\,dA = \int_0^h yb\,\frac{h - y}{h}\,dy = \frac{b}{h}\int_0^h (hy - y^2)\,dy$$

$$= \frac{b}{h}\left[h\frac{y^2}{2} - \frac{y^3}{3} \right]_0^h \qquad Q_x = \frac{1}{6}bh^2$$

(b) Ordinate of Centroid. Recalling the first of Eqs. (A.4) and observing that $A = \frac{1}{2}bh$, we have

$$Q_x = A\bar{y} \qquad \frac{1}{6}bh^2 = (\tfrac{1}{2}bh)\bar{y}$$

$$\bar{y} = \tfrac{1}{3}h$$

Fig. A.7

Fig. A.8

A.2. DETERMINATION OF THE FIRST MOMENT AND CENTROID OF A COMPOSITE AREA

Consider an area *A*, such as the trapezoidal area shown in Fig. A.9, which may be divided into simple geometric shapes. As we saw in the preceding section, the first moment Q_x of the area with respect to the *x* axis is represented by the integral $\int y\,dA$, which extends over the entire area *A*. Dividing *A* into its component parts A_1, A_2, A_3, we write

$$Q_x = \int_A y\,dA = \int_{A_1} y\,dA + \int_{A_2} y\,dA + \int_{A_3} y\,dA$$

or, recalling the second of Eqs. (A.3),

$$Q_x = A_1\bar{y}_1 + A_2\bar{y}_2 + A_3\bar{y}_3$$

where $\bar{y}_1, \bar{y}_2,$ and $\bar{y}_3$ represent the ordinates of the centroids of the component areas. Extending this result to an arbitrary number of compo-

Fig. A.9

nent areas, and noting that a similar expression may be obtained for Q_y, we write

$$Q_x = \sum_i A_i \bar{y}_i \qquad Q_y = \sum_i A_i x_i \qquad \text{(A.5)}$$

To obtain the coordinates $\bar{X}$ and $\bar{Y}$ of the centroid C of the composite area A, we substitute $Q_x = A\bar{Y}$ and $Q_y = A\bar{X}$ into Eqs. (A.5). We have

$$A\bar{Y} = \sum_i A_i \bar{y}_i \qquad A\bar{X} = \sum_i A_i \bar{x}_i$$

Solving for $\bar{X}$ and $\bar{Y}$ and recalling that the area A is the sum of the component areas A_i, we write

$$\bar{X} = \frac{\displaystyle\sum_i A_i \bar{x}_i}{\displaystyle\sum_i A_i} \qquad \bar{Y} = \frac{\displaystyle\sum_i A_i \bar{y}_i}{\displaystyle\sum_i A_i} \qquad \text{(A.6)}$$

EXAMPLE A.02

Locate the centroid C of the area A shown in Fig. A.10.

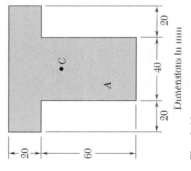

Dimensions in mm

Fig. A.10

Selecting the coordinate axes shown in Fig. A.11, we note that the centroid C must be located on the y axis, since this axis is an axis of symmetry; thus, $\bar{X} = 0$.

Dividing A into its component parts A_1 and A_2, we use the second of Eqs. (A.6) to determine the ordinate $\bar{Y}$ of the centroid. The actual computation is best carried out in tabular form.

Dimensions in mm

Fig. A.11

	Area, mm²	$\bar{y}_i$, mm	$A_i\bar{y}_i$, mm³
A_1	(20)(80) = 1600	70	112×10^3
A_2	(40)(60) = 2400	30	72×10^3
	$\displaystyle\sum_i A_i = 4000$		$\displaystyle\sum_i A_i\bar{y}_i = 184 \times 10^3$

$$\bar{Y} = \frac{\displaystyle\sum_i A_i\bar{y}_i}{\displaystyle\sum_i A_i} = \frac{184 \times 10^3 \text{ mm}^3}{4 \times 10^3 \text{ mm}^2} = 46 \text{ mm}$$

EXAMPLE A.03

Referring to the area A of Example A.02, we consider the horizontal x' axis through its centroid C. (Such an axis is called a *centroidal axis*.) Denoting by A' the portion of A located above that axis (Fig. A.12), determine the first moment of A' with respect to the x' axis.

Fig. A.12

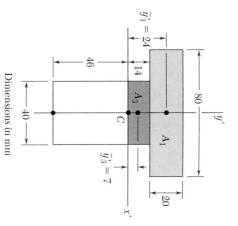

Dimensions in mm

Fig. A.13

Solution. We divide the area A' into its components A_1 and A_3 (Fig. A.13). Recalling from Example A.02 that C is located 46 mm above the lower edge of A, we determine the ordinates $\bar{y}'_1$ and $\bar{y}'_3$ of A_1 and A_3 and express the first moment $Q'_{x'}$ of A' with respect to x' as follows:

$$Q'_{x'} = A_1\bar{y}'_1 + A_3\bar{y}'_3$$
$$= (20 \times 80)(24) + (14 \times 40)(7) = 42.3 \times 10^3 \text{ mm}^3$$

Alternative Solution. We first note that since the centroid C of A is located on the x' axis, the first moment $Q_{x'}$ of the *entire area* A with respect to that axis is zero:

$$Q_{x'} = A\bar{y}' = A(0) = 0$$

Denoting by A'' the portion of A located below the x' axis and by $Q''_{x'}$ its first moment with respect to that axis, we have therefore

$$Q_{x'} = Q'_{x'} + Q''_{x'} = 0 \quad \text{or} \quad Q'_{x'} = -Q''_{x'}$$

which shows that the first moments of A' and A'' have the same magnitude and opposite signs. Referring to Fig. A.14, we write

$$Q''_{x'} = A_4\bar{y}'_4 = (40 \times 46)(-23) = -42.3 \times 10^3 \text{ mm}^3$$

and

$$Q'_{x'} = -Q''_{x'} = +42.3 \times 10^3 \text{ mm}^3$$

Dimensions in mm

Fig. A.14

A.3. SECOND MOMENT, OR MOMENT OF INERTIA, OF AN AREA; RADIUS OF GYRATION

Consider again an area A located in the xy plane (Fig. A.1) and the element of area dA of coordinates x and y. The *second moment, or moment of inertia,* of the area A with respect to the x axis, and the second moment, or moment of inertia, of A with respect to the y axis are defined, respectively, as

$$I_x = \int_A y^2 \, dA \qquad I_y = \int_A x^2 \, dA \qquad (A.7)$$

Fig. A.1 (repeated)

These integrals are referred to as *rectangular moments of inertia*, since they are computed from the rectangular coordinates of the element dA. While each integral is actually a double integral, it is possible in many applications to select elements of area dA in the shape of thin horizontal or vertical strips, and thus reduce the computations to integrations in a single variable. This is illustrated in Example A.04.

We now define the *polar moment of inertia* of the area A with respect to point O (Fig. A.15) as the integral

$$J_O = \int_A \rho^2 \, dA \qquad (A.8)$$

Fig. A.15

where ρ is the distance from O to the element dA. While this integral is again a double integral, it is possible in the case of a circular area to select elements of area dA in the shape of thin circular rings, and thus reduce the computation of J_O to a single integration (see Example A.05).

We note from Eqs. (A.7) and (A.8) that the moments of inertia of an area are positive quantities. If SI units are used, moments of inertia are expressed in m^4 or mm^4; if U.S. customary units are used, they are expressed in ft^4 or in^4.

An important relation may be established between the polar moment of inertia J_O of a given area and the rectangular moments of inertia I_x and I_y of the same area. Noting that $\rho^2 = x^2 + y^2$, we write

$$J_O = \int_A \rho^2 \, dA = \int_A (x^2 + y^2) \, dA = \int_A y^2 \, dA + \int_A x^2 \, dA$$

or

$$J_O = I_x + I_y \qquad (A.9)$$

The *radius of gyration* of an area A with respect to the x axis is defined as the quantity r_x, that satisfies the relation

$$I_x = r_x^2 A \qquad (A.10)$$

where I_x is the moment of inertia of A with respect to the x axis. Solving Eq. (A.10) for r_x, we have

$$r_x = \sqrt{\frac{I_x}{A}} \qquad (A.11)$$

In a similar way, we define the radii of gyration with respect to the y axis and the origin O. We write

$$I_y = r_y^2 A \qquad r_y = \sqrt{\frac{I_y}{A}} \qquad (A.12)$$

$$J_O = r_O^2 A \qquad r_O = \sqrt{\frac{J_O}{A}} \qquad (A.13)$$

Substituting for J_O, I_x, and I_y in terms of the corresponding radii of gyration in Eq. (A.9), we observe that

$$r_O^2 = r_x^2 + r_y^2 \qquad (A.14)$$

EXAMPLE A.04

For the rectangular area of Fig. A.16, determine (a) the moment of inertia I_x of the area with respect to the centroidal x axis, (b) the corresponding radius of gyration r_x.

(a) *Moment of Inertia I_x.*   We select as an element of area a horizontal strip of length b and thickness dy (Fig. A.17). Since all the points within the strip are at the same distance y from the x axis, the moment of inertia of the strip with respect to that axis is

$$dI_x = y^2\, dA = y^2(b\, dy)$$

Integrating from $y = -h/2$ to $y = +h/2$, we write

$$I_x = \int_A y^2\, dA = \int_{-h/2}^{+h/2} y^2(b\, dy) = \tfrac{1}{3}b[y^3]_{-h/2}^{+h/2}$$

$$= \tfrac{1}{3}b\!\left(\frac{h^3}{8} + \frac{h^3}{8}\right)$$

or

$$I_x = \tfrac{1}{12}bh^3$$

Fig. A.16

Fig. A.17

(b) *Radius of Gyration r_x.*   From Eq. (A.10), we have

$$I_x = r_x^2 A \qquad \tfrac{1}{12}bh^3 = r_x^2(bh)$$

and, solving for r_x,

$$r_x = h/\sqrt{12}$$

For the circular area of Fig. A.18, determine (a) the polar moment of inertia J_O, (b) the rectangular moments of inertia I_x and I_y.

Fig. A.18

Fig. A.19

(a) Polar Moment of Inertia. We select as an element of area a ring of radius ρ and thickness $d\rho$ (Fig. A.19). Since all the points within the ring are at the same distance ρ from the origin O, the polar moment of inertia of the ring is

$$dJ_O = \rho^2 \, dA = \rho^2 (2\pi\rho \, d\rho)$$

Integrating in ρ from 0 to c, we write

$$J_O = \int_A \rho^2 \, dA = \int_0^c \rho^2(2\pi\rho \, d\rho) = 2\pi \int_0^c \rho^3 \, d\rho$$

$$J_O = \tfrac{1}{2}\pi c^4$$

(b) Rectangular Moments of Inertia. Because of the symmetry of the circular area, we have $I_x = I_y$. Recalling Eq. (A.9), we write

$$J_O = I_x + I_y = 2I_x \qquad \tfrac{1}{2}\pi c^4 = 2I_x$$

and, thus,

$$I_x = I_y = \tfrac{1}{4}\pi c^4$$

The results obtained in the preceding two examples, and the moments of inertia of other common geometric shapes, are listed in a table inside the back cover of this book.

A.4. PARALLEL-AXIS THEOREM

Consider the moment of inertia I_x of an area A with respect to an arbitrary x axis (Fig. A.20). Denoting by y the distance from an element of area dA to that axis, we recall from Sec. A.3 that

$$I_x = \int_A y^2 \, dA$$

Fig. A.20

Let us now draw the *centroidal x' axis*, i.e., the axis parallel to the x axis which passes through the centroid C of the area. Denoting by y' the distance from the element dA to that axis, we write $y = y' + d$, where d is the distance between the two axes. Substituting for y in the integral representing I_x, we write

$$I_x = \int_A y^2 \, dA = \int_A (y' + d)^2 dA$$

$$I_x = \int_A y'^2 \, dA + 2d \int_A y' \, dA + d^2 \int_A dA \qquad (A.15)$$

The first integral in Eq. (A.15) represents the moment of inertia $\bar{I}_{x'}$ of the area with respect to the centroidal x' axis. The second integral represents the first moment $Q_{x'}$ of the area with respect to the x' axis and is equal to zero, since the centroid C of the area is located on that axis. Indeed, we recall from Sec. A.1 that

$$Q_{x'} = A\bar{y}' = A(0) = 0$$

Finally, we observe that the last integral in Eq. (A.15) is equal to the total area A. We have, therefore,

$$I_x = \bar{I}_{x'} + Ad^2 \qquad (A.16)$$

This formula expresses that the moment of inertia I_x of an area with respect to an arbitrary x axis is equal to the moment of inertia $\bar{I}_{x'}$ of the area with respect to the centroidal x' axis parallel to the x axis, *plus* the product Ad^2 of the area A and of the square of the distance d between the two axes. This result is known as the *parallel-axis theorem*. It makes it possible to determine the moment of inertia of an area with respect to a given axis, when its moment of inertia with respect to a centroidal axis of the same direction is known. Conversely, it makes it possible to determine the moment of inertia $\bar{I}_{x'}$ of an area A with respect to a centroidal axis x', when the moment of inertia I_x of A with respect to a parallel axis is known, by *subtracting* from I_x the product Ad^2. We should note that the parallel-axis theorem may be used *only if one of the two axes involved is a centroidal axis.*

A similar formula may be derived, which relates the polar moment of inertia J_O of an area with respect to an arbitrary point O and the polar moment of inertia $\bar{J}_C$ of the same area with respect to its centroid C. Denoting by d the distance between O and C, we write

$$J_O = \bar{J}_C + Ad^2 \qquad (A.17)$$

A.5. DETERMINATION OF THE MOMENT OF INERTIA OF A COMPOSITE AREA

Consider a composite area A made of several component parts A_1, A_2, and so forth. Since the integral representing the moment of inertia of A may be subdivided into integrals extending over A_1, A_2, and so forth, the moment of inertia of A with respect to a given axis will be obtained by adding the moments of inertia of the areas A_1, A_2, and so forth, with respect to the same axis. The moment of inertia of an area made of several of the common shapes shown in the table inside the back cover of this book may thus be obtained from the formulas given in that table. Before adding the moments of inertia of the component areas, however, the parallel-axis theorem should be used to transfer each moment of inertia to the desired axis. This is shown in Example A.06.

EXAMPLE A.06

Determine the moment of inertia $\bar{I}_x$ of the area shown with respect to the centroidal x axis (Fig. A.21).

Location of Centroid. The centroid C of the area must first be located. However, this has already been done in Example A.02 for the given area. We recall from that example that C is located 46 mm above the lower edge of the area A.

Computation of Moment of Inertia. We divide the area A into the two rectangular areas A_1 and A_2 (Fig. A.22), and compute the moment of inertia of each area with respect to the x axis.

Rectangular Area A_1. To obtain the moment of inertia $(I_x)_1$ of A_1 with respect to the x axis, we first compute the moment of inertia of A_1 *with respect to its own centroidal axis* x'. Recalling the formula derived in part a of Example A.04 for the centroidal moment of inertia of a rectangular area, we have

$$(\bar{I}_{x'})_1 = \tfrac{1}{12}bh^3 = \tfrac{1}{12}(80 \text{ mm})(20 \text{ mm})^3 = 53.3 \times 10^3 \text{ mm}^4$$

Using the parallel-axis theorem, we transfer the moment of inertia of A_1 from its centroidal axis x' to the parallel axis x:

$$(I_x)_1 = (\bar{I}_{x'})_1 + A_1 d_1^2 = 53.3 \times 10^3 + (80 \times 20)(24)^2$$
$$= 975 \times 10^3 \text{ mm}^4$$

Rectangular Area A_2. Computing the moment of inertia of A_2 with respect to its centroidal axis x'', and using the parallel-axis theorem to transfer it to the x axis, we have

$$(\bar{I}_{x''})_2 = \tfrac{1}{12}bh^3 = \tfrac{1}{12}(40)(60)^3 = 720 \times 10^3 \text{ mm}^4$$
$$(I_x)_2 = (\bar{I}_{x''})_2 + A_2 d_2^2 = 720 \times 10^3 + (40 \times 60)(16)^2$$
$$= 1334 \times 10^3 \text{ mm}^4$$

Entire Area A. Adding the values computed for the moments of inertia of A_1 and A_2 with respect to the x axis, we obtain the moment of inertia $\bar{I}_x$ of the entire area:

$$\bar{I}_x = (I_x)_1 + (I_x)_2 = 975 \times 10^3 + 1334 \times 10^3$$
$$\bar{I}_x = 2.31 \times 10^6 \text{ mm}^4$$

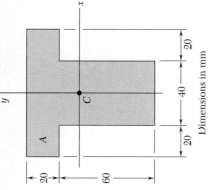

Dimensions in mm

Fig. A.21

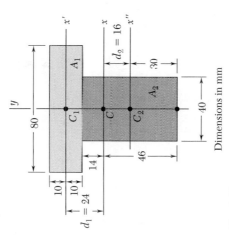

Dimensions in mm

Fig. A.22

Appendix B. Typical Properties of Selected Materials Used in Engineering[1,5] (U.S. Customary Units)

Material	Specific Weight, lb/in³	Ultimate Strength Tension, ksi	Ultimate Strength Compression,[2] ksi	Ultimate Strength Shear, ksi	Yield Strength[3] Tension, ksi	Yield Strength[3] Shear, ksi	Modulus of Elasticity, 10⁶ psi	Modulus of Rigidity, 10⁶ psi	Coefficient of Thermal Expansion, 10⁻⁶/°F	Ductility, Percent Elongation in 2 in.
Steel										
Structural (ASTM-A36)	0.284	58			36	21	29	11.2	6.5	21
High-strength-low-alloy										
ASTM-A709 Grade 50	0.284	65			50		29	11.2	6.5	21
ASTM-A913 Grade 65	0.284	80			65		29	11.2	6.5	17
ASTM-A992 Grade 50	0.284	65			50		29	11.2	6.5	21
Quenched & tempered ASTM-A709 Grade 100	0.284	110			100		29	11.2	6.5	18
Stainless, AISI 302										
Cold-rolled	0.286	125			75		28	10.8	9.6	12
Annealed	0.286	95			38		28	10.8	9.6	50
Reinforcing Steel										
Medium strength	0.283	70			40		29	11	6.5	
High strength	0.283	90			60		29	11	6.5	
Cast Iron										
Gray Cast Iron 4.5% C, ASTM A-48	0.260	25	95	35			10	4.1	6.7	0.5
Malleable Cast Iron 2% C, 1% Si, ASTM A-47	0.264	50	90	48	33	22	24	9.3	6.7	10
Aluminum										
Alloy 1100-H14 (99% Al)	0.098	16		10	14	8	10.1	3.7	13.1	9
Alloy 2014-T6	0.101	66		40	58	33	10.9	3.9	12.8	13
Alloy 2024-T4	0.101	68		41	47		10.6		12.9	19
Alloy 5456-H116	0.095	46		27	33	19	10.4		13.3	16
Alloy 6061-T6	0.098	38		24	35	20	10.1	3.7	13.1	17
Alloy 7075-T6	0.101	83		48	73		10.4	4	13.1	11
Copper										
Oxygen-free copper (99.9% Cu)										
Annealed	0.322	32		22	10		17	6.4	9.4	45
Hard-drawn	0.322	57		29	53		17	6.4	9.4	4
Yellow Brass (65% Cu, 35% Zn)										
Cold-rolled	0.306	74		43	60	36	15	5.6	11.6	8
Annealed	0.306	46		32	15	9	15	5.6	11.6	65
Red Brass (85% Cu, 15% Zn)										
Cold-rolled	0.316	85		46	63		17	6.4	10.4	3
Annealed	0.316	39		31	10		17	6.4	10.4	48
Tin bronze (88 Cu, 8Sn, 4Zn)	0.318	45		21	21		14		10	30
Manganese bronze (63 Cu, 25 Zn, 6 Al, 3 Mn, 3 Fe)	0.302	95		48	48		15		12	20
Aluminum bronze (81 Cu, 4 Ni, 4 Fe, 11 Al)	0.301	90	130	40	40		16	6.1	9	6

(Table continued on page 748)

Appendix B. Typical Properties of Selected Materials Used in Engineering[1,5] (SI Units)

Material	Density kg/m³	Ultimate Strength Tension, MPa	Ultimate Strength Compression, MPa	Ultimate Strength Shear, MPa	Yield Strength[3] Tension, MPa	Yield Strength[3] Shear, MPa	Modulus of Elasticity, GPa	Modulus of Rigidity, GPa	Coefficient of Thermal Expansion, $10^{-6}/°C$	Ductility, Percent Elongation in 50 mm
Steel										
Structural (ASTM-A36)	7860	400			250	145	200	77.2	11.7	21
High-strength-low-alloy										
ASTM-A709 Grade 345	7860	450			345		200	77.2	11.7	21
ASTM-A913 Grade 450	7860	550			450		200	77.2	11.7	17
ASTM-A992 Grade 345	7860	450			345		200	77.2	11.7	21
Quenched & tempered										
ASTM-A709 Grade 690	7860	760			690		200	77.2	11.7	18
Stainless, AISI 302										
Cold-rolled	7920	860			520		190	75	17.3	12
Annealed	7920	655			260	150	190	75	17.3	50
Reinforcing Steel										
Medium strength	7860	480			275		200	77	11.7	
High strength	7860	620			415		200	77	11.7	
Cast Iron										
Gray Cast Iron 4.5% C, ASTM A-48	7200	170	655	240			69	28	12.1	0.5
Malleable Cast Iron 2% C, 1% Si, ASTM A-47	7300	345	620	330	230		165	65	12.1	10
Aluminum										
Alloy 1100-H14 (99% Al)	2710	110		70	95	55	70	26	23.6	9
Alloy 2014-T6	2800	455		275	400	230	75	27	23.0	13
Alloy 2024-T4	2800	470		280	325		73		23.2	19
Alloy 5456-H116	2630	315		185	230	130	72		23.9	16
Alloy 6061-T6	2710	260		165	240	140	70	26	23.6	17
Alloy 7075-T6	2800	570		330	500		72	28	23.6	11
Copper										
Oxygen-free copper (99.9% Cu)										
Annealed	8910	220		150	70		120	44	16.9	45
Hard-drawn	8910	390		200	265		120	44	16.9	4
Yellow-Brass (65% Cu, 35% Zn)										
Cold-rolled	8470	510		300	410	250	105	39	20.9	8
Annealed	8470	320		220	100	60	105	39	20.9	65
Red Brass (85% Cu, 15% Zn)										
Cold-rolled	8740	585		320	435		120	44	18.7	3
Annealed	8740	270		210	70		120	44	18.7	48
Tin bronze (88 Cu, 8Sn, 4Zn)	8800	310			145		95		18.0	30
Manganese bronze (63 Cu, 25 Zn, 6 Al, 3 Mn, 3 Fe)	8360	655			330		105		21.6	20
Aluminum bronze (81 Cu, 4 Ni, 4 Fe, 11 Al)	8330	620	900		275		110	42	16.2	6

(Table continued on page 749)

Appendix B. Typical Properties of Selected Materials Used in Engineering[1,5]
(U.S. Customary Units)

Continued from page 746

Material	Specific Weight, lb/in^3	Ultimate Strength Tension, ksi	Compression[2], ksi	Shear, ksi	Yield Strength[3] Tension, ksi	Shear, ksi	Modulus of Elasticity, 10^6 psi	Modulus of Rigidity, 10^6 psi	Coefficient of Thermal Expansion, 10^{-6}/°F	Ductility, Percent Elongation in 2 in.
Magnesium Alloys										
Alloy AZ80 (Forging)	0.065	50		23	36		6.5	2.4	14	6
Alloy AZ31 (Extrusion)	0.064	37		19	29		6.5	2.4	14	12
Titanium										
Alloy (6% Al, 4% V)	0.161	130			120		16.5		5.3	10
Monel Alloy 400 (Ni-Cu)										
Cold-worked	0.319	98			85	50	26		7.7	22
Annealed	0.319	80			32	18	26		7.7	46
Cupronickel (90% Cu, 10% Ni)										
Annealed	0.323	53			16		20	7.5	9.5	35
Cold-worked	0.323	85			79		20	7.5	9.5	3
Timber, air dry[4]										
Douglas fir	0.017	15	7.2	1.1			1.9	.1	Varies 1.7 to 2.5	
Spruce, Sitka	0.015	8.6	5.6	1.1			1.5			
Shortleaf pine	0.018		7.3	1.4			1.7	.07		
Western white pine	0.014		5.0	1.0			1.5			
Ponderosa pine	0.015	8.4	5.3	1.1			1.3			
White oak	0.025		7.4	2.0			1.8			
Red oak	0.024		6.8	1.8			1.8			
Western hemlock	0.016	13	7.2	1.3			1.6			
Shagbark hickory	0.026		9.2	2.4			2.2			
Redwood	0.015		6.1	0.9			1.3			
Concrete										
Medium strength	0.084		4.0				3.6		5.5	
High strength	0.084		6.0				4.5		5.5	
Plastics										
Nylon, type 6/6, (molding compound)	0.0412	11	14		6.5		0.4		80	50
Polycarbonate	0.0433	9.5	12.5		9		0.35		68	110
Polyester, PBT (thermoplastic)	0.0484	8	11		8		0.35		75	150
Polyester elastomer	0.0433	6.5		5.5			0.03		500	500
Polystyrene	0.0374	8	13				0.45		70	2
Vinyl, rigid PVC	0.0520	6	10		6.5		0.45		75	40
Rubber	0.033	2							90	600
Granite (Avg. values)	0.100	3	35	5			10	4	4	
Marble (Avg. values)	0.100	2	18	4			8	3	6	
Sandstone (Avg. values)	0.083	1	12	2			6	2	5	
Glass, 98% silica	0.079	7					9.6	4.1	44	

[1] Properties of metals vary widely as a result of variations in composition, heat treatment, and mechanical working.
[2] For ductile metals the compression strength is generally assumed to be equal to the tension strength.
[3] Offset of 0.2 percent.
[4] Timber properties are for loading parallel to the grain.
[5] See also Marks' *Mechanical Engineering Handbook*, 10th ed., McGraw-Hill, New York, 1996; *Annual Book of ASTM*, American Society for Testing Materials, Philadelphia, Pa.; *Metals Handbook*, American Society for Metals, Metals Park, Ohio; and *Aluminum Design Manual*, The Aluminum Association, Washington, DC.

Appendix B. Typical Properties of Selected Materials Used in Engineering[1,5]
(SI Units)

Continued from page 747

Material	Density kg/m³	Ultimate Strength			Yield Strength[3]		Modulus of Elasticity, GPa	Modulus of Rigidity, GPa	Coefficient of Thermal Expansion, 10⁻⁶/°C	Ductility, Percent Elongation in 50 mm
		Tension, MPa	Compression,[2] MPa	Shear, MPa	Tension, MPa	Shear, MPa				
Magnesium Alloys										
Alloy AZ80 (Forging)	1800	345	160		250		45	16	25.2	6
Alloy AZ31 (Extrusion)	1770	255	130		200		45	16	25.2	12
Titanium										
Alloy (6% Al, 4% V)	4430	900			830		115		9.5	10
Monel Alloy 400(Ni-Cu)										
Cold-worked	8830	675			585	345	180		13.9	22
Annealed	8830	550			220	125	180		13.9	46
Cupronickel (90% Cu, 10% Ni)										
Annealed	8940	365			110		140	52	17.1	35
Cold-worked	8940	585			545		140	52	17.1	3
Timber, air dry										
Douglas fir	470	100	50	7.6			13	0.7	Varies	
Spruce, Sitka	415	60	39	7.6			10	0.5	3.0 to 4.5	
Shortleaf pine	500		50	9.7			12			
Western white pine	390		34	7.0			10			
Ponderosa pine	415	55	36	7.6			9			
White oak	690		51	13.8			12			
Red oak	660		47	12.4			12			
Western hemlock	440	90	50	10.0			11			
Shagbark hickory	720		63	16.5			15			
Redwood	415	65	42	6.2			9			
Concrete										
Medium strength	2320		28				25		9.9	
High strength	2320		40				30		9.9	
Plastics										
Nylon, type 6/6, (molding compound)	1140	75	95		45		2.8		144	50
Polycarbonate	1200	65	85		35		2.4		122	110
Polyester, PBT (thermoplastic)	1340	55	75		55		2.4		135	150
Polyester elastomer	1200	45	40				0.2		125	500
Polystyrene	1030	55	90		55		3.1		125	2
Vinyl, rigid PVC	1440	40	70		45		3.1		135	40
Rubber	910	15							162	600
Granite (Avg. values)	2770	20	240	35			70	4	7.2	
Marble (Avg. values)	2770	15	125	28			55	3	10.8	
Sandstone (Avg. values)	2300	7	85	14			40	2	9.0	
Glass, 98% silica	2190	50	50				65	4.1	80	

[1] Proprties of metals very widely as a result of variations in composition, heat treatment, and mechanical working.

[2] For ductile metals the compression strength is generally assumed to be equal to the tension strength.

[3] Offset of 0.2 percent.

[4] Timber proprties are for loading parallel to the grain.

[5] See *Marks' Mechanical Engineering Handbook*, 10th ed., McGraw-Hill, New York, 1996; *Annual Book of ASTM*, American Society for Testing Materials, Philadelphia, Pa.; *Metals Handbook*, American Society of Metals, Metals Park, Ohio; and *Aluminum Design Manual*, The Aluminum Association, Washington, DC.

Appendix C. Properties of Rolled-Steel Shapes
(U.S. Customary Units)

W Shapes
(Wide-Flange Shapes)

Designation†	Area A, in²	Depth d, in.	Flange Width b_f, in.	Flange Thickness t_f, in.	Web Thickness t_w, in.	Axis X-X I_x, in⁴	Axis X-X S_x, in³	Axis X-X r_x, in.	Axis Y-Y I_y, in⁴	Axis Y-Y S_y, in³	Axis Y-Y r_y, in.
W36 × 300	88.3	36.74	16.655	1.680	0.945	20300	1110	15.2	1300	156	3.83
135	39.7	35.55	11.950	0.790	0.600	7800	439	14.0	225	37.7	2.38
W33 × 201	59.1	33.68	15.745	1.150	0.715	11500	684	14.0	749	95.2	3.56
118	34.7	32.86	11.480	0.740	0.550	5900	359	13.0	187	32.6	2.32
W30 × 173	50.8	30.44	14.985	1.065	0.655	8200	539	12.7	598	79.8	3.43
99	29.1	29.65	10.450	0.670	0.520	3990	269	11.7	128	24.5	2.10
W27 × 146	42.9	27.38	13.965	0.975	0.605	5630	411	11.4	443	63.5	3.21
84	24.8	26.71	9.960	0.640	0.460	2850	213	10.7	106	21.2	2.07
W24 × 104	30.6	24.06	12.750	0.750	0.500	3100	258	10.1	259	40.7	2.91
68	20.1	23.73	8.965	0.585	0.415	1830	154	9.55	70.4	15.7	1.87
W21 × 101	29.8	21.36	12.290	0.800	0.500	2420	227	9.02	248	40.3	2.89
62	18.3	20.99	8.240	0.615	0.400	1330	127	8.54	57.5	13.9	1.77
44	13.0	20.66	6.500	0.450	0.350	843	81.6	8.06	20.7	6.36	1.26
W18 × 106	31.1	18.73	11.200	0.940	0.590	1910	204	7.84	220	39.4	2.66
76	22.3	18.21	11.035	0.680	0.425	1330	146	7.73	152	27.6	2.61
50	14.7	17.99	7.495	0.570	0.355	800	88.9	7.38	40.1	10.7	1.65
35	10.3	17.70	6.000	0.425	0.300	510	57.6	7.04	15.3	5.12	1.22
W16 × 77	22.6	16.52	10.295	0.760	0.455	1110	134	7.00	138	26.9	2.47
57	16.8	16.43	7.120	0.715	0.430	758	92.2	6.72	43.1	12.1	1.60
40	11.8	16.01	6.995	0.505	0.305	518	64.7	6.63	28.9	8.25	1.57
31	9.12	15.88	5.525	0.440	0.275	375	47.2	6.41	12.4	4.49	1.17
26	7.68	15.69	5.500	0.345	0.250	301	38.4	6.26	9.59	3.49	1.12
W14 × 370	109	17.92	16.475	2.660	1.655	5440	607	7.07	1990	241	4.27
145	42.7	14.78	15.500	1.090	0.680	1710	232	6.33	677	87.3	3.98
82	24.1	14.31	10.130	0.855	0.510	882	123	6.05	148	29.3	2.48
68	20.0	14.04	10.035	0.720	0.415	723	103	6.01	121	24.2	2.46
53	15.6	13.92	8.060	0.660	0.370	541	77.8	5.89	57.7	14.3	1.92
43	12.6	13.66	7.995	0.530	0.305	428	62.7	5.82	45.2	11.3	1.89
38	11.2	14.10	6.770	0.515	0.310	385	54.6	5.87	26.7	7.88	1.55
30	8.85	13.84	6.730	0.385	0.270	291	42.0	5.73	19.6	5.82	1.49
26	7.69	13.91	5.025	0.420	0.255	245	35.3	5.65	8.91	3.54	1.08
22	6.49	13.74	5.000	0.335	0.230	199	29.0	5.54	7.00	2.80	1.04

†A wide-flange shape is designated by the letter W followed by the nominal depth in inches and the weight in pounds per foot.

(Table continued on page 752)

Appendix C.   Properties of Rolled-Steel Shapes
(SI Units)

W Shapes
(Wide-Flange Shapes)

Designation†	Area A, mm²	Depth d, mm	Flange Width b_f, mm	Flange Thickness t_f, mm	Web Thickness t_w, mm	Axis X-X I_x 10⁶ mm⁴	Axis X-X S_x 10³ mm³	Axis X-X r_x mm	Axis Y-Y I_y 10⁶ mm⁴	Axis Y-Y S_y 10³ mm³	Axis Y-Y r_y mm
W920 × 446	57000	933	423	42.70	24.0	8470	18200	385	540	2550	97.3
201	25600	903	304	20.10	15.2	3250	7200	356	94.4	621	60.7
W840 × 299	38100	855	400	29.20	18.2	4790	11200	355	312	1560	90.5
176	22400	835	292	18.80	14.0	2460	5890	331	78.2	536	59.1
W760 × 257	32600	773	381	27.10	16.6	3420	8850	324	250	1310	87.6
147	18700	753	265	17.00	13.2	1660	4410	298	52.9	399	53.2
W690 × 217	27700	695	355	24.80	15.4	2340	6730	291	185	1040	81.7
125	16000	678	253	16.30	11.7	1190	3510	273	44.1	349	52.5
W610 × 155	19700	611	324	19.00	12.7	1290	4220	256	108	667	74.0
101	13000	603	228	14.90	10.5	764	2530	242	29.5	259	47.6
W530 × 150	19200	543	312	20.30	12.7	1010	3720	229	103	660	73.2
92	11800	533	209	15.60	10.2	552	2070	216	23.8	228	44.9
66	8370	525	165	11.40	8.9	351	1340	205	8.57	104	32.0
W460 × 158	20100	476	284	23.90	15.0	796	3340	199	91.4	644	67.4
113	14400	463	280	17.30	10.8	556	2400	196	63.3	452	66.3
74	9450	457	190	14.50	9.0	333	1460	188	16.6	175	41.9
52	6630	450	152	10.80	7.6	212	942	179	6.34	83.4	30.9
W410 × 114	14600	420	261	19.30	11.6	462	2200	178	57.2	438	62.6
85	10800	417	181	18.20	10.9	315	1510	171	18.0	199	40.8
60	7580	407	178	12.80	7.7	216	1060	169	12.1	136	40.0
46.1	5890	403	140	11.20	7.0	156	774	163	5.14	73.4	29.5
38.8	4990	399	140	8.80	6.4	127	637	160	4.04	57.7	28.5
W360 × 551	70100	455	418	67.60	12.0	2260	9930	180	825	3950	108
216	27600	375	394	27.70	17.3	712	3800	161	283	1440	101
122	15500	363	257	21.70	13.0	365	2010	153	61.5	479	63.0
101	12900	357	255	18.30	10.5	302	1690	153	50.6	397	62.6
79	10100	354	205	16.80	9.4	227	1280	150	24.2	236	48.9
64	8140	347	203	13.50	7.7	178	1030	148	18.9	186	48.2
57.8	7220	358	172	13.10	7.9	161	899	149	11.1	129	39.2
44	5730	352	171	9.80	6.9	122	693	146	8.18	95.7	37.8
39	4980	353	128	10.70	6.5	102.0	578	143	3.75	58.6	27.4
32.9	4170	349	127	8.50	5.8	82.7	474	141	2.91	45.8	26.4

†A wide-flange shape is designated by the letter W followed by the nominal depth in millimeters and the mass in kilograms per meter.

(Table continued on page 753)
(Table continued on page 753)

Appendix C. Properties of Rolled-Steel Shapes
(U.S. Customary Units)
Continued from page 750

W Shapes
(Wide-Flange Shapes)

Designation†	Area A, in²	Depth d, in.	Flange Width b_f, in.	Flange Thickness t_f, in.	Web Thickness t_w, in.	I_x in⁴	S_x in³	r_x, in.	I_y in⁴	S_y in³	r_y, in.
W12 × 96	28.2	12.71	12.160	0.900	0.550	833	131	5.44	270	44.4	3.09
72	21.1	12.25	12.040	0.670	0.430	597	97.4	5.31	195	32.4	3.04
50	14.7	12.19	8.080	0.640	0.370	394	64.7	5.18	56.3	13.9	1.96
40	11.8	11.94	8.005	0.515	0.295	310	51.9	5.13	44.1	11.0	1.93
35	10.3	12.50	6.560	0.520	0.300	285	45.6	5.25	24.5	7.47	1.54
30	8.79	12.34	6.520	0.440	0.260	238	38.6	5.21	20.3	6.24	1.52
26	7.65	12.22	6.490	0.380	0.230	204	33.4	5.17	17.3	5.34	1.51
22	6.48	12.31	4.030	0.425	0.260	156	25.4	4.91	4.66	2.31	0.847
16	4.71	11.99	3.990	0.265	0.220	103	17.1	4.67	2.82	1.41	0.773
W10 × 112	32.9	11.36	10.415	1.250	0.755	716	126	4.66	236	45.3	2.68
68	20.0	10.40	10.130	0.770	0.470	394	75.7	4.44	134	26.4	2.59
54	15.8	10.09	10.030	0.615	0.370	303	60.0	4.37	103	20.6	2.56
45	13.3	10.10	8.020	0.620	0.350	248	49.1	4.32	53.4	13.3	2.01
39	11.5	9.92	7.985	0.530	0.315	209	42.1	4.27	45.0	11.3	1.98
33	9.71	9.73	7.960	0.435	0.290	170	35.0	4.19	36.6	9.20	1.94
30	8.84	10.47	5.810	0.510	0.300	170	32.4	4.38	16.7	5.75	1.37
22	6.49	10.17	5.750	0.360	0.240	118	23.2	4.27	11.4	3.97	1.33
19	5.62	10.24	4.020	0.395	0.250	96.3	18.8	4.14	4.29	2.14	0.874
15	4.41	9.99	4.000	0.270	0.230	68.9	13.8	3.95	2.89	1.45	0.810
W8 × 58	17.1	8.75	8.220	0.810	0.510	228	52.0	3.65	75.1	18.3	2.10
48	14.1	8.50	8.110	0.685	0.400	184	43.3	3.61	60.9	15.0	2.08
40	11.7	8.25	8.070	0.560	0.360	146	35.5	3.53	49.1	12.2	2.04
35	10.3	8.12	8.020	0.495	0.310	127	31.2	3.51	42.6	10.6	2.03
31	9.13	8.00	7.995	0.435	0.285	110	27.5	3.47	37.1	9.27	2.02
28	8.25	8.06	6.535	0.465	0.285	98.0	24.3	3.45	21.7	6.63	1.62
24	7.08	7.93	6.495	0.400	0.245	82.8	20.9	3.42	18.3	5.63	1.61
21	6.16	8.28	5.270	0.400	0.250	75.3	18.2	3.49	9.77	3.71	1.26
18	5.26	8.14	5.250	0.330	0.230	61.9	15.2	3.43	7.97	3.04	1.23
15	4.44	8.11	4.015	0.315	0.245	48.0	11.8	3.29	3.41	1.70	0.876
13	3.84	7.99	4.000	0.255	0.230	39.6	9.91	3.21	2.73	1.37	0.843
W6 × 25	7.34	6.38	6.080	0.455	0.320	53.4	16.7	2.70	17.1	5.61	1.52
20	5.87	6.20	6.020	0.365	0.260	41.4	13.4	2.66	13.3	4.41	1.50
16	4.74	6.28	4.030	0.405	0.260	32.1	10.2	2.60	4.43	2.20	0.966
12	3.55	6.03	4.000	0.280	0.230	22.1	7.31	2.49	2.99	1.50	0.918
9	2.68	5.90	3.940	0.215	0.170	16.4	5.56	2.47	2.19	1.11	0.905
W5 × 19	5.54	5.15	5.030	0.430	0.270	26.2	10.2	2.17	9.13	3.63	1.28
16	4.68	5.01	5.000	0.360	0.240	21.3	8.51	2.13	7.51	3.00	1.27
W4 × 13	3.83	4.16	4.060	0.345	0.280	11.3	5.46	1.72	3.86	1.90	1.00

†A wide-flange shape is designated by the letter W followed by the nominal depth in inches and the weight in pounds per foot.

Appendix C.   Properties of Rolled-Steel Shapes
(SI Units)
Continued from page 751

W Shapes
(Wide-Flange Shapes)

Designation[†]	Area A, mm²	Depth d, mm	Flange Width b_f, mm	Flange Thickness t_f, mm	Web Thickness t_w, mm	Axis X-X I_x 10^6 mm⁴	Axis X-X S_x 10^3 mm³	Axis X-X r_x mm	Axis Y-Y I_y 10^6 mm⁴	Axis Y-Y S_y 10^3 mm³	Axis Y-Y r_y mm
W310 × 143	18200	323	309	22.9	14.0	348	2150	138	113	731	78.8
107	13600	311	306	17.0	10.9	248	1590	135	81.2	531	77.3
74	9480	310	205	16.3	9.4	165	1060	132	23.4	228	49.7
60	7590	303	203	13.1	7.5	129	851	130	18.3	180	49.1
52	6670	318	167	13.2	7.6	119	748	134	10.3	123	39.3
44.5	5690	313	166	11.2	6.6	99.2	634	132	8.55	103	38.8
38.7	4940	310	165	9.7	5.8	85.1	549	131	7.27	88.1	38.4
32.7	4180	313	102	10.8	6.6	65.0	415	125	1.92	37.6	21.4
23.8	3040	305	101	6.7	5.6	42.7	280	119	1.16	23.0	19.5
W250 × 167	21300	289	265	31.8	19.2	300	2080	119	98.8	746	68.1
101	12900	264	257	19.6	11.9	164	1240	113	55.5	432	65.6
80	10200	256	255	15.6	9.4	126	984	111	43.1	338	65.0
67	8580	257	204	15.7	8.9	101	809	110	22.2	218	51.0
58	7420	252	203	13.5	8.0	87.3	693	108	18.8	185	50.3
49.1	6250	247	202	11.0	7.4	70.6	572	106	15.1	150	49.2
44.8	5720	266	148	13.0	7.6	71.1	535	111	7.03	95.0	35.1
32.7	4180	258	146	9.1	6.1	48.9	379	108	4.73	64.8	33.7
28.4	3630	260	102	10.0	6.4	40.0	308	105	1.78	34.9	22.1
22.3	2850	254	102	6.9	5.8	28.9	228	101	1.23	24.1	20.8
W200 × 86	11000	222	209	20.6	13.0	94.7	853	92.4	31.4	300	53.2
71	9100	216	206	17.4	10.2	76.6	709	91.7	25.4	247	52.8
59	7560	210	205	14.2	9.1	61.1	582	89.9	20.4	199	51.9
52	6660	206	204	12.6	7.9	52.7	512	89.0	17.8	175	51.7
46.1	5860	203	203	11.0	7.2	45.5	448	87.9	15.3	151	51.1
41.7	5310	205	166	11.8	7.2	40.9	399	87.8	9.01	109	41.2
35.9	4580	201	165	10.2	6.2	34.4	342	86.7	7.64	92.6	40.8
31.3	4000	210	134	10.2	6.4	31.4	299	88.6	4.1	61.2	32.0
26.6	3390	207	133	8.4	5.8	25.8	249	87.2	3.3	49.6	31.2
22.5	2860	206	102	8.0	6.2	20.0	194	83.6	1.42	27.8	22.3
19.3	2480	203	102	6.5	5.8	16.6	164	81.8	1.15	22.5	21.5
W150 × 37.1	4730	162	154	11.6	8.1	22.2	274	68.5	7.07	91.8	38.7
29.8	3790	157	153	9.3	6.6	17.2	219	67.4	5.56	72.7	38.3
24.0	3060	160	102	10.3	6.6	13.4	168	66.2	1.83	35.9	24.5
18.0	2290	153	102	7.1	5.8	9.17	120	63.3	1.26	24.7	23.5
13.5	1730	150	100	5.5	4.3	6.87	91.6	63.0	0.918	18.4	23.0
W130 × 28.1	3580	131	128	10.9	6.9	10.9	166	55.2	3.81	59.5	32.6
23.8	3010	127	127	9.1	6.1	8.80	139	54.1	3.11	49.0	32.1
W100 × 19.3	2480	106	103	8.8	7.1	4.77	90.0	43.9	1.61	31.3	25.5

†A wide-flange shape is designated by the letter W followed by the nominal depth in millimeters and the mass in kilograms per meter.

Appendix C. Properties of Rolled-Steel Shapes
(U.S. Customary Units)

S Shapes
(American Standard Shapes)

Designation†	Area A, in²	Depth d, in.	Flange Width b_f, in.	Flange Thick-ness t_f, in.	Web Thick-ness t_w, in.	Axis X-X I_x, in⁴	S_x, in³	r_x, in.	Axis Y-Y I_y, in⁴	S_y, in³	r_y, in.
S24 × 121	35.6	24.50	8.050	1.090	0.800	3160	258	9.43	83.3	20.7	1.53
106	31.2	24.50	7.870	1.090	0.620	2940	240	9.71	77.1	19.6	1.57
100	29.3	24.00	7.245	0.870	0.745	2390	199	9.02	47.7	13.2	1.27
90	26.5	24.00	7.125	0.870	0.625	2250	187	9.21	44.9	12.6	1.30
80	23.5	24.00	7.000	0.870	0.500	2100	175	9.47	42.2	12.1	1.34
S20 × 96	28.2	20.30	7.200	0.920	0.800	1670	165	7.71	50.2	13.9	1.33
86	25.3	20.30	7.060	0.920	0.660	1580	155	7.89	46.8	13.3	1.36
75	22.0	20.00	6.385	0.795	0.635	1280	128	7.62	29.8	9.32	1.16
66	19.4	20.00	6.255	0.795	0.505	1190	119	7.83	27.7	8.85	1.19
S18 × 70	20.6	18.00	6.251	0.691	0.711	926	103	6.71	24.1	7.72	1.08
54.7	16.1	18.00	6.001	0.691	0.461	804	89.4	7.07	20.8	6.94	1.14
S15 × 50	14.7	15.00	5.640	0.622	0.550	486	64.8	5.75	15.7	5.57	1.03
42.9	12.6	15.00	5.501	0.622	0.411	447	59.6	5.95	14.4	5.23	1.07
S12 × 50	14.7	12.00	5.477	0.659	0.687	305	50.8	4.55	15.7	5.74	1.03
40.8	12.0	12.00	5.252	0.659	0.462	272	45.4	4.77	13.6	5.16	1.06
35	10.3	12.00	5.078	0.544	0.428	229	38.2	4.72	9.87	3.89	0.980
31.8	9.35	12.00	5.000	0.544	0.350	218	36.4	4.83	9.36	3.74	1.00
S10 × 35	10.3	10.00	4.944	0.491	0.594	147	29.4	3.78	8.36	3.38	0.901
25.4	7.46	10.00	4.661	0.491	0.311	124	24.7	4.07	6.79	2.91	0.954
S8 × 23	6.77	8.00	4.171	0.425	0.441	64.9	16.2	3.10	4.31	2.07	0.798
18.4	5.41	8.00	4.001	0.425	0.271	57.6	14.4	3.26	3.73	1.86	0.831
S6 × 17.25	5.07	6.00	3.565	0.359	0.465	26.3	8.77	2.28	2.31	1.30	0.675
12.5	3.67	6.00	3.332	0.359	0.232	22.1	7.37	2.45	1.82	1.09	0.705
S5 × 10	2.94	5.00	3.004	0.326	0.214	12.3	4.92	2.05	1.22	0.809	0.643
S4 × 9.5	2.79	4.00	2.796	0.293	0.326	6.79	3.39	1.56	0.903	0.646	0.569
7.7	2.26	4.00	2.663	0.293	0.193	6.08	3.04	1.64	0.764	0.574	0.581
S3 × 7.5	2.21	3.00	2.509	0.260	0.349	2.93	1.95	1.15	0.586	0.468	0.516
5.7	1.67	3.00	2.330	0.260	0.170	2.52	1.68	1.23	0.455	0.390	0.522

†An American Standard Beam is designated by the letter S followed by the nominal depth in inches and the weight in pounds per foot.

Appendix C. Properties of Rolled-Steel Shapes
(SI Units)

S Shapes
(American Standard Shapes)

Designation†	Area A, mm²	Depth d, mm	Flange Width b_f, mm	Flange Thickness t_f, mm	Web Thickness t_w, mm	Axis X-X I_x 10⁶ mm⁴	Axis X-X S_x 10³ mm³	Axis X-X r_x mm	Axis Y-Y I_y 10⁶ mm⁴	Axis Y-Y S_y 10³ mm³	Axis Y-Y r_y mm
S610 × 180	22900	622	204	27.7	20.3	1320	4240	240	34.9	341	39.0
158	20100	622	200	27.7	15.7	1230	3950	247	32.5	321	39.9
149	19000	610	184	22.1	18.9	995	3260	229	20.2	215	32.3
134	17100	610	181	22.1	15.9	938	3080	234	19.0	206	33.0
119	15200	610	178	22.1	12.7	878	2880	240	17.9	198	34.0
S510 × 143	18200	516	183	23.4	20.3	700	2710	196	21.3	228	33.9
128	16400	516	179	23.4	16.8	658	2550	200	19.7	216	34.4
112	14200	508	162	20.2	16.1	530	2090	193	12.6	152	29.5
98.3	12500	508	159	20.2	12.8	495	1950	199	11.8	145	30.4
S460 × 104	13300	457	159	17.6	18.1	385	1685	170	10.4	127	27.5
81.4	10400	457	152	17.6	11.7	333	1460	179	8.83	113	28.8
S380 × 74	9500	381	143	15.6	14.0	201	1060	145	6.65	90.8	26.1
64	8150	381	140	15.8	10.4	185	971	151	6.15	85.7	27.1
S310 × 74	9480	305	139	16.7	17.4	126	826	115	6.69	93.2	26.1
60.7	7730	305	133	16.7	11.7	113	741	121	5.73	83.6	26.8
52	6650	305	129	13.8	10.9	95.3	625	120	4.19	63.6	24.8
47.3	6010	305	127	13.8	8.9	90.5	593	122	3.97	61.1	25.3
S250 × 52	6670	254	126	12.5	15.1	61.2	482	95.8	3.59	55.7	22.9
37.8	4820	254	118	12.5	7.9	51.1	402	103	2.86	47.5	24.1
S200 × 34	4370	203	106	10.8	11.2	26.8	264	78.3	1.83	33.8	20.2
27.4	3500	203	102	10.8	6.9	23.9	235	82.6	1.60	30.6	21.1
S150 × 25.7	3270	152	91	9.1	11.8	10.8	142	57.5	1.00	21.3	17.2
18.6	2370	152	85	9.1	5.8	9.11	120	62.0	0.782	18.0	18.0
S130 × 15	1890	127	76	8.3	5.4	5.07	79.8	51.8	0.513	13.2	16.3
S100 × 14.1	1800	102	71	7.4	8.3	2.82	55.3	39.6	0.383	10.5	14.4
11.5	1460	102	68	7.4	4.9	2.53	49.6	41.6	0.328	9.41	14.8
S75 × 11.2	1430	76	64	6.6	8.9	1.20	31.6	29.0	0.254	7.72	13.1
8.5	1070	76	59	6.6	4.3	1.03	27.1	31.0	0.190	6.44	13.3

†An American Standard Beam is designated by the letter S followed by the nominal depth in millimeters and the mass in kilograms per meter.

Appendix C. Properties of Rolled-Steel Shapes
(U.S. Customary Units)

C Shapes
(American Standard Channels)

Designation†	Area A, in²	Depth d, in.	Flange Width b_f, in.	Flange Thickness t_f, in.	Web Thickness t_w, in.	Axis X-X I_x, in⁴	Axis X-X S_x, in³	Axis X-X r_x, in.	Axis Y-Y I_y, in⁴	Axis Y-Y S_y, in³	Axis Y-Y r_y, in.	Axis Y-Y $\bar{x}$, in.
C15 × 50	14.7	15.00	3.716	0.650	0.716	404	53.8	5.24	11.0	3.78	0.867	0.798
40	11.8	15.00	3.520	0.650	0.520	349	46.5	5.44	9.23	3.37	0.886	0.777
33.9	9.96	15.00	3.400	0.650	0.400	315	42.0	5.62	8.13	3.11	0.904	0.787
C12 × 30	8.82	12.00	3.170	0.501	0.510	162	27.0	4.29	5.14	2.06	0.763	0.674
25	7.35	12.00	3.047	0.501	0.387	144	24.1	4.43	4.47	1.88	0.780	0.674
20.7	6.09	12.00	2.942	0.501	0.282	129	21.5	4.61	3.88	1.73	0.799	0.698
C10 × 30	8.82	10.00	3.033	0.436	0.673	103	20.7	3.42	3.94	1.65	0.649	0.649
25	7.35	10.00	2.886	0.436	0.526	91.2	18.2	3.52	3.36	1.48	0.676	0.617
20	5.88	10.00	2.739	0.436	0.379	78.9	15.8	3.66	2.81	1.32	0.692	0.606
15.3	4.49	10.00	2.600	0.436	0.240	67.4	13.5	3.87	2.28	1.16	0.713	0.634
C9 × 20	5.88	9.00	2.648	0.413	0.448	60.9	13.5	3.22	2.42	1.17	0.642	0.583
15	4.41	9.00	2.485	0.413	0.285	51.0	11.3	3.40	1.93	1.01	0.661	0.586
13.4	3.94	9.00	2.433	0.413	0.233	47.9	10.6	3.48	1.76	0.962	0.669	0.601
C8 × 18.75	5.51	8.00	2.527	0.390	0.487	44.0	11.0	2.82	1.98	1.01	0.599	0.565
13.75	4.04	8.00	2.343	0.390	0.303	36.1	9.03	2.99	1.53	0.854	0.615	0.553
11.5	3.38	8.00	2.260	0.390	0.220	32.6	8.14	3.11	1.32	0.781	0.625	0.571
C7 × 12.25	3.60	7.00	2.194	0.366	0.314	24.2	6.93	2.60	1.17	0.703	0.571	0.525
9.8	2.87	7.00	2.090	0.366	0.210	21.3	6.08	2.72	0.968	0.625	0.581	0.540
C6 × 13	3.83	6.00	2.157	0.343	0.437	17.4	5.80	2.13	1.05	0.642	0.525	0.514
10.5	3.09	6.00	2.034	0.343	0.314	15.2	5.06	2.22	0.866	0.564	0.529	0.499
8.2	2.40	6.00	1.920	0.343	0.200	13.1	4.38	2.34	0.693	0.492	0.537	0.511
C5 × 9	2.64	5.00	1.885	0.320	0.325	8.90	3.56	1.83	0.632	0.450	0.489	0.478
6.7	1.97	5.00	1.750	0.320	0.190	7.49	3.00	1.95	0.479	0.378	0.493	0.484
C4 × 7.25	2.13	4.00	1.721	0.296	0.321	4.59	2.29	1.47	0.433	0.343	0.450	0.459
5.4	1.59	4.00	1.584	0.296	0.184	3.85	1.93	1.56	0.319	0.283	0.449	0.457
C3 × 6	1.76	3.00	1.596	0.273	0.356	2.07	1.38	1.08	0.305	0.268	0.416	0.455
5	1.47	3.00	1.498	0.273	0.258	1.85	1.24	1.12	0.247	0.233	0.410	0.438
4.1	1.21	3.00	1.410	0.273	0.170	1.66	1.10	1.17	0.197	0.202	0.404	0.436

†An American Standard Channel is designated by the letter C followed by the nominal depth in inches and the weight in pounds per foot.

Appendix C.   Properties of Rolled-Steel Shapes
(SI Units)

C Shapes
(American Standard Channels)

Designation[†]	Area A, mm²	Depth d, mm	Flange Width b_f, mm	Flange Thickness t_f, mm	Web Thickness t_w, mm	Axis X-X I_x 10⁶ mm⁴	S_x 10³ mm³	r_x mm	Axis Y-Y I_y 10⁶ mm⁴	S_y 10³ mm³	r_y mm	$\bar{x}$ mm
C380 × 74	9480	381	94	16.5	18.2	167	877	133	4.54	61.5	21.9	20.2
60	7570	381	89	16.5	13.2	144	756	138	3.79	54.7	22.4	19.7
50.4	6430	381	86	16.5	10.2	134	688	143	3.34	50.5	22.8	19.9
C310 × 45	5690	305	80	12.7	13.0	67.2	441	109	2.09	33.2	19.2	17.0
37	4720	305	77	12.7	9.8	59.7	391	112	1.83	30.5	19.7	17.0
30.8	3920	305	74	12.7	7.2	53.4	350	117	1.57	27.7	20.0	17.4
C250 × 45	5670	254	76	11.1	17.1	42.7	336	86.8	1.58	26.5	16.7	16.3
37	4750	254	73	11.1	13.4	37.9	298	89.3	1.38	24.0	17.0	15.6
30	3780	254	69	11.1	9.6	32.6	257	92.9	1.14	21.2	17.4	15.3
22.8	2880	254	65	11.1	6.1	27.7	218	98.1	0.912	18.5	17.8	15.8
C230 × 30	3800	229	67	10.5	11.4	25.4	222	81.8	0.997	19.1	16.2	14.7
22	2840	229	63	10.5	7.2	21.2	185	86.4	0.796	16.5	16.7	14.9
19.9	2530	229	61	10.5	5.9	19.8	173	88.5	0.708	15.4	16.7	15.0
C200 × 27.9	3560	203	64	9.9	12.4	18.2	179	71.5	0.817	16.4	15.1	14.3
20.5	2660	203	59	9.9	7.7	14.9	147	75.7	0.620	13.7	15.4	13.9
17.1	2170	203	57	9.9	5.6	13.4	132	78.6	0.538	12.6	15.7	14.4
C180 × 18.2	2310	178	55	9.3	8.0	10.0	112	65.8	0.470	11.2	14.3	13.1
14.6	1850	178	53	9.3	5.3	8.83	99.2	69.1	0.400	10.2	14.7	13.7
C150 × 19.3	2450	152	54	8.7	11.1	7.11	93.6	53.9	0.420	10.2	13.1	12.9
15.6	1980	152	51	8.7	8.0	6.21	81.7	56.0	0.347	9.01	13.2	12.5
12.2	1540	152	48	8.7	5.1	5.35	70.4	58.9	0.276	7.82	13.4	12.7
C130 × 13	1710	127	48	8.1	8.3	3.70	58.3	46.5	0.264	7.37	12.4	12.2
10.4	1310	127	47	8.1	4.8	3.25	51.2	49.8	0.229	6.74	13.2	13.0
C100 × 10.8	1370	102	43	7.5	8.2	1.90	37.3	37.2	0.172	5.44	11.2	11.4
8.0	1020	102	40	7.5	4.7	1.61	31.6	39.7	0.130	4.56	11.3	11.5
C75 × 8.9	1130	76.2	40	6.9	9.0	0.850	22.3	27.4	0.122	4.25	10.4	11.3
7.4	936	76.2	37	6.9	6.6	0.751	19.7	28.3	0.0948	3.62	10.1	10.8
6.1	765	76.2	35	6.9	4.3	0.671	17.6	29.6	0.0765	3.16	10.0	10.8

†An American Standard Channel is designated by the letter C followed by the nominal depth in millimeters and the mass in kilograms per meter.

Appendix C. Properties of Rolled-Steel Shapes
(U.S. Customary Units)

Angles
Equal Legs

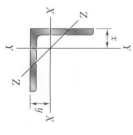

Size and Thickness, in.	Weight per Foot, lb/ft	Area, in²	Axis X-X and Axis Y-Y				Axis Z-Z r, in.
			I, in⁴	S, in³	r, in.	x or y, in.	
L8 × 8 × 1	51.0	15.0	89.0	15.8	2.44	2.37	1.56
¾	38.9	11.4	69.7	12.2	2.47	2.28	1.58
½	26.4	7.75	48.6	8.36	2.50	2.19	1.59
L6 × 6 × 1	37.4	11.0	35.5	8.57	1.80	1.86	1.17
¾	28.7	8.44	28.2	6.66	1.83	1.78	1.17
⅝	24.2	7.11	24.2	5.66	1.84	1.73	1.18
½	19.6	5.75	19.9	4.61	1.86	1.68	1.18
⅜	14.9	4.36	15.4	3.53	1.88	1.64	1.19
L5 × 5 × ¾	23.6	6.94	15.7	4.53	1.51	1.52	0.975
⅝	20.0	5.86	13.6	3.86	1.52	1.48	0.978
½	16.2	4.75	11.3	3.16	1.54	1.43	0.983
⅜	12.3	3.61	8.74	2.42	1.56	1.39	0.990
L4 × 4 × ¾	18.5	5.44	7.67	2.81	1.19	1.27	0.778
⅝	15.7	4.61	6.66	2.40	1.20	1.23	0.779
½	12.8	3.75	5.56	1.97	1.22	1.18	0.782
⅜	9.8	2.86	4.36	1.52	1.23	1.14	0.788
¼	6.6	1.94	3.04	1.05	1.25	1.09	0.795
L3½ × 3½ × ½	11.1	3.25	3.64	1.49	1.06	1.06	0.683
⅜	8.5	2.48	2.87	1.15	1.07	1.01	0.687
¼	5.8	1.69	2.01	0.794	1.09	0.968	0.694
L3 × 3 × ½	9.4	2.75	2.22	1.07	0.898	0.932	0.584
⅜	7.2	2.11	1.76	0.833	0.913	0.888	0.587
¼	4.9	1.44	1.24	0.577	0.930	0.842	0.592
L2½ × 2½ × ½	7.7	2.25	1.23	0.724	0.739	0.806	0.487
⅜	5.9	1.73	0.984	0.566	0.753	0.762	0.487
¼	4.1	1.19	0.703	0.394	0.769	0.717	0.491
³⁄₁₆	3.07	0.902	0.547	0.303	0.778	0.694	0.495
L2 × 2 × ⅜	4.7	1.36	0.479	0.351	0.594	0.636	0.389
¼	3.19	0.938	0.348	0.247	0.609	0.592	0.391
⅛	1.65	0.484	0.190	0.131	0.626	0.546	0.398

Appendix C.   Properties of Rolled-Steel Shapes
(SI Units)

Angles
Equal Legs

Size and Thickness, mm	Mass per Meter, Kg/m	Area, mm²	Axis X-X and Axis Y-Y				Axis Z-Z
			I 10⁶ mm⁴	S 10³ mm³	r mm	x or y mm	r mm
L203 × 203 × 25.4	75.9	9670	36.9	258	61.8	60.0	39.7
19.0	57.9	7350	28.9	199	62.7	57.8	40.0
12.7	39.3	4990	20.2	137	63.6	55.5	40.4
L152 × 152 × 25.4	55.7	7080	14.6	139	45.4	47.2	29.5
19.0	42.7	5420	11.6	108	46.3	44.9	29.7
15.9	36.0	4580	10.0	92.5	46.7	43.9	29.9
12.7	29.2	3700	8.22	75.2	47.1	42.7	30.0
9.5	22.2	2800	6.34	57.4	47.6	41.5	30.2
L127 × 127 × 19.0	35.1	4470	6.54	74.0	38.3	38.6	24.7
15.9	29.8	3790	5.66	63.2	38.6	37.5	24.8
12.7	24.1	3060	4.68	51.7	39.1	36.5	25.0
9.5	18.3	2320	3.63	39.6	39.6	35.3	25.1
L102 × 102 × 19.0	27.5	3520	3.23	46.3	30.3	32.3	19.9
15.9	23.4	2990	2.81	39.7	30.7	31.3	19.9
12.7	19.0	2430	2.34	32.6	31.0	30.2	19.9
9.5	14.6	1850	1.83	25.1	31.5	29.0	20.0
6.4	9.8	1260	1.29	17.4	32.0	28.0	20.3
L89 × 89 × 12.7	16.5	2100	1.52	24.5	26.9	26.9	17.4
9.5	12.6	1600	1.19	18.8	27.3	25.8	17.4
6.4	8.6	1100	0.845	13.1	27.7	24.6	17.6
L76 × 76 × 12.7	14.0	1770	0.915	17.5	22.7	23.6	14.8
9.5	10.7	1350	0.725	13.6	23.2	22.5	14.9
6.4	7.3	932	0.517	9.50	23.6	21.4	15.0
L64 × 64 × 12.7	11.4	1460	0.524	12.1	18.9	20.6	12.5
9.5	8.7	1130	0.419	9.40	19.3	19.4	12.5
6.4	6.1	778	0.302	6.62	19.7	18.4	12.6
4.8	4.6	591	0.235	5.09	19.9	17.8	12.7
L51 × 51 × 9.5	7.0	879	0.202	5.80	15.2	16.2	9.95
6.4	4.7	612	0.147	4.09	15.5	15.1	9.94
3.2	2.4	316	0.0806	2.17	16.0	13.9	10.1

Appendix C. Properties of Rolled-Steel Shapes
(U.S. Customary Units)

Angles
Unequal Legs

Size and Thickness, in.	Weight per Foot, lb/ft	Area, in²	Axis X-X				Axis Y-Y				Axis Z-Z	
			I_x, in⁴	S_x, in³	r_x, in.	y, in.	I_y, in⁴	S_y, in³	r_y, in.	x, in.	r_z, in.	tan α
L8 × 6 × 1	44.2	13.0	80.8	15.1	2.49	2.65	38.8	8.92	1.73	1.65	1.28	0.543
¾	33.8	9.94	63.4	11.7	2.53	2.56	30.7	6.92	1.76	1.56	1.29	0.551
½	23.0	6.75	44.3	8.02	2.56	2.47	21.7	4.79	1.79	1.47	1.30	0.558
L6 × 4 × ¾	23.6	6.94	24.5	6.25	1.88	2.08	8.68	2.97	1.12	1.08	0.860	0.428
½	16.2	4.75	17.4	4.33	1.91	1.99	6.27	2.08	1.15	0.987	0.870	0.440
⅜	12.3	3.61	13.5	3.32	1.93	1.94	4.90	1.60	1.17	0.941	0.877	0.446
L5 × 3 × ½	12.8	3.75	9.45	2.91	1.59	1.75	2.58	1.15	0.829	0.750	0.648	0.357
⅜	9.8	2.86	7.37	2.24	1.61	1.70	2.04	0.888	0.845	0.704	0.654	0.364
¼	6.6	1.94	5.11	1.53	1.62	1.66	1.44	0.614	0.861	0.657	0.663	0.371
L4 × 3 × ½	11.1	3.25	5.05	1.89	1.25	1.33	2.42	1.12	0.864	0.827	0.639	0.543
⅜	8.5	2.48	3.96	1.46	1.26	1.28	1.92	0.866	0.879	0.782	0.644	0.551
¼	5.8	1.69	2.77	1.00	1.28	1.24	1.36	0.599	0.896	0.736	0.651	0.558
L3½ × 2½ × ½	9.4	2.75	3.24	1.41	1.09	1.20	1.36	0.760	0.704	0.705	0.534	0.486
⅜	7.2	2.11	2.56	1.09	1.10	1.16	1.09	0.592	0.719	0.660	0.537	0.496
¼	4.9	1.44	1.80	0.755	1.12	1.11	0.777	0.412	0.735	0.614	0.544	0.506
L3 × 2 × ½	7.7	2.25	1.92	1.00	0.924	1.08	0.672	0.474	0.546	0.583	0.428	0.414
⅜	5.9	1.73	1.53	0.781	0.940	1.04	0.543	0.371	0.559	0.539	0.430	0.428
¼	4.1	1.19	1.09	0.542	0.957	0.993	0.392	0.260	0.574	0.493	0.435	0.440
L2½ × 2 × ⅜	5.3	1.55	0.912	0.547	0.768	0.831	0.514	0.363	0.577	0.581	0.420	0.614
¼	3.62	1.06	0.654	0.381	0.784	0.787	0.372	0.254	0.592	0.537	0.424	0.626

Appendix C. Properties of Rolled-Steel Shapes
(SI Units)

Angles
Unequal Legs

Size and Thickness, mm	Mass per Meter kg/m	Area mm²	Axis X-X				Axis Y-Y				Axis Z-Z	
			I_x 10^6 mm⁴	S_x 10^3 mm³	r_x mm	y mm	I_y 10^6 mm⁴	S_y 10^3 mm³	r_y mm	x mm	r_z mm	tan α
L203 × 152 × 25.4	65.5	8370	33.5	247	63.3	67.4	16.0	145	43.7	41.9	32.4	0.541
19.0	50.1	6380	26.2	190	64.1	65.1	12.7	113	44.6	39.6	32.7	0.551
12.7	34.1	4350	18.4	131	65.0	62.7	8.96	78.1	45.4	37.3	33.0	0.556
L152 × 102 × 19.0	35.0	4470	10.1	102	47.5	52.5	3.65	49.0	28.6	27.5	21.9	0.435
12.7	24.0	3060	7.20	70.8	48.5	50.3	2.64	34.4	29.4	25.3	22.2	0.446
9.5	18.2	2320	5.56	54.0	49.0	49.1	2.06	26.4	29.8	24.1	22.4	0.452
L127 × 76 × 12.7	19.0	2420	3.93	47.6	40.3	44.4	1.06	18.6	20.9	19.0	16.3	0.355
9.5	14.5	1840	3.06	36.6	40.8	43.3	0.841	14.5	21.4	17.8	16.6	0.362
6.4	9.8	1260	2.14	25.2	41.2	42.1	0.598	10.1	21.8	16.6	16.8	0.369
L102 × 76 × 12.7	16.4	2100	2.12	31.1	31.8	33.9	1.00	18.1	21.8	20.9	16.2	0.536
9.5	12.6	1600	1.66	24.0	32.2	32.8	0.792	14.1	22.2	19.8	16.3	0.545
6.4	8.6	1100	1.17	16.6	32.6	31.6	0.564	9.83	22.6	18.6	16.5	0.552
L89 × 64 × 12.7	13.9	1780	1.36	23.3	27.6	30.6	0.581	12.7	18.1	18.1	13.7	0.491
9.5	10.7	1360	1.07	18.0	28.0	29.5	0.463	9.83	18.5	16.9	13.8	0.503
6.4	7.3	938	0.759	12.5	28.4	28.3	0.333	6.91	18.8	15.8	13.9	0.512
L76 × 51 × 12.7	11.5	1450	0.795	16.4	23.4	27.4	0.283	7.84	14.0	14.9	10.9	0.420
9.5	8.8	1120	0.632	12.7	23.8	26.2	0.228	6.11	14.3	13.7	10.9	0.434
6.4	6.1	772	0.453	8.90	24.2	25.1	0.166	4.32	14.7	12.6	11.1	0.446
L64 × 51 × 9.5	7.9	1000	0.388	9.10	19.5	21.3	0.217	5.99	14.7	14.8	10.8	0.610
6.4	5.4	695	0.280	6.39	20.1	20.2	0.158	4.24	15.1	13.7	10.8	0.621

Appendix. D. Beam Deflections and Slopes

	Beam and Loading	Elastic Curve	Maximum Deflection	Slope at End	Equation of Elastic Curve
1			$-\dfrac{PL^3}{3EI}$	$-\dfrac{PL^2}{2EI}$	$y = \dfrac{P}{6EI}(x^3 - 3Lx^2)$
2			$-\dfrac{wL^4}{8EI}$	$-\dfrac{wL^3}{6EI}$	$y = -\dfrac{w}{24EI}(x^4 - 4Lx^3 + 6L^2x^2)$
3			$-\dfrac{ML^2}{2EI}$	$-\dfrac{ML}{EI}$	$y = -\dfrac{M}{2EI}x^2$
4			$-\dfrac{PL^3}{48EI}$	$\pm\dfrac{PL^2}{16EI}$	For $x \le \tfrac{1}{2}L$: $y = \dfrac{P}{48EI}(4x^3 - 3L^2x)$
5			For $a > b$: $-\dfrac{Pb(L^2 - b^2)^{3/2}}{9\sqrt{3}EIL}$ at $x_m = \sqrt{\dfrac{L^2 - b^2}{3}}$	$\theta_A = -\dfrac{Pb(L^2 - b^2)}{6EIL}$ $\theta_B = +\dfrac{Pa(L^2 - a^2)}{6EIL}$	For $x < a$: $y = -\dfrac{Pb}{6EIL}[x^3 - (L^2 - b^2)x]$ For $x = a$: $y = -\dfrac{Pa^2b^2}{3EIL}$
6			$-\dfrac{5wL^4}{384EI}$	$\pm\dfrac{wL^3}{24EI}$	$y = -\dfrac{w}{24EI}(x^4 - 2Lx^3 + L^3x)$
7			$-\dfrac{ML^2}{9\sqrt{3}EI}$	$\theta_A = +\dfrac{ML}{6EI}$ $\theta_B = -\dfrac{ML}{3EI}$	$y = -\dfrac{M}{6EIL}(x^3 - L^2x)$

APPENDIX

E

Fundamentals of Engineering Examination

Engineers are required to be licensed when their work directly affects the public health, safety, and welfare. The intent is to ensure that engineers have met minimum qualifications, involving competence, ability, experience, and character. The licensing process involves an initial exam, called the *Fundamentals of Engineering Examination*, professional experience, and a second exam, called the *Principles and Practice of Engineering*. Those who successfully complete these requirements are licensed as a *Professional Engineer*. The exams are developed under the auspices of the *National Council of Examiners for Engineering and Surveying*.

The first exam, the *Fundamentals of Engineering Examination*, can be taken just before or after graduation from a four-year accredited engineering program. The exam stresses subject material in a typical undergraduate engineering program, including *Mechanics of Materials*. The topics included in the exam cover much of the material in this book. The following is a list of the main topic areas, with references to the appropriate sections in this book. Also included are problems that can be solved to review this material.

Stresses (1.3–1.8; 1.11–1.12)
Problems: 1.2, 1.9, 1.31, 1.39

Strains (2.2–2.3; 2.5–2.6; 2.8–2.11; 2.14–2.15)
Problems: 2.3, 2.17, 2.39, 2.50, 2.64, 2.69

Torsion (3.2–3.6; 3.13)
Problems: 3.5, 3.22, 3.38, 3.52, 3.140, 3.142

Bending (4.2–4.6; 4.12–4.14)
Problems: 4.9, 4.27, 4.43, 4.46, 4.121, 4.127

Shear and Bending-Moment Diagrams (5.2–3)
Problems: 5.5, 5.17, 5.47, 5.50

Normal Stresses in Beams (5.1–3)
Problems: 5.23, 5.24, 5.62, 5.67

Shear (6.2–6.4; 6.6–6.7)
Problems: 6.1, 6.10, 6.31, 6.39

Transformation of Stresses and Strains (7.2–7.4; 7.7–7.9)
Problems: 7.8, 7.19, 7.31, 7.37, 7.81, 7.83, 7.98, 7.107

Deflection of Beams (9.2–9.4; 9.7)
Problems: 9.6, 9.15, 9.67, 9.77

Columns (10.2–10.4)
Problems: 10.12, 10.21, 10.25

Strain Energy (11.2–11.4)
Problems: 11.12, 11.15, 11.22

Photo Credits

Index

Answers to Problems

Answers to problems with a number set in straight type are given on this and the following pages. Answers to problems with a number set in italic are not listed.

CHAPTER 1

1.1 (a) 84.9 MPa. (b) −96.8 MPa.
1.2 $d_1 = 22.6$ mm; $d_2 = 40.2$ mm.
1.3 (a) 17.93 ksi. (b) 22.6 ksi.
1.4 $d_1 = 1.059$ in.; $d_2 = 0.714$ in.
1.7 (a) 72.2 MPa. (b) −83.3 MPa.
1.8 (a) 101.6 MPa. (b) −21.7 MPa.
1.9 (a) 14.64 ksi. (b) −9.96 ksi.
1.10 (a) −281 psi. (b) 107.1 psi.
1.13 (a) 17.86 kN. (b) −41.4 MPa.
1.14 (a) 12.73 MPa. (b) −4.77 MPa.
1.15 308 mm.
1.16 43.4 mm.
1.17 1.633 in.
1.18 12.57 kips.
1.19 178.6 mm.
1.20 321 mm.
1.23 (a) 80.8 MPa. (b) 127.0 MPa. (c) 203 MPa.
1.24 (a) 31.1 MPa. (b) 48.8 MPa. (c) 78.1 MPa.
1.25 (a) 7.28 ksi. (b) 18.30 ksi.
1.26 (a) 8.92 ksi. (b) 22.4 ksi. (c) 11.21 ksi.
1.29 $\sigma = 565$ kPa; $\tau = 206$ kPa.
1.30 (a) 5.31 kN. (b) 182.0 kPa.
1.31 (a) 3290 lb. (b) 75.5 psi.
1.32 $\sigma = 55.1$ psi; $\tau = 65.7$ psi.
1.35 $\sigma = −37.1$ MPa; $\tau = 17.28$ MPa.
1.36 337 kN.
1.37 3.60.
1.38 27.8 mm.
1.39 3.64.
1.40 4.55 kips.
1.43 3.23.
1.44 216 mm.
1.45 2.87.
1.46 0.8125 in.
1.49 3.72 kN.
1.50 3.97 kN.
1.51 589 lb.
1.52 938 lb.
1.53 2.42.
1.54 2.05.
1.57 (a) 362 kg. (b) 1.718.
1.58 (a) 629 lb. (b) 1.689.
1.59 −54.0 MPa.
1.60 1.084 ksi.
1.62 9.22 kN.
1.63 (a) 10.84 ksi. (b) 5.11 ksi.
1.65 1.800.
1.66 4.49 kips.
1.68 277 lb.
1.69 $22.8° \leq \theta \leq 32.1°$.
1.C2 (c) 16 mm $\leq d \leq 22$ mm. (d) 18 mm $\leq d \leq 22$ mm.
1.C3 (c) 0.70 in. $\leq d \leq 1.10$ in. (d) 0.85 in. $\leq d \leq 1.25$ in.
1.C4 (b) For $\beta = 38.66°$, $\tan \beta = 0.8$; BD is perpendicular to BC.
       (c) F.S. = 3.58 for $\alpha = 26.6°$; **P** is perpendicular to line AC.
1.C5 (b) Member of Fig. P 1.29, for $\alpha = 70°$:
       (1) 565 kPa; (2) 206 kPa; (3) 2.23; (4) 7.29; (5) 2.23.
       Member of Fig. P 1.32, for $\alpha = 40°$:
       (1) 55.1 psi; (2) 65.7 psi; (3) 2.72; (4) 3.26; (5) 2.72.
1.C6 (d) $P_{all} = 5.79$ kN; stress in links is critical.

CHAPTER 2

2.1 (a) 9.96 mm. (b) 109.1 MPa.
2.2 (a) 0.0303 in. (b) 15.28 ksi.
2.3 (a) 9.09 ksi. (b) 1.760.
2.4 (a) 5.32 mm. (b) 1.750 m.
2.5 (a) 11.31 kN. (b) 400 MPa.
2.7 (a) 0.546 mm. (b) 36.3 MPa.
2.9 48.4 kips.
2.10 10.70 mm.
2.11 1.988 kN.
2.12 0.429 in.
2.15 (a) 9.53 kips. (b) 1.254×10^{-3} in.
2.16 (a) 32.8 kN. (b) 0.0728 mm.
2.17 (a) 0.01819 mm. (b) −0.0909 mm.
2.19 5.74 kips.
2.21 50.4 kN.
2.22 2.11 mm; 2.03 mm.
2.23 (a) 0.1767 in. (b) 0.1304 in.
2.26 (a) −0.0302 mm. (b) 0.01783 mm.
2.27 0.1095 mm.
2.28 3.51 kips.
2.29 (a) $pgL^2/2E$. (b) $W/2$.

775

2.30 $pgh^2/16E$.
2.33 (a) 47.5 MPa. (b) 0.1132 mm.
2.34 (a) 75.9 kN. (b) 120 MPa.
2.35 $\sigma_s = -8.34$ ksi; $\sigma_c = -1.208$ ksi.
2.38 695 kips.
2.39 (a) 0.0762 mm.
2.40 (a) 26.2 ksi.
(b) -9.82 ksi.
2.41 (a) $\mathbf{R}_A = 11.92$ kips ←; $\mathbf{R}_D = 20.1$ kips ←.
(b) 3.34×10^{-3} in.
2.42 (a) $\mathbf{R}_A = 76.6$ kN →; $\mathbf{R}_D = 64.6$ kN ←.
(b) -0.0394 mm.
2.43 (a) $\mathbf{R}_A = 47.3$ kN →; $\mathbf{R}_D = 35.3$ kN ←.
(b) -0.00605 mm.
2.44 177.4 lb.
2.47 $P_A = 0.525P$; $P_B = 0.200P$; $P_C = 0.275P$.
2.48 $P_A = 0.1P$; $P_B = 0.2P$; $P_C = 0.3P$; $P_D = 0.4P$.
2.49 (a) -98.3 MPa. (b) -38.3 MPa.
2.50 (a) 1.567 ksi. (b) -0.298 ksi.
2.51 75.4° C.
2.52 $\sigma_s = -9.47$ MPa; $\sigma_c = 0.391$ MPa.
2.53 (a) $\sigma_{AB} = -21.1$ ksi; $\sigma_{BC} = -6.50$ ksi.
(b) 0.00364 in.
2.54 (a) $\sigma_{AB} = -44.4$ MPa; $\sigma_c = -100$ MPa.
(b) 0.500 mm ↓.
2.57 (a) 217 kN. (b) 0.2425 mm.
2.58 (a) 98.6° C. (b) 450.0266 mm.
2.59 (a) -22.1 ksi. (b) 0.01441 in.
2.60 (a) -7.55 ksi. (b) 10.00467 in.
2.63 (a) 0.00780 in. (b) -0.000256 in.
2.64 216 MPa; 74.5 MPa; 0.451.
2.65 (a) -2.43 mm. (b) 0.0961 mm. (c) 0.00400 mm.
2.68 (a) 1.324×10^{-3} in. (b) -99.3×10^{-6} in.
(c) -12.41×10^{-6} in. (d) -12.41×10^{-6} in.^{2}
2.69 (a) 352×10^{-6} in. (b) 82.8×10^{-6} in.
(c) 307×10^{-6} in.
2.70 0.0269%.
2.71 (a) 0.0754 mm. (b) 0.1028 mm. (c) 0.1222 mm.
2.72 (a) -0.0724 mm. (b) -0.01531 mm.
2.77 10.26 MPa.
2.78 6.17×10^3 kN/m.
2.79 0.0187 in.
2.80 0.818 in.; 2.42 in.
2.83 (a) 242×10^{-6}; 18.40 mm^3.
(b) 519×10^{-6}; 39.4 mm^3.
2.84 (a) 16.55×10^{-6} in^3. (b) 16.54×10^{-6} in^3
2.85 (a) 588×10^{-6} in. (b) 33.2×10^{-3} in^3. (c) 0.0294%.
2.86 (a) -0.0746 mm; -143.9 mm^3.
(b) -0.0306 mm; -521 mm^3.
2.87 3.46 kips.
2.88 6.11.
2.93 (a) 14.04 ksi. (b) 18.2 ksi.
2.94 5.71 kips.
2.95 58.3 kN. (b) 64.3 kN.
2.96 (a) 87.0 MPa. (b) 75.2 MPa. (c) 73.9 MPa.

2.99 (a) 50 kN. (b) 110%.
2.100 55 kN.
2.101 (a) 310.5 kN; 0.705 mm.
(b) 310.5 kN; 4.205 mm.
2.102 (a) 310.5 kN; 7.81 mm.
(b) 310.5 kN; 10.81 mm.
2.103 2.65 kips; 0.1117 in.
2.104 3.68 kips; 0.1552 in.
2.107 (a) 23.9 kN. (b) 250 MPa. (c) 0.
2.108 (a) 23.9 kN. (b) 250 MPa. (c) 0.310 mm.
2.111 (a) 112.1 kips. (b) 82.9 ksi. (c) 0.00906 in.
2.112 (a) 0.0309 in.
(b) 64 ksi.
(c) 0.00387 in.
2.113 (a) 0.1042 mm. (b) -65.2 MPa.
2.114 (a) 0.00778 mm. (b) -6.06 MPa.
2.115 8 ksi.
2.116 18.78 ksi.
2.117 (a) -250 MPa. (b) 117.4 MPa.
2.118 (a) 140.0° C. (b) 260.1° C.
2.121 (a) $\sigma_{AD} = 250$ MPa; $\sigma_{BE} = 124.3$ MPa.
(b) 0.622 mm.
2.122 (a) $\sigma_{BE} = 250$ MPa; $\sigma_{AD} = 233$ MPa.
(b) 1.322 mm.
2.123 (a) $\sigma_{AD} = -4.70$ MPa; $\sigma_{BE} = 19.34$ MPa.
(b) 0.0967 mm.
2.124 1.219 in.
2.125 (a) $\sigma_{tube} = 67.9$ MPa; $\sigma_{rod} = -55.6$ MPa.
(b) $\delta_{tube} = 0.2425$ mm; $\delta_{rod} = -0.1325$ mm.
2.128 0.0455 in; 8.51° ▽.
2.131 $\delta_A = 0.237$ mm ←; $\delta_B = 0.296$ mm →;
$\delta_C = 2.43$ mm →.
2.132 (a) -57.7° C.
(b) $\delta_C = 0.0329$ mm →; $\delta_B = 0.0263$ mm →.
2.133 (a) 9 mm. (b) 62 kN.
2.135 (a) 0.00310 mm. (b) -0.001104 in.
2.C1 Prob. 2.18: (a) 2.95 mm. (b) 5.29 mm.
2.C4 Prob. 2.109: (a) 0.292 mm (b) $\sigma_{AB} = 250$ MPa;
$\sigma_{BC} = -307$ MPa. (c) 0.027 mm.
2.C6 (a) -0.40083. (b) -0.10100. (c) -0.00405.

CHAPTER 3

3.1 641 N · m.
3.2 87.3 MPa.
3.3 12.44 ksi.
3.4 4.12 kip · in.
3.5 (a) 70.5 MPa. (b) 55.8 mm.
3.6 (a) 10.74 kN · m (b) 22.8 kN · m.
3.9 (a) 56.6 MPa. (b) 36.6 MPa.
3.10 39.8 mm.
3.11 (a) BC. (b) 9.66 ksi.
3.12 (a) AB. (b) 10.06 ksi.
3.15 9.16 kip · in.
3.16 (a) 1.503 in. (b) 1.853 in.

3.17 3.18 kN · m.
3.19 (a) 50.3 mm.
    (b) 63.4 mm.
3.21 (a) 72.5 MPa. (b) 68.7 MPa.
3.22 (a) 59.6 mm. (b) 43.9 mm.
3.23 (a) 1.158 in; (b) 0.835 in.
3.25 1.189 kip · in.
3.26 (a) 37.1 mm. (b) 31.7 mm.
3.27 515 N · m.
3.29 (a) $T/w = (\tau_m/2\rho g)[c_1^2 + c_2^2)/c_2]$.
    (b) $T/w = (T/w)_0[1 + (c_1/c_2)^2]$.
3.30 1.0; 1.025; 1.120; 1.200; 1.0.
3.31 (a) 4.21°. (b) 5.25°.
3.32 (a) 2.83 kip · in. (b) 13.00°.
3.33 9.38 ksi.
3.35 (a) 2.53°. (b) 3.42°.
3.37 (a) 0.741°. (b) 1.573°.
3.38 (a) 24°. (b) 75.6°.
3.39 7.94°.
3.40 4.52°.
3.41 3.12°.
3.44 1.749°.
3.45 (a) 82.1 mm. (b) 109.4 mm.
3.46 1.392 in.
3.47 62.9 mm.
3.48 42.1 mm.
3.52 (a) 73.6 MPa. (b) 34.4 MPa. (c) 5.07°.
3.53 4.13°.
3.54 (a) 4.72 ksi. (b) 7.08 ksi. (c) 4.35°.
3.55 7.37°.
3.56 $\tau_{AB} = 39.6$ MPa; $\tau_{CD} = 31.7$ MPa.
3.57 $\tau_{AB} = 68.8$ MPa; $\tau_{CD} = 14.75$ MPa.
3.58 $\tau_{AB} = 10.34$ MPa; $\tau_{CD} = 48.6$ MPa.
3.61 12.24 MPa.
3.62 0.241 in.
3.65 (a) $T/2\pi r_1^2$ u/ $r - r_1$.
3.66 (a) 73.7 MPa. (b) 0.510°.
3.67 $d = 21.4$ mm.
3.68 $d = 0.265$ in.
3.69 $d = 1.528$ in.
3.70 $d = 6.69$ mm.
3.71 $d_2 = 58.2$ mm.
3.73 25.6 kW.
3.74 1.89 mm.
3.76 8 mm.
3.77 (a) 5926 psi. (b) 1.940°.
3.78 (a) 0.685 in. (b) 5.38°.
3.81 (a) 16.02 Hz. (b) 27.7 Hz.
3.82 $d = 74.0$ mm.
3.83 4.91 Hz.
3.84 934 rpm.
3.87 313 kW.
3.88 268 kW.
3.89 0.225 in.
3.90 2076 rpm.
3.91 (a) 21.6 MPa. (b) 17.9 MPa.

3.94 (a) 129.4 MPa; 27 mm.
    (b) 145 MPa; 23.4 mm.
3.95 (a) 18.11 ksi; 0.75 in.
    (b) 21 ksi; 0.369 in.
3.96 21.2 N · m.
3.97 47.7 ksi.
3.100 (a) 283 N · m. (b) 12.95 mm.
3.101 145 MPa; 19.75°.
3.102 (a) $1.126 \, \phi_Y$. (b) $1.587 \, \phi_Y$.
    (c) $2.15 \, \phi_Y$.
3.103 21 ksi; 21.0°.
3.104 (a) 11.71 kN · m; 3.44°.
    (b) 14.12 kN · m; 4.82°.
3.105 (a) 8.04°. (b) 14.89 kN · m.
3.111 2.32 kN · m.
3.112 2.26 kN · m.
3.113 44.9 MPa.
3.114 11.75 ksi.
3.115 (a) 40.5 MPa. (b) 2.09°.
3.118 5.77°.
3.119 (b) $0.221\tau_Y c^3$.
3.120 3.11°.
3.121 9.45 kip · in.
3.122 (a) 10.40 ksi; 9.31°.
    (b) 8.65 ksi; 6.77°.
3.123 (a) 1.730 kip · in; 8.95°.
    (b) 2.08 kip · in; 7.82°.
3.124 (a) 30.8 MPa; 0.535°.
    (b) 37.9 MPa; 0.684°.
3.125 (a) 1.300 kN · m; 0.869°.
    (b) 1.055 kN · m; 0.902°.
3.126 (a) 1.193 in. (b) 1.170 in. (c) 0.878 in.
3.129 0.944.
3.130 1.223 mm.
3.131 1.356.
3.132 1.198.
3.133 (a) 157.0 kN · m.
    (b) 8.72°.
3.136 (a) 925 N · m. (b) 5.79°.
3.137 8.47 MPa.
3.138 18.67 MPa.
3.139 6.79 ksi; 4.53 ksi.
3.140 7.86 ksi; 5.24 ksi.
3.141 8.45 N · m.
3.142 16.85 N · m.
3.145 22.6 ksi; 15.04 ksi.
3.146 (a) 12.73 MPa. (b) 5.40 N · m.
3.149 (a) $1 + t^2/4c_m^2$.
    (b) 0.25%; 1%; 4%.
3.151 0.587°.
3.154 3.79°.
3.155 211 N · m.
3.156 (a) 2.16 kip · in. (b) 2.07 kip · in.
    (c) 1.92 kip · in.
3.157 1.221.
3.159 (a) 0.347 in. (b) 37.2°.

3.161 39.4 MPa.
3.C1 *Prob. 3.150:* (a) 8.55 ksi. (b) 9.13 ksi.
3.C2 *Prob. 3.42:* 1.914°.
3.C4 *Prob. 3.160:* (a) 2.78 ksi at $\rho = 0.5$ in. (b) 5.68°.
3.C5 (a) −3.282%. (b) −0.853%. (c) −0.138%.
    (d) −0.0055%.
3.C6 (a) −1.883%. (b) −0.484%. (c) −0.078%.
    (d) −0.0031%.

CHAPTER 4

4.1 (a) −2.38 ksi. (b) −0.650 ksi.
4.2 (a) −116.4 MPa. (b) −87.3 MPa.
4.3 243 kN · m.
4.4 56.3 kN · m.
4.5 (a) 1.405 kip · in. (b) 3.19 kip · in.
4.6 2.38 kN · m.
4.7 259 kip · in.
4.8 187.1 kip · in.
4.9 67.8 MPa; −81.8 MPa.
4.12 (a) 83.7 MPa. (b) −146.4 MPa.
    (c) 14.67 MPa.
4.15 2.22 kips.
4.16 2.05 kips.
4.17 1.957 kips.
4.18 2.57 kips.
4.19 7.67 kN · m.
4.20 3.79 kN · m.
4.21 20.4 kip · in.
4.24 1.372 kip · in.
4.25 (a) 105.5 lb · in. (b) 113.3 in.
4.26 (a) 145 ksi. (b) 384 lb · in.
4.27 (a) 193.3 ksi. (b) 0.0483 lb · in.
4.30 (a) 53.6 MPa; 379 m.
    (b) 158.9 MPa; 127.5 m.
4.31 (a) 9.17 kN · m. (b) 10.24 kN · m.
4.32 0.950.
4.33 0.949.
4.34 (a) $\sqrt{2}$. (b) $\sqrt{3}$.
4.35 (a) 1007 in. (b) 3470 in. (c) 0.01320°.
4.36 (a) 334 ft. (b) 0.0464°.
4.37 (a) 139.6 m. (b) 481 m.
4.38 (a) $(\delta_x)_{max}(y^2 - c^2)/2\,pc$.
4.39 1.240 kN · m.
4.40 1.043 kN · m.
4.43 330 kip · in.
4.44 685 kip · in.
4.45 (a) −56.0 MPa. (b) 66.4 MPa.
4.46 (a) −56.0 MPa. (b) 68.4 MPa.
4.47 (a) −1.179 ksi. (b) 9.94 ksi.
4.48 (a) −1.526 ksi. (b) 17.67 ksi.
4.49 8.70 m.
4.50 8.59 m.
4.53 11.73 kN.
4.54 9.50 kN.

4.55 (a) 330 MPa. (b) −26.0 MPa.
4.56 (a) 292 MPa. (b) −21.3 MPa.
4.57 (a) 29.0 ksi. (b) −1.163 ksi.
4.58 32.4 kip · ft.
4.62 (a) $\sigma_a = 6.86$ ksi; $\sigma_b = 6.17$ ksi; $\sigma_s = 4.11$ ksi.
    (b) 151.9 ft.
4.63 (a) $\sigma_s = 8.96$ ksi; $\sigma_a = 1.792$ ksi; $\sigma_b = 0.896$ ksi.
    (b) 349 ft.
4.64 (a) −22.5 ksi. (b) +17.78 ksi.
4.65 (a) 52.3 MPa. (b) 132.1 MPa.
4.66 (a) 40.8 MPa. (b) 145.2 MPa.
4.67 (a) 6.15 MPa. (b) −8.69 MPa.
4.69 (a) 219 MPa. (b) 176 MPa.
4.70 (a) 128 N · m. (b) 142 N · m.
4.71 (a) 6.79 kip · in. (b) 5.59 kip · in.
4.72 (a) 4.71 ksi. (b) 5.72 ksi.
4.75 (a) 144 N · m. (b) 208 N · m.
4.76 (a) 115.2 N · m. (b) 171.2 N · m.
4.77 (a) 284 lb · in. (b) 410 lb · in.
4.79 8.94 ft.
4.80 (a) 5.87 mm. (b) 2.09 m.
4.81 (a) 21.9 mm. (b) 7.81 m.
4.82 (a) 1260 lb · in. (b) 1304 lb · in.
4.83 (a) 5.76 kN · m. (b) 8.32 kN · m.
4.85 (a) 1759 kip · in. (b) 2650 kip · in.
4.86 (a) 3340 kip · in. (b) 4730 kip · in.
4.88 (a) 8.16 kN · m. (b) 1.443.
4.89 (a) 2840 kip · in. (b) 1.611.
4.90 (a) 4820 kip · in. (b) 1.443.
4.91 19.01 kN · m.
4.92 2.03 kN · m.
4.93 20.7 kip · in.
4.94 212 kip · in.
4.96 48.6 kN · m.
4.97 120 MPa.
4.98 106.4 MPa
4.99 25.7 ksi.
4.100 18.62 ksi.
4.101 106.7 MPa. (b) ±20.8 mm; 0. (c) 16.08 m.
4.102 (a) 99.6 MPa. (b) ±21.2 mm; 0. (c) 15.77 m.
4.106 (a) $0.707\rho_y$. (b) $6.09\rho_y$.
4.108 (a) 292 MPa. (b) 7.01 mm.
4.109 (a) 4.69 m. (b) 7.29 kN · m.
4.111 (a) 43 ksi. (b) 10.75 kip · in.
4.115 (a) −8.33 MPa; −8.33 MPa.
    (b) −15.97 MPa; 4.86 MPa.
4.117 (a) −102.8 MPa. (b) 80.6 MPa.
4.118 (a) 13.52 ksi. (b) −11.52 ksi.
4.119 (a) 16.34 ksi. (b) −13.78 ksi.
4.121 (a) −37.5 MPa.
    (b) −11.62 MPa.
4.123 10.83 mm.
4.124 (a) −0.75 ksi. (b) −2.00 ksi. (c) −1.50 ksi.
4.125 94.8 kN ≤ P ≤ 177.3 kN.
4.126 94.8 kN ≤ P ≤ 177.3 kN.
4.127 3.07 kips.

4.129 (a) 270 kips. (b) 196.0 kips.
4.130 (a) 2.54 kN.
(b) 3.89 mm to the right of C.
4.131 (a) 40.3 kN.
(b) 56.3 mm from left edge.
4.132 (a) 69.6 kN.
(b) 41.9 mm from left edge.
4.133 (a) 6.18 ksi. (b) −7.87 ksi.
(c) 0.448 in. above point D.
4.137 9.00 kN.
4.138 $P = 44.2$ kips; $Q = 57.3$ kips.
4.139 $P = 9.21$ kips; $Q = 48.8$ kips.
4.140 (a) 30 mm. (b) 94.5 kN · m.
4.143 (a) 75 mm. (b) 40 MPa.
4.144 (a) 150 mm. (b) −10 MPa.
4.145 (a) 9.86 ksi. (b) −2.64 ksi. (c) −9.86 ksi.
4.146 (a) −3.37 MPa. (b) −18.60 MPa. (c) 3.37 MPa.
4.147 (a) −17.16 ksi. (b) 6.27 ksi. (c) 17.16 ksi.
4.148 (a) 0.321 ksi. (b) −0.107 ksi. (c) 0.427 ksi.
4.149 (a) 55.3 MPa. (b) −141.2 MPa. (c) −55.3 MPa.
4.150 (a) 8.12 ksi. (b) −7.98 ksi. (c) 3.65 ksi.
4.151 (a) 57.3°. (b) 75.1 MPa.
4.152 (a) 18.2°. (b) 80.9 MPa.
4.153 (a) 19.2°. (b) 11.21 ksi.
4.155 (a) 27.5°. (b) 8.44 ksi.
4.157 10.46 ksi.
4.158 113.0 MPa.
4.160 (a) 633 kPa. (b) −233 kPa.
(c) 146.2 mm from point A.
4.162 (a) 4.09 ksi; −1.376 ksi.
(b) 3.74 in. above point A.
4.163 (a) 2.98 ksi; 1.161 ksi.
(b) 2.22 in. to the right of B.
4.164 36.8 mm.
4.166 91.3 kN.
4.167 17.11 mm.
4.168 29.1 kip · in.
4.170 733 N · m.
4.171 1.323 kN · m.
4.172 6.19 ksi.
4.173 900 N · m.
4.174 4.7°.
4.180 (a) −45.0 MPa. (b) −42.2 MPa.
4.181 (a) −43.2 MPa. (b) 33.0 MPa.
4.182 (a) −5.15 ksi. (b) −3.72 ksi.
4.183 −4.34 ksi; 2.64 ksi.
4.186 512 lb.
4.187 881 lb.
4.188 −148.6 MPa.
4.189 60.9 mm.
4.192 (a) −64.1 MPa. (b) 65.2 MPa.
4.193 (a) −106.1 MPa. (b) 38.9 MPa.
4.194 107.8 N · m.
4.195 (a) 6.74 ksi. (b) −3.45 ksi.
4.196 1.584 in.
4.197 (a) −45.2 MPa. (b) 17.40 MPa.

4.198 (a) −43.3 MPa. (b) 14.43 MPa.
4.201 (a) 63.9 MPa. (b) −52.6 MPa.
4.202 84.7 MPa.
4.203 (a) −3.65 ksi. (b) 3.72 ksi.
4.204 (a) −8.51 ksi. (b) 2.10 ksi.
4.207 $\frac{1}{2}(\bar{r} + \sqrt{r^2 - c^2})$.
4.208 $\frac{1}{2}h^2(b_1 + b_2)/[(r_2b_1 - r_1b_2)\ln(r_2/r_1) - h(b_1 - b_2)]$.
4.209 $\frac{1}{2}h/[(r_2/h)\ln(r_2/r_1) - 1]$.
4.211 (a) 152.3 kips. (b) $x = 0.595$ in.; $z = 0.571$ in.
(c) 8.70 ksi.
4.212 (a) −70.9 MPa. (b) −14.17 MPa. (c) 25.4 m.
4.214 (a) 9.33 ksi. (b) 8.00 ksi.
4.217 (a) 46.6° ⤢. (b) 79.6 MPa.
4.218 (a) 56.7 kN · m. (b) 20 m.
4.219 (a) $\sigma_A = \frac{1}{2}\sigma_I$; $\sigma_B = \sigma_I$; $\sigma_C = -\sigma_I$; $\sigma_D = \frac{1}{2}\sigma_I$.
(b) 1.333 ρ_I.
4.221 (a) 112.5 MPa. (b) 12.80 m.
4.C1 $a = 4$ mm: σ_a = 50.6 MPa, σ_s = 107.9 MPa;
$a = 14$ mm: σ_a = 89.7 MPa, σ_s = 71.8 MPa.
(a) 111.6 MPa. (b) 6.61 mm.
4.C3 $\beta = 30°$: σ_A = −7.83 ksi, σ_B = −5.27 ksi,
σ_C = 7.19 ksi, σ_D = 5.91 ksi;
$\beta = 120°$: σ_A = 1.557 ksi, σ_B = 6.01 ksi,
σ_C = −2.67 ksi, σ_D = −4.89 ksi.
4.C4 r_1/h = 0.529 for 50% increase in σ_{max}.
4.C5 Prob. 4.10. 8.82 ksi, −14.71 ksi.
Prob. 4.11: 15.40 ksi, −10.38 ksi.
4.C6 y_y = 0.8 in.: 76.9 kip · in., 552 in.;
y_y = 0.2 in.: 95.5 kip · in., 138.1 in.
4.C7 $a = 0.2$ in.: −7.27 ksi, $a = 0.8$ in.: −6.61 ksi
For $a = 0.625$ in., σ = −6.51 ksi.

CHAPTER 5

5.1 $M_{max} = Pab/L$.
5.2 $M_{max} = wL^2/8$.
5.4 $|M|_{max} = wL^2/2$.
5.5 $M_{max} = w(L^2 - 4a^2)/8$.
5.7 From A to B: $V = Pb/L$, $M = Pbx/L$.
From B to C: $V = -Pa/L$, $M = Pa(L - x)/L$.
5.8 $V = w(L - 2x)/2$; $M = wx(L - x)/2$.
5.10 $V = w(L - x)$; $M = -w(L - x)^2/2$.
5.11 From A to B: $V = w(L - 2a)/2$, $M = w(L - 2a)x/2$.
From B to C: $V = w(L - 2x)/2$,
$M = w[(L - 2a)x - (x - a)^2]/2$.
From C to D: $V = -w(L - 2a)/2$,
$M = w(L - 2a)(L - x)/2$.
5.13 (a) 2.15 kN. (b) 150 N · m.
5.14 (a) 300 N. (b) 67.5 N · m.
5.15 (a) 30 kips. (b) 90 kip · ft.
5.16 (a) 23 kips. (b) 140 kip · ft.
5.17 (a) 3.45 kN. (b) 1125 N · m.
5.19 (a) 18 kN. (b) 12.15 kN · m.
5.21 10.89 MPa.
5.22 950 psi.

5.23 128.6 MPa.

5.24 11.56 ksi.

5.27 $V_{max} = 24$ kN; $|M|_{max} = 32$ kN·m; $\sigma_{max} = 31.1$ MPa.

5.28 $V_{max} = 5.6$ kN; $|M|_{max} = 1.92$ kN·m;
$\sigma_{max} = 38.7$ MPa.

5.29 $V_{max} = 27.5$ kips; $|M|_{max} = 45$ kip·ft;
$\sigma_{max} = 14.14$ ksi.

5.30 $V_{max} = 1.5$ kips; $|M|_{max} = 3$ kip·ft; $\sigma_{max} = 2.11$ ksi.

5.33 (a) 1.371 m. (b) 26.6 MPa.

5.34 (a) 866 mm. (b) 5.74 MPa.

5.35 (a) 2.67 kips. (b) 1.404 ksi.

5.36 (a) 1.26 ft. (b) 7.20 ksi.

5.37 $|V|_{max} = 342$ N; $M_{max} = 51.6$ N·m;
$\sigma_{max} = 17.19$ MPa.

5.39 (a) 33.3 mm. (b) 6.66 mm.

5.51 (a) 140 N. (b) 33.6 N·m.

5.52 (a) 600 N. (b) 180 N·m.

5.53 10.89 MPa.

5.54 950 psi.

5.55 128.6 MPa.

5.56 11.56 ksi.

5.57 (a) $V = -w_0 x + w_0 x^2/2L + w_0 L/3$;
$M = -w_0 x^2/2 + w_0 x^3/6L + w_0 Lx/3$.

5.58 (b) $0.06415\, w_0 L^2$.

5.61 (a) $V = (w_0 L/\pi)\cos(\pi x/L)$; $M = (w_0 L^2/\pi^2)\sin(\pi x/L)$.
(b) $w_0 L^2/\pi^2$.

5.62 $V_{max} = 20.7$ kN; $M_{max} = 9.75$ kN·m;
$\sigma_{max} = 73.5$ MPa.

5.63 $V_{max} = 15$ kips; $M_{max} = 37.5$ kip·ft; $\sigma_{max} = 9$ ksi.

5.64 $V_{max} = 8$ kips; $M_{max} = 16$ kip·ft; $\sigma_{max} = 6.98$ ksi.

5.67 $|V|_{max} = 200$ kN; $|M|_{max} = 300$ kN·m;
$\sigma_{max} = 136.4$ MPa.

5.68 $|V|_{max} = 5.6$ kN; $|M|_{max} = 1.92$ kN·m;
$\sigma_{max} = 38.7$ MPa.

5.70 $|V|_{max} = 30$ lb; $|M|_{max} = 24$ lb·ft; $\sigma_{max} = 6.95$ ksi.

5.71 $|V|_{max} = 1150$ N; $|M|_{max} = 221$ N·m; $P = 500$ N;
$Q = 250$ N.

5.72 (a) $|V|_{max} = 6$ kips; $|M|_{max} = 12$ kip·ft.
(b) 6.21 ksi.

5.74 (a) $|V|_{max} = 1440$ N; $|M|_{max} = 360$ N·m.
(b) 25.0 MPa.

5.75 173.2 mm.

5.76 361 mm.

5.77 6.67 in.

5.78 2.95 in.

5.81 W 410 × 60.

5.82 W 250 × 28.4.

5.83 W 21 × 44.

5.84 W 18 × 50.

5.85 S 310 × 47.3.

5.86 S 510 × 98.3.

5.89 C 180 × 14.6.

5.90 C 230 × 22.

5.91 3/8 in.

5.92 3/8 in.

5.93 W 610 × 101.

5.94 S 24 × 80.

5.95 7.48 kN.

5.96 7.32 kN.

5.99 (a) 1.485 kN/m. (b) 1.935 m.

5.100 (a) 4.01 kN. (b) 3.27 m.

5.101 (a) S 15 × 42.9. (b) W 27 × 84.

5.102 (a) S 20 × 66. (b) W 27 × 84.

5.105 W 27 × 84.

5.106 23.2%.

5.107 383 mm.

5.108 336 mm.

5.109 (a) $V = 2P/3 - P\langle x - a\rangle^0$; $M = 2Px/3 - P\langle x - a\rangle$.
(b) $Pa/3$.

5.110 (a) $V = -w_0 x + w_0\langle x - 2a\rangle^1 + 3w_0 a/2$;
$M = -w_0 x^2/2 + w_0\langle x - 2a\rangle^2/2 + 3w_0 ax/2$.
(b) $w_0 a^2/2$.

5.111 (a) $V = -w_0\langle x - a\rangle^1 - 3w_0 a/4 + (15w_0 a/4)\langle x - 2a\rangle^0$;
$M = -(w_0/2)\langle x - a\rangle^2 - 3w_0 ax/4 +$
$(15w_0 a/4)\langle x - 2a\rangle^1$.
(b) $-w_0 a^2/2$.

5.112 (a) $V = -w_0 x + w_0\langle x - a\rangle^1$;
$M = -w_0 x^2/2 + (w_0/2)\langle x - a\rangle^2$.
(b) $-3w_0 a^2/2$.

5.113 (a) $V = -w_0 x^2/2a + w_0\langle x - a\rangle^1 + (w_0/2a)\langle x - a\rangle^2$;
$M = -w_0 x^3/6a + (w_0/2)\langle x - a\rangle^2 +$
$(w_0/6a)\langle x - a\rangle^3$.
(b) $-2w_0 a^2/3$.

5.114 (a) $V = -w_0 x^2/2a + w_0\langle x - a\rangle^1 - (w_0/2a)\langle x - a\rangle^2$;
$M = -w_0 x^3/6a + (w_0/2)\langle x - a\rangle^2 - (w_0/6a)\langle x - a\rangle^3$.
(b) $-5w_0 a^2/6$.

5.117 (a) $V = 62.5 - 25\langle x - 0.6\rangle^1 + 25\langle x - 2.4\rangle^1 -$
$40\langle x - 0.6\rangle^0 - 40\langle x - 2.4\rangle^0$ kN;
$M = 62.5x - 12.5\langle x - 0.6\rangle^2 + 12.5\langle x - 2.4\rangle^2 -$
$40\langle x - 0.6\rangle^1 - 40\langle x - 2.4\rangle^1$ kN·m.
(b) 47.6 kN·m.

5.118 (a) $V = -1.25x + 1.25\langle x - 1.2\rangle^1 - 1.25\langle x - 2.4\rangle^1 +$
$2.4 - 1.8\langle x - 1.8\rangle^0$ kN;
$M = -0.625x^2 + 0.625\langle x - 1.2\rangle^2 -$
$0.625\langle x - 2.4\rangle^2 + 2.4x - 1.8\langle x - 1.8\rangle^1$ kN·m.
(b) 2.52 kN·m.

5.119 (a) $V = 40 - 20\langle x - 2\rangle^1 - 20\langle x - 4\rangle^0 -$
$20\langle x - 6\rangle^0$ kips;
$M = 40x - 20\langle x - 2\rangle^1 - 20\langle x - 4\rangle^1 -$
$20\langle x - 6\rangle^1$ kip·ft.
(b) 120 kip·ft.

5.121 (a) $V = -10 + 25\langle x - 15\rangle^0 - 25\langle x - 35\rangle^0 +$
$20\langle x - 60\rangle$ kips;
$M = -10x + 25\langle x - 15\rangle^1 - 25\langle x - 35\rangle^1 +$
$20\langle x - 60\rangle^1$ kip·ft.
(b) 20.4 ksi.

5.123 (a) 225.6 kN·m; 3.625 m.
(b) 60.6 MPa.

5.124 (a) 28.3 kN·m; 1.938 m.
(b) 45.2 MPa.

5.C2 (a) 173.2 mm. (b) 48.0 mm. (c) 203 mm.

5.C4 For $x = 13.5$ ft: $M_1 = 131.25$ kip·ft; $M_2 = 156.25$ kip·ft; $M_C = 150.0$ kip·ft.

CHAPTER 6

6.1 60 mm.

6.2 2 kN.

6.3 (a) 31.5 lb. (b) 43.2 psi.

6.4 (a) 372 lb. (b) 64.4 psi.

6.5 180.6 kN.

6.6 209 kN.

6.9 (a) 7.40 ksi. (b) 6.70 ksi.

6.10 (a) 3.17 ksi. (b) 2.40 ksi.

6.11 (a) 920 kPa. (b) 765 kPa.

6.12 (a) 114.1 MPa. (b) 96.9 MPa.

6.15 (a) Acceptable. (b) Not acceptable; $h = 379$ mm.

6.16 (a) Acceptable. (b) Acceptable.

6.17 (a) Acceptable. (b) Acceptable.

6.18 (a) Acceptable. (b) Acceptable.

6.21 (a) 12.55 MPa. (b) 18.82 MPa.

6.22 (a) 1.745 ksi. (b) 2.81 ksi.

6.23 19.61 MPa.

6.24 3.21 ksi.

6.25 356 kN; 254 kN.

6.26 (a) Horizontal diameter. (b) 1.333.

6.29 336 N.

6.30 (a) 239 N. (b) 549 N.

6.31 1835 lb.

6.32 2410 lb.

6.35 (a) 95.2 MPa. (b) 112.9 MPa.

6.36 (a) 101.6 MPa. (b) 79.9 MPa.

6.37 (a) 40.5 psi. (b) 55.2 psi.

6.38 (a) 2.67 in. (b) 41.6 psi.

6.39 (a) 41.4 MPa. (b) 41.4 MPa.

6.40 (a) 18.23 MPa. (b) 14.59 MPa. (c) 46.2 MPa.

6.43 255 kN.

6.44 9.05 mm.

6.45 7.19 ksi.

6.46 7.19 ksi.

6.47 266 kN/m.

6.48 $\tau_a = 10.76$ MPa; $\tau_b = 0$; $\tau_c = 11.21$ MPa; $\tau_d = 22.0$ MPa; $\tau_e = 9.35$ MPa.

6.51 (a) 2.08. (b) 2.10.

6.52 (a) 2.25. (b) 2.12.

6.53 1.422 in.

6.54 (a) $(V \sin \theta)/(\pi r_m t)$.

6.59 (a) 1.323 ksi. (b) 1.329 ksi.

6.60 (a) 0.888 ksi. (b) 1.453 ksi.

6.61 1.25 a.

6.62 $3(b^2 - a^2)/(6a + 6b + h)$.

6.63 $10a/29$.

6.64 $5a/7$.

6.65 (a) 48.7 mm.
(b) 0 at A; 15.95 and 26.6 MPa at B: 73.1 and 43.9 MPa at D; 60.5 MPa at midpoint of DE.

5.125 (a) 121.5 kip·ft; 6 ft.
(b) W 16 × 40.

5.126 (a) 122.7 kip·ft; 6.5 ft.
(b) W 16 × 40.

5.129 $|V|_{max} = 20.7$ kN at $x = 0.9$ m;
$|M|_{max} = 9.72$ kN·m at $x = 2.7$ m.

5.130 $|V|_{max} = 89.0$ kN for $0 < x < 2$ m;
$|M|_{max} = 178.0$ kN·m at $x = 2$ m.

5.131 $|V|_{max} = 15.30$ kips at $x = 0$;
$|M|_{max} = 38.0$ kip·ft at $x = 5.5$ ft.

5.132 $|V|_{max} = 12.75$ kips for 4 ft $< x < 6$ ft;
$|M|_{max} = 39.0$ kip·ft at $x = 6$ ft.

5.133 (a) $|V|_{max} = 13.80$ kN at $x - 5$ m;
$|M|_{max} = 16.16$ kN·m at $x = 2.84$ m.
(b) 83.3 MPa.

5.134 (a) $|V|_{max} = 40.0$ kN at $x = 2$ m;
$|M|_{max} = 30.0$ kN·m at $x = 4$ m.
(b) 40.0 MPa.

5.137 (a) $h_0(x/L)^{3/2}$.
(b) 150 mm.

5.138 (a) $h = h_0 \left[\dfrac{(\pi x/2L) - \sin(\pi x/2L)}{(\pi/2) - 1} \right]^{1/2}$
(b) 176.7 mm.

5.139 (a) $h_0 x/L$.
(b) 30 kips.

5.140 (a) $h_0[(x/L)(1 - x/L)]^{1/2}$.
(b) 6.67 kip/in.

5.143 1.800 m.

5.144 1.900 m.

5.145 6.00 ft; 4.00 ft.

5.146 5.00 ft; 2.50 ft.

5.147 (a) $d_0 \sqrt[3]{y/L}$. (b) 5.09 kN.

5.148 (a) $d_0(y/L)^{2/3}$. (b) 50.3 mm.

5.151 (a) 152.6 MPa. (b) 133.6 MPa.

5.152 (a) 155.2 MPa. (b) 147.4 MPa

5.153 (a) 25.0 ksi. (b) 18.03 ksi.

5.154 (a) 24.0 ksi. (b) 29.3 ksi.

5.155 (a) 4.49 m. (b) 211 mm.

5.158 125.4 kN.

5.159 (a) $x = 0.400$ m. (b) 156.3 MPa.

5.160 (a) $r = 0.24$ m. (b) 150 MPa.

5.163 (a) $x = 30$ in. (b) 12.80 kips.

5.164 (a) $x = 15$ in. (b) 320 lb/in.

5.165 (a) 18.00 kips. (b) 72.0 kip·ft.

5.166 136.0 MPa.

5.167 (a) 34.6 in. (b) 13.72 ksi.

5.169 (a) 27.8 kip·ft. (b) 13.53 ksi.

5.170 (a) $V = -12 + 35\langle x - 0.9\rangle^0 - 24\langle x - 2.1\rangle^0 - 12\langle x - 3.3\rangle^0$ kN;
$M = -12x + 35\langle x - 0.9\rangle^1 - 24\langle x - 2.1\rangle^1 - 12\langle x - 3.3\rangle^1$ kN·m.
(b) 16.80 kN·m.

5.172 342 mm.

5.173 W 310 × 38.7.

5.175 (a) $d_0[(4x/L)(1 - x/L)]^{1/3}$.
(b) 32.1 mm.

6.66 (a) 41.8 mm.
(b) 0 at A; 13.31 and 7.99 MPa at B; 46.8 and 78.0 MPa at D; 91.9 MPa at midpoint of DE.
6.69 0.727 in.
6.70 0.433 in.
6.71 20.2 mm.
6.72 9.64 mm.
6.75 3.75 in.
6.76 3.00 in.
6.77 21.7 mm.
6.78 74.2 mm.
6.81 (a) 1200 N: 22.9 N · m.
(b) 25.4 MPa.
6.82 (a) 1200 N: 22.9 N · m.
(b) 35.3 MPa.
6.83 12 ksi at corner.
6.84 16 ksi at 2 in. below corner.
6.89 81.5 mm.
6.90 193.5 mm.
6.92 (a) 2.73 ksi. (b) 1.665 ksi.
6.93 728 N.
6.94 (a) 59.9 psi. (b) 79.8 psi.
6.95 20.1 ksi.
6.98 1.265 in.
6.99 40 mm.
6.C1 (a) $h = 173.2$ mm. (b) $h = 379$ mm.
6.C2 (a) $L = 37.5$ in.; $b = 1.250$ in.
(b) $L = 70.3$ in.; $b = 1.172$ in.
(c) $L = 59.8$ in.; $b = 1.396$ in.
6.C4 (a) $\tau_{max} = 2.02$ ksi; $\tau_B = 1.80$ ksi. (b) 194 psi.
6.C5 Prob. 6.68: (a) 0.794 in. (b) $\tau_A = 0$; $\tau_B = 6.62$ ksi and 3.31 ksi; $\tau_C = 7.72$ ksi.
6.C6 Prob. 6.100: $e = 2.37$ in.

CHAPTER 7

7.1 $\sigma = 9.46$ ksi; $\tau = 1.013$ ksi ↘
7.2 $\sigma = 10.93$ ksi; $\tau = 0.536$ ksi ↘
7.3 $\sigma = -0.521$ MPa; $\tau = 56.4$ MPa ↘
7.4 $\sigma = 32.9$ MPa; $\tau = 71.0$ MPa ↙
7.7 (a) −37.0°; 53.0°.
(b) −13.60 MPa ($\theta = 53.0°$); $\sigma_{min} = -86.4$ MPa.
7.8 (a) −31.0°; 59.0°.
(b) $\sigma_{max} = 52$ MPa ($\theta = -31.0°$); $\sigma_{min} = -84$ MPa.
7.9 (a) −31.0°; 59.0°.
(b) 17 ksi; (c) 3 ksi.
7.10 (a) −26.6°; 63.4°.
(b) 5 ksi. (c) 6 ksi.
7.11 (a) 8.0°; 98.0°.
(b) 36.4 MPa. (c) −50 MPa.
7.13 (a) −37.5 MPa; 57.5 MPa; −25.4 MPa.
(b) −30.1 MPa; 50.1 MPa; 35.9 MPa.
7.15 (a) 9.02 ksi; −13.02 ksi; 3.80 ksi.
(b) 5.34 ksi; −9.34 ksi; −9.06 ksi.
7.16 (a) −4.80 ksi; 20.8 ksi; 0.30 ksi.
(b) 3.90 ksi; 12.10 ksi; 12.13 ksi.

7.17 (a) −0.300 MPa. (b) −2.92 MPa.
7.18 (a) 346 psi. (b) −200 psi.
7.19 (a) 18.4°. (b) 16.67 ksi.
7.22 (a) 14.3°. (b) 117.3 MPa.
7.23 $\sigma_{max} = 35.4$ MPa; $\sigma_{min} = -35.4$ MPa; $\tau_{max} = 35.4$ MPa.
7.24 $\sigma_{max} = 12.18$ MPa; $\sigma_{min} = -48.7$ MPa; $\tau_{max} = 30.5$ MPa.
7.25 $\sigma_{max} = 25.1$ ksi; $\sigma_{min} = -0.66$ ksi; $\tau_{max} = 12.88$ ksi.
7.26 $\sigma_{max} = 5.12$ ksi; $\sigma_{min} = -1.64$ ksi; $\tau_{max} = 3.38$ ksi.
7.27 $\sigma_{max} = 40.9$ ksi.
7.29 15 MPa $\leq \sigma_x \leq$ 135 MPa.
7.30 (a) −2.89 MPa.
(b) 12.77 MPa; 1.23 MPa.
7.32 For 7.12: (a) 14.0°; 104.0°.
(b) 68 MPa; (c) −16 MPa.
7.53 (a) −8.66 MPa.
(b) 17 MPa; −3 MPa.
7.55 $\sigma_{max} = 72.9$ MPa ($\theta_p = 24.6°$ ↘);
$\sigma_{min} = 27.1$ MPa ($\theta_p = 114.6°$ ↗).
7.58 $\sigma_{max} = 12$ ksi ($\theta_p = 33.7°$ ↘);
$\sigma_{min} = -1$ ksi ($\theta_p = 123.7°$ ↗).
7.59 (a) 33.7°; 123.7°.
(b) 18 ksi. (c) 6.5 ksi.
7.60 −5.2° $\leq \theta \leq$ 132.0°; 174.9° $\leq \theta \leq$ 312.0°.
7.61 16.5° $\leq \theta \leq$ 110.1°; 196.5° $\leq \theta \leq$ 290.1°.
7.62 −45° $\leq \theta \leq$ 8.13°; 45° $\leq \theta \leq$ 98.13°.
7.63 −120 MPa $\leq \tau_{xy} \leq$ 120 MPa.
−141.4 MPa $\leq \tau_{xy} \leq$ 141.4 MPa.
7.65 (b) $\tau_{xy} = \pm\sqrt{\sigma_x\sigma_y - \sigma_{max}\sigma_{min}}$
7.66 (a) 100 MPa. (b) 110 MPa.
7.67 (a) 94.3 MPa. (b) 105.3 MPa.
7.69 (a) 10 ksi. (b) 12.5 ksi.
7.70 (a) 85 MPa. (b) 85 MPa. (c) 95 MPa.
7.71 (a) 97.5 MPa. (b) 85 MPa. (c) 120 MPa.
7.72 (a) 6.5 ksi. (b) 7.5 ksi. (c) 6.5 ksi.
7.74 (a) no yielding: 1.287.
(b) no yielding: 1.018.
(c) yielding.
7.75 1 ksi; 7.8 ksi.
7.76 56.9 MPa; −130 MPa.
7.77 (a) ±6 ksi. (b) ±11.24 ksi.
7.80 (a) 45.7 MPa. (b) 92.9 MPa.
7.81 ±60 ksi.
7.82 (a) no yielding: 1.119.
(b) yielding. (c) yielding.
7.83 (a) no yielding: 1.083.
(b) yielding. (c) yielding.
7.85 717 N · m.
7.87 8.19 kip · in.
7.88 9.46 kip · in.
7.89 no rupture.
7.90 no rupture.
7.91 no rupture.
7.93 ±49.1 MPa.

7.95 1.457 kip $\cdot$ in.

7.96 ± 50 MPa.

7.97 (a) 8 ksi. (b) 6.67 ksi. (c) 8.89 ksi.

7.98 166.5 psi.

7.99 $\sigma_{max} = 8.61$ ksi; $\tau_{max} = 4.31$ ksi (out of plane).

7.100 5.04.

7.101 7.74 mm.

7.104 $\sigma_{max} = 37.9$ MPa; $\tau_{max} = 18.93$ MPa (out of plane).

7.106 $\sigma_{max} = 11.82$ MPa; $\tau_{max} = 5.91$ ksi (out of plane).

7.107 $\sigma_{max} = 136$ MPa; $\tau_{max} = 68$ MPa (out of plane).

7.108 $\sigma_{max} = 7.58$ ksi; $\tau_{max} = 3.79$ ksi (out of plane).

7.110 2.95 MPa.

7.111 3.41 MPa.

7.112 (a) 3750 psi. (b) 1079 psi.

7.113 $-26.8° \le \beta \le 26.8°, 63.2° \le \beta \le 116.8°$.

7.115 56.8°.

7.116 2.84 MPa.

7.118 $\sigma_{max} = 68.6$ MPa; $\tau_{max} = 34.2$ MPa (out of plane).

7.120 (a) 21.4 MPa. (b) 14.17 MPa.

7.121 (a) 11.00 MPa. (b) 21.5 MPa.

7.123 $\sigma_{max} = 13.09$ ksi; $\tau_{max} = 6.54$ ksi (out of plane).

7.124 (a) 28.0 MPa. (b) 1.400 MPa.

7.125 (a) 36.0 MPa. (b) 1.800 MPa.

7.126 $\epsilon_{x'} = -670\ \mu$; $\epsilon_{y'} = -50\ \mu$; $\epsilon_{z} = -474\ \mu$.

7.127 $\epsilon_{x'} = 36.7\ \mu$; $\epsilon_{y'} = 283\ \mu$; $\epsilon_{z} = 227\ \mu$.

7.128 $\epsilon_{x'} = -653\ \mu$; $\epsilon_{y'} = 303\ \mu$; $\epsilon_{z} = -829\ \mu$.

7.130 $\epsilon_{x'} = -670\ \mu$; $\epsilon_{y'} = -50\ \mu$; $\epsilon_{z} = -474\ \mu$.

7.131 $\epsilon_{x'} = -36.7\ \mu$; $\epsilon_{y'} = 283\ \mu$; $\epsilon_{z} = 227\ \mu$.

7.132 $\epsilon_{x'} = -653\ \mu$; $\epsilon_{y'} = 303\ \mu$; $\epsilon_{z} = -829\ \mu$.

7.134 (a) $\theta_p = 21.6° \downarrow$; $\epsilon_a = 279\ \mu$; $\epsilon_b = -599\ \mu$; $\epsilon_c = 160\ \mu$.
(b) $877\ \mu$. (c) $877\ \mu$.

7.135 (a) $\theta_p = 33.7° \downarrow$; $\epsilon_a = 100\ \mu$; $\epsilon_b = -420\ \mu$; $\epsilon_c = 160\ \mu$.
(b) $520\ \mu$. (c) $580\ \mu$.

7.136 (a) $\theta_p = 25.1° \downarrow$; $\epsilon_a = 985\ \mu$; $\epsilon_b = -265\ \mu$;
$\epsilon_c = -360\ \mu$.
(b) $1250\ \mu$. (c) $1345\ \mu$.

7.137 (a) $\theta_p = 30.1° \downarrow$; $\epsilon_a = -149.2\ \mu$; $\epsilon_b = -3.51\ \mu$;
$\epsilon_c = 250\ \mu$.
(b) $202\ \mu$. (c) $601\ \mu$.

7.140 (a) $+513\ \mu$. at $31.0° \downarrow$; $+87.5\ \mu$ at $59.0° \downarrow$.
(b) $425\ \mu$. (c) $513\ \mu$.

7.141 (a) $+243\ \mu$ at $7.8° \downarrow$; $+56.6\ \mu$ at $97.8° \uparrow$.
(b) $186.8\ \mu$. (c) $243\ \mu$.

7.142 (a) $\epsilon_{max} = 734\ \mu$; $\epsilon_{min} = -84.3\ \mu$.
(b) $819\ \mu$.

7.143 380×10^{-6} in/in.

7.144 (a) $-300\ \mu$.
(b) $\epsilon_{max} = 435\ \mu$; $\epsilon_{min} = -315\ \mu$; $\gamma_{max} = 750\ \mu$.

7.145 415×10^{-6} in/in.

7.148 $\theta_p = 28.2° \downarrow$; $\epsilon_a = 820\ \mu$; $\epsilon_b = -1521\ \mu$; $\epsilon_c = 299\ \mu$.

7.149 $\theta_p = 57.9° \uparrow$; $\sigma_{max} = -4.53$ ksi; $\sigma_{min} = -10.40$ ksi.

7.150 367 psi.

7.152 5.84 kN $\cdot$ m.

7.153 4.33 kN $\cdot$ m.

7.154 87.0 kips; 37.9 kips.

7.156 (a) 1.50. (b) 1.688.

7.157 (a) 36.9 MPa. (b) 18.44 MPa.
(c) $129\ \mu$.

7.159 (a) $\theta_p = 55.8° \uparrow$ and $34.2° \uparrow$;
$\sigma_{max} = 18$ ksi; $\sigma_{min} = -1$ ksi.
(b) 9.5 ksi.

7.160 (a) $\sigma_{max} = 45.1$ MPa; $\tau_{max} = 9.40$ MPa.
(b) $\sigma_{max} = 45.1$ MPa; $\tau_{max} = 7.49$ MPa.

7.162 (a) 5650 psi. (b) 1970 psi.

7.164 (a) $\theta_p = 50.7° \uparrow$ and $39.3° \uparrow$;
(b) $\sigma_{max} = 16.20$ ksi; $\sigma_{min} = -4.20$ ksi.
(c) 10.20 ksi.

7.165 (a) 0.775 MPa. (b) -2.69 MPa.

7.166 250 psi.

7.C1 *Prob. 7.14:* (a) $+24.0$ MPa, -104.0 MPa, -1.498 *MPa*.
(b) -19.51 MPa, -60.5 MPa, -60.7 MPa.

7.C3 *Prob. 7.84:* (a) No; F.S. = 1.228. Yes.
(b) No; F.S. = 1.098. (c) Yes.

7.C4 *Prob. 7.92:* Yes. ($\sigma_a = 5.85$ ksi, $\sigma_b = -18.88$ ksi.)
Prob. 7.94: 3.67 ksi.

7.C5 *Prob. 7.129:* $\epsilon_{x'} = -450\ \mu$; $\epsilon_{y'} - 200\ \mu$;
$\gamma_{x'y'} = 375\ \mu$.

7.C7 *Prob. 7.138:* (a) $37.4\ \mu$ at $37.5° \uparrow$, $-187.6\ \mu$ at $52.5° \downarrow$. (b) $155.2\ \mu$. (c) $187.6\ \mu$.
Prob. 7.139: (a) $388\ \mu$ at $11.3° \uparrow$, $62.5\ \mu$ at $101.3° \uparrow$.
(b) $325\ \mu$. (c) $388\ \mu$.

CHAPTER 8

8.1 (a) 144.3 MPa. (b) 262 MPa.
(c) Not acceptable.

8.2 (a) 144.3 MPa. (b) 178.0 MPa.
(c) Acceptable.

8.3 (a) 28.8 ksi. (b) 28.5 ksi.
(c) Acceptable.

8.4 (a) 28.8 ksi. (b) 29.7 ksi.
(c) Not acceptable.

8.5 (a) W 690×125.
(b) $\sigma_m = 128.2$ MPa; $\tau_m = 47.3$ MPa;
$\sigma_{max} = 124.0$ MPa.

8.6 (a) W 690×125.
(b) $\sigma_m = 117.5$ MPa; $\tau_m = 34.7$ MPa;
$\sigma_{max} = 121.7$ MPa.

8.9 (a) 152.8 MPa. (b) 147.2 MPa.

8.10 (a) 131.3 MPa. (b) 135.5 MPa.

8.11 (a) 18.06 ksi. (b) 17.32 ksi.

8.12 (a) 18.22 ksi. (b) 17.39 ksi.

8.13 (a) 17.80 ksi. (b) 16.98 ksi.

8.14 (a) 19.33 ksi. (b) 20.6 ksi.

8.18 184.4 MPa.

8.19 873 lb.

8.20 402 lb.

8.22 (a) $\tau_H = 6880$ psi; $\tau_K = 6760$ psi.
(b) $\tau_H = 7420$ psi; $\tau_K = 7010$ psi.

8.25 57.7 mm.

8.26 54.3 mm.

8.29 46.5 mm.

8.30 45.9 mm.

8.31 (a) $\sigma_a = 3.28$ MPa; $\tau_a = 0$;
(b) $\sigma_b = -3.92$ MPa; $\tau_b = 0$;
(c) $\sigma_c = -0.320$ MPa; $\tau_c = 0.600$ MPa.

8.32 (a) -11.07 ksi; 0.
(b) 2.05 ksi; 2.15 ksi.
(c) 15.17 ksi; 0.

8.33 (a) 11.87 ksi; 0.
(b) 2.05 ksi; 2.15 ksi.
(c) -7.78 ksi; 0.

8.35 (a) -37.9 MPa; 14.06 MPa.
(b) -131.6 MPa; 0.

8.37 (a) $\sigma_{\max} = 0.592$ ksi at $\theta = 13.3°\nwarrow$;
$\sigma_{\min} = -10.59$ ksi at $\theta = 103.3°\nwarrow$.
(b) $\tau_{\max} = 5.59$ ksi.

8.38 (a) $\sigma_{\max} = 0.150$ ksi at $\theta = 4.6°\nwarrow$;
$\sigma_{\min} = -23.5$ ksi at $\theta = 94.6°\nwarrow$.
(b) $\tau_{\max} = 11.82$ ksi.

8.39 -3.96 ksi; 0.938 ksi.

8.40 -14.98 MPa; 17.29 MPa.

8.43 $\sigma_{\max} = 1.69$ MPa at $\theta = -23.0$ ksi; $\tau_{\max} = 12.34$ ksi.

8.44 (a) $\sigma_{\max} = 3.34$ ksi; $\sigma_{\min} = -3.34$ ksi; $\tau_{\max} = 3.34$ ksi.
(b) $\sigma_{\max} = 0.89$ ksi; $\sigma_{\min} = -9.69$ ksi; $\tau_{\max} = 5.29$ ksi.

8.45 $\sigma_{\max} = 55$ MPa at $\theta = 45°\nwarrow$;
$\sigma_{\min} = 55$ MPa at $\theta = 45°\swarrow$.

8.46 $\tau_{\max} = 55$ MPa.

8.47 (a) $\sigma_{\max} = 24.1$ MPa; $\sigma_{\min} = -78.7$ MPa;
$\tau_{\max} = 51.4$ MPa.

8.48 (a) $\sigma_{\max} = 3.79$ ksi at $\theta = 33.7°\nwarrow$;
$\sigma_{\min} = -8.50$ ksi at $\theta = 123.7°\nwarrow$.
(b) 6.15 ksi.

8.51 (a) $\sigma_{\max} = 4.3$ MPa at $\theta = 12.1°\nwarrow$;
$\sigma_{\min} = -93.4$ MPa at $\theta = 102.1°\nwarrow$.
(b) 48.9 MPa.

8.52 $\sigma_{\max} = 0.413$ ksi; $\sigma_{\min} = -0.0378$ ksi;
$\tau_{\max} = 0.225$ ksi.

8.53 $\sigma_{\max} = 1.798$ ksi; $\sigma_{\min} = -0.006$ ksi;
$\tau_{\max} = 0.902$ ksi.

8.54 $\tau_{\max} = 2830$ psi.

8.55 -14.35 MPa; 9.26 MPa.

8.56 -54.9 MPa; 4.17 MPa.

8.57 $\sigma_{\max} = 1506$ psi at $\theta = 31.1°\nwarrow$;
$\sigma_{\min} = -4150$ psi at $\theta = 121.1°\nwarrow$;
$\tau_{\max} = 2830$ psi.

8.58 $\sigma_{\max} = 140$ psi at $\theta = 12.7°\nwarrow$;
$\sigma_{\min} = -2780$ psi at $\theta = 102.7°\nwarrow$; $\tau_{\max} = 1462$ psi.

8.61 $\sigma_{\max} = 25.2$ MPa; $\sigma_{\min} = -0.87$ MPa;
$\tau_{\max} = 13.06$ MPa.

8.62 $\sigma_{\max} = 34.6$ MPa; $\sigma_{\min} = -10.18$ MPa;
$\tau_{\max} = 22.4$ MPa.

Point a: $\sigma_{\max} = 86.5$ MPa; $\sigma_{\min} = 0$; $\tau_{\max} = 43.3$ MPa.
Point b: $\sigma_{\max} = 58.5$ MPa; $\sigma_{\min} = -1.53$ MPa;
$\tau_{\max} = 30.0$ MPa.
Point d: $\sigma_{\max} = 42.4$ MPa; $\sigma_{\min} = -0.19$ MPa;
$\tau_{\max} = 21.3$ MPa.
Point e: $\sigma_{\max} = 12.74$ MPa; $\sigma_{\min} = 0$; $\tau_{\max} = 6.37$ MPa.

8.63 $\sigma_{\max} = 5.58$ ksi at $\theta = 34.1°\swarrow$;
$\sigma_{\min} = -12.23$ ksi at $\theta = 55.9°\nwarrow$; $\tau_{\max} = 8.91$ ksi.

8.64 $\sigma_{\max} = 5.55$ ksi at $\theta = 30.1°\swarrow$;
$\sigma_{\min} = -16.47$ ksi at $\theta = 59.9°\nwarrow$; $\tau_{\max} = 11.01$ ksi.

8.67 (a) $6Pl(\cos \beta/h - \sin \beta/b)/bh$.
(b) $\tan^{-1}(b/h)$.

8.68 (a) 51.4 kN. (b) 39.7 kN.

8.69 $\sigma_{\max} = 3470$ psi; $\sigma_{\min} = -3470$ psi; $\tau_{\max} = 3470$ psi.

8.71 $\sigma_{\max} = 77.7$ MPa; $\sigma_{\min} = -12.04$ MPa;
$\tau_{\max} = 44.9$ MPa.

8.73 (a) W310 × 38.7.
(b) $\sigma_m = 147.3$ MPa; $\tau_m = 18.3$ MPa;
$\sigma_{\max} = 139.7$ MPa.

8.74 15.86 ksi, 19.76 ksi.

8.75 39.5 mm.

8.77 1.822 in.

8.79 $\sigma = -20.2$ MPa; $\tau = 2.82$ MPa.

8.81 $\sigma_a = -16.41$ ksi, $\tau_a = 0$; $\sigma_b = -15.63$ ksi, $\tau_b = 47$ psi;
$\sigma_c = -7.10$ ksi, $\tau_c = 1.256$ ksi.

8.82 $\sigma_{\max} = 30.1$ MPa at $\theta = 8.2°\nwarrow$;
$\sigma_{\min} = -0.62$ MPa at $\theta = 81.8°\swarrow$; $\tau_{\max} = 15.37$ MPa.

8.84 (a) $\sigma_{\max} = 31.9$ MPa; $\sigma_{\min} = -6.99$ MPa;
$\tau_{\max} = 19.42$ MPa.
(b) $\sigma_{\max} = 14.21$ MPa; $\sigma_{\min} = -19.18$ MPa;
$\tau_{\max} = 16.70$ MPa.

8.C3 *Prob.* 8.76: 1.578 in.

8.C5 *Prob.* 8.50: $\sigma = 1.792$ ksi; $\theta_p = 0.1042$ ksi.

8.C6 *Prob.* 8.35: (a) $\sigma = -37.9$ MPa, $\tau = 14.06$ MPa.
(b) $\sigma = -131.6$ MPa, $\tau = 0$.

8.C7 *Prob.* 8.72a: $\sigma_{\max} = 12.89$ ksi, $\theta_p = -8.9°$;
-0.319 ksi, $\theta_p = 81.1°$; $\tau_{\max} = 6.61$ ksi.

CHAPTER 9

9.1 (a) $y = M_0(L - x)^2/2EI$.
(b) $M_0L^2/2EI\uparrow$, (c) $M_0L/EI\swarrow$.

9.2 (a) $y = -Px^2(3L - x)/6EI$.
(b) $PL^3/3EI\downarrow$, (c) $PL^2/2EI\swarrow$.

9.3 (a) $y = -w_0(x^4 - 4L^3x + 3L^4)/24EI$.
(b) $wL^4/8EI\downarrow$, (c) $wL^3/6EI\swarrow$.

9.4 (a) $y = -w_0(L^3x^2/6 - Lx^4/12 + x^5/120)/EIL$.
(b) $11u_0L^4/120EI\downarrow$, (c) $w_0L^3/8EI\swarrow$.

9.5 (a) $y = w(L^3x/16 - x^4/24 - L^3x/48)/EI$.
(b) $wL^3/48EI\downarrow$. (c) 0.

9.6 (a) $y = w(Lx^3/10 - x^4/24 - L^2x^2/20 - L^3x/120)/EI$.
(b) $13 \, wL^4/1920 \, EI\downarrow$, (c) $wL^3/120 \, EI\swarrow$.

9.9 (a) $0.00652 \, w_0L^4/EI\downarrow$; 0.519 L.
(b) 0.229 in. $\downarrow$.

9.10 (a) $0.00652 \, w_0L^4/EI\downarrow$; 0.481 L.
(b) 0.229 in. $\downarrow$.

9.11 (a) $0.423 L$; $0.0642 M_0L^2/EI\uparrow$.
(b) 45.3 kN · m.

9.12 (a) $0.211 L$; $0.01604 M_0L/EI$.
(b) 6.09 m.

9.15 (a) $y = P(ax^2/2 - aLx/2 + a^3/6)/EI$.
(b) 1.976 mm $\downarrow$.

9.16 (a) $y = w(x^4/24 - L^3x/54)/EI$.
(b) 0.397 mm ↓.
9.17 3 $wL/8$ ↑.
9.18 $3M_0/2L$ ↑.
9.20 $11\,w_0L/40$ ↑.
9.21 3 kips ↑.
9.23 $\mathbf{R}_A = 5P/16$ ↑; $M_C = 5PL/32$; $M_B = -3PL/16$.
9.24 $\mathbf{R}_A = 7wL/128$ ↑; $M_C = 0.0273wL^2$.
$M_B = -0.0703wL^2$; $M_{max} = 0.0288wL^2$.
9.25 $\mathbf{R}_A = 9w_0L/640$ ↑; $M_C = 0.00703w_0L^2$;
$M_B = -0.0276w_0L^2$, $M_{max} = 0.00814w_0L^2$.
9.26 $\mathbf{R}_B = 9M_0/8L$ ↑; $M = -7M_0/16$ to left of C.
9.29 $4M_0/3L$ ↑; $2M_0L^2/243\,EI$ ↑.
9.30 $4P/27$ ↑; $11PL^3/2187\,EI$ ↓.
9.31 $\mathbf{R}_A = P/2$ ↑; $\mathbf{M}_A = PL/8$ ↱; $M_C = PL/8$.
9.32 $\mathbf{R}_A = w_0L/4$ ↑; $\mathbf{M}_A = 5w_0L^2/96$ ↱; $M_C = w_0L^2/32$.
9.33 (a) $y = w_0(x^6/90 - Lx^5/30 + L^3x^3/18 - L^5x/30)/EIL^2$.
$w_0L^3/30\,EI$ ↘. (c) $61\,w_0L^4/15760\,EI$ ↓.
9.34 (a) $y = 2w_0L^4[-8\cos(\pi x/2L) - \pi^2x^2/L^2 + 2\pi(\pi-2)x/L + \pi(4-\pi)]/\pi^4EI$.
(b) $0.1473\,w_0L^3/EI$ ↘. (c) $0.1089\,w_0L^4/EI$ ↓.
9.35 (a) $y = P\{bx^3 - L\langle x-a\rangle^3 - b(L^2 - b^2)x\}/6EIL$.
(b) $Pb(L^2 - b^2)/6EIL$ ↘. (c) $Pa^2b^2/3EIL$ ↓.
9.36 (a) $y = w\{Lx^3/48 - \langle x-L/2\rangle^4/24 - 7L^3x/384\}/EI$.
(b) $7wL^3/384\,EI$ ↘. (c) $5wL^4/768\,EI$ ↓.
9.37 (a) $y = M_0\{x^3 - 3L\langle x-a\rangle^2 + (3b^2 - L^2)x\}/6EIL$.
(b) $M_0(3b^2 - L^2)/6EIL$ ↙. (c) $M_0ab(b-a)/3EIL$ ↑.
9.38 (a) $y = w_0\{-\langle x-L/2\rangle^5/60 + L^2x^3/144 - 37L^4x/5760\}/EIL$.
(b) $37w_0L^3/5760\,EI$ ↘. (c) $3w_0L^4/1280\,EI$ ↓.
9.41 (a) $y = w\{-3L^2x^2/16 + Lx^3/12 - \langle x-L/2\rangle^4/24\}/EI$.
(b) $7wL^4/192\,EI$ ↘. (c) $41wL^4/384\,EI$ ↓.
9.42 (a) $y = w\{ax^3/6 - 3L\langle x-a\rangle^2 + (3b^2 - L^2)x\}/6EIL$.
(b) $M_0(3b^2 - L^2)/6EIL$ ↙. (c) $M_0ab(b-a)/3EIL$ ↑.
9.43 (a) $y = w_0\{-\langle x-L/2\rangle^5/60 + L^2x^3/144 - 37L^4x/5760\}/EIL$.
9.44 (a) $y = w\{-x^4/24 + L\langle x-L/2\rangle^3/8 + \langle x-L/2\rangle^4/24 - \langle x-L\rangle^4/24 + 17L^3x/384 - 5L^4/256\}/EI$.
(b) $5wL^4/256\,EI$ ↓. (c) $wL^4/768\,EI$ ↑.
9.46 (a) 2.48×10^{-3} rad ↘. (b) 1.071 mm ↓.
9.48 (a) 0.873×10^{-3} rad ↘. (b) 1.641 mm ↓.
9.49 14.00×10^{-3} rad ↘ at A; 12.32×10^{-3} rad ↙ at D; 0.341 in. ↓ maximum.
9.50 0.873×10^{-3} rad ↘ at A; 1.035×10^{-3} rad ↙ at B; 1.642 mm ↓ maximum.
9.51 (a) $5P/16$ ↑. (b) $7\,PL^3/168\,EI$ ↓.
9.52 (a) $9\,M_0/8L$ ↓. (b) $M_0L^2/128\,EI$ ↑.
9.53 (a) $2P/3$ ↑. (b) $5\,PL^3/486\,EI$ ↓.
9.54 (a) $41\,wL/128$ ↑. (b) $19\,wL^4/6144\,EI$ ↓.
9.57 41.3 kN ↑. (b) 0.710 mm ↓.
9.58 11.54 kN ↑. (b) 4.18 mm ↓.
9.59 $3M_0/2L$ ↑; $M_0/4$ ↱.
(b) $M_0L/16\,EI$ ↙.
9.60 (a) $1.280\,wa$ ↑; $1.333\,wa^2$ ↱.
(b) $0.907\,wa^4/EI$ ↓.

9.61 0.1619 in. ↓; 3.74 ft from A.
9.62 1.393 mm ↓; 0.857 m from A.
9.63 0.341 in. ↓; 3.34 ft from A.
9.64 1.648 mm ↓; 2.86 m from A.
9.67 (a) $8PL^3/243\,EI$ ↓. (b) $19PL^2/162EI$ ↘.
9.68 (a) $PL^3/486\,EI$ ↓. (b) $PL^2/81\,EI$ ↘.
9.69 (a) $wL^4/128\,EI$ ↓. (b) $wL^3/72\,EI$ ↘.
9.70 (a) $19\,Pa^3/6\,EI$ ↓. (b) $5\,Pa^2/2\,EI$ ↘.
9.73 $PL^2/24\,EI$ ↙; $PL^3/48\,EI$ ↓.
9.74 $wL^3/48\,EI$ ↙; $wL^4/384\,EI$ ↑.
9.75 (a) 2.58×10^{-3} rad ↘. (b) 2.70 mm ↓.
9.76 (a) 0.601×10^{-3} rad ↘. (b) 3.67 mm ↓.
9.77 7.91×10^{-3} rad ↙; 0.340 in. ↓.
9.78 6.98×10^{-3} rad ↙; 0.1571 in. ↓.
9.81 (a) $\mathbf{R}_A = 4P/3$ ↑; $\mathbf{M}_A = PL/3$ ↱. (b) $\mathbf{R}_B = 2P/3$ ↑.
9.82 (a) $\mathbf{R}_A = 41\,wL/128$ ↑.
(b) $\mathbf{R}_B = 23wl/128$ ↑; $\mathbf{M}_B = 7wL^2/128$ ↓.
9.83 $\mathbf{R}_A = 7P/32$ ↑; $\mathbf{R}_C = 33P/16$ ↑; $\mathbf{R}_E = 23P/32$ ↑.
9.84 $\mathbf{R}_A = 2M_0/L$ ↑; $\mathbf{R}_B = 3M_0/L$ ↓; $\mathbf{R}_C = M_0/L$ ↑.
9.87 (a) 5.06×10^{-3} rad ↘. (b) 47.7×10^{-3} in. ↓.
9.88 121.5 N/m.
9.89 (a) 10.38 mm ↓. (b) 23.1 mm ↓.
9.90 $y_B = -0.210$ in. ↓; $y_C = -0.171$ in. ↓.
9.92 (a) 0.00937 mm ↓. (b) 229 N ↑.
9.93 5.63 kN ↓.
9.95 0.1975 in. ↓.
9.96 (a) 31.2 mm ↓. (b) 17.89 mm ↑.
9.97 (a) $wL^3/6EI$ ↙. (b) $wL^4/8EI$ ↓.
9.99 (a) $5\,Pa^2/2EI$ ↘. (b) $7\,Pa^3/2EI$ ↓.
9.100 (a) $3Pa^2/2EI$ ↘. (b) $11Pa^3/6EI$ ↓.
9.102 (a) $7wL^3/48EI$ ↘. (b) $7wL^4/192EI$ ↓.
9.103 (a) 5.22×10^{-3} rad ↙. (b) 10.88 mm ↓.
9.104 (a) 9.67×10^{-3} rad ↙. (b) 0.814 in. ↓.
9.106 (a) 5.84×10^{-3} rad ↙. (b) 0.300 in. ↓.
9.107 $y_D = 25M_0a^2/12EI$ ↓; $\theta_D = 11M_0a/6EI$ ↘.
9.108 (a) $7wa^4/36EI$ ↓. (b) $47wa^4/72EI$ ↓.
9.110 (a) 6.10×10^{-3} rad ↙. (b) 6.03 mm ↓.
9.111 (a) $PL^2/16EI$ ↘. (b) $PL^3/48EI$ ↓.
9.112 (a) $PL^2/32EI$ ↘. (b) $PL^3/128EI$ ↓.
9.114 (a) $wa^2(3L - 2a)/12EI$ ↘.
(b) $wa^2(3L - 2a^2)/48EI$ ↓.
9.115 (a) 3.67×10^{-3} rad ↘. (b) 5.63 mm ↓.
9.117 (a) $5Pa^2/8EI$ ↘. (b) $3Pa^3/4EI$ ↓.
9.118 (a) $11Pa^2/6EI$ ↘. (b) $35Pa^3/18EI$ ↓.
9.119 (a) 2.12×10^{-3} rad ↘. (b) 0.1133 in. ↓.
9.120 (a) 5.17×10^{-3} rad ↙. (b) 21.0 mm ↓.
9.121 (a) 1.735×10^{-3} rad ↘. (b) 0.0639 in. ↓.
9.122 (a) 3.13×10^{-3} rad ↙. (b) 1.016 mm ↑.
9.124 3.55 kN.
9.125 $0.223\,L$.
9.126 $0.211\,L$.
9.127 (a) $2M_0L^2/81\,EI$ ↑. (b) $M_0L/18EI$ ↙.
9.128 (a) $4PL^3/243EI$ ↑. (b) $14PL^2/81EI$ ↙.
9.129 (a) $5PL^3/768EI$ ↓. (b) $3PL^2/128EI$ ↙.
9.130 (a) $17w_0L^4/2916EI$ ↓. (b) $7w_0L^3/360EI$ ↘.
9.132 (a) 8.70×10^{-3} rad ↘. (b) 15.03 mm ↓.

9.133 (a) 3.14×10^{-3} rad ↘, (b) 4.15 mm ↓.
9.137 (a) 1.042×10^{-3} rad ↘, (b) 2.71 mm ↓.
9.138 (a) 1.231×10^{-3} rad ↘, (b) 0.221 in. ↓.
9.139 (a) 0.481×10^{-3} rad ↗, (b) 0.141 in. ↓.
9.140 (a) 3.61×10^{-3} rad ↘, (b) 0.960 mm ↑.
9.141 $3M_0L/16EI$ ↘, (b) $M_0L/8EI$ ↙.
9.142 (a) $5PL^2/96\,EI$ ↘, (b) $PL^2/24EI$ ↙.
(c) $M_0L^2/24\,EI$ ↙.
9.143 (c) $PL^3/64EI$ ↓.
$\sqrt{3}\,M_0L^2/27EI$ ↑, at $\sqrt{3}L/3$ from A.
9.144 $\sqrt{3}\,PL^3/256EI$ ↓, at $\sqrt{3}L/4$ from A.
9.145 $0.00652w_0L^4/EI$ ↓, at $0.519\,L$ from A.
9.146 1.511 mm ↓, at 2.81 m from A.
9.147 4.15 mm ↓, at 2.14 m from A.
9.150 $2PL^3/3EI_0$ ↓.
9.151 $3M_0/2L$ ↑.
9.153 $7\,wL/128$ ↑.
9.154 $0.275\,w_0L^3$ ↑.
9.155 $R_A = M_0/4L$ ↑; $R_B = 5M_0/4L$ ↑, $R_C = 3M_0/2L$ ↓.
9.156 $R_A = wL/6$ ↓; $R_B = 3wL/4$ ↑, $R_C = 5wL/12$ ↑.
9.157 $R_A = 37$ kN ↑; $M_C = 55.6$ kN · m;
$M_B = -73.3$ kN · m.
9.159 $144\,EI/L^3$.
9.160 $48\,EI/7L^3$.
9.162 (a) $5M_0L/32EI$ ↘, (b) $3M_0L^2/64EI$ ↓.
9.164 $29\,w_0L^4/960\,EI$ ↓.
9.165 (a) 3.47 kips ↑, (b) 0.0372 in. ↓.
9.167 (a) 5.94 mm ↓. (b) 6.75 mm ↓.
9.168 2.45 kips.
9.169 (a) $17\,PL^3/972\,EI$ ↓, (b) $19\,PL^3/972\,EI$ ↓.
9.170 $0.0210\,PL^3/EI$ at $0.537\,L$ from A.
9.172 (a) 6.95 mm. (b) 46.3 kN.
9.C1 Prob. 9.79; 6.30×10^{-3} rad ↘; 5.53 mm ↓;
Prob. 9.80; 5.56×10^{-3} rad ↘; 2.5 mm ↓.
9.C2 $a = 6\,ft$; (a) 3.14×10^{-3} rad ↘, 0.292 in. ↓;
(b) 0.397 in. ↓ at 11.27 ft to the right of A.
9.C3 $x = 1.6\,m$; (a) 7.90×10^{-3} rad ↘, 8.16 mm ↓;
(b) 6.05×10^{-3} rad ↘, 5.79 mm ↓;
(c) 1.021×10^{-3} rad ↘, 0.314 mm ↓.
9.C4 Prob. 9.14; $x = 0.75$ m; 4.01×10^{-3} rad ↘;
4.03 mm ↓.
9.C5 (a) $a = 3\,ft$; 1.586×10^{-3} rad ↘; 0.1369 in. ↓;
(b) $a = 1.0\,m$; 0.293×10^{-3} rad ↘; 0.479 mm ↓.
9.C6 Prob. 9.116; $x = 18$ in.; 4.47×10^{-3} rad ↘;
0.1341 in. ↓.
9.C7 $x = 2.5\,m$; 5.31 mm ↓; $x = 5.0\,m$; 12.28 mm ↓.

CHAPTER 10

10.1 K/L.
10.2 kL.
10.3 $kL/4$.
10.4 K/L.
10.5 $ka^2/2l$.
10.8 $d > 392$ mm.
10.9 (a) 6.65 lb. (b) 21.0 lb.
10.10 305 kN.
10.11 (a) 13.06 kN. (b) 22.9 mm.
(c) 40.6%.
10.12 (a) 12.04 kips. (b) 6.25%.
10.15 168.4 kN.
10.16 2.125.
10.17 335 kips.
10.18 (a) 94.8 kN. (b) 449 kN.
10.21 (1) 2.64 kN; (2) 0.661 kN; (3) 10.57 kN; (4) 5.39 kN;
(5) 2.64 kN.
10.22 (a) 2.29. (b) (2) 1.768 in.; (3) 1.25 in.; (4) 1.046 in.
10.23 8.61 kN.
10.24 (a) $L_{BC} = 1.96$ m; $L_{CD} = 0.49$ m. (b) 23.1 kN.
10.25 (a) 1/2. (b) $d = 1.144$ in.; $b = 0.572$ in.
10.28 12.08 m.
10.29 (a) 0.1451 in. (b) 5.99 ksi.
10.30 (a) 13.24 mm. (b) 79.0 MPa.
10.31 (a) 0.247 in. (b) 12.94 ksi.
10.32 (a) 1.658 mm. (b) 78.9 MPa.
10.35 (a) 13.29 kips. (b) 15.50 ksi.
10.36 (a) 235 kN. (b) 149.6 MPa.
10.37 (a) 151.6 kN. (b) 109.5 MPa.
10.38 (a) 368 kN. (b) 103.8 MPa.
10.39 (a) 223 kN. (b) 62.8 MPa.
10.40 (a) 67.6 kips. (b) 14.3 ksi.
10.43 (a) 189 kN. (b) 229 kN.
10.44 (a) 147 kN. (b) 174 kN.
10.45 (a) 38.6 kips. (b) 0.628.
10.46 (a) 54.9 kips. (b) 0.397.
10.49 (a) 4.54 m. (b) 2.41 m.
10.50 (a) 8.31 m. (b) 2.54 m.
10.51 $d = 2.125$ in.
10.52 2.625 in.
10.55 2.95.
10.56 3.00.
10.57 (a) 368 kN. (b) 916 kN.
10.58 (a) 87.6 kips. (b) 87.6 kips.
10.59 (a) 61.0 kips. (b) 31.6 kips.
10.60 (a) 1568 kN. (b) 633 kN.
10.61 45.4 kN.
10.64 319 kN.
10.66 174.3 kips.
10.68 292 kips.
10.69 117.5 kN.
10.70 156.2 kN.
10.72 0.884 in.
10.73 9 mm.
10.74 4.01 in.
10.75 121 mm.
10.77 W 360 × 216.
10.78 W 200 × 46.1.
10.81 (a) 30.1 mm. (b) 33.5 mm.
10.82 $\frac{1}{2}$ in.-thickness.
10.83 $\frac{3}{8}$ in.-thickness.
10.84 6.4 mm.

10.85 300 kN
10.86 (a) 73.7 kips; 54.6 kips.
    (b) 183.9 kips; 136.2 kips.

10.87 $\frac{5}{16}$ in.
10.88 W 310 × 74.
10.89 17.48 kN.
10.90 34.7 kN.
10.91 11.91 kips.
10.92 9.72 kips.
10.95 42.0 in.
10.96 47.6 in.
10.97 4.81 m.
10.98 5.48 m.
10.101 (a) 11.76 mm. (b) 6.43 mm.
10.102 7.75 mm.
10.105 70.9 mm.
10.106 81.5 mm.
10.107 83.9 mm.
10.108 87.6 mm.
10.109 70.0 mm.
10.110 80.4 mm.
10.111 W 200 × 59.
10.112 W 250 × 58.
10.113 W 12 × 40.
10.115 12 mm.
10.117 79.0 kN.
10.119 $\pi^2 b^2/12L^2\alpha$.
10.120 4.00 kN.
10.121 0.384 in.
10.124 W 250 × 67.
10.125 (a) 26.4 kN. (b) 32.2 kN.
10.127 16.46 ft.
10.128 1.894 in
10.C1 $r = 8\,mm$: 9.07 kN
    $r = 16\,mm$: 70.4 kN
10.C2 $b = 1.0\,in.$: 3.85 kips
    $b = 1.375\,in.$: 6.07 kips
10.C3 $h = 5.0\,m$: 9819 kg
    $h = 7.0\,m$: 13 255 kg
10.C4 $l' = 35\,kips$. (a) 0.086 in., (b) 4.69 ksi.
    $P = 55\,kips$: (a) 0.146 in.; (b) 7.65 ksi.

CHAPTER 11

11.1 (a) 43.1 in · lb/in³. (b) 72.8 in · lb/in³.
    (c) 172.4 in · lb/in³.
11.2 (a) 21.6 kJ/m³. (b) 336 kJ/m³.
    (c) 142.0 kJ/m³.
11.3 (a) 177.9 kJ/m³. (b) 712 kJ/m³.
    (c) 160.3 kJ/m³.
11.5 (a) 1296 kJ/m³. (b) 90 MJ/m³.
11.7 (a) 150 kJ/m³. (b) 63 MJ/m³.
11.8 (a) 221 in · lb/in³. (b) 8900 in · lb/in³.
11.9 (a) 176.2 in · lb.
    (b) 11.72 in · lb/in³; 5.65 in · lb/in³.

11.11 (a) 168.8 in · lb.
    (b) 0.882 in · lb/in³; 5.65 in · lb/in³.
11.12 (a) 12.18 J.
    (b) 15.83 kJ/m³; 38.6 kJ/m³.
11.15 724 in · lb.
11.16 13.73 mm.
11.17 4.98.
11.18 6.04.
11.20 $0.24856\,P^2L/EA_{min}$; −0.575%.
11.21 102.7 in · lb.
11.22 $3.81\,P^2l/EA$.
11.24 $2.37\,P^2l/EA$.
11.25 $0.233\,P^2l/EA$.
11.26 1015 J.
11.28 $P^2a^2(a + L)/6EI$.
11.29 $P^2a^2(3L - 4a)/6EI$.
11.31 $w^2L^5/240\,EI$.
11.34 894 J.
11.35 662 J.
11.36 894 in · lb.
11.38 14.70 J.
11.40 2.45×10^6 in · lb.
11.41 9.33 ksi.
11.43 -11.39 ksi $\leq \sigma_y \leq 19.39$ ksi.
11.45 (a) no yield.
    (b) yield.
11.46 -2.65 MPa $\leq \sigma_z \leq 122.7$ MPa.
11.47 $2M_0^2l(1 + 3Ed^2/10GL^2)/Ebd^3$
11.49 $Q^2\ln(R_2/R_1)/4\pi GL$.
11.50 9.12 lb.
11.51 25.5 ft/s.
11.52 0.997 kg.
11.54 4.76 kg.
11.56 11.50 in.
11.57 7.87 in.
11.58 (a) 21.0 kN. (b) 171.7 MPa.
    (c) 8.58 mm.
11.59 (a) 7.67 kN. (b) 312 MPa.
    (c) 23.5 mm.
11.60 8.50 ft/s.
11.62 (a) 15.63 mm. (b) −83.8 N · m.
    (c) 208 MPa.
11.63 (a) 23.6 mm. (b) −64.4 N · m.
    (c) 157.6 MPa.
11.64 (a) 0.1451 in. (b) 28.0 ksi.
11.66 0.0357 in.
11.67 0.0335 in.
11.69 7.12.
11.70 (b) $y'_m = 13.25$ mm;
    $|(y'_m - y_m)/y_m| = 0.152$; $y'_m/2h = 0.166$.
11.71 $Pa^2b^2/3EIL$ ↓.
11.72 $Pa^2(a + L)/3EI$ ↓.
11.73 $M_0(a^3 + b^3)/3EIL^2$ ↶.
11.74 $M_0(L + 3a)/3EI$ ↶.
11.75 $3PL^3/16EI$ ↓.
11.77 $5M_0L/16EI$ ↶.

11.79 11.57 mm ↓.

11.80 3.12°.

11.84 $3.375\ Pl/EA$ →.

11.86 1.030 mm →.

11.87 0.0298 in. →.

11.89 0.366 in. ↓.

11.90 (a) and (b) $(P^2L^2 + 3M_0PL + 3M_0^2)L/6EI$.

11.92 (a) and (b) $(P^2L^2 - 6M_0PL + 16M_0^2)L/96EI$.

11.94 $7P^2L^3/24EI$.

11.95 $M_0^2L/2EI$.

11.96 $5PL^3/48EI$ ↓.

11.97 $17wL^4/384EI$ ↓.

11.98 $3PL^2/8EI$ ∠.

11.99 $7wL^3/48EI$ ∠.

11.100 $Pab(3La - 2a^2 + 2b^2)/6EIL^2$ ∢.

11.102 $wL^4/768EI$ ↑.

11.103 $PaL^2/16EI$ ↑.

11.105 $PaL/24EI$ ∠.

11.106 5.08 mm ↓.

11.107 0.329 in. ↓.

11.108 0.987 in. ↓.

11.109 6.98×10^{-3} rad ∢.

11.113 0.1124 mm ←.

11.114 0.233 in. ↓.

11.115 0.1504 in. →.

11.116 0.1459 mm ↓.

11.118 (a) $PR^3/2EI$ →. (b) $\pi PR^3/4EI$ ↓.

11.119 (a) $2Pl^3/3EI$ →. (b) $Pl^2/6EI$ ∠.

11.120 (a) $5Pl^3/3EI$ →. (b) $2Pl^2/EI$ ∢.

11.122 (a) $PL^3/6EI$ ↓, (b) $0.1443\ PL^3/EI$ →.

11.124 $5P/16$ ↑; $M_C = 5PL/32$.

11.125 $3M_0/2L$; $M_A = M_0/2$.

11.127 $7wL/128$ ↑; $M_B = -0.0703wL^2$.

11.128 $14P/27$ ↑; $M_B = -0.1481\ wL$.

11.130 $Pl/(1 + 2\cos^3\phi)$.

11.131 $3P/4$.

11.133 0.652 P.

11.135 $2P/3$.

11.136 0.807 in. ↓.

11.137 1.64 mm →.

11.140 4.84°.

11.141 (a) 225 kJ/m³. (b) 180 MPa.

11.143 47.5 mm ↓.

11.144 52.0 in.

11.146 1.111 mm ↓.

11.147 $\{3Emv_0^2c^2/IL\}^{1/2}$.

11.C2 (a) $a = 15$ in.: $\sigma_D = 17.19$ ksi, $\sigma_C = 21.0$ ksi;
$a = 45$ in.: $\sigma_D = 36.2$ ksi, $\sigma_C = 14.74$ ksi.
(b) $a = 18.34$ in., $\sigma = 20.67$ ksi.

11.C3 (a) $L = 200$ mm: $h = 2.27$ mm;
$L = 800$ mm: $h = 1.076$ mm.
(b) $L = 440$ mm: $h = 3.23$ mm.

11.C4 $a = 300$ mm: 1.795 mm, 179.46 MPa;
$a = 600$ mm: 2.87 mm, 179.59 MPa.

11.C5 $a = 2$ m: (a) 30.0 J; (b) 7.57 mm, 60.8 J.
$a = 4$ m: (a) 21.9 J; (b) 8.87 mm, 83.4 J.

11.C6 $a = 20$ in: (a) 13.26 in.; (b) 99.5 kip · in.;
(c) 803 lb.
$a = 50$ in: (a) 9.46 in.; (b) 93.7 kip · in.;
(c) 996 lb.

		$\bar{x}$	$\bar{y}$	Area
Triangular area			$\dfrac{h}{3}$	$\dfrac{bh}{2}$
Quarter-circular area		$\dfrac{4r}{3\pi}$	$\dfrac{4r}{3\pi}$	$\dfrac{\pi r^2}{4}$
Semicircular area		0	$\dfrac{4r}{3\pi}$	$\dfrac{\pi r^2}{2}$
Semiparabolic area		$\dfrac{3a}{8}$	$\dfrac{3h}{5}$	$\dfrac{2ah}{3}$
Parabolic area		0	$\dfrac{3h}{5}$	$\dfrac{4ah}{3}$
Parabolic spandrel		$\dfrac{3a}{4}$	$\dfrac{3h}{10}$	$\dfrac{ah}{3}$
Circular sector		$\dfrac{2r\sin\alpha}{3\alpha}$	0	αr^2
Quarter-circular arc		$\dfrac{2r}{\pi}$	$\dfrac{2r}{\pi}$	$\dfrac{\pi r}{2}$
Semicircular arc		0	$\dfrac{2r}{\pi}$	πr
Arc of circle		$\dfrac{r\sin\alpha}{\alpha}$	0	$2\alpha r$